T0212755

Lecture Notes in Artificial Intelligence 9346

Subseries of Lecture Notes in Computer Science

More information about this series at http://www.springer.com/series/1244

Toby Walsh (Ed.)

Algorithmic Decision Theory

4th International Conference, ADT 2015
Lexington, KY, USA, September 27–30, 2015
Proceedings

 Springer

Editor
Toby Walsh
University of New South Wales
Kensington
Australia

ISSN 0302-9743   ISSN 1611-3349   (electronic)
Lecture Notes in Artificial Intelligence
ISBN 978-3-319-23113-6   ISBN 978-3-319-23114-3   (eBook)
DOI 10.1007/978-3-319-23114-3

Library of Congress Control Number: 2015946998

LNCS Sublibrary: SL7 – Artificial Intelligence

Springer Cham Heidelberg New York Dordrecht London

Printed on acid-free paper

Springer International Publishing AG Switzerland is part of Springer Science+Business Media
(www.springer.com)

Preface

The 4th International Conference on Algorithmic Decision Theory (ADT 2015) brought together researchers and practitioners coming from diverse areas such as artificial intelligence, database systems, operations research, decision theory, discrete mathematics, game theory, multiagent systems, computational social choice, and theoretical computer science with the goal of improving the theory and practice of modern decision support. Previous conferences were held in Venice (2009), Piscataway (2011), and Brussels (2013).

Some of the scientific challenges facing the ADT community include big preference data, combinatorial structures, partial and/or uncertain information, distributed decision making, and large user bases. Such challenges occur in realworld decision making in domains such as electronic commerce, recommender systems, network optimization (communication, transport, energy), risk assessment and management, and e-government.

This volume contains the papers presented at ADT 2015. The conference itself was held during September 27–30 in Lexington, Kentucky. The meeting was co-located with the 13th International Conference on Logic Programming and Non-monotonic Reasoning (LPNMR 2015). A joint session and invited speaker were shared with this conference.

Each submission to ADT 2015 was reviewed by at least three Program Committee members. Reviewing was double blind. The committee decided to accept 32 papers out of the 70 submissions, giving an acceptance rate of 45 %. The program also includes six peer-reviewed papers written for the associated doctorial consortium.

I would like to thank Judy Goldsmith for her tireless efforts as local chair, the three invited speakers (Steve Brams, Jerome Lang, and Brent Venable), the LPNMR 2015 chairs for their help and cooperation, as well as the many people who volunteered to help in some way or other to make the meeting happen. I also thank the authors for trusting us with their latest research results, as well as the Program Committee and all their additional reviewers for their help with selecting and preparing the conference program.

I end by thanking my wife, Andrea, and my daughter, Bronte, who gave me the time and space to be a program chair yet again. Thank you.

July 2015   Toby Walsh

Organization

Program Committee

Haris Aziz	NICTA and University of New South Wales, Australia
Dorothea Baumeister	Universität Duesseldorf, Germany
Craig Boutilier	University of Toronto, Canada
Felix Brandt	Technische Universität München, Germany
Markus Brill	Duke University, USA
Alexander Brodsky	George Mason University, USA
Ioannis Caragiannis	University of Patras, Greece
Edith Elkind	University of Oxford, UK
Markus Endres	Universität Augsburg, Germany
Ulle Endriss	ILLC, University of Amsterdam, The Netherlands
Gabor Erdelyi	Universität Siegen, Germany
Piotr Faliszewski	AGH University of Science and Technology, Poland
Helene Fargier	IRIT-CNRS, France
Felix Fischer	University of Cambridge, UK
Umberto Grandi	University of Padova, Italy
Lane A. Hemaspaandra	University of Rochester, USA
Christian Klamler	University of Graz, Austria
Ron Lavi	The Technion – Israel Institute of Technology, Israel
Thierry Marchant	Universiteit Gent, Belgium
Reshef Meir	Harvard University, USA
Nina Narodytska	University of Toronto and University of New South Wales, Canada/Australia
Rolf Niedermeier	TU Berlin, Germany
Svetlana Obraztsova	NTUA, Greece
Sasa Pekec	Duke University, USA
Patrice Perny	LIP6
Marc Pirlot	Université de Mons, Belgium
Jörg Rothe	Universität Duesseldorf, Germany
Alexis Tsoukias	CNRS - LAMSADE, France
Paolo Viappiani	CNRS and LIP6, Université Pierre et Marie Curie
Toby Walsh	NICTA and UNSW, Australia
Mark Wilson	University of Auckland, New Zealand
Lirong Xia	RPI

Additional Reviewers

Airiau, Stéphane
Brandl, Florian
Bredereck, Robert
Chen, Jiehua
Darmann, Andreas
De Keijzer, Bart
Destercke, Sébastien
Dickerson, John P.
Doutre, Sylvie
Ehrgott, Matthias
Fitzsimmons, Zack
Geist, Christian

Hazon, Noam
Hemaspaandra, Edith
Lackner, Martin
Lee, Hooyeon
Lesca, Julien
Ma, Hongyao
Neugebauer, Daniel
Neumann, Frank
Nguyen, Nhan-Tam
Perotto, Filipo
Perrault, Andrew
Pfandler, Andreas

Rabinovich, Zinovi
Reger, Christian
Rey, Anja
Rey, Lisa
Ruzika, Stefan
Schadrack, Hilmar
Schend, Lena
Schnoor, Henning
Selker, Ann-Kathrin
Staněk, Rostislav
Talmon, Nimrod
Turrini, Paolo

Contents

Social Choice

Allocation and Other Problems

Doctoral Consortium

Preferences

Beyond Theory and Data in Preference Modeling: Bringing Humans into the Loop

Thomas E. Allen[1], Muye Chen[2], Judy Goldsmith[1], Nicholas Mattei[3(✉)], Anna Popova[4], Michel Regenwetter[2], Francesca Rossi[5], and Christopher Zwilling[2]

[1] University of Kentucky, Lexington, KY, USA
{teal223,goldsmit}@cs.uky.edu
[2] University of Illinois, Urbana–Champaign, IL, USA
{mchen67,regenwet,zwillin1}@illinois.edu
[3] NICTA and UNSW, Sydney, Australia
nicholas.mattei@nicta.com.au
[4] Dell Research Labs, Austin, TX, USA
anna_popova@dell.com
[5] University of Padova, Padova, Italy
frossi@math.unipd.it

Abstract. Many mathematical frameworks aim at modeling human preferences, employing a number of methods including utility functions, qualitative preference statements, constraint optimization, and logic formalisms. The choice of one model over another is usually based on the assumption that it can accurately describe the preferences of humans or other subjects/processes in the considered setting and is computationally tractable. Verification of these preference models often leverages some form of real life or domain specific data; demonstrating the models can predict the series of choices observed in the past. We argue that this is not enough: to evaluate a preference model, humans must be brought into the loop. Human experiments in controlled environments are needed to avoid common pitfalls associated with exclusively using prior data including introducing bias in the attempt to clean the data, mistaking correlation for causality, or testing data in a context that is different from the one where the data were produced. Human experiments need to be done carefully and we advocate a multi-disciplinary research environment that includes experimental psychologists and AI researchers. We argue that experiments should be used to validate models. We detail the design of an experiment in order to highlight some of the significant computational, conceptual, ethical, mathematical, psychological, and statistical hurdles to testing whether decision makers' preferences are consistent with a particular mathematical model of preferences.

1 Introduction

In the AI world of preference modeling, researchers often test their preference framework, particularly in the realm of recommendation systems and other decision support systems. However, most of the testing focuses on usability and functionality. Almost none that we are aware of looks at whether humans actually act

© Springer International Publishing Switzerland 2015
T. Walsh (Ed.): ADT 2015, LNAI 9346, pp. 3–18, 2015.
DOI: 10.1007/978-3-319-23114-3_1

the way a certain preference model states; i.e., test the underlying assumptions of the model itself. Interest in testing preference models proposed in computer science began, for us, when thinking about conditional preference networks (CP-nets) [6]. Although there are many hundreds of papers on CP-nets, none that we know of has looked at actually eliciting CP-nets from non-computer scientists, nor done choice-based tests to see if people act in a manner consistent with having an underlying CP-net preference structure. In this paper we describe both the process and the challenges that go into designing and implementing a human subjects experiment to test, for instance, the validity of CP-nets. We argue that human subjects experiments are an important opportunity for both interdisciplinary collaboration as well as extending the scope and impact of preference research in computer science.

Even within the work on preference elicitation, we have noticed a focus on optimization (see, e.g., [7]) to make the process fast and not too invasive for the user. While we celebrate the increasing libraries of preference data available, such as PrefLib [44], we also have concerns about the efficacy of using those data alone for validating preference models. In particular, we see many models validated on the Sushi Dataset [30], e.g. [25], which was generated for a very particular scenario and yet is now exploited for tests in fundamentally different settings. When we generalize or attempt to switch the domain of some data we introduce bias, which can potentially lead to spurious conclusions about the methods under study [53]. There are usability studies for preference elicitation software (e.g., [9,52]) and humans are being brought into the loop in recommender systems (e.g., [28,66]). These studies are crucial steps and the efforts should be rewarded and expanded within the broader communities that work with preferences. Running good tests with human subjects is necessary and nontrivial.

When we say that we advocate for studies with human subjects, by this we do not mean tests involving introspection. There is an urban legend in AI that the early work on chess involved asking chess players to introspect, and that this destroyed their intuitive processes. This likely refers to De Groots' work on chess:

> *"The only way of working with 'systematic introspection' would have been to interrupt the process after, say, every two minutes in order to have the subject introspect, and then continue. A few preliminary trials, however, with the author as subject showed this technique to be relatively ineffective as well as extraordinarily troublesome. After each interruption one feels disturbed and cannot continue normally. Apart from being unpleasant for the subject the technique is highly artificial in that it disrupts the unity of the thought process [16, pp. 80–81]."*

If we were to ask athletes to pay active attention to every body movement during peak performance, quite plausibly they would either disregard our instructions or fall short of peak performance due to a lack of focus. This is why athletes have coaches who monitor them. Likewise, asking decision makers to divert attention and memory resources away from their task in order to monitor their decision making introspectively likely interferes with the very process we are studying, making introspection an ineffective method for eliciting preferences or thought processes [50,51,70]. Indeed, without actively allocating cognitive resources to commit information to memory, there is no reason to expect that a decision maker can accurately recall the deliberations underlying his decision afterwards. This is why psychologists run laboratory experiments where human actions in a controlled environment are observed, rather than asking people how they think. They draw inferences about latent preferences from observable quantities such as choice proportions, buying or selling prices, reaction times, eye movements, all of which need not reveal one single consistent picture [39,71]. The challenge is to model the relationship between theoretical constructs (e.g., preferences) and observed data (e.g., choices) [59].

2 Preferences in Computer Science

Preference handling in artificial intelligence is a robust and well developed discipline with its own working groups and specialized workshops [23]. Often, much of the work takes the form of creating or defining models and then analyzing the computational complexity of various reasoning tasks within these models [8,19]. One such model that has gained prominence since its introduction in 2004 is the CP-net [6]. A CP-net is a formal model able to capture conditional preference statements (cp-statements) such as, "For dinner, if I have beef, I prefer fruit to ice cream for dessert, but if I have fish, I prefer ice cream to fruit for dessert" and "I prefer beef over fish for dinner."

Formally, a CP-net [6] consists of a directed graph $G = \langle V, E \rangle$, where the nodes V represent *variables* (sometimes called *features*) of an object, each with its own finite domain or set of *values*. For each variable V_i in the graph there is a possibly empty set of parent variables $Pa(V_i)$. For each variable, an ordinal preference relation over its values is specified by a collection of cp-statements, called a *conditional preference table (CPT)*. The assignment of values to $Pa(V_i)$ can affect the preference relation over V_i. For example, in Fig. 1 the variable PROTEIN can take the values *beef* or *fish* and *beef* is preferred. As PROTEIN has no parents, there is only one cp-statement. However, $Pa(\text{DESSERT}) = \text{PROTEIN}$ and therefore, depending on the assignment to PROTEIN either *fruit* is preferred to *ice cream* or vice versa.

A CP-net is a compact representation of the preference graph on *outcomes*. An outcome is a complete assignment of values to variables. The outcome graph $G_O = \langle V_O, E_O \rangle$ has nodes representing each possible set of values for the feature variables, and a directed edge between any two nodes that differ on exactly one feature value. The direction of the edge is determined by the preference over that feature, conditioned on the (otherwise fixed) values of its parent variables. The transitive closure of the preference graph gives the partial order over outcomes specified by the CP-net. A *sequence of worsening flips* is a directed path from an outcome o to o' through the outcome graph. This flipping sequence, if it exists, proves that outcome o is preferred to o'. We call this relation *dominance* and it is NP-hard to compute in the general case. Part of the difficulty of computing dominance arises because the outcome graph is exponentially larger than the CP-net graph, and improving flipping sequences can be exponentially long [6]. Tractable subproblems exist (such as when the graph has the shape of a directed tree), as well as computational heuristics for determining dominance [34]. Understanding the CP-net as a human decision making tool may help us formalize other cases where reasoning with CP-nets is tractable, such as when the preference graph has low degree or the dominance relation has short flipping sequences [2].

Preference representation and reasoning plays a key role in many other areas broadly included under the umbrella of Artificial Intelligence (AI). For example, within the area of *constraint reasoning* [62], the annual MAX-SAT solver competition[1] includes problem instances that encode both hard constraints and soft preferences for domains such as scheduling, time-tabling, and facility location. Both traditional *social choice* and *computational social choice* [12] are fields that actively work with choice data and are beginning the transition towards working with repurposed data explicitly [44,63]. However, these fields are still primarily focused on worst case assumptions, not behavioral or explanatory models of human decision making. Various forms of weighted logic representations such as *penalty logics* [17], *possibilistic logics* [20], and *answer set programs with soft constraints* [72] all explicitly rank states (assignments to all parameters) of the world. This ordered set of states is often interpreted as preferences over the states themselves. These fields primarily focus on algorithms for and quantifying the complexity of reasoning with choice data and preference models; testing these models and theories against human choice behavior is not a central focus.

The data focused fields of *machine learning* and *data mining* investigate preferences both implicitly and explicitly when working with, for example, large volumes of customer data [27]. This is perhaps most obvious in the sub-fields of *recommender systems* [61] and *preference learning* [22]. The objective in both of these areas is to learn and interpret observed choices (data) in order to make tangible recommendations or predictions, e.g. algorithms for the Netflix Prize Challenge [5] or Amazon product recommendations [35]. These areas are well developed preference handling fields where data are readily available from both academic sources [3] and as part of a number of industrial or commercial licenses or competitions (e.g. Kaggle, and the Yelp Academic Datasets). However, often these systems are only evaluated on their ability to minimize an error or loss function [5,61] when compared to held out choice data (i.e., data not in the training set). The explicit goal of these systems is not to understand the features of a user's internal preference reasoning or if the system itself can affect the user's explicit choice.

Humans are being brought into the loop in more and more areas of computer science, often leading to important and exciting impacts. More researchers are focusing on understanding *how* users reason internally and *why* users implement recommendations [18,56,66]. Trust building through explanation in recommendation systems is becoming standard practice due to its increased effectiveness

---

[1] http://maxsat.ia.udl.cat/introduction/.

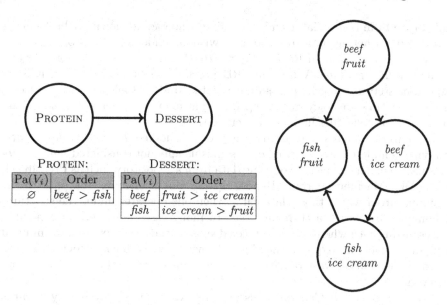

Fig. 1. A simple CP-net (left) and the induced preference graph over outcomes (right).

in leading to *implemented* recommendations [55]. Additional experiments with human subjects have also helped to validate activities like learning preferences through click tracking [28] and understanding the cognitive burden of asking certain preference queries [9]. The field of *computer-human interaction (CHI)* often performs studies of human behavior, validating models with laboratory experiment. Indeed, the most recent ACM Computer-Human Interaction conference (ACM:CHI 2014) provided courses on survey design and performing human studies [40,47]. However, in the broad set of communities that deal with preferences in AI, the human element is still often misunderstood.

The omission of human centered testing bypasses both a host of practical considerations and formal verification of preference models. These problems require controlled human subjects experiments and offer exciting opportunities for cross disciplinary research. There are over 800 references to the original CP-nets paper with not a single human subjects study to investigate whether a CP-net is a model of human choice, nor any testing of the model for consistency with respect to the way individuals reason about individual preference.

3 Legal Considerations in Human Subjects Research

Collecting data from human participants involves a panoply of challenges including important legal and ethical considerations that are sometimes poorly understood or considered by researchers. We have encountered colleagues in the US[2]

---

[2] Human-subjects standards vary from country to country and are also sometimes imposed by international academic societies.

and abroad from non-social science disciplines who had unknowingly broken laws by illegally gathering data from humans without undergoing appropriate prior review by an *Institutional Review Board (IRB)* and without undergoing legally required *ethics training*. While the IRB process can be cumbersome, it is an important step in using human subjects data. Modern tools such as Amazon's Mechanical Turk are a key resource [14,43] that many in preference handling are embracing for collecting human subjects data [41].

Data from humans may or may not be considered *human subjects data*. Studying completely anonymized data sets is usually not considered *human subjects research*; hence, the ease of using data from a repository such as PrefLib [44] or the UCI Machine Learning Repository [3]. But if one can link data to the individual from whom those data came, then one operates under the strictures of human subjects research regulations. In many cases, however, an expedited process is in place when an IRB officer deems a study exempt, due to minimal risk, and waives the requirement of a full review by the board. Reviews by the board will evaluate a vast range of considerations or requirements, of which we review a few.

INCENTIVIZATION: Generally, research in experimental psychology rewards participants in one of two major ways. Most experiments recruit undergraduate psychology students in exchange for course credit, this is commonly referred to as the "subject pool." Other experiments pay participants with cash or other rewards. Decision making experiments in this category often give some of the chosen options as real rewards in order to motivate participants to invest cognitive effort and reveal true preferences. Generally, behavioural and experimental economists disregard studies that do not link rewards to performance as being "insufficiently incentivized" [29]. There are also considerations of "over-incentivization" in that very large rewards can be blocked by some IRBs for being coercive. Another consideration is whether participants are allowed to receive payments, e.g., based on age, legal, or immigration status.

INFORMED CONSENT & DECEPTION: Since the infamous "Milgram experiments" [45] in which participants were led to believe that they were torturing others, ethical issues in human subjects research have been discussed in great detail. Many protections have been put in place to protect participants in psychology, economics, and medical experiments from being harmed. Scientists and lab personnel are required to undergo extensive training, e.g., the *Collaborative Institutional Training Initiative (CITI,* https://www.citiprogram.org/), before they may carry out research on humans. Ethical issues of informed consent emerged prominently in the mass media recently when it came to light that Facebook carried out social and emotional experiments on some of its users without clear-cut informed consent [68]. In behavioural and experimental economics, outright deception is often frowned upon [29].

CONFIDENTIALITY: It is straightforward that the protection of human subjects, besides avoiding immediate bodily or psychological harm, starts with proper confidentiality assurances. Considerations of what constitutes "anonymized" data is a growing concern in computer science and other disciplines. High-profile cases in recent years have shown that even a sequence of movie rental dates can be enough to discern personally identifiable information from a supposedly anonymized dataset [49]. Besides the obvious concerns about data trails from scheduling participants, time-stamped electronic data collection, and accounting records of payments, the use of cloud-based tools, such as storage or email, where servers may reside outside the country, or with commercial providers, threatens confidentiality.

4   Perspectives from Mathematical Psychology

Let C be a finite set of choice alternatives, and let $\succ$ denote pairwise preference, i.e., $x \succ y$ with $x, y \in C$ denotes that a person strictly prefers x to y. Many models of preferences, including CP-nets, require $\succ$ to be transitive.

How would one test whether decision makers' preferences are transitive? Psychologists differentiate between *theoretical constructs* and *observables*. A binary preference relation, a real valued utility function, a CP-net, are theoretical constructs that we cannot observe directly, just as a physicist cannot observe gravity itself. In decision making, actual choices made by actual people are observables that are presumably related to the latent construct, just as an apple falling is an observable manifestation of gravity.

A major conceptual, mathematical, and experimental challenge for testing theories about preferences comes from the fact that decision makers experience uncertainty in what to choose when faced with multi-attribute options in which attributes trade off in complex ways. Experimentally, we observe substantial amounts of variability between people and even within a single person over repeated choices among the same options. It is not uncommon for a decision maker to choose x over y on 70 % of occasions, and y over x otherwise, even within a one-hour study. This led economists and psychologists to mathematically model uncertainty and variability in choice. Arguably, the most natural way to model uncertainty in choice is via probabilistic models [4,10,36–38,67].

There are two major classes of probabilistic choice models. One assumes that the theoretical construct of preference is deterministic but choices are probabilistic, the other assumes that the theoretical construct itself is probabilistic. For transitivity, the first model type assumes that each decision maker has one fixed deterministic preference $\succ$ over the course of the experiment, whereas the latter model casts preferences as a probability distribution over a set of transitive preferences. For CP-nets the analogue is to distinguish two major possibilities:

1. the decision maker uses one single fixed CP-net, but makes probabilistic *errors* in revealing this CP-net in overt choices;
2. the choice probabilities are *induced* by an unknown probability distribution over a collection of CP-nets.

An *error* specification may assume that the decision maker has unknown preference $\succ$, and if $x \succ y$ then she is more likely to pick x than y, formally and more precisely,

$$P_{xy} > \tau, \text{ with a bound on error rates of } 1 - \tau \leq \frac{1}{2}. \tag{1}$$

A *random preference specification* considers a (finite) collection $\mathcal{R}$ of permissible preference relations (e.g., transitive relations, CP-nets, etc.), a probability distribution $\mathbb{P}$ on $\mathcal{R}$, and models the binary choice probability P_{xy} as a marginal probability

$$P_{xy} = \sum_{\succ \in \mathcal{R}} \mathbb{P}(x \succ y). \tag{2}$$

Characterizing the binary choice probabilities that are consistent with a *random preference specification* (2) can be mathematically and computationally prohibitive. In the case that $\mathcal{R}$ is the collection of all strict linear orders over a finite set $\mathcal{C}$, the binary choice probabilities (2) form a convex polytope known as the *linear ordering polytope* [21,24,33]. The mathematical structure of this polytope is known only for small sizes of $\mathcal{C}$ and finding a complete minimal description in terms of *facet-defining inequalities* is computationally hard [42]. Building a random preference model in which the collection of permissible preferences $\mathcal{R}$ is composed of CP-nets would require that we understand the permissible binary choice probabilities (2). The currently standard approach would be to employ methods from polyhedral combinatorics, by defining and studying appropriate *CP-net-polytopes*, in which probability distributions over CP-nets are conceptualized as convex combinations of deterministic CP-nets.

We have sketched the conceptual and mathematical challenge of defining uncertain choices induced by theoretical preferences that form CP-nets, using probabilities. The next challenge is that those probabilities $\{P_{xy} \mid x \neq y; x, y \in \mathcal{C}\}$, in turn, are theoretical constructs. If we are to study CP-nets in the laboratory and if we are to allow different decision makers to use CP-nets differently, then we need to draw inferences about probabilities from finite samples using appropriate statistical tools. Both the error models (1) and the random preference models (2) impose multiple simultaneous order-constraints on the parameters of joint Bernoulli processes. This causes serious challenges in maximum-likelihood methods because point estimates may lie on the boundary of the parameter space (e.g., on a face of a convex polytope) where standard likelihood theory breaks down. Frequentist and Bayesian order-constrained likelihood-based inference methods have only become available recently [15,32,48,59]. Some of the algorithms, e.g., for computing Bayes Factors between two competing convex polytopes, are computationally expensive, with current researchers sometimes using thousands of CPU-hours per Bayes Factor.[3]

The task, then, for a quantitative test of CP-nets in individual decision makers, includes: the development of "probabilistic specifications" that represent the uncertainty experienced by the decision maker, the adoption of suitable statistical tools, and the design and implementation of an experiment that generates

---

[3] For an example of the complexities involved in testing transitivity of preferences, including a critical review of the prior literature, see, e.g. [11,57,58,60].

data suitable for either testing the mathematical model as a hypothesis or for selecting between the model for CP-nets and alternative theoretical proposals. Both the mathematical characterization and the statistical inference involve significant mathematical and computational challenges.

5   Other Considerations for Laboratory Experiments

In defining a laboratory experiment on human decision making, attention must also be give to the following issues which, while not legal or ethical in nature, can affect the design and implementation of an experiment.

DATA BIAS: Statistical inferences from finite sample data generally require repeated observations either from multiple people or from a given participant. In order to eliminate potential biases and the effect of irrelevant variables, a decision-making experiment asking participants to decide among choice options can implement a variety of "cross-balancing" precautions. These include, e.g., showing a given choice option randomly in different locations on a display to compensate for attentional biases and making different stimuli "equally complex" to balance cognitive load. Statistical tests and analyses often assume independent and identically distributed observations. These assumptions affect the experimental design itself, e.g., separating repeated observations through decoys to attenuate violations of independence.

CORRELATION VS. CAUSALITY: This has important implications for selecting experimental methods over data mining or other approaches. If one wishes to make causal attributions that values in one variable "cause" outcomes in another variable, one needs to use random assignment to experimental conditions (e.g., placebo versus treatment).

FALSIFIABILITY, DIAGNOSTICITY, AND PARSIMONY: According to these principles, theoretical predictions motivate what stimuli to use and hence precede data collection. Epistemologically, restrictive theories are favored because they lead to falsifiable predictions [54]. There are at least three major ways in which behavioral scientists use statistical inference.

1. Many scholars support a theoretical claim by statistically rejecting a *null hypothesis* of "no effect," a practice that has come under intense criticism [13,46,69].
2. Others, similar to data mining methods, formulate mathematical models and use statistics to estimate parameters through data fitting, then interpret the inferred parameter values in terms of scientific primitives. Oftentimes the validity or replicability of the findings are assessed through goodness-of-fit on hold-out samples or through predictions about future data.
3. More and more behavioral scientists use Bayesian methods to carry out competitions among theories that vary in their parsimony, by weighing prior beliefs with empirical evidence, and penalizing flexible models [31,48].

Several disciplines within social science are currently engaged in a major debate about replicability,[4] publication bias, and scientific integrity [26,64,65].

---

[4] See, e.g., http://psychfiledrawer.org/.

Most social science journals only consider novel findings for publication, leading some researchers to draw scientific conclusions from very slight statistical effects, and several high-profile scholars have been accused and/or found guilty of faking their data. The practical consequence of the recent debate is that researchers must take care to ensure that their models and experiments stem from rigorous theories, which make precise predictions that can be tested in a laboratory setting through the use of appropriately applied statistics.

6   Case Study: The CP-net Experiment

We have recently completed data collection on an experiment to test whether decision makers subjectively represent preferences in a way that is consistent with a mathematical CP-net representation. We have incorporated the considerations above with many additional practical and logistic constraints.

A good rule of thumb for running a first experiment in a given domain is to start simple. Since there is no prior empirical work on actual CP-nets of actual people, we needed to design the study without having to hypothesize too many details about CP-nets that are suitable for the domain under consideration. Otherwise, were we to conclude that "our" CP-net is not descriptive of our participants, we would not learn much about the general descriptive validity of CP-nets. If we allow *all* CP-nets on a given set of choice options as potential preference states, then we need to limit the number of CP-nets that are possible. We do not want to make the CP-nets trivial but we also cannot make them overly complex as it will lead to intractable experiments. Therefore we limit ourselves to acyclic dependency graphs with four binary nodes/variables. This means that our CP-nets have 16 choice alternatives. This permits a rich set of preference states, exactly 481,776 distinct CP-nets (computed as all possible non-degenerate boolean functions on $n = 4$ binary variables [1]).

The next major set of considerations is to decide on actual stimuli that are both interesting and that may tell us something about everyday decision making, at least at face value. Furthermore, at least some of the stimuli need to be 'deliverable' as real prizes while the other ones need to be 'cross-balanced.' We therefore selected two domains, restaurant menu choices (since they are common hypothetical illustrations in the CP-net literature) and choices among retail goods or services. For example, for the restaurant menu options we chose "appetizer" versus "dessert" as one attribute, "chicken" versus "shrimp" as another attribute, etc.

Since we incentivized our participants by offering them some of their choices as real rewards, team members spent significant time contacting retail and restaurant managers to find ways to purchase rewards through university purchase orders and to ensure that a person will be given the exact reward we specify (as opposed to being able to use, say, a gift certificate in a fungible way). Likewise, multiple team members agonized over finding a sufficiently rich set of 'comparable' stimuli, even for trials that are not used to determine real rewards. For example, all stimuli need to be credible as potential rewards of a comparable value and payable by a federal grant. Over two domains with 60 participants the experiment distributed rewards of $4992.00 USD. The distributed rewards consisted of $2400.00 of prescribed meals at a local restaurant, $220.00 of video rentals, and $2372.00 of merchandise at the university bookstore.

There are also major tradeoffs between practical, logistical and statistical prerogatives: Ultimately, participants need to make sufficiently many choices among sufficiently many options to allow statistical estimation, hypothesis testing, or model selection. We decided to make each "trial" of the experiment a *ternary paired comparison*, i.e., two meals are presented and the decision maker can either express a preference for one, the other, or express "no preference." Statistically, this means that each "trial" provides an observation for a trinomial random variable. In order to obtain repeated observations, we needed to show each pair several times, at the risk of making the experiment laborious and repetitive. Hence, we substituted different "instantiations" of a given "choice" on different trials, "chicken" on one trial could be "Chicken Marsala" and on another trial could be "Chicken della Nonna." However, this means that we may have introducing many unintended variables that we are not modeled in the CP-net. Therefore, when showing a participant two meals that share the value "chicken" for the variable "main dish," we showed them the identical chicken dish in both options, so as to make it impossible for them to have a pairwise preference on a variable that we model as having identical values in both options.

There is a tradeoff between the number of times we ask a user to make a decision and the statistical tests we can then employ to perform reliable statistical analysis. Since we are asking users to choose between two meals, with 16 total meals, that gives 120 trials or pairwise comparisons that we must elicit from each user, and each of these trials must be repeated. Some of our models are convex (polytopes), in which case we can pool data across subjects even if there are individual differences between them. Some of our (error) models are not convex and should best be evaluated separately for each individual. Advanced statistical methods that do not require asymptotic statistics can get by with fewer than 10 trials per user per question.

The tasks involved in preparing for this experiment include: computing a list of all 4-variable CP-nets; developing the initial set of variables for each domain; negotiating agreements with the Institutional Review Board for human-subject research; getting agreement from vendors to provide specified rewards (and dealing with the video rental business going out of business before some of the long term rewards could be redeemed); creating multiple equivalent wordings of the same reward (e.g., 6 T-shirts in one trial and a half dozen short sleeved shirts in another); developing and testing the GUIs and interface functionality on iPads for the experiment. These tasks took about 350–400 person hours. As members of the team are extremely experienced with navigating the bureaucracy of IRB approval and negotiating non-fungible rewards with outside vendors, this number likely underestimates the time required for a first time experimenter.

For data collection, we consider both the participants' time and the cost of running the study: 2 sessions for each of the 60 participants and about 90 min per session. There are 5 iPads, but scheduling is complicated, so we ran about

50 experimental sessions to get the data from those 60 participants. The person overseeing each experimental session spent about an hour for each experimental session distributing and collecting the informed consent paperwork, making sure the app was running on the iPads and ready to use, making sure the iPads were charged, introducing people to the study, answering questions, explaining the payment, scheduling their second session, making sure the results were uploaded, making sure each participant's payment was provided confidentially in a separate room, making sure the post-test questionnaire was filled out, etc. Thus, the number of person-hours for running the experiment (about 180 person-hours of subjects' time, plus about 100 hours of experimenters' time) was slightly less than the time spent preparing the experiment itself.

7   Conclusion

In this paper we have highlighted some of the key pitfalls and challenges associated with human subjects experiments within preference model testing. Ideally, experimentalists in AI can use this as both a call to action and as a starting point for conducting their own experiments both in human subjects labs and leveraging the power of online tools such as Mechanical Turk [41,43]. We have only skimmed the surface of the relevant literatures in computer science and psychology. There is a vast literature on experimental studies in other fields including decision sciences, experimental economics, medical, and other cognitive studies areas. We hope this article serves as a jumping off point into the literature.

Data are available in large quantities, but we should resist the temptation to rely on past data alone when testing a preference modeling framework. Human experimentation should be part of the testing process. However, in doing this, we need to pay attention to several conceptual, mathematical, statistical, computational, legal and ethical considerations, as well as tackle many practical and logistic complications. We believe that AI and psychology researchers should work together in this endeavor. For AI researchers, understanding the functions and limitations of human decision making can lead to the development of more accurate models and heuristics in the multitude of areas that engage with humans and preferences. For psychologists, understanding the computational burden of reasoning with various preference models can inform new experiments and processes.

We have just completed data collection at the time of acceptance of this manuscript, after clearing all the significant development and logistical hurdles we have outlined in this paper. Proper analysis of this data will take months; discrepancies in the publication culture of computer science and psychology means we must target psychology journal submissions first (as data must be novel for publication). This will give us the first real experiment which will contemplate the question of whether or not subjects' preferences over two domains (retail and food) are at least noisily consistent with CP-net models and whether or not, given adequate instruction, the subjects can write these preferences down in a way that is consistent with their previous choices.

References

1. Allen, T.E., Goldsmith, J., Mattei, N.: Counting, ranking, and randomly generating CP-nets. In: 8th Workshop on Advances in Preference Handling (MPREF 2014), AAAI-14 Workshop Series (2014)
2. Allen, T.E.: CP-nets with indifference. In: 2013 51st Annual Allerton Conference on Communication, Control, and Computing (Allerton), pp. 1488–1495. IEEE (2013)
3. Bache, K., Lichman, M.: UCI machine learning repository, School of Information and Computer Sciences, University of California, Irvine (2013). http://archive.ics.uci.edu/ml
4. Becker, G., DeGroot, M., Marschak, J.: Stochastic models of choice behavior. Behav. Sci. **8**, 41–55 (1963)
5. Bennett, J., Lanning, S.: The Netflix prize. In: Proceedings of the KDD Cup and Workshop (2007)
6. Boutilier, C., Brafman, R., Domshlak, C., Hoos, H., Poole, D.: CP-nets: A tool for representing and reasoning with conditional ceteris paribus preference statements. J. Artif. Intell. Res. **21**, 135–191 (2004)
7. Boutilier, C., Patrascu, R., Poupart, P., Schuurmans, D.: Constraint-based optimization and utility elicitation using the minimax decision criterion. Artif. Intell. **170**(8), 686–713 (2006)
8. Brafman, R.I., Domshlak, C.: Preference handling–an introductory tutorial. AI Mag. **30**(1), 58 (2009)
9. Braziunas, D., Boutilier, C.: Assessing regret-based preference elicitation with the utpref recommendation system. In: Proceedings of the 11th ACM Conference on Electronic Commerce (EC), pp. 219–228. ACM (2010)
10. Carbone, E., Hey, J.: Which error story is best? J. Risk Uncertain. **20**, 161–176 (2000)
11. Cavagnaro, D., Davis-Stober, C.: Transitive in our preferences, but transitive in different ways: an analysis of choice variability. Decision **1**(1), 102–122 (2014)
12. Chevaleyre, Y., Endriss, U., Lang, J., Maudet, N.: Preference handling in combinatorial domains: from AI to social choice. AI Mag. **29**(4), 37–46 (2008)
13. Cohen, J.: The earth is round ($p < .05$). Am. Psychol. **49**(12), 997 (1994)
14. Crump, M.J.C., McDonnell, J.V., Gureckis, T.M.: Evaluating Amazon's mechanical Turk as a tool for experimental behavioral research. PloS one **8**(3), e57410 (2013)
15. Davis-Stober, C.: Analysis of multinomial models under inequality constraints: applications to measurement theory. J. Math. Psychol. **53**, 1–13 (2009)
16. De Groot, A.D.: Thought and Choice in Chess (Psychological Studies), 2nd edn. Mouton de Gruyter, The Hague (1978)
17. De Saint-Cyr, F.D., Lang, J., Schiex, T.: Penalty logic and its link with Dempster-Shafer theory. In: Proceedings of the UAI, pp. 204–211 (1994)
18. Dodson, T., Mattei, N., Guerin, J., Goldsmith, J.: An English-language argumentation interface for explanation generation with Markov decision processes in the domain of academic advising. ACM Trans. Interact. Intell. Syst. (TiiS) **3**(3), 18 (2013)
19. Domshlak, C., Hüllermeier, E., Kaci, S., Prade, H.: Preferences in AI: an overview. Artif. Intell. **175**(7), 1037–1052 (2011)
20. Dubois, D., Lang, J., Prade, H.: A brief overview of possibilistic logic. In: Kruse, R., Siegel, P. (eds.) Symbolic and Quantitative Approaches to Uncertainty. LNCS, vol. 548, pp. 53–57. Springer, Heidelberg (1991)

21. Fiorini, S.: Determining the automorphism group of the linear ordering polytope. Discret. Appl. Math. **112**, 121–128 (2001)
22. Fürnkranz, J., Hüllermeier, E.: Preference Learning. Springer, New York (2010)
23. Goldsmith, J., Junker, U.: Preference handling for artificial intelligence. AI Mag. **29**(4), 9 (2009)
24. Grötschel, M., Jünger, M., Reinelt, G.: Facets of the linear ordering polytope. Math. Program. **33**, 43–60 (1985)
25. Guo, S., Sanner, S., Bonilla, E.V.: Gaussian process preference elicitation. In: Advances in Neural Information Processing Systems, pp. 262–270 (2010)
26. Ioannidis, J.: Why most published research findings are false. PLoS Med. **2**(8), e124 (2005)
27. Jawaheer, G., Weller, P., Kostkova, P.: Modeling user preferences in recommender systems: a classification framework for explicit and implicit user feedback. ACM Trans. Interact. Intell. Syst. (TiiS) **4**(2), 8 (2014). http://doi.acm.org/10.1145/2512208
28. Joachims, T., Granka, L., Pan, B., Hembrooke, H., Radlinski, F., Gay, G.: Evaluating the accuracy of implicit feedback from clicks and query reformulations in web search. ACM Trans. Inf. Syst. (TOIS) **25**(2), 7 (2007)
29. Kagel, J., Roth, A.: The Handbook of Experimental Economics. Princeton University Press, Princeton (1995)
30. Kamishima, T.: Nantonac collaborative filtering: recommendation based on order responses. In: The 9th International Conference on Knowledge Discovery and Data Mining (KDD), pp. 583–588 (2003)
31. Kass, R., Raftery, A.: Bayes factors. J. Am. Stat. Assoc. **90**(430), 773–795 (1995)
32. Klugkist, I., Hoijtink, H.: The Bayes factor for inequality and about equality constrained models. Comput. Stat. Data Anal. **51**, 6367–6379 (2007)
33. Koppen, M.: Random utility representation of binary choice probabilities: critical graphs yielding critical necessary conditions. J. Math. Psychol. **39**, 21–39 (1995)
34. Li, M., Vo, Q.B., Kowalczyk, R.: Efficient heuristic approach to dominance testing in CP-nets. In: Proceedings of the AAMAS, pp. 353–360, Richland, SC, USA (2011)
35. Linden, G., Smith, B., York, J.: Amazon.com recommendations: item-to-item collaborative filtering. IEEE Internet Comput. **7**(1), 76–80 (2003)
36. Loomes, G., Sugden, R.: Testing different stochastic specifications of risky choice. Economica **65**, 581–598 (1998)
37. Luce, R.: Individual Choice Behavior: A Theoretical Analysis. Wiley, New York (1959)
38. Luce, R.: Four tensions concerning mathematical modeling in psychology. Annu. Rev. Psychol. **46**, 1–26 (1995)
39. Luce, R.: Joint receipt and certainty equivalents of gambles. J. Math. Psychol. **39**, 73–81 (1995)
40. MacKenzie, I.S., Castellucci, S.J.: Empirical research methods for human-computer interaction. In: Proceedings of the CHI, pp. 1013–1014 (2014)
41. Mao, A., Procaccia, A.D., Chen, Y.: Better human computation through principled voting. In: Proceedings of the 27th AAAI Conference on Artificial Intelligence (AAAI) (2013)
42. Martí, R., Reinelt, G.: The Linear Ordering Problem: Exact and Heuristic Methods in Combinatorial Optimization. Applied Mathematical Science, vol. 175. Springer, Heidelberg (2011)
43. Mason, W., Suri, S.: Conducting behavioral research on Amazon's Mechanical Turk. Behav. Res. Methods **44**(1), 1–23 (2012)

44. Mattei, N., Walsh, T.: Preflib: a library for preferences http://www.preflib.org. In: Perny, P., Pirlot, M., Tsoukiàs, A. (eds.) ADT 2013. LNCS, vol. 8176, pp. 259–270. Springer, Heidelberg (2013)
45. Milgram, S.: Behavioral study of obedience. J. Abnorm. Soc. Psychol. **67**(4), 371 (1963)
46. Morey, R., Rouder, J., Verhagen, J., Wagenmakers, E.: Why hypothesis tests are essential for psychological science: a comment on Cumming. Psychol. Sci. **25**(6), 1289–1290 (2014)
47. Müller, H., Sedley, A., Ferrall-Nunge, E.: Designing unbiased surveys for HCI research. In: Proceedings of the CHI, pp. 1027–1028 (2014)
48. Myung, J., Karabatsos, G., Iverson, G.: A Bayesian approach to testing decision making axioms. J. Math. Psychol. **49**, 205–225 (2005)
49. Narayanan, A., Shmatikov, V.: Robust de-anonymization of large sparse datasets. In: IEEE Symposium on Security and Privacy, pp. 111–125 (2008)
50. Nisbett, R.E., Wilson, T.D.: Telling more than we can know: verbal reports on mental processes. Psychol. Rev. **84**(3), 231 (1977)
51. Nordgren, L.F., Dijksterhuis, A.: The devil is in the deliberation: thinking too much reduces preference consistency. J. Consum. Res. **36**(1), 39–46 (2009)
52. Pommeranz, A., Broekens, J., Wiggers, P., Brinkman, W.P., Jonker, C.M.: Designing interfaces for explicit preference elicitation: a user-centered investigation of preference representation and elicitation process. User Model. User-Adap. Inter. **22**(4–5), 357–397 (2012)
53. Popova, A., Regenwetter, M., Mattei, N.: A behavioral perspective on social choice. Ann. Math. Artif. Intell. **68**(1–3), 5–30 (2013)
54. Popper, K.: The Logic of Scientific Discovery. Hutchinson, London (1959)
55. Pu, P., Chen, L.: Trust building with explanation interfaces. In: Proceedings of the 11th International Conference on Intelligent User Interfaces (IUI), pp. 93–100 (2006)
56. Pu, P., Chen, L., Hu, R.: Evaluating recommender systems from the user's perspective: survey of the state of the art. User Model. User-Adap. Inter. **22**(4–5), 317–355 (2012)
57. Regenwetter, M., Dana, J., Davis-Stober, C.P.: Transitivity of preferences. Psychol. Rev. **118**, 42–56 (2011)
58. Regenwetter, M., Dana, J., Davis-Stober, C.P.: Testing transitivity of preferences on two-alternative forced choice data. Front. Psychology **1**, 148 (2010). doi:10. 3389/fpsyg.2010.00148
59. Regenwetter, M., Davis-Stober, C.P., Lim, S.H., Guo, Y., Popova, A., Zwilling, C., Cha, Y.C., Messner, W.: QTest: quantitative testing of theories of binary choice. Decision **1**(1), 2–34 (2014)
60. Regenwetter, M., Davis-Stober, C.: Behavioral variability of choices versus structural inconsistency of preferences. Psychol. Rev. **119**(2), 408–416 (2012)
61. Ricci, F., Rokach, L., Shapira, B., Kantor, P.B. (eds.): Recommender Systems Handbook. Springer, New York (2011)
62. Rossi, F., Beek, P.V., Walsh, T.: Handbook of Constraint Programming. Elsevier, Amsterdam (2006)
63. Rossi, F., Venable, K.B., Walsh, T.: A short introduction to preferences: between artificial intelligence and social choice. Synth. Lect. Artif. Intell. Mach. Learn. **5**(4), 1–102 (2011)
64. Schooler, J.: Unpublished results hide the decline effect. Nature **470**(7335), 437 (2011)

65. Simmons, J., Nelson, L., Simonsohn, U.: False-positive psychology: undisclosed flexibility in data collection and analysis allows presenting anything as significant. Psychol. Sci. **22**(11), 1359–1366 (2011)
66. Tintarev, N., Masthoff, J.: Designing and evaluating explanations for recommender systems. In: Ricci, F., Rokach, L., Shapira, B., Kantor, P.B. (eds.) Recommender Systems Handbook, pp. 479–510. Springer, London (2011)
67. Tversky, A.: Intransitivity of preferences. Psychol. Rev. **76**, 31–48 (1969)
68. Waldman, K.: Facebook's unethical experiment. Slate (2014). http://www. slate.com/articles/health_and_science/science/2014/06/facebook_unethical_ experiment_it_made_news_feeds_happier_or_sadder_to_manipulate.html
69. Wetzels, R., Matzke, D., Lee, M., Rouder, J., Iverson, G., Wagenmakers, E.: Statistical evidence in experimental psychology an empirical comparison using 855 t tests. Perspect. Psychol. Sci. **6**(3), 291–298 (2011)
70. Wilson, T.D., Schooler, J.W.: Thinking too much: introspection can reduce the quality of preferences and decisions. J. Pers. Soc. Psychol. **60**(2), 181 (1991)
71. von Winterfeldt, D., Chung, N.K., Luce, R., Cho, Y.: Tests of consequence monotonicity in decision making under uncertainty. J. Exper. Psychol. Learn. Mem. Cogn. **23**, 406–426 (1997)
72. Zhu, Y., Truszczynski, M.: On optimal solutions of answer set optimization problems. In: Cabalar, P., Son, T.C. (eds.) LPNMR 2013. LNCS, vol. 8148, pp. 556–568. Springer, Heidelberg (2013)

Reasoning with Preference Trees over Combinatorial Domains

Xudong Liu$^{(\boxtimes)}$ and Miroslaw Truszczynski

Department of Computer Science, University of Kentucky,
Lexington, KY, USA
{liu,mirek}@cs.uky.edu

Abstract. Preference trees, or *P-trees* for short, offer an intuitive and often concise way of representing preferences over combinatorial domains. In this paper, we propose an alternative definition of P-trees, and formally introduce their compact representation that exploits occurrences of identical subtrees. We show that P-trees generalize lexicographic preference trees and are strictly more expressive. We relate P-trees to answer-set optimization programs and possibilistic logic theories. Finally, we study reasoning with P-trees and establish computational complexity results for the key reasoning tasks of comparing outcomes with respect to orders defined by P-trees, and of finding optimal outcomes.

1   Introduction

Preferences are essential in areas such as constraint satisfaction, decision making, multi-agent cooperation, Internet trading, and social choice. Consequently, preference representation languages and algorithms for reasoning about preferences have received substantial attention [8]. When there are only a few objects (or *outcomes*) to compare, it is both most direct and feasible to represent preference orders by their explicit enumerations. The situation changes when the domain of interest is *combinatorial*, that is, its elements are described in terms of combinations of values of *issues*, say $x_1, \ldots, x_n$ (also called *variables* or *attributes*), with each issue x_i assuming values from some set D_i — its *domain.*

Combinatorial domains appear commonly in applications. Since their size is exponential in the number of issues, they are often so large as to make explicit representations of preference orders impractical. Therefore, designing languages to represent preferences on elements from combinatorial domains in a concise and intuitive fashion is important. Several such languages have been proposed including penalty and possibilistic logics [4], conditional preference networks (CP-nets) [2], lexicographic preference trees (LP-trees) [1], and answer-set optimization (ASO) programs [3].

In this paper, we focus our study on combinatorial domains with *binary* issues. We assume that each issue x has the domain $\{x, \neg x\}$ (we slightly abuse the notation here, overloading x to stand both for an issue and for one of the elements of its domain). Thus, outcomes in the combinatorial domain determined by the set $\mathcal{I} = \{x_1, \ldots, x_n\}$ of binary issues are simply complete and consistent

© Springer International Publishing Switzerland 2015
T. Walsh (Ed.): ADT 2015, LNAI 9346, pp. 19–34, 2015.
DOI: 10.1007/978-3-319-23114-3_2

sets of literals over $\mathcal{I}$. We denote the set of all such sets of literals by $CD(\mathcal{I})$. We typically view them as truth assignments (interpretations) of the propositional language over the vocabulary $\mathcal{I}$. This allows us to use propositional formulas over $\mathcal{I}$ as concise representations of sets of outcomes over $\mathcal{I}$. Namely, each formula φ represents the set of outcomes that satisfy φ (make φ true).

For example, let us consider preferences on possible ways to arrange a vacation. We assume that vacations are described by four binary variables:

1. *activity* (x_1) with values *water sports* (x_1) and *hiking* ($\neg x_1$),
2. *destination* (x_2) with *Florida* (x_2) and *Colorado* ($\neg x_2$),
3. *time* (x_3) with *summer* (x_3) and *winter* ($\neg x_3$), and
4. the mode of *travel* (x_4) could be *car* (x_4) and *plane* ($\neg x_4$).

A complete and consistent set of literals $\neg x_1 \neg x_2 x_3 x_4$ represents the hiking vacation in Colorado in the summer to which we travel by car.

To describe sets of vacations we can use formulas. For instance, vacations that take place in the summer (x_3) or involve water sports (x_1) can be described by the formula $x_3 \vee x_1$, and vacations in Colorado ($\neg x_2$) that we travel to by car (x_4) by the formula $\neg x_2 \wedge x_4$.

Explicitly specifying strict preference orders on $CD(\mathcal{I})$ becomes impractical even for combinatorial domains with as few as 7 or 8 issues. However, the setting introduced above allows us to specify total preorders on outcomes in terms of desirable properties outcomes should have. For instance, a formula φ might be interpreted as a definition of a total preorder in which outcomes satisfying φ are preferred to those that do not satisfy φ (and outcomes within each of these two groups are equivalent). More generally, we could see an expression (a sequence of formulas)

$$\varphi_1 > \varphi_2 > \ldots > \varphi_k$$

as a definition of a total preorder in which outcomes satisfying φ_1 are preferred to all others, among which outcomes satisfying φ_2 are preferred to all others, etc., and where outcomes not satisfying any of the formulas φ_i are least preferred. This way of specifying preferences is used (with minor modifications) in possibilistic logic [4] and ASO programs [3]. In our example, the expression

$$x_3 \wedge x_4 > \neg x_3 \wedge \neg x_2$$

states that we prefer summer vacations (x_3) where we drive by car (x_4) to vacations in winter ($\neg x_3$) in Colorado ($\neg x_2$), with all other vacations being the least preferred.

This linear specification of preferred formulas is sometimes too restrictive. An agent might prefer outcomes that satisfy a property φ to those that do not. Within the first group that agent might prefer outcomes satisfying a property ψ_1 and within the other a property ψ_2. Such *conditional* preference can be naturally captured by a form of a decision tree presented in Fig. 1. Leaves, shown as boxes, represent sets of outcomes satisfying the corresponding conjunctions of formulas ($\varphi \wedge \psi_1$, $\varphi \wedge \neg \psi_1$, etc.).

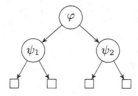

Fig. 1. A preference tree

Trees such as the one in Fig. 1 are called *preference trees*, or *P-trees*. They were introduced by Fraser [5,6], who saw them as a convenient way to represent conditional preferences. Despite their intuitive nature they have not attracted much interest in the preference research in AI. In particular, they were not studied for their relationship to other preference formalisms. Further, the issue of compact representations received only an informal treatment by Fraser (P-trees in their full representation are often impractically large), and the algorithmic issues of reasoning with P-trees were also only touched upon.

In this paper, we propose an alternative definition of preference trees, and formally define their compact representation that exploits occurrences of identical subtrees. P-trees are reminiscent of LP-trees [1]. We discuss the relation between the two concepts and show that P-trees offer a much more general, flexible and expressive way of representing preferences. We also discuss the relationship between P-trees and ASO preferences, and between P-trees and possibilistic logic theories. We study the complexity of the problems of comparing outcomes with respect to orders defined by preference trees, and of problems of finding optimal outcomes.

Our paper is organized as follows. In the next section, we formally define P-trees and a compact way to represent them. In the following section we present results comparing the language of P-trees with other preference formalisms. We then move on to study the complexity of the key reasoning tasks for preferences captured by P-trees and, finally, conclude by outlining some future research directions.

2   Preference Trees

In this section, we define preference trees and discuss their representation. Let $\mathcal{I}$ be a set of binary issues. A *preference tree* (*P-tree*, for short) over $\mathcal{I}$ is a binary tree with all nodes other than leaves labeled with propositional formulas over $\mathcal{I}$. Each P-tree T defines a natural strict order $\succ_T$ on the set of its leaves, the order of their enumeration from left to right.

Given an outcome $M \in CD(\mathcal{I})$, we define the *leaf of M in T* as the leaf reached by starting at the root of T and proceeding downwards. When at a node t labeled with φ, if $M \models \varphi$, we descend to the left child of t; otherwise, we descend to the right child of t. We denote the leaf of M in T by $l_T(M)$.

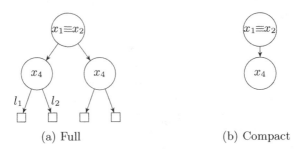

Fig. 2. P-trees on vacations

We use the concept of the leaf of an outcome M in a P-tree T to define a total preorder on $CD(\mathcal{I})$. Namely, for outcomes $M, M' \in CD(\mathcal{I})$, we set $M \succeq_T M'$ (M is *preferred* to M'), if $l_T(M) \succeq_T l_T(M')$, and $M \succ_T M'$ (M is *strictly preferred* to M'), if $l_T(M) \succ_T l_T(M')$.[1] We say that M is *equivalent* to M', $M \approx_T M'$, if $l_T(M) = l_T(M')$. Finally, M is *optimal* if there exists no M' such that $M' \succ_T M$.

Let us come back to the vacation example and assume that an agent prefers vacations involving water sports in Florida or hiking in Colorado over the other options. This preference is described by the formula $(x_1 \wedge x_2) \vee (\neg x_1 \wedge \neg x_2)$ or, more concisely, as an equivalence $x_1 \equiv x_2$. Within each of the two groups of vacations (satisfying the formula and not satisfying the formula), driving (x_4) is the preferred transporting mode. These preferences can be captured by the P-tree in Fig. 2a. We note that in this example, the preferences at the second level are *unconditional*, that is, they do not depend on preferences at the top level.

To compare two outcomes, $M = \neg x_1 \neg x_2 \neg x_3 x_4$ and $M' = x_1 x_2 x_3 \neg x_4$, we walk down the tree and find that $l_T(M) = l_1$ and $l_T(M') = l_2$. Thus, we have $M \succ_T M'$ since l_1 precedes l_2.

The key property of P-trees is that they can represent any total preorder on $CD(\mathcal{I})$.

Proposition 1. *For every set $\mathcal{I}$ of binary issues, for every set $D \subseteq CD(\mathcal{I})$ of outcomes over $\mathcal{I}$, and for every total preorder $\succeq$ on D into no more than 2^n clusters of equivalent outcomes, there is a P-tree T of depth at most n such that the preorder determined by T on $CD(\mathcal{I})$ when restricted to D coincides with $\succeq$ (that is, $\succeq_{T|D} = \succeq$).*

Proof. Let $\succeq$ be a total preorder on a subset $D \subseteq CD(\mathcal{I})$ of outcomes over $\mathcal{I}$, and let $D_1 \succ D_2 \succ \ldots \succ D_m$ be the corresponding strict ordering of clusters of equivalent outcomes, with $m \leq 2^n$. If $m = 1$, a single-leaf tree (no decision nodes, just a box node) represents this preorder. This tree has depth 0 and so, the assertion holds. Let us assume then that $m > 1$, and let us define $D' = D_1 \cup \ldots \cup D_{\lceil m/2 \rceil}$ and $D'' = D \setminus D'$. Let $\varphi_{D'}$ be a formula such that models of D' are

---

[1] We overload the symbols $\succeq_T$ and $\succ_T$ by using them both for the order on the leaves of T and the corresponding preorder on the outcomes from $CD(\mathcal{I})$.

precisely the outcomes in D' (such a formula can be constructed as a disjunction of conjunctions of literals, each conjunction representing a single outcome in D'). If we place $\varphi_{D'}$ in the root of a P-tree, that tree represents the preorder with two clusters, D' and D'', with D' preceding D''. Since each of D' and D'' has no more than 2^{n-1} clusters, by induction, the preorders $D_1 \succ \ldots \succ D_{\lceil m/2 \rceil}$ and $D_{\lceil m/2 \rceil + 1} \succ \ldots \succ D_m$ can each be represented as a P-tree with depth at most $n - 1$. Placing these trees as the left and the right subtrees of $\varphi_{D'}$ respectively results in a P-tree of depth at most n that represents $\succeq$. $\square$

Compact Representation of P-trees. Proposition 1 shows P-trees to have high expressive power. However, the construction described in the proof has little practical use. First, the P-tree it produces may have a large size due to the large sizes of labeling formulas that are generated. Second, to apply it, one would need to have an explicit enumeration of the preorder to be modeled, and that explicit representation in practical settings is unavailable.

However, preferences over combinatorial domains that arise in practice typically have structure that can be elicited from a user and exploited when constructing a P-tree representation of the preferences. First, decisions at each level are often based on considerations involving only very few issues, often just one or two and very rarely more than that. Moreover, the subtrees of a node that order the "left" and the "right" outcomes are often identical or similar.

Exploiting these features often leads to much smaller representations. A *compact P-tree over $\mathcal{I}$* is a tree such that

1. every node is labeled with a Boolean formula over $\mathcal{I}$, and
2. every non-leaf node t labeled with φ has either two outgoing edges, with the left one meant to be taken by outcomes that satisfy φ and the right one by those that do not (Fig. 3a), or one outgoing edge pointing
   - straight-down (Fig. 3b), which indicates that the two subtrees of t are *identical* and the formulas labeling every pair of corresponding nodes in the two subtrees are the *same*,
   - left (Fig. 3c), which indicates that right subtree of t is a leaf, or
   - right (Fig. 3d), which indicates that left subtree of t is a leaf.

The P-tree in Fig. 2a can be collapsed as both subtrees of the root are the same (including the labeling formulas). This leads to a tree in Fig. 2b with a

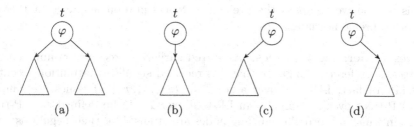

Fig. 3. Compact P-trees

straight-down edge. We note that we drop box-labeled leaves in compact representations of P-trees, as they no longer have an interpretation as distinct clusters.

Empty Leaves in P-trees. Given a P-tree T one can prune it so that all sets of outcomes corresponding to its leaves are non-empty. However, keeping empty clusters may lead to compact representations of much smaller (in general, even exponentially smaller) size.

A full P-tree T in Fig. 4a uses labels $\varphi_1 = \neg x_1 \vee x_3$, $\varphi_2 = x_2 \vee \neg x_4$, and $\varphi_3 = x_2 \wedge x_3$. We check that leaves l_1, l_2 and l_3 are empty, that is, the conjunctions $\varphi_1 \wedge \neg \varphi_2 \wedge \varphi_3$, $\neg \varphi_1 \wedge \varphi_2 \wedge \varphi_3$ and $\neg \varphi_1 \wedge \neg \varphi_2 \wedge \varphi_3$ are unsatisfiable. Pruning T one obtains a compact tree T' (Fig. 4b) that is smaller compared to T, but larger than T'' (Fig. 4c), another compact representation of T, should we allow empty leaves and exploit the structure of T.

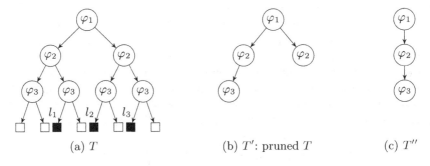

(a) T (b) T': pruned T (c) T''

Fig. 4. P-trees with empty leaves

That example generalizes and leads to the question of finding small-sized representations of P-trees (we conjecture that the problem in its decision version asking about the existence of a compact representation of size at most k is NP-complete). From now on, we assume that P-trees are given in their compact representation.

3   P-trees and Other Formalisms

In this section we compare the preference representation language of P-trees with other preference languages.

P-trees Generalize LP-trees. As stated earlier, P-trees are reminiscent of LP-trees, a preference language that has received significant attention recently [1,10,11]. In fact, LP-trees over a set $\mathcal{I} = \{x_1, \ldots, x_n\}$ of issues are simply special P-trees over $\mathcal{I}$. Namely, an LP-tree over $\mathcal{I}$ can be defined as a P-tree over $\mathcal{I}$, in which all formulas labeling nodes are atoms x_i or their negations $\neg x_i$, depending on whether x_i or $\neg x_i$ is preferred, and every path from the root to a leaf has all atoms x_i appear on it exactly once. Clearly, LP-trees are full binary trees of depth n (assuming they have an implicit extra level of "non-decision" nodes representing outcomes) and determine strict *total orders* on outcomes in $CD(\mathcal{I})$ (no indifference between different outcomes). An example of an LP-tree over $\{x_1, x_2, x_3, x_4\}$ for our vacation example is given in Fig. 5.

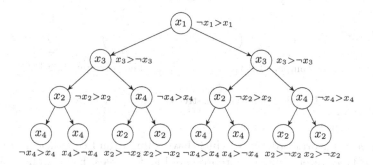

Fig. 5. A full LP-tree on vacations

In general representing preferences by LP-trees is impractical. The size of the representation is of the same order as that of an explicit enumeration of the preference order. However, in many cases preferences on outcomes have structure that leads to LP-trees with similar subtrees. That structure can be exploited, as in P-trees, to represent LP-trees compactly. Figure 6a shows a compact representation of the LP-tree in Fig. 5. We note the presence of conditional preference tables that make up for the lost full binary tree structure. Together with the simplicity of the language, compact representations are essential for the practical usefulness of LP-trees. The compact representations of LP-trees translate into compact representations of P-trees, in the sense defined above. This matter is not central to our discussion and we simply illustrate it with an example. The compactly represented P-tree in Fig. 6b is the counterpart to the compact LP-tree in Fig. 6a, where $\varphi = (x_2 \wedge x_4) \vee (\neg x_2 \wedge \neg x_4)$.

The major drawback of LP-trees is that they can capture only a very small fraction of preference orders. One can show that the number, say $G(n)$, of LP-trees over n issues is

$$G(n) = \prod_{k=0}^{n-1} (n-k)^{2^k} \cdot 2^{2^k}$$

and is asymptotically much smaller than $L(n) = (2^n)!$, the number of all preference orders of the corresponding domain of outcomes. In fact, one can show that

$$\frac{G(n)}{L(n)} < \frac{1}{2^{(2^n \cdot (n - \log n - 2))}}.$$

This is in stark contrast with Proposition 1, according to which every total preorder can be represented by a P-tree.

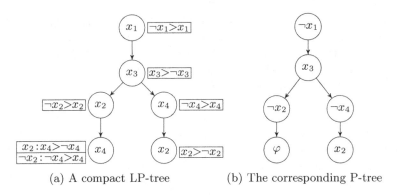

(a) A compact LP-tree   (b) The corresponding P-tree

Fig. 6. A compact LP-tree as a compact P-tree

Even very natural orderings, which have simple (and compact) representations by P-trees often cannot be represented as LP-trees. For instance, there is no LP-tree on $\{x_1, x_2\}$ representing the order $00 \succ 11 \succ 01 \succ 10\}$. However, the P-trees (both full and compact) in Fig. 2 do specify it.

P-trees Extend ASO-Rules. The formalism of ASO-rules [3] provides an intuitive way to express preferences over outcomes as total preorders. An ASO-rule partitions outcomes into ordered clusters according to the semantics of the formalism. Formally, an ASO-rule r over $\mathcal{I}$ is a preference rule of the form

$$C_1 > \ldots > C_m \leftarrow B, \qquad (1)$$

where all C_i's and B are propositional formulas over $\mathcal{I}$. For each outcome M, rule r of the form (1) determines its *satisfaction degree*. It is denoted by $SD_r(M)$ and defined by

$$SD_r(M) = \begin{cases} 1, & M \models \neg B \\ m+1, & M \models B \wedge \bigwedge_{1 \leq i \leq m} \neg C_i \\ min\{i : M \models C_i\}, & \text{otherwise.} \end{cases}$$

We say that an outcome M is weakly preferred to an outcome M' ($M \succeq_r M'$) if $SD_r(M) \leq SD_r(M')$. Thus, the notion of the satisfaction degree (or, equivalently, the preference r) partitions outcomes into (in general) $m + 1$ clusters.[2]

Let us consider the domain of vacations. An agent may prefer hiking in Colorado to water sports in Florida if she is going on a summer vacation. Such preference can be described as an ASO-rule:

$$\neg x_1 \wedge \neg x_2 > x_1 \wedge x_2 \leftarrow x_3.$$

Under the semantics of ASO, this preference rule specifies that the most desirable vacations are summer hiking vacations to Colorado and all winter vacations, the

---

[2] This definition is a slight adaptation of the original one.

next preferred vacations are summer water sports vacations to Florida, and the least desirable vacations are summer hiking vacations to Florida and summer water sports vacations to Colorado.

It is straightforward to express ASO-rules as P-trees. For an ASO-rule r of form (1), we define a P-tree T_r as shown in Fig. 7. That is, every node in T_r has the right child only (the left child is a leaf representing an outcome and is not explicitly shown). Moreover, the labels of nodes from the root down are defined as follows: $\varphi_1 = \neg B \vee C_1$, and $\varphi_i = C_i$ ($2 \leq i \leq m$).

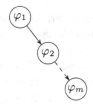

Fig. 7. A P-tree T_r

Theorem 1. *Given an ASO-rule r, the P-tree T_r has size linear in the size of r, and for every two outcomes M and M'*

$$M \succeq_r^{ASO} M' \text{ iff } M \succeq_{T_r} M'$$

Proof. The P-tree T_r induces a total preorder $\succeq_{T_r}$ where outcomes satisfying φ_1 are preferred to outcomes satisfying $\neg\varphi_1 \wedge \varphi_2$, which are then preferred to outcomes satisfying $\neg\varphi_1 \wedge \neg\varphi_2 \wedge \varphi_3$, and so on. The least preferred are the ones satisfying $\bigwedge_{1 \leq i \leq m} \neg\varphi_i$. Clearly, the order $\succeq_{T_r}$ is precisely the order $\succeq_r^{ASO}$ given by the ASO rule r.   □

There are other ways of translating ASO-rules to P-trees. For instance, it might be beneficial if the translation produced a more balanced tree. Keeping the definitions of φ_i, $1 \leq i \leq m$, as before and setting $\varphi_{m+1} = B \wedge \neg C_1 \wedge \ldots \wedge \neg C_m$, we could proceed as in the proof of Proposition 1.

For example, if $m = 6$, we build the P-tree T_r^b in Fig. 8, where $\psi_1 = \varphi_1 \vee \varphi_2 \vee \varphi_3 \vee \varphi_4$, $\psi_2 = \varphi_1 \vee \varphi_2$, $\psi_3 = \varphi_1$, $\psi_4 = \varphi_3$, $\psi_5 = \varphi_5 \vee \varphi_6$, and $\psi_6 = \varphi_5$. The indices i's of the formulas ψ_i's indicate the order in which the corresponding formulas are built recursively.

This P-tree representation of a preference r of the form (1) is balanced with the height $\lceil \log_2(m + 1) \rceil$. Moreover, the property in Theorem 1 also holds for the balanced tree T_r^b. The size of T_r^b is in $O(s_r \log s_r)$, where s_r is the size of rule r. It is then larger by the logarithmic factor than T_r but has a smaller depth.

Representing P-trees as RASO-Theories. Preferences represented by compact P-trees cannot in general be captured by ASO preferences without a significant (in some cases, exponential) growth in the size of the representation.

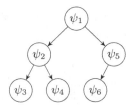

Fig. 8. T_r^b when $m = 6$

However, any P-tree can be represented as a set of *ranked* ASO-rules, or an RASO-theory [3], aggregated by the Pareto method.

We first show how Pareto method is used to order outcomes with regard to a set of *unranked* ASO-rules. Let M and M' be two outcomes. Given a set P of unranked ASO-rules, M is weakly preferred to M' with respect to P, $M \succeq_P^u M'$, if $SD_r(M) \leq SD_r(M')$ for every $r \in P$. Moreover, M is strictly preferred to M', $M \succ_P^u M'$, if $M \succeq_P^u M'$ and $SD_r(M) < SD_r(M')$ for some $r \in P$. Finally, M is equivalent to M', $M \approx_P^u M'$, if $SD_r(M) = SD_r(M')$ for every $r \in P$.

In general, the resulting preference relation is not total. However, by ranking rules according to their importance total preorders can in some cases be obtained. Let us assume $P = \{P_1, \ldots, P_g\}$ is a collection of ranked ASO preferences divided into g sets P_i, with each set P_i consisting of ASO-rules of rank d_i so that $d_1 < d_2 < \ldots < d_g$. We assume that a lower rank of a preference rule indicates its higher importance. We define $M \succeq_P^{rk} M'$ w.r.t P if for every i, $1 \leq i \leq g$, $M \approx_{P_i}^u M'$, or if for some i, $1 \leq i \leq g$, $M \succ_{P_i}^u M'$, and $M \approx_{P_j}^u M'$ for every j, $j < i$.

Given a P-tree T, we construct an RASO-theory Φ_T as follows. We start with $\Phi_T = \emptyset$. For every node t_i in a P-tree T, we update $\Phi_T = \Phi_T \cup \{\varphi_i \overset{d_i}{\leftarrow} conditions\}$, where φ_i is the formula labeling node t_i, d_i, the rank of the ASO-rule, is the depth of node t_i, and *conditions* is the conjunction of formulas φ_j or $\neg\varphi_j$ labeling all nodes t_j that have two children and that are ancestors of t_i in T. We use φ_j in the conjunction if the path from the root to t_i descends from t_j to its left child. Otherwise, we use $\neg\varphi_j$.

For instance, the P-tree T in Fig. 6b gives rise to the following RASO-theory:

$$\neg x_1 \overset{1}{\leftarrow}.$$
$$x_3 \overset{2}{\leftarrow}.$$
$$\neg x_2 \overset{3}{\leftarrow} x_3. \qquad \neg x_4 \overset{3}{\leftarrow} \neg x_3.$$
$$(x_2 \wedge x_4) \vee (\neg x_2 \wedge \neg x_4) \overset{4}{\leftarrow} x_3. \qquad x_2 \overset{4}{\leftarrow} \neg x_3.$$

Theorem 2. *For every P-tree T, the RASO-theory Φ_T has size polynomial in the size of T, and for every two outcomes M and M'*

$$M \succeq_{\Phi_T}^{RASO} M' \text{ iff } M \succeq_T M'$$

Proof. The claim concerning the size of Φ_T is evident from the construction.

($\Leftarrow$) Let us assume $M \succeq_T M'$. Denote by $(\varphi_{i_1}, \ldots, \varphi_{i_j})$ the order of formulas labeling the path determined by M from the root to a leaf. Let φ_{i_k}, $1 \leq k \leq j$, be the first formula that M and M' evaluate differently. Then, $M \models \varphi_{i_k}$ and $M' \not\models \varphi_{i_k}$. Denote by d the depth of φ_{i_k} in T. Based on the construction of Φ_T, for every RASO-rule r of rank less than d, we have $M \approx_r^{ASO} M'$. For every RASO-rule r of rank d, we have $M \succ_r^{ASO} M'$ if r comes from φ_{i_k}, and we have $M \approx_r^{ASO} M'$ for other rules of rank d (in fact, the satisfaction degrees of M and M' on all these other rules are equal to 1). Thus, $M \succeq_{\Phi_T}^{RASO} M'$. If M and M' evaluate all formulas φ_{i_k}, $1 \leq k \leq j$, the same, then $M \approx_{\Phi_T}^{RASO} M'$ and so, $M \succeq_{\Phi_T}^{RASO} M'$, too.

($\Rightarrow$) Towards a contradiction, let us assume that $M \succeq_{\Phi_T}^{RASO} M'$ and $M' \succ_T M$ hold. We again denote by $(\varphi_{i_1}, \ldots, \varphi_{i_j})$ the order of formulas labeling the path determined by M from the root to a leaf. There must exist some formula φ_{i_k}, $1 \leq k \leq j$, such that $M' \models \varphi_{i_k}$, $M \not\models \varphi_{i_k}$, and all formulas φ_ℓ, $1 \leq \ell \leq k - 1$, are evaluated in the same way by M and M'. Based on RASO ordering, we have $M' \succ_{\Phi_T}^{RASO} M$, contradiction. $\square$

Hence, the relationship between P-trees and ASO preferences can be summarized as follows. Every ASO preference rule can be translated into a P-tree, and every P-tree into a theory of ranked ASO preference rules. In both cases, the translations have size polynomial in the size of the input. Examining the inverse direction, the size of the ASO rule translated from a P-tree could be exponential, and the orders represented by ranked ASO theories *strictly include* the orders induced by P-trees, as RASO-theories describe *partial* preorders in general.

P-trees Extend Possibilistic Logic. A possibilistic logic theory Π over a vocabulary $\mathcal{I}$ is a set of *preference pairs*

$$\{(\phi_1, a_1), \ldots, (\phi_m, a_m)\},$$

where every ϕ_i is a Boolean formula over $\mathcal{I}$, and every a_i is a real number such that $1 \geq a_1 > \ldots > a_m \geq 0$ (if two formulas have the same importance level, they can be replaced by their conjunction). Intuitively, a_i represents the importance of ϕ_i, with larger values indicating higher importance.

The *tolerance degree* of outcome M with regard to preference pair (ϕ, a), $TD_{(\phi,a)}(M)$, is defined by

$$TD_{(\phi,a)}(M) = \begin{cases} 1, & M \models \phi \\ 1 - a, & M \not\models \phi \end{cases}$$

Based on that, the tolerance degree of outcome M with regard to a *set* Π of preference pairs, $TD_\Pi(M)$, is defined by

$$TD_\Pi(M) = min\{TD_{(\phi_i, a_i)}(M) : 1 \leq i \leq m\}.$$

The larger $TD_\Pi(M)$, the more preferred M is.

For example, for the domain of vacations, we might have the following set of preference pairs $\{(\neg x_1 \wedge x_3, 0.8), (x_2 \wedge x_4, 0.5)\}$. According to the possibilistic logic interpretation, vacations satisfying both preferences are the most preferred, those satisfying $\neg x_1 \wedge x_3$ but falsifying $x_2 \wedge x_4$ are next in the preference order, and those falsifying $\neg x_1 \wedge x_3$ are the worst.

Similarly as for ASO-rules, we can apply different methods to encode a possibilistic logic theory in P-trees. Here we discuss one of them. We define T_Π to be an unbalanced P-tree shown in Fig. 7 with labels φ_i defined as follows: $\varphi_1 = \bigwedge_{1 \leq i \leq m} \phi_i$, $\varphi_2 = \bigwedge_{1 \leq i \leq m-1} \phi_i \wedge \neg \phi_m$, $\varphi_3 = \bigwedge_{1 \leq i \leq m-2} \phi_i \wedge \neg \phi_{m-1}$, and $\varphi_m = \phi_1 \wedge \neg \phi_2$.

Theorem 3. *For every possibilistic theory Π, the P-tree T_Π has size polynomial in the size of Π, and for every two outcomes M and M'*

$$M \succeq_\Pi^{Poss} M' \ \ iff \ \ M \succeq_{T_\Pi} M'.$$

Proof. It is clear that the size of the P-tree T_Π is polynomial in the size of Π. Let $mi(M, \Pi)$ denote the maximal index j such that M satisfies all $\phi_1, \ldots, \phi_j$ in Π. (If M falsifies all formulas in Π, we have $mi(M, \Pi) = 0$.) One can show that $M \succeq_\Pi^{Poss} M'$ if and only if $mi(M, \Pi) \geq mi(M', \Pi)$, and $mi(M, \Pi) \geq mi(M', \Pi)$ if and only if $M \succeq_{T_\Pi} M'$. Therefore, the theorem follows. $\square$

4   Reasoning Problems and Complexity

In this section, we study decision problems on reasoning about preferences described as P-trees, and provide computational complexity results for the three reasoning problems defined below.

Definition 1. *Dominance-testing (*DomTest*): given a P-tree T and two distinct outcomes M and M', decide whether $M \succeq_T M'$.*

Definition 2. *Optimality-testing (*OptTest*): given a P-tree T and an outcome M of T, decide whether M is optimal.*

Definition 3. *Optimality-with-property (*OptProp*): given a P-tree T and some property α expressed as a Boolean formula over the vocabulary of T, decide whether there is an optimal outcome M that satisfies α.*

Our first result shows that P-trees support efficient dominance testing.

Theorem 4. *The* DomTest *problem can be solved in time linear in the height of the P-tree T.*

Proof. The DomTest problem can be solved by walking down the tree. The preference between M and M' is determined at the first non-leaf node n where M and M' evaluate φ_n differently. If such node does not exist before arriving at a leaf, $M \approx_T M'$. $\square$

An interesting reasoning problem not mentioned above is to decide whether there exists an optimal outcome with respect to the order given by a P-tree. However, this problem is trivial as the answer simply depends on whether there is any outcome at all. However, optimality *testing* is a different matter. Namely, we have the following result.

Theorem 5. *The* OPTTEST *problem is coNP-complete.*

Proof. We show that the complementary problem, testing non-optimality of an outcome M, is NP-complete. The membership is obvious. A witness of non-optimality of M is any outcome M' such that $M' \succ_T M$, a property that can be verified in linear time (cf. Theorem 4). NP-hardness follows from a polynomial time reduction from SAT [7]. Given a CNF formula $\Phi = c_1 \wedge \ldots \wedge c_n$ over a set of variables $V = \{X_1, \ldots, X_m\}$, we construct a P-tree T and an outcome M as follows.

1. We choose $X_1, \ldots, X_m, unsat$ as issues, where $unsat$ is a new variable;
2. we define the P-tree T_Φ (cf. Fig. 9) to consist of a single node labeled by $\Psi = \Phi \wedge \neg unsat$;
3. we set $M = \{unsat\}$.

We show that $M = \{unsat\}$ is not an optimal outcome if and only if $\Phi = c_1 \wedge \ldots \wedge c_n$ is satisfiable.

($\Rightarrow$) Assume that $M = \{unsat\}$ is not an optimal outcome. Since $M \not\models \Psi$, M belongs to the right leaf and there must exist an outcome M' such that $M' \succ M$. This means that $M' \models \Phi \wedge \neg unsat$. Thus, Φ is satisfiable.

($\Leftarrow$) Let M' be a satisfying assignment to Φ over $\{X_1, \ldots, X_m\}$. Since no $c_i \in \Phi$ mentions $unsat$, we can assume $unsat \notin M'$. So $M' \models \Psi$ and M' is optimal. Thus, $M = \{unsat\}$ is not optimal. $\square$

Fig. 9. The P-tree T_Φ

Theorem 6. *The* OPTPROP *problem is* Δ_2^P-*complete.*

Proof. (Membership) The problem is in the class Δ_2^P. Let T be a given preference tree. To check whether there is an optimal outcome that satisfies a property α, we start at the root of T and move down. As we do so, we maintain the information about the path we took by updating a formula ψ, which initially is set to $\top$ (a generic tautology). Each time we move down to the left from a node t, we update ψ to $\psi \wedge \varphi_t$, and when we move down to the right, to $\psi \wedge \neg\varphi_t$. To decide whether to move down left or right form a node t, we check if $\varphi_t \wedge \psi$ is satisfiable by making a call to an NP oracle for deciding satisfiability. If $\varphi_t \wedge \psi$

is satisfiable, we proceed to the left subtree and, otherwise, to the right one. We then update t to be the node we moved to and repeat. When we reach a leaf of the tree (which represents a cluster of outcomes), this cluster is non-empty, consists of all outcomes satisfying ψ and all these outcomes are optimal. Thus, returning YES, if $\psi \wedge \alpha$ is satisfiable and NO, otherwise, correctly decides the problem. Since the number of oracle calls is polynomial in the size of the tree T, the problem is in the class Δ_2^P.

(Hardness) The maximum satisfying assignment (MSA) problem[3] [9] is Δ_2^P-complete. We first show that MSA remains Δ_2^P-hard if we restrict the input to Boolean formulas that are satisfiable and have models other than the all-false model (i.e., $\neg x_1 \ldots \neg x_n$).

Lemma 1. *The MSA problem is Δ_2^P-complete under the restriction to formulas that are satisfiable and have models other than the all-false model.*

Proof. The membership in Δ_2^P is evident. Given a Boolean formula Φ over $\{x_1, \ldots, x_n\}$, we define $\Psi = \Phi \vee (x_0 \wedge \neg x_1 \wedge \ldots \wedge \neg x_n)$ over $\{x_0, x_1, \ldots, x_n\}$. It is clear that Ψ is satisfiable, and has at least one model other than the all-false one. Let M be a lexicographically maximum assignment satisfying Φ and assume that M has $x_n = 1$. Extending M by $x_0 = 1$ yields a lexicographically maximum assignment satisfying Ψ and this assignment obviously satisfies $x_n = 1$, too. Conversely, if M is a lexicographically maximum assignment satisfying Ψ and $x_n = 1$ holds in M, then it follows that $M \models \Phi$. Thus, M restricted to $\{x_1, \ldots, x_n\}$ is a lexicographically maximal assignment satisfying Φ and $x_n = 1$. Thus, the unrestricted problem has a polynomial reduction to the restricted one. That proves Δ_2^P-hardness.   □

We now show the hardness of the OPTPROP problem by a reduction from this restricted version of the MSA problem. Let Φ be a satisfiable propositional formula over variables $x_1, \ldots, x_n$ that has at least one model other than the all-false one. We construct an instance of the OPTPROP problem as follows. We define the P-tree T_Φ as shown in Fig. 10, where every node is labeled by formula $\Phi \wedge x_i$, and we set $\alpha = x_n$.

Fig. 10. The P-tree T_Φ

---

[3] Given a Boolean formula Φ over $\{x_1, \ldots, x_n\}$, the maximum satisfying assignment (MSA) problem is to decide whether $x_n = 1$ in the lexicographically maximum satisfying assignment for Φ. (If Φ is unsatisfiable, the answer is *no*.)

Our P-tree T_Φ induces a total preorder consisting of a sequence of singleton clusters, each containing an outcome satisfying Φ, followed by a single cluster comprising all outcomes that falsify Φ and the all-false model. By our assumption on Φ, the total preorder has at least two non-empty clusters. Moreover, all singleton clusters preceding the last one are ordered lexicographically. Thus, the optimal outcome of T_Φ satisfies α if and only if the lexicographical maximum satisfying outcome of Φ satisfies x_n.   □

5   Conclusion and Future Work

We investigated the qualitative preference representation language of *preference trees*, or *P-trees*. This language was introduced in early 1990s (cf. [5,6]), but have not received a substantial attention as a formalism for preference representation in AI. We studied formally the issue of compact representations of P-trees, established its relationship to other preference languages such as lexicographic preference trees, possibilistic logic and answer-set optimization. For several preference reasoning problems on P-trees we derived their computational complexity.

P-trees are quite closely related to possibilistic logic theories or preference expressions in answer-set optimization. However, they allow for much more structure among formulas appearing in these latter two formalisms (arbitrary trees as opposed to the linear structure of preference formulas in the other two formalisms). This structure allows for representations of conditional preferences. P-trees are also more expressive than lexicographic preference trees. This is the case even for P-trees in which every node is labeled with a formula involving just two issues, as we illustrated with the $00 \succ 11 \succ 01 \succ 01$ example. Such P-trees are still simple enough to correspond well to the way humans formulate hierarchical models of preferences, with all their decision conditions typically restricted to one or two issues.

Our paper shows that P-trees form a rich preference formalism that deserves further studies. Among the open problems of interest are those of learning P-trees and their compact representations, aggregating P-trees coming from different sources (agents), and computing optimal consensus outcomes. These problems will be considered in the future work.

References

1. Booth, R., Chevaleyre, Y., Lang, J., Mengin, J., Sombattheera, C.: Learning conditionally lexicographic preference relations. In: ECAI, pp. 269–274 (2010)
2. Boutilier, C., Brafman, R., Domshlak, C., Hoos, H., Poole, D.: CP-nets: a tool for representing and reasoning with conditional ceteris paribus preference statements. J. Artif. Intell. Res. **21**, 135–191 (2004)
3. Brewka, G., Niemelä, I., Truszczynski, M.: Answer set optimization. In: IJCAI, pp. 867–872 (2003)
4. Dubois, D., Lang, J., Prade, H.: A brief overview of possibilistic logic. In: ECSQARU, pp. 53–57 (1991)

5. Fraser, N.M.: Applications of preference trees. In: Proceedings of IEEE Systems Man and Cybernetics Conference, pp. 132–136. IEEE (1993)
6. Fraser, N.M.: Ordinal preference representations. Theory Decis. **36**(1), 45–67 (1994)
7. Garey, M.R., Johnson, D.S.: Computers and Intractability: A Guide to the Theory of NP-Completeness. W. H. Freeman & Co., New York (1979)
8. Kaci, S.: Working with Preferences: Less Is More. Cognitive Technologies. Springer, Berlin (2011)
9. Krentel, M.W.: The complexity of optimization problems. J. Comput. Syst. Sci. **36**(3), 490–509 (1988)
10. Lang, J., Mengin, J., Xia, L.: Aggregating conditionally lexicographic preferences on multi-issue domains. In: CP, pp. 973–987 (2012)
11. Liu, X., Truszczynski, M.: Aggregating Conditionally lexicographic preferences using answer set programming solvers. In: Perny, P., Pirlot, M., Tsoukiàs, A. (eds.) ADT 2013. LNCS, vol. 8176, pp. 244–258. Springer, Heidelberg (2013)

Group Activity Selection from Ordinal Preferences

Andreas Darmann[(✉)]

University of Graz, Graz, Austria
andreas.darmann@uni-graz.at

Abstract. We consider the situation in which group activities need to be organized for a set of agents when each agent can take part in at most one activity. The agents' preferences depend both on the activity and the number of participants in that activity. In particular, the preferences are given by means of strict orders over such pairs "(activity, group size)", including the possibility "do nothing". Our goal will be to assign agents to activities on basis of their preferences, the minimum requirement being that no agent prefers doing nothing, i.e., not taking part in any activity at all. We take two different approaches to establish such an assignment: (i) by use of k-approval scores; (ii) considering stability concepts such as Nash and core stability.

For each of these approaches, we analyse the computational complexity involved in finding a desired assignment. Particular focus is laid on two natural special cases of agents' preferences which allow for positive complexity results.

1 Introduction

In many situations activities need to be organized for a set of agents, with the agents having preferences over the activities. However, often the preferences of the agents do not depend solely on the activity itself, but also on the number of participants in the activity (see also [7]). We consider such a scenario and assume that each agent can be assigned to at most one activity. E.g., consider a company which would like to provide free sports classes in order to achieve a high employee satisfaction [10], or the organizer of a social or business event (such as a workshop), who wants to arrange social activities for the free afternoon [7]. In the former case, for cost reasons the company might allow each employee to take part in at most one activity; in the latter case, since the activities take place at the same time, each agent can take part in at most one activity. Now, often the agents have preferences not only over the available activities, but also over the number of attendees of the activity. For example, one would be willing to take a sauna with up to 5 attendees, but does not wish to take part if the sauna is more crowded. On the other hand, for activities connected with costs that need to be shared by the attendees, an agent might only take part if a high number of attendees joins, while the desired numbers of attendees may be different for each agent. These examples already indicate two natural special cases we will

© Springer International Publishing Switzerland 2015
T. Walsh (Ed.): ADT 2015, LNAI 9346, pp. 35–51, 2015.
DOI: 10.1007/978-3-319-23114-3_3

consider in this paper: vaguely speaking, in the case of decreasing preferences the agents want the number of other agents joining an activity to be as low as possible; in the case of increasing preferences, the agents would like as many agents as possible to join the same activity.

Thus, in this work we consider a setting in which the agents have preferences over pairs of the form "(activity, group size)". In these pairs, we include the possibility "do nothing" to which we refer as the void activity $a_\emptyset$. Note that $a_\emptyset$ allows the agents to express which pairs "(activity, group size)" they are not really happy with by ranking them below $a_\emptyset$ (they would rather do nothing than join the respective activity with the corresponding total number of attendees). Throughout this paper, we assume the agents' preferences to be strict orders over such pairs (including $a_\emptyset$).

The goal, of course, would be a "good" assignment of agents to activities. As a minimum requirement, the assignment should be *individually rational*, i.e., no agent should be forced to take part in an activity with a total number of attendees such that she would prefer doing nothing. Taking into account the special cases of increasing and decreasing preferences, we follow two different approaches to find a "good" assignment, and provide computational complexity results for each of them:

(i) By use of k-approval scores derive a group decision from the agents' rankings (Sect. 3); (ii) Find stable assignments with respect to different stability concepts such as Nash or core stability (Sect. 4).

Finally, note that, as already pointed out in [7], avoiding to take into account the number of participants may lead to rather unsatisfactory assignments. One might, e.g., externally impose constraints on the number of participants instead, or simply neglect the number of participants. In the latter case, imagine 80 agents assigned to take a sauna. In the first case, for instance, in an activity (such as a bus trip) whose costs need to be shared the minimum number of participants such that an agent is actually willing to take part in the activity (i.e., such that an agent prefers joining the activity to doing nothing) might highly vary.

1.1 Related Work

The most closely related work is [7]. There, the general *group activity selection problem (GASP)* is introduced, where the agents' preferences are weak orders over the pairs "(activity, group size)". The authors analyse computational complexity aspects of the special case of a-GASP, where the agent's preferences are not strict orders but trichotomous; i.e., each agent partitions the set of pairs "(activity,group size)" into the following three clusters: pairs approved by an agent (i.e., pairs that are preferred to the void activity), the void activity itself, and pairs that are disapproved by the agents (i.e., pairs which the void activity is preferred to). In that framework, the focus is laid on maximum individually rational assignments (i.e., an individually rational assignment with the maximum number of agents assigned to a non-void activity) and Nash stable assignments, while the adaption of further stability concepts is briefly discussed. In their

Table 1. Computational complexity with respect to stability in o-GASP

	General preferences	Increasing preferences	Decreasing preferences
Nash stable	NP-complete	NP-complete	in P
Individually stable	NP-complete	NP-complete	in P
Contractually ind. stable	in P		
Core stable	NP-complete	NP-complete	in P

computational study, several special cases including increasing and decreasing preferences are considered.

In our work, we focus on another obvious variant of the group activity selection problem, in which the agents' preferences are strict orders. We apply the stability concepts and the notions of increasing/decreasing preferences of a-GASP to our framework and, in Sect. 4, analyse the computational complexity involved in finding such stable assignments with respect to increasing/decreasing preferences. An overview over the results settled in Sect. 4 is given in Table 1, for strict orders in general and the special cases of increasing resp. decreasing preferences (NP-completeness results refer to the decision problem if there exists an assignment satisfying the respective stability notion; membership in P indicates that such an assignment always exists and can be found in polynomial time).

The model presented in this paper is closely related to that of *hedonic games* [2,5], and in particular *anonymous hedonic games* [2]. In the latter setting, the goal is to find a "good" partition of the set of agents into several groups, while the agents have preferences over the size of the possible group they are part of. In contrast, in our setting the agents' preferences depend on the group size and the specific activity they would like to join.

In general (i.e., non-anonymous) hedonic games, the agents' preferences are over the possible groups (i.e., the composition of the group rather than only its size) they are part of. By introducing dummy agents for the activities and suitable preferences, the general group activity selection problem – and hence our model – can be embedded in the general hedonic game framework (see [7] for a detailed description of that representation). As already pointed out by [7], note that our setting has useful structural properties that distinguish it from a hedonic game though – e.g., it allows for a succinct representation of agents' preferences. In addition, in our model two natural special cases are inherent that admit efficient algorithms for finding good outcomes.

Both anonymous and non-anonymous hedonic games have been studied by [1] from a computational viewpoint. In particular, in [1] it is shown that deciding whether there is an outcome that is core stable, Nash stable, individually stable or contractually individually stable is NP-complete for both anonymous and non-anonymous hedonic games. While these results translate to the general group activity selection problem introduced in [7], they do not directly imply similar hardness results for the setting considered in this work.

Table 2. Computational complexity of maximizing k-approval scores in o-GASP

	General preferences	Increasing preferences	Decreasing preferences
1-approval	in P		in P
2-approval	NP-complete	in P	in P
3-approval	NP-complete		in P
k-approval, $k \geq 4$	NP-complete		NP-complete

Further related work includes the works of [9,11]. In [9], the stable invitation problem is introduced, both in an anonymous and non-anonymous version. In the stable invitation problem, the goal is to find a set of agents to be invited to an event. In the anonymous stable invitation problem, the agents have preferences over the number of invitees; in the non-anonymous case, the agents additionally specify a set of accepted agents and a set of rejected agents. [9] provide a number of complexity results with respect to stability, and also consider strategic behaviour of the agents.

In the work of [11], agents have preferences over pairs made up of a role and a coalition; the role refers to the actual role the agent takes in the coalition, and the coalition specifies the composition of roles that make up the coalition. [11] provide a number of computational complexity results with respect to different notions of stability.

In Sect. 3, we apply k-approval scores used in voting theory (see [6] for a survey) to find a desirable outcome. Such scores have been analysed outside of their classical framework, e.g., with respect to computational complexity in fair division problems (see, for instance, [3,8]). In this work, we provide computational complexity results for finding an individually rational assignment maximizing k-approval scores, for any fixed $k \in \mathbb{N}$. We point out that approval scores in general take back the setting to the one of a-GASP analysed in [7] (see Sect. 2.1 for details); an analysis of k-approval scores, however, is not provided in [7]. Table 2 summarizes the results of Sect. 3 (NP-completeness results refer to the decision problem version).

2   Formal Framework

Definition 1. *An instance* (N, A, P) *of the* Group activity selection problem with ordinal preferences *(o-GASP) is given as follows.* N *is a set of agents with* $n = |N|$; *unless stated otherwise, the agents are denoted by* $N = \{1, \ldots, n\}$. $A = A^* \cup \{a_\emptyset\}$ *is a set of activities, where* $A^* = \{a_1, \ldots, a_m\}$; $X = (A^* \times \{1, \ldots, n\}) \cup \{a_\emptyset\}$ *is the set of alternatives. Finally, the profile* $P = \langle V_1, \ldots, V_n \rangle$ *consists of* n *votes, one for each agent. For agent* i, *a vote* V_i *(also denoted by* $\succ_i$*) is a strict order over* X; *the set* $S_i \subseteq X$ *such that for each* $x \in S_i$ *we have* $x \succ_i a_\emptyset$, *is the* induced approval vote *of agent* i. *We say that agent* i *approves of the alternatives in* S_i.

Definition 2. *Given an instance* (N, A, P) *of* o-GASP, *we say that agent* i *has* increasing preferences *if for each* $a \in A^*$, $(a, k) \succ_i (a, k-1)$ *holds for each* $k \in \{2, \ldots, n\}$. *An agent* i *has* decreasing preferences *if for each* $a \in A^*$, $(a, k-1) \succ_i (a, k)$ *holds for each* $k \in \{2, \ldots, n\}$.

An instance (N, A, P) *of* o-GASP *has* increasing (decreasing) *preferences, if each agent* $i \in N$ *has increasing (decreasing) preferences.*

Definition 3. *Given an instance* (N, A, P) *of* o-GASP, *a mapping* $\pi : N \to A$ *is called* assignment. *The set* $\pi^a := \{i \in N | \pi(i) = a\}$ *denotes the set of agents assigned to* $a \in A$. *The set* $\pi_i := \{j \in N | \pi(j) = \pi(i)\}$ *denotes the set of agents assigned to the same activity as agent* $i \in N$.

An assignment is said to be individually rational *if for every* $a \in A^*$ *and every agent* $i \in \pi^a$ *it holds that* $(a, |\pi^a|) \succ_i a_\emptyset$.

As done in [7], we consider individual rationality as a minimum fairness and stability requirement an assignment must satisfy: If an assignment is not individually rational, then there is an agent that would rather join the void activity (i.e., do nothing) than taking part in the assigned activity (and hence this agent wants to deviate from that activity).

Note that the assignment π defined by $\pi(i) = a_\emptyset$ for all $i \in N$ is always individually rational; therewith, an individually rational assignment always exists. Let $\#(\pi) = |\{i \in N \mid \pi(i) \neq a_\emptyset\}|$ denote the number of agents assigned to a non-void activity. Finally, π is *maximum individually rational* if π is individually rational and $\#(\pi) \geq \#(\pi')$ for every individually rational assignment π'.

In this paper, we consider individually rational assignments only. In particular, in any instance (N, A, P) of o-GASP we will restrict the attention to the part of the profile which excludes alternatives ranked below $a_\emptyset$. In addition, throughout the paper we assume that $S_i \neq \emptyset$ holds for each agent i, since otherwise in any individually rational assignment i can only participate in the void activity. Finally, in this paper several proofs are omitted due to space constraints.

2.1 k-approval Scores

In an instance of o-GASP, a scoring function f maps an assignment to a nonnegative real number by means of $f(\pi) := \sum_{i \in N} f_i(\pi(i), |\pi_i|)$ with $f_i : X \to \mathbb{R}_0^+$. In *approval scores*, for $i \in N$, let $f_i(x) = 1$ for $x \in S_i$ and $f_i(x) = 0$ for $x \notin S_i$. k-*approval scores*, $k \in \mathbb{N}$, correspond to approval scores in the case that $|S_i| = k$ holds for all $i \in N$. The value $f(\pi)$ is called *score of* π.

Our first task will be to find an individually rational assignment that maximizes k-approval scores (Sect. 3).

Note that in general, approval scores take back the setting to the one of a-GASP considered in [7], and an individually rational assignment with maximum approval score corresponds to a maximum individually rational assignment. For the problem of finding such an assignment a number of computational complexity results are given in [7]. For instance, it is shown that finding a maximum individually rational assignment is NP-hard both for decreasing and increasing preferences; on the positive side, if there is only one activity, a maximum individually rational assignment can be found in polynomial time.

However, the complexity involved in finding a maximum individually rational assignment in the case that $|S_i| = k$ for each agent i, i.e., the problem of finding an individually rational assignment of maximum k-approval score, $k \in \mathbb{N}$, was not analysed. We close this gap by giving a detailed analysis of that problem in Sect. 3.

2.2 Stability Concepts

A different approach is to investigate upon individually rational assignments that satisfy classical stability concepts such as, for instance, Nash and core stability. The stability concepts considered in this paper are defined as follows.

A Nash stable assignment requires that no agent has the wish to deviate from her assigned activity. Abusing notation, for any $k \in \mathbb{N}$, we associate $(a_\emptyset, k)$ with $a_\emptyset$, i.e., we define $(a_\emptyset, k) := a_\emptyset$.

Definition 4. *Given an instance* (N, A, P) *of* o-GASP, *an assignment* $\pi : N \to A$ *is* Nash stable *if it is individually rational and there is no agent* $i \in N$ *and no* $a \in A^* \setminus \{\pi(i)\}$ *such that* $(a, |\pi^a| + 1) \succ_i (\pi(i), |\pi_i|)$.

Note that in the above definition of Nash stability an agent is allowed to deviate in favour of activity a even if the agents currently assigned to a are opposed to this. In contrast, in the notion of *individual stability* an agent is allowed to join such a group of agents only if none of the group members objects. *Contractual individual stability* additionally requires that none of the agents assigned to the activity the agent leaves are worse off.

Definition 5. *Given an instance* (N, A, P) *of* o-GASP, *an individually rational assignment* $\pi : N \to A$ *is* individually stable, *if there is no agent* $i \in N$ *and no* $a \in A^* \setminus \pi(i)$ *such that* $(a, |\pi^a| + 1) \succ_i (\pi(i), |\pi_i|)$, *and for all* $i' \in \pi^a$ *it holds that* $(a, |\pi^a| + 1) \succ_{i'} (a, |\pi^a|)$.

Definition 6. *Given an instance* (N, A, P) *of* o-GASP, *an individually rational assignment* $\pi : N \to A$ *is* contractually individually stable, *if there is no agent* $i \in N$ *and no* $a \in A^* \setminus \pi(i)$ *such that (i)* $(a, |\pi^a| + 1) \succ_i (\pi(i), |\pi_i|)$, *(ii) for all* $i' \in \pi^a$ *it holds that* $(a, |\pi^a| + 1) \succ_{i'} (a, |\pi^a|)$, *and (iii) there is no* $h \in \pi_i$ *with* $(\pi(i), |\pi_i| - 1) \prec_h (\pi(i), |\pi_i|)$.

In a *weak core stable* assignment π, there is no subset E of agents such that, by deviating from π, all members of E can strictly improve their situation. In a *strong core stable* assignment π, there is no subset E of agents such that, by deviating from π, at least one member of E can strictly improve her situation while none of the other members is worse off. Generalizing the notion of individual stability, both stability notions require the deviating group of agents to use an activity to which only members of the group are assigned.

Definition 7. *Given an instance* (N, A, P) *of* o-GASP, *an individually rational assignment* $\pi : N \rightarrow A$ *is* weak core stable *if there is no* $E \subseteq N$ *and no* $a \in A^*$ *with* $\pi^a \subset E$ *such that* $(a, |E|) \succ_i (\pi(i), |\pi_i|)$ *for all* $i \in E$.

Definition 8. *Given an instance* (N, A, P) *of* o-GASP, *an individually rational assignment* $\pi : N \rightarrow A$ *is* strong core stable *if there is no* $E \subseteq N$ *and no* $a \in A^*$ *with* $\pi^a \subset E$ *such that* $(a, |E|) \succ_i (\pi(i), |\pi_i|)$ *holds for at least one* $i \in E$ *while there is no* $i \in E$ *with* $(a, |E|) \prec_i (\pi(i), |\pi_i|)$.

Relations Between the Concepts. Clearly, a strong core stable assignment is also weak core stable. In particular, the concepts of weak core stability and strong core stability coincide.

Proposition 1. *An assignment* π *is weak core stable if and only if* π *is strong core stable.*

Proof. Clearly, the if-part holds. For the only-if-part, consider a weak core stable assignment π. If π is not strong core stable, then there exist a subset $E \subseteq N$ and an activity $a \in A^*$ with $\pi^a \subset E$ such that $(\pi(h), |\pi_h|) \nsucc_h (a, |E|)$ for all $h \in E$ and

$$(a, |E|) \succ_i (\pi(i), \pi_i) \tag{1}$$

for at least one agent $i \in E$. Because π is weak core stable, there must be an agent $j \in E$ with $(a, |E|) \nsucc_j (\pi(j), |\pi_j|)$. However, with $(\pi(j), |\pi_j|) \nsucc_j (a, |E|)$ this means that $\pi(j) = a$ (thus, $\pi_j = \pi^a$) and $|\pi_j| = |\pi^a| = |E|$ hold. Now, (1) and $|\pi^a| = |E|$ imply that for agent i we get $i \in E \setminus \pi^a$. In turn, this implies that $\pi^a \setminus E \neq \emptyset$ holds. This contradicts to $\pi^a \subset E$. $\square$

Thus, in what follows we use the notion *core stability* instead of weak/strong core stability.

Summing up, we can observe the following relations between the concepts (here, $a \Rightarrow b$ means if assignment π satisfies a, then it satisfies b):

Nash stable $\Rightarrow$ individually stable $\Rightarrow$ contractually individually stable; core stable $\Rightarrow$ individually stable.

In addition, in the case of *increasing preferences*, Nash stable $\Leftrightarrow$ individually stable holds. In the case of *decreasing preferences*, the following relations hold:

core stable $\Leftrightarrow$ individually stable $\Leftrightarrow$ contractually individually stable (the first equivalence is stated in Proposition 2 on page 15).

3   Maximizing *k*-approval Scores

Theorem 1. *In* o-GASP *with 1-approval scores, an individually rational assignment with maximum score can be found in polynomial time.*

Proof. Clearly, it is sufficient to find, for each $a \in A$, the maximum $h \in N$ such that $(a, h) \in S_i$ holds for at least h agents and assign h of these agents to a. Obviously, this can be done in polynomial time. $\square$

Theorem 2. *Given $\kappa \in \mathbb{N}$, for* o-GASP *with 2-approval scores it is* NP-*complete to decide if there is an individually rational assignment with score at least κ.*

Proof. The proof proceeds by a reduction from the NP-complete problem E3-OCC-MAX-2SAT [4]. □

Theorem 3. *For* o-GASP *with k-approval scores, in the case of increasing preferences an individually rational assignment of maximum score can be found in polynomial time, for any fixed $k \in \mathbb{N}$.*

Proof. If $k \geq \frac{n}{2}$, then obviously n is bounded by $2k$ and thus by a constant. Since the number of possible assignments of agents to activities is bounded by $(m+1)^n$, it follows that there are at most $(m+1)^{2k}$ possible assignments. Thus, an individually rational assignment of maximum score can be determined in time exponential in k. Since k is fixed, we can therefore find an individually rational assignment of maximum score in polynomial time.

If $k < \frac{n}{2}$, then by increasing preferences we know that for each activity a, either no agent or at least $\left\lceil \frac{n}{2} \right\rceil + 1$ agents are assigned to a in an individually rational assignment. Thus, in any individually rational assignment there is at most one activity to which at least one agent is assigned. Hence, if there is an $h \in \mathbb{N}$ for which there is an activity a such that $(a, h) \in S_i$ holds for at least h agents, then it is sufficient to find the maximum such h. If such an $h \in \mathbb{N}$ does not exist, then the only individually rational assignment is $\pi(i) = a_\emptyset$ for each $i \in N$. Clearly, this can be checked in polynomial time. □

Theorem 4. *In* o-GASP *with 2-approval scores, in the case of decreasing preferences an individually rational assignment of maximum score can be found in polynomial time.*

Proof. By decreasing preferences, an agent cannot approve of (a, h) for any $h \geq 3$ and $a \in A^*$. In addition, $(a, 2) \in S_i$ implies $(a, 1) \in S_i$. Let A_2 be the set of activities $a \in A^*$ for which $(a, 2)$ is approved by at least two agents. Clearly, for each $a \in A_2$ at most two of the agents that approve of a can be assigned to activity a (and hence to any activity of A^*) in any individually rational assignment. Note that to any $a \in A^* \setminus A_2$, at most one agent can be assigned in any individually rational assignment.

We now construct an individually rational assignment of maximum score in two steps. In the first step, we arbitrarily assign two of the agents who approve of $a \in A_2$ to a. In a second step, for the remaining activities we apply a maximum cardinality matching in a bipartite graph to derive our desired assignment.

1. For each $a \in A_2$, assign two arbitrary agents approving of $(a, 2)$ to a.
2. Let $A' := A^* \setminus A_2$. Let $G = (V, E)$ with $V = N \cup A'$, $E = \{\{i, a\} | i \in N, a \in A', (a, 1) \in S_i\}$. Compute a maximum cardinality matching M in G. For $a \in A'$, assign agent i to a if $\{i, a\} \in M$.

It is not hard to verify that the resulting assignment π is in fact an individually rational assignment of maximum score. □

Theorem 5. *For any $k \leq 3$, in o-GASP with k-approval scores, in the case of decreasing preferences an individually rational assignment of maximum score can be found in polynomial time.*

Proof (sketch). The cases $k = 1$ and $k = 2$ are covered by Theorems 1 and 4[1].

Let $k = 3$. Analogously to the case $k = 2$, by decreasing preferences an agent cannot approve of (a, h) for any $h \geq 4$ and $a \in A^*$. In addition, $(a, 3) \in S_i$ implies $(a, 2) \in S_i$ and $(a, 1) \in S_i$. Analogously to A_2, let A_j be the set of activities $a \in A^*$ for which (a, j) is approved by at least j agents, $j \in \{1, 3\}$. Analogously to the case $k = 2$, we construct an individually rational assignment of maximum score in two steps. Before dealing with these, we state a simple observation.

Observation. There is an individually rational assignment of maximum score such that to each $a \in A_3$ exactly three agents are assigned, and to each activity of $A_2 \setminus A_3$ exactly two agents are assigned.

Therefore, in the first step for each $a \in A_3$ we arbitrarily assign to a three of the agents who approve of $(a, 3)$. Clearly, in the second step it hence suffices to find an assignment of maximum score for the remaining activities and agents. This will be done by computing a minimum cost flow in a dedicated graph. In what follows, we introduce that graph and argue that the resulting assignment is in fact an individually rational assignment of maximum total score. Note that for the second step, we restrict our attention to activities in $(A_1 \cup A_2) \setminus A_3$, since to each activity of A_3 agents have already been assigned in the first step.

Let $A' := (A_1 \cup A_2) \setminus A_3$. We define the directed graph $G = (V, E)$ with $V = \{s, t\} \cup N \cup \{a, a^1, \tilde{a}^1, a^2 | a \in A'\}$ as follows:

- for each $a \in A'$,
  - introduce the edges (s, a), (a, a^1), $(a^1, \tilde{a}^1)$, (a, a^2) of zero cost each, and (a^1, t) of cost -3
  - for each $i \in N$, introduce
    * the edge $(\tilde{a}^1, i)$ of cost -4 if $(a, 1) \in S_i$
    * the edge (a^2, i) of cost -9 if $(a, 2) \in S_i$ and $a \in A_2$
- for each $i \in N$ introduce the edge (i, t) of zero cost
- the capacity of each edge entering in $\tilde{a}^1 \in A'$, $i \in N$ or t is 1; for $a \in A'$, the capacity of an edge entering in a, a^1 or a^2 is 2.

Note that for $a \in A'$, an edge emanating from a^2 is only contained in G if at least two agents approve of $(a, 2)$.

Let f be an integer min cost flow from s to t (over all flow sizes $1, \ldots, 2m$). The proof continues by showing that f induces an individually rational assignment $\pi_f : N \to A'$ of maximum score. This can be proven by the use of the following properties:

(i) for each $a \in A_2 \setminus A_3$, f sends two units of flow along (a, a^2); (ii) each assignment $\pi : N \to A'$ induces an integer flow from s to t in G.   □

---

[1] In fact, the case of 2-approval scores can be embedded in the 3-approval setting. The 2-approval case, however, allows for a very intuitive determination of an assignment of maximum score as described in the proof of Theorem 4.

Theorem 6. *Let $\kappa \in \mathbb{N}$. For any fixed $k \geq 4$, under k-approval scores in o-GASP it is NP-complete to decide if there is an individually rational assignment with score at least κ, even when all agents have decreasing preferences.*

Proof. The proof proceeds by a reduction from E3-OCC-MAX-2SAT [4].   □

4   Finding Stable Assignments

In this section, we consider the problem of finding a stable assignment w.r.t. the different notions of stability defined in Sect. 2.2.

4.1   Nash Stability

The first stability concept we consider is Nash stability. In the setting of a-GASP, it is NP-complete to decide whether a Nash stable assignment exists; however, in the case of increasing or decreasing preferences, a Nash stable assignment always exists and can be found in polynomial time [7]. In contrast, as Example 1 shows, in the setting of o-GASP even when all agents have increasing preferences a Nash stable assignment does not necessarily exist. In particular, we show that in o-GASP (i) it is NP-complete to decide whether a Nash stable assignment exists even if all agents have increasing preferences, while (ii) in the case of decreasing preferences, a Nash stable assignment always exists and can be found in polynomial time.

Example 1. Let $N = \{1, 2, 3, 4, 5, 6\}$ and $A^* = \{a, b, c\}$. Let P be given as follows:

1	2	3	4	5	6
$(b, 6)$	$(a, 6)$	$(c, 6)$	$(b, 6)$	$(a, 6)$	$(c, 6)$
$(b, 5)$	$(a, 5)$	$(c, 5)$	$(b, 5)$	$(a, 5)$	$(c, 5)$
$(b, 4)$	$(a, 4)$	$(c, 4)$	$(b, 4)$	$(a, 4)$	$(c, 4)$
$(b, 3)$	$(a, 3)$	$(c, 3)$	$(b, 3)$	$(a, 3)$	$(c, 3)$
$(a, 6)$	$(a, 2)$	$(b, 6)$	$(b, 2)$	$(c, 6)$	$(c, 2)$
$(a, 5)$	$a_\emptyset$	$(b, 5)$	$a_\emptyset$	$(c, 5)$	$a_\emptyset$
$(a, 4)$		$(b, 4)$		$(c, 4)$	
$(a, 3)$		$(b, 3)$		$(c, 3)$	
$(a, 2)$		$(b, 2)$		$(c, 2)$	
$(a, 1)$		$(b, 1)$		$(c, 1)$	
$a_\emptyset$		$a_\emptyset$		$a_\emptyset$	

Clearly, all agents have increasing preferences. Note that for each activity $\alpha \in A^*$ there are exactly three agents who rank (α, k) above $a_\emptyset$ for some k.

Assume that there is a Nash stable assignment π. Clearly, π must assign agent 1 to a non-void activity, because otherwise 1 would like to join a. Analogously, agents 3 and 5 must be assigned to a non-void activity.

Assume $\pi(1) = a$. Then $\pi(2) = a$ must hold since π is Nash-stable. For the same reason, $\pi(5) = a$ must hold as well. Recall that also agent 3 has to be assigned to a non-void activity, i.e., $\pi(3) \in \{b, c\}$ holds. $\pi(3) = c$ would imply that three agents are assigned to c, which is impossible due to $\pi(5) = a$. Thus,

$\pi(3) = b$ holds. Clearly, this implies $\pi(4) = b$ because otherwise 4 would like to join activity b in contradiction with Nash stability. But then 1 would like to join b, which violates Nash stability.

Therefore, $\pi(1) = b$ must hold, implying $\pi(3) = \pi(4) = b$. Since agent 5 must be assigned to a non-void activity, $\pi(5) = c$ follows ($\pi(5) = a$ is impossible because this would require $\pi(1) = \pi(2) = a$), which in turn implies $\pi(6) = c$. But now agent 3 would like to join c which violates Nash stability.

Thus, a Nash stable assignment does not exist. Note that removing any agent from the above instance results in a new instance which admits a Nash stable assignment. E.g., if agent 6 is removed, then the assignment π' with $\pi'(2) = a_\emptyset$, $\pi'(1) = \pi'(3) = \pi'(4) = b$ and $\pi'(5) = c$ is Nash stable.

Theorem 7. *It is* NP*-complete to decide whether* o-GASP *admits a Nash stable assignment, even when all agents have increasing preferences.*

Proof. The proof proceeds by a reduction from EXACT COVER BY 3-SETS. □

On the positive side, for the case of decreasing preferences, by means of Algorithm 1 we propose a polynomial-time algorithm that computes a Nash stable assignment. Given an individually rational assignment π, if i wants to deviate from $\pi(i)$ we denote by $fc(i, \pi)$ agent i's *favourite choice*, i.e., the activity $a \in A^* \setminus \{\pi(i)\}$ such that $(a, |\pi^a| + 1)$ is the best-ranked among all alternatives in the ranking $\succ_i$ that satisfy $(a', |\pi^{a'}| + 1) \succ_i (\pi(i), |\pi_i|)$.

Starting with all agents being assigned to the void activity, in each step Algorithm 1 assigns an agent who wishes to deviate from the current assignment to her favourite choice b. If another agent assigned to b opposes to this, i.e., wants to deviate after i joins b, that agent is assigned to her favourite choice in the subsequent step.

Theorem 8. *Given an instance of* o-GASP *with decreasing preferences, a Nash stable assignment always exists and can be determined in polynomial time.*

Proof. We show that Algorithm 1 terminates in polynomial time, with the resulting assignment π being Nash stable. Note that for each $b \in A^*$, during the execution of Algorithm 1, the number of agents assigned to b, i.e., π^b is non-decreasing.

Complexity. First, we show that in the execution of Algorithm 1, for any $i \in N$, $b \in A^*$ and $1 \le k \le n$, b is at most once i's favourite choice with currently $k - 1$ agents assigned to b, i.e., at most once we have $\pi(i) = a_\emptyset$, $fc(i, \pi) = b$ and $|\pi^b| = k - 1$. Assume the opposite. That means, during one loop of the algorithm i is assigned to b (let loop ℓ denote the first such loop), and in a later loop i is assigned to $a_\emptyset$ again. Note that we must have

$$(b, k) \succ_i a_\emptyset \tag{2}$$

and

$$(b, k) \succ_j (d, |\pi^d| + 1) \tag{3}$$

Algorithm 1. Nash stable assignment for decreasing preferences

1: $\pi(h) := a_\emptyset$ for all $h \in N$. $C := \emptyset$. $i := 1$. $loop := 0$.
2: **while** $C \subset N$ **do**
3:     $loop := loop + 1$.   // counts the number of executions of the while-loop
4:     **if** $\nexists a \in A^* \setminus \{\pi(i)\}$ such that $(a, |\pi^a| + 1) \succ_i (\pi(i), |\pi_i|)$ **then**
5:         $C := C \cup \{i\}$
6:         $i := \min (N \setminus C)$
7:     **else**
8:         $b := fc(i, \pi)$
9:         **if** there is a $j \in \pi^b$ such that
            $- a_\emptyset \succ_j (b, |\pi^b| + 1)$ or
            $- (d, |\pi^d| + 1) \succ_j (b, |\pi^b| + 1)$ for some $d \in A^* \setminus \{b\}$
            then
10:             $C := (C \cup \{i\}) \setminus \{j\}$
11:             $\pi(i) := b$
12:             $\pi(j) := a_\emptyset$
13:             $i := j$
14:         **else**
15:             $C := C \cup \{i\}$
16:             $\pi(i) := b$

---

for all $d \in A^* \setminus \{b\}$ after loop ℓ. Clearly, directly *after* execution of loop ℓ, $|\pi^b| \in \{k - 1, k\}$ holds, depending on whether i was assigned to b according to lines 10–13 or 15–16. In the latter case $|\pi^b| = k$ holds. Thus, b with $|\pi^b| = k - 1$ cannot be any agent's favourite choice in a later stage because the number of agents assigned to b is non-decreasing during the execution of the algorithm.

Thus, we must have $|\pi^b| = k-1$ after loop ℓ. To enable that in a loop $g > \ell$, b is again i's favourite choice, i must be assigned to $a_\emptyset$ in a loop ℓ' with $\ell < \ell' < g$. This is only possible if either (i) $a_\emptyset \succ_i (b, |\pi^b|+1)$ or (ii) $(d, |\pi^d|+1) \succ_i (b, |\pi^b|+1)$ for some $d \in A^* \setminus \{b\}$ holds in loop ℓ'. Clearly, since for each activity the number of agents assigned to the activity in loop ℓ' is at least as high as in loop ℓ, there are at least $k - 1$ agents assigned to b when entering loop ℓ'. Assume we have $|\pi^b| = k - 1$ at the beginning of loop ℓ'. Obviously, (i) cannot hold in this case due to (2). Analogously, (3) contradicts with (ii), because the agents have decreasing preferences and the number of agents assigned to d is at least as high at the beginning of loop ℓ' as after loop ℓ.

As a consequence, at the beginning of loop ℓ' we must have $|\pi^b| \geq k$. This, however, contradicts with the fact that, in loop g, b with $|\pi^b| = k - 1$ is i's favourite choice, because the number of agents assigned to b in loop g is at least as high as in loop ℓ'.

Hence, for each $i \in N$, $b \in A^*$ and $1 \leq k \leq n$, b is at most once i's favourite choice with currently $k - 1$ agents assigned to b. With $m = |A^*|$ that means that the algorithm terminates after at most $n \cdot mn = mn^2$ loops. In each loop, the highest computational effort is the execution of line 9, which, in the worst case, requires checking $(d, |\pi^d| + 1) \succ_j (b, |\pi^b| + 1)$ for all agents j and activities $d \in A^* \setminus \{b\}$; this can be bounded by $\mathcal{O}(m^2 n^2)$. Thus, the running time of the algorithm can be bounded by $\mathcal{O}(mn^2 \cdot m^2 n^2)$, i.e., by $\mathcal{O}(m^3 n^4)$.

Correctness. After termination of the algorithm, we have $C = N$; let π^* denote the final assignment π, i.e., the assignment π after the final loop. It is easy to verify that π^* is individually rational. Let $i \in N$, with $\pi^*(i) = a \in A^* \cup \{a_\emptyset\}$. Let ℓ denote the loop in which i was put into set C for the last time. In that loop, i was assigned to her favourite choice. Thus, right after loop ℓ, i did not want to deviate from $\pi(i)$ in order to join another activity $b \in A^* \cup \{a_\emptyset\}$. Since the number of agents assigned to an activity is non-decreasing and each agent has decreasing preferences, i does not want to deviate from $\pi^*(i)$ and join an activity to which agents are assigned by π^*. □

4.2 (Contractual) Individual Stability

In the case of increasing preferences individual stability coincides with Nash stability. Theorem 7 hence implies the following corollary.

Corollary 1. *It is NP-complete to decide whether o-GASP admits an individually stable assignment, even when all agents have increasing preferences.*

In contrast, for the case of decreasing preferences we already know that a Nash stable assignment can be found in polynomial time by use of Algorithm 1 (see Theorem 8). Since Nash stability implies individual stability, we can conclude that in case of decreasing preferences an individually stable assignment can be determined in polynomial time. However, below we propose a very simple algorithm which efficiently determines an individually stable assignment.

Algorithm 2. Individually stable assignment for decreasing preferences

---

1: $\pi(h) := a_\emptyset$ for all $h \in N$. $N' := N$. $A' := A^*$.
           // A' is the subset of activities to which no agent is assigned
2: **while** $N' \neq \emptyset$ **do**
3:      $i := \min N'$
4:      **if** $\exists\, a \in A'$ such that $(a, 1) \succ_i a_\emptyset$ **then**
5:          assign i to $a \in A'$ with $(a, 1) \succ_i (a', 1)$ for all $a' \in A' \setminus \{a\}$
6:          $A' := A' \setminus \{a\}$
7:          $N' := N \setminus \{i\}$

Theorem 9. *Given an instance of o-GASP with decreasing preferences, an individually stable assignment always exists and can be determined in polynomial time.*

Proof. Algorithm 2 runs in $\mathcal{O}(n^2 m)$ time. For correctness, observe that an agent assigned to an activity does not want (i) another agent to join due to the agent's decreasing preferences and (ii) does not want to join another activity to which no agent is assigned by the choice of a in line 5. Again by line 5, for an agent i assigned to $a_{\emptyset}$ there is no activity to which no agent is assigned that i would like to join; for activities $a^* \in A^*$ with $|\pi^{a^*}| \geq 1$, each agent assigned to a^* objects to i joining a^*. □

Algorithm 3. Contractually individually stable assignment

1: $\pi(h) := a_{\emptyset}$ for all $h \in N$. $N' := N$.
2: **while** $N' \neq \emptyset$ **do**
3:      $i := \min N'$
4:      **if** $\exists a \in A^* \setminus \pi(i)$ such that
        I. $(a, |\pi^a| + 1) \succ_i (\pi(i), |\pi_i|)$,
        II. for all $i' \in \pi^a$ it holds that $(a, |\pi^a| + 1) \succ_{i'} (a, |\pi^a|)$, and
        III. there is no $h \in \pi_i$ with $(\pi(i), |\pi_i| - 1) \prec_h (\pi(i), |\pi_i|)$
     then
5:         take the best-ranked such a in the ranking $\succ_i$
6:         $\pi(i) := a$
7:         $N' := N$
8:      **else**
9:         $N' := N' \setminus \{i\}$

On the other hand, a contractually individually stable assignment always exists and can be found efficiently. Algorithm 3 computes such an assignment. In each step, the algorithm picks an agent i and checks if i wants to join another activity a (I), such that no agent currently assigned to the same activity as i is worse off (III), and all agents currently assigned to a are better off (II); i is assigned to the best-ranked of such activities, if such an activity exists. Note that if i wishes to deviate from a non-void activity, (III) means that each agent currently assigned to the same activity as i is better off when i leaves. The algorithm proceeds as long as there is an agent i for which such an improvement is possible.

Theorem 10. *Given an instance of* o-GASP, *a contractually individually stable assignment always exists and can be determined in polynomial time.*

Proof. The correctness of the algorithm is obvious. Clearly, the algorithm terminates since each time an agent is assigned to an activity in line 6, at least one agent is strictly better off while no agent is worse off. For the running time of the algorithm, note that each agent i can deviate from its current assigment at most $n \cdot m$ times. The running time of each execution of lines 4 -7 (i.e., the if-part) can very roughly be bounded by $\mathcal{O}((nm)^2)$. Thus, the overall running time of the algorithm is bounded by a polynomial in n and m. □

4.3   Core Stability

For the setting of a-GASP, a weak core stable assignment always exists and can be determined efficiently [7]. While in o-GASP an analogous result holds for decreasing preferences as well (see Corollary 2), in general this is not the case. In particular, our results show that deciding whether o-GASP admits a core stable assignment is computationally hard even in the case of increasing preferences.

As the following example emphasizes, even in the case of increasing preferences a core stable assignment does not always exist.

Example 2. Let us again consider instance (N, A, P) of Example 1. Assume that there is a core stable assignment π. Suppose $\pi(1) = a_\emptyset$. Then $\pi^a = \emptyset$ follows, and the set $E = \{1\}$ would prefer joining a, which violates core stability. Thus, agent 1 must be assigned to a non-void activity; analogously, agents 3 and 5 must be assigned to a non-void activity.

If $\pi(1) = a$, then $\pi(2) = \pi(5) = a$ follows (otherwise $|\pi^a| \leq 2$ holds and hence each member of $E = \{1, 2, 5\} \supset \pi^a$ would be better off with all members of E being assigned to a). Recall that agent 3 is assigned to a non-void activity. By individual rationality of π, $\pi(5) \neq c$ hence implies that $\pi(3) = b$ holds. However, by $\pi(1) = a$ it follows that $|\pi^b| \leq 2$ holds; but then each member of $E = \{1, 3, 4\} \supset \pi^b$ would be better off with all members of E being assigned to b. This violates core stability.

Therewith, $\pi(1) = b$ holds, which means $\pi(3) = \pi(4) = b$. Since agent 5 is assigned to a non-void activity, $\pi(5) = c$ must hold due to $\pi(1) \neq a$. By $\pi(3) \neq c$ we can conclude that $|\pi^c| \leq 2$ holds. But then each member of $E = \{3, 5, 6\} \supset \pi^c$ is better off with all members of E being assigned to c, which violates core stability.

Thus, a core stable assignment does not exist. However, note that if agent 6 is removed, then the assignment $\pi'(1) = \pi'(3) = \pi'(4) = b$ and $\pi'(5) = c$ is core stable.

Theorem 11. *It is NP-complete to decide whether o-GASP admits a core stable assignment, even when all agents have increasing preferences.*

Proof. The hardness-proof proceeds by a reduction from EXACT COVER BY 3-SETS. For membership in NP, observe that a core stable assignment π serves as certificate: For each $a \in A^*$, by scanning the whole profile for each value $k > |\pi^a|$ in polynomial time we can check if there exists a set $E \supset \pi^a$ of size $|E| = k$ such that $(a, |E|) \succ_i (\pi(i), |\pi_i|)$ for all $i \in E$. Thus, overall in polynomial time we can verify if π is indeed core stable.   □

In contrast, in the case of decreasing preferences a core stable assignment always exists and can be determined in polynomial time. This is a consequence of the following proposition.

Proposition 2. *In any instance of o-GASP with decreasing preferences, an assignment π is core stable if and only if π is individually stable.*

Proof. The only-if part is satisfied for general instances of o-GASP. For the if-part, assume that π is individually stable but not core stable. I.e., there are $E \subseteq N$ and $a \in A^*$ with $\pi^a \subset E$ such that $(a, |E|) \succ_i (\pi(i), |\pi_i|)$ for all $i \in E$. For $j \in \pi^a$, $(a, |E|) \succ_j (a, |\pi_a|)$ follows, which by $|E| > |\pi^a|$ and decreasing preferences implies $\pi^a = \emptyset$. Take an arbitrary $i \in E \backslash \pi^a$, i.e., $i \in E$. By decreasing preferences, $(a, |E|) \succ_i (\pi(i), |\pi_i|)$ implies $(a, |\pi^a| + 1) \succ_i (\pi(i), |\pi_i|)$. Trivially, $(a, |\pi^a| + 1) \succ_j (a, |\pi^a|)$ is satisfied for all $j \in \pi^a = \emptyset$. Thus, π is not individually stable in contradiction with our assumption. $\quad\square$

Corollary 2. *Given an instance of* o-GASP *with decreasing preferences, a core stable assignment always exists and can be found in polynomial time.*

Proof. The proof follows from Proposition 2 and Theorem 9. $\quad\square$

5   Conclusion

In this paper, we have analysed the computational complexity involved in finding solutions to the problem of assigning agents to activities on basis on their preferences over pairs made up of an activity and the size of the group of agents participating in that activity. A number of solution concepts are provided for this problem. With respect to the two natural special cases of decreasing and increasing preferences, computational complexity results are provided for the task of finding such a solution. A related research direction would be to find an individually rational assignment maximizing other types of positional scores. Another interesting direction for future research would be to investigate the complexity involved in the considered problems if the number of activities is bounded.

Acknowledgments. The author would like to thank Christian Klamler for helpful comments and Jérôme Lang for pointing towards research directions considered in this paper. In addition, the author would like to thank the anonymous referees for valuable comments that helped to improve the paper. Andreas Darmann was supported by the Austrian Science Fund (FWF): [P 23724-G11] "Fairness and Voting in Discrete Optimization".

References

1. Ballester, C.: NP-completeness in hedonic games. Games Econ. Behav. **49**, 1–30 (2004)
2. Banerjee, S., Konishi, H., Sönmez, T.: Core in a simple coalition formation game. Soc. Choice Welfare **18**, 135–153 (2001)
3. Baumeister, D., Bouveret, S., Lang, J., Nguyen, N., Nguyen, T., Rothe, J., Saffidine, A.: Axiomatic and computational aspects of scoring allocation rules for indivisible goods. In: 5th International Workshop on Computational Social Choice (COMSOC), Pittsburgh, USA, pp. 1–14 (2014)
4. Berman, P., Karpinski, M.: Improved approximation lower bounds on small occurrence optimization. Electron. Colloquium Comput. Complex. (ECCC) **10**(8), 1–19 (2003)

5. Bogomolnaia, A., Jackson, M.: The stability of hedonic coalition structures. Games Econ. Behav. **38**, 201–230 (2002)
6. Brams, S.J., Fishburn, P.C.: Voting procedures. In: Arrow, K.J., Sen, A.K., Suzumura, K. (eds.) Handbook of Social Choice and Welfare, vol. 1, pp. 173–236. Elsevier, Amsterdam (2002)
7. Darmann, A., Elkind, E., Kurz, S., Lang, J., Schauer, J., Woeginger, G.: Group activity selection problem. In: Goldberg, P. (ed.) WINE 2012. LNCS, vol. 7695, pp. 156–169. Springer, Heidelberg (2012)
8. Darmann, A., Schauer, J.: Maximizing Nash product social welfare in allocating indivisible goods. In: 5th International Workshop on Computational Social Choice (COMSOC), Pittsburgh, USA, pp. 1–14 (2014)
9. Lee, H., Shoham, Y.: Stable invitations. In: Proceedings of AAAI 2015 (2015)
10. Skowron, P., Faliszewski, P., Slinko, A.: Fully proportional representation as resource allocation: approximability results. In: IJCAI 2013, Proceedings of the 23rd International Joint Conference on Artificial Intelligence, Beijing, China, 3–9 August 2013, pp. 353–359 (2013)
11. Spradling, M., Goldsmith, J., Liu, X., Dadi, C., Li, Z.: Roles and teams hedonic game. In: Perny, P., Pirlot, M., Tsoukiàs, A. (eds.) ADT 2013. LNCS, vol. 8176, pp. 351–362. Springer, Heidelberg (2013)

Towards Decision Making via Expressive Probabilistic Ontologies

Erman Acar[✉], Camilo Thorne, and Heiner Stuckenschmidt

Data and Web Science Group, Universität Mannheim, Mannheim, Germany
{erman,camilo,heiner}@informatik.uni-mannheim.de

Abstract. We propose a framework for automated multi-attribute decision making, employing the probabilistic non-monotonic description logics proposed by Lukasiewicz in 2008. Using this framework, we can model artificial agents in decision-making situation, wherein background knowledge, available alternatives and weighted attributes are represented via probabilistic ontologies. It turns out that extending traditional utility theory with such description logics, enables us to model decision-making problems where probabilistic ignorance and default reasoning plays an important role. We provide several decision functions using the notions of expected utility and probability intervals, and study their properties.

1   Introduction

Preference representation and its link to decision support systems is an ongoing research problem in artificial intelligence, gaining more attention every day. This interest has led on the one hand to the analysis of decision-theoretic problems using methods common in A.I. and knowledge representation, and on the other hand to apply methods from classical decision theory to improve decision support systems. In this regard there has been a growing interest over the last decade in the use of logics to model preferences, see [1,3,14–19].

Description Logics (DLs) are a family of knowledge representation languages that are based on (mostly) decidable fragments of first order logic. They were designed as formal languages for knowledge representation becoming one of the major formalisms in this field over the last decade. Alongside this and from a more practical point of view, they formally underpin semantic web OWL Web Ontology Language[1], the semantic web key representation and ontology standard (defined by the World Wide Web Consortium).

In this work, we propose a formal framework which is based on expressive probabilistic DLs [13], viz., the non-monotonic P-$\mathcal{SHOIN}(\mathbf{D})$ family of DL languages, designed to model uncertainty and uncertain, non-monotonic reasoning.

In such languages one can express objective (statistical) uncertainty (terminological knowledge concerning concepts), as well as subjective (epistemic) uncertainty (assertional knowledge concerning individuals). Furthermore, due to their non-monotonicity, one can represent and reason with default knowledge.

---

[1] http://www.w3.org/TR/owl-features/.

© Springer International Publishing Switzerland 2015
T. Walsh (Ed.): ADT 2015, LNAI 9346, pp. 52–68, 2015.
DOI: 10.1007/978-3-319-23114-3_4

Also, their probabilistic component employs imprecise probabilities to model uncertainty, which in turn allows to model probabilistic ignorance with considerable flexibility, in contrast to classical probability theory.

We show that our framework can represent decision-theoretic problems and solve them using DL inference services, taking advantage of imprecise probabilities and background knowledge (as represented by ontologies) to compute expected utilities in a fine-grained manner that goes beyond traditional decision theory. One reason why this is possible within a DL-based decision making framework, is because one can express the various dependency relations between attributes/decision criteria with rich DL concept hierarchies and evaluate thereafter *alternatives* in terms of their logical implications.

Our framework can be interpreted as modeling the behavior of an agent, or as a model for systems that support group decisions. In this work, we pursue the former interpretation and focus on modeling artificial agents where each attribute/decision criterion has an independent local utility value (weight). We consider *available alternatives* in the form of DL individuals, and attributes in the form of DL concepts. Finally, we represent the agent's background knowledge and beliefs via a probabilistic DL knowledge base.

In this work, we present several decision functions in order to model agents with different characteristics. Furthermore, the employed logic's use of imprecise probabilities to model uncertainty, allows considerable expressive power to model non-standard decision behaviour that violate the axioms of (classical) expected utility e.g., Ellsberg paradox. Using the framework, we show that it is straightforward to provide decision functions which model ambiguity averse decision-making. In so doing, we investigate the various properties of such decision functions as well as their connection to ontological knowledge.

2   Preliminaries

Preferences and Utility. Traditional utility theory [10] models the behavior of rational agents, by quantifying their available choices in terms of their utility, modeling preference (and eventual courses of action) in terms of the induced partial orders and utility maximization.

Let $A = \{a_1, \ldots, a_n\}$ be a set of alternatives, and a (rational) *preference* is a complete and transitive binary relation $\succeq$ on A. Then, for any $a_i, a_j \in A$ for $i, j \in \{1, \ldots, n\}$, *strict preference* and *indifference* is defined as follows: $a_i \succ a_j$ iff $a_i \succeq a_j$ and $a_j \not\succeq a_i$ (*Strict preference*), $a_i \sim a_j$ iff $a_i \succeq a_j$ and $a_j \succeq a_i$ (*Indifference*).

It is said that, a_i is *weakly preferred*[2] (*strictly preferred*) to a_j whenever $a_i \succeq a_j$ ($a_i \succ a_j$), a_i is indifferent to a_j whenever $a_i \sim a_j$. Moreover, a_i, a_j are *incomparable* iff $a_i \parallel a_j \iff a_i \not\succeq a_j$ and $a_j \not\succeq a_i$, which implies that $\succeq$ is a partial ordering.

---

[2] It is also called *preference-indifference* relation, since it is the union of strict preference and indifference relation.

In order to represent the preference relation numerically one introduces the notion of *utility*, which is is a function that maps an alternative to a positive real number reflecting its degree of desire. For a decision theoretic framework, two questions are essential; given a (finite) set of alternatives (i) which alternative is the best one(s)? (ii) How does the whole preference relation look like i.e., a complete list of order of alternatives (e.g., $a_1 \succ a_3 \succeq \ldots$). Throughout the paper, these two main questions will also be of our concern, along with a restriction to single (non-sequential) decisions.

Formally, given a finite set of alternatives $A = \{a_1, \ldots, a_n\}$, and preference $\succeq$ on A, $u : A \to \mathbb{R}$ is a *utility function* iff for any $a_i, a_j \in A$ with $i, j \in \{1, \ldots, n\}$, $a_i \succ a_j \iff u(a_i) > u(a_j)$, $a_i \succeq a_j \iff u(a_i) \geq u(a_j)$, $a_i \sim a_j \iff u(a_i) = u(a_j)$.

For the proof that such a function exists, we refer the reader to the so-called representation theorems in [7].

The basic principle in utility theory is that a rational agent should always try to maximize its utility, or *should take the choice with the highest utility*. Utility functions modeling behaviours based on more than one attribute (i.e., n-ary) are called multi-attribute utility functions. Let $X = \{X_1, \ldots, X_n\}$ be a set of attributes where $n \geq 2$, and $\Omega = X_1 \times \cdots \times X_n$ be the set of outcomes over which the agent's preference relation is defined. An alternative/outcome is a tuple $(x_1, \ldots, x_n) \in \Omega$. Let $\succeq$ be the preference relation defined over X, then u is a *multi-attribute utility function* representing $\succeq$ iff for all $(x_1, \ldots, x_n), (y_1, \ldots, y_n) \in \Omega$, $(x_1, \ldots, x_n) \succeq (y_1, \ldots, y_n) \iff u(x_1, \ldots, x_n) \geq u(y_1, \ldots, y_n)$. For an introductory text on multi-attribute utility theory, see [10].

Moreover, a utility function u is said to be *unique up to affine transformation* iff for any real numbers $m > 0$ and c, $u(x) \geq u(x')$ iff $m \cdot u(x) + c \geq m \cdot u(x') + c$.

Along the paper, we will use two running examples to point out two important limitations of traditional decision theory that we will overcome with description logics. theoretic: *Ellsberg's Paradox* and a a *touristic agent* example.

Ellsberg's Paradox. Assume that there is an urn, full with three different colours of balls, namely red, blue and yellow. You know only that 1/3 of the balls are red, and the blue and yellow balls together make up the remaining 2/3. However, it is possible that either there is no single blue ball (that is all of them are yellow) or that all of them are blue. Now, before randomly picking up a ball from the urn, you are asked to make a guess, choosing red or blue with the following two gambles:

1st Gamble: If you guess correctly, you get the prize.
2nd Gamble: If you guess correctly, or the ball is yellow, you get the prize.

If you prefer to choose red to blue (i.e., *red* $\succ$ *blue*) in *Gamble A*, then following the *sure-thing principle*, you are supposed to also have *red* $\succ$ *blue* in *Gamble B*, since

$$U(red) \cdot Pr(red) > U(blue) \cdot Pr(blue)$$
$$\Longrightarrow$$
$$U(red) \cdot (Pr(red) + Pr(yellow)) > U(blue) \cdot (Pr(blue) + Pr(yellow)).$$

However, people usually choose $red \succ blue$ in the first gamble, and $blue \succ red$ in the second. This particular situation is called *Ellsberg's paradox* [6] and is not compatible with the preferential predictions that ensue from subjective utility theory (which we will not mention here (see [7] for details)), and arises in the presence of *ambiguity* in probabilities [6].

The Tourist Example. Imagine a tourist trying to decide in which hotel to stay. He/she would rather stay at a 5 star hotel, rather than a 4 star hotel (among other features he may desire). But what if the hotel suggested in his trip has a bad reputation? Intuitively, *we* know that this has a negative impact, but how do we factor in *background knowledge* about, say, hotels? This example will be used to motivate the importance of using structured knowledge (e.g., ontologies, concept hierarchies) in decision making in order to perform logical reasoning.

The P-$\mathcal{SHOIN}$(D) Probabilistic DL. Lukasiewicz's probabilistic description logics (DLs), see [13], extend classical DLs with probabilistic, non-monotonic reasoning. DLs are logics –typically fragments of first order logic– specifically designed to represent and reason on structured knowledge, where domains of

Table 1. Syntax and semantics of the DL $\mathcal{SHOIN}$(D). Notice that **D** refers to concrete domains. The first block introduces *individuals*. The second block recursively defines *concepts* (others can be introduced by explicit definition), while the third does it with roles. The fourth formally introduces terminological statements, resp., concept (ISA) and role *inclusion* statements. Finally, the fifth block introduces assertional facts a.k.a. membership assertions, resp. concept and role *membership* assertions. A TBox $\mathcal{T}$ is a set of terminological statements, an ABox $\mathcal{A}$ is a set of assertions, and a KB is a pair $T = (\mathcal{T}, \mathcal{A})$. Entailment and satisfiability are defined in the usual way. The syntax and semantics of P-$\mathcal{SHOIN}$(D) extend this definition.

Syntax	Semantics w.r.t. classical interpretation $\mathcal{I} = (\Delta^{\mathcal{I}}, \cdot^{\mathcal{I}})$
i	$i^{\mathcal{I}} \in \Delta^{\mathcal{I}}$
A	$A^{\mathcal{I}} \subseteq \Delta^{\mathcal{I}}$
D	$D^{\mathcal{I}} \subseteq \mathbf{D} = \text{NUM} \cup \text{STRING}$
$OneOf(i_1, \ldots, i_n)$	$(OneOf(i_1, \ldots, i_n))^{\mathcal{I}} := \{i_1, \ldots, i_n\}$
$\neg\phi$	$(\neg\phi)^{\mathcal{I}} := \Delta^{\mathcal{I}} \setminus \phi^{\mathcal{I}}$
$\exists r.\phi$	$(\exists r.\phi)^{\mathcal{I}} := \{d \mid \text{exists } e \text{ s.t. } (d, e) \in r^{\mathcal{I}} \text{ and } e \in \phi^{\mathcal{I}}\}$
$\exists_{\leq k} r$	$(\exists_{\leq k} r)^{\mathcal{I}} := \{d \mid \text{exists at most } k \text{ es s.t. } (d, e) \in r^{\mathcal{I}}\}$
$\phi_1 \sqcap \phi_2$	$(\phi_1 \sqcap \phi_2)^{\mathcal{I}} := \phi^{\mathcal{I}} \cap \phi'^{\mathcal{I}}$
p	$p^{\mathcal{I}} \subseteq \Delta^{\mathcal{I}} \times \Delta^{\mathcal{I}}$
r^-	$(r^-)^{\mathcal{I}} := \{(d, e) \mid (d, e) \in r^{\mathcal{I}}\}$
$Tr(r)$	$(Tr(r))^{\mathcal{I}} :=$ the transitive closure of r in $\Delta^{\mathcal{I}} \times \Delta^{\mathcal{I}}$
$\phi_1 \sqsubseteq \phi_2$	$\mathcal{I} \models \phi_1 \sqsubseteq \phi_2 \text{iff} \phi_1^{\mathcal{I}} \subseteq \phi_2^{\mathcal{I}}$
$r_1 \sqsubseteq r_2$	$\mathcal{I} \models r_1 \sqsubseteq r_2 \text{iff} r_1^{\mathcal{I}} \subseteq r_2^{\mathcal{I}}$
$\phi(i)$	$\mathcal{I} \models \phi(i) \text{iff} i^{\mathcal{I}} \in \phi^{\mathcal{I}}$
$r(i, j)$	$\mathcal{I} \models r(i, j) \text{iff} (i, j)^{\mathcal{I}} \in r^{\mathcal{I}}$

interest are represented as composed of *objects* structured into: (i) *concepts*, corresponding to classes, denoting sets of objects; (ii) *roles*, corresponding to (binary) relationships, denoting binary relations on objects. Knowledge is predicated through so-called *assertions*, i.e., logical axioms, organized into an intensional component (called TBox, for "terminological box"), and an extensional one (called ABox, for "assertional box"), viz. the former consists of a set of universal statements and the latter of a set of atomic facts. A DL *knowledge base* (KB) is then defined as the combination of a TBox and an ABox.

For simplicity, we restrict the discussion in this paper to the P-$\mathcal{SHOIN}$(**D**) family of probabilistic logics, which is an extension of the known $\mathcal{SHOIN}$(**D**) DL whose syntax and semantics we briefly recall in Table 1. $\mathcal{SHOIN}$(**D**) underpins the OWL-DL fragment of OWL (in the OWL 1.1 standard).

Example 1. DLs KBs can be used to formally model domain knowledge, and formally reason over it. Consider the hotel domain. Consider now the KB with TBox T = {*OneStarHotel* $\sqsubseteq$ *Hotel* $\sqcap$ $\exists hasService.ExtendedBreakfast$}, which states that *every one star hotel is an hotel and there is an extended breakfast service*, and ABox A = {*OneStarHotel(tapir)*}, which says that *Tapir is a one star hotel*.[3] Following $\mathcal{SHOIN}$(**D**) semantics, we will conclude that *Tapir is a hotel and it has an extended breakfast service* (T, A) $\models$ *Hotel* $\sqcap$ $\exists hasService.ExtendedBreakfast(tapir)$. ♣

Given that the semantics of the P-$\mathcal{SHOIN}$(**D**) family is very rich, we avoid giving a full description of it (which would go beyond the scope of this paper), and provide, rather a basic overview of their syntax and semantics, and cover its main properties (on which our results rely) in a succinct Appendix. For its full definition and properties, we refer the reader to [13]. A general remark is that the framework that we present here is (w.l.o.g.) independent from a particular choice of P-DL, provided they cover numeric domains (more in general, data types).

Syntax. The P-$\mathcal{SHOIN}$(**D**) family extends the syntax of $\mathcal{SHOIN}$(**D**) with the language of conditional constraints defined as follows: $\mathbf{I}_P$ is the set of *probabilistic individuals o*, disjoint from classical individuals $\mathbf{I}_C = \mathbf{I} \backslash \mathbf{I}_P$, $\mathcal{C}$ is a finite nonempty set of basic classification concepts or basic *c-concepts*, which are (not necessarily atomic) concepts in $\mathcal{SHOIN}$(**D**) that are free of individuals from $\mathbf{I}_P$. Informally, they are the DL concepts relevant for defining probabilistic relationships. In what follows we overload the notation for concepts with that of c-concepts.

In addition to probabilistic individuals, TBoxes and ABoxes can be extended in P-$\mathcal{SHOIN}$(**D**) to *probabilistic* TBoxes (PTBoxes P) and ABoxes (PABoxes P_o), via so-called *conditional constraints*, expressing (or encoding) uncertain, default knowledge about domains of interest. A PTBox *conditional constraint* is an expression $(\psi|\phi)[l, u]$, where ψ and ϕ are c-concepts, and $l, u \in [1, 0]$. Informally, $(\psi|\phi)[l, u]$ encodes that the probability of ψ given ϕ lies, *by default*, within $[l, u]$. A PABox constraint $(\psi|\phi)[l, u] \in P_o$ however, relativizes constraint $(\psi|\phi)[l, u]$ to the individual o.

---

[3] By convention, objects are written with lower case.

A probabilistic KB $\mathcal{K} := (T, P, (P_o)_{o \in \mathbf{I}_P})$ consists of T a classical KB[4], P a PTBox (a set of conditional constraints), and a collection of PABoxes, each of which is a (possibly empty) set of relativized conditional constraints for each probabilistic $o \in \mathbf{I}_P$.

Semantics. A *world* I is a finite set of basic c-concepts $\phi \in \mathcal{C}$ such that $\{\phi(i) \mid \phi \in I\} \cup \{\neg\phi(i) \mid \phi \in \mathcal{C}\backslash I\}$ is satisfiable, where i is a new individual (intuitively worlds specify an individual unique up to identity), whereas $\mathcal{I}_C$ is the set of all worlds relative to $\mathcal{C}$. $I \models T$ iff $T \cup \{\phi(i) \mid \phi \in I\} \cup \{\neg\phi(i) \mid \phi \in \mathcal{C}\backslash I\}$ is satisfiable. $I \models \phi$ iff $\phi \in I$. $I \models \neg\phi$ iff $I \models \phi$ does not hold. For c-concepts ϕ and ψ, $I \models \psi \sqcap \phi$ iff $I \models \psi$ and $I \models \phi$. Note that above notion of satisfiability based on worlds is compatible with the satisfiability of classical knowledge bases, that is, there is a classical interpretation $\mathcal{I} = (\Delta^{\mathcal{I}}, \cdot^{\mathcal{I}})$ that satisfies T iff there is a world $I \in \mathcal{I}_C$ that satisfies T.[5]

A *probabilistic interpretation* Pr is a probability function $Pr : \mathcal{I}_C \rightarrow [0,1]$ with $\sum_{I \in \mathcal{I}_C} Pr(I) = 1$. $Pr \models T$, iff $I \models T$ for every $I \in \mathcal{I}_C$ such that $Pr(I) > 0$. The *probability* of a c-concept ϕ in Pr is defined as $Pr(\phi) = \sum_{I \models \phi} Pr(I)$. For c-concepts ϕ and ψ with $Pr(\phi) > 0$, we write $Pr(\psi|\phi)$ to abbreviate $Pr(\psi \sqcap \phi)/Pr(\phi)$. For a conditional constraint $(\psi|\phi)[l,u]$, $Pr \models (\psi|\phi)[l,u]$ iff $Pr(\phi) = 0$ or $Pr(\psi|\phi) \in [l,u]$. For a set of conditional constraints $\mathcal{F}$, $Pr \models \mathcal{F}$ iff $Pr \models F$ for all $F \in \mathcal{F}$. Notice that T has a satisfying classical interpretation $\mathcal{I} = (\Delta^{\mathcal{I}}, \cdot^{\mathcal{I}})$ iff $Pr \models T$.[6] We provide further technical details in the Appendix.

Satisfaction and entailment in $\mathcal{SHOIN}(\mathbf{D})$ can be extended to probabilistic interpretations Pr, see the Appendix. More important for our purposes are the *defeasible* entailment relations induced by P-$\mathcal{SHOIN}(\mathbf{D})$, viz., *lexicographic entailment* $\|\!\sim^{lex}$ and *tight lexicographic entailment* $\|\!\sim^{lex}_{tight}$. Probabilistic KBs in general and conditional constraints in particular encode as we said probable, default knowledge, and *tolerate* to some degree inconsistency (w.r.t. classical knowledge). Lexicographic entailment supports such tolerance by intuitively: (i) partitioning P (ii) selecting the lexicographically least set in such partition consistent with T. See the Appendix for the technicalities.

Reasoning Problems. A reasoning problem that will be of our interest is *probabilistic membership* PCMEM (probabilistic concept membership): given a consistent probabilistic KB $\mathcal{K}$, a probabilistic individual $o \in \mathbf{I}_P$, and a c-concept ψ, compute $l, u \in [0,1]$ such that $\mathcal{K} \|\!\sim^{lex}_{tight} (\psi|\top)[l,u]$ for o.

3 Representing Decision Making Problems

In this section we introduce probabilistic DL decision bases. Regarding notation, we will try to stick to that in [13] as much as possible, to give the reader easy access to the referred paper.

---

[4] Note that T is not used to denote a classical TBox anymore but rather the whole classical knowledge base, TBox and ABox.

[5] See Proposition 4.8 in [13].

[6] See Proposition 4.9 in [13].

Attributes and Preferences. We define the non-empty *set of attributes* as a subset of c-concepts derived from basic c-concepts $\mathcal{C}$. Informally, every world I determines a subset of attributes that is to be satisfied. We will assume that the set of attributes X possibly contains redundancies.

Decision Base. We define a *decision base* that models an agent in a decision situation; *background knowledge* of the agent is modelled by a probabilistic knowledge base, the finite set of available *alternatives* are modelled by a set of individuals, and a *weight function* that is defined over the set of *attributes* which will be used to derive the preference relation of the agent.

Definition 1 (Decision Base). *A probabilistic description logic decision base is a triple* $\mathcal{D} := (\mathcal{K}, \mathcal{A}, \mathcal{U})$ *where:*

- $\mathcal{K} = (T, P, (P_o)_{o \in I_P})$ *is a* consistent *probabilistic KB encoding background knowledge,*
- $\mathcal{A} \subseteq \boldsymbol{I}$ *is the set of alternatives,*
- $\mathcal{U}$ *is UBox, that is a finite graph of a bounded real-valued function* $w : X \longrightarrow \mathbb{R}^+$ *with* $w(\bot) = 0$. †

Informally, the role of $\mathcal{K}$ is to provide assertional information about the alternatives at hand, along with the general terminological knowledge information that the agent may require to reason further over alternatives; indeed X is the set of concepts ϕ such that $\mathcal{K}$ logically entails $\phi(a)$. Moreover, $\mathcal{U}$ can be defined to include negative weights as well, (i.e., $w : X \longrightarrow \mathbb{R}$ instead or $\mathbb{R}^+$) to model undesirable outcomes or punishments.[7] However, for the sake of brevity, we will consider here only positive weights.

Alternatives with Classical Knowledge. In this particular setting, we assume we are in possession of *certain* information about the alternatives, and consider only the *certain* subsumption relations between concepts. We do this by providing a value function for alternatives, defined over the classical component of the framework (i.e., the classical DL KB T in the decision base).

Definition 2 (Utility of an Alternative). *Given a decision base* $\mathcal{D} = (\mathcal{K}, \mathcal{A}, \mathcal{U})$, *the* utility *of an alternative* $a \in \mathcal{A}$ *is,*

$$U(a) := \sum \{w(\phi) \mid T \models \phi(a) \wedge \phi \in X\}. \tag{1}$$

where $\mathcal{K} = (T, P, (P_o)_{o \in I_P})$ *and* $a \in \boldsymbol{I_C}$. †

---

[7] Alternatively, $\mathcal{U}$ can be studied in two partition, that is, the set of pairs with non-negative (denoted $\mathcal{U}^+$) and negative weights (denoted $\mathcal{U}^-$). In extreme cases, $\mathcal{U} = \mathcal{U}^+$ when $\mathcal{U}^- = \emptyset$ (similarly for $\mathcal{U} = \mathcal{U}^+$).

In this work, for the sake of simplicity, we define U as a summation. Note however that U can be potentially any utility function, such as, e.g., $U(a) = 2(p_1 w(\phi) \cdot p_2 w(\psi)) + p_3 \exp(w(\gamma)) + c$, were a to satisfy $\psi, \phi, \gamma \in X$, where $p_i, c \in \mathbb{R}$, for $i = 1, 2, 3$. Furthermore, we assume that w is defined without knowing the exact knowledge base and its transitive closure on subsumption, without having *complete knowledge* about the ontological relations between attributes.

Notice that each alternative corresponds to an outcome. Using U, we define the preference relation $\succ$ over alternatives $A = \{a_1, \dots, a_n\}$: $a_i \succ a_j$ iff $U(a_i) > U(a_j)$, for $i, j \in \{1, \dots, n\}$; $\succeq$ and $\sim$ are defined similarly.

Definition 3 (Optimal Choice). *Given a decision base* $\mathcal{D} = (\mathcal{K}, \mathcal{A}, \mathcal{U})$, *the optimal choice w.r.t.* $\mathcal{D}$ *is,*

$$Opt(\mathcal{A}) := \arg\max_{a \in \mathcal{A}} U(a) \qquad (2)$$

That is, an alternative gets a reward for satisfying each attribute independently.

†

Intuitively, the function U measures the value of an alternative with respect to the concepts (possibly deduced) that it belongs. The following proposition is an immediate result of that.

Proposition 1. *Let* T *be a classical part of the knowledge base of* $\mathcal{D}$ *and* $a_1, a_2 \in A$ *be any two alternatives. If for every* $\phi \in X$ *with* $T \models \phi(a_1)$, *there is a* $\psi \in X$ *with* $T \models \psi(a_2)$ *such that* $T \models \phi \sqsubseteq \psi$, *then* $a_1 \succeq a_2$.

Proof. ψ be any basic c-concept such that $T \models \psi(a_2)$ and $(\psi, w(\psi)) \in \mathcal{U}$, then $U(a_2) \geq w(\psi)$. By assumption, there is a $\phi \in X$ such that $T \models \phi \sqsubseteq \psi$ and $T \models \phi(a)$, hence $T \models \psi(a)$. It follows that $U(a_1) \geq w(\psi)$, therefore $a_1 \succeq a_2$. $\square$

Intuitively, *ceteris paribus* (everything else remains the same) any thing that belongs to a subconcept should be at least as desirable as something that belongs to a superconcept; for instance, a *new sport car* is at least as desirable as a *sport car* (since anything that is a *new sport car* is a *sport car* i.e., *new sport car* $\sqsubseteq$ *sport car*).[8] The following results says that two alternatives are of same desirability if they belong to exactly the same concepts.

Corollary 1. *Let* $\mathcal{D}$ *be decision base with a classical knowledge base* T *and a set of alternatives* $\mathcal{A}$. *Then for any two alternatives* $a, a' \in \mathcal{A}$, $a \sim a'$ *iff* $\{\psi \mid \psi \in X, T \models \psi(a)\} = \{\phi \mid \phi \in X, T \models \phi(a)\}$.

Proof. By applying Proposition 1 in both directions (i.e., $a \sim a' \implies a \succeq a'$ and $a' \succeq a$). $\square$

---

[8] Recall that we concern ourselves with desirable attributes, i.e., weights are non-negative.

The intuitive explanation for Corollary 1 is that we measure the desirability (and non-desirability) of things, according to what they are, or to which concepts they belong. This brings forward the importance of reasoning, since it might not be obvious at all that two alternatives actually belong to exactly the same concepts w.r.t attributes.

Example 2. Consider the following decision base about choosing a trip:

$$T=\{\ hasHotel(trip1, merdan), hasHotel(trip2, armada), FiveStarHotel(meridian),$$
$$Expensive \sqsubseteq \neg Economic, \exists hasHotel.FiveStarHotel \sqsubseteq Expensive,$$
$$\exists hasHotel.ThreeStarHotel \sqsubseteq Economic, ThreeStarHotel(armada)\}$$
$$\mathcal{U}=\{\ (Expensive, 10), (Economic, 15)\} \qquad \mathcal{A} = \{trip1, trip2\}$$

Here $U(trip1) = 10$, since the agent knows that *trip1* has a five star hotel, it is an *Expensive* trip. Similarly, $U(trip2) = 15$, therefore $trip2 \succ trip1$. $Opt(\mathcal{A}) = trip2$. ♣

Properties of the Utility Function. Since every individual is corresponds to a subset of attributes that it satisfies, in this section we will treat U as if it was formally defined over the set of attributes X rather than that of individuals so that we can discuss some common properties of U following the definitions given in [3].

Proposition 2. *Suppose that U is a value function. Then U is (a) normalized, (b) non-negative, (c) is monotone, (d) concave, (e) sub-additive, (f) unique up to positive affine transformation.*

Proof. We deal with each property separately:

(a) This holds when the individual does not satisfy any attributes, whence $U(\emptyset) = 0$.
(b) Follows from Proposition 1 and property (a).
(c) Follows from Proposition 1.
(d) Let $Y, Z, T \subseteq X$ with $Z \subseteq Y$. Since the classical part of the logic is monotonic and weights are positive, whenever $\mathcal{I} \models Y$, $\mathcal{I} \models Z$, which implies $U(X \cup Y) - U(Y) \leq U(X \cup Z) - U(Z)$.
(e) Follows from (d).
(f) Let $Y, Z \subseteq X$ with $U(Y) \geq U(Z)$ and $M(x) = ax + b$ with $a > 0$ and b,

$$M(U(Y)) \geq M(U(Z)) \iff a \cdot \textstyle\sum_{\psi \in Y} w(\psi) + b \geq a \cdot \textstyle\sum_{\psi \in Z} w(\psi) + b$$
$$\iff \textstyle\sum_{\psi \in Y} w(\psi) \qquad \geq \textstyle\sum_{\psi \in Z} w(\psi). \qquad \square$$

Decisions with Default Ontological Reasoning. In many situations, preferential statements that are done by human agents are not meant to be strict statements, say, as in the formal sciences, nor do they take full ontological knowledge into account. When someone asserts that she prefers a *suite* to a *standard room* (i.e., *suite* $\succ$ *standard room*), it is often the case that the statement is not meant to hold for every suite e.g., a *a burned suite* (*burned suite* $\not\succ$ *standard room*). We would like to model such preferential statements in our framework, but they potentially violate Proposition 1. Indeed, the decision rule for classical ontologies (Definition 3) cannot deal with such cases. To do this we need to go beyond classical KBs, and consider full P-$\mathcal{SHOIN}(\mathbf{D})$ KBs and their reasoning techniques.

Example 3. (cont'd) Now consider the previous example extended with the following knowledge P and T:

$$T = \{BadFamedFiveStarHotel \sqsubseteq FiveStarHotel\}$$
$$\mathcal{P} = \{(Desirable|\exists hasHotel.FiveStarHotel)[1,1],$$
$$(\neg Desirable|\exists hasHotel.BadFamedFiveStarHotel)[1,1]\}$$
$$\mathcal{U} = \{(Desirable, 10)\}$$

which encodes the following knowledge: *A bad famed five star hotel is a five star hotel. Generally, a trip which has a five star hotel is desirable. Generally, a trip which has a bad famed five star hotel is undesirable.*   ♣

With classical $\mathcal{SHOIN}(\mathbf{D})$ reasoning, it follows that *any trip that has a bad famed five star hotel is a trip that has five star hotel*, in symbols $\exists hasHotel.BadFamedFiveStarHotel \sqsubseteq \exists hasHotel.FiveStarHotel$. Note that in the light of this information, it is entailed that *trip*1 is desirable. However, if the agent also learns (added to its knowledge base) that *meridian is a bad famed five star hotel*, then *trip*1 will not be desirable anymore[9].

Decisions with Ontological Probabilistic Reasoning. In this section, we will generalize our previously introduced choice functions with probabilities, that will result in different behavioral characteristics in the presence of uncertainty. Those behavioral characteristics can be interpreted as different types of agents (optimistic, pessimistic etc.), or a decision support system that orders alternatives with respect to different criteria (best possible uncertain outcome, worst possible uncertain outcome etc.) and user preferences.

A remark on notation before defining expected utility intervals: we will use the notation $[\text{PCMEM}(\mathcal{K}, a, \phi)]$ to denote the tight interval $[l, u]$ that is the answer to the query PCMEM, with regard to knowledge base $\mathcal{K}$, individual $a \in I_{\mathbf{P}}$ and c-concept ϕ. Moreover, $l = \llcorner\text{PCMEM}(\mathcal{K}, a, \phi)\lrcorner$ and $r = \ulcorner\text{PCMEM}(\mathcal{K}, a, \phi)\urcorner$

As we have a set of probability functions instead of a single probability function which results in probability intervals, we get an interval of the expected utilities. That is, $EU(a) = \sum_{\phi \in X} \Pr(\phi) \cdot w(\phi)$ is the expected utility of an alternative a w.r.t. Pr, and EI is the expected utility interval defined as follows.

---

[9] This is done via Lehmann's lexicographic entailment; in this particular example z-partition is (P_0, P_1) where $P_0 = \{(\neg Desirable|\exists hasHotel.FiveStarHotel)[1,1]\}$ and $P_1 = \{(Desirable|\exists hasHotel.FiveStarHotel)[1,1]\}$ that is, $(T, P) \cup BadFamedFiveStarHotel(meridian) \mid\hspace{-0.3em}\sim^{lex} \neg Desirable(trip1)$.

Definition 4 (Expected Utility Interval of an Alternative). *Given a decision base* $\mathcal{D}$, *the* expected utility interval *of an alternative* $a \in \mathcal{A}$ *is,*

$$EI(a) := [\sum_{\phi \in X} \llcorner \text{PCMEM}(\mathcal{K}, a, \phi) \lrcorner \cdot w(\phi), \sum_{\phi \in X} \ulcorner \text{PCMEM}(\mathcal{K}, a, \phi) \urcorner \cdot w(\phi)] \quad (3)$$

Notice that each element in the interval is an expected utility, defined via a (potentially) distinct probability distribution. For simplicity we will denote by $\underline{EI}(a)$ *the infimum and* $\overline{EI}(a)$ *the supremum of* $EI(a)$ *(i.e., the extrema of* $EI(a)$*).* †

Now, using expected utility intervals we will define some decision functions (mainly from the literature of imprecise probabilities) which generalize the notion of choices by maximum expected utility. A *decision function* δ maps non-empty sets of alternatives A to a subset of A where $a \in \delta(\mathcal{A})$ iff $a \succeq a'$ for every $a' \in \mathcal{A}$[10].

We proceed to define decision functions characterizing different kinds of rational agents. In terms of their use of intervals, they are similar to the Γ-*maximax*, Γ-*minimax, Interval Dominance* and *E-admissibility* in the literature of imprecise probabilities [8].

Definition 5 (Optimistic, Pessimistic Choices). *Given a decision base* $\mathcal{D} = (\mathcal{K}, \mathcal{A}, \mathcal{U})$, *and* $EI(a) = [\underline{EI}(a), \overline{EI}(a)]$ *for any* $a \in \mathcal{A}$ *w.r.t.* $\mathcal{D}$, *then* δ *is, resp.* optimistic *or* pessimistic *iff*

$$\overline{Opt}(\mathcal{A}) := \arg\max_{a \in \mathcal{A}} \overline{EI}(a) \qquad or \qquad \underline{Opt}(\mathcal{A}) := \arg\max_{a \in \mathcal{A}} \underline{EI}(a). \quad (4)$$

We denote the preference order w.r.t. optimistic and pessimistic choice with $\succeq^{\overline{opt}}$, $\succeq_{\underline{opt}}$ *respectively. Strict orders are defined accordingly.* †

Definition 6 (Cautious Choice). *The decision function* δ *is said to be cautious iff* $\delta_{id}(\mathcal{A}) := \{a \in \mathcal{A} \mid \underline{EI}(a) \geq \overline{EI}(a') \text{ for all } a' \in \mathcal{A}\}$. †

We will denote the preference ordering of cautious choices with $\succ_{id}$ (*id* for interval dominance). Interval dominance offers a formalisation for incomparability; that is, if two alternatives a and a' have neither overlapping expected utility intervals (i.e., $EI(a) \neq EI(a')$), nor dominate each other (which means that an agent cannot decide between them), then $a_1 \parallel a_2$. Notice that $\succeq_{id}$ is a partial weak order whereas $\succeq_{\underline{opt}}$ and $\succeq^{\overline{opt}}$ are total-weak orders.

Example 4. Consider an hotel choosing agent and KB $\mathcal{K}$ where

$$T = \{GoodHotel \sqsubseteq \neg BadHotel, Hotel \sqsubseteq FourStarHotel \sqcup OneStarHotel\}$$
$$\mathcal{P} = \{((GoodHotel|FourStarHotel)[1,1], (BadHotel|OneStarHotel)[1,1]\}$$
$$P_{ritz} = \{(FourStarHotel|\top)[0.5, 0.7], (Hotel|\top)[1,1]\}$$
$$P_{tivoli} = \{(FourStarHotel|\top)[0.3, 0.1](Hotel|\top)[1,1]\}$$
$$P_{holiday} = \{(OneStarHotel|\top)[0.1, 0.3], (Hotel|\top)[1,1]\}$$

---

[10] Note that this definition essentially coincides with that choice functions in the imprecise probability literature [8], with the exception that it is allowed to return an empty set.

According to $\mathcal{K}$ our agent knows that *a good hotel is not a bad hotel* (symmetrically), that *usually a four star hotel is a good hotel, and usually one star hotel is a bad hotel,* that *tivoli* is a four star hotel with a probability of at least 0.3, and so on. For the sake of simplicity, it is further assumed that an hotel is either four star or one star hotel. Assume that the agent can choose among three courses of action, viz., among alternatives $\mathcal{A} = \{ritz, tivoli, holiday\}$, relatively to the UBox $\mathcal{U} = \{(GoodHotel, 10), (BadHotel, 0)\}$. If the agent is pessimistic, she will choose *holiday* since it is a Good Hotel with probability of at least 0.7, (his preference being *holiday* $\succ_{opt}$ *ritz* $\succ_{opt}$ *tivoli*). An optimistic agent will instead choose *tivoli* (i.e., *tivoli* $\succ^{\overline{opt}}$ *ritz* $\succ^{\overline{opt}}$ *holiday*). Finally, a cautious agent prefers *holiday* to *ritz*. However, in general it cannot make a choice, since *tivoli* $\|$ *ritz* and *tivoli* $\|$ holiday. ♣

Notice that interval dominance is a very strict restriction that is not very helpful in *normative* settings. We give a less strict version based on Levi's notion of E-admissibility in [8,12] (E for expected).

Definition 7 (E-Admissible Choice). *An alternative $a \in \mathcal{A}$ is E-admissible $(a \in \delta_e(\mathcal{A}))$ iff for every $\phi \in X$, there is a $Pr(\phi) \in [l, u]$ s.t. $K \parallel\!\!\sim^{lex}_{tight} a : \phi[l, u]$, and for every $a' \in \mathcal{A}\backslash\{a\}$ and for every $Pr'(\phi) \in [l', u']$ s.t. $K \parallel\!\!\sim^{lex}_{tight} a' : \phi[l', u']$, $Pr(\phi) > Pr'(\phi)$ holds. We denote the preference relation with $\succeq_e$.* †

Informally, δ_e looks for a probability distribution that lets an alternative weakly dominates every other.

Example 5. Consider alternatives $\mathcal{A} = \{a_1, a_2, a_3\}$ with expected utility intervals on a single attribute, that are $[5, 7], [1, 10]$ and $[1, 8]$. Assume that there are two distributions Pr and Pr' such that expected utility of each alternatives w.r.t. Pr is 5, 7, 6, and 6, 7, 8 w.r.t. Pr'. Also assume that there is no Pr'' such that $EU(a_1) \geq EU(a_2)$ and $EU(a_1) \geq EU(a_3)$. Then, $\delta_e(\mathcal{A}) = \{a_2, a_3\}$, that is, $a_3 \|_e a_2$ and $a_2 \succeq_e a_1$ as well as $a_3 \succeq_e a_1$.

Proposition 3. *The following statements hold: (i) $\succeq^{\overline{opt}} \subseteq \succeq_e$ and, on the other hand, (ii) $\succeq_{id} \subseteq \succeq^{\overline{opt}} \cap \succeq_{opt}$.*

Proof. We prove each condition separately:

(a) Let $(a, a') \in \succeq^{\overline{opt}}$. By definition of $\overline{Opt}(\mathcal{A})$, there is a $Pr(\phi) \in [l, u]$, (indeed $Pr(\phi) = u$) such that, on the one hand $\overline{EI}(\overline{Opt}(\mathcal{A})) = Pr(\phi) \cdot w(\phi)$, and $\overline{EI}(\overline{Opt}(\mathcal{A})) \geq \overline{EI}(\overline{Opt}(\mathcal{A}\backslash\overline{Opt}(\mathcal{A})))$ on the other hand. These fact together imply that $(a, a') \in \succeq_e$.

(b) Let $(a, a') \in \succ_i$; then $\underline{EI}(a) \geq \underline{EI}(a')$, which means (i) $a = \underline{Opt}(\mathcal{A})$ and (ii) $a' = \overline{Opt}(\mathcal{A})$, whence $(a, a') \in \succeq^{\overline{opt}} \cap \succeq_{opt}$. □

Modeling Ambiguity Averse Decisions. As it is commonly motivated by imprecise probability literature, the classical theory of probability is not able make distinctions between different layers of uncertainty. One such common example is that under complete ignorance.

In this section, we will encode the Ellsberg example in our framework and show that it is possible to model ambiguity averse decisions.

One popular interpretation for the behaviour explained in preliminary section is that, human agents tend to prefer more precise outcomes to less precise ones. That is, one feels safer where one has an idea about *risk* (one is less ignorant about the outcomes). The theory of imprecise probabilities offers a straightforward representation of the problem.

Definition 8 (Ellsberg-like Choice). *Given alternatives* $a, a' \in \mathcal{A}$, $a \succ_{ebg} a'$ *holds iff* $(\underline{EI}(a) + \overline{EI}(a))/2 = (\underline{EI}(a') + \overline{EI}(a'))/2$, *and* $(\overline{EI}(a) - \underline{EI}(a)) < (\overline{EI}(a') - \underline{EI}(a'))$. *We will denote the corresponding decision function as* δ_{ebg} *and call it an* Ellsberg-like choice.   †

Informally, such a function chooses a tighter interval where means are the same. Th reader is invited to verify that the preference relation *Ellsberg-dominates* denoted $\succ_{ebg}$, behaves accordingly to the experiment scenario given in Preliminaries (Sect. 2).

Example 6. One possible encoding of the problem is as follows. For convenience, we will give $l, u \in \mathbb{Q}$.

$$T = \{\, Yellow \sqsubseteq \neg Blue, Blue \sqsubseteq \neg Red, Yellow \sqsubseteq \neg Red$$
$$P = \{(Red|\top)[1/3, 1/3]\}, \mathcal{A} = \{choosered, chooseblue\}$$
$$P_{choosered} = \{(ChosenRed|\top)[1,1], (ChosenBlue|\top)[0,0]\}$$
$$P_{chooseblue} = \{(ChosenBlue|\top)[1,1], (ChosenRed|\top)[0,0]\}$$
$$\mathcal{U} = \{(Red \sqcap ChosenRed, 1), (Blue \sqcap ChosenBlue, 1)\}$$

Notice that agent only knows that red balls are one third of the domain (as well as red, blue and yellow are distinct). The framework automatically infers that yellow balls are of between 0 and 2/3 (as well as for red), and yellow or blue are 2/3 exactly. Given this information, it is easy to verify that *choosered* $\succ_{ebg}$ *chooseblue*. Now modifying UBox, i.e., replacing $(Red \sqcap ChosenRed, 1)$ with $((Red \sqcup Yellow) \sqcap ChosenRed, 1)$, and replacing $(Blue \sqcap ChosenBlue, 1)$ with $((Blue \sqcup Yellow) \sqcap ChosenBlue, 1)$, agent has the preference *chooseblue* $\succ_{ebg}$ *choosered* (since $EI(choosered) = [1/3, 1]$ and $EI(chooseblue) = [2/3, 2/3]$).

Note that it is still too strict, which one may not expect to hold often. Below, we will give a more tolerant form of this function.

Definition 9 (Ambiguity Averse Opportunist Choice). *Given alternatives* $a, a' \in \mathcal{A}$, $a \succ_{ag} a'$ *iff* $\overline{EI}(a) \leq \overline{EI}(a')$, $\underline{EI}(a) \geq \underline{EI}(a')$ *and* $\underline{EI}(a) - \underline{EI}(a') \geq \overline{EI}(a') - \overline{EI}(a)$. *We call the induced choice function an* ambiguity averse choice.   †

Intuitively, it brings an extra condition such that the mean needs to be greater or equal. The following result shows that $\succ_{ebg}$ is a special case of $\succ_{ag}$.

Proposition 4. *Let a, a' be two alternatives. Then, $a \succ_{ebg} a'$ implies $a \succ_{ag} a'$.*

Proof. Assume that (i) $(\underline{EI}(a) + \overline{EI}(a))/2 = (\underline{EI}(a') + \overline{EI}(a'))/2$ and also (ii) $(\overline{EI}(a) - \underline{EI}(a)) < (\overline{EI}(a') - \underline{EI}(a'))$. Then by (i), it follows that $(\underline{EI}(a) + \overline{EI}(a)) = (\underline{EI}(a') + \overline{EI}(a'))$ (iii), that is $(\underline{EI}(a) - \underline{EI}(a')) = (\overline{EI}(a') - \overline{EI}(a))$, hence $(\underline{EI}(a) - \underline{EI}(a'))/(\overline{EI}(a') - \overline{EI}(a)) = 1$. We know that $\overline{EI}(a) \geq \underline{EI}(a)$ and $\overline{EI}(a') \geq \underline{EI}(a')$. By (ii), $(\underline{EI}(a) - \underline{EI}(a')) < (\overline{EI}(a') - \overline{EI}(a))$ (iv), and by (iii) and (iv), $(\overline{EI}(a') - \overline{EI}(a)) \geq 0$, hence $\overline{EI}(a') \geq \overline{EI}(a)$. Similarly for $\underline{EI}(a) \geq \underline{EI}(a')$ □

In a loose sense, one can combine them with the previously mentioned functions (e.g., δ_{e+ag}) in order to model more complex behaviours. However, we leave their compositions and compatibilities, along with subtle connections to the probabilistic ontologies to future work.

4 Related Work

Our framework can be seen as a part of the literature on weighted logics for representing preferences [3,11], with an emphasis on agent modeling. Our notion of UBox to generate utility functions was for instance partially derived from the notion of goal bases (occasionally defined in terms of multi-sets) as understood in the literature of propositional languages for preferences [11,19]. There is also a substantial tradition on defeasible reasoning over preferences, see [2,4,5,9], on which we have leveraged.

On the DL side, several weighted DL languages have been proposed, albeit without covering uncertainty over instances [16,17]. In them, constructs similar to goal bases are used, called "preference sets", and elements of multi-attribute utility theory are partially incorporated into their settings.

Further recent works which can be considered to be loosely related (as *sensu stricto* non utility-theoretic) recent approaches include: an application of DL-based ontologies to CP-Nets, see [15], and a probabilistic logic-based setting [14] based on Markov Logics (precise probabilities) and using Markov networks to model and reason over preferences.

An uncertainty-based approach which attempts to focus on multi-criteria decision making (MCDM) problems is [18]; it is mainly based on the application of general fuzzy logic to MCDM problems. Although the terms *utility* and *preference* are not explicitly used, it refers to preferences implicitly.

5 Conclusions and Further Work

We have introduced a description logic based framework, to effectively express and solve non-sequential decision-making problems with multiple attributes.

As the major part of decision theory literature takes uncertainty into account, we based our approach on Lukasiewicz' P-$\mathcal{SHOIN}(\mathbf{D})$ family of probabilistic description logics ([13]). We have shown that it is straightforward to define decision functions representing ambiguity aversion; a case that violates the axioms of expected utility. In so doing, one can define preference relations and decision functions that we believe model decisions by rational (human) agents much better.

Another major direction is to investigate the value of information (structured knowledge in this context) in different ontological frameworks, viz., to explore in which ways and how much prior knowledge influences decisions about to be taken by agents.

Furthermore, it would be interesting to extend the framework to sequential decisions (e.g., a $\mathcal{D}_i \to \mathcal{D}_{i+1}$ sequence of decision bases). This is possible, since the language extensively uses conditional constraints. Once a sequential extension is defined, one can express strategies and game-theoretic issues. Furthermore, it would be interesting to apply the framework or an appropriate modification, to common problems such as *fair division, voting, preference aggregation* etc.

We are currently working on the implementation of the framework as a Protégé[11] plug-in. The development of our Protégé plugin is motivated by the idea to demonstrate the benefits of our approach to a set of different application scenarios where decision making is involved.

Appendix

Consistency, Lexicographic and Logical Consequence. A probabilistic interpretation Pr *verifies* a conditional constraint $(\psi|\phi)[l, u]$ iff $Pr(\phi) = 1$ and $Pr(\psi) \models (\psi|\phi)[l, u]$. Moreover, Pr *falsifies* $(\psi|\phi)[l, u]$ iff $Pr(\phi) = 1$ and $Pr(\psi) \not\models (\psi|\phi)[l, u]$. A set of conditional constraints $\mathcal{F}$ *tolerates* a conditional constraint $(\psi|\phi)[l, u]$ under a classical knowledge base T, iff there is model Pr of $T \cup \mathcal{F}$ that verifies $(\psi|\phi)[l, u]$ (i.e., $Pr \models T \cup \mathcal{F} \cup \{(\psi|\phi)[l, u], (\phi|\top)[1, 1]\}$). A PTBox $PT = (T, P)$ is *consistent* iff T is satisfiable, and there exists an ordered partition $(P_0, \ldots, P_k)$ of P such that each P_i (where $i \in \{0, \ldots, k\}$) is the set of all $F \in P\backslash(P_0 \cup \cdots \cup P_{i-1})$ that are tolerated under T by $P\backslash(P_0 \cup \cdots \cup P_{i-1})$. Following [13], we note that such ordered partition of PT is unique if it exists, and is called *z-partition*. A probabilistic knowledge base $KB = (T, P, (P_o)_{o \in \mathbf{I}_P})$ is consistent iff $PT = (T, P)$ is consistent, and for every probabilistic individuals $o \in \mathbf{I}_P$, there is a Pr such that $Pr \models T \cup P_o$.

For probabilistic interpretations Pr and Pr', Pr is *lexicographically preferable* (or *lex-preferable*) to Pr' iff there exists some $i \in \{0, \ldots, k\}$ such that $|\{F \in P_i \mid Pr \models F\}| > |\{F \in P_i|Pr' \models F\}|$ and $|\{F \in P_j \mid Pr \models F\}| = |\{F \in P_j \mid Pr' \models F\}|$ for all $i < j \leq k$. A probabilistic interpretation Pr is a *lexicographically minimal* (or *lex-minimal*) model of $T \cup \mathcal{F}$ iff $Pr \models T \cup \mathcal{F}$ and there is no

---

[11] http://protege.stanford.edu/.

Pr' such that $Pr' \models T \cup \mathcal{F}$ and Pr' is lex-preferable to Pr. A conditional constraint $(\psi|\phi)[l, u]$ is a *lexicographic consequence* (or *lex-consequence*) of a set of conditional constraints $\mathcal{F}$ under a PTBox PT (or $\mathcal{F} \;\|\!\!\sim^{lex} (\psi|\phi)[l, u]$) under PT, iff $Pr(\psi) \in [l, u]$ for every lex-minimal model Pr of $T \cup \mathcal{F} \cup \{(\phi|\top)[1, 1]\}$. Moreover, $PT \;\|\!\!\sim^{lex} F$, iff $\emptyset \;\|\!\!\sim^{lex} F$ under PT. Note that the notion of lex-consequence faithfully generalizes the classical class subsumption. That is, given a consistent PTBox $PT = (T, P)$, a set of conditional constraints $\mathcal{F}$, and c-concepts ϕ and ψ, if $T \models \phi \sqsubseteq \psi$, then $\mathcal{F} \;\|\!\!\sim^{lex} (\psi|\phi)[1, 1]$ under PT.

Furthermore, we say that $(\psi|\phi)[l, u]$ is a *tight lexicographic consequence* (or *tight lex-consequence*) of $\mathcal{F}$ under PT, denoted $F \;\|\!\!\sim^{lex}_{tight} (\psi|\phi)[l, u]$ under PT, iff $l = \inf\{Pr(\psi) \mid Pr \;\|\!\!\sim^{lex} T \cup \mathcal{F} \cup \{(\phi|\top)[1, 1]\}$ and $u = \sup\{Pr(\psi) \mid Pr \;\|\!\!\sim^{lex} T \cup \mathcal{F} \cup \{(\phi|\top)[1, 1]\}$. Moreover, $PT \;\|\!\!\sim^{lex}_{tight} F$ iff $\emptyset \;\|\!\!\sim^{lex} F$. Note that $[l, u] = [1, 0]$ (empty interval) when there is no such model. For a probabilistic knowledge base $KB = (T, P, (P_o)_{o \in \mathbf{I}_P})$, $KB \;\|\!\!\sim^{lex} F$ where F is a conditional constraint for $o \in \mathbf{I}_P$ iff $P_o \;\|\!\!\sim^{lex} F$ under (T, P). Moreover, $KB \;\|\!\!\sim^{lex}_{tight} F$ iff $P_o \;\|\!\!\sim^{lex}_{tight} F$ under (T, P). A conditional constraint $(\psi|\phi)[l, u]$ is a *logical consequence* of $T \cup \mathcal{F}$ (i.e., $T \cup F \models (\psi|\phi)[l, u]$) iff each model of $T \cup \mathcal{F}$ is also a model of $(\psi|\phi)[l, u]$. Furthermore, $(\psi|\phi)[l, u]$ is a *tight logical consequence* of $T \cup F$ (i.e., $T \cup \mathcal{F} \models_{tight} (\psi|\phi)[l, u]$, iff $l = \inf\{Pr(\psi|\phi) \mid Pr \models T \cup \mathcal{F}$ and $Pr(\phi) > 0\}$ and $u = \sup\{Pr(\psi|\phi) \mid Pr \models T \cup \mathcal{F}$ and $Pr(\phi) > 0\}$. Given a PTBox $PT = (T, P)$, $Q \subseteq P$ is *lexicographically preferable* (or *lex-preferable*) to $Q' \subseteq P$ iff there exists some $i \in 0, \ldots, k$ such that $|Q \cap P_i| > |Q' \cap P_i|$ and $|Q \cap P_j| = |Q' \cap P_j|$ for all $i < j \leq k$, where $(P_0, \ldots, P_k)$ is the z-partition of PT. Q is *lexicographically minimal* (or *lex-minimal*) in a set S of subsets of P iff $Q \in S$ and no $Q' \in S$ is lex-preferable to Q. Furthermore, let $\mathcal{F}$ be a set of conditional constraints, and ϕ and ψ be two concepts, then a set $\mathcal{Q}$ of lexicographically minimal subsets of P exists such that $F \;\|\!\!\sim^{lex} (\psi|\phi)[l, u]$ under PT iff $T \cup Q \cup \mathcal{F} \cup (\phi|\top)[1, 1] \models (\psi|\top)[l, u]$ for all $Q \in \mathcal{Q}$. This is extended to tight case lex-consequence.

References

1. Bienvenu, M., Lang, J., Wilson, N.: From preference logics to preference languages, and back. In: Proceedings of the International Conference on Principles and Knowledge Representation and Reasoning KR (2010)
2. Boutilier, C.: Toward a logic for qualitative decision theory. In: Proceedings of the International Conference on Principles and Knowledge Representation and Reasoning KR (1994)
3. Chevaleyre, Y., Endriss, U., Lang, J.: Expressive power of weighted propositional formulas for cardinal preference modeling. In: Proceedings of the International Conference on Principles of Knowledge Representation and Reasoning, KR (2006)
4. Delgrande, J.P., Schaub, T.: Expressing preferences in default logic. Artif. Intell. **123**(1–2), 41–87 (2000)
5. Delgrande, J.P., Schaub, T., Tompits, H., Wang, K.: A classification and survey of preference handling approaches in nonmonotonic reasoning. Comput. Intell. **20**(2), 308–334 (2004)
6. Ellsberg, D.: Risk, ambiguity, and the savage axioms. Q. J. Econ. **75**, 643–669 (1961)

7. Fishburn, P.C.: Utility Theory for Decision Making. Robert E. Krieger Publishing Co., Huntington, New York (1969)

8. Huntley, N., Hable, R., Troffaes, M.C.M.: Decision making. In: Augustin, T., Coolen, F.P.A., de Cooman, G., Troffaes, M.C.M. (eds.) Introduction to Imprecise Probabilities, pp. 190–206. Wiley, Chichester (2014)

9. Kaci, S., van der Torre, L.: Reasoning with various kinds of preferences: logic, non-monotonicity, and algorithms. Ann. OR **163**(1), 89–114 (2008)

10. Keeney, R.L., Raiffa, H.: Decisions with Multiple Objectives: Preferences and Value Tradeoffs. Wiley, New York (1976)

11. Lafage, C., Lang, J.: Logical representation of preferences for group decision making. In: Proceedings of the International Conference on Principles and Knowledge Representation and Reasoning KR, San Francisco (2000)

12. Levi, I.: The Enterprise of Knowledge. MIT Press, Cambridge, MA (1980)

13. Lukasiewicz, T.: Expressive probabilistic description logics. Artif. Intell. **172**(6–7), 852–883 (2008)

14. Lukasiewicz, T., Martinez, M.V., Simari, G.I.: Probabilistic preference logic networks. In: Proceedings of the European Conference on Artificial Intelligence ECAI (2014)

15. Di Noia, T., Lukasiewicz, T.: Combining CP-nets with the power of ontologies. In: AAAI (Late-Breaking Developments) (2013)

16. Ragone, A., Di Noia, T., Donini, F.M., Di Sciascio, E., Wellman, M.P.: Computing utility from weighted description logic preference formulas. In: Baldoni, M., Bentahar, J., van Riemsdijk, M.B., Lloyd, J. (eds.) DALT 2009. LNCS, vol. 5948, pp. 158–173. Springer, Heidelberg (2010)

17. Ragone, A., Di Noia, T., Donini, F.M., Di Sciascio, E., Wellman, M.P.: Weighted description logics preference formulas for multiattribute negotiation. In: Godo, L., Pugliese, A. (eds.) SUM 2009. LNCS, vol. 5785, pp. 193–205. Springer, Heidelberg (2009)

18. Straccia, U.: Multi criteria decision making in fuzzy description logics: a first step. In: Velásquez, J.D., Ríos, S.A., Howlett, R.J., Jain, L.C. (eds.) KES 2009, Part I. LNCS, vol. 5711, pp. 78–86. Springer, Heidelberg (2009)

19. Uckelman, J., Chevaleyre, Y., Endriss, U., Lang, J.: Representing utility functions via weighted goals. Math. Log. Q. **55**(4), 341–361 (2009)

Manipulation

Manipulation of k-Approval in Nearly Single-Peaked Electorates

Gábor Erdélyi[1]($\boxtimes$), Martin Lackner[2]($\boxtimes$), and Andreas Pfandler[1,2]($\boxtimes$)

[1] School of Economic Disciplines, University of Siegen, Siegen, Germany
erdelyi@wiwi.uni-siegen.de
[2] Institute of Information Systems, TU Wien, Vienna, Austria
{lackner,pfandler}@dbai.tuwien.ac.at

Abstract. For agents it can be advantageous to vote insincerely in order to change the outcome of an election. This behavior is called manipulation. The Gibbard-Satterthwaite theorem states that in principle every non-trivial voting rule with at least three candidates is susceptible to manipulation. Since the seminal paper by Bartholdi, Tovey, and Trick in 1989, (coalitional) manipulation has been shown NP-hard for many voting rules. However, under single-peaked preferences – one of the most influential domain restrictions – the complexity of manipulation often drops from NP-hard to P.

In this paper, we investigate the complexity of manipulation for the k-approval and veto families of voting rules in nearly single-peaked elections, exploring the limits where the manipulation problem turns from P to NP-hard. Compared to the classical notion of single-peakedness, notions of nearly single-peakedness are more robust and thus more likely to appear in real-world data sets.

1 Introduction

Elections are a useful framework for preference aggregation with many applications in both human societies and in multiagent systems. Well-known examples are political elections in human societies and the design of recommender systems [16], planning [10], and machine learning [22] in multiagent systems, just to name a few.

Informally, an election is given by a set of candidates and a set of voters who have to express their preferences over the set of candidates. A voting rule describes how to aggregate the voters' preferences in order to determine the winners of a given election. In computational social choice, a central research topic is to study computational questions regarding insincere behavior in elections. A prominent example is *coalitional manipulation*. Coalitional manipulation deals with situations in which a group of voters casts their votes strategically in order to alter the outcome of an election. (If the coalition has size one, the problem is called *single manipulation*). The famous Gibbard-Satterthwaite theorem says that, in principle, every reasonable voting rule for at least three candidates is susceptible to manipulation [17,20].

© Springer International Publishing Switzerland 2015
T. Walsh (Ed.): ADT 2015, LNAI 9346, pp. 71–85, 2015.
DOI: 10.1007/978-3-319-23114-3_5

Manipulability is considered to be an undesirable property for a voting rule. In their seminal paper Bartholdi, Tovey, and Trick suggested that although voting rules are manipulable, the manipulator's task of successfully manipulating the election can still be computationally hard, i.e., NP-hard [1]. Indeed, since the paper by Bartholdi, Tovey, and Trick, (coalitional) manipulation has been shown NP-hard for many voting rules.

In contrast, under domain restrictions, the computational complexity of manipulation drops from NP-hard to P for many voting rules. One popular model of domain restriction in elections is the model of single-peaked preferences introduced by Black [2]. Unfortunately, the concept of single-peakedness is fragile and is unlikely to appear in real-world data sets. To overcome this limitation, recent research has established notions of nearly single-peaked preferences which are more robust [6,7,11,14].

To the best of our knowledge, the only paper investigating (coalitional) manipulation in nearly single-peaked elections is the work by Faliszewski, Hemaspaandra, and Hemaspaandra [14] (a detailed comparison to our paper can be found in the Related Work Section). Our paper follows this new line of research and extends it with the following contributions:

- In our complexity analysis, we provide dichotomy results for constructive coalitional weighted manipulation under k-approval in the voter deletion model. The voter deletion model assumes that at most ℓ voters are not single-peaked with respect to the linear axis. Our results pinpoint the border between P membership and NP-completeness with respect to the number of approved candidates and the distance to single-peakedness.
- For veto we show how the complexity of constructive coalitional weighted manipulation behaves under seven notions of nearly single-peakedness that have been recently introduced [11,14]. Our dichotomies show that constructive coalitional weighted manipulation in nearly single-peaked electorates is either trivial (and therefore in P) or NP-complete depending on the distance to single-peakedness.

Related Work. Our work continues the line of research on manipulation of elections. The first paper investigating manipulation in elections is the seminal paper of Bartholdi, Tovey, and Trick [1], where they studied the single manipulation problem with unweighted voters and proved the problem to be solvable in polynomial-time for all scoring rules.

Constructive coalitional weighted manipulation (CCWM, for short) was first introduced by Conitzer, Sandholm, and Lang [5]. Later, Hemaspaandra and Hemaspaandra provided a dichotomy result for the CCWM problem for scoring rules [18]. In particular, they showed that CCWM is easy for plurality, but is NP-hard for all other k-approval and k-veto rules. Procaccia and Rosenschein have extended this line of research by studying the average-case complexity of manipulation [19].

Walsh was the first who studied the complexity of manipulation in single-peaked elections, especially with a view to answer the question whether the complexity of manipulation changes under single-peaked elections [21]. In particular, he demonstrated that the complexity of CCWM under single transferable vote remains NP-hard even for single-peaked elections. Faliszewski et al. proved that for single-peaked profiles, CCWM for m-candidate 3-veto elections is NP-complete for $m = 5$ and is in P for all other m [13]. Furthermore, they showed that for single-peaked profiles, CCWM for veto is in P and they completely characterized which scoring rules have easy CCWM problems and which scoring rules have hard CCWM problems for three-candidate elections. Brandt et al. generalized the latter result for m-candidate scoring rules [3].

The present paper was mostly motivated by Faliszewski, Hemaspaandra, and Hemaspaandra [14]: Amongst others, they investigated the complexity of the CCWM problem under veto elections in nearly single-peaked societies, where they used the nearly single-peaked notion of ℓ-Voter Deletion (which is called ℓ-Maverick in their paper). We extend their results to seven common notions of nearly single-peakedness that were recently discussed in the literature [11,14].

Two recent publications have studied the complexity of computing the distance to single-peaked electorates. Erdélyi, Lackner, and Pfandler [11] have focused on the single-peaked domain whereas Bredereck, Chen, and Woeginger [4] considered distances to a larger number of domain restrictions. Both papers mostly contain NP-hardness results with a few notable exceptions such as that the candidate deletion distance is computable in polynomial time. For a practical use of nearly single-peaked preferences, it would be desirable to have efficient algorithms to compute distances. This line of research has been initiated by Elkind and Lackner [8], where several approximation and fixed-parameter algorithms have been presented.

Organization. The remainder of the paper is organized as follows. In Sect. 2, we recap some voting theory basics. Section 3 gives an overview on the nearly single-peakedness notions and their relations handled in this paper. Our results on manipulation are presented in Sect. 4. Section 5 provides some conclusions and future directions.

2   Preliminaries

Let C be a finite set of *candidates*, V be a finite set of *voters*, and let $\succ$ be a *vote* (i.e., a total order) on C. Without loss of generality let $V = \{1, \ldots, n\}$. Let $\mathcal{P} = (\succ_1, \ldots, \succ_n)$ be a *(preference) profile*, i.e., a collection of votes. For simplicity, we will write for each voter $i \in V$, $c_1 c_2 \ldots c_m$ instead of $c_1 \succ_i c_2 \succ_i \ldots \succ_i c_m$. For two preference profiles on the same set of candidates $\mathcal{P} = (\succ_1, \ldots, \succ_n)$ and $\mathcal{L} = (\succ_{n+1}, \ldots, \succ_s)$, let $(\mathcal{P}, \mathcal{L}) = (\succ_1, \ldots, \succ_s)$ define the *union* of the two preference profiles. An *election* is defined as a triple $E = (C, V, \mathcal{P})$, where C is the set of candidates, V the set of voters, and $\mathcal{P}$ a preference profile over C. Throughout the paper let m denote the number of candidates and n the number of votes.

A *voting correspondence (or voting rule)* $\mathcal{F}$ is a mapping from a given election $E = (C, V, \mathcal{P})$ to a non-empty subset $W \subseteq C$; we call the candidates in W the *winners* of the election E. A prominent class of voting rules is the class of scoring rules, which are defined using a *scoring vector* $\alpha = (\alpha_1, \ldots, \alpha_m)$, $\alpha_i \in \mathbb{N}$, $\alpha_1 \geq \cdots \geq \alpha_m$. In an *m-candidate scoring rule* each voter has to specify a tie-free linear ordering of all candidates and gives α_i points to the candidate ranked in position i. The winners of the election are the candidates with the highest overall score. *k-approval* is an m-candidate scoring rule with $\alpha_1 = \cdots = \alpha_k = 1$ and $\alpha_{k+1} = \cdots = \alpha_m = 0$. *veto* is the scoring rule defined by the scoring vector $\alpha_1 = \cdots = \alpha_{m-1} = 1$ and $\alpha_m = 0$.

In the case of k-approval we say that the first k candidates in a given ranking have been *approved* whereas the others have been *disapproved*. For k-approval and veto, preferences actually do not have to be full rankings but *dichotomous preferences* suffice. Dichotomous preferences only distinguish between approved and disapproved candidates. In this paper we often do not give full rankings but rather the set of approved candidates. Strictly speaking, we use this notation to describe some total order that ranks the approved candidates above the disapproved candidates; all such total orders are equivalent from the perspective of k-approval.

Definition 1. *Let an* axis *A be a total order on C denoted by $<$. Furthermore, let $\succ$ be a vote with c as its highest ranked candidate. The vote $\succ$ is* single-peaked *with respect to A if for any $x, y \in C$, if $x < y < c$ or $c < y < x$ then $c \succ y \succ x$ has to hold. A preference profile $\mathcal{P}$ is said to be* single-peaked *with respect to an axis A if each vote is single-peaked with respect to A. A preference profile $\mathcal{P}$ is said to be* single-peaked consistent *if there exists an axis A such that $\mathcal{P}$ is single-peaked with respect to A.*

Note that, given a set of approved candidates and an axis A, there exists a single-peaked total order that corresponds to these approved candidates if and only if the candidates form an interval on A. Thus, for dichotomous preferences, one could also define single-peakedness in terms of intervals on an axis. We remark that recently several other domain restrictions specifically for dichotomous preferences have been proposed and studied [9].

To establish NP-hardness results we will reduce from the well-known NP-complete problem PARTITION (see, e.g., [15]), which is defined as follows.

PARTITION

Given: A finite multiset $S = \{x_1, \ldots, x_s\}$ of positive integers with $\sum_{i=1}^{s} x_i = 2X$ for some positive integer X.

Question: Is there a subset $S' \subset S$ such that the sum of the elements in S' is exactly X?

---

3 Nearly Single-Peakedness

As we build upon the notions of nearly single-peakedness which were studied by Erdélyi, Lackner, and Pfandler [11], we briefly recapitulate the relevant definitions and results. All these notions have been previously introduced and defined in the literature [11, 12, 14].

In the following, let $E = (C, V, \mathcal{P})$ be an election and ℓ a positive integer. Also, by $\mathcal{P}[C']$ we denote the profile $\mathcal{P}$ restricted to the candidates in C'. Analogously if A is an axis over C, we denote by $A[C']$ the axis A restricted to candidates in C'.

Voter Deletion: A profile $\mathcal{P}$ is ℓ-*Voter Deletion single-peaked consistent* if by removing at most ℓ votes from $\mathcal{P}$ one can obtain a preference profile $\mathcal{P}'$ that is single-peaked consistent. (We remark that this notion is also referred to as ℓ-maverick-SP [14] and as ℓ-maverick single-peaked consistent [11]).

Candidate Deletion: A profile $\mathcal{P}$ is ℓ-*Candidate Deletion single-peaked consistent* if we can obtain a set $C' \subseteq C$ by removing at most ℓ candidates from C such that the preference profile $\mathcal{P}[C']$ is single-peaked consistent.

Local Candidate Deletion: Let A be an axis over C. A vote $\succ$ on a candidate set $C' \subset C$ is called a partial vote. A partial vote on C' is said to be *single-peaked with respect to A* if it is single-peaked with respect to $A[C']$. A profile $\mathcal{P}$ is ℓ-*Local Candidate Deletion single-peaked consistent* if there exists an axis A such that by removing at most ℓ candidates from each vote we obtain a partial profile $\mathcal{P}'$ that is single-peaked with respect to A.

Additional Axes: A profile $\mathcal{P}$ is ℓ-*Additional Axes single-peaked consistent* if there is a partition $V_1, \ldots, V_{\ell+1}$ of the voter set V such that the corresponding preference profiles $\mathcal{P}_1, \ldots, \mathcal{P}_{\ell+1}$ are single-peaked consistent.

Global Swaps: A profile $\mathcal{P}$ is ℓ-*Global Swaps single-peaked consistent* if $\mathcal{P}$ can be made single-peaked by performing at most ℓ swaps of consecutive candidates in the profile. (Note that these swaps can be performed wherever we want – we can have ℓ swaps in only one vote, or one swap each in ℓ votes.)

Local Swaps: A profile $\mathcal{P}$ is ℓ-*Local Swaps single-peaked consistent* if $\mathcal{P}$ can be made single-peaked consistent by performing no more than ℓ swaps of consecutive candidates per vote.

Candidate Partition: A profile $\mathcal{P}$ is ℓ-*Candidate Partition single-peaked consistent* if the set of candidates C can be partitioned into at most ℓ disjoint sets $C_1, \ldots, C_\ell$ with $C_1 \cup \ldots \cup C_\ell = C$ such that the profiles $\mathcal{P}[C_1], \ldots, \mathcal{P}[C_\ell]$ are single-peaked consistent.

We denote by $VD(\mathcal{P})/CD(\mathcal{P})/LCD(\mathcal{P})/AA(\mathcal{P})/GS(\mathcal{P})/LS(\mathcal{P})/CP(\mathcal{P})$ the smallest ℓ such that $\mathcal{P}$ is ℓ-Voter Deletion/ℓ-Candidate Deletion/ℓ-Local Candidate Deletion/ℓ-Additional Axes/ℓ-Global Swaps/ℓ-Local Swaps/ℓ-Candidate Partition single-peaked consistent.

Theorem 2. (cf. [11]) *Let $\mathcal{P}$ be a preference profile. Then the following inequalities hold:*

(1) $LS(\mathcal{P}) \leq GS(\mathcal{P})$.   *(4) $LCD(\mathcal{P}) \leq LS(\mathcal{P})$.*   *(7) $CP(\mathcal{P}) \leq CD(\mathcal{P}) + 1$.*

(2) $LCD(\mathcal{P}) \leq CD(\mathcal{P})$.   *(5) $VD(\mathcal{P}) \leq GS(\mathcal{P})$.*   *(8) $CP(\mathcal{P}) \leq LS(\mathcal{P}) + 1$.*

(3) $CD(\mathcal{P}) \leq GS(\mathcal{P})$.   *(6) $AA(\mathcal{P}) \leq VD(\mathcal{P})$.*

This list is complete in the following sense: Inequalities that are not listed here and that do not follow from transitivity do not hold in general. The resulting partial order with respect to $\leq$ is displayed in Fig. 1 as a Hasse diagram.

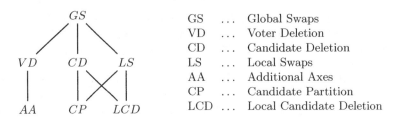

Fig. 1. Hasse diagram of the partial order described in Theorem 2.

Finally, let us summarize the complexity results concerning the detection of nearly single-peaked elections. Here, the question is whether a given election is ℓ-X single-peaked consistent. This problem is polynomial-time solvable for the Candidate Deletion distance and NP-complete for all $X \in \{$Voter Deletion, Local Candidate Deletion, Additional Axes, Global Swaps, Local Swaps$\}$ [4,11].

4 Manipulation

In what follows we investigate the computational complexity of coalitional manipulation in scoring rules under the assumption that the underlying elections are nearly single-peaked. For this we have to formally define the coalitional weighted manipulation problem in general and for nearly single-peaked electorates. Let $\mathcal{F}$ be a voting correspondence.

$\mathcal{F}$-CONSTRUCTIVE COALITIONAL WEIGHTED MANIPULATION ($\mathcal{F}$-CCWM)

Given: An election $(C, V, \mathcal{P})$, where C is a set of candidates, V a set of nonmanipulative voters, and $\mathcal{P} = (P_1, \ldots, P_h)$ a preference profile; in addition, a set of manipulative voters S with $V \cap S = \emptyset$, a weight function w from $V \cup S$ to $\mathbb{N}$, and a distinguished candidate $p \in C$.

Question: Is there a preference profile $\mathcal{L} = (L_1, \ldots, L_s)$ for the manipulative voters in S such that p is a (co-)winner in $(C, V \cup S, (\mathcal{P}, \mathcal{L}))$ with respect to the voting correspondence $\mathcal{F}$?

In this paper we study $\mathcal{F}$-CCWM for nearly single-peaked preferences. For a fixed, non-negative integer ℓ and $X \in \{$Voter Deletion, Candidate Deletion, Local Candidate Deletion, Additional Axes, Global Swaps, Local Swaps, Candidate Partition$\}$, we define $\mathcal{F}$-ℓ-X-CCWM to be $\mathcal{F}$-CCWM restricted to profiles that are ℓ-X single-peaked consistent with respect to an axis A. Note that the combined election $(C, V \cup S, (\mathcal{P}, \mathcal{L}))$ has to be ℓ-X single-peaked consistent. In addition, we assume that this axis is part of the input. In the case of the additional axes distance, we assume that all axes are part of the input; in the case of the candidate partition distance, we assume that the actual partition is part of the input. To be more precise, $\mathcal{F}$-ℓ-X-CCWM is defined as follows:

$\mathcal{F}$-ℓ-X-CCWM

Given: An $\mathcal{F}$-CCWM instance, an axis A, additional axes $A_1, \ldots, A_\ell$ if X is Additional Axes, a partition of the candidate set C if X is Candidate Partition.

Question: Is there a preference profile $\mathcal{L} = (L_1, \ldots, L_s)$ for the manipulative voters in S such that (i) p is a (co-)winner in $(C, V \cup S, (\mathcal{P}, \mathcal{L}))$ with respect to the voting correspondence $\mathcal{F}$ and (ii) $(C, V \cup S, (\mathcal{P}, \mathcal{L}))$ is ℓ-X single-peaked consistent?

Remark. As mentioned earlier, it is NP-hard to verify whether an election is ℓ-X single-peaked consistent for all notions of distance X considered in this paper except for the Candidate Deletion distance (for which this problem is in P) and for the Candidate Partition distance (for which its complexity is not known) [4,11]. These NP-hardness results, however, do not influence the complexity of $\mathcal{F}$-ℓ-X-CCWM due to our assumption that the axis is part of the input. Given a fixed axis A and $X \in \{$Voter Deletion, Candidate Deletion, Local Candidate Deletion, Global Swaps, Local Swaps$\}$, it requires only polynomial time to verify that an election is ℓ-X single-peaked with respect to A. The same holds for Additional Axis if all axes are given and for Candidate Partition if the partition of the candidates is given. Consequently, the complexity of $\mathcal{F}$-ℓ-X-CCWM can be studied separately from the complexity of deciding ℓ-X single-peaked consistency.

Let us start with our first result on CCWM. Following the notation of Faliszewski, Hemaspaandra, and Hemaspaandra [14], let $(\alpha_1, \alpha_2, \alpha_3)$ *elections* denote three-candidate scoring rule elections with scoring-vector $\alpha = (\alpha_1, \alpha_2, \alpha_3)$. In that paper it was proven that for each $\alpha_1 \geq \alpha_2 > \alpha_3$, $(\alpha_1, \alpha_2, \alpha_3)$-1-VOTER DELETION-CCWM is NP-complete. This result implies that VETO-1-VOTER DELETION-CCWM for three-candidate elections is NP-complete. In contrast, VETO-CCWM is in P for single-peaked societies. The following proposition makes use of Theorem 2 and shows that the same holds for all other notions of distance studied in this paper.

Proposition 3. *Let* $X \in \{$Candidate Deletion, Local Candidate Deletion, Additional Axes, Global Swaps, Local Swaps$\}$. *For each* $\alpha_1 \geq \alpha_2 > \alpha_3$, *the problems* $(\alpha_1, \alpha_2, \alpha_3)$-1-$X$-CCWM *and* $(\alpha_1, \alpha_2, \alpha_3)$-2-CANDIDATE PARTITION-CCWM *are* NP-*complete.*

Proof. Faliszewski, Hemaspaandra, and Hemaspaandra [14] show NP-completeness of the $(\alpha_1, \alpha_2, \alpha_3)$-1-VOTER DELETION-CCWM problem. We now show that a three candidate, 1-voter deletion single-peaked consistent election is also 1-X single-peaked consistent for all X and 2-candidate partition single-peaked consistent. It is easy to see that every election E over three candidates is 1-candidate deletion, 1-local candidate deletion, and 2-candidate partition single-peaked consistent. From Theorem 2, Inequality (6), follows that E is also 1-additional axes single-peaked consistent.

Let $C = \{a, b, c\}$ be the set of candidates, and without loss of generality assume that E is 1-voter deletion single-peaked consistent along the axis $a < b < c$. Note that there are only two possible non-single-peaked votes, acb and cab. In both votes, swapping the last two candidates leaves us with single-peaked votes with respect to axis $a < b < c$. Thus, E is 1-global swaps single-peaked consistent. From Theorem 2, Inequality (1), follows that E is 1-local swaps single-peaked consistent. □

4.1 Manipulation for k-Approval

We continue by investigating the computational complexity of manipulation for the k-approval voting rule in ℓ-Voter Deletion single-peaked societies. CCWM for k-approval is known to be NP-complete in general [18]. This holds even for single-peaked elections in many settings [3]. We extend these results to ℓ-Voter Deletion single-peaked societies. More concretely, we show a dichotomy result: k-APPROVAL-ℓ-VOTER DELETION-CCWM is in P if and only if $\ell < \frac{2k-m}{m-k}$, and NP-complete otherwise. This gives a complete picture for k-approval with respect to ℓ-Voter Deletion single-peakedness. Both the P membership and NP-hardness result are generalizations of the results for veto elections [14] and our proofs can be seen as refinements of the corresponding proofs.

Theorem 4. *Let $m \geq 3, k > 1, \ell \geq 1$ be fixed integers such that $k < m$ and $\ell \geq \frac{2k-m}{m-k}$. Then k-APPROVAL-ℓ-VOTER DELETION-CCWM for elections with m candidates is NP-complete.*

Proof. Membership in NP is trivial. We reduce from PARTITION with a sum of $2X$. Let b be a positive integer such that $\max(1, 2k - m) \leq b \leq k - 1$. Note that such a b always exists since $1 < k < m$. Let $C = \{x, y, p, c_1, \ldots, c_{m-3}\}$, where p is the distinguished candidate. To construct the votes in $\mathcal{P}$, we split the sequence $pc_1 \cdots c_{b-1}$ into consecutive blocks of size $m - k$. (If necessary the last block is filled with candidates from $c_b, \ldots, c_{m-3}$.) Let $d = \left\lceil \frac{b}{m-k} \right\rceil$. These blocks give d sets of candidates $D_1, \ldots, D_d$. Furthermore, let $e = \frac{m-3-b}{2}$. We fix the axis A to $c_{b+\lceil e \rceil} < \cdots < c_{m-3} < x < p < c_1 < \cdots < c_{b-1} < y < c_b < \cdots < c_{b+\lceil e \rceil -1}$. The profile $\mathcal{P}$ comprises the following votes:

– $\mathcal{P}$ contains d votes of weight X: For each $i \in \{1, \ldots, d\}$,

$$V_i : C \setminus D_i \text{ is approved}, w(V_i) = X.$$

Note that all these votes are not single-peaked.
– $\ell - d$ votes of weight 1: For $i \in \{d + 1, \ldots, \ell\}$,

$$V_i : \{x, y, p, c_1, \ldots, c_{k-4}, c_{m-4}\}, w(V_i) = 1.$$

Also these votes are not single-peaked and hence we have exactly ℓ non-single-peaked votes. Consequently, we force the manipulators to cast single-peaked votes.

Let the set $\mathcal{L}$ consist of s manipulators with weights $x_1, \ldots, x_s$.

At this point, candidates have the following scores: The candidates x and y are approved by all votes and hence they have a score of $d \cdot X + (\ell - d)$. Candidate p is approved by all votes except V_1 since $p \in D_1$ and thus has a score of $(d-1) \cdot X + (\ell - d)$. The candidates $c_1, \ldots, c_{b-1}$ have a score of at most $(d-1) \cdot X + (\ell - d)$ since they are contained in at least one D_i. The candidates $c_b, \ldots, c_{m-3}$ have a score of at most $d \cdot X + (\ell - d)$.

Since the manipulators can cast only single-peaked votes and they want to approve p but not both x and y, the manipulators approve an interval on A of length k that contains p and either x or y.

Intuitively, the manipulators can only make p a winner if they manage to give $2X$ points to p and X points to x and y such that x, y, and p are tied. Note that in a single-peaked vote where p is approved, also either x or y has to be approved. We claim that there is a subset $S' \subset S$ such that the elements in S' sum to X if and only if p can be made a winner of the election by constructive coalitional weighted manipulation.

"$\Rightarrow$": Suppose there is a subset $S' \subset S$ such that the elements in S' sum to X. Let all the manipulators whose weight is in S' approve

$$\{x, p, c_1, \ldots, c_{b-1}, c_{m-k+b-1}, \ldots, c_{m-3}\},$$

i.e., they approve a "block" of length k that starts at c_{b-1} and goes to the left on axis A. All the manipulators whose weight is in $S \setminus S'$ approve

$$\{p, y, c_1, \ldots, c_{k-2}\},$$

i.e., they approve a "block" of length k that starts at p and goes to the right. In both cases the vote is single-peaked. It is easy to see that p gains $2X$, whereas x and y only gain X points. Note that the maximum score where p ties with x and y is $(d+1) \cdot X + (\ell - d)$ as x and y obtain $d \cdot X + (\ell - d)$ points from the nonmanipulators and the manipulators can distribute a score of $2X$, but never approve x and y together. Note that none of the other candidates can surpass the score of x, y, and p. Hence, x, y, and p are among the winners tied for first place making the distinguished candidate p a winner.

"$\Leftarrow$": Suppose that p can be made a winner of the election by constructive coalitional weighted manipulation. Note that according to the scores given by the nonmanipulators, p is missing X points to be tied with x and y. The only way p can gain X points on these two candidates is if the manipulators can be divided into two groups, both weighing X points. The first group approves p, x, and suitable candidates from $\{c_1, \ldots, c_{m-3}\}$. The second group approves p, y, and suitable candidates from $\{c_1, \ldots, c_{m-3}\}$. Thereby, p gains $2X$ points, whereas x and y gain only X points each. Thus, there is a subset $S' \subset S$ such that the elements in S' sum to X. $\qquad\square$

Theorem 5. *If $\ell < \frac{2k-m}{m-k}$, then k-APPROVAL-ℓ-VOTER DELETION-CCWM is in P.*

Proof. Without loss of generality let $A : c_1 < c_2 < \cdots < c_m$. Every vote disapproves $m - k$ candidates. Consequently, the ℓ non-single-peaked voters disapprove at most $\ell \cdot (m - k) < 2k - m$ candidates. Therefore, there is at least one candidate contained in $\{c_{m-k+1}, \ldots, c_k\}$ that is approved by all non-single-peaked voters, since $|\{c_{m-k+1}, \ldots, c_k\}| = 2k - m$. (Note that $0 \le \ell < \frac{2k-m}{m-k}$ implies $2k - m > 0$). Single-peaked voters may disapprove only candidates in $\{c_1, \ldots, c_{m-k}, c_{k+1}, \ldots, c_m\}$. Thus, candidates in $\{c_{m-k+1}, \ldots, c_k\}$ approved by all non-single-peaked voters are also approved by all single-peaked voters. Since there exists at least one candidate that is approved by all voters (including the manipulators), p is a winner if and only if it is approved by all voters (both manipulators and nonmanipulators). $\square$

The following corollary shows how Theorem 4 carries over to VETO-ℓ-VOTER DELETION-CCWM.

Corollary 6. (also shown in [14]) *Let $m, \ell \in \mathbb{N}$ be fixed such that $\ell > m - 3$. Then* VETO-ℓ-VOTER DELETION-CCWM *is NP-complete. Otherwise,* VETO-ℓ-VOTER DELETION-CCWM *is in* P.

4.2 Manipulation for Veto

In this section we study the complexity of constructive coalitional weighted manipulation in nearly single-peaked societies under the veto rule. For veto, CCWM is NP-complete in general [18], whereas the problem is in P for single-peaked elections [13]. In contrast to the previous section, we study here only a single rule, veto, but consider a variety of notions for nearly single-peakedness. Table 1 summarizes the complexity results regarding VETO-ℓ-X-CCWM under several notions of nearly single-peakedness. Note that all results in this table yield dichotomies.

Table 1. Complexity results regarding VETO-ℓ-X-CCWM under several notions of nearly single-peakedness, assuming $m \ge 3$.

X	in P	NP-complete	Reference
Voter Deletion	$\ell \le m - 3$	$\ell > m - 3$	[14] & Cor. 6
Candidate Deletion	$\ell \le m - 3$	$\ell > m - 3$	Thm. 7
Local Candidate Del.	$\ell = 0$	$\ell \ge 1$	Prop. 8
Global Swaps	$m = 2k$: $\ell \le k^2 - k - 1$	$\ell > k^2 - k - 1$	Thm. 10
	$m = 2k - 1$: $\ell \le k^2 - 2k$	$\ell > k^2 - 2k$	Thm. 10
Local Swaps	$\ell < \lfloor \frac{m-1}{2} \rfloor$	$\ell \ge \lfloor \frac{m-1}{2} \rfloor$	Thm. 11
Candidate Partition	$\ell < \frac{m}{2}$	$\ell \ge \frac{m}{2}$	Thm. 12
Additional Axes	$\ell < \frac{m}{2} - 1$	$\ell \ge \frac{m}{2} - 1$	Thm. 13

In the following we will prove each entry of this table. We assume throughout this section that there are at least three candidates, since for less than three candidates VETO-CCWM is in P [5,18].

All P membership proofs in this section follow the same reasoning as the proof of Theorem 5. More specifically, we show that there is at least one candidate that is never vetoed. As a consequence, a candidate p can only be amongst the winners if p is never vetoed (both by the nonmanipulators and the manipulators). Clearly, it is possible in polynomial time to determine whether a candidate is approved by all nonmanipulative voters and to construct the manipulator's votes that approve p. In the following P membership proofs we only argue that there is indeed at least one candidate that is never vetoed and omit the remainder of the argument.

Theorem 7. *Let $m \geq 3$. For each $\ell \geq 0$, if $\ell \leq m - 3$ VETO-ℓ-CANDIDATE DELETION-CCWM is in P and NP-complete otherwise.*

Proof. We are first handling the $\ell \leq m-3$ case. Let A be the axis along which the election is nearly single-peaked and let c_l and c_r be the leftmost and rightmost candidates in A, respectively. Note that in a veto election over a single-peaked society, only c_l and c_r can be vetoed. In an ℓ-Candidate Deletion single-peaked society there are at most ℓ additional candidates vetoed in those votes not consistent with the axis A. Thus, there are at most $\ell + 2 \leq m - 1$ candidates that are vetoed. Consequently, there has to be at least one candidate who never got vetoed.

We now turn to the case where $\ell > m - 3$. In this case, hardness follows immediately from the fact that every profile is $(m-2)$-candidate deletion single-peaked consistent and VETO-CCWM is an NP-hard problem [5,18,19].   □

In the following proposition we require that $\ell \geq 1$. The $\ell = 0$ case would mean that the election is single-peaked, for which Brandt et al. [3] proved that constructive coalitional weighted manipulation under the veto rule is in P.

Proposition 8. *For each $m \geq 3$ and $\ell \geq 1$, VETO-ℓ-LOCAL CANDIDATE DELETION-CCWM is NP-complete.*

Proof. The crucial observation here is that with $\ell \geq 1$ every candidate can be vetoed, since the vetoed candidate can be the one that is locally deleted. Thus, this problem is equivalent to VETO-CCWM, which is NP-complete for each $m \geq 3$ [5,18,19].   □

In the following, for any two candidates $c_1, c_2 \in C$ let $d_A(c_1, c_2)$ be the *distance of two candidates on the axis A*. For example, for the axis $A = c_1 < c_3 < c_5 < c_4 < c_2 < c_6$ the distance $d_A(c_1, c_2) = 4$.

Lemma 9. *Let $E = (C, V, \mathcal{P})$ be a single-peaked election along the axis A, where c_l and c_r are the leftmost and rightmost candidates, respectively. The number of swaps required to make a candidate $c \in C$ the lowest-ranked candidate in a vote $v \in V$ is at least $min(d_A(c, c_r), d_A(c, c_l))$.*

Proof. Without loss of generality assume that c is closer to c_r in A than to c_l (i.e., $d_A(c, c_r) < d_A(c, c_l)$). If a vote v coincides with the axis A then clearly exactly $min(d_A(c, c_r), d_A(c, c_l)) = d_A(c, c_r)$ swaps are needed to make c the candidate who gets vetoed in v.

If v does not coincide with A, we have to distinguish three cases. First, let c be the peak of v. In this case it is clear that c has to be swapped with all other candidates to get vetoed and thus we need exactly $d_A(c, c_r) + d_A(c, c_l) \geq min(d_A(c, c_r), d_A(c, c_l))$ swaps. Second, let c be left from v's peak on axis A. This means that, according to the definition of single-peakedness, all the candidates left from c on axis A are ranked lower than c in v. To swap c through to the last position, we will have to make at least $d_A(c, c_l) > min(d_A(c, c_r), d_A(c, c_l))$ swaps. Finally, let c be right from v's peak on axis A. This means that all the candidates on axis A right from c are ranked lower than c in v. To swap c through to the last position, we will have to make at least $d_A(c, c_r) = min(d_A(c, c_r), d_A(c, c_l))$ swaps. □

Using Lemma 9, the following two theorems can be shown.

Theorem 10. *Let $k \geq 2$ be a positive integer.*

1. *Let the number of candidates be $m = 2k$. For each $\ell \geq 0$, VETO-ℓ-GLOBAL SWAPS-CCWM is in P if $\ell \leq k^2 - k - 1$ and NP-complete otherwise.*
2. *Let the number of candidates be $m = 2k-1$. For each $\ell \geq 0$, VETO-ℓ-GLOBAL SWAPS-CCWM is in P if $\ell \leq k^2 - 2k$ and NP-complete otherwise.*

Proof. Without loss of generality, let $A : c_1 < \cdots < c_m$ be the axis for which the election is nearly single-peaked. Let us consider case (1) first, i.e., $m = 2k$. We count the number of candidates who can be vetoed. These are the two candidates c_1 and c_m, and those candidates that can be "swapped" to the last position with at most ℓ swaps. Observe that it requires at least one swap each for swapping c_2 and c_{m-1} to the lowest position in a vote (cf. Lemma 9). For c_3 and c_{m-2} at least three swaps are required, etc. We make at most $k^2 - k - 1 = -1 + 2\sum_{i=0}^{k-1} i$ swaps and consequently less than m different candidates can be swapped to a last position in some vote (cf. Lemma 9). Thus, there is at least one candidate who is never vetoed. In case (2), i.e., $m = 2k - 1$, note that $k^2 - 2k = -1 + (k-1) + 2\sum_{i=0}^{k-2} i$ and hence less than m can be vetoed.

To show hardness we reduce from PARTITION. Let $m = 2k$ and $\ell \geq k^2 - k$. (The case that $m = 2k-1$ works analogously). Given a multiset $S = \{x_1, \ldots, x_s\}$ of s integers that sum to $2X$, define the following instance of VETO-ℓ-GLOBAL SWAPS-CCWM. Let $C = \{p, c_l, c_r, c_1, \ldots, c_{m-3}\}$ be the set of candidates and let p be the distinguished candidate. Let A be the axis for which the election is nearly single-peaked and let candidates c_l and c_r be the leftmost and rightmost candidates in A. Let the nonmanipulative voters consist of $m-2$ voters, each with weight X such that for every candidate $c \in C \setminus \{c_l, c_r\}$ there is a nonmanipulative voter who ranks c last but otherwise the votes are identical with the axis A or its reverse $\overline{A}$ (if c is closer to c_l on A, then we choose the axis as vote which ranks c_l last). Note that in this case we need $2\sum_{i=0}^{k-1} i = k^2 - k$ global swaps to make the profile single-peaked which is still less or equal ℓ. Let $\mathcal{L}$ consist of s manipulators with weights $x_1, \ldots, x_s$.

We claim that there is a subset $S' \subset S$ such that the elements in S' sum to X if and only if p can be made a winner of the election by constructive coalitional weighted manipulation.

"$\Rightarrow$": Suppose there is a subset $S' \subset S$ such that the elements in S' sum to X. Let all the manipulators whose weight is in S' vote identically to the axis A, and all the manipulators whose weight is in $S \setminus S'$ vote reverse. It is easy to see that now all candidates tie for first place and, thus, the distinguished candidate p is a winner.

"$\Leftarrow$": Suppose that p can be made a winner of the election by constructive coalitional weighted manipulation. Note that p ties (with $c_1, \ldots, c_{m-3}$) for the third place X points behind both candidates c_l and c_r. The only way p can gain X points on these two candidates is if the manipulators can be divided into two groups, both weighing X points and vetoing candidates c_l and c_r, respectively. Thus, there is a subset $S' \subset S$ such that the elements in S' sum to X. $\square$

Theorem 11. *Let $m \geq 3$ denote the number of candidates. For each $\ell \geq 0$, VETO-ℓ-LOCAL SWAPS-CCWM is in P if $\ell < \lfloor \frac{m-1}{2} \rfloor$ and NP-complete otherwise.*

Proof. Let A be the axis along the election is nearly single-peaked, and let c_l and c_r be the leftmost and rightmost candidates in A, respectively. Observe that there is a candidate on A with distance at least $\lfloor \frac{m-1}{2} \rfloor$ to both c_l or c_r. Thus, for $\ell < \lfloor \frac{m-1}{2} \rfloor$, there is a candidate that is never vetoed.

For showing hardness, note that when we start with the single-peaked votes $c_1 \succ c_2 \succ \cdots \succ c_m$ or $c_m \succ \cdots \succ c_2 \succ c_1$, $\lfloor \frac{m-1}{2} \rfloor$ swaps suffice to make any candidate rank last. Thus, every candidate can be vetoed and VETO-ℓ-LOCAL SWAPS-CCWM for $\ell \geq \lfloor \frac{m-1}{2} \rfloor$ is equivalent to VETO-CCWM, which is NP-complete for $m \geq 3$ [5, 18, 19]. $\square$

Theorem 12. *Let $m \geq 3$ be the number of candidates in an election E. For each $\ell \geq 1$, VETO-ℓ-CANDIDATE PARTITION-CCWM is in P if $\ell < \frac{m}{2}$ and NP-complete otherwise.*

Proof. In the $\ell < \frac{m}{2}$ case we again count the number of candidates who can be vetoed. As we can veto at most two candidates per partition and we have ℓ single-peaked partitions, there can be at most 2ℓ different candidates being vetoed. Since $2\ell < m$, there has to be at least one candidate who is never vetoed.

For the other case, $\ell \geq \frac{m}{2}$, note that since there are at least $\frac{m}{2}$ partitions, all candidates can be vetoed while preserving candidate partition single-peakedness. Hardness for this case follows from the result for the general case [5, 18, 19]. $\square$

Finally, we turn to VETO-ℓ-ADDITIONAL AXES-CCWM.

Theorem 13. *Let $m \geq 3$. For each $\ell \geq 0$, VETO-ℓ-ADDITIONAL AXES-CCWM is in P if $\ell < \frac{m}{2} - 1$ and NP-complete otherwise.*

Proof. The proof is similar to the candidate partition case (Theorem 12). The important observation is that for this purpose candidate partition and alternative axes provide the same freedom: In both cases at most two candidates per axis or partition can be vetoed. Note that the -1 in the bound on ℓ comes from the fact that ℓ *additional* axes give us $\ell + 1$ axes in total.   □

5   Conclusions and Open Questions

We have investigated the computational complexity of manipulation in nearly single-peaked elections, where we focused on k-approval and veto. For veto we have studied seven notions of nearly single-peakedness that were recently studied in the literature [11,14]. In contrast, for k-approval, we have explored how k influences the complexity if we consider voter deletion as notion for nearly single-peakedness. In both cases we proved dichotomies that exactly pinpoint the border of tractability. These results give insight into the sources of hardness and reveal in which settings we can hope for computationally hard instances.

There are several ways to continue with this direction of research. Extending our results to k-approval (or even arbitrary scoring rules) for all notions of nearly single-peakedness is certainly an important direction to go. Another way is to consider other notions of strategic behavior such as bribery and control in the light of nearly single-peakedness.

Acknowledgments. We thank the anonymous ADT-2015 referees for their very helpful comments and suggestions. This work was supported by the Austrian Science Fund (FWF): P25518, Y698, and the German Research Foundation (DFG): ER 738/2-1.

References

1. Bartholdi, J., Tovey, C., Trick, M.: The computational difficulty of manipulating an election. Soc. Choice Welfare **6**(3), 227–241 (1989)
2. Black, D.: On the rationale of group decision making. J. Polit. Econ. **56**(1), 23–34 (1948)
3. Brandt, F., Brill, M., Hemaspaandra, E., Hemaspaandra, L.A.: Bypassing combinatorial protections: polynomial-time algorithms for single-peaked electorates. In: Proceedings of the 24th AAAI Conference on Artificial Intelligence (AAAI 2010), pp. 715–722. AAAI Press (2010)
4. Bredereck, R., Chen, J., Woeginger, G.J.: Are there any nicely structured preference profiles nearby? In: Proceedings of the 23rd International Joint Conference on Artificial Intelligence (IJCAI 2013), pp. 62–68. AAAI Press (2013)
5. Conitzer, V., Sandholm, T., Lang, J.: When are elections with few candidates hard to manipulate? J. ACM vol. 54(3), Article 14 (2007)
6. Cornaz, D., Galand, L., Spanjaard, O.: Bounded single-peaked width and proportional representation. In: Proceedings of the 20th European Conference on Artificial Intelligence (ECAI 2012). Frontiers in Artificial Intelligence and Applications, vol. 242, pp. 270–275. IOS Press (2012)

7. Cornaz, D., Galand, L., Spanjaard, O.: Kemeny elections with bounded single-peaked or single-crossing width. In: Proceedings of the 23rd International Joint Conference on Artificial Intelligence (IJCAI 2013). pp. 76–82. AAAI Press (2013)
8. Elkind, E., Lackner, M.: On detecting nearly structured preference profiles. In: Proceedings of the 28th AAAI Conference on Artificial Intelligence (AAAI 2014). pp. 661–667. AAAI Press (2014)
9. Elkind, E., Lackner, M.: Structure in dichotomous preferences. In: Proceedings of the 24th International Joint Conference on Artificial Intelligence (IJCAI 2015). AAAI Press (to appear 2015)
10. Ephrati, E., Rosenschein, J.: A heuristic technique for multi-agent planning. Ann. Math. Artif. Intell. **20**(1–4), 13–67 (1997)
11. Erdélyi, G., Lackner, M., Pfandler, A.: Computational aspects of nearly single-peaked electorates. In: Proceedings of the 27th AAAI Conference on Artificial Intelligence (AAAI 2013), pp. 283–289. AAAI Press (July 2013)
12. Escoffier, B., Lang, J., Öztürk, M.: Single-peaked consistency and its complexity. In: Proceedings of the 18th European Conference on Artificial Intelligence (ECAI 2008). Frontiers in Artificial Intelligence and Applications, vol. 178, pp. 366–370. IOS Press (2008)
13. Faliszewski, P., Hemaspaandra, E., Hemaspaandra, L.A., Rothe, J.: The shield that never was: societies with single-peaked preferences are more open to manipulation and control. Inf. Comput. **209**(2), 89–107 (2011)
14. Faliszewski, P., Hemaspaandra, E., Hemaspaandra, L.A.: The complexity of manipulative attacks in nearly single-peaked electorates. Artif. Intell. **207**, 69–99 (2014)
15. Garey, M., Johnson, D.: Computers and Intractability: A Guide to the Theory of NP-Completeness. W. H. Freeman and Company, New York (1979)
16. Ghosh, S., Mundhe, M., Hernandez, K., Sen, S.: Voting for movies: The anatomy of recommender systems. In: Proceedings of the 3rd Annual Conference on Autonomous Agents (AGENTS 1999), pp. 434–435. ACM Press (1999)
17. Gibbard, A.: Manipulation of voting schemes. Econometrica **41**(4), 587–601 (1973)
18. Hemaspaandra, E., Hemaspaandra, L.: Dichotomy for voting systems. J. Comput. Syst. Sci. **73**(1), 73–83 (2007)
19. Procaccia, A.D., Rosenschein, J.S.: Junta distributions and the average-case complexity of manipulating elections. J. Artif. Intell. Res. (JAIR) **28**, 157–181 (2007)
20. Satterthwaite, M.: Strategy-proofness and Arrow's conditions: existence and correspondence theorems for voting procedures and social welfare functions. J. Econ. Theor. **10**(2), 187–217 (1975)
21. Walsh, T.: Uncertainty in preference elicitation and aggregation. In: Proceedings of the 22nd AAAI Conference on Artificial Intelligence (AAAI 2007). pp. 3–8. AAAI Press (2007)
22. Xia, L.: Designing social choice mechanisms using machine learning. In: Proceedings of the 12th International Conference on Autonomous Agents and Multiagent Systems (AAMAS 2013), pp. 471–474. International Foundation for Autonomous Agents and Multiagent Systems (2013)

Manipulation and Bribery When Aggregating Ranked Preferences

Ying Zhu and Miroslaw Truszczynski[⊠]

Department of Computer Science, University of Kentucky,
Lexington, KY 40506, USA
mirek@cs.uky.edu

Abstract. Manipulation and bribery have received much attention from the social choice community. We study these concepts for preference formalisms that identify a set of optimal outcomes rather than a single winning outcome. We assume that preferences may be ranked (differ in importance), and we use the Pareto principle adjusted to the case of ranked preferences as the preference aggregation rule. For two important classes of preferences, representing the extreme ends of the spectrum, we provide characterizations of situations when manipulation and bribery is possible, and establish the complexity of the problems to decide that.

1 Introduction

In a common *preference reasoning* scenario, a group of agents is presented with a set of *configurations* or *outcomes*. These outcomes come from a *combinatorial* domain, that is, they are characterized by several multivalued attributes and are represented as tuples of attribute values. Each agent has her individual *preferences* on the outcomes. The problem is to *aggregate* these preferences, that is, to define a "group" preference relation or, at the very least, to identify outcomes that could be viewed by the entire group as good consensus choices. This scenario has received much attention in the AI and decision theory communities [12,19,20].

The two key questions are: how to represent preferences, and how to reason about them. Modeling preferences over combinatorial domains is challenging as the sheer number of outcomes makes explicit representations infeasible. To circumvent the problem of size, one resorts to implicit representation languages that aim to provide concise and intuitive approximations to agents' true preferences. The survey by Domshlak et al. [12] and the monograph by Kaci [20] discuss several of them.

The other aspect of the scenario above is *preference aggregation*. The goal is to *aggregate* diverse preferences of a group of agents into a single *consensus preference ordering* on outcomes or, for some applications, into a set of *consensus optimal outcomes*. Preference aggregation may cause agents to behave *strategically*. They may misrepresent their true preferences, or coerce others to do so, in order to secure consensus preference aggregation outcomes that are more favorable to them. This is the problem that we study in our paper.

© Springer International Publishing Switzerland 2015
T. Walsh (Ed.): ADT 2015, LNAI 9346, pp. 86–102, 2015.
DOI: 10.1007/978-3-319-23114-3_6

The preference aggregation scenario and the problem of strategic misrepresentation of preferences are similar to problems in social choice theory [1,2]. There the concern is to determine a *winner* (sometimes, a strict ordering of the candidates) based on the *votes* cast by a group of *voters*. If we think of voters as agents, of candidates as options, and of votes as preferences, the connection is evident and was noted before [11]. However, the distinguishing aspect of the voting problem considered in social choice is that the number of options (candidates in an election) is *small* and preferences (votes) are specified *explicitly*. The main research goals are to design *voting rules* (procedures to determine a winner or winners based on votes), to identify socially desirable properties that voting rules should have, and to determine which voting rules have which properties. Most of the common voting rules proposed and studied in social choice rely on some form of quantitative scoring [5]. They are typically quite strong. That is, when the number of voters is sufficiently large they rarely return multiple winners.

In contrast, in research of preferences over combinatorial domains the number of candidates is *large*. Thus, the primary objective is to design languages to represent preferences in an intuitive and concise way [12,20]. Collections of preferences in the language, in some cases ranked, to reflect a varying importance of agents, are called *preference theories*. The semantics of the language, that is, a mapping assigning to a *preference theory* a *set* of *preferred* objects from the domain plays the role of a preference aggregation method. To identify preferred elements, quantitative methods similar to basic voting rules have been considered. However, the primary focus has been on qualitative principles such as the *Pareto rule* and its *ranked* versions [9]. Such rules are in general weaker (select more outcomes as optimal) than the quantitative ones. In fact, it is one of the reasons why they are of interest. Namely, in the context of combinatorial domains, concise representations of preferences are likely to introduce errors, being only *approximations* to actual preferences. Strongly discriminating rules may fail to return an outcome optimal with respect to the true preferences of the agents. Weaker formalisms, which select more outcomes, offer a better chance that a true optimal outcome will not be missed.

The problem we study in this paper, misrepresenting preferences by agents to influence preference aggregation to their advantage, has its roots in *strategic voting* studied in social choice [2,18,22]. Strategic voting comes in two flavors. *Manipulation* consists of a voter misrepresenting her vote to secure a better outcome for herself [18,22]. *Bribery* consists of coercing *other* voters to vote against their preferences [13].

Manipulation, arguably the more fundamental of the two and being around longer, has received by far more attention so far. The classical work of Gibbard and Satterthwaite [18,22] established the main impossibility result stating that no rule in a certain broad class is robust to manipulation (or *strategy proof*). However, some researchers argued that one of the key desiderata on the class of rules considered in the Gibbard-Satterthwaite result, the requirement that a rule be *resolute* (that is, always returning a single winner) is in many cases unreasonable [17,21,23] and at odds with socially desirable requirements of equal treatment of candidates and voters [23]. This critique opened the door to research of strategy proofness of voting rules that are *irresolute*, that is, may return several winners. Early results identifying situations in which multi-winner rules are strategy proof, as well as those when impossibility results similar to that of Gibbard and Satterthwaite still hold, were found by Gärdenfors [17] and Kelly [21]. Additional results along these lines are surveyed by Taylor [23] and Barberà [3]. More recently strategy proofness of irresolute rules have been studied by Brandt [6], Brandt and Brill [7], and Brandt and Geist [8]. It turns out that how preferences on candidates are extended to preferences on *sets* is essential for the possibility of strategic behaviors. In particular, Brandt et al. found several strategy proof irresolute rules for the so-called Kelly, Fishburn and Gärdenfors extensions.

The way we view preference reasoning corresponds to the setting of irresolute rules in social choice research. We model agents' preferences as total *preorders* on the space D of outcomes. We assign ranks to preferences as, in real settings, agents will have a hierarchical structure and some will be more important than others. We select a ranked version of *Pareto efficiency* as the principle of preference aggregation. We define the manipulation and bribery in this setting, and establish conditions under which manipulation and bribery are possible.

In each case, the key question is whether misrepresenting preferences can improve for a particular agent the quality of the *collection* of all preferred outcomes resulting from preference aggregation. To be able to decide this question, we have to settle on a way to compare *subsets* of D based on that agent's preference preorder on *elements* of D, an issue that underlies all research on strategy proofness of irresolute rules. In this paper, we focus on four natural extensions of a total preorder on D to a total preorder on the power set $\mathcal{P}(D)$. For each of these extensions we characterize a possibility of manipulation or bribery under the Pareto rule (or ranked Pareto rule, for ranked theories). These results apply directly to the setting of social choice as they do not depend on any preference representation language. Since in many cases strategy proofness cannot be assured, following the well established research in computational social choice [4,13,14,16], we turn attention to study the complexity of deciding whether manipulation or bribery are possible. Indeed, the intractability of computing deviations from true preferences to improve the outcome for an agent may serve as a barrier against strategic behaviors. We look at these questions in the setting of combinatorial domains, the setting not covered by the earlier results. We use our characterizations results as the main tool in this part of our work.

2   Technical Preliminaries

A *preference* on D is a total preorder on D, that is, a binary relation on D that is reflexive, transitive and total. Each such relation, say $\succeq$, determines two associated relations: *strict preference*, denoted $\succ$, where $x \succ y$ if and only if $x \succeq y$ and $y \not\succeq x$, and *indifference*, denoted $\approx$, where $x \approx y$ if and only if $x \succeq y$ and $y \succeq x$. The indifference relation $\approx$ is an equivalence relation on D and partitions D into equivalence classes, $D_1, \ldots, D_m$, which we always enumerate from the most to the least preferred. Using this notation, we can describe a total preorder $\succeq$ by the expression

$$\succeq: \quad D_1 \succ D_2 \succ \cdots \succ D_m.$$

For example, a total preorder $\succeq$ on $D = \{a, b, c, d, e, f\}$ such that $a \approx d$, $b \approx e \approx f$ and $a \succ b \succ c$ (these identities uniquely determine $\succeq$) is specified by an expression

$$\succeq: \quad a, d \succ b, e, f \succ c.$$

(we omit braces from the notation specifying sets of outcomes to keep the notation simple). For every $x \in D$, we define the *quality degree* of x in $\succeq$, written $q_\succeq(x)$, as the unique i such that $x \in D_i$.

Let us consider a group $\mathcal{A}$ of N agents each with her own preference on D and with its rank in the set. We denote these agents by integers from $\{1, \ldots, N\}$, their ranks by $r_1, \ldots, r_N$ (lower rank values imply higher importance), and their preferences by $\succeq_1, \ldots, \succeq_N$. Sometimes, we write $\succeq_{i,r_i}$ for the preference of an agent i, indicating at the same time, the rank of the agent (the rank of her preference). We write $D_1^i, \ldots, D_{m_i}^i$ for the equivalence classes of the relation $\approx_i$ enumerated, as above, from the most to the least preferred with respect to $\succeq_i$. We call the sequence $(\succeq_{1,r_1}, \ldots, \succeq_{N,r_N})$ of preferences of agents in $\mathcal{A}$ a (preference) *profile* of $\mathcal{A}$. For instance,

$$\succeq_{1,1}: \quad f \succ a, c, e \succ b, d$$
$$\succeq_{2,1}: \quad a, c \succ d, e, f \succ b$$
$$\succeq_{3,2}: \quad a \succ b, c \succ d \succ e, f.$$

is a profile of agents $1, 2$ and 3. The preferences of agents 1 and 2 are equally ranked and more important than the preferences of agent 3.

Let $\mathcal{A}$ be a set of N agents with a profile $P = (\succeq_{1,r_1}, \ldots, \succeq_{N,r_N})$. We say that $a \in D$ is *Pareto preferred* in P to $b \in D$ (more formally, Pareto-preferred by a group $\mathcal{A}$ of agents with profile P), written $a \succeq_P b$, if for every $i \in \mathcal{A}$ such that $b \succ_i a$, there is $j \in \mathcal{A}$ such that $r_j < r_i$ and $a \succ_j b$. Similarly, $a \in D$ is *strictly Pareto-preferred* in P to $b \in D$, written $a \succ_P b$, if $a \succeq_P b$ and $b \not\succeq_P a$, that is, precisely when there is a rank r such that for *every* $i \in \mathcal{A}$ with $r_i \leq r$, $a \succeq_i b$, and for at least one $i \in \mathcal{A}$ with $r_i \leq r$, $a \succ_i b$. Finally, $a \in D$ is *Pareto optimal* in P if there is no $b \in D$ such that $b \succ_P a$. We denote the set of all elements in D that are Pareto-optimal in P by $Opt(P)$. Virtually all preference aggregation techniques select "group optimal" elements from those that are Pareto-optimal. From now on, we omit the term "Pareto" when speaking about the preference relation $\succeq_P$ on D and optimal elements of D determined by this relation, as we do not consider any other preference aggregation principles.

Let P be the profile given above. Considering the preferences of agents 1 and 2, a and c are indifferent, no outcome can strictly dominate a, c or f, and outcomes b, d, e are strictly dominated by a and c. According to the preference of agent 3, a is strictly better than c. Thus, $Opt(P) = \{a, f\}$. It is interesting to note that for each of the first two agents, the set $Opt(P)$ contains at least one of her "top-rated" outcomes. This is an instance of a general *fairness* property of the Pareto principle.

Theorem 1. *For every profile P of a set $\mathcal{A}$ of agents, and for every top-ranked agent $i \in \mathcal{A}$, the set $Opt(P)$ of optimal outcomes for P contains at least one outcome most preferred by i.*

Proof. Let us pick any outcome $w \in D$ that is optimal for i (that is, $w \in D_1^i$). Clearly, there is $v \in Opt(P)$ such that $v \succeq_P w$. In particular, $v \succeq_i w$. Thus, $v \in D_1^i$ and $v \in Opt(P)$. □

Coming back to our example, it is natural to ask how satisfied agent 3 is with the result of preference aggregation and what means might she have to influence the result. If she submits a different ("dishonest") preference, say

$$\succeq'_{3,2}: \quad a, c \succ b \succ d \succ e, f$$

then, writing P' for the profile $(\succeq_{1,1}, \succeq_{2,1}, \succeq'_{3,2})$, $Opt(P') = \{a, c, f\}$. It may be that agent 3 would prefer $\{a, c, f\}$ to $\{a, f\}$, for instance, because the new set contains an additional highly preferred outcome for her. Thus, agent 3 may have an incentive to misrepresent her preference to the group. We will call such behavior *manipulation*. Similarly, agent 3 might keep her preference unchanged but convince agent 1 to replace his preference with

$$\succeq'_{1,1}: \quad b \succ f \succ a, c, e \succ d.$$

Denoting the resulting profile $(\succeq'_{1,1}, \succeq_{2,1}, \succeq_{3,2})$ by P'', $Opt(P'') = \{a, b, f\}$ and, because of the same reason as above, this collection of outcomes may also be preferred to $\{a, f\}$ by agent 3. Thus, agent 3 may have an incentive to try to coerce other agents to change their preference. We will call such behavior *simple bribery*.

We now formally define *manipulation* and *simple bribery*. For a profile $P = (\succeq_{1,r_1}, \ldots, \succeq_{N,r_N})$ and a preference $\succeq'_{i,r_i}$, we write $P_{\succeq_{i,r_i}/\succeq'_{i,r_i}}$ for the profile obtained from P by replacing the preference $\succeq_{i,r_i}$ of the agent i with the preference $\succeq'_{i,r_i}$. Let now $\mathcal{A}$ be a group of N agents with a profile $P = (\succeq_{1,r_1}, \ldots, \succeq_{N,r_N})$, and let $\succeq'_{i,r_i}$ be a preference of agent i on *subsets* of D.

Manipulation: An agent i can *manipulate* preference aggregation if there is a preference $\succeq'_{i,r_i}$ such that $Opt(P_{\succeq_{i,r_i}/\succeq'_{i,r_i}}) \succ'_i Opt(P)$.

Simple Bribery: An agent t is a target for *bribery* by an agent i, if there is a preference $\succeq'_{t,r_t}$ such that $Opt(P_{\succeq_{t,r_t}/\succeq'_{t,r_t}}) \succ'_i Opt(P)$.[1]

Clearly, when deciding whether to manipulate (or bribe), agents must be able to compare sets of outcomes and not just single outcomes. This is why we

---

[1] Bribery is traditionally understood as an effort by an *external* agent to bribe a group of voters to obtain a more satisfying result. To stress the difference between this notion and the notion we consider in the paper, we use the term simple bribery.

assumed that the agent i has a preorder $\succeq'_i$ on $\mathcal{P}(D)$. However, even when D itself is not a combinatorial domain, $\mathcal{P}(D)$ is. Thus, explicit representations of that preorder may be infeasible.

The question then is whether the preorder $\succeq'_i$ of $\mathcal{P}(D)$, which parameterizes the definitions of manipulation and bribery, can be expressed in terms of the preorder $\succeq_i$ on D, as the latter clearly imposes some strong constraints on the former. This problem has received attention from the social choice and AI communities [3,10,15,17,21] and it turns out to be far from trivial. The difficulty comes from the fact that there are several ways to "lift" a preorder from D to the power set of D, none of them fully satisfactory (cf. impossibility theorems [3]). In this paper, we sidestep this issue and simply select and study several most direct and natural "liftings" of preorders on sets to preorders on power sets. We introduce them below. We write X and Y for subsets of D and $\succeq$ for a total preorder on D that we seek to extend to a total preorder on $\mathcal{P}(D)$.

Compare Best: $X \succeq^{cb} Y$ if there is $x \in X$ such that for every $y \in Y$, $x \succeq y$.

Compare Worst: $X \succeq^{cw} Y$ if there is $y \in Y$ such that for every $x \in X$, $x \succeq y$.

For the next two definitions, we assume that $\succeq$ partitions D into strata $D_1, \ldots, D_m$, as discussed above.

Lexmin: $X \succeq^{lmin} Y$ if for every i, $1 \le i \le m$, $|X \cap D_i| = |Y \cap D_i|$, or if for some i, $1 \le i \le m$, $|X \cap D_i| > |Y \cap D_i|$ and, for every $j \le i - 1$, $|X \cap D_j| = |Y \cap D_j|$.

Average-Rank:[2] $X \succeq^{ar} Y$ if $ar_\succeq(X) \le ar_\succeq(Y)$, where for a set $Z \subseteq D$, $ar_\succeq(Z)$ denotes the average rank of an element in Z and is defined by $ar_\succeq(Z) = \sum_{i=1}^{m} i \frac{|Z \cap D_i|}{|Z|}$.

Finally, we describe the classes of profiles that we focus on here. Namely, as the setting of ranked preferences is rich, we restrict attention to the two "extreme" cases. In the first one, all agents are equally ranked. In such case, the Pareto principle makes many outcomes optimal as pairs of outcomes are often incomparable. Nevertheless, all practical aggregation techniques, can be understood as simply refining the set of Pareto-optimal outcomes. Thus, improving the quality of the Pareto-optimal set is a desirable objective as it increases a chance of a more favorable outcome once a refinement is applied. In the second setting, we assume all agents have distinct ranks. In such case, the Pareto principle is natural and quite effective, resulting in a total preorder refining the one of the most important agent by breaking ties based on preferences of lower ranked agents.

3   Equally Ranked Preferences

In this section, we discuss the manipulation and simple bribery problems in the case where all preferences are equally ranked, and study them with respect

---

[2] This method is well defined only if both sets to compare are non-empty. This is not a strong restriction because our aggregation method returns only non-empty sets of optimal outcomes.

to each of the four extensions of total preorders on D to $\mathcal{P}(D)$ defined above. An *equally ranked preference profile* is a profile $P = (\succeq_{1,r_1}, \ldots, \succeq_{N,r_N})$, where $r_1 = \cdots = r_N$. To simplify the notation, we write it as $P = (\succeq_1, \ldots, \succeq_N)$.

3.1 Manipulation

Given a set $\mathcal{A}$ of N agents and a profile $P = (\succeq_1, \ldots, \succeq_N)$, the manipulation problem is to determine whether an agent i can find a total preorder $\succeq$ such that $Opt(P_{\succeq_i/\succeq}) \succ_i' Opt(P)$ where $\succeq_i'$ is the total preorder agent i uses to compare subsets of D.

Theorem 2. *Manipulation is impossible for* compare best *and* compare worst *on profiles of equally ranked preferences.*

Proof. Let $\mathcal{A}$ be a set of N agents $1, \ldots, N$ with a profile of equally ranked preferences $P = (\succeq_1, \ldots, \succeq_N)$. We want to show that for every $i \in \mathcal{A}$ and every total preorder $\succeq$, $Opt(P) \succeq_i^{cb} Opt(P_{\succeq_i/\succeq})$ and $Opt(P) \succeq_i^{cw} Opt(P_{\succeq_i/\succeq})$.

For *compare best*, let $v \in Opt(P)$ be an outcome that is also optimal for i (such a v exists by Theorem 1). It follows that for every $w \in D$, $v \succeq_i w$. Thus, $v \succeq_i w$, for every $w \in Opt(P_{\succeq_i/\succeq})$. By the definition of $\succeq_i^{cb}$, $Opt(P) \succeq_i^{cb} Opt(P_{\succeq_i/\succeq})$.

For *compare worst*, let us assume that there is a total preorder $\succeq$ such that $Opt(P_{\succeq_i/\succeq}) \succ_i^{cw} Opt(P)$. It follows from the definition of $\succeq_i^{cw}$ that there is $w' \in Opt(P)$ such that for every $w \in Opt(P_{\succeq_i/\succeq})$, $w \succ_i w'$. Thus, $w' \notin Opt(P_{\succeq_i/\succeq})$ and, consequently, there is $v \in Opt(P_{\succeq_i/\succeq})$ such that $v \succ_{P_{\succeq_i/\succeq}} w'$. It follows that $v \succeq_j w'$, for every agent $j \neq i$. Since by an earlier observation, $v \succ_i w'$, we obtain $v \succ_P w'$, a contradiction with $w' \in Opt(P)$. $\square$

On the other hand, manipulation is possible for every agent using the *lexmin* comparison rule precisely when not every outcome in D is optimal. The reason is that by changing her preference an agent can cause a Pareto-nonoptimal outcome become Pareto-optimal, while keeping the optimality status of every other outcome unchanged.

Theorem 3. *Let $\mathcal{A}$ be a set of N agents $1, \ldots, N$ with a profile of equally ranked preferences $P = (\succeq_1, \ldots, \succeq_N)$ and let $i \in \mathcal{A}$. There exists a total preorder $\succeq$ such that $Opt(P_{\succeq_i/\succeq}) \succ_i^{lmin} Opt(P)$ if and only if $Opt(P) \neq D$.*

Proof. ($\Leftarrow$) Let us assume that $\succeq_i$ is given by

$$\succeq_i: \quad D_1^i \succ_i \ldots \succ_i D_{m_i}^i.$$

Let ℓ be the smallest k such that $D_k^i \setminus Opt(P) \neq \emptyset$ and let $a \in D_\ell^i \setminus Opt(P)$. We will now construct a preference $\succeq$ for agent i so that $Opt(P) \cup \{a\} = Opt(P_{\succeq_i/\succeq})$. For that preference, we have $Opt(P_{\succeq_i/\succeq}) \succ_i^{lmin} Opt(P)$, which demonstrates that i can manipulate preference aggregation in P.

To construct $\succeq$, we first note that since $a \notin Opt(P)$, there is $w \in Opt(P)$ such that $w \succ_P a$. Since $w \succeq_i a$ and $a \in D_\ell^i$, $w \in D_j^i$, for some $j \leq \ell$. Without loss of generality, we may assume that this w is chosen so that to minimize j.

In the remainder of the proof, we write P^{-i} for the profile obtained from P by removing the preference of agent i. To simplify the notation, we also write P' for $P_{\succeq_i/\succeq}$.

Case 1: $w \approx_{P^{-i}} a$. Since $w \succ_P a$, we have $w \succ_i a$, that is, $j < \ell$. Let us define $\succeq$ as follows:

$$\succeq: \quad D'_1 \succ \ldots \succ D'_{m_i},$$

where $D'_j = D^i_j \cup \{a\}$, $D'_\ell = D^i_\ell \setminus \{a\}$, and $D'_k = D^i_k$, for the remaining $k \in [1..m_i]$. Thus $a \approx_{P'} w$. We also have that for every $w', w'' \in D \setminus \{a\}$, $w' \succeq_{P'} w''$ if and only if $w' \succeq_P w''$ (the degrees of quality of outcomes other than a remain the same when we move from P to P'). Finally, for every $w' \in D$, $a \succeq_{P'} w'$ if and only if $w \succeq_{P'} w'$. These observations imply that $Opt(P') = Opt(P) \cup \{a\}$.

Case 2: $w \succ_{P^{-i}} a$. Let us define $\succeq$ as follows:

$$\succeq: \quad D'_1 \succ \ldots \succ D'_{m_i+1},$$

where $D'_k = D^i_k$, for $k < j$, $D'_j = \{a\}$, $D'_{\ell+1} = D^i_\ell \setminus \{a\}$, and $D'_k = D^i_{k-1}$, for every $k \in \{j+1, \ldots, m_i + 1\}$ such that $k \neq \ell + 1$. Informally, $\succeq$ is obtained by pulling a from D^i_ℓ, and inserting it as a singleton cluster directly before D^i_j. Since a is the only outcome moved, for every $w', w'' \in D \setminus \{a\}$, $w' \succeq_{P'} w''$ if and only if $w' \succeq_P w''$ (and similarly, for the derived relation $\succ_{P'}$).

Let us observe that $a \in Opt(P')$. Indeed, if for some $w' \in D$, $w' \succeq_{P'} a$, then $w' \in D^i_k$, for some $k < j$. It follows that $w' \in Opt(P)$ and $w' \succ_i a$. Consequently, $w' \succ_P a$, contrary to our choice of w.

Let $w' \in Opt(P)$ and let us assume that $w'' \succ_{P'} w'$ for some $w'' \in D$. Since $a \notin Opt(P)$, $w' \neq a$. If $w'' \neq a$, then $w'' \succ_P w'$ (indeed, a is the only outcome whose relation to other outcomes changes when we move from P to P'). This is a contradiction with $w' \in Opt(P)$. Thus, $w'' = a$. Consequently, $w'' \succ_{P'} w'$ implies $a \succeq_{P^{-i}} w'$ and $a \succeq w'$. By the construction of $\succeq$, the latter property implies that $w \succeq_i w'$. Since $w \succ_{P^{-i}} a \succeq_{P^{-i}} w'$, $w \succ_P w'$, a contradiction. It follows that $w' \in Opt(P')$ and, consequently, we have $Opt(P) \cup \{a\} \subseteq Opt(P')$.

Conversely, let us consider $w' \in Opt(P')$ such that $w' \neq a$. Let us assume that for some $w'' \in Opt(P)$, $w'' \succ_P w'$. If $w'' \neq a$, we can get $w'' \succ_{P'} w'$, a contradiction. If $w'' = a$, we can get $a \succeq_k w'$ for every $k \in \mathcal{A}$ and $k \neq i$ from $a \succ_P w'$ and $a \succ w'$. Thus $a \succ_{P'} w'$ which contradicts the property that $w' \in Opt(P')$. It follows that $w' \in Opt(P)$. Thus, $Opt(P') \subseteq Opt(P) \cup \{a\}$. Consequently, $Opt(P) \cup \{a\} = Opt(P')$.

($\Rightarrow$) If $Opt(P) = D$, then there is no set S such that $S \succ_i^{lmin} Opt(P)$. □

For the *average-rank* preorder for comparing sets, an agent can manipulate the result to her advantage if there are Pareto-nonoptimal outcomes that are highly preferred by the agent, or when there are Pareto-optimal outcomes that are low in the preference of that agent, as the former can be made optimal and the latter made non-optimal without changing the Pareto-optimality status of other outcomes.

Theorem 4. *Let $\mathcal{A}$ be a set of N agents $1, \ldots, N$ with a profile of equally ranked preferences $P = (\succeq_1, \ldots, \succeq_N)$ and let $i \in \mathcal{A}$. There exists a total preorder $\succeq$ such that $Opt(P_{\succeq_i/\succeq}) \succ_i^{ar} Opt(P)$ if and only if:*

1. *For some $j < ar_{\succeq_i}(Opt(P))$, there exists $a' \in D_j^i$ such that $a' \notin Opt(P)$; or*
2. *For some $j > ar_{\succeq_i}(Opt(P))$, there are $a' \in Opt(P) \cap D_j^i$ and $a'' \in Opt(P)$ such that $a' \neq a''$, and $a'' \succeq_k a'$, for every $k \in \mathcal{A}$, $k \neq i$.*

Proof. ($\Leftarrow$) Let us assume that the first condition holds. Let ℓ be the smallest k such that $D_k^i \setminus Opt(P) \neq \emptyset$, and let $a' \in D_\ell^i \setminus Opt(P)$. Reasoning as in the proof of the previous theorem, we can construct a total preorder $\succeq$ such that $Opt(P') = Opt(P) \cup \{a'\}$ (where P' denotes $P_{\succeq_i/\succeq}$). Clearly, $ar_{\succeq_i}(Opt(P')) < ar_{\succeq_i}(Opt(P))$ and so, $Opt(P') \succ_i^{ar} Opt(P)$ (i can manipulate).

If the second condition is satisfied then, let us assume that $a'' \in D_{j'}^i$. Then, we have $j' \geq j$ (otherwise, $a'' \succ_P a'$, contradicting optimality of a' in P). Let us construct $\succeq$ as in the previous argument, but substituting a'' for a' (and, as before, we write P' for $P_{\succeq_i/\succeq}$). Without loss of generality, we may select a'' so that j' be minimized.

We know that $a'' \in Opt(P')$. Moreover, by the definition, $a'' \succ a'$. Thus, $a'' \succ_{P'} a'$ and so, $a' \notin Opt(P')$.

Next, if $w \in Opt(P)$ and $w \succ_i a'$, then $w \in Opt(P')$. To show this, let us assume that there is $w' \in Opt(P')$ such that $w' \succ_{P'} w$. Since $w \succ_i a'$, $w \neq a''$ and $w \succ a''$. The latter implies that $w' \neq a'$. Thus, $w' \succ_P w$, a contradiction.

Finally, if $w \notin Opt(P)$ and $a' \succeq_i w$, $w \notin Opt(P')$. Indeed, it is clear that if $w' \succ_P w$ then $w' \succ_{P'} w$.

Since $j > ar_{\succeq_i}(Opt(P))$, these observations imply that $ar_{\succeq_i}(Opt(P')) < ar_{\succeq_i}(Opt(P))$.

($\Rightarrow$) We set $x = ar_{\succeq_i}(Opt(P))$. By the assumption, there is a total preorder $\succeq$ on D such that $ar_{\succeq_i}(Opt(P_{\succeq_i/\succeq})) < x$. Let us set $O = Opt(P_{\succeq_i/\succeq})$ and let D_1 be the set of all elements $w \in D$ such that $q_{\succeq_i}(w) < x$. If $D_1 \setminus Opt(P) \neq \emptyset$, then the condition (1) holds. Thus, let us assume that $D_1 \subseteq Opt(P)$. We denote by O' the set obtained by

1. removing from $Opt(P)$ every element $w \in D_1 \setminus O$
2. removing from $Opt(P)$ every element $w \notin O$ such that $q_{\succeq_i}(w) = x$
3. including every element $w \in O \setminus Opt(P)$ such that $q_{\succeq_i}(w) = x$.

We have $ar_{\succeq_i}(O') \geq x$. Moreover, O' differs from O (if at all) only on elements w such that $q_{\succeq_i}(w) > x$. If O contains every element $w \in Opt(P)$ such that $q_{\succeq_i}(w) > x$, then $ar_{\succeq_i}(O) \geq ar_{\succeq_i}(O')$ and so, $ar_{\succeq_i}(O) \geq x$, a contradiction. Thus, there is an element $w \in Opt Opt(P)$ such that $q_{\succeq_i}(w) > x$ and $w \notin O$. Since $O = Opt(P_{\succeq_i/\succeq})$, it is only possible if the condition (2) holds. $\square$

The main message of these theorems is that when the result of preference aggregation is a *set* of optimal outcomes, then even the most elementary aggregation rule, Pareto principle, may be susceptible to manipulation. Whether it is or is not depends on how agents measure the quality of a set. If the comparison is based on the best or worst outcomes, manipulation is not possible (a positive result). However, under less simplistic rules such as *lexmin* or *average-rank* the possibility for manipulation emerges (a negative result that, in some settings, we later moderate by means of the complexity barrier).

3.2 Simple Bribery

In the same setting, the simple bribery problem is to decide whether an agent i can find an agent t, $t \neq i$, and a total preorder $\succeq$ such that $Opt(P_{\succeq_t/\succeq}) \succ_i' Opt(P)$. Our results on bribery are similar to those we obtained for manipulation, with one notable exception, and show that whether bribery is possible depends on how agents measure the quality of sets of outcomes.

Theorem 5. *Simple bribery is impossible for* compare best *on profiles of equally ranked preferences.*

The result can be proved in the same way as Theorem 2. We stress that it states that no agent using *compare best* preorder on sets can successfully bribe *any other* agent.

The situation changes if agents are interested in maximizing the worst outcomes in a set. Unlike in the case of manipulation, simple bribery may now be possible. Given a set $X \subseteq D$ and a total preorder $\succeq$, by $Min_{\succeq}(X)$ we denote the set of all "worst" elements in X, that is the set that contains every element $x \in X$ such that for every $y \in X$, $y \succeq x$.

Theorem 6. *Let $\mathcal{A}$ be a set of N agents $1, \ldots, N$ with a profile of equally ranked preferences $P = (\succeq_1, \ldots, \succeq_N)$ and let $i \in \mathcal{A}$. There exist $t \in \mathcal{A}$, $t \neq i$, and a total preorder $\succeq$ such that $Opt(P_{\succeq_t/\succeq}) \succ_i^{cw} Opt(P)$ if and only if for every $a \in Min_{\succeq_i}(Opt(P))$, there is $a' \in D$ such that $a' \succ_i a$, and $a' \succeq_k a$, for every $k \in \mathcal{A}$, $k \neq t$.*

Proof. ($\Leftarrow$) To define $\succeq$, we modify the total preorder $\succeq_t$ as follows. For every $a \in Min_{\succeq_i}(Opt(P))$, we move a' (the element satisfying $a' \succ_i a$, and $a' \succeq_k a$, for every $k \in \mathcal{A}$, $k \neq t$, whose existence is given by the assumption) from its cluster in $\succeq_t$ to the cluster of $\succeq_t$ containing a.

First, we note that for every $a \in Min_{\succeq_i}(Opt(P))$, $a' \succ_{P_{\succeq_t/\succeq}} a$. Second, the only change when moving from P to $P_{\succeq_t/\succeq}$ is in the profile of agent t, and that profile changes by *promoting* elements a' (indeed, for every $a \in Min_{\succeq_i}(Opt(P))$, $a \succ_t a'$; otherwise, we would have $a' \succ_P a$, contrary to $a \in Opt(P)$). Thus, some of these elements might become optimal but their degrees of quality in $\succeq_i$ are better than those of their corresponding elements a. Finally, other elements than a's cannot become optimal. These three observations imply that $Opt(P_{\succeq_t/\succeq}) \succ_i^{cw} Opt(P)$.

($\Rightarrow$) Let an agent $t \neq i$ and a total preorder $\succeq$ satisfy $Opt(P_{\succeq_t/\succeq}) \succ_i^{cw} Opt(P)$. To simplify notation, we set $Q = P_{\succeq_t/\succeq}$.

Let us consider $a \in Min_{\succeq_i}(Opt(P))$. Since $Opt(Q) \succ_i^{cw} Opt(P)$, $a \notin Opt(Q)$. It follows that there is $a' \in Opt(Q)$ such that $a' \succ_Q a$. Thus $a' \succ_i a$ (otherwise, we would have $Opt(P) \succeq_i Opt(Q)$, a contradiction). Moreover, for every $k \in \mathcal{A}$, $k \neq t$, $a' \succeq_k a$. □

Simple bribery is also possible when *lexmin* or *average-rank* methods are used by agents to extend a preorder on D to a preorder on $\mathcal{P}(D)$. Similarly to Theorem 5, the following two theorems are literal generalizations of the earlier results on manipulation and we omit the proofs.

Theorem 7. *Let $\mathcal{A}$ be a set of N agents $1, \ldots, N$ with a profile of equally ranked preferences $P = (\succeq_1, \ldots, \succeq_N)$ and let $i, t \in \mathcal{A}$, $t \neq i$. There exists a total preorder $\succeq$ such that $Opt(P_{\succeq_t / \succeq}) \succ_i^{lmin} Opt(P)$ if and only if $Opt(P) \neq D$.*

Theorem 8. *Let $\mathcal{A}$ be a set of N agents $1, \ldots, N$ with a profile of equally ranked preferences $P = (\succeq_1, \ldots, \succeq_N)$ and let $i \in \mathcal{A}$. There exist $t \in \mathcal{A}$, $t \neq i$, and a total preorder $\succeq$ such that $Opt(P_{\succeq_t / \succeq}) \succ_i^{ar} Opt(P)$ if and only if:*

1. *For some $j < ar_{\succeq_i}(Opt(P))$, there exists $a' \in D_j^i$ such that $a' \notin Opt(P)$; or*
2. *For some $j > ar_{\succeq_i}(Opt(P))$, there are $a' \in Opt(P) \cap D_j^i$, and $a'' \in Opt(P)$ such that $a' \neq a''$ and $a'' \succeq_k a'$, for every $k \in \mathcal{A}$, $k \neq t$.*

Theorems 6, 7 and 8 show that a possibility for simple bribery may arise when *compare worst*, *lexmin* and *average-rank* are used to compare sets of outcomes. There is, however, a difference between *lexmin* and the other two methods. For the former, if simple bribery is possible, then every agent can be the target (can be used as t in the theorem). This is not the case for the other two methods.

4 Strictly Ranked Preferences

In this section, we discuss the manipulation and simple bribery problems in the setting in which all agents have distinct ranks and so, can be seen as strictly ranked. A strictly ranked preference profile can be written as $P = (\succeq_{1,1}, \ldots, \succeq_{N,N})$ (after possibly relabeling agents). In this section, we will write such profiles as $P = (\succeq_1, \ldots, \succeq_N)$. Such a preference formalism generates a total preorder over outcomes. Moreover, all optimal outcomes are indifferent and share the same quality degree for every preference. In general, proceeding from the most important preference to the least, the relation between two outcomes is decided by the first preference, where they have different quality degrees.

Our first two results in this section concern the manipulation problem.

Theorem 9. *Manipulation is impossible for* compare best, compare worst *and* average-rank *on profiles of strictly ranked preferences.*

Proof. Let $\mathcal{A}$ be a set of N agents $1, \ldots, N$ with a profile $P = (\succeq_1, \ldots, \succeq_N)$ and $i \in \mathcal{A}$. Let us assume the preference of an agent $i \in \mathcal{A}$ is

$$\succeq_i: \quad D_1^i \succ_i D_2^i \succ_i \cdots \succ_i D_{m_i}^i.$$

Since all optimal outcomes are indifferent, we can assume $a \in D_j^i$ for every $a \in Opt(P)$. Let a be any optimal outcome. According to the definitions of *compare-best, compare-worst* and *average-rank*, if $Opt(P')$ is better than $Opt(P)$ based on the corresponding extension of $\succeq_i$, then there exists $a' \in Opt(P')$ such that $a' \succ_i a$. Since $a' \notin Opt(P)$, $a \succ_P a'$. And because of $a' \succ_i a$, $a \succ_j a'$ for some $j < i$. This can not be changed no matter how i changes her preference. Thus $a \succ_{P'} a'$ and $a' \notin Opt(P')$.   □

For *lexmin*, agent i can get a better result by making non-optimal outcomes equivalent or worse to currently optimal outcomes become optimal. The precise description of the conditions when it is possible is given below.

Theorem 10. *Let $\mathcal{A}$ be a set of N agents $1, \ldots, N$ with a profile of strictly ranked preferences $P = (\succeq_1, \ldots, \succeq_N)$. For every $i \in \mathcal{A}$, manipulation is possible for* lexmin *if and only if at least one of the following conditions holds:*

1. *There exists $a' \in D \setminus Opt(P)$, such that for every $a'' \in Opt(P)$, $a'' \approx_{P/i} a'$* [3]
2. *There exists $a' \in D \setminus Opt(P)$, such that*

$$|\{a : a \in D, a \approx_P a'\}| > |Opt(P)|,$$

   and for every $a'' \in Opt(P)$ and for every $l \leq i$, $a'' \approx_l a'$
3. *There exists $a' \in D \setminus Opt(P)$, such that*

$$|\{a : a \in D, a \approx_P a'\}| = |Opt(P)|,$$

   for every $a'' \in Opt(P)$ and for every $l \leq i$, $a'' \approx_l a'$, and for some $w \in D$, $w \approx_{P/i} a'$ and $w' \not\approx_i a'$.

We will now consider simple bribery. There are rather intuitive conditions describing when simple bribery is possible for the *compare best, compare worst* and *average-rank* set comparison methods, and somewhat more complicated ones for *lexmin*.

Theorem 11. *Let $\mathcal{A}$ be a set of N agents $1, \ldots, N$ with a profile of strictly ranked preferences $P = (\succeq_1, \ldots, \succeq_N)$. For every $i \in \mathcal{A}$, simple bribery is possible for* compare best, compare worst *and* average-rank *if and only if there exists $a' \in D$ such that $a' \succ_i a$ for every $a \in Opt(P)$.*

Since all optimal answer sets in $Opt(P)$ are indifferent, if $Opt(P') \succ_i^{cb/cw/ar} Opt(P)$, then there exists $a' \in Opt(P')$ such that $a' \succ_i a$. Thus, it is clear that if there is no $a' \in D$ such that $a' \succ_i a$ for every $a \in Opt(P)$, $Opt(P)$ cannot be dominated by any set of optimal answer sets and simple bribery is impossible. If such a' exists, the agent i can bribe the agent on the top level to modify her preference by putting a' at the first place. Then $Opt(P') = \{a'\}$ and $Opt(P') \succ_i^{cb/cw/ar} Opt(P)$. If agent i is at the top level, her top choice must be in $Opt(P)$ and such a' does not exist.

Theorem 12. *Let $\mathcal{A}$ be a set of N agents $1, \ldots, N$ with a profile of strictly ranked preferences $P = (\succeq_1, \ldots, \succeq_N)$. For every $i \in \mathcal{A}$, simple bribery is possible for* lexmin *if and only if at least one of the following three conditions holds:*

---

[3] $a'' \approx_{P/i} a'$ means $a'' \approx_P a'$ except for $\succeq_i$.

1. *There is $a' \in D$ such that for all $a'' \in Opt(P)$, $a' \succ_i a''$*
2. *There is $a' \in D \setminus Opt(P)$ such that for some $t \in \mathcal{A}$, $t \neq i$, and for every $a'' \in Opt(P)$, $a' \approx_{P/t} a''$*
3. *There is $a' \in D$ and $t, j \in \mathcal{A}$ such that $t \leq j$, $t \neq i$,*

$$|\{a : a \in D, a \approx_{P/t} a'\}| > |Opt(P)|,$$

for every $a'' \in Opt(P)$, $a' \approx_i a''$, $a'' \succ_j a'$, and $a'' \approx_l a'$, for every $l < j$.

5 Complexity

So far we studied the problems of manipulation and simple bribery ignoring the issue of how preferences (total preorders) on D are represented. In this section, we will establish the complexity of deciding whether manipulation or simple bribery are possible. For this study, we have to fix a preference representation schema.

First, let us assume that preference orders on elements of D are represented explicitly as sequences $D_1, \ldots, D_m$ of the indifference strata, enumerating them from the most preferred to the least preferred. For this representation, the characterizations we presented in the previous section imply that the problems of the existence of manipulation and bribery can be solved in polynomial time. Thus, in the "explicit representation" setting, computational complexity cannot serve as a barrier against them.

However, for combinatorial domains explicit representations are not feasible. We now take for D a common combinatorial domain given by a set U of binary attributes. We view elements of U as propositional variables and assume that each element of U can take a value from the domain $\{true, false\}$. In this way, we can view D as the set of all truth assignments on U. Following a common convention, we identify a truth assignment on U with the subset of U consisting of elements that are true under the assignment. Thus, we can think of D as the power set $\mathcal{P}(U)$ of U.

By taking this perspective, we can use a formula φ over U as a concise implicit representation of the set $M(\varphi) = \{X \subseteq U : X \models \varphi\}$ of all interpretations of U (subsets of U) that satisfy φ, and we can use sequences of formulas to define total preorders on $\mathcal{P}(U)$ $(= D)$.

A *preference statement* over U is an expression

$$\varphi_1 > \varphi_2 > \cdots > \varphi_m, \tag{1}$$

where all φ_is are formulas over U and $\varphi_1 \vee \cdots \vee \varphi_m$ is a tautology. A preference statement $p = \varphi_1 > \varphi_2 > \cdots > \varphi_m$ determines a sequence $(D_1, \ldots, D_m)$ of subsets of $\mathcal{P}(U)$, where, for every $i = 1, \ldots, m$,

$$D_i = \{X \subseteq U : X \models \varphi_i\} \setminus (D_1 \cup \cdots \cup D_{i-1}).$$

These subsets are disjoint and cover the entire domain $\mathcal{P}(U)$ (the latter by the fact that $\varphi_1 \vee \cdots \vee \varphi_m$ is a tautology). It follows that if $X \subseteq U$, then there is a unique i_X such that $X \in D_{i_X}$. The relation $\succeq_p$ defined so that $X \succeq_p Y$ precisely when $i_X \leq i_Y$ is a total preorder on $\mathcal{P}(U)$. We say that the preference expression p *represents* the preorder $\succeq_p$. [4]

This form of modeling preferences (total preorders) is quite common. Preference statements were considered by Brewka, Niemelä and Truszczynski [9] as elements of preference modules in answer-set optimization programs. [5] Furthermore, modulo slight differences in the notation, preference statements can also be viewed as preference theories of the possibilistic logic [20].

We will now study the complexity of the existence of manipulation and simple bribery when preferences are given in terms of preference statements. That is, we assume that the input to these problems consists of N ranked preferences $\succeq_{1,r_1}, \ldots, \succeq_{N,r_N}$. We will denote by $(D_1^i, \ldots, D_{m_i}^i)$ the sequence of indifference strata determined by $\succeq_{i,r_i}$, as defined above. We refer to these two problems as the *existence-of-manipulation* (EM) problem and the *existence-of-simple-bribery* (ESB) problem, respectively. These problems are parameterized by the method used to compare sets. We denote the methods by cb (*compare best*), cw (*compare worst*), $lmin$ (*lexmin*) and ar (*average-rank*).

For equally ranked preference statements, since for the *compare best* and *compare worst* methods for comparing sets manipulation is impossible, the problems are (trivially) in P. Similarly, the problem of deciding whether simple bribery is possible for *compare best* is in P, too. The summary of the complexity results for all the cases is given by the following theorem.

Theorem 13. *The complexity of deciding whether manipulation and simple bribery are possible for equally ranked preferences with four ways to lift preorders on outcomes to preorders on sets of outcomes is as follows:*

	cb	cw	$lmin$	ar
EM	P	P	NP-comp	Σ_2^P-hard, in PSPACE
ESB	P	Σ_2^P-hard, Π_2^P-hard, in Δ_3^P	NP-comp	Σ_2^P-hard, in PSPACE

For strictly ranked preferences, for the *compare best, compare worst* and *average-rank* methods, the manipulation is impossible and so, deciding its existence is trivially in P. On the other hand, simple bribery is possible for all set comparison methods. The complete complexity results are given by Theorem 14.

Theorem 14. *The complexity of deciding whether manipulation and simple bribery is possible for strictly ranked preferences with four ways to lift preorders on outcomes to preorders on sets of outcomes is as follows:*

---

[4] The partition of D into strata that is determined by $\succeq_p$ is not always $(D_1, \ldots, D_m)$ as some sets D_i may be empty.

[5] The original definition [9] allows for more general preference statements. However, they all can be effectively expressed as preference statements we defined here.

	cb	cw	$lmin$	ar
EM	P	P	Δ_2^P-$hard$	P
ESB	Δ_2^P-$comp$	Δ_2^P-$comp$	Δ_2^P-$hard$	Δ_2^P-$comp$

6   Conclusions and Future Work

We studied manipulation and simple bribery problems arising when one aggregates sets of ranked preferences. As a preference aggregation method we used the Pareto rule. We considered two extreme cases of that general setting. In one of them, all preferences are equally ranked, in the other one the preferences are strictly ranked. In the scenario we investigated, agents submit preferences on elements of the space of outcomes but, when considering manipulation and simple bribery, they need to assess the quality of *sets* of such elements. In the paper, we considered several natural ways in which a total preorder on a space of outcomes can be lifted to a total preorder on the space of sets of outcomes. For each of these "liftings", we found conditions characterizing situations when manipulation (simple bribery) are possible. These characterizations show that in many cases it is impossible for any agent to strategically misrepresent preferences (*compare best* and *compare worst* for manipulation, in both equally ranked and strictly ranked settings; *compare best* for simple bribery in the equally ranked setting; and, somewhat surprisingly, *average-rank* for manipulation in the strictly ranked setting). In those cases, the Pareto principle is "strategy-proof".

However, in all other cases, it is no longer the case. Manipulation and simple bribery cannot be *a priori* excluded. To study whether computational complexity may provide a barrier against strategic misrepresentation of preferences, we considered a simple logical preference representation language closely related to possibilistic logic and answer-set optimization. For sets of preferences given in this language (in the settings of equally ranked or strictly ranked preferences) and for each way of lifting preorders from sets to power sets for which manipulation and simple bribery are possible, we proved that deciding the existence of manipulation or simple bribery is intractable.

Our work leaves several interesting open problems. First, methods to lift preorders from sets to power sets can be defined axiomatically in terms of properties for the lifted preorders to satisfy. Are there general results characterizing the existence of manipulation (simple bribery) for lifted preorders specified only by axioms they satisfy? Second, we do not know the exact complexity of the problems EB^{cw}, EM^{ar} and EB^{ar} for the equally ranked preferences, nor for EM^{lmin} and EB^{lmin} for the strictly ranked preferences (the superscript indicates the set comparison method used). Finally, in the setting of equally ranked preferences, most aggregation rules of practical significance properly extend the Pareto one. We conjecture that at least for some of these rules, one can derive results on existence of manipulation and simple bribery from our results concerning the Pareto rule.

Acknowledgments. The authors wish to thank the anonymous reviewers for useful comments and pointers to relevant literature.

References

1. Arrow, K.: Social Choice and Individual Values. Cowles Foundation Monographs Series. Yale University Press, New Haven (1963)
2. Arrow, K., Sen, A., Suzumura, K. (eds.): Handbook of Social Choice and Welfare. Elsevier, North-Holland (2002)
3. Barberà, S., Bossert, W., Pattanaik, P.K.: Ranking sets of objects. Springer, Newyork (2004)
4. Bartholdi, J.J.I., Tovey, C., M, Trick: The computational difficulty of manipulating an election. Soc. Choice Welfare **6**(3), 227–241 (1989)
5. Brams, S., Fishburn, P.: Voting procedures. In: Arrow, K., Sen, A., Suzumura, K. (eds.) Handbook of Social Choice and Welfare, pp. 173–206. Elsevier, Amsterdam (2002)
6. Brandt, F.: Group-strategyproof irresolute social choice functions. In: Walsh, T. (ed.) Proceedings of the 22nd International Joint Conference on Artificial Intelligence, IJCAI 2011. pp. 79–84. IJCAI/AAAI (2011)
7. Brandt, F., Brill, M.: Necessary and sufficient conditions for the strategyproofness of irresolute social choice functions. In: Apt, K.R. (ed.) Proceedings of the 13th Conference on Theoretical Aspects of Rationality and Knowledge, TARK-2011. pp. 136–142. ACM (2011)
8. Brandt, F., Geist, C.: Finding strategyproof social choice functions via SAT solving. In: Bazzan, A.L.C., Huhns, M.N., Lomuscio, A., Scerri, P. (eds.) International Conference on Autonomous Agents and Multi-Agent Systems, AAMAS 2014, pp. 1193–1200. IFAAMAS/ACM (2014)
9. Brewka, G., Niemelä, I., Truszczynski, M.: Answer set optimization. In: IJCAI. pp. 867–872 (2003)
10. Brewka, G., Truszczynski, M., Woltran, S.: Representing preferences among sets. In: Proceedings of AAAI 2010 (2010)
11. Chevaleyre, Y., Endriss, U., Lang, J., Maudet, N.: Preference handling in combinatorial domains: from AI to social choice. AI Mag. **29**(4), 37–46 (2008)
12. Domshlak, C., Hüllermeier, E., Kaci, S., Prade, H.: Preferences in AI: an overview. Artif. Intell. **175**(7–8), 1037–1052 (2011)
13. Faliszewski, P., Hemaspaandra, E., Hemaspaandra, L.A.: The complexity of bribery in elections. Proc. AAAI **6**, 641–646 (2006)
14. Faliszewski, P., Hemaspaandra, E., Hemaspaandra, L.A.: Using complexity to protect elections. Commun. ACM **53**(11), 74–82 (2010)
15. Fishburn, P.: Even-chance lotteries in social choice theory. Theor. Decis. **3**, 18–40 (1972)
16. Fitzsimmons, Z., Hemaspaandra, E., Hemaspaandra, L.A.: Control in the presence of manipulators: Cooperative and competitive cases. In: Proceedings of the Twenty-Third International Joint Conference on Artificial Intelligence, pp. 113–119. IJCAI 2013, AAAI Press (2013)
17. Gärdenfors, P.: Manipulation of social choice functions. J. Econ. Theor. **13**(2), 217–228 (1976)
18. Gibbard, A.: Manipulation of voting schemes: a general result. Econometrica **41**(4), 587–601 (1973)

19. Goldsmith, J., Junker, U.: Special issue on preferences. AI Mag. **29**(4), 37–46 (2008)
20. Kaci, S.: Working with Preferences: Less Is More. Cognitive Technologies. Springer, Heidelberg (2011)
21. Kelly, J.: Strategy-proofness and social choice functions without single-valuedness. Econometrica **45**(2), 439–446 (1977)
22. Satterthwaite, M.A.: Strategy-proofness and Arrow's conditions: existence and correspondence theorems for voting procedures and social welfare functions. J. Econ. Theor. **10**, 187–217 (1975)
23. Taylor, A.: Social Choice and the Mathematics of Manipulation. Cambridge University Press, Cambridge (2005)

Complexity of Manipulative Actions
When Voting with Ties

Zack Fitzsimmons[1][✉] and Edith Hemaspaandra[2]

[1] College of Computing and Information Sciences,
Rochester Institute of Technology, Rochester, NY 14623, USA
zmf6921@rit.edu
[2] Department of Computer Science, Rochester Institute of Technology,
Rochester, NY 14623, USA
eh@cs.rit.edu

Abstract. Most of the computational study of election problems has assumed that each voter's preferences are, or should be extended to, a total order. However in practice voters may have preferences with ties. We study the complexity of manipulative actions on elections where voters can have ties, extending the definitions of the election systems (when necessary) to handle voters with ties. We show that for natural election systems allowing ties can both increase and decrease the complexity of manipulation and bribery, and we state a general result on the effect of voters with ties on the complexity of control.

1 Introduction

Elections are commonly used to reach a decision when presented with the preferences of several agents. This includes political domains as well as multiagent systems. In an election agents can have an incentive to cast a strategic vote in order to affect the outcome. An important negative result from social-choice theory, the Gibbard-Satterthwaithe theorem, states that every reasonable election system is susceptible to strategic voting (a.k.a. manipulation) [16,26].

Although every reasonable election system can be manipulated, it may be computationally infeasible to determine if a successful manipulation exists. Bartholdi et al. introduced the notion of exploring the computational complexity of the manipulation problem [1]. They expanded on this work by introducing and analyzing the complexity of control [2]. Control models the actions of an election organizer, referred to as the chair, who has control over the structure of the election (e.g., the voters) and wants to ensure that a preferred candidate wins. Faliszewski et al. introduced the model of bribery [9]. Bribery is closely related to manipulation, but instead of asking if voters can cast strategic votes to ensure a preferred outcome, bribery asks if a subcollection of the voters can be paid to change their vote to ensure a preferred outcome.

It is important that we understand the complexity of these election problems on votes that allow ties, since in practical settings voters often have ties between some of the candidates. This is seen in the online preference repository

© Springer International Publishing Switzerland 2015
T. Walsh (Ed.): ADT 2015, LNAI 9346, pp. 103–119, 2015.
DOI: 10.1007/978-3-319-23114-3_7

PREFLIB, which contains several election datasets containing votes with ties, ranging from political elections to elections created from rating data [23]. Most of the computational study of election problems for partial votes has assumed that each voter's preferences should be extended to a total order (see e.g., the possible and necessary winner problems [21]). However an agent may view two options as explicitly equal and it makes sense to view these preferences as votes with ties, instead of as partial rankings that can be extended.

Election systems are sometimes even explicitly defined for voters with ties. Both the Kemeny rule [20] and the Schulze rule [27] are defined for votes that contain ties. Also, there exist variants of the Borda count that are defined for votes that contain ties [8].

The computational study of the problems of manipulation, control, and bribery has largely been restricted to elections that contain voters with tie-free votes. Important recent work by Narodytska and Walsh [25] studies the computational complexity of the manipulation problem for top orders, i.e., votes where the candidates ranked last are all tied and are otherwise total orders. The manipulation results in this paper can be seen as an extension of the work by Narodytska and Walsh. We consider orders that allow a voter to state ties at each position of his or her preference order, i.e., weak orders. We mention that in contrast to the work by Narodytska and Walsh [25], we give an example of a natural case where manipulation becomes hard when given votes with ties, while it is in P for total orders. Additionally, we are the first to study the complexity of the standard models of control and bribery for votes that contain ties. However, we mention here that Baumeister et al. consider a different version of bribery called extension bribery, for top orders (there called top-truncated votes) [3].

The organization of this paper is as follows. In Sect. 2 we state the formal definitions and problem statements needed for our results. The results in Sect. 3 are split into three sections, each showing a different behavior of voting with ties. In Sect. 3.1 we give examples of election systems where the problems of manipulation, bribery, and control increase in complexity from P to NP-complete. Conversely, in Sect. 3.2 we give examples of election systems where the complexity of manipulation and bribery becomes easier, and state a general result about the complexity of control. In Sect. 3.3 we solve an open question from Narodytska and Walsh [25] and give examples of election systems whose manipulation complexities are unaffected by voters with ties. Additionally, we completely characterize 3-candidate Copeland$^\alpha$ coalitional weighted manipulation for rational and irrational voters with ties. We discuss related work in Sect. 4 and our general conclusions and open directions in Sect. 5.

2   Preliminaries

An *election* consists of a finite set of candidates C and a collection of voters V (also referred to as a preference profile). Each voter in V is specified by its preference order. We consider voters with varying amounts of ties in their preferences. A *total order* is a linear ordering of all of the candidates from most to least preferred. A *weak order* is a transitive, reflexive, and antisymmetric ordering where the indifference relation ("$\sim$") is transitive. In general, a weak order can be viewed as a total order with ties. As usual, we will colloquially refer to indifference as ties throughout this paper since the indifference relation specifies the preference for two elements being equal. A *top order* is a weak order with all tied candidates ranked last, and a *bottom order* is a weak order with all tied candidates ranked first. In Example 1 below we present examples of each of the orders examined in this paper.

Example 1. *Given the candidate set $\{a, b, c, d\}$, $a > b \sim c > d$ is a weak order, $a \sim b > c > d$ is a bottom order, $a > b > c \sim d$ is a top order, and $a > b > c > d$ is a total order. Notice that every bottom order and every top order is also a weak order, and that every total order is also a top, bottom, and weak order.*

An *election system*, $\mathcal{E}$, maps an election, i.e., a finite candidate set C and a collection of voters V, to a set of winners, where the winner set can be any subset of the candidate set. The voters in an election can sometimes have an associated weight where a voter with weight w counts as w unweighted voters.

We examine two important families of election systems, the first being scoring rules. A scoring rule uses a vector of the form $\langle s_1, \ldots, s_m \rangle$, where m denotes the number of candidates, to determine each candidate's score when given a preference profile. When the preferences are all total orders, a candidate at position i in the preference order of a voter receives a score of s_i from that voter. The candidate(s) with the highest total score win. We consider the following three scoring rules.

Plurality: with scoring vector $\langle 1, 0, \ldots, 0 \rangle$.
Borda: with scoring vector $\langle m - 1, m - 2, \ldots, 1, 0 \rangle$.
t-**Approval:** with scoring vector $\langle \underbrace{1, \ldots, 1}_{t}, 0, \ldots, 0 \rangle$.

To properly handle voters with ties in their preference orders we define several natural extensions which generalize the extensions from Baumeister et al. [3] and Narodytska and Walsh [25].

Write a preference order with ties as $G_1 > G_2 > \cdots > G_r$ where each G_i is a set of tied candidates. For each set G_i, let $k_i = \sum_{j=1}^{i-1} \|G_j\|$ be the number of candidates strictly preferred to every candidate in the set. See the caption of Table 1 for an example.

We now introduce the following scoring-rule extensions, which as stated above, generalize previously used scoring-rule extensions [3, 25]. In Table 1 we present an example of each of these extensions for Borda.

Min: Each candidate in G_i receives a score of $s_{k_i + \|G_i\|}$.
Max: Each candidate in G_i receives a score of $s_{k_i + 1}$.
Round down: Each candidate in G_i receives a score of s_{m-r+i}.

Average: Each candidate in G_i receives a score of

$$\frac{\sum_{j=k_i+1}^{k_i+\|G_i\|} s_j}{\|G_i\|}.$$

Table 1. The score of each candidate for preference order $a > b \sim c > d$ using Borda with each of our scoring-rule extensions. We write this order as $\{a\} > \{b, c\} > \{d\}$, i.e., $G_1 = \{a\}$, $G_2 = \{b, c\}$, and $G_3 = \{d\}$. Note that $k_1 = 0$, $k_2 = 1$, and $k_3 = 3$

Borda	score(a)	score(b)	score(c)	score(d)
Min	3	1	1	0
Max	3	2	2	0
Round down	2	1	1	0
Average	3	1.5	1.5	0

The optimistic and pessimistic models from the work by Baumeister et al. [3] are the same as our max and min extensions respectively, for top orders. All of the scoring-rule extensions for top orders found in the work by Narodytska and Walsh [25] can be realized by our definitions above, with our round-down and average extensions yielding the same scores for top orders as their round-down and average extensions. With the additional modification that $s_m = 0$ our min scoring-rule extension yields the same scores for top orders as round up in the work by Narodytska and Walsh [25].

Notice that plurality using the max scoring-rule extension for bottom orders is the same as approval voting, where each voter indicates either approval or disapproval of each candidate and the candidate(s) with the most approvals win. For example, given the set of candidates $\{a, b, c, d\}$, an approval vector that approves of a and c, and a preference order $a \sim c > b > d$ yield the same scores for approval and plurality using max respectively.

In addition to scoring rules, elections can be defined by the pairwise majority elections between the candidates. One important example is Copeland$^\alpha$ [7] (where α is a rational number between 0 and 1), which is scored as follows. Each candidate receives one point for each pairwise majority election he or she wins and receives α points for each tie. We also mention that Copeland^1 is often referred to, and will be throughout this paper, as Llull [17]. We apply the definition of Copeland$^\alpha$ to weak orders in the obvious way (as was done for top orders in [3, 25]).

We sometimes look at voters whose preferences need not be rational and we refer to those voters as "irrational." This simply means that for every unordered pair a, b of distinct candidates, the voter has $a > b$ or $b > a$. For example, a voter's preferences could be $(a > b, b > c, c > a)$. We also look at irrational votes with ties.

When discussing elections defined by pairwise majority elections we sometimes refer to the *induced majority graph* of a preference profile. A preference profile V where each voter has preferences over the set of candidates C induces the majority graph with a vertex set equal to the candidate set and an edge set defined as follows. For every $a, b \in C$ the graph contains the edge $a \rightarrow b$ if more voters have $a > b$ than $b > a$.

2.1 Election Problems

We examine the complexity of the following election problems.

The coalitional manipulation problem (where a coalition of manipulators seeks to change the outcome of the election) for weighted voters, first studied by Conitzer et al. [6], is described below.

Name: $\mathcal{E}$-CWCM
Given: A candidate set C, a collection of nonmanipulative voters V where each voter has a positive integral weight, a preferred candidate $p \in C$, and a collection of manipulative voters W.
Question: Is there a way to set the votes of the manipulators such that p is an $\mathcal{E}$ winner of the election $(C, V \cup W)$?

Electoral control is the problem of determining if it is possible for an election organizer with control over the structure of an election, whom we refer to as the election chair, to ensure that a preferred candidate wins [2]. We formally define the specific control action of constructive control by adding voters (CCAV) below. CCAV is one of the most natural cases of electoral control and it models scenarios such as targeted voter registration drives where voters whose votes will ensure the goal of the chair are added to the election.

Name: $\mathcal{E}$-CCAV
Given: A candidate set C, a collection of voters V, a collection of unregistered voters U, a preferred candidate $p \in C$, and an add limit $k \in \mathbb{N}$.
Question: Is there a subcollection of the unregistered voters $U' \subseteq U$ such that $\|U'\| \leq k$ and p is an $\mathcal{E}$ winner of the election $(C, V \cup U')$?

Bribery is the problem of determining if it is possible to change the votes of a subcollection of the voters, within a certain budget, to ensure that a preferred candidate wins [9]. The case for unweighted voters is defined below, but we also consider the case for weighted voters.

Name: $\mathcal{E}$-Bribery
Given: A candidate set C, a collection of voters V, a preferred candidate $p \in C$, and a bribe limit $k \in \mathbb{N}$.
Question: Is there a way to change the votes of at most k of the voters in V so that p is an $\mathcal{E}$ winner?

2.2 Computational Complexity

We use the following NP-complete problems in our proofs of NP-completeness.

Name: Exact Cover by 3-Sets
Given: A nonempty set of elements $B = \{b_1, \ldots, b_{3k}\}$ and a collection $\mathcal{S} = \{S_1, \ldots, S_n\}$ of 3-element subsets of B.
Question: Does there exist a subcollection $\mathcal{S}'$ of $\mathcal{S}$ such that every element of B occurs in exactly one member of $\mathcal{S}'$?

Name: Partition
Given: A nonempty set of positive integers $k_1, \ldots, k_t$ such that $\sum_{i=1}^{t} k_i = 2K$.
Question: Does there exist a subset A of $k_1, \ldots, k_t$ such that $\sum A = K$?[1]

Some of our results utilize the following variation of Partition, referred to as Partition', for which we prove NP-completeness by a reduction from Partition.

Name: Partition'
Given: A nonempty set of positive even integers $k_1, \ldots, k_t$ and a positive even integer $\widehat{K}$.
Question: Does there exist a partition (A, B, C) of $k_1, \ldots, k_t$ such that $\sum A = \sum B + \widehat{K}$?

Theorem 1. *Partition' is NP-complete.*

Proof. The construction here is similar to the first part of the reduction to a different version of Partition from Faliszewski et al. [9].

Given $k_1, \ldots, k_t$ such that $\sum_{i=1}^{t} k_i = 2K$, corresponding to an instance of Partition, we construct the following instance $k_1', \ldots, k_t', \ell_1', \ldots, \ell_t', \widehat{K}$ of Partition'. Let $k_i' = 4^i + 4^{t+1}k_i$, $\ell_i' = 4^i$, and $\widehat{K} = 4^{t+1}K + \sum_{i=1}^{t} 4^i$. (Note that in Faliszewski et al. [9] "3"s were used, but we use "4"s here so that when we add a subset of $k_1', \ldots, k_t', \ell_1', \ldots, \ell_t', \widehat{K}$, we never have carries in the last $t+1$ digits base 4, and we set the last digit to 0 to ensure that all numbers are even.)

If there exists a partition (A, B, C) of $k_1', \ldots, k_t', \ell_1', \ldots, \ell_t'$ such that $\sum A = \sum B + \widehat{K}$, then $\forall i, 1 \leq i \leq t$, $\lfloor (\sum A)/4^i \rfloor \bmod 4 = \lfloor (\sum B + \widehat{K})/4^i \rfloor \bmod 4$. Note that $\lfloor (\sum A)/4^i \rfloor \bmod 4 = \|A \cap \{k_i', \ell_i'\}\|$, $\lfloor (\sum B)/4^i \rfloor \bmod 4 = \|B \cap \{k_i', \ell_i'\}\|$, and $\lfloor \widehat{K}/4^i \rfloor \bmod 4 = 1$. So, $\|A \cap \{k_i', \ell_i'\}\| = \|B \cap \{k_i', \ell_i'\}\| + 1$. It follows that exactly one of k_i' or ℓ_i' is in A and neither is in B. Since this is the case for every i, it follows that $B = \emptyset$. Now look at all k_i such that k_i' is in A. That set will add up to K, and so our original Partition instance is a positive instance.

For the converse, it is immediate that a subset D of $k_1, \ldots, k_t$ that adds up to K can be converted into a solution for our Partition' instance, namely, by putting k_i' in A for every k_i in D, putting ℓ_i' in A for every k_i not in D, letting $B = \emptyset$, and putting all other elements of $k_1', \ldots, k_t', \ell_1', \ldots, \ell_t'$ in C. □

3   Results

3.1   Complexity Goes up

The related work on the complexity of manipulation of top orders [25] did not find a natural case where manipulation complexity increases when moving from total orders to top orders. We will show such cases in this section.

---

[1] Here and elsewhere we write $\sum A$ to denote $\sum_{a \in A} a$.

Single-peakedness is a restriction on the preferences of the voters introduced by Black [4]. Given a total order A over the candidates, referred to as an axis, a collection of voters is single-peaked with respect to A if each voter has preferences that strictly increase to a peak and then strictly decrease, only strictly increase, or only strictly decrease with respect to A.

For our purposes we consider the model of top order single-peakedness introduced by Lackner [22] where given an axis A, a collection of voters is single-peaked with respect to A if no voter has preferences that strictly decrease and then strictly increase with respect to A. Notice that for total orders, if a preference profile is single-peaked with respect to Black's model [4] it is also single-peaked with respect to Lackner's model [22].

For single-peaked preferences we follow the model of manipulation from Walsh [28] where the axis is given and both the nonmanipulators and the manipulators all cast votes that are single-peaked with respect to the given axis. 3-candidate Borda CWCM is known to be in P for single-peaked voters [12].

Theorem 2. *[12] 3-candidate Borda CWCM for single-peaked total orders is in* P.

We now consider the complexity of 3-candidate Borda CWCM for top orders that are single-peaked. In all of our reductions the axis is $a <_A p <_A b$. Single-peakedness with respect to this axis allows the following top order votes: $a > p > b$, $a \sim p \sim b$, $a > p \sim b$, $p > a > b$, $p > b > a$, $p > a \sim b$, $b > p > a$, and $b > p \sim a$. It does not allow $a > b > p$ or $b > a > p$.

Theorem 3. *3-candidate Borda CWCM for single-peaked top orders using max is* NP-*complete.*

Proof. Given a nonempty set of positive integers $k_1, \ldots, k_t$ such that $\sum_{i=1}^{t} k_i = 2K$ we construct the following instance of manipulation.

Let the set of candidates be $C = \{a, b, p\}$. We have two nonmanipulators with the following weights and votes.

- One weight $3K$ nonmanipulator voting $a > p \sim b$.
- One weight $3K$ nonmanipulator voting $b > p \sim a$.

From the nonmanipulators, score$(p) = 6K$, while score(a) and score(b) are both $9K$.

Let there be t manipulators, with weights $k_1, \ldots, k_t$. Without loss of generality, all of the manipulators put p first. Then p receives a score of $10K$ overall. However, a and b can score at most K each from the votes of the manipulators, for p to be a winner. So the manipulators must split their votes so that a subcollection of manipulators with weight K votes $p > a > b$ and a subcollection with weight K votes $p > b > a$. Notice that these are the only votes possible to ensure that p wins and that the manipulators cannot simply all vote $p > a \sim b$ since both a and b receive a point from that vote (since we are using max) and we have no points to spare. $\square$

The above argument for max does not immediately apply to the other scoring-rule extensions. In particular, for min the optimal vote for the manipulators is always to rank p first and to rank the remaining candidates tied and less preferred than p (as in Proposition 3 of Narodytska and Walsh [25]). So that case is in P, with an optimal manipulator vote of $p > a \sim b$.

It is not hard to modify the proof to show that the reduction of the proof of Theorem 3 also works for the round-down case.

Theorem 4. *3-candidate Borda CWCM for single-peaked top orders using round down is NP-complete.*

The average scoring-rule extension case is more complicated since it is less close to Partition than the previous cases. We will still be able to show NP-completeness, but we have to reduce from the special, restricted version of Partition that we defined previously in Sect. 2.2 as Partition'.[2]

Theorem 5. *3-candidate Borda CWCM for single-peaked top orders using average is NP-complete.*

Proof. Let $k_1, \ldots, k_t, \widehat{K}$ be an instance of Partition'. We are asking whether there exists a partition (A, B, C) of $k_1, \ldots, k_t$ such that $\sum A = \sum B + \widehat{K}$. Recall that all numbers involved are even. Let $k_1, \ldots, k_t$ sum to $2K$. Without loss of generality, assume that $\widehat{K} \leq 2K$.

Let the candidates $C = \{a, b, p\}$. We have two nonmanipulators with the following weights and votes.

– One weight $6K + \widehat{K}$ nonmanipulator voting $a > p \sim b$.
– One weight $6K - \widehat{K}$ nonmanipulator voting $b > p \sim a$.

From the nonmanipulators, score(p) is $6K$, score(a) + score(b) = $30K$ and score(a) – score(b) = $3\widehat{K}$.

Let there be t manipulators, with weights $3k_1, \ldots, 3k_t$.

First suppose there exists a partition (A, B, C) of $k_1, \ldots, k_t$ such that $\sum A = \sum B + \widehat{K}$. For every $k_i \in A$, let the weight $3k_i$ manipulator vote $p > b > a$. For every $k_i \in B$, let the weight $3k_i$ manipulator vote $p > a > b$. For every $k_i \in C$, let the weight $3k_i$ manipulator vote $p > a \sim b$. Notice that after this manipulation that score(p) = $18K$, score(a) = score(b), and score(a) + score(b) = $30K + 6K$. It follows that score(p) = score(a) = score(b) = $18K$.

For the converse, suppose that p can be made a winner. Without loss of generality, assume that p is ranked uniquely first by all manipulators. Then score(p) = score(a) = score(b) = $18K$. Let A' be the set of manipulator weights that vote $p > b > a$, let B' be the set of manipulator weights that vote $p > a > b$, and let C' be the set of manipulator weights that vote $p > a \sim b$. No

---

[2] A similar situation occurred in the proof of Proposition 5 in Narodytska and Walsh [25], where a (very different) specialized version of Subset Sum was constructed to prove that 3-candidate Borda CWCM (in the non-single-peaked case) for top orders using average remained NP-complete.

other votes are possible. Let $A = \{k_i \mid 3k_i \in A'\}$, $B = \{k_i \mid 3k_i \in B'\}$, and $C = \{k_i \mid 3k_i \in C'\}$. Therefore (A, B, C) corresponds to a partition of $k_1, \ldots, k_t$. Note that $\sum A = \sum B + \widehat{K}$. $\square$

We now consider cases where the complexity of control can increase when moving from total order votes to votes with ties. We examine the complexity of CCAV, which is one of the most natural models of control and known to be in P for plurality for total orders [2].

Theorem 6. *[2] Plurality CCAV for total orders is in P.*

However below we show two cases where CCAV for plurality is NP-complete for bottom orders and weak orders.

As mentioned in the Preliminaries, plurality using max for bottom orders is the same as approval voting. So the theorem below immediately follows from the proof of Theorem 4.43 from Hemaspaandra et al. [19].

Theorem 7. *Plurality CCAV for bottom orders and weak orders using max is NP-complete.*

We now show that the case of plurality for bottom orders and weak orders using average is NP-complete.

Theorem 8. *Plurality CCAV for bottom orders and weak orders using average is NP-complete.*

Proof. Let $B = \{b_1, \ldots, b_{3k}\}$ and a collection $\mathcal{S} = \{S_1, \ldots S_n\}$ of 3-element subsets of B be an instance of Exact Cover by 3-Sets, where each $S_j = \{b_{j_1}, b_{j_2}, b_{j_3}\}$. Without loss of generality let k be divisible by 4 and let $\ell = 3k/4$. We construct the following instance of control by adding voters.

Let the candidates $C = \{p\} \cup B$. Let the addition limit be k. Let the collection of registered voters consist of the following $(3k^2 + 9k)/4 + 1$ voters. (When "$\cdots$" appears at the end of a vote the remaining candidates from C are ranked lexicographically. For example, given the candidate set $\{a, b, c, d\}$, the vote $b > \cdots$ denotes the vote $b > a > c > d$.)

- For each i, $1 \leq i \leq \ell$, $k + 3$ voters voting $b_i \sim b_{i+\ell} \sim b_{i+2\ell} \sim b_{i+3\ell} > \cdots$.
- One voter voting $p > \cdots$.

Let the collection of unregistered voters consist of the following n voters.

- For each $S_j \in \mathcal{S}$, one voter voting $p \sim b_{j_1} \sim b_{j_2} \sim b_{j_3} > \cdots$.

Notice that from the registered voters, the score of each b_i candidate is $(k-1)/4$ greater than the score of p. Thus the chair must add voters from the collection of unregistered voters so that no b_i candidate receives more than $1/4$ more points, while p must gain $k/4$ points. Therefore the chair must add the voters that correspond to an exact cover. $\square$

We now present a case where the complexity of bribery goes from P for total orders to NP-complete for votes with ties.

Theorem 9. *[9] Unweighted bribery for plurality for total orders is in* P.

The proof that bribery for plurality for bottom orders and weak order using max is NP-complete immediately follows from the proof of Theorem 4.2 from Faliszewski et al. [9], which showed bribery for approval to be NP-complete.

Theorem 10. *Unweighted bribery for plurality for bottom orders and weak orders using max is* NP-*complete.*

3.2   Complexity Goes down

Narodytska and Walsh [25] show that the complexity of coalitional manipulation can go down when moving from total orders to top orders. In particular, they show that the complexity of coalitional manipulation (weighted or unweighted) for Borda goes from NP-complete to P for top orders using round-up. This is because in round-up an optimal manipulator vote is to put p first and have all other candidates tied for last.

In contrast, notice that the complexity of a (standard) control action cannot decrease when more lenient votes are allowed. This is because the votes that create hard instances of control are still able to be cast when more general votes are possible. The election chair is not able to directly change votes, except in a somewhat restricted way in candidate control cases, but it is clear to see how this does not affect the statement below.

Observation 11. *If a (standard) control problem is hard for a type of vote with ties, it remains hard for votes that allow more ties.*

What about bribery? Bribery can be viewed as a two-phase action consisting of control by deleting voters followed by manipulation. Hardness for a bribery problem is typically caused by hardness of the corresponding deleting voters problem or the corresponding manipulation problem. If the deleting voters problem is hard, this problem remains hard for votes that allow ties, and it is likely that the bribery problem remains hard as well. Our best chance of finding a bribery problem that is hard for total orders and easy for votes with ties is a problem whose manipulation problem is hard, but whose deleting voters problem is easy. Such problems exist, e.g., all weighted m-candidate t-approval systems except plurality and triviality.[3]

Theorem 12. *[9] Weighted bribery for m-candidate t-approval for all $t \geq 2$ and $m > t$ is* NP-*complete.*

For m-candidate t-approval elections (except plurality and triviality) the corresponding weighted manipulation problem was shown to be NP-complete by Hemaspaandra and Hemaspaandra [18] and the corresponding deleting voters problem was shown to be in P by Faliszewski et al. [10].

---

[3] By triviality we mean a scoring rule with a scoring vector that gives each candidate the same score.

Theorem 13. *Weighted bribery for m-candidate t-approval for weak orders and for top orders using min is in* P.

Proof sketch. To perform an optimal bribery, we cannot simply perform an optimal deleting voter action followed by an optimal manipulation action. For example, if the score of b is already at most the score of p, it does not make sense to delete a voter with vote $b > p \sim a$. But in the case of bribery, we would change this voter to $p > a \sim b$, which could be advantageous.

However, the weighted constructive control by deleting voters (WCCDV) algorithm from [10] still basically works. Since m is constant, there are only a constant number of different votes possible. And we can assume without loss of generality that we bribe only the heaviest voters of each vote-type and that each bribed voter is bribed to put p first and have all other candidates tied for last. In order to find out if there exists a successful bribery of k voters, we look at all the ways we can distribute this k among the different types of votes. We then manipulate the heaviest voters of each type to put p first and have all other candidates tied for last, and see if that makes p a winner.   □

3.3   Complexity Remains the Same

Narodytska and Walsh [25] show that 4-candidate Copeland$^{0.5}$ CWCM remains NP-complete for top orders. They conjecture that this is also the case for 3 candidates and point out that the reduction that shows this for total orders from Faliszewski et al. [13] won't work. We will prove their conjecture, with a reduction similar to the proof of Theorem 5.[4]

Theorem 14. *3-candidate Copeland$^{\alpha}$ CWCM remains NP-complete for top orders, bottom orders, and weak orders, for all rational $\alpha \in [0,1)$ in the nonunique winner case (our standard model).*

Proof. Let $k_1, \ldots, k_t$ and $\widehat{K}$ be an instance of Partition', which asks whether there exists a partition (A, B, C) of $k_1, \ldots, k_t$ such that $\sum A = \sum B + \widehat{K}$.

Let $k_1, \ldots, k_t$ sum to $2K$ and without loss of generality assume that $\widehat{K} \leq 2K$. We now construct an instance of CWCM. Let the candidate set $C = \{a, b, p\}$ and let the preferred candidate be p. Let there be two nonmanipulators with the following weights and votes.

- One weight $K + \widehat{K}/2$ nonmanipulator voting $a > b > p$.
- One weight $K - \widehat{K}/2$ nonmanipulator voting $b > a > p$.

From the votes of the nonmanipulators, score$(a) = 2$, score$(b) = 1$, and score$(p) = 0$. In the induced majority graph, there is the edge $a \to b$ with weight $\widehat{K}$, the edge $a \to p$ with weight $2K$, and the edge $b \to p$ with weight $2K$. Let there be t manipulators with, weights $k_1, \ldots, k_t$.

---

[4] Menon and Larson independently proved the top order case of the following theorem [24].

Suppose that there exists a partition of $k_1, \ldots, k_t$ into (A, B, C) such that $\sum A = \sum B + \widehat{K}$. Then for each $k_i \in A$, have the manipulator with weight k_i vote $p > b > a$, for each $k_i \in B$, have the manipulator with weight k_i vote $p > a > b$, and for each $k_i \in C$ have the manipulator with weight k_i vote $p > a \sim b$. From the votes of the nonmanipulators and manipulators, $\mathrm{score}(a) = \mathrm{score}(b) = \mathrm{score}(p) = 2\alpha$.

For the other direction, suppose that p can be made a winner. When all of the manipulators put p first then $\mathrm{score}(p) = 2\alpha$ (the highest score that p can achieve). Since $\alpha < 1$, the manipulators must have voted such that a and b tie. This means that a subcollection of the manipulators with weight K voted $p > b > a$, a subcollection with weight $K - \widehat{K}$ voted $p > a > b$, and a subcollection with weight $\widehat{K}$ voted $p > a \sim b$. No other votes would cause b and a to tie. Notice that the weights of the manipulators in the three different subcollections form a partition (A, B, C) of $k_1, \ldots, k_t$ such that $\sum A = \sum B + \widehat{K}$.   □

3-candidate Copeland$^\alpha$ CWCM is unusual in that the complexity can be different if we look at the unique winner case instead of the nonunique winner case (our standard model). We can prove that the only 3-candidate Copeland CWCM case that is hard for the unique winner model remains hard using a very similar approach.

Theorem 15. *3-candidate Copeland^0 CWCM remains NP-complete for top orders, bottom orders, and weak orders, in the unique winner case.*

Proof. Let $k_1, \ldots, k_t$ and $\widehat{K}$ be an instance of Partition′, which asks whether there exists a partition (A, B, C) of $k_1, \ldots, k_t$ such that $\sum A = \sum B + \widehat{K}$.

Let $k_1, \ldots k_t$ sum to $2K$ and without loss of generality assume that $\widehat{K} \leq 2K$. We now construct an instance of CWCM. Let the candidate set $C = \{a, b, p\}$. Let the preferred candidate be $p \in C$. Let there be two nonmanipulators with the following weights and votes.

- One weight $K + \widehat{K}/2$ nonmanipulator voting $a > p > b$.
- One weight $K - \widehat{K}/2$ nonmanipulator voting $b > a > p$.

From the votes of the nonmanipulators $\mathrm{score}(a) = 2$, $\mathrm{score}(b) = 0$, and $\mathrm{score}(p) = 1$. The induced majority graph contains the edge $a \to b$ with weight $\widehat{K}$, the edge $a \to p$ with weight $2K$, and the edge $p \to b$ with weight $\widehat{K}$. Let there be t manipulators, with weights $k_1, \ldots, k_t$.

Suppose that there exists a partition of $k_1, \ldots, k_t$ into (A, B, C) such that $\sum A = \sum B + \widehat{K}$. Then for each $k_i \in A$ have the manipulator with weight k_i vote $p > b > a$, for each $k_i \in B$ have the manipulator with weight k_i vote $p > a > b$, and for each $k_i \in C$ have the manipulator with weight k_i vote $p > a \sim b$. From the votes of the nonmanipulators and the manipulators $\mathrm{score}(p) = 1$ and $\mathrm{score}(a) = \mathrm{score}(b) = 0$.

For the other direction, suppose that p can be made a unique winner. When all of the manipulators put p first then $\mathrm{score}(p) = 1$. So the manipulators must have voted so that a and b tie, since otherwise either a or b would tie with p and

p would not be a unique winner. Therefore a subcollection of the manipulators with weight K voted $p > b > a$, a subcollection with weight $K - \hat{K}$ voted $p > a > b$, and a subcollection with weight $\hat{K}$ voted $p > a \sim b$. No other votes would cause a and b to tie. □

Theorem 16. *3-candidate Copeland$^\alpha$ CWCM remains in* P *for top orders, bottom orders, and weak orders, for $\alpha = 1$ for the nonunique winner case and for all rational $\alpha \in (0, 1]$ in the unique winner case.*

The proof of this theorem follows using the same arguments as the proof of the case without ties from Faliszewski et al. [13].

Tournament Result. We now state a general theorem on two-voter tournaments for votes with ties. See Brandt et al. [5] for related work on tournaments constructed from a fixed number of voters with total orders.

Theorem 17. *A majority graph can be induced by two weak orders if and only if it can be induced by two total orders.*

Proof sketch. Given two weak orders v_1 and v_2 that describe preferences over a candidate set C, we construct two total orders, v_1' and v_2' iteratively as follows.

For each pair of candidates $a, b \in C$ and $i \in \{1, 2\}$, if $a > b$ in v_i then set $a > b$ in v_i'.

For each pair of candidates $a, b \in C$, if $a > b$ in v_1 (v_2) and $a \sim b$ in v_2 (v_1) then the majority graph induced by v_1 and v_2 contains the edge $a \to b$. To ensure that the majority graph induced by v_1' and v_2' contains the edge $a \to b$ we must set $a > b$ in v_2' (v_1').

After performing the above steps there may still be a set of candidates $C' \subseteq C$ such that v_1 and v_2 are indifferent between each pair of candidates in C'. For each pair of candidates $a, b \in C'$, $a \sim b$ in v_1 and v_2, which implies the majority graph does not contain and edge between a and b. To ensure that majority graph induced by v_1' and v_2' does not contain an edge between a and b, without loss of generality set v_1' to strictly prefer the lexicographically smaller to the lexicographically larger candidate and the reverse in v_2'.

The process described above constructs two orders v_1' and v_2' and ensures that the majority graph induced by v_1 and v_2 is the same as the majority graph induced by v_1' and v_2'. Since for each pair of candidates $a, b \in C$ and $i \in \{1, 2\}$ we consider each possible case where $a \sim b$ is in v_i and set either $a > b$ or $b > a$ in the corresponding order v_i', it is clear that v_1' and v_2' are total orders. □

Observe that as a consequence of Theorem 17 we get a transfer of NP-hardness from total orders to weak orders for two manipulators when the result depends only on the induced majority graph. The proofs for Copeland$^\alpha$ unweighted manipulation for two manipulators for all rational α for total orders depend only on the induced majority graph [13,14], so we can state the following corollary to Theorem 17.

Corollary 18. *Copeland$^\alpha$ unweighted manipulation for two manipulators for all rational $\alpha \neq 0.5$ for weak orders is* NP-complete.

Irrational Voter Copeland Results. As mentioned in the preliminaries, another way to give more flexibility to voters is to let the voters be irrational. A voter with irrational preferences can state preferences that are not necessarily transitive and as mentioned in Faliszewski et al. [11] a voter is likely to posses preferences that are not transitive when making a decision based on multiple criteria.

Additionally, the preferences of voters can include ties as well as irrationality. When voters are able to state preferences that can contain irrationality and ties they can represent all possible pairwise preferences that they may have over all of the candidates.

It is known that unweighted Copeland$^\alpha$ manipulation is NP-complete for total orders for all rational α except 0.5 [13,14]. For irrational voters, this problem is in P for $\alpha = 0$, 0.5, and 1, and NP-complete for all other α [14]. *Weighted manipulation for Copeland$^\alpha$ has not been studied for irrational voters. We will do so here.*

Theorem 19. *3-candidate Copeland$^\alpha$ CWCM remains in P for irrational voters with or without ties, for $\alpha = 1$ for the nonunique winner case and for all rational $\alpha \in (0,1]$ in the unique winner case.*

Theorem 20. *3-candidate Copeland$^\alpha$ CWCM remains NP-complete for irrational voters with or without ties, for $\alpha = 0$ in the unique winner case and for all rational $\alpha \in [0,1)$ in the nonunique winner case.*

The proofs of the above two theorems follow from the arguments in the proofs of the corresponding rational cases, i.e., the proofs of Theorem 4.1 and 4.2 from Faliszewski et al. [13] for the case of voters without ties and the proofs of Theorems 14, 15, and 16 above for the case of voters with ties.

When $\alpha = 1$, also known as Llull, interesting things happen. It is known that 4-candidate Llull CWCM is in P for the unique and nonunique winner cases [15]. For larger fixed numbers of candidates, this is open. Though it is known that unweighted manipulation for Llull (with an unbounded number of candidates) is NP-complete in the nonunique winner case [14]. In contrast, we will show now that for irrational voters, all these problems are in P.

Theorem 21. *Llull CWCM is in P for irrational voters with or without ties, in the nonunique winner case and in the unique winner case.*

Proof. Given a set of candidates C, a collection of voters V, k manipulators, and a preferred candidate $p \in C$, the preferences of the manipulators will always contain $p > a$ for all candidates $a \neq p$. This determines the score of p. In addition, let the initial preferences of the manipulators be $a > b$ for each pair of candidates $a, b \in C - \{p\}$ such that a defeats b in V or such that a ties b in V and a is lexicographically smaller than b. Note that, if $k > 0$, there are no pairwise ties in the election with the manipulators set in this way and that the manipulators all have strict preferences between every pair of candidates (i.e., no ties in their preferences). For every $a \neq p$, let $score_0(a)$ be the score of a with the manipulators set as above.

Construct the following flow network. The nodes are: a source s, a sink t, and all candidates other than p. For every $a \in C - \{p\}$, add an edge with capacity $\text{score}_0(a)$ from s to a and add an edge with capacity $\text{score}(p)$ from a to t. For every $a, b \in C - \{p\}$, add an edge from candidate a to candidate b with capacity 1 if, when all manipulators set $b > a$, the score of a decreases by 1 (and the score of b increases by 1).

If there is manipulation such that p is a winner, then for every candidate $a \in C - \{p\}$, $\text{score}(a) \leq \text{score}(p)$ so there is a network flow that saturates all edges that go out from s.

If there is a network flow that saturates all edges that go out from s then for every $a, b \in C - \{p\}$ such that there is a unit of flow from a to b, change $a > b$ to $b > a$ in all manipulators.

This construction can be adapted to the unique winner case by letting the capacity of the edge from a to t be $\text{score}(p) - 1$ instead of $\text{score}(p)$.   □

4   Related Work

The recent work by Narodytska and Walsh [25] studied the complexity of manipulation for top orders and is very influential to our computational study of more general votes with ties. Baumeister et al. [3] and Narodytska and Walsh [25] studied several extensions for election systems for top orders, which we further extend for weak orders.

Most of the related work in the computational study of election problems assumes that the partial or tied preferences of the voters must be extended to total orders. We mention the important work on partial orders by Konczak and Lang [21] that introduces the possible and necessary winner problems. Given a preference profile of partial votes, a possible winner is a candidate that wins in at least one extension of the votes to total orders, while a necessary winner wins in every extension [21].

Baumeister et al. [3] also look at the possible winner problem and in their case they examine the problem given different types of incomplete votes, i.e., top truncated, bottom truncated, and top and bottom truncated. Baumeister et al. also introduced the problem of extension bribery, where given voters with preferences that are top truncated, voters are paid to extend their vote to ensure that a preferred candidate wins [3]. We do not consider the problem of extension bribery, but instead we use the standard model of bribery introduced by Faliszewski et al. [9]. In this model the briber can set the entire preferences of a subcollection of voters to ensure that a preferred candidate wins [9].

5   Conclusions and Future Work

We examined the computational complexity of the three most commonly studied manipulative attacks on elections when voting with ties. We found a natural case for manipulation where the complexity increases for voters with ties, whereas it is easy for total orders. For bribery we found examples where the complexity increases and where it decreases. We examined the complexity of Copeland$^\alpha$ elections for voters with ties and even irrational votes with and without ties. It would be interesting to see how the complexity of other election problems are affected by voters with ties, specifically weak orders, which we consider to be a natural model for preferences in practical settings.

Acknowledgments. The authors thank Aditi Bhatt, Kimaya Kamat, Matthew Le, David Narváez, Amol Patil, Ashpak Shaikh, and the anonymous referees for their helpful comments. This work was supported in part by NSF grant no. CCF-1101452 and a National Science Foundation Graduate Research Fellowship under NSF grant no. DGE-1102937.

References

1. Bartholdi III, J., Tovey, C., Trick, M.: The computational difficulty of manipulating an election. Soc. Choice Welf. **6**(3), 227–241 (1989)
2. Bartholdi III, J., Tovey, C., Trick, M.: How hard is it to control an election? Math. Comput. Model. **16**(8/9), 27–40 (1992)
3. Baumeister, D., Faliszewski, P., Lang, J., Rothe, J.: Campaigns for lazy voters: truncated ballots. In: Proceedings of the 11th International Conference on Autonomous Agents and Multiagent Systems, pp. 577–584, June 2012
4. Black, D.: On the rationale of group decision-making. J. Polit. Econ. **56**(1), 23–34 (1948)
5. Brandt, F., Harrenstein, P., Kardel, K., Seedig, H.: It only takes a few: on the hardness of voting with a constant number of agents. In: Proceedings of the 12th International Conference on Autonomous Agents and Multiagent Systems, pp. 375–382, May 2013
6. Conitzer, V., Sandholm, T., Lang, J.: When are elections with few candidates hard to manipulate? J. ACM **54**(3), 1–33 (2007). Article 14
7. Copeland, A.: A "reasonable" social welfare function. Mimeographed notes from a Seminar on Applications of Mathematics to the Social Sciences, University of Michigan (1951)
8. Emerson, P.: The original Borda count and partial voting. Soc. Choice Welf. **40**(2), 352–358 (2013)
9. Faliszewski, P., Hemaspaandra, E., Hemaspaandra, L.: How hard is bribery in elections? J. Artif. Intell. Res. **35**, 485–532 (2009)
10. Faliszewski, P., Hemaspaandra, E., Hemaspaandra, L.: Weighted electoral control. In: Proceedings of the 12th International Conference on Autonomous Agents and Multiagent Systems, pp. 367–374, May 2013
11. Faliszewski, P., Hemaspaandra, E., Hemaspaandra, L., Rothe, J.: Llull and Copeland voting computationally resist bribery and constructive control. J. Artif. Intell. Res. **35**, 275–341 (2009)
12. Faliszewski, P., Hemaspaandra, E., Hemaspaandra, L., Rothe, J.: The shield that never was: societies with single-peaked preferences are more open to manipulation and control. Inf. Comput. **209**(2), 89–107 (2011)
13. Faliszewski, P., Hemaspaandra, E., Schnoor, H.: Copeland voting: ties matter. In: Proceedings of the 7th International Conference on Autonomous Agents and Multiagent Systems, pp. 983–990, May 2008

14. Faliszewski, P., Hemaspaandra, E., Schnoor, H.: Manipulation of Copeland elections. In: Proceedings of the 9th International Conference on Autonomous Agents and Multiagent Systems, pp. 367–374, May 2010

15. Faliszewski, P., Hemaspaandra, E., Schnoor, H.: Weighted manipulation for four-candidate Llull is easy. In: Proceedings of the 20th European Conference on Artificial Intelligence, pp. 318–323, August 2012

16. Gibbard, A.: Manipulation of voting schemes. Econometrica **41**(4), 587–601 (1973)

17. Hägele, G., Pukelsheim, F.: The electoral writings of Ramon Llull. Stud. Lulliana **41**(97), 3–38 (2001)

18. Hemaspaandra, E., Hemaspaandra, L.: Dichotomy for voting systems. J. Comput. Syst. Sci. **73**(1), 73–83 (2007)

19. Hemaspaandra, E., Hemaspaandra, L., Rothe, J.: Anyone but him: the complexity of precluding an alternative. Artif. Intell. **171**(5–6), 255–285 (2007)

20. Kemeny, J.: Mathematics without numbers. Daedalus **88**, 577–591 (1959)

21. Konczak, K., Lang, J.: Voting procedures with incomplete preferences. In: Proceedings of the IJCAI-05 Multidisciplinary Workshop on Advances in Preference Handling, pp. 124–129, July/August 2005

22. Lackner, M.: Incomplete preferences in single-peaked electorates. In: Proceedings of the 28th AAAI Conference on Artificial Intelligence, pp. 742–748, July 2014

23. Mattei, N., Walsh, T.: PREFLIB: a library for preferences. In: Perny, P., Pirlot, M., Tsoukiàs, A. (eds.) ADT 2013. LNCS, vol. 8176, pp. 259–270. Springer, Heidelberg (2013)

24. Menon, V., Larson, K.: Complexity of manipulation in elections with partial votes. Technical report. arXiv:1505.05900 [cs.GT], arXiv.org, May 2015

25. Narodytska, N., Walsh, T.: The computational impact of partial votes on strategic voting. In: Proceedings of the 21st European Conference on Artificial Intelligence, pp. 657–662, August 2014

26. Satterthwaite, M.: Strategy-proofness and arrow's conditions: existence and correspondence theorems for voting procedures and social welfare functions. J. Econ. Theor. **10**(2), 187–217 (1975)

27. Schulze, M.: A new monotonic and clone-independent, reversal symmetric, and condorcet-consistent single-winner election method. Soc. Choice Welf. **36**(2), 267–303 (2011)

28. Walsh, T.: Uncertainty in preference elicitation and aggregation. In: Proceedings of the 22nd AAAI Conference on Artificial Intelligence, pp. 3–8, July 2007

Learning and Other Issues

Bayesian Network Structure Learning with Messy Inputs: The Case of Multiple Incomplete Datasets and Expert Opinions

Shravan Sajja and Léa A. Deleris[✉]

IBM Research - Ireland, Dublin, Ireland
lea.deleris@ie.ibm.com

Abstract. In this paper, we present an approach to build the structure of a Bayesian network from multiple disparate inputs. Specifically, our method accepts as input multiple partially overlapping datasets with missing data along with expert opinions about the structure of the model and produces an associated directed acyclic graph representing the graphical layer of a Bayesian network. We provide experimental results where we compare our algorithm with an application of Structural Expectation Maximization. We also provide a real world example motivating the need for combining disparate sources of information even when noisy and not fully aligned with one another.

Keywords: Bayesian network · Expert opinions · Multiple incomplete datasets

1 Introduction

Decision and Risk Analysis make extensive uses of influence diagrams, at the heart of which lies the probabilistic model known as a Bayesian network. Bayesian networks (BNs) [14] are a compact representation of the relationships among random variables. They provide an intuitive graphical representation of the variables dependence and independence relations along with an efficient way to perform inference queries. A Bayesian network is composed of two main elements: (i) the structure of the network captured by the directed acyclic graph and representing dependence and independence relations among the variables and (ii) the parameters of the network in the form of conditional probability tables, representing conditional distributions of the variables given all possible scenarios of their parents. Building a Bayesian network typically involves first determining the structure of the network and then estimating the parameters. We focus in this paper on the first step, the determination of the directed acyclic graph underlying the network.

In the past, it has been customary to assume that one would have access to one single clean input (one set of experts or one dataset) to determine the structure of the network. This assumption needs to be revisited as radical technology

© Springer International Publishing Switzerland 2015
T. Walsh (Ed.): ADT 2015, LNAI 9346, pp. 123–138, 2015.
DOI: 10.1007/978-3-319-23114-3_8

changes from the past 30 years have modified the context in which Bayesian networks are built. While the availability of information and experts has significantly increased, the quality of this knowledge may not have improved in general. Technology makes it simpler to assemble in a few clicks a distributed team of experts, and through web-based techniques, to elicit information from them in a minimally disruptive way. However, web-based elicitation can also result in experts being less engaged and consequently less focused and reliable. Having access to a distributed team of experts also increases the chance of conflicting opinions while the asynchronous approach to elicitation makes it difficult to resolve conflict directly. Similarly, Big Data does not necessarily imply better data. In fact, Big Data is mainly a euphemism to talk about overwhelming data: undeniably in large volumes but mostly unstructured, incomplete, evolving and noisy.

A critical objective for decision and risk analysis is thus to be able to make use of messy inputs in a rigorous way. By messy inputs, we mean *multiple disparate and noisy* inputs including multiple overlapping and incomplete datasets, conflicting expert opinions, facts extracted from the literature or from statistical sources such as census information, and constraints derived from domain knowledge such as reference ontologies. In this paper, we address one such subproblem by providing an approach to build the *structure* of a Bayesian network from multiple, incomplete and partially overlapping datasets and expert opinions. Our focus is on combining the work that has been carried out in the Artificial Intelligence community around structure learning and whenever relevant to adapt it to our purpose.

2 Related Work

The problem of building a Bayesian network structure from multiple datasets has been studied for some years, thus there is a fair amount of literature upon which to build on. Although it is possible to collate the datasets together and apply the Structural Expectation Maximization algorithm [7], this is not satisfying as the algorithm assumes that data is missing at random, which would not apply in this context [3].

One stream of research has thus explored how multitask learning methods (i.e. simultaneously learning multiple structures from multiple datasets) could be extended to learning one common structure. However, these approaches do not generalize simply to cases where datasets have only partial overlap in their attributes [12]. In addition, learning a single representative structure may lead to negative transfer of knowledge as the number of datasets increase [13].

Another stream of research deals with the fusion of Bayesian networks learnt from individual datasets. This means that a Bayesian network structure is generated from each dataset and then a consensus Bayesian network is defined [9,10,15]. These methods depend on the prior knowledge of node ordering of the final target Bayesian network and they fuse two networks at a time which further leads to questions about the bias in selecting the order in which multiple networks are fused. This problem was also addressed by [3,19]. The ION algorithm proposed by [3] learns an equivalence class of structures as an output from a set of input structures learnt from the individual datasets. However, ION algorithm was not capable of resolving contradictions between multiple input structures, which may arise because of statistical errors in the conditional independence tests. This issue was resolved in [17,18] by using Fisher's method [6] of combining p-values to resolve conflicts in the measurement of conditional independence relations. However, [17,18] do not address the issues of either missing data in partially overlapping datasets or the incorporation of expert opinions.

Finally, a third stream of research extends constraint based algorithm for structure learning to cases with multiple datasets. An extension of PC-algorithm for multiple datasets was proposed by [8], though the authors assume that they have completely overlapping datasets as their inputs. To perform conditional independence testing required by the PC algorithm, they suggest a heuristic based on the combination of mutual information measured between variables from different datasets. We follow a similar approach in this paper but we (i) extend to partially overlapping datasets with missing entries, (ii) incorporate expert opinions into the structure building algorithm and (iii) borrow the method from [17] for combining conditional independence tests, namely Fisher's method. Grounded in statistics, it has been broadly validated in meta-analysis and in that respect is a more plausible solution to the problem of combining statistical independence tests from multiple datasets than the heuristic defined by [8].

3 Problem Set Up: Learning a Bayesian Network from Multiple Sources

The main goal of this paper is to learn a single Bayesian network from multiple sources of information, which we define more formally in this section after introducing some notations. The next section will discuss the structure building algorithm.

3.1 Preliminary Definitions

Let $\mathcal{V} = \{X_1, X_2, \ldots, X_n\}$ be a set of random variables. A Bayesian network $\mathcal{BN} = \langle \mathcal{G}, \Theta \rangle$ is defined by

- a directed acyclic graph (DAG) $\mathcal{G} = \langle \mathcal{V}, \mathcal{E} \rangle$ where $\mathcal{V}$ represents the set of nodes (one node for each variable) and $\mathcal{E}$ the set of edges representing the relationships between the variables,
- parameters $\Theta = \{\theta_{ijk}\}_{1 \leq i \leq n, 1 \leq j \leq q_i, 1 \leq k \leq r_i}$, the set of conditional probability tables of each node X_i knowing its parents state $Pa(\mathcal{G}, X_i)$ (with r_i and q_i as respective cardinalities of X_i and $Pa(\mathcal{G}, X_i)$).

In this paper we consider directed graphs containing only directed edges ($\rightarrow$), undirected graphs containing only undirected edges ($-$) and partially directed graphs containing directed and/or undirected edges.

The skeleton of a Bayesian network graph $\mathcal{G}$ over $\mathcal{V}$ is an undirected graph over $\mathcal{V}$ that contains an edge $\{X, Y\}$ for every edge (X, Y) in $\mathcal{G}$. Two nodes of a graph are said to be *adjacent* if there is an edge connecting the two nodes and we assume that there is at most one edge between any two nodes. The set of all adjacent nodes for a node X_i in a graph $\mathcal{G}$ is denoted by $ADJ(\mathcal{G}, X_i)$. Thus if there is an edge from X_i to X_j, then $X_i \in ADJ(\mathcal{G}, X_j)$, however, if $X_i \to X_j$, then X_i is called a *parent* of X_j and X_j is called the *child* of X_i. The set of all such parents of node X_i in a graph $\mathcal{G}$ is denoted by $Pa(\mathcal{G}, X_i)$ and the set of children of X_i by $Ch(\mathcal{G}, X_i)$.

Let $\mathcal{G} = \langle \mathcal{V}, \mathcal{E} \rangle$ be a graph. An *ordering* of the nodes $X_1, \ldots, X_n$ is a topological ordering relative to $\mathcal{G}$ if, whenever we have $X_i \to X_j$, then $i < j$.

Definition 1 (Conditional Independence). *Let $\mathcal{V}$ be a set of random variables with $\mathbf{X}, \mathbf{Y}, \mathbf{Z} \subseteq \mathcal{V}$ disjoint sets of random variables, and let $\mathbf{P}$ be a joint probability distribution defined on $\mathcal{V}$, then $\mathbf{X}$ is said to be conditionally independent of $\mathbf{Y}$ given $\mathbf{Z}$, denoted by $\mathbf{X} \perp\!\!\!\perp \mathbf{Y} | \mathbf{Z}$, if*

$$P(X|Y, Z) = P(X|Z).$$

There can be more than one possible directed acyclic graph (DAG) to capture a set of conditional dependence and independence statements. In that sense, DAGs constitute an equivalence class where two DAGs are equivalent if and only if they have the same skeleton and the same v-structures [20]. To describe each equivalence class, we use partially directed acyclic graphs (PDAGs), in which some arcs are directed and others are undirected. In fact, we use a subset of PDAG, the completed-PDAGs (CPDAGs) which is defined as the set of PDAGs that have only undirected arcs and unreversible directed arcs. CPDAGs are also referred to as maximally oriented graphs.

3.2   Defining Input Datasets

The set of input datasets is denoted by $\mathbf{D} = \{D_1, D_2, \ldots, D_m\}$. Given any two datasets D_i and D_j, we assume that we are in a non trivial case where there is at least one overlapping variable within each dataset, i.e. $\forall i, \exists j : Supp(D_i) \cap Supp(D_j) \neq \phi$, where support $Supp$ of a dataset is defined as the set of all variables measured in a dataset. We assume that for each dataset are composed of s_j i.i.d. samples, $D_j = \{D_j[1], D_j[2], \ldots, D_j[s_j]\}$. We further assume that there can be some missing values in each dataset (assumed to be missing at random). Finally we assume that there are no hidden variables and that datasets collectively cover all variables, i.e., $\bigcup_j Supp(D_j) = \mathcal{V}$. We also assume that the effort of pre-processing the datasets to match and link entities has been completed: variables have been matched so there is no ambiguity in mapping variables among the datasets, all variables are categorical, and matching variables have the same set of possible values across the datasets. For instance, assume that a variable *Smoker* appears in multiple datasets then all instances have been identified as representing the same concept and their states have been unified and take the same values (for instance $\{Never, Former, Current\}$) in all those datasets.

3.3 Defining Input Expert Opinions

We consider prior domain knowledge in the form of conditional independence statements from multiple experts, and also direct dependence information such as forbidden arcs and required arcs.

Conditional Independence (CI) Statements are of the form $\mathbf{X} \perp\!\!\!\perp \mathbf{Y} | \mathbf{Z}$ where $\mathbf{X}, \mathbf{Y}, \mathbf{Z} \subseteq \mathcal{V}$ are disjoint sets of random variables. In order to resolve conflicts among multiple CI statements, we assume that a confidence measure is attached to each CI statement in the form a p-value. Such p-values come naturally when performing statistical tests. In the context of expert-provided opinions, they need to be further defined. If the expert derives his/her conclusion from statistical analysis, as is common for instance in the medical literature, then a p-value is readily available assuming it is reported. In other cases, when the statements of conditional dependence or independence of the expert come from his/her accumulated expertise, then we refer to the definition of the concept of p-value namely the probability of obtaining the observed result (i.e. whether the expert says "D" or "I") from the sample (the expert knowledge) given that the null hypothesis (dependence) is actually true. In mathematical terms, if the expert says "D" then the p-value is $P("D"|D)$ and if the expert says "I", then the p-value is $P("I"|D)$. This means that we need to evaluate $P("I"|D)$ (and $P("D"|D)$ if we allow for a "does not know" answer) which represents a somewhat subjective estimation of the expert abilities. Such measures are not unusual, however, they have been used in models for expert opinion aggregation using Bayesian updating [1] though there remain ample room in the field of expert opinion aggregation for deriving practical estimation procedures of the likelihood model.

Forbidden and Required Arcs. Let F be the set of directed edges which are forbidden, R is the set of directed edges which are required. According to [11], background knowledge described by $\mathcal{K} = \langle F, R \rangle$ is consistent with graph $\mathcal{G}$ if and only if there exists a graph $\mathcal{G}'$ which is consistent DAG extension of $\mathcal{G}$ such that

(i) all of the edges in R are oriented correctly in $\mathcal{G}'$ and
(ii) no edge $A \to B$ in F is oriented as such in $\mathcal{G}$.

For background described by $\mathcal{K} = \langle F, R \rangle$, we assume that we have a single expert opinion (or a consensus of multiple expert opinions). This assumption doesn't diminish the generality of our work, since forbidden and required arcs are generally used to model temporal constraints in dynamic Bayesian networks and there is seldom any conflict in the expert opinions regarding temporal constraints.

4 Structure Learning Method

As we mentioned previously, the structure learning method used in this paper is a modified version of the constraint-based method: PC-algorithm [16]. The original PC-algorithm was designed for learning a CPDAG from a single complete dataset. It consists of three main stages. In the first stage, the algorithm begins with a complete undirected graph and learns a skeleton by recursively deleting the edges based on conditional independence tests performed on the dataset. In the second stage, some edges are oriented by adding the set of v-structures, obtaining a PDAG. In the final stage, some other undirected edges are oriented to form a CPDAG, which represents the equivalence class of DAGs associated with the input datasets. One known issue associated with the original PC-algorithm is that its iterative approach makes the generation process dependent on the order of variables used to construct the initial graph $\mathcal{G}$. To address that problem, we implement the modifications proposed by [2] to make it order independent. This order-independent version of PC-algorithm was termed PC-Stable.

In the following sections, we consider each stage of the original PC-algorithm and discuss the modifications made to extend it to simultaneously handle multiple partially-overlapping and incomplete datasets together with background knowledge from multiple experts. We named our algorithm PCFS where F stands for Fischer, as we used Fischer's method for combining conditional independence tests and S stands for Stable as we implement the order independent version of the PC-algorithm. A sketch of our algorithm is given in Fig. 1.

4.1   Preliminary Step: Addressing Data Incompleteness

The PC-algorithm, and in general all constraint-based algorithms, use information about conditional independence obtained by performing statistical tests on the data. However, constraint-based methods are not equipped to handle missing data in the datasets [4]. Hence as a preliminary step, we perform data imputation for all datasets using Structural Expectation Maximization (EM) [7]. We use EM algorithm to construct individual Bayesian networks $\mathcal{BN}_i$ for each incomplete dataset D_i. These Bayesian networks will be used to estimate the missing data in their respective datasets using forward sampling with available data as evidence.

4.2   Step 1: Skeleton Generation

The PC-algorithm starts from the complete undirected graph associated with the variable set from which edges are recursively deleted to obtain the skeleton. We make a first modification here by making PCFS start with an initial graph which is obtained after removing edges corresponding to variables that are never jointly measured in any two datasets. In other words, our starting point is the union of the complete graphs associated with the variables in each dataset. This new initialization allows us to reduce false positive errors that indicate an edge in the learned model that does not appear in the true model.

The algorithm then proceeds as follows: For every possible pair of variables X_i and X_j and for increasing value of $|\mathbf{S}|$, where $\mathbf{S}$ denote the conditioning subset, $\mathbf{S} \subseteq a(X_i) - \{X_j\}$, it tests whether $X_i \perp\!\!\!\perp X_j | \mathbf{S}$. The term $a(X_i)$ represents adjacency of X_i in the initial graph. Contrary to changing adjacencies in the original PC-algorithm, the order independent version requires $a(X_i)$ to be determined at the onset of each iteration on $|\mathbf{S}|$ [2].

Several statistical methods are available to test conditional independence between any variables X, Y given $\mathbf{Z} \subseteq \mathcal{V}$ from each dataset and these test results can be combined to obtain a common CI decision [8]. It was empirically shown in [17] that using the PC-algorithm based on evaluating p-values (for the hypothesis $X \perp\!\!\!\perp Y | \mathbf{Z}$) from each dataset and then combining these p-values using Fisher's method [6] was outperforming many existing methods. Another important advantage of using p-values is that they do not depend on the shapes of distributions across the different datasets.

Suppose we have k independent tests for $(X \perp\!\!\!\perp Y | \mathbf{Z})$, with p values given by $p_1, p_2, \ldots, p_k$. These p-values are obtained from the k datasets (out of m available datasets) where X, Y and $\mathbf{Z}$ are jointly measured. Fisher proposed a method to combine p-values from multiple sources of data using the test statistic

$$T_F = -2\Sigma_{i=1}^{k}\log(p_i),$$

where T_F has a χ^2 distribution with $2k$ degrees of freedom under the null hypothesis. T_F is tested against $F_{\chi^2_{2k}}^{-1}(1 - \alpha)$, where α is the significance level. In [17], Fisher's method was applied for p-values that were obtained from multiple datasets, however, in this paper we use Fisher's formula to further combine with multiple expert opinions in the form of CI statements.

We assume that there are ℓ expert opinions available for the null hypothesis $X \perp\!\!\!\perp Y | \mathbf{Z}$ and the p-values corresponding to these expert opinions are given by $p_1^e, p_2^e, \ldots, p_\ell^e$. Then based on Fisher's method we combine p-values from multiple sources of information using the test statistic

$$T_F = -2 \left(\Sigma_{i=1}^{k}\log(p_i) + \Sigma_{i=1}^{\ell}\log(p_i^e) \right),$$

where T_F has a χ^2 distribution with $2(k + \ell)$ degrees of freedom under the null hypothesis. Thus T_F is tested against $F_{\chi^2_{2(k+\ell)}}^{-1}(1-\alpha)$. In essence, we assume that each expert has as much influence on the aggregated result than one dataset.

In the original PC algorithm, the conditioning set for which the independence holds is recorded in the *SepSet* for use when orienting the arcs. PCFS also records the combined p-values corresponding to the edges that have been removed. This information is used at the very last stage of the algorithm when we search for admissible CPDAG to the skeleton obtained.

4.3   Steps 2 and 3: Orienting the Edges

The second and third stages of the PC-Algorithm focus on transforming the skeleton obtained into a CPDAG. The second stage begins with orienting v-structures in the skeleton. V-structures are instances of the pattern $X_i \rightarrow X_j \leftarrow X_k$ in the DAG. In order to determine v-structures, the algorithm considers all the unshielded triples in the skeleton, i.e. $X_i - X_j - X_k$, and orients an unshielded triple as $X_i \rightarrow X_j \leftarrow X_k$ if and only if $X_j \notin Sepset(X_i, X_k)$. Once all v-structures have been determined, the third stage seeks to orients some of the remaining undirected edges using rules $\mathbf{R_1}$, $\mathbf{R_2}$ and $\mathbf{R_3}$ (also known as Meek's rules) to produce a CPDAG [11].

As mentioned before, we implement order-independent variations of those two steps. In addition, we adapt the third stage of the algorithm to incorporate domain knowledge in the form of Required and Forbidden arcs. Specifically, we implement the methods proposed by Meek to account for prior background knowledge [11]. Meek's algorithm compares PDAG with the domain knowledge and if some information is conflicting, the algorithm produces an error. Otherwise, it iteratively adds the prior knowledge (directed edges) not present in the PDAG and again applies the orientation rules $\mathbf{R_1}$, $\mathbf{R_2}$ and $\mathbf{R_3}$ recursively with another rule $\mathbf{R_4}$. These orientation rules are necessary to infer all the new edge orientations induced by the addition of the prior knowledge and they produce a PDAG consistent with $\mathcal{K} = \langle F, R \rangle$ [11, Theorem 4].

Note that the approach mentioned above produces no output CPDAG (and an error) whenever there is a conflict between the input PDAG and domain knowledge.

4.4   Step 4: Obtaining a DAG

Following these orientation rules, we apply the DAGEXTENSION algorithm to convert the output CPDAG to a DAG [5]. It is possible that the output CPDAG does not admit a consistent DAG extension. In those cases, and also in cases where there were no output CPDAG after the third step, we use a greedy search algorithm to modify the skeleton.

Specifically, this algorithm, called GSEARCH, adds back some of the edges previously removed during the skeleton generation process following a chosen heuristic. Our reason for revisiting the output skeleton from step 1 and searching the space of its neighbors is that the skeleton generation process is based on imperfect statistical tests while the orientation of arcs is deterministic. For each new candidate skeleton, we re-apply the orientation steps to determine whether there exist an admissible output. This approach leads to alternation between the space of skeletons and the space of PDAGs. This alternation is terminated in two scenarios:

- if the function DAGEXTENSION produces a DAG as an output or
- if the function DAGEXTENSION produces an empty set and the greedy search heuristic exhausts all the deleted edges from the skeleton generation process.

In the first case, we have a successful completion of our method. The second scenario indicates that there is no consistent explanation of the datasets **D** together with domain knowledge in terms of a Bayesian network for the given significance value α. In such situations we perform a search over the space of different possible values of α. This adaptive adjustment of α is not addressed in this paper.

There are several possible heuristics that can be used for thickening the skeleton (adding previously removed edges). In our implementation of GSEARCH, we choose the subset of deleted edges which corresponds to the largest p-value associated to the set of edges removed during the skeleton generation phase.

A randomly selected edge from this subset is then added to the skeleton to go back to the orientation stage and we update the *Sepset* accordingly. Upon failure to produce to a DAG output from the orientation stage, the previously added edge is removed and another edge from the subset is added to the skeleton. Once, all the edges in the subset corresponding to the largest p-value are exhausted, it proceeds to second largest p-value and so on. This process is continued until an admissible output is found or the GSEARCH function exhausts all the deleted edges. Other variations of this greedy search involve addition of more than one edge at a time.

5 Experimental Results

We present in this section two types of experiments that we have undertaken. The first set of experiments is aimed at validating our approach to build Bayesian networks from multiple partially overlapping datasets by comparing it with a baseline approach, namely applying Structural Expectation Maximization to the stacked datasets. We seek to determine whether the additional sophistication of our approach is useful compared with naive methods. The second set of experiments present the application of our method to a real world context, namely vulnerability modeling for eldercare. Our aim in those experiments is to illustrate the approach but also to explore the benefits of combining multiple datasets and expert opinions in building more reliable models.

5.1 Validation Experiments

Set Up: Validation experiments are based on synthetic data generated from a known Bayesian network, specifically, the four-node Sprinkler network represented on Fig. 2. In this experiment, we look at creating a Bayesian network structure from two partially overlapping datasets of the same size (ranging from 10 to 1000 entries) and for different overlap sizes (ranging from 1 to 4). For each overlap size, we generated all possible combinations of variables overlap. For each such configuration, we generated then generated the associated datasets by sampling from our reference network. In fact, we generate 10 iterations for each dataset size and overlap variable configuration. We create network structures using PCFS for several values of α, specifically {0.001, 0.01, 0.05, 0.1, 0.2} but also using Structural Expectation Maximization on the stacked datasets. We evaluate the distance of each output structure to our ground truth (the reference network) using Graphical Edit Distance (GED). The GED between two DAGs corresponds to the minimal sequence of operations needed to transform the one DAG into another. Operations considered are edge-insertion, edge-deletion and edge-reversal.

Results: Fig. 3 presents the results of our experiments grouped by overlap size of the input datasets. We start with general observations. First it appears that edit distance generally decreases across all models with higher level of overlap.

Algorithm 4.1. PCFS $(\mathbf{D}, C, F, R, \alpha)$

Require: Datasets $\mathbf{D} = \{D_1, \ldots, D_m\}$,

Require: Constraint maps C, Forbidden arcs F,

Required arcs R, Significance value α.
Ensure: DAG G'

$G \leftarrow$ *Initial undirected graph*
$Sepset \leftarrow$ *Empty set*
$RPvalue \leftarrow$ *Empty matrix*
for each $X \in \mathcal{V}$
  do $a(X) = ADJ(X, G)$
$i \leftarrow 0$
repeat
 for each $X \in \mathcal{V}$

$$\mathbf{do} \begin{cases} \textbf{for each } Y \in ADJ(X) \\ \mathbf{do} \begin{cases} \textbf{for each } \mathbf{S} \subseteq a(X) - \{Y\}, |\mathbf{S}| = i \\ \mathbf{do} \begin{cases} [decision, cpvalue] = \mathbf{ComPvalue}(\mathbf{D}, C, X, Y, \mathbf{S}, \alpha) \\ \textbf{if } (decision) \\ \textbf{then} \begin{cases} \text{Remove the edge } X - Y \text{ from } G; \\ \text{Save } \mathbf{S} \text{ in } Sepset; \\ \text{Save } cpvalue \text{ in } RPvalue; \end{cases} \end{cases} \end{cases} \end{cases}$$

$i \leftarrow i + 1$
until $|ADJ(X)| \le i, \forall X$
$G' \leftarrow \mathbf{orient}(G, Sepset, F, R)$
while $isempty(G')$

$$\mathbf{do} \begin{cases} \textbf{if } isempty(RPvalue) \\ \textbf{then} \begin{cases} G' \leftarrow Empty set \\ \textbf{return } (G') \\ \textbf{Comment: } \text{There is No consistent DAG for this value of } \alpha \end{cases} \\ \textbf{else } \{[G', RPvalue] = \mathbf{Gsearch}(G, Sepset, RPvalue, F, R) \end{cases}$$

return (G')

Fig. 1. PCFS algorithm

We also note that the sample sizes that we have considered have limited effects on the distances, as for each overlap size, the patterns repeat across the sample sizes. While we expected that higher overlap led to better models, we are surprised that whether the datasets have 10 or 1000 entries - a difference of two orders of magnitude - does not seem to affect performance.

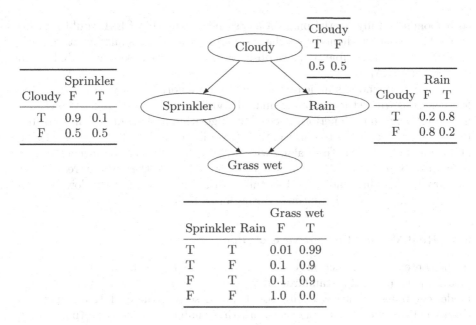

| | | Cloudy |
	T	F
	0.5	0.5

| | Sprinkler | |
Cloudy	F	T
T	0.9	0.1
F	0.5	0.5

| | Rain | |
Cloudy	F	T
T	0.2	0.8
F	0.8	0.2

| | | Grass wet | |
Sprinkler	Rain	F	T
T	T	0.01	0.99
T	F	0.1	0.9
F	T	0.1	0.9
F	F	1.0	0.0

Fig. 2. Reference Bayesian network for the validation experiments

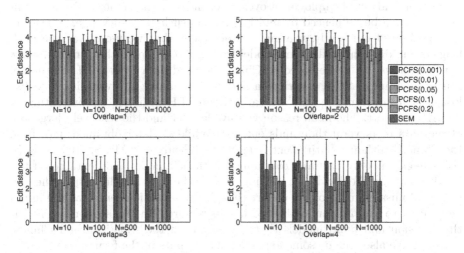

Fig. 3. Edit distance between outputs of various structure learning methods and the original network for different overlap sizes.

We now compare the performance of the different models. For overlap of 1, all PCFS variations outperform SEM, sometimes significantly. For overlap of 2 and 3, SEM is already performing quite well and only PCFS algorithm with a judiciously chosen value of α display slightly better performances. For overlap of 4 this pattern is even clearer. This is not surprising as this latter situation corresponds to fully overlapping datasets and means that SEM would be thus applied in a context aligned with the assumption that data is missing at random. However, for some sample sizes, we observe that there are cases where PCFS still outperforms SEM.

Overall, our findings indicate that the PCFS algorithm may be better suited for situations where the overlap is small. In addition it seems that the choice of α has an influence on the performance of the algorithm and should be investigated further. Our objective is to perform further similar experiments, in fact exploring all four-node DAG structures along with larger networks, so as to have a stronger base to draw conclusions from. We feel however that the current results are encouraging as they indicate that there are situations, here the low overlap cases, in which PCFS significantly outperforms SEM.

5.2   Real-World Illustrative Example

In the context of a project on social care for the elderly, we have built a model to provide person-specific standardized vulnerability assessment of patients. This model combines a Markov chain model that describes the evolution of persons through states with a Bayesian network that enables to customize the Markov chain parameters. For instance, in a different social care context, the evolution of a person in and out of employment over time, and thus the vulnerability to unemployment, can be influenced by multiple factors such as gender, age, education. At the time of our study, the main impediment that we faced was the challenge of building the underlying model from limited information. In fact, while our system was to be deployed in China, we used data from the Longitudinal Study on Aging (LSOA, description and data available at www.cdc.gov/nchs/lsoa/lsoa2.htm) from the US National Institute of Health, hence representing an American cohort. The main reason was availability and the relatively large size of the dataset (several thousands entries). While we later obtained data from the China Health and Retirement Longitudinal Study (CHARLS, http://charls.ccer.edu.cn), it only partially overlapped with our model and contained a few hundred entries. We decided at the time to keep the initial model despite the geographical misalignment of the cohort though this situation served as the initial motivation to the problem addressed in this research. In this section, we revisit this decision and build models that leverage both the American and Chinese datasets. We also include some expert opinion inputs in the form of conditional dependence statements that were extracted from the literature.

Set Up: For our illustrative example, we selected a subset of the variables used in the original eldercare model. Specifically our aim is to estimate the risk of being unable to live independently, measured through the level of difficulty experienced by the person in terms of walking and eating independently (Variables *Difficulty Walking, Difficulty Eating*). Those represent a subset of the Activity of Daily Living (ADLs) which are broadly used in the social care domain. Based on the more extensive model, we selected 13 variables to predict the ability of elderly to walk and eat independently. Those were: *Age, Difficulty in Doing Light Housework, Difficulty Finger Grasping (US), Difficulty Picking Coin (China), Difficulty in Reaching OverHead, Difficulty Stooping, Difficulty Walking Quarter Mile (US), Difficulty Walking 1 km (China), Gender, Has Diabetes, Marital Status, Smoker* and *Veteran (US)*. For the ones that are not common to both datasets, we have indicated in parenthesis for which country they are reported. Both datasets have missing values. The Chinese dataset contains 463 entries and the American one 3604 entries.

In terms of expert opinions, we relied on the search feature provided by the PubMed service (http://www.ncbi.nlm.nih.gov/pubmed/) to identify possible dependence or independence relations among the variables in our model. We do not claim to be exhaustive in this search. We found evidence of dependence between *Smoker* and *Difficulty Walking* (3 papers: PubMed IDs 1576581, 23741513, 18394514), *Gender* and *Difficulty Walking* (1 paper: PubMed ID 18394514) and between *Has Diabetes* and *Difficulty Walking* (1 paper: PubMed ID 23741513). Finally we add logical dependence constraints between the following pairs of variables: *Difficulty Walking Quarter Mile* and *Difficulty Walking*, *Difficulty Walking Quarter Mile* and *Difficulty Walking 1 km*, *Difficulty Walking* and *Difficulty Walking 1 km*, *Difficulty Finger Grasping* and *Difficulty Picking Coin*. As we did not have p-values for the above opinions, we used a somewhat arbitrary number representing experts reliability, namely $P(``D"|D) = 0.95$ thus $P(``I"|D) = 0.05$.

Using this information, we built several Bayesian network models: Two using PCFS with $\alpha = 0.05$, one including expert opinions and one without. We also built three models using Structural Expectation Maximization (SEM): the first one using only the Chinese data, the second using the American data and the third one using both datasets stacked together.

We do not have access to a ground truth in this case. To compare the models to one another, we evaluate how accurate they are at predicting *Difficulty Walking* and *Difficulty Eating*. This means that we need to learn the parameters of the Bayesian networks for the ones built with PCFS (SEM learns structure and parameters at once). For that task we used the parameter Expectation Maximization using the stacked datasets. To avoid overfitting, we set aside 10 % of each dataset for testing and used only the remaining 90 % for training.

Results: We present the accuracy of each model on each cohort on Fig. 4.
Our first observation pertains to the comparison of performance between PCFS and PCFS with constraints. For both cohorts and both prediction tasks (*Difficulty Eating* and *Difficulty Walking*), we see no difference in performance between the two models. Thus in this case, the information brought by expert opinions had no influence on the model. This may be due to the fact that our expert opinions consisted only in dependence statements which are somewhat neutral from a Bayesian network perspective. Indeed, adding an unwanted arc is problematic as it increases the size of the conditional probability table, and thus the difficulty of the parameter elicitation task, but it does not rule out expressing

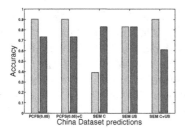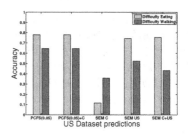

Fig. 4. Predictions of *Difficulty Walking, Difficulty Eating* using different Bayesian networks

independence at the parameter level. By contrast, the absence of an arc irremediably constrains the nature of the relations among the variables. Furthermore, it is possible that the relations from the expert opinion pool are already strongly captured in the datasets, even with limited number of samples (as in the Chinese cohort).

We now analyze of the predictions for the Chinese cohort, on the left side of the figure. The results in that figure are based on about 40 samples. Looking at the prediction for *Difficulty Eating*, PCFS outperforms two SEM models (China and US) but its performance is similar to the model from SEM applied on the combined datasets. Interestingly, the SEM model trained on the US cohort performs better than the one trained on the Chinese cohort. The most likely explanation for this result is the difference in size between the datasets. Looking at the prediction for *Difficulty Walking*, we observe unexpected results. The best models are the SEM models with either the Chinese or American datasets and both have similar performance. In addition the SEM model trained on the combined datasets performs poorly, the worst of all our 5 models. In other words the model trained on American input only works as well as the one trained on Chinese data with an accuracy of about 80 %, but when combined through SEM they lead to an accuracy of about 60 %, much below the combination through PCFS which provides an accuracy of 70 %. We do not have an explanation why combining the datasets leads to lower performance than each separately but we observe that PCFS appears in this instance to be a better approach to combining knowledge.

Looking at the American cohort, on the right side of Fig. 4, we observe that for *Difficulty Eating*, predictions based on PCFS are on par, though slightly better than those with SEM on the American or combined datasets, slightly below 80 % but that the predictions made with the Chinese dataset are much lower (around 10 % accuracy). Here again the relative sizes of the datasets plays a role. The results suggest that combining knowledge has limited value. In fact, as the SEM model based on US data is also quite useful for predicting difficulties with eating on the Chinese cohort, it seems that for that specific prediction task, the dataset captures sufficient knowledge and does not benefit from external sources of information. For *Difficulty Walking*, PCFS outperforms the SEM algorithms with an accuracy of about 60 %. As in the Chinese cohort, it appears to be a more useful approach to knowledge combination than SEM for which the model based on the combined dataset leads to an accuracy of about 40 % which is lower than the model based on US data only which provides an accuracy of approximately 50 %.

6 Conclusion

This paper presents an algorithm, named PCFS, which adapts the PC-Algorithm, Fischer's method and Meek's approach to incorporating forbidden and required arcs so as to be able to build a Bayesian network structure from multiple partially overlapping incomplete datasets and expert opinions in the form of conditional independence statements and domain knowledge. We focus in this paper on learning the structure of the model, not on the parameters. While we present validation based on the four-node Sprinkler network, further experiments are warranted to understand both the influence of the significance parameter along with the benefits of PCFS over Structural Expectation Maximization. Our findings from the real world case study are encouraging. They did validate that combining knowledge could lead to better models and that in such cases PCFS appeared as a valuable alternative to Structural Expectation Maximization.

References

1. Clemen, R.T., Winkler, R.L.: Combining probability distributions from experts in risk analysis. Risk Anal. **19**(2), 187–203 (1999). http://dx.doi.org/10.1111/j.1539-6924.1999.tb00399.x
2. Colombo, D., Maathuis, M.H.: Order-independent constraint-based causal structure learning (2012). arXiv preprint, arXiv:1211.3295
3. Danks, D., Glymour, C., Tillman, R.E.: Integrating locally learned causal structures with overlapping variables. In: Advances in Neural Information Processing Systems, pp. 1665–1672 (2009)
4. Dash, D., Druzdzel, M.J.: Robust independence testing for constraint-based learning of causal structure. In: Proceedings of the Nineteenth Conference on Uncertainty in Artificial Intelligence, pp. 167–174. Morgan Kaufmann Publishers Inc. (2002)
5. Dor, D., Tarsi, M.: A simple algorithm to construct a consistent extension of a partially oriented graph. Technicial report R-185, Cognitive Systems Laboratory, UCLA (1992)
6. Fisher, R.A.: Statistical Methods for Research Workers, 11th edn. Oliver and Boyd, London (1950)
7. Friedman, N.: The Bayesian structural em algorithm. In: Proceedings of the Fourteenth Conference on Uncertainty in Artificial Intelligence, pp. 129–138. Morgan Kaufmann Publishers Inc. (1998)
8. Luis, R., Sucar, L.E., Morales, E.F.: Inductive transfer for learning Bayesian networks. Mach. Learn. **79**(1–2), 227–255 (2010)

9. Matzkevich, I., Abramson, B.: The topological fusion of Bayes nets. In: Proceedings of the Eighth International Conference on Uncertainty in Artificial Intelligence, pp. 191–198. Morgan Kaufmann Publishers Inc. (1992)
10. Matzkevich, I., Abramson, B.: Some complexity considerations in the combination of belief networks. In: Proceedings of the Ninth International Conference on Uncertainty in Artificial Intelligence, pp. 152–158. Morgan Kaufmann Publishers Inc. (1993)
11. Meek, C.: Causal inference and causal explanation with background knowledge. In: Proceedings of the Eleventh Conference on Uncertainty in Artificial Intelligence, pp. 403–410. Morgan Kaufmann Publishers Inc. (1995)
12. Niculescu-Mizil, A., Caruana, R.: Inductive transfer for Bayesian network structure learning. J. Mach. Learn. Res. Proc. Track **2**, 339–346 (2007)
13. Oyen, D., Lane, T.: Leveraging domain knowledge in multitask Bayesian network structure learning. In: AAAI (2012)
14. Pearl, J.: Probabilistic Reasoning in Intelligent Systems: Networks of Plausible Inference. Morgan Kaufmann Publishers Inc., San Francisco (1988)
15. Peña, J.M.: Finding consensus Bayesian network structures. J. Artif. Intell. Res. (JAIR) **42**, 661–687 (2011)
16. Spirtes, P., Glymour, C.N., Scheines, R.: Causation, Prediction, and Search, vol. 81. MIT press, Cambridge (2000)
17. Tillman, R.E.: Structure learning with independent non-identically distributed data. In: Proceedings of the 26th Annual International Conference on Machine Learning, pp. 1041–1048. ACM (2009)
18. Tillman, R.E., Spirtes, P.: Learning equivalence classes of acyclic models with latent and selection variables from multiple datasets with overlapping variables. In: International Conference on Artificial Intelligence and Statistics, pp. 3–15 (2011)
19. Tsamardinos, I., Triantafillou, S., Lagani, V.: Towards integrative causal analysis of heterogeneous data sets and studies. J. Mach. Learn. Res. **98888**(1), 1097–1157 (2012)
20. Verma, T., Pearl, J.: Equivalence and synthesis of causal models. In: Proceedings of the Sixth Annual Conference on Uncertainty in Artificial Intelligence, pp. 255–270. Elsevier Science Inc. (1990)

Reducing the Number of Queries in Interactive Value Iteration

Hugo Gilbert[1]([✉]), Olivier Spanjaard[1], Paolo Viappiani[1], and Paul Weng[2,3]

[1] CNRS, LIP6 UMR 7606, Sorbonne Universités, UPMC Univ Paris 06, Paris, France
{hugo.gilbert,olivier.spanjaard,paolo.viappiani}@lip6.fr
[2] SYSU-CMU Joint Institute of Engineering, Guangzhou, China
[3] SYSU-CMU Shunde International Joint Research Institute, Shunde, China
paweng@cmu.edu

Abstract. To tackle the potentially hard task of defining the reward function in a Markov Decision Process (MDPs), a new approach, called Interactive Value Iteration (IVI) has recently been proposed by Weng and Zanuttini (2013). This solving method, which interweaves elicitation and optimization phases, computes a (near) optimal policy without knowing the precise reward values. The procedure as originally presented can be improved in order to reduce the number of queries needed to determine an optimal policy. The key insights are that (1) asking queries should be delayed as much as possible, avoiding asking queries that might not be necessary to determine the best policy, (2) queries should be asked by following a priority order because the answers to some queries can enable to resolve some other queries, (3) queries can be avoided by using heuristic information to guide the process. Following these ideas, a modified IVI algorithm is presented and experimental results show a significant decrease in the number of queries issued.

1 Introduction

In problems of sequential decision-making under uncertainty, an agent has to repeatedly choose according to her current state an action whose consequences are uncertain in order to maximize a certain criterion in the long run. Such problems can be represented as Markov Decision Processes (MDPs) [9]. In this model, the outcome of each action is stochastic and numeric rewards are granted each time an action is performed. The numerical values of rewards are defined either by the environment or by a human user and the goal of the agent is to choose a policy (which specifies which action to take in every state) such as to maximize the expectation of the discounted sum of future rewards. The latter case, where a human must specify the reward values, represents a difficulty when using MDP methods as this task can be cognitively hard, even for an expert user. And yet, it is well known that the optimal policy is extremely sensitive to the numerical reward values. This problem has motivated much work aiming at mitigating the burden of defining precisely the reward function. In the literature, four main approaches can be distinguished, although their boundaries may be blurry.

© Springer International Publishing Switzerland 2015
T. Walsh (Ed.): ADT 2015, LNAI 9346, pp. 139–152, 2015.
DOI: 10.1007/978-3-319-23114-3_9

Robust Approach. In the first approach, the parameters of the MDP (i.e., rewards and possibly also probabilities) are assumed to be imprecisely known. A natural way [2,7] to handle such situation is to search for robust solutions, i.e., solutions that are as good as possible even in the worst case. However, this method often leads to solutions that are too pessimistic. A better approach is based on minimax regret [20]. Here, one tries to minimize the gap between the value of the best policy (after the true reward values are revealed) and that of the chosen policy. However, this leads to NP-hard problems.

Non Standard Decision Criterion. A second approach is to change the decision criterion optimized by the agent. Different criteria have been proposed. For instance, Delage and Mannor [6] proposed to use a criterion based on quantiles (on distributions over history values). Unfortunately, by changing the decision criterion, one loses all the nice properties satisfied by the standard criterion (e.g., existence of an optimal stationary deterministic policy). Weng [16,17] proposed two new decision criteria that could be used when only the order over rewards, but not the exact values, is known. In both cases, all or some of the nice properties of the standard criterion are preserved. However, those approaches have not yet been experimentally evaluated.

Preference Learning. Another approach, which has been mainly developed for reinforcement learning (i.e., a context more general than MDP), aims at learning the reward values, either from demonstrations (live [1] or from recorded logs [8]) or from interactions with a human tutor [15]. One drawback of this approach is that it generally assumes that demonstrations from human tutors can be easily translated into the state/action representation of the learning agent, which may be difficult as humans and agents evolve in different state/action spaces.

Preference Elicitation. A final approach assumes a human tutor is present and the agent may query her to get more precise information about reward values. In a series of papers, Regan and Boutilier [10–13] show how to compute policies which optimize minmax regret with respect to all candidate functions, and discuss how this criterion can be used to generate informative queries to ask the tutor about the true reward function. Iteratively issuing such queries is shown to allow convergence to the optimal policy for this function. However the problem of computing such robust policy reveals to be NP-hard [20] and their algorithm issues bound queries which can be cognitively difficult to answer. Following the same line of research, Weng and Zanuttini [18], revisited a well known algorithm for solving MDPs, *Value Iteration*, by incorporating the elicitation process in the solving procedure. In this new algorithm called *Interactive Value Iteration* (IVI), a human tutor is queried about multi-sets of rewards when the information acquired so far does not allow to continue solving the MDP. This procedure is appealing as answering comparison queries is much less cognitively demanding than giving the reward function to optimize. However, the original IVI procedure does not try to explicitly minimize the number of queries issued and may lead to a prohibitive effort for the tutor.

In this paper, we address the problem of modifying IVI in order to reduce the number of queries issued. To that aim, we propose a variation of IVI based on the ideas that (1) delaying the queries open the possibility that some of them meanwhile become unnecessary, (2) the order in which the queries are asked matters, and this order should be optimized, (3) queries can be avoided by using heuristic information to guide the iteration process. We show empirically that these combined techniques can greatly reduce the number of queries issued.

The paper is organized as follows. After recalling the main features of interactive value iteration (Sect. 2), we present our proposed optimizations (Sect. 3). Finally, we provide the results of numerical tests that show a significant decrease in the number of queries issued (Sect. 4).

2   Background

2.1   Markov Decision Process

A *Markov Decision Process* (MDP) [9] is defined by a tuple $\mathcal{M} = (S, A, p, r, \gamma)$ where S is a finite set of states, A is a finite set of actions, $p : S \times A \to \mathcal{P}(S)$ is a transition function with $\mathcal{P}(S)$ being the set of probability distributions over states, $r : S \times A \to \mathbb{R}$ is a reward function and $\gamma \in [0,1[$ is a discount factor.

A (stationary, deterministic) *policy* $\pi : S \to A$ associates an action to each state. Such a policy is evaluated by a *value function* $v^\pi : S \to \mathbb{R}$ and a *Q-function*, $Q^\pi : S \times A \to \mathbb{R}$ defined as follows:

$$v^\pi(s) = r(s, \pi(s)) + \gamma \sum_{s' \in S} p(s, \pi(s), s') v^\pi(s') \tag{1}$$

$$Q^\pi(s, a) = r(s, a) + \gamma \sum_{s' \in S} p(s, a, s') v^\pi(s') \tag{2}$$

Then a preference relation is defined over policies by: $\pi \succsim \pi' \Leftrightarrow \forall s \in S, v^\pi(s) \geq v^{\pi'}(s)$. A solution to an MDP is a policy, called *optimal policy*, that ranks the highest with respect to $\succsim$. Such a policy can be found by solving the *Bellman equations*.

$$v^*(s) = \max_{a \in A} r(s, a) + \gamma \sum_{s' \in S} p(s, a, s') v^*(s') \tag{3}$$

As can be seen, the preference relation $\succsim$ over policies is directly induced by the reward function r. In a setting where the reward function is not known with certainty, taking the maximum over actions (as done in (3)) can be problematic. In this paper we will query ordinal information from a tutor to unveil the maximal actions.

2.2   Ordinal Reward MDP

In this paper, while the rewards' numerical values are assumed to be unknown, we suppose that the order over rewards is given. Such situation can be represented as an *Ordinal Reward MDP* (ORMDP) [16] defined by a tuple $(S, A, p, \hat{r}, \gamma)$ where the reward function $\hat{r} : S \times A \to E$ takes its values in a set $E = \{r_1 < r_2 \ldots < r_k\}$ of unknown ordered rewards.

In order to count the number of each unknown reward obtained by a policy, an ORMDP can be reformulated as a Vector Reward MDP (VMDP) $(S, A, p, \overline{r}, \gamma)$ where $\overline{r}(s, a)$ is the vector in $\mathbb{R}^k$ whose i^{th} component is 1 for $\hat{r}(s, a) = r_i$, and 0 on the other components. Like in standard MDPs, in such a VMDP we can define the value function $\overline{v}^\pi$ and the Q-function $\overline{Q}^\pi$ of a policy π by:

$$\overline{v}^\pi(s) = \overline{r}(s, \pi(s)) + \gamma \sum_{s' \in S} p(s, \pi(s), s') \overline{v}^\pi(s') \tag{4}$$

$$\overline{Q}^\pi(s, a) = \overline{r}(s, a) + \gamma \sum_{s' \in S} p(s, a, s') \overline{v}^\pi(s') \tag{5}$$

where additions and multiplications are componentwise. The i-th component of a vector $v \in \mathbb{R}^k$ can be interpreted as the expected number of unknown reward r_i.

Therefore, a value function in a state can be interpreted as a multi-set or a bag of elements of E. Now comparing policies amount to comparing vectors. Interactive value iteration [18] is a procedure inspired from value iteration to find the optimal policy according to the true unknown reward function by querying when needed an expert for comparisons of two bags of elements of E. We present in the next section how this algorithm works.

2.3   Interactive Value Iteration

In order to find an (approximate) optimal policy for an ORMDP with an initially unknown preference relation over vectors, Weng and Zanuttini [18] developed a variant of value iteration, named Interactive Value Iteration (IVI), where the agent may ask a tutor which of two sequences of rewards should be preferred. An example of query is "$r_1 + r_3 \geq 2r_2$?", meaning "is receiving r_1 and r_3 as good as receiving $2r_2$?". As the tutor answers queries, the set of admissible reward functions shrinks and more vectors can be compared without querying the tutor.

At the beginning of the process, the agent only knows the order over rewards, i.e., $r_1 < r_2 < \ldots < r_k$, and knows that she can set without loss of generality r_1 to 0 and r_k to 1 [17]. This initial knowledge is represented by a set $\mathcal{K}$ of linear constraints. When having to choose the best of two value vectors $\overline{v}$ and $\overline{v}'$ in a given state (see Algorithm 2), function $StDom(\overline{v}, \overline{v}')$ compares the two vectors with respect to the following dominance relation (which is analogous to stochastic dominance):

$$\forall j = 1, \ldots, k \quad \sum_{i=j}^{k} \overline{v}_i \geq \sum_{i=j}^{k} \overline{v}'_i \tag{6}$$

In case no dominance is found, the next step is to check whether one of the two vectors is necessarily preferred to the other given the constraints in $\mathcal{K}$. The function $KDom(\overline{v}, \overline{v}')$ checks whether $\overline{v}$ is not less preferred than $\overline{v}'$ (denoted by $\overline{v} \succeq \overline{v}'$) by solving the following linear program:

$$z_K^* = \min \ (\overline{v} - \overline{v}') \cdot r \tag{7}$$
$$\text{s.t.} \quad r \in \mathcal{C}(\mathcal{K}) \tag{8}$$

where the dot in (7) denotes the inner product and $\mathcal{C}(\mathcal{K})$ the set of reward functions $r = (r_1, \ldots, r_k)$ satisfying all constraints in $\mathcal{K}$. We distinguish the following three cases: (1) A non-negative optimal value for the objective function z_K^* implies that for any possible reward function, $\overline{v} \succeq \overline{v}'$; $\overline{v}$ is then said to KDominate $\overline{v}'$. (2) In case of negativity of the optimal value z_K^*, $KDom(\overline{v}', \overline{v})$ is called to check if $\overline{v}' \succeq \overline{v}$ for all reward function satisfying constraints in $\mathcal{K}$. (3) If the two vectors can still not be compared, the tutor is asked which of $\overline{v}$ or $\overline{v}'$ should be preferred. A query is a 2-subset $\{\overline{v}, \overline{v}'\}$ from which the tutor picks her preferred element (if indifferent, she arbitrarily picks one of them). If she picks $\overline{v}$, then $\overline{v} \succeq \overline{v}'$, otherwise $\overline{v}' \succeq \overline{v}$. If $\overline{v} \succeq \overline{v}'$ (resp. $\overline{v}' \succeq \overline{v}$), then the new constraint $(\overline{v} - \overline{v}').r \geq 0$ (resp. $(\overline{v}' - \overline{v}).r \geq 0$) is added to $\mathcal{K}$.

Algorithm 1 summarizes the IVI procedure. It uses the functions *Init*, that returns the initial set $\mathcal{K}$ of linear constraints induced by $r_1 < \ldots < r_k$, and *getBest* (Algorithm 2) that returns the best of two vectors. The function *getBest* calls first function *StDom*, then function *KDom*, and finally function query (Algorithm 3) if no dominance is found. Note that $StDom(\overline{v}, \overline{v}') = KDom(\overline{v}, \overline{v}')$ for initial set $\mathcal{K}$, and that $StDom(\overline{v}, \overline{v}') \Rightarrow KDom(\overline{v}, \overline{v}')$ in all cases. Nevertheless, the IVI procedure takes advantage of the computational efficiency of *StDom* to save some calls to *KDom*.

Weng and Zanuttini [18] showed that the number of queries used by IVI is polynomial in the size of the MDP. However, as the tutor is queried each time two vectors cannot be compared, it seems that a more sophisticated approach could greatly reduce the number of queries needed by the algorithm. In the next section, we propose a modified version of IVI that integrates several techniques in order to reduce the number of queries.

3   Modified Interactive Value Iteration

Interactive Value Iteration is appealing for users as answering comparison queries is significantly less cognitively demanding than directly defining the reward function. And yet to be a reasonable alternative, the number of queries needs to be as low as possible. As stated in the introduction, several ideas are investigated to address this problem: (1) delaying queries in order to avoid asking some unnecessary queries; (2) prioritizing queries in order to ask the most informative ones first; (3) allowing small mistakes in the dominance tests in the early stages in order to anticipate the shrinking of the set of admissible reward functions.

Algorithm 1. IVI [18]

Data: $S, A, p, \hat{r}, \gamma, E, \epsilon$
Result: $\overline{v}_t$
1 $t \leftarrow 0$
2 compute $\overline{r}$ from $\hat{r}$
3 $\mathcal{K} \leftarrow \text{Init}(E)$
4 for $s \in S$ do $\overline{v}_0(s) \leftarrow (0, \ldots, 0)$
5 repeat
6 $\quad t \leftarrow t + 1$
7 $\quad$ for $s \in S$ do
8 $\qquad best \leftarrow (0, \ldots, 0)$
9 $\qquad$ for $a \in A$ do
10 $\qquad\quad \overline{v} \leftarrow \overline{r}(s, a) + \gamma \sum_{s' \in S} p(s, a, s') \overline{v}_{t-1}(s')$
11 $\qquad\quad (best, \mathcal{K}) \leftarrow getBest(best, \overline{v}, \mathcal{K})$
12 $\qquad \overline{v}_t(s) \leftarrow best$
13 until $\|\overline{v}_t - \overline{v}_{t-1}\| < \epsilon$
14 return $\overline{v}_t$

Algorithm 2. getBest

Data: $\overline{v}, \overline{v}', \mathcal{K}$
Result: $(v, \mathcal{K})$
if $StDom(\overline{v}, \overline{v}')$ then return $(\overline{v}, \mathcal{K})$
if $StDom(\overline{v}', \overline{v})$ then return $(\overline{v}', \mathcal{K})$
if $KDom(\overline{v}, \overline{v}', \mathcal{K})$ then return $(\overline{v}, \mathcal{K})$
if $KDom(\overline{v}', \overline{v}, \mathcal{K})$ then return $(\overline{v}', \mathcal{K})$
$(best, \mathcal{K}) \leftarrow query(\overline{v}, \overline{v}', \mathcal{K})$
return $(best, \mathcal{K})$

Algorithm 3. query

Data: $\overline{v}, \overline{v}', \mathcal{K}$
Result: $(v, \mathcal{K})$
Ask query $\{\overline{v}, \overline{v}'\}$
if $\overline{v} \succeq \overline{v}'$ then
$\quad$ return $(\overline{v}, \mathcal{K} \cup \{(\overline{v} - \overline{v}').r \geq 0\})$
else
$\quad$ return $(\overline{v}', \mathcal{K} \cup \{(\overline{v}' - \overline{v}).r \geq 0\})$

3.1 Delaying the Queries

In Algorithm 1, the tutor is queried whenever two vectors $\overline{v}$ and $\overline{v}'$ cannot be compared. The intuition behind the improvement that we propose is that by delaying the query phase after all vectors (each vector refers to a possible action in a given state) are generated we might benefit from the fact that new dominance relations might appear later on. Indeed when trying to assess which of three vectors $\overline{v}_1, \overline{v}_2$, and $\overline{v}_3$ is the best, we may be able to find that $\overline{v}_3$ dominates both $\overline{v}_1$ and $\overline{v}_2$; this is enough for us, even if we ignore which is better between $\overline{v}_1$ and $\overline{v}_2$. Therefore querying which of $\overline{v}_1$ and $\overline{v}_2$ is best, as would do Algorithm 1, is unnecessary. Thus, delaying the querying step from the loop over actions will prevent asking some unnecessary queries. For this purpose, we replace Lines 8 to 13 in Algorithm 1 by the following lines:

for $s \in S$ do
$\quad \overline{Q}(s) \leftarrow \emptyset$
$\quad$ for $a \in A$ do
$\qquad \overline{Q}(s, a) \leftarrow \overline{r}(s, a) + \gamma \sum_{s' \in S} p(s, a, s') \overline{v}_{t-1}(s')$
$\qquad \overline{Q}(s) \leftarrow \overline{Q}(s) \cup \{\overline{Q}(s, a)\}$
$\quad \overline{Q}(s) \leftarrow StDFilter(\overline{Q}(s))$

$\overline{Q}(s) \leftarrow KDFilter(\overline{Q}(s))$
while $|\overline{Q}(s)| > 1$ do
$\quad$ Let $\{\overline{v}, \overline{v}'\} \subseteq \overline{Q}(s)$
$\quad (\_, \mathcal{K}) \leftarrow query(\overline{v}, \overline{v}', \mathcal{K})$
$\quad \overline{Q}(s) \leftarrow KDFilter(\overline{Q}(s))$
$\overline{v}_t(s) \leftarrow$ unique vector of $\overline{Q}(s)$

where primitives $StDFilter$ and $KDFilter$ are given in Algorithms 4 and 5: the former checks the dominance described by Eq. 6 for each pair of vectors $\overline{v}, \overline{v}'$ in $\overline{Q}$ and returns the set of undominated vectors; the latter does the same thing for K-dominance.

Basically, $\overline{Q}(s)$ is a set of vectors (discounted collections of unknown rewards) associated to a state s while $\overline{Q}(s, a)$ is related to the value of taking an action

a in state s. As in the original IVI, we filter out dominated and K-dominated vectors before starting the querying phase. Finally, in the **while** loop we query the user about pairs of non-dominated vectors in $\overline{Q}(s)$ until $\overline{Q}(s)$ contains a single element, that is assigned to $\overline{v}_t(s)$ in the last line. This idea of delaying the queries can be pushed further by delaying the querying phase out of the loop over states or by waiting several time steps before asking queries. We will return to this point when discussing future works in Sect. 5.

Algorithm 4. StDFilter

Data: $\overline{Q}$
Result: Filtered $\overline{Q}$
for $\overline{v} \in \overline{Q}$ do
    for $\overline{v}' \in \overline{Q}$ with $v \neq v'$ do
        if $StDom(\overline{v}, \overline{v}')$ then
            $\overline{Q} \leftarrow \overline{Q} \setminus \{\overline{v}'\}$

return $\overline{Q}$

Algorithm 5. KDFilter

Data: $\overline{Q}$
Result: Filtered $\overline{Q}$
for $\overline{v} \in \overline{Q}$ do
    for $\overline{v}' \in \overline{Q}$ with $v \neq v'$ do
        if $KDom(\overline{v}, \overline{v}')$ then
            $\overline{Q} \leftarrow \overline{Q} \setminus \{\overline{v}'\}$

return $\overline{Q}$

3.2 Prioritizing the Queries

By delaying the queries out of the loop over actions and even out of the loop over states, one can choose to ask queries in a different order than the sequential one induced by the original IVI procedure. Certainly this order will count as queries are not all equally informative. Thus queries that we presume might solve many others should be asked first. Defining a relevance score to guide the querying process seems to be a promising technique to curb the number of queries necessary to solve the MDP. For this purpose, we now replace Lines 8 to 13 in Algorithm 1 by the following lines:

for $s \in S$ do
    $\overline{Q}(s) \leftarrow \emptyset$
    for $a \in A$ do
        $\overline{Q}(s,a) \leftarrow \overline{r}(s,a) + \gamma \sum_{s' \in S} p(s,a,s')\overline{v}_{t-1}(s')$
        $\overline{Q}(s) \leftarrow \overline{Q}(s) \cup \{\overline{Q}(s,a)\}$
    $\overline{Q}(s) \leftarrow$ StDFilter$(\overline{Q}(s))$
    $\overline{Q}(s) \leftarrow$ KDFilter$(\overline{Q}(s))$
$Queries \leftarrow \bigcup_s \{\{\overline{v}, \overline{v}'\} \subseteq \overline{Q}(s) : \overline{v} \neq \overline{v}'\}$

while $Queries \neq \emptyset$ do
    $\{\overline{v}, \overline{v}'\} \leftarrow \arg\max\{\mathcal{X}\text{-score}(\{\overline{v}, \overline{v}'\}) : \{\overline{v}, \overline{v}'\} \subseteq Queries\}$
    $(\_, \mathcal{K}) \leftarrow$ query$(\overline{v}, \overline{v}', \mathcal{K})$
    for $s \in S$ do $\overline{Q}(s) =$ KDFilter$(\overline{Q}(s))$
    $Queries \leftarrow \bigcup_s \{\{\overline{v}, \overline{v}'\} \subseteq \overline{Q}(s) : \overline{v} \neq \overline{v}'\}$

where function $\mathcal{X}$-score (with $\mathcal{X} \in \{\mathcal{Q},\mathcal{K},\mathcal{S}\}$) is one of the priority functions described below, the value of which is intended to reflect how informative a query is, and $Queries$ is the set of all unsolved queries over S. Note that here the best vector returned by query$(\overline{v}, \overline{v}', \mathcal{K})$ does not need to be saved.

To define priority functions evaluating the informative value of a query, several methods could be considered. Here we propose two types of heuristics. The first type aims at reducing as much as possible the set of remaining unsolved queries (*i.e.*, focusing on the impact on the cardinality of *Queries*, the set of unsolved queries) while the second type tries to reduce as much as possible the set of admissible reward functions (*i.e.*, focusing on the impact on the polytope $\mathcal{C}(\mathcal{K})$).

Strategies Aiming at Reducing the Cardinality of Set Queries. Let $\{\overline{v}, \overline{v}'\}$ be a query. If we know that, for instance, $\overline{v}$ is preferred to $\overline{v}'$, this induces an additional constraint that reduces the polytope $\mathcal{C}(\mathcal{K})$. This new, smaller, polytope may induce new relations of K-dominance — for instance, we might be able to check that for all $r \in \mathcal{C}(\mathcal{K} \cup \{\overline{v} \succeq \overline{v}'\})$ a vector $\overline{v}_1$ is preferred to another vector $\overline{v}_2$ — in other words the information carried in an answer to a query can generalize to other queries. A natural idea to evaluate the information value of a query is then to count the number of queries that can be resolved if $\overline{v} \succeq \overline{v}'$, the number of queries resolved if, on the other hand, $\overline{v}' \succeq \overline{v}$ and to take the minimum between the two values. This relevance score will be called $\mathcal{Q}$-score ($\mathcal{Q}$ for *Queries*). Let $Q_{val}(\overline{v}, \overline{v}')$ be the number of queries decided if $\overline{v} \succeq \overline{v}'$; then:

$$\mathcal{Q}\text{-score}(\{\overline{v}, \overline{v}'\}) = \min\{Q_{val}(\overline{v}, \overline{v}'), Q_{val}(\overline{v}', \overline{v})\}. \tag{9}$$

This idea is simple and natural but comes at the cost of solving $4(N-1)$ linear programs (one for each call to KDom) in the worst case if N is the number of queries.

Strategies Aiming at Directly Reducing Polytope $\mathcal{C}(\mathcal{K})$. Another idea is to use the optimal objective values given by procedure KDom. Consider a query $\{\overline{v}, \overline{v}'\} \in$ *Queries*. Let $\mathcal{K}_{val}(\overline{v}, \overline{v}')$ be the optimal value of the linear program described by Eqs. 7–8, normalized by $||\overline{v} - \overline{v}'||$. Since neither $\overline{v}$ nor $\overline{v}'$ could be filtered, both $\mathcal{K}_{val}(\overline{v}, \overline{v}')$ and $\mathcal{K}_{val}(\overline{v}', \overline{v})$ are necessarily negative. We define the priority of query $\{\overline{v}, \overline{v}'\}$ as

$$\mathcal{K}\text{-score}(\{\overline{v}, \overline{v}'\}) = \min\{|\mathcal{K}_{val}(\overline{v}, \overline{v}')|, |\mathcal{K}_{val}(\overline{v}', \overline{v})|\}. \tag{10}$$

The idea of $\mathcal{K}$-score is that $\mathcal{K}_{val}(\overline{v}, \overline{v}')$ and $\mathcal{K}_{val}(\overline{v}', \overline{v})$ give us an approximation of the volume of the polytope on both sides of the constraint defined by the query. Thus it is likely that a high $\mathcal{K}$-score query will reduce largely the polytope no matter the answer of the tutor. This alternative has the benefits of its algorithmic simplicity. Indeed the computation of $\mathcal{K}_{val}(\overline{v}, \overline{v}')$ and $\mathcal{K}_{val}(\overline{v}', \overline{v})$ only requires to solve small linear programs.

Alternatively, following the same idea, we sampled rewards in the admissible reward space (using a Gibbs sampler [5]). Let $\mathcal{SR}$ be the set of samples generated and consider a query $\{\overline{v}, \overline{v}'\} \in$ *Queries*. Let $\mathcal{S}_{val}(\overline{v}, \overline{v}') = |\{r \in \mathcal{SR} : (\overline{v} - \overline{v}').r \geq 0\}|$. We define the priority of query $\{\overline{v}, \overline{v}'\}$ as its $\mathcal{S}$-score ($\mathcal{S}$ for sampling):

$$\mathcal{S}\text{-score}(\{\overline{v}, \overline{v}'\}) = \min\{\mathcal{S}_{val}(\overline{v}, \overline{v}'), \mathcal{S}_{val}(\overline{v}', \overline{v})\}. \tag{11}$$

The numbers of samples on each side of the hyperplane defined by the query $\{\overline{v}, \overline{v}'\}$ gives us an approximation of the volumes of the polytope on each side of the query. Hence a query which has roughly 50 % of samples on both sides will approximately cut the polytope in two equal volumes and it will be deemed a very informative query according to this strategy. A similar idea was used by Rosenthal and Veloso [14] in a context where rewards are a weighted sum of known subrewards and the elicitation procedure searches for the unknown weights.

3.3 Allowing Small Mistakes in the Early Stages

The modification of IVI we propose in this subsection aims at making a compromise between the number of iterations of the procedure and the number of queries issued. The idea is to take the "risk" of slowering convergence by avoiding as much as possible to ask queries in the early stages. For this purpose, the condition $z_K^* \geq 0$ in function $KDom$ (we recall that z_K^* corresponds to the optimal value of program (7–8)) is loosened: a vector $\overline{v}'$ is considered to be dominated if $z_K^* \geq -\text{err}(t)$, where $\text{err}(t) \geq 0$ is a function that decreases to 0 with t. Clearly, this trick has the potentiality to avoid many queries during the early stage with the possible drawback of temporarily driving IVI towards a misleading direction. To insure that this modification will not prevent the algorithm to converge towards the optimal value function, we modify the main loop of IVI so that the algorithm will keep running until $\text{err}(t) \leq \delta$ with $\delta << 1$.

3.4 Synthesis

Algorithm 6 synthesizes our modifications to IVI. The initialization of the algorithm (lines 1 to 4) is unchanged. The main loop (lines 5 to 22) iterates until the value function converges to the optimal value function and the $\text{err}(t)$ function converges to 0. From line 7 to line 13, we fill the sets of possible value vectors for each state by computing and appending the Q-values of the corresponding state-action pairs (lines 7 to 11) and then filter out the vectors (lines 12 and 13) that we already know are dominated by using functions $StDFilter$ (Algorithm 4) and $KDFilter$ (Algorithm 5). This latter algorithm now takes an extra parameter (i.e., $\text{err}(t)$) for implementing the idea presented in the previous subsection. In the second part of the loop we consider the set of all unsolved queries $Queries$ composed of pairs of non-dominated vectors of a same state. While there exists unsolved queries we select and issue the most informative query (lines 16 and 17) using the priority score, $\mathcal{X}$-score ($\mathcal{X} \in \{\mathcal{Q}, \mathcal{K}, \mathcal{S}\}$). Once the tutor answered the query, the acquired information may enable to filter other vectors (line 18) thus reducing the number of unsolved queries (line 19). Once all the queries are solved, each set $\overline{Q}(s)$ is composed of a single element, corresponding to $\overline{v}_t(s)$ (the value vector of s for the next time step). The optimal value function for $\mathcal{M} = (S, A, p, \hat{r}, \gamma)$ is returned. By using standard bookkeeping techniques, we could return the optimal policy as well.

Algorithm 6. Modified IVI

Data: $S, A, p, \hat{r}, \gamma, E, \epsilon,$ err
Result: $\bar{v}_t$

1   $t \leftarrow 0$
2   compute $\bar{r}$ from $\hat{r}$
3   $\mathcal{K} \leftarrow \text{Init}(E)$
4   **for** $s \in S$ **do** $\bar{v}_0(s) \leftarrow (0, \ldots, 0)$
5   **repeat**
6     $t \leftarrow t + 1$
7   **for** $s \in S$ **do**
8       $\overline{Q}(s) \leftarrow \emptyset$
9   **for** $a \in A$ **do**
10         $\overline{Q}(s,a) \leftarrow \bar{r}(s,a) +$ $\gamma \sum_{s' \in S} p(s,a,s')\bar{v}_{t-1}$
11         $\overline{Q}(s) \leftarrow$ $\overline{Q}(s) \cup \{\overline{Q}(s,a)\}$
12       $\overline{Q}(s) \leftarrow$ $\text{StDFilter}(\overline{Q}(s))$
13       $\overline{Q}(s) \leftarrow \text{KDFilter}(\overline{Q}(s),$ $\text{err}(t))$

14   $Queries \leftarrow \bigcup_s \{\{\bar{v}, \bar{v}'\} \subseteq$ $\overline{Q}(s) : \bar{v} \neq \bar{v}'\}$
15   **while** $Queries \neq \emptyset$ **do**
16     $\{\bar{v}, \bar{v}'\} \leftarrow \arg\max\{\mathcal{X}\text{-}$ $\text{score}(\{\bar{v}, \bar{v}'\}) : \{\bar{v}, \bar{v}'\} \subseteq$ $Queries\}$
17     $(\_, \mathcal{K}) \leftarrow \text{query}(\bar{v}, \bar{v}', \mathcal{K})$
18   **for** $s \in S$ **do** $\overline{Q}(s) =$ $\text{KDFilter}(\overline{Q}(s), \text{err}(t))$
19     $Queries \leftarrow \bigcup_s \{\{\bar{v}, \bar{v}'\} \subseteq$ $\overline{Q}(s) : \bar{v} \neq \bar{v}'\}$
20   /*each $\overline{Q}(s)$ is now a singleton*/
21   **for** $s \in S$ **do** $\bar{v}_t = \overline{Q}(s)$
22 **until** $\|\bar{v}_t - \bar{v}_{t-1}\| < \epsilon$ and err$(t) < \delta$
23 **return** $\bar{v}_t$

---

4   Numerical Tests

We tested our approach on three different domains: randomly generated MDPs, autonomic computing [4] and a simulated setting of personalized assistance to impaired people (Coach domain [3])[1]. The original IVI and the improved version described in Sect. 3 were coded in python using Gurobi 6.0 as an LP solver. The discount factor γ is set to 0.95, ϵ to 10^{-3}, δ to 10^{-7} and err(t) = e^{-t}. The number of samples used by the $\mathcal{S}$-score is 5000. All numeric results are averaged over 20 runs.

Random MDPs. We first compared IVI and different variations of our improved IVI on randomly generated MDPs; Given fixed n, m, k (the numbers of states, of actions and of different types of rewards), we randomly generate the transition function assuming that each pair (s, a) has $\lfloor \log_2(n) \rfloor$ successors (chosen uniformly from the set of states) and transition probabilities are obtained by sampling between 0 and 1 and then normalizing. The type of reward of each pair (s, a) is picked from the uniform categorical distribution $r_1, \ldots, r_k$; the numerical values are randomly generated in interval $[0, 1]$ and reordered in order to be consistent.

In Figs. 1 (the number of queries asked as a function of n; $k = |E|$ is fixed to 10 and m to 5) and 2 (the number of queries asked as a function of the

---

[1] In both the autonomic computing domain and in Coach, we randomly generated the transition values and the rewards in such a way to satisfy the constraints imposed by the problem domain.

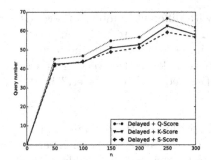

Fig. 1. Number of queries vs number of states for each priority scores.

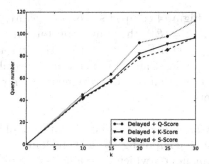

Fig. 2. Number of queries vs number of rewards for each priority scores.

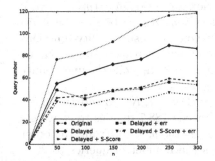

Fig. 3. Number of queries vs number of states for different query strategies.

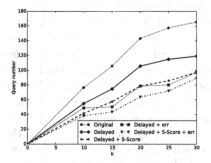

Fig. 4. Number of queries vs number of rewards for different query strategies.

number $|E|$ of different ordinal rewards; n is fixed to 50 and m to 5) we compare the different priority scores that can be used to choose the next query to ask; the graph shows that all three techniques (Q-score, K-score and S-score) are similarly effective with S-score performing best.

In Fig. 3 we compare the impact of the different improvements that we described in Subsects. 3.1–3.3 with the performance of the original IVI ($k = |E|$ is fixed to 10 and m to 5). For the queries' priority we focus on the S-score since this seemed to be the best performing strategy. As expected, IVI asks the highest number of queries (around 80 with 100 states; 120 with 300 states); delaying the moment of asking queries already gives a very significant advantage (around 60 and 85 queries for 100, 300 states). If we additionally ask queries according to the priority induced by the S-score, the number of queries reduces even more (around 45 and 55 queries for 100, 300 states); but surprisingly this improvement is less than the improvement obtained by the combination of delaying the queries and the heuristic of allowing small errors in the initial iterations. The best results are obtained by combining the S-score with the "error" heuristic (about 40 queries with 300 states). Interestingly, with our improvements, the number of queries asked by modified IVI grows very slowly with respect to the number of states.

Figure 4 compares the same strategies but with different numbers of rewards (i.e., $k = |E|$), the number of states n being fixed to 50 and m to 5. By comparing Fig. 3 with Fig. 4 we can see that parameter k impacts much more the number of queries than parameter n; especially when considering the version of IVI including all our improvements (denoted as "Delayed + $\mathcal{S}$-score + err" in the plots).

Autonomic Computing. We also applied our algorithm on the domain of autonomic computing [4]. In this domain, we assume there are κ application server elements on which N available resources have to be assigned. A feasible allocation is an integer vector $\overline{a} = (a_1, \ldots, a_\kappa)$ with $\sum_{i=1}^{\kappa} a_i \leq N$. The client's demand (changing over time) is an integer vector $\overline{d} = (d_1, \ldots, d_\kappa)$ representing κ levels of demand in $\{1, \ldots, D\}$. A state of the MDP is a vector $(\overline{a}, \overline{d})$ defining current allocation and demand. An action is the adoption of a new allocation $\overline{m} = (m_1, \ldots, m_\kappa)$. The reward of taking action $\overline{m}$ in state $(\overline{a}, \overline{d})$ is $r((\overline{a}, \overline{d}), \overline{m}) = u(\overline{a}, \overline{d}) - c(\overline{a}, \overline{d}, \overline{m})$ where $u(\overline{a}, \overline{d}) = \sum_{i=1}^{\kappa} u(n_i, d_i)$ is the sum of non-decreasing utility-functions $u(a_i, d_i)$ and $c(\overline{a}, \overline{d}, \overline{m})$ is the sum of the costs for removing a resource unit from the server. An action deterministically sets the next allocation, while the uncertainty about demands is stochastic and exogenous.

We ran IVI and modified IVIs on instances with $\kappa = 2$, $N = 3$ and $D = 3$ (90 states and 10 actions). While the original IVI needs 188.7 queries to converge, by delaying the queries we reduce this number to 108.9. Prioritizing the order in which the queries are asked further reduces the number of queries to 86.9. Finally by using the heuristic that allows small mistakes in the first iterations of the algorithm, we only need to ask 66.3 queries.

Coach. Finally, we present our experimental results on the "Coach" domain [3]. In this problem, we provide assistance to a person with dementia accomplishing a daily-life activity (e.g., handwashing) that is decomposed into $T = \{0, \ldots, l\}$ phases. Different types of aids are available, modeled by actions $A = \{0, \ldots, m\}$; $a \in A$ is a form of assistance, each associated with a different level of intrusiveness between 0 and m; 0 represents no prompt (no aid is given), $m-1$ represents the most intrusive prompt and m means that a caregiver has to be called. The goal is to aid the person in completing the task, with enough aid but avoiding being too intrusive.

A state in the MDP is described as a tuple (t, d, f) where $t \in T$ is the current timestep, $d \in D = \{0, \ldots, 5\}$ is the delay (time already spent in the current phase of the task) and $f \in A$ is the last prompt used. Transitions model the chance of "success", i.e., the probability that the person moves to the next phase. To model the effectiveness of the aid, at each phase $t < l$ of the task, the probability of success is increasing with the level of intrusiveness of the action a; however the probability is decreasing with d. The reward associated to taking action a in state (t, d, f) is defined by $r((t, d, f), a) = r_{goal}(t) + r_{progress}(d) + r_{delay}(d) + r_{prompt}(a)$ where $r_{goal}(t)$ gives a large reward when the final phase is reached and 0 otherwise, $r_{progress}(d)$ is a small reward when passing to the next phase with no delay and 0 otherwise, $r_{delay}(d)$ and $r_{prompt}(a)$ are increasing cost functions.

We ran IVI and our improved versions of IVI on instances with $l = 14$ and $m = 6$ (630 states and 7 actions). The original version of IVI needed 169.9 queries to converge. By delaying the queries we reduced this number to 114.7. Prioritizing the order in which the queries where issued curbed this score to 77.9. Lastly by allowing small mistakes at the beginning of the algorithm the number of queries issued decreased to 71.2.

5   Conclusion and Future Works

IVI is an appealing procedure that mitigates the burden of defining the reward function of an MDP by interweaving the elicitation and resolution phases. In order to find an optimal policy for an MDP, this procedure queries the tutor about comparisons of multisets of rewards when needed. This paper presents modifications to the original algorithm that are shown to reduce substantially the number of queries issued. The main ideas of the paper are that we can avoid unnecessary queries by delaying the querying phase, and reasoning about the order in which we ask the query.

A natural extension of our work would be to explore new priority scores to guide the querying process. For instance, a strategy alternative to the ones proposed in Sect. 3.1 would be to work in the space of differences of value vectors $\overline{v} - \overline{v}'$. Points $\overline{v} - \overline{v}'$ for which $\overline{v} \succeq \overline{v}'$ (resp. $\overline{v}' \succeq \overline{v}$) would be labelled $+$ (resp. $-$) and an SVM method would be used to find the hyperplane (going through point $\overline{0}$) best separating $+$ and $-$ labels. The vector orthogonal to this hyperplane can be interpreted as the most likely reward function given $\mathcal{K}$. The next query would then be the unsolved query closest to this hyperplane.

Additionally, we intend to delay even more the querying phase by waiting several time steps before asking queries. In this setting (similar to the one of multiobjective MDPs [19]), sets $\overline{Q}(s)$ would not reduce to a singleton at the end of each iteration. Our preliminary results in this direction (where we delay over 3 time steps) are promising and lead to an even more important reduction of the number of queries. Indeed only ≈ 25 queries are needed to solve a random MDP with 50 states, 5 actions and 10 ordinal rewards (results averaged on 20 runs). However, the number of possible value vector for each state can easily explode in this setting and we need to adapt our algorithm to prevent this from happening.

Acknowledgments. Work supported by the French National Research Agency through the Idex Sorbonne Universites, ELICIT project under grant ANR-11-IDEX-0004-02.

References

1. Abbeel, P., Ng, A.: Apprenticeship learning via inverse reinforcement learning. In: Proceedings of Twenty-First International Conference on Machine Learning, ICML 2004. ACM, New York (2004)

2. Bagnell, J., Ng, A., Schneider, J.: Solving uncertain markov decision processes. Technical report, CMU (2001)
3. Boger, J., Hoey, J., Poupart, P., Boutilier, C., Fernie, G., Mihailidis, A.: A planning system based on markov decision processes to guide people with dementia through activities of daily living. IEEE Trans. Inf. Technol. Biomed. **10**, 323–333 (2006)
4. Boutilier, C., Das, R., Kephart, J.O., Tesauro, G., Walsh, W.E.: Cooperative negotiation in autonomic systems using incremental utility elicitation. In: Proceedings of the Nineteenth Conference on Uncertainty in Artificial Intelligence, pp. 89–97 (2003)
5. Casella, G., George, E.I.: Explaining the gibbs sampler. Am. Stat. **46**, 167–174 (1992)
6. Delage, E., Mannor, S.: Percentile optimization in uncertain Markov decision processes with application to efficient exploration. In: ICML, pp. 225–232 (2007)
7. Givan, R., Leach, S., Dean, T.: Bounded-parameter Markov decision process. Artif. Intell. **122**(1–2), 71–109 (2000)
8. Piot, B., Geist, M., Pietquin, O.: Boosted and reward-regularized classification for apprenticeship learning. In: International conference on Autonomous Agents and Multi-Agent Systems, AAMAS 2014, Paris, France, 5–9 May 2014, pp. 1249–1256 (2014)
9. Puterman, M.: Markov Decision Processes: Discrete Stochastic Dynamic Programming, 1st edn. Wiley, New York (1994)
10. Regan, K., Boutilier, C.: Regret-based reward elicitation for Markov decision Processes. In: Proceedings of the Twenty-Fifth Conference on Uncertainty in Artificial Intelligence, UAI 2009, pp. 444–451. AUAI Press, Arlington (2009)
11. Regan, K., Boutilier, C.: Robust policy computation in reward-uncertain MDPS using nondominated policies. In: Fox, M., Poole, D. (eds.) AAAI. AAAI Press (2010)
12. Regan, K., Boutilier, C.: Eliciting additive reward functions for Markov decision processes. In: Proceedings of the Twenty-Second International Joint Conference on Artificial Intelligence, IJCAI 2011, vol. 3, pp. 2159–2164. AAAI Press (2011)
13. Regan, K., Boutilier, C.: Robust online optimization of reward-uncertain MDPs. In: Proceedings of the Twenty-Second International Joint Conference on Artificial Intelligence, IJCAI 2011, vol. 3, pp. 2165–2171. AAAI Press (2011)
14. Rosenthal, S., Veloso, M.M.: Monte carlo preference elicitation for learning additive reward functions. In: RO-MAN, pp. 886–891. IEEE (2012)
15. Thomaz, A., Hoffman, G., Breazeal, C.: Real-time interactive reinforcement learning for robots. In: AAAI Workshop Human Comprehensible Machine Learning, pp. 9–13 (2005)
16. Weng, P.: Markov decision processes with ordinal rewards: reference point-based preferences. In: Proceedings of the 21st International Conference on Automated Planning and Scheduling, ICAPS 2011, Freiburg, Germany, 11–16 June 2011 (2011)
17. Weng, P.: Ordinal decision models for Markov decision processes. In: ECAI 2012–20th European Conference on Artificial Intelligence. Including Prestigious Applications of Artificial Intelligence (PAIS 2012) System Demonstrations Track, pp. 828–833, Montpellier, France, 27–31 August 2012 (2012)
18. Weng, P., Zanuttini, B.: Interactive value iteration for Markov decision processes with unknown rewards. In: Rossi, F. (ed.) IJCAI. IJCAI/AAAI (2013)
19. White, D.J.: Multi-objective infinite-horizon discounted Markov decision processes. J. Math. Anal. Appl. **89**(2), 639–647 (1982)
20. Xu, H., Mannor, S.: Parametric regret in uncertain Markov decision processes. In: CDC, pp. 3606–3613. IEEE (2009)

Learning the Parameters of a Non Compensatory Sorting Model

Olivier Sobrie[1,2]([⊠]), Vincent Mousseau[1], and Marc Pirlot[2]

[1] LGI, CentraleSupelec, Grande Voie des Vignes, 92295 Châteney-Malabry, France
`olivier.sobrie@gmail.com`, `vincent.mousseau@ecp.fr`
[2] Faculté Polytechnique, Université de Mons, 9, Rue de Houdain, 7000 Mons,
Belgium
`marc.pirlot@umons.ac.be`

Abstract. We consider a multicriteria sorting procedure based on a majority rule, called MR-Sort. This procedure allows to sort each object of a set, evaluated on multiple criteria, in a category selected among a set of pre-defined and ordered categories. With MR-Sort, the ordered categories are separated by profiles which are vectors of performances on the different attributes. Using the MR-Sort rule, an object is assigned to a category if it is at least as good as the category lower profile and not better than the category upper profile. To determine whether an object is as good as a profile, the weights of the criteria on which the object performances are better than the profile performances are summed up and compared to a threshold. If the sum of weights is at least equal to the threshold, then the object is considered at least as good as the profile. In view of increasing the expressiveness of the model, we substitute additive weights by a capacity to represent the power of coalitions of criteria. This corresponds to the Non-Compensatory Sorting model characterized by Bouyssou and Marchant. In the paper we describe a mixed integer program and a heuristic algorithm that enable to learn the parameters of this model from assignment examples.

1 Introduction

In Multiple Criteria Decision Analysis (MCDA), the "sorting problem setting" (or ordered classification) consists in assigning each alternative of a set, evaluated on several monotone criteria, in a category selected among a set of pre-defined and ordered categories. Several MCDA methods are designed to handle sorting problems. In this paper, we consider a sorting model that satisfies the requirements of the non-compensatory sorting models characterized in [1,2]. The model is a generalization of MR-Sort [3,4]. In MR-Sort, categories are separated by profiles which are vectors of performances on the different criteria. Each criterion of the model is associated a weight representing its importance or its voting power. Using this model, without veto, we assign an alternative to a category if it is considered at least as good as the category lower profile and not at least as good as the category upper profile. An alternative is considered as good as a profile if

© Springer International Publishing Switzerland 2015
T. Walsh (Ed.): ADT 2015, LNAI 9346, pp. 153–170, 2015.
DOI: 10.1007/978-3-319-23114-3_10

its performances are at least as good as the profile performances on a weighted majority of criteria. In MR-Sort, the weighted majority of criteria is reached if the sum of weights of criteria on which the alternative is at least as good as the profile is greater than a threshold.

Such a model contrasts with utility based models such as UTADIS [5,6]. It belongs to a class of decision models referred as noncompensatory in the literature [7,8], because it just takes into account whether or not an evaluation is above the profile value, not by how much it passes or misses this profile value. These methods are well suited to criteria assessed on ordinal scales.

Consider a MR-Sort model involving 4 criteria (c_1, c_2, c_3 and c_4) and 2 ordered categories ($C_2 \succ C_1$), separated by a profile b_1. Using this model, an alternative is assigned to the "good" category (C_2) iff its performances are as good as the profile b_1 on at least one of the four following minimal criteria coalitions: $c_1 \wedge c_2$, $c_3 \wedge c_4$, $c_1 \wedge c_4$ and $c_2 \wedge c_4$. A coalition of criteria is said to be minimal if removing any criterion is enough to reject the assertion "alternative a is as good as profile b". With an MR-Sort model, this can be achieved by selecting, for instance, the following weights and majority threshold: $w_1 = 0.3$, $w_2 = 0.2$, $w_3 = 0.1$, $w_4 = 0.4$ and $\lambda = 0.5$. We have $w_1 + w_2 = \lambda$, $w_3 + w_4 = \lambda$, $w_1 + w_4 > \lambda$ and $w_2 + w_4 > \lambda$. All the other coalitions of criteria, which are not supersets of the four minimal coalitions listed above, are not sufficient to be considered as good as b_1 (e.g. $w_1 + w_3 < \lambda$).

Assume that we want a model for which the two minimal sufficient criteria coalitions are: $c_1 \wedge c_2$ and $c_3 \wedge c_4$. Modeling this classification rule with an MR-Sort model is impossible. To model these rules, we have to choose weights w_i, $i = 1, \ldots, 4$, summing up to 1, such that $w_1 + w_2 \geq \lambda$ and $w_3 + w_4 \geq \lambda$. Summing these two inequalities yields $1 \geq 2\lambda$. If we want these coalitions to be the only minimal sufficient ones, we must also have : $w_1 + w_3 < \lambda$, $w_1 + w_4 < \lambda$, $w_2 + w_3 < \lambda$ and $w_2 + w_4 < \lambda$. Summing these four inequalities yields $2 < 4\lambda$. Hence, there exist no weights and majority threshold for which the 2 above coalitions are the only two minimal sufficient coalitions. In view of being able to represent such a type of rule, we consider in this paper an extension of MR-Sort allowing to model interactions between criteria. This formulation expresses the majority rule of MR-Sort by using a capacity like in the Choquet Integral [9]. This model is called the Non Compensatory Sorting Model (NCS model). It was introduced and characterized in [1,2].

In this paper, we aim at studying the additional descriptive ability of the NCS model as compared to MR-Sort. We assess this experimentally on real datasets. The paper is organized as follows. The next section describes formally what is a non compensatory sorting model. Section 3 recalls previous work dealing with learning the parameters of MR-Sort models from assignment examples. The next two sections describe respectively a Mixed Integer Program and a heuristic algorithm that allow to learn the parameters of a NCS model. Some experimental results are finally presented.

2   MR-Sort and NCS Models

2.1   MR-Sort Model

MR-Sort is a method for assigning objects to ordered categories. It is a simplified version of ELECTRE TRI, another MCDA method [10, 11].

The MR-Sort rule works as follows. Formally, let X be a set of objects evaluated on n ordered attributes (or criteria), $F = \{1, ..., n\}$. We assume that X is the Cartesian product of the criteria scales, $X = \prod_{j=1}^{n} X_j$. An object $a \in X$ is a vector $(a_1, \ldots, a_j, \ldots, a_n)$, where $a_j \in X_j$ for all j. The ordered categories which the objects are assigned to by the MR-Sort model are denoted by C_h, with $h = 1, \ldots, p$. Category C_h is delimited by its lower limit profile b_{h-1} and its upper limit profile b_h, which is also the lower limit profile of category C_{h+1} (provided $0 < h < p$). The profile b_h is the vector of criterion values $(b_{h,1}, \ldots, b_{h,j}, \ldots, b_{h,n})$, with $b_{h,j} \in X_j$ for all j. We denote by $P = \{1, ..., p\}$ the list of category indices. By convention, the best category, C_p, is delimited by a fictive upper profile, b_p, and the worst one, C_1, by a fictive lower profile, b_0. It is assumed that the profiles dominate one another, i.e.: $b_{h-1,j} \leq b_{h,j}$, for $h = \{1, \ldots, p\}$ and $j = \{1, \ldots, n\}$.

Using the MR-Sort procedure, an object is assigned to a category if its criterion values are at least as good as the category lower profile values on a weighted majority of criteria while this condition is not fulfilled when the object's criterion values are compared to the category upper profile values. In the former case, we say that the object is *preferred* to the profile, while, in the latter, it is not. Formally, if an object $a \in X$ is *preferred* to a profile b_h, we denote this by $a \succcurlyeq b_h$. Object a is preferred to profile b_h whenever the following condition is met:

$$a \succcurlyeq b_h \Leftrightarrow \sum_{j: a_j \geq b_{h,j}} w_j \geq \lambda, \tag{1}$$

where w_j is the nonnegative weight associated with criterion j, for all j and λ sets a majority level. The weights satisfy the normalization condition $\sum_{j \in F} w_j = 1$; λ is called the *majority threshold*; it satisfies $\lambda \in [1/2, 1]$.

The preference relation $\succcurlyeq$ defined by (1) is called an *outranking* relation without veto or a *concordance* relation ([11]; see also [12, 13] for an axiomatic description of such relations). Consequently, the condition for an object $a \in X$ to be assigned to category C_h reads:

$$\sum_{j: a_j \geq b_{h-1,j}} w_j \geq \lambda \quad \text{and} \quad \sum_{j: a_j \geq b_{h,j}} w_j < \lambda. \tag{2}$$

The MR-Sort assignment rule described above involves $pn + 1$ parameters, i.e. n weights, $(p - 1)n$ profiles evaluations and one majority threshold.

A *learning set* A is a subset of objects $A \subseteq X$ for which an assignment is known. For $h \in P$, A_h denotes the subset of objects $a \in A$ which are assigned to category C_h. The subsets A_h are disjoint; some of them may be empty.

2.2 NCS Model

Limitation of MR-Sort. Before describing the NCS model, we show the limits of MR-Sort. As an illustration, consider an application in which a committee for a higher education program has to decide about the admission of students on the basis of their evaluations in 4 courses: math, physics, chemistry and history. To be accepted in the program, the committee considers that a student should have a sufficient majority of evaluations above 10/20. From the committee point of view, courses (criteria) coalitions don't have the same importance. The strength of a coalition of courses varies as a function of the courses belonging to the coalition. The committee stated that the following subsets are the minimal coalitions of courses in which the evaluation should be above 10/20 in order to be accepted: {math, physics}, {math, chemistry} and {chemistry, history}. To illustrate this rule, Table 1 shows evaluations of several students and, for each student, whether he is accepted or refused.

Representing these assignments by using a MR-Sort model with profile fixed at 10/20 in each course is impossible. There are no additive weights allowing to model such rules. MR-Sort is not adapted to handle such type of problems since it does not allow to model attribute interactions. In view of taking criterion interactions into account, we modify the definition of the global outranking relation, $a \succcurlyeq b_h$, given in (1).

Capacity. The new model described hereafter uses capacities. A capacity is a function $\mu : 2^F \to [0, 1]$ such that:

- $\mu(B) \geq \mu(A)$, for all $A \subseteq B \subseteq F$ (monotonicity);
- $\mu(\emptyset) = 0$ and $\mu(F) = 1$ (normalization).

The Möbius transform allows to express the capacity in another form:

$$\mu(A) = \sum_{B \subseteq A} m(B) \quad \forall A \subseteq F \quad \text{with } m(B) = \sum_{C \subseteq B} (-1)^{|B|-|C|} \mu(C).$$

The value $m(B)$ can be interpreted as the weight that is exclusively allocated to B as a whole. A capacity can be defined directly by its Möbius transform

Table 1. Evaluation of students and their acceptance/refusal status

	Math	Physics	Chemistry	History	A/R
James	11	11	9	9	A
Marc	11	9	11	9	A
Robert	9	9	11	11	A
John	11	9	9	11	R
Paul	9	11	9	11	R
Pierre	9	11	11	9	R

also called Möbius interaction. A Möbius interaction or Möbius mass m is a set function $m : 2^F \rightarrow [-1, 1]$ satisfying the following conditions:

$$\sum_{j \in K \subseteq J \cup \{j\}} m(K) \geq 0 \quad \forall j \in F, J \subseteq F \backslash \{i\} \quad \text{and} \quad \sum_{K \subseteq F} m(K) = 1. \tag{3}$$

If m is a Möbius interaction, the set function defined by $\mu(A) = \sum_{B \subseteq A} m(B)$ is a capacity. Conditions (3) guarantee that μ is monotone [14].

NCS Model. Using a capacity to express the weight of the coalition in favor of an object, we transform the outranking rule (1) as follows:

$$a \succcurlyeq b_h \Leftrightarrow \mu(A) \geq \lambda \text{ with } A = \{j \in F : a_j \geq b_{h,j}\}$$

$$\text{and } \mu(A) = \sum_{B \subseteq A} m(B) \tag{4}$$

Computing the value of $\mu(A)$ with the Möbius transform requires the evaluation of $2^{|A|}$ parameters. In a model involving n criteria, this implies the elicitation of 2^n parameters, with $\mu(\emptyset) = 0$ and $\mu(F) = 1$. To reduce the number of parameters to elicit, we use a 2-additive capacity in which all the interactions involving more than 2 criteria are equal to zero. Inferring a 2-additive capacity for a model having n criteria requires the determination of $\frac{n(n+1)}{2} - 1$ parameters.

Finally, the condition for an object $a \in X$ to be assigned to category C_h can be expressed as follows:

$$\mu(F_{a \geq b_{h-1}}) \geq \lambda \quad \text{and} \quad \mu(F_{a \geq b_h}) < \lambda \tag{5}$$

with $F_{a \geq b_{h-1}} = \{j \in F : a_j \geq b_{h-1,j}\}$ and $F_{a \geq b_h} = \{j \in F : a_j \geq b_{h,j}\}$.

This model fits with the definition of a NCS model given in [1,2]. We note that MR-Sort is a special case of a NCS model in which a simple additive capacity is used.

3   Learning the Parameters of a MR-Sort Model

Learning the parameters of MR-Sort and ELECTRE TRI models has been already studied in several articles [3,4,15–21]. In this section, we recall how to learn the parameters of an MR-Sort model using respectively an exact method [3] and a heuristic algorithm [4].

3.1   Mixed Integer Programming

Learning the parameters of a MR-Sort model using linear programming techniques has been proposed in [3]. The paper describes a Mixed Integer Program (MIP) taking a set of assignment examples and their vector of performances as input and finding the parameters of a MR-Sort model such that the largest possible number of examples are restored by the inferred model. We recall in this subsection the main steps to obtain the MIP formulation.

The condition for an object x to be assigned to category C_h (Equation (2)) can be written as follows:

$$a \in C_h \iff \begin{cases} \sum_{j=1}^n c_{a,j}^{h-1} \geq \lambda & \text{with } c_{a,j}^{h-1} = \begin{cases} w_j & \text{if } a_j \geq b_{h-1,j} \\ 0 & \text{otherwise} \end{cases} \\ \sum_{j=1}^n c_{a,j}^h < \lambda & \text{with } c_{a,j}^h = \begin{cases} w_j & \text{if } a_j \geq b_{h,j} \\ 0 & \text{otherwise} \end{cases} \end{cases}$$

To linearize these constraints, we introduce for each value $c_{a,j}^l$, with $l = \{h-1, h\}$, a binary variable $\delta_{a,j}^l$ that is equal to 1 when the performance of the object a is at least as good as or better than the performance of the profile b_l on criterion j and 0 otherwise. To obtain the value of $\delta_{a,j}^l$, we add the following constraints, where M is an arbitrary large positive constant:

$$M(\delta_{a,j}^l - 1) \leq a_j - b_{l,j} < M \cdot \delta_{a,j}^l \tag{6}$$

By using the value $\delta_{a,j}^l$, the values of $c_{a,j}^l$ are obtained as follows:

$$\begin{cases} c_{a,j}^l \geq 0 \\ c_{a,j}^l \leq w_j \end{cases} \qquad \begin{cases} c_{a,j}^l \leq \delta_{a,j}^l \\ c_{a,j}^l \geq \delta_{a,j}^l - 1 + w_j \end{cases}$$

The objective function of the MIP consists in maximizing the number of examples compatible with the learned model, i.e. minimizing the 0/1 loss function. In order to model this, new binary variables, γ_a for all $a \in A$, are introduced. The value of γ_a is equal to 1 if object a is assigned to the expected category, i.e. the category it is assigned to in the learning set, and equal to 0 otherwise. To obtain the correct value of γ_a variables, two additional constraints are added:

$$\begin{cases} \sum_{j=1}^n c_{a,j}^{h-1} \geq \lambda + M(\gamma_a - 1) \\ \sum_{j=1}^n c_{a,j}^h < \lambda - M(\gamma_a - 1) \end{cases}$$

The objective function chosen for the linear program consists in maximizing the number of examples compatible with the model. Formally it reads: $\max \sum_{a \in A} \gamma_a$. Finally, the combination of all the constraints leads to the MIP given in Appendix A.

3.2 A Heuristic Algorithm

The MIP presented in the previous section is not suitable for large data sets because of the high computing time that is required to infer the MR-Sort parameters. In view of learning MR-Sort models in the context of large data sets, a heuristic algorithm has been proposed in [4]. As in the MIP, the heuristic algorithm takes as input a set of assignment examples and their vectors of performances. The algorithm returns the parameters of a MR-Sort model.

$$\min \sum_{a \in A} (x'_a + y'_a)$$

s.t.

$$\sum_{j:a_j \geq b_{h-1,j}} w_j - x_a + x'_a = \lambda \qquad\qquad \forall a \in A_h, h = \{2, ..., p\}$$

$$\sum_{j:a_j \geq b_{h,j}} w_j + y_a - y'_a = \lambda - \epsilon \qquad\qquad \forall a \in A_h, h = \{1, ..., p-1\}$$

$$\sum_{j=1}^{n} w_j = 1$$

$$w_j \in [0;1] \qquad\qquad \forall j \in F$$

$$\lambda \in [0.5;1]$$

$$x_a, y_a, x'_a, y'_a \in \mathbb{R}_0^+$$

ε a small positive number.

$$(7)$$

The heuristic algorithm proposed in [4] works as follows. First a population of MR-Sort models is initialized. After the initialization, the two following steps are repeated iteratively on each model in the population:

1. A linear program optimizes the weights and the majority threshold on the basis of assignment examples and fixed profiles.
2. Given the inferred weights and the majority threshold, a heuristic adjusts the profiles of the model on the basis of the assignment examples.

After applying these two steps to all the models in the population, the $\lfloor \frac{n}{2} \rfloor$ models restoring the least numbers of examples are reinitialized. These steps are repeated until the heuristic finds a model that fully restores all the examples or after a number of iterations specified a priori.

The linear program designed to learn the weights and the majority threshold is given by (7). It minimizes a sum of slack variables, x'_a and y'_a, that is equal to 0 when all the objects are correctly assigned, i.e. assigned to the category defined in the input data set. We remark that the objective function of the linear program does not explicitly minimize the 0/1 loss but a sum of slacks. This implies that compensatory effects might appear, with undesirable consequences on the 0/1 loss. However in this heuristic, we consider that these effects are acceptable. The linear program doesn't involve binary variables. Therefore, the computing time remains reasonable when the size of the problem increases.

The objective function of the heuristic varying the profiles maximizes the number of examples compatible with the model. To do so, it iterates over each profile h and each criterion j and identifies a set of candidate moves for the profile, which correspond to the performances of the examples on criterion j located between profiles $h-1$ and $h+1$. Each candidate move is evaluated as a function of the probability to improve the classification accuracy of the model. To evaluate if a candidate move is likely or unlikely to improve the classification of one or several objects, the examples which have an evaluation on criterion

j located between the current value of the profile, $b_{h,j}$, and the candidate move, $b_{h,j} + \delta$ (resp. $b_{h,j} - \delta$), are grouped in different subsets:

$V_{h,j}^{+\delta}$ (**resp.** $V_{h,j}^{-\delta}$): the sets of objects misclassified in C_{h+1} instead of C_h (resp. C_h instead of C_{h+1}), for which moving the profile b_h by $+\delta$ (resp. $-\delta$) on j results in a correct assignment.

$W_{h,j}^{+\delta}$ (**resp.** $W_{h,j}^{-\delta}$): the sets of objects misclassified in C_{h+1} instead of C_h (resp. C_h instead of C_{h+1}), for which moving the profile b_h by $+\delta$ (resp. $-\delta$) on j strengthens the criteria coalition in favor of the correct classification but will not by itself result in a correct assignment.

$Q_{h,j}^{+\delta}$ (**resp.** $Q_{h,j}^{-\delta}$): the sets of objects correctly classified in C_{h+1} (resp. C_{h+1}) for which moving the profile b_h by $+\delta$ (resp. $-\delta$) on j results in a misclassification.

$R_{h,j}^{+\delta}$ (**resp.** $R_{h,j}^{-\delta}$): the sets of objects misclassified in C_{h+1} instead of C_h (resp. C_h instead of C_{h+1}), for which moving the profile b_h by $+\delta$ (resp. $-\delta$) on j weakens the criteria coalition in favor of the correct classification but does not induce misclassification by itself.

$T_{h,j}^{+\delta}$ (**resp.** $T_{h,j}^{-\delta}$): the sets of objects misclassified in a category higher than C_h (resp. in a category lower than C_{h+1}) for which the current profile evaluation weakens the criteria coalition in favor of the correct classification.

A formal definition of these sets can be found in [4]. The evaluation of the candidate moves is done by aggregating the number of elements in each subset. Finally, the choice to move or not the profile on the criterion is determined by comparing the candidate move evaluation to a random number drawn uniformly. These operations are repeated multiple times on each profile and each criterion.

4   Mixed Integer Program to Learn a 2-Additive NCS Model

As compared to a MR-Sort model, a NCS model involves more parameters. In a standard MR-Sort model, a weight is associated to each criterion, which makes overall n parameters to elicit. With a NCS model limited to two-additive capacities, the computation of the strength of a coalition of criteria involves the weights of the criteria in the coalition and the pairwise interactions (Möbius coefficients) between these criteria. Overall there are $\frac{n(n+1)}{2} - 1$ coefficients. In the two-additive case, let us denote by m_j the weights of criterion j and by $m_{j,k}$ the Möbius interactions between criteria j and k. The capacity $\mu(A)$ of a subset of criteria is obtained as: $\mu(A) = \sum_{j \in A} m_j + \sum_{\{j,k\} \subseteq A} m_{j,k}$. The constraints (3) on the interaction read:

$$m_j + \sum_{k \in J} m_{j,k} \geq 0 \qquad \forall j \in F, \forall J \subseteq F \backslash \{j\} \qquad (8)$$

and $\sum_{j \in F} m_j + \sum_{\{j,k\} \subseteq F} m_{j,k} = 1$.

The number of monotonicity constraints evolves exponentially as a function of the number of criteria, n. In [22], two other formulations are proposed in order to reduce significantly the number of constraints ensuring the monotonicity of the capacities. The first formulation reduces the number of constraints to $2n^2$ but leads to a non linear program. The second formulation reduces the number of constraints to $n^2 + 1$ without introducing non linearities but adds n^2 extra variables.

With a 2-additive MR-Sort model, the constraints for an alternative a to be assigned to a category h (5) can also be expressed as follows:

$$\begin{cases} \sum_{j=1}^{n} c_{a,j}^{h-1} + \sum_{j=1}^{n} \sum_{k=1}^{j} c_{a,j,k}^{h-1} & \geq \lambda + M(\gamma_a - 1) \\ \sum_{j=1}^{n} c_{a,j}^{h} + \sum_{j=1}^{n} \sum_{k=1}^{j} c_{a,j,k}^{h} & < \lambda - M(\gamma_a - 1) \end{cases} \tag{9}$$

with:

- $c_{a,j}^{h-1}$ (resp. $c_{a,j}^{h}$) equals m_j if the performance of alternative a is at least as good as the performance of profile b_{h-1} (resp. b_h) on criterion j, and equals 0 otherwise;
- $c_{a,j,k}^{h-1}$ (resp. $c_{a,j,k}^{h}$) equals $m_{j,k}$ if the performance of alternative a is at least as good as the performance of profile b_{h-1} (resp. b_h) on criteria j and k, and equals 0 otherwise.

For all $a \in A$, $j \in F$ and $l \in P$, constraints (8) imply that $c_{a,j}^l \geq 0$ and that $c_{a,j,k}^l \in [-1, 1]$. The values of $c_{a,j}^{h-1}$ and $c_{a,j}^{h}$ are obtained in a similar way as it is done for learning the parameters of a standard MR-Sort model by replacing the weights with the corresponding Möbius coefficients (10).

$$\begin{cases} c_{a,j}^l & \geq 0 \\ c_{a,j}^l & \leq m_j \end{cases} \qquad \begin{cases} c_{a,j}^l & \leq \delta_{a,j}^l \\ c_{a,j}^l & \geq \delta_{a,j}^l - 1 + m_j \end{cases} \tag{10}$$

However it is not the case for the variables $c_{a,j,k}^{h-1}$ and $c_{a,j,k}^{h}$, because they involve two criteria. To linearize the formulation, we introduce new binary variables, $\Delta_{a,j,k}^l$ equal to 1 if alternative a has better performances than profile b_l on criteria j and k and equal to 0 otherwise. We obtain the value of $\Delta_{a,j,k}^l$ thanks to the conjunction of constraints given in (6) and the following constraints:

$$2\Delta_{a,j,k}^l \leq \delta_{a,j}^l + \delta_{a,j}^k \leq \Delta_{a,j,k}^l + 1$$

In order to obtain the value of $c_{a,j,k}^l$, which can be either positive or negative, for all $l \in P$, we decompose the variable in two parts, $\alpha_{a,j,k}^l$ and $\beta_{a,j,k}^l$ such that $c_{a,j,k}^l = \alpha_{a,j,k}^l - \beta_{a,j,k}^l$ with $\alpha_{a,j,k}^l \geq 0$ and $\beta_{a,j,k}^l \geq 0$. The same is done for $m_{j,k}$ which is decomposed as follows: $m_{j,k} = m_{j,k}^+ - m_{j,k}^-$ with $m_{j,k}^+ \geq 0$ and $m_{j,k}^- \geq 0$. The values of $\alpha_{a,j,k}^l$ and $\beta_{a,j,k}^l$ are obtained thanks to the following constraints:

$$\begin{cases} \alpha_{a,j,k}^l & \leq \Delta_{a,j,k}^l \\ \alpha_{a,j,k}^l & \leq m_{j,k}^+ \\ \alpha_{a,j,k}^l & \geq \Delta_{a,j,k}^l - 1 + m_{j,k}^+ \end{cases} \qquad \begin{cases} \beta_{a,j,k}^l & \leq \Delta_{a,j,k}^l \\ \beta_{a,j,k}^l & \leq m_{j,k}^- \\ \beta_{a,j,k}^l & \geq \Delta_{a,j,k}^l - 1 + m_{j,k}^- \end{cases}$$

Finally, we obtain the MIP displayed in Appendix B.

$$\min \sum_{a \in A}(x'_a + y'_a)$$
s.t.

$$\sum_{j:a_j \geq b_{h-1,j}}^{n} \left(m_j + \sum_{k:a_k \geq b_{h-1,k}}^{j} m_{j,k} \right) - x_a + x'_a = \lambda \qquad \forall a \in A_h,$$

$$h = \{2, ..., p\}$$

$$\sum_{j:a_j \geq b_{h,j}}^{n} \left(m_j + \sum_{k:a_k \geq b_{h,k}}^{j} m_{j,k} \right) + y_a - y'_a = \lambda - \varepsilon \, \forall a \in A_h,$$

$$h = \{1, ..., p-1\}$$

$$\sum_{j=1}^{n} m_j + \sum_{j=1}^{n} \sum_{k=1}^{j} m_{j,k} = 1$$

$$m_j + \sum_{k \in J} m_{j,k} \geq 0 \qquad\qquad \forall j \in F, \forall J \subseteq F \backslash \{j\}$$

$$\lambda \in [0.5; 1]$$
$$m_j \in [0, 1] \qquad\qquad \forall j \in F$$
$$m_{j,k} \in [-1, 1] \qquad\qquad \forall j \in F, \forall k \in F, k < j$$
$$x_a, y_a, x'_a, y'_a \in \mathbb{R}_0^+ \qquad\qquad a \in A$$
$$\varepsilon \text{ a small positive number.}$$

$$(11)$$

---

5   A Heuristic Algorithm to Learn a 2-Additive NCS Model

The MIP described in the previous section requires a lot of binary variables and is therefore not well-suited for large problems. In the present section, we describe an adaptation of the heuristic described in Subsect. 3.2 in view of learning the parameters of a NCS model. Like for the MIP in the previous section, we limit the model to 2-additive capacities in order to reduce the number of coefficients as compared to a model with a general capacity.

One of the components that needs to be adapted in the heuristic in order to be able to learn a 2-additive NCS model is the linear program that infers the weights and the majority threshold (7). Like in the MIP described in the previous section, we use the Möbius transform to express capacities. In view of inferring Möbius coefficients, m_j and $m_{j,k}$, $\forall j, \forall k$ with $k < j$, we modify the linear program as shown in (11).

The value of $x_a - x'_a$ (resp. $y_a - y'_a$) represents the difference between the capacity of the criteria belonging to the coalition in favor of $a \in A_h$ w.r.t. b_{h-1} (resp. b_h) and the majority threshold. If both $x_a - x'_a$ and $y_a - y'_a$ are positive, then object a is assigned to the correct category. In order to try to maximize the number of examples correctly assigned by the model, the objective function of the linear program minimizes the sum of x'_a and y'_a, i.e. the objective function is $\min \sum_{a \in A}(x'_a + y'_a)$.

The heuristic adjusting the profile also needs some adaptations in view of taking capacities into account. More precisely, the formal definition of the sets in which objects are classified for computing the candidate move evaluation should be adapted. The semantics of the sets, recalled in Sect. 3.2 remains identical. The formal definitions of these sets have to be adapted to take into account the capacity. The rest of the algorithm remains unchanged.

6 Experiments

The use of the MIP for learning a NCS model is limited because of the large number of binary variables involved. It contains more binary variables than the MIP learning the parameters of a simple additive MR-Sort model. Experiments reported in [3] have demonstrated that the computing time required to learn the parameters of a standard MR-Sort model having a small number of criteria and categories from a small set of assignment examples becomes quickly prohibitive. Therefore we cannot expect to be able to treat large problems using the MIP for learning NCS models.

In view of assessing the performance of the heuristic algorithm designed for learning the parameters of a NCS model, we use it to learn NCS models from several real data sets presented in Table 2. These data sets, available at http://www.uni-marburg.de/fb12/kebi/research/repository/monodata, have been already used to assess other algorithms (e.g. [4,23]). They involve from 120 to 1728 instances, from 4 to 8 monotone attributes and from 2 to 36 categories. In our experiments, categories have been binarized by thresholding at the median.

In our first experiment, we use 50 % of the alternatives in the data sets as learning set and the rest as test set. We learn MR-Sort and NCS models using both heuristics. We repeat this procedure for 100 random splits of the data sets

Table 2. Data sets

Data set	#Instances	#Attributes	#Categories
DBS	120	8	2
CPU	209	6	4
BCC	286	7	2
MPG	392	7	36
ESL	488	4	9
MMG	961	5	2
ERA	1000	4	4
LEV	1000	4	5
CEV	1728	6	4

Table 3. Average and standard deviation of the classification accuracy of the test set when using 50 % of the examples as learning set and the rest as test set

Data set	Heuristic MR-Sort	Heuristic NCS
DBS	0.8377 ± 0.0469	0.8312 ± 0.0502
CPU	0.9325 ± 0.0237	0.9313 ± 0.0272
BCC	0.7250 ± 0.0379	0.7328 ± 0.0345
MPG	0.8219 ± 0.0237	0.8180 ± 0.0247
ESL	0.8996 ± 0.0185	0.8970 ± 0.0173
MMG	0.8268 ± 0.0151	0.8335 ± 0.0138
ERA	0.7944 ± 0.0173	0.7944 ± 0.0156
LEV	0.8408 ± 0.0122	0.8508 ± 0.0188
CEV	0.9064 ± 0.0119	0.9118 ± 0.0263

Table 4. Average and standard deviation of the classification accuracy of the learning set when using the MR-Sort and NCS models learned on the whole data set

Data set	Heuristic MR-Sort	Heuristic NCS
DBS	0.9268 ± 0.0096	0.9326 ± 0.0087
CPU	0.9643 ± 0.0048	0.9703 ± 0.0091
BCC	0.7605 ± 0.0147	0.7761 ± 0.0085
MPG	0.8419 ± 0.0099	0.8389 ± 0.0069
ESL	0.9164 ± 0.0033	0.9168 ± 0.0042
MMG	0.8419 ± 0.0099	0.8409 ± 0.0091
ERA	0.8035 ± 0.0052	0.8027 ± 0.0053
LEV	0.8501 ± 0.0082	0.8643 ± 0.0038
CEV	0.9005 ± 0.0141	0.9172 ± 0.0101

in learning and test sets. We observe from Table 3 that the classification accuracy obtained with the NCS heuristic is on average comparable to the one obtained with the MR-Sort heuristic. The use of a more expressive model does not help much to improve the classification accuracy of the test set.

In a second experiment, we check the ability of MR-Sort and NCS to restore the whole data set. To do so, we run both heuristics 100 times. The average classification accuracy and standard deviation of the learning set are given in Table 4. The NCS heuristic does not always give better results than the MR-Sort one in restoring the learning set examples. Except for the MPG data set, we observe a slight advantage (of the order of one standard deviation) in favor of NCS when the number of attributes is at least 6. There is almost no difference for the data sets described by 4 or 5 attributes and for MPG (7 attributes).

Table 5. Average computing time (in seconds) required to find a solution with MR-Sort and NCS heuristics when using all the examples as learning set

Data set	Heuristic MR-Sort	Heuristic NCS
DBS	3.0508	6.9547
CPU	3.1646	5.2069
BCC	3.3700	7.7545
MPG	4.4136	9.9294
ESL	3.8466	7.2495
MMG	6.1481	13.4848
ERA	5.9689	14.4875
LEV	5.8986	13.2356
CEV	11.1122	31.7042

Average computing times of the results in Table 4 are displayed in Table 5. Learning a NCS model can take up to almost 3 times as much as learning a simple MR-Sort model.

The above experiments on benchmark data sets available in the literature failed to show a clear advantage at using NCS rather than MR-Sort. This raises the following question. Which type of data set would reveal a gain of expressivity provided by NCS over MR-Sort? We investigate this question in the next section.

7   Potential Gain in Descriptive Power with the NCS Model

Among NCS assignment rules, some can be exactly represented by additive weights and a threshold (the MR-Sort rules), while the others require a non-additive capacity and a threshold. We call the latter *non-additive* NCS rules. These are not MR-Sort rules but they can be *approximated* by a MR-Sort model. The experiment described below aims at assessing how well a non-additive NCS rule can be approximated by a MR-Sort rule.

Consider a NCS model assigning alternatives in two categories, C_1 and C_2. For a given profile, the set of all possible alternatives can be partitioned in 2^n subsets, where n is the number of criteria. Each of these subsets is characterized by one of the 2^n relative positions of an alternative w.r.t. the profile. On each criterion, the performance of an alternative is either at least as good as the profile or worse. Due to the ordinal nature of the NCS rule, all alternatives that share the same relative position w.r.t. the profile (i.e. all alternatives in the same class of the partition in 2^n subsets) are assigned to the same category. If we assume that the evaluations of the alternatives on all criteria range in the $[0, 1]$ interval, we can set the profile values to 0.5 on all criteria. The set of n-dimensional Boolean vectors is composed of exactly one example of each possible relative position w.r.t. the profile.

Our experiments are conducted as follows.

1. We modify the MIP described in Sect. 3.1 to learn only the weights and the majority threshold of a MR-Sort model on the basis of fixed profiles and assignment examples. The objective function of the MIP remains the minimization of the 0/1 loss.
2. We generate all possible NCS rules for $n = 4, 5, 6$ criteria. For more detail about how this can be done, see [24]; the list of all non-equivalent NCS rules is available at http://olivier.sobrie.be/shared/mbfs/. Each non-additive NCS rule, is used to assign the set of n-dimensional Boolean vectors to one of the two categories (using the 0.5 constant profile). These sets of representative alternatives constitute our learning sets.
3. The modified MIP is used to learn the weights and majority threshold of a MR-Sort model, which restores as well as possible the assignments made by the non-additive NCS rule.

The results of the experimentation are displayed in Table 6. Each row of the table contains the results for a given number of criteria, $n = 4, 5, 6$. The second column shows the percentage of non-additive NCS rules among all possible rules for each given number of criteria. The last three columns contain the min, max and average percentage of the 2^n examples assigned by non-additive rules that cannot be restored by a simple additive model.

Table 6. Average, minimum and maximum 0/1 loss of the learning sets after learning additive weights and the majority threshold of a MR-Sort model

n	% Non-additive	MR-Sort		
		Min.	Max.	Avg.
4	11 %	6.2 %	6.2 %	6.2 %
5	57 %	3.1 %	9.4 %	3.9 %
6	97 %	1.6 %	12.5 %	4.8 %

We observe that a MR-Sort model on 4 criteria is, in the worst case, not able to restore 6.2 % of the examples in the learning set (1 example out of 16). With 5 and 6 criteria, the maximum 0/1 loss increases respectively to 9.4 % (3 examples out of 32) and 12.5 % (8 examples out of 64).

Note that these proportions were obtained using learning sets in which each type of relative position w.r.t. the profile is represented exactly once. Therefore these conclusions should be valid for learning sets in which all types of relative positions are approximately equally represented. On a test set, the difference in classification performance between a non-additive NCS rule and its approximation by a MR-Sort rule can be amplified, or, on the contrary, can fade, depending on the proportion of the test alternatives belonging to the various types of relative positions w.r.t. the profile.

Table 6 reveals another important information. The proportion of non-additive NCS rules among all NCS rules quickly grows with the number of attributes: from 11 % of 2-additive NCS rules for $n = 4$ to 97 % for $n = 6$. It hence becomes more and more likely that a NCS rule is not a MR-Sort one when n grows.

The results in Table 6 could help to better understand the relatively poor gains observed in the previous section when comparing the heuristic algorithm for learning a 2-additive NCS model and a MR-Sort model. We noticed that the classification accuracy of the learned NCS rule tended to be slightly better for the data sets involving at least 6 attributes. The lack of an advantage for data sets involving 4 attributes might be due to the relative scarcity of non-additive NCS rules for $n = 4$ (11 %). When a gain is obtained, it is tiny, which might result from the fact that the approximation of a non-additive NCS rule by a MR-Sort rule is relatively good, at least up to $n = 6$. Investigating the NCS for $n \geq 7$ model in a systematic way, using the same method as we did in our last experiments, is almost impossible due to the extremely fast growth of the number of possible NCS rules (see [24]). It is however arguable that non-additive NCS rules could be at an advantage, as compared to MR-Sort rules, when the number of attributes is at least as large as 6.

A MIP Learning the Parameters of a MR-Sort Model

$$\max \sum_{a \in A} \gamma_a$$
s.t.
$$\sum_{j=1}^{n} c_{a,j}^{h-1} \geq \lambda + M(\gamma_a - 1) \ \forall a \in A_h, h = \{2,...,p\}$$
$$\sum_{j=1}^{n} c_{a,j}^{h} < \lambda - M(\gamma_a - 1) \ \forall a \in A_h, h = \{1,...,p-1\}$$

$$
\begin{aligned}
a_j - b_{l,j} &< M \cdot \delta_{a,j}^l &&\forall j \in F, \forall a \in A_h, \forall h \in P, l = \{h-1, h\} \\
a_j - b_{l,j} &\geq M(\delta_{a,j}^l - 1) &&\forall j \in F, \forall a \in A_h, \forall h \in P, l = \{h-1, h\} \\
c_{a,j}^l &\leq \delta_{a,j}^l &&\forall j \in F, \forall a \in A_h, \forall h \in P, l = \{h-1, h\} \\
c_{a,j}^l &\leq w_j &&\forall j \in F, \forall a \in A_h, \forall h \in P, l = \{h-1, h\} \quad (12) \\
c_{a,j}^l &\geq \delta_{a,j}^l - 1 + w_j &&\forall j \in F, \forall a \in A_h, \forall h \in P, l = \{h-1, h\} \\
b_{h,j} &\geq b_{h-1,j} &&\forall j \in F, h = \{2,...,p-1\}
\end{aligned}
$$

$$\sum_{j=1}^{n} w_j = 1$$

$$
\begin{aligned}
\delta_{a,j}^l &\in \{0,1\} &&\forall j \in F, \forall a \in A_h, \forall h \in P, l = \{h-1, h\} \\
c_{a,j}^l &\in [0,1] &&\forall j \in F, \forall a \in A_h, \forall h \in P, l = \{h-1, h\} \\
b_{h,j} &\in \mathbb{R} &&\forall j \in F, \forall h \in P \\
\gamma_a &\in \{0,1\} &&\forall a \in A \\
w_j &\in [0,1] &&\forall j \in F \\
\lambda &\in [0.5,1]
\end{aligned}
$$

B   MIP Learning the Parameters of a 2-Additive NCS Model

$$\max \sum_{a\in A} \gamma_a$$

s.t.

$$\sum_{j=1}^{n}\left(c_{a,j}^{h-1} + \sum_{k=1}^{j}\alpha_{a,j,k}^{h-1} - \sum_{k=1}^{j}\beta_{a,j,k}^{h-1}\right) \geq \lambda + M(\gamma_a - 1)\ \forall a \in A_h,$$

$$h = \{2,...,p\}$$

$$\sum_{j=1}^{n}\left(c_{a,j}^{h} + \sum_{k=1}^{j}\alpha_{a,j,k}^{h} - \sum_{k=1}^{j}\beta_{a,j,k}^{h}\right) < \lambda - M(\gamma_a - 1)\ \forall a \in A_h,$$

$$h = \{1,\cdots,p-1\}$$

$$m_j + \sum_{k\in J}(m_{j,k}^{+} - m_{j,k}^{-}) \geq 0 \qquad\qquad \forall j \in F, \forall J \subseteq F\backslash\{j\}$$

$$\sum_{j=1}^{n}m_j + \sum_{j=1}^{n}\sum_{k=1}^{j}(m_{j,k}^{+} - m_{j,k}^{-}) = 1$$

$$b_{h,j} \geq b_{h-1,j} \qquad \forall j \in F, h = \{2,...,p\}$$

$$c_{a,j}^{l} \leq \delta_{a,j}^{l} \qquad \forall j \in F, \forall a \in A_h, \forall h \in P, l = \{h-1,h\}$$

$$c_{a,j}^{l} \leq m_j \qquad \forall j \in F, \forall a \in A_h, \forall h \in P, l = \{h-1,h\}$$

$$c_{a,j}^{l} - m_j \geq \delta_{a,j}^{l} - 1 \qquad \forall j \in F, \forall a \in A_h, \forall h \in P, l = \{h-1,h\}$$

$$a_j - b_{l,j} < M\cdot\delta_{a,j}^{l} \qquad \forall j \in F, \forall a \in A_h, \forall h \in P, l = \{h-1,h\}$$

$$a_j - b_{l,j} \geq M(\delta_{a,j}^{l} - 1) \qquad \forall j \in F, \forall a \in A_h, \forall h \in P, l = \{h-1,h\}$$

$$\delta_{a,j}^{l} + \delta_{a,k}^{l} \geq 2\Delta_{a,j,k}^{l} \qquad \forall\{j,k\} \in F : k < j, \forall a \in A_h, \forall h \in P, l = \{h-1,h\}$$

$$\delta_{a,j}^{l} + \delta_{a,k}^{l} \leq \Delta_{a,j,k}^{l} + 1 \qquad \forall\{j,k\} \in F : k < j, \forall a \in A_h, \forall h \in P, l = \{h-1,h\}$$

$$\alpha_{a,j,k}^{l} \leq \Delta_{a,j,k}^{l} \qquad \forall\{j,k\} \in F : k < j, \forall a \in A_h, \forall h \in P, l = \{h-1,h\}$$

$$\alpha_{a,j,k}^{l} \leq m_{j,k}^{+} \qquad \forall\{j,k\} \in F : k < j, \forall a \in A_h, \forall h \in P, l = \{h-1,h\}$$

$$\alpha_{a,j,k}^{l} + m_{j,k}^{+} \geq \Delta_{a,j,k}^{l} - 1 \qquad \forall\{j,k\} \in F : k < j, \forall a \in A_h, \forall h \in P, l = \{h-1,h\}$$

$$\beta_{a,j,k}^{l} \leq \Delta_{a,j,k}^{l} \qquad \forall\{j,k\} \in F : k < j, \forall a \in A_h, \forall h \in P, l = \{h-1,h\}$$

$$\beta_{a,j,k}^{l} \leq m_{j,k}^{-} \qquad \forall\{j,k\} \in F : k < j, \forall a \in A_h, \forall h \in P, l = \{h-1,h\}$$

$$\beta_{a,j,k}^{l} - m_{j,k}^{-} \geq \Delta_{a,j,k}^{l} - 1 \qquad \forall\{j,k\} \in F : k < j, \forall a \in A_h, \forall h \in P, l = \{h-1,h\}$$

$$c_{a,j}^{l} \in [0,1] \qquad \forall j \in F, \forall a \in A_h, \forall h \in P, l = \{h-1,h\}$$

$$\delta_{a,j}^{l} \in \{0,1\} \qquad \forall j \in F, \forall a \in A_h, \forall h \in P, l = \{h-1,h\}$$

$$\alpha_{a,j,k}^{l}, \beta_{a,j,k}^{l} \in [0,1] \qquad \forall\{j,k\} \in F : k < j, \forall a \in A_h, \forall h \in P, l = \{h-1,h\}$$

$$\Delta_{a,j,k}^{l} \in \{0,1\} \qquad \forall\{j,k\} \in F : k < j, \forall a \in A_h, \forall h \in P, l = \{h-1,h\}$$

$$m_j \in [0,1] \qquad \forall j \in F$$

$$m_{j,k}^{+}, m_{j,k}^{-} \in [0,1] \qquad \forall j \in F, \forall k \in F, k < j$$

$$b_{h,j} \in \mathbb{R} \qquad \forall j \in F, \forall h \in P$$

$$\gamma_a \in \{0,1\} \qquad \forall a \in A$$

$$\lambda \in [0,1]$$

$$(13)$$

References

1. Bouyssou, D., Marchant, T.: An axiomatic approach to noncompensatory sorting methods in MCDM, I: the case of two categories. Eur. J. Oper. Res. **178**(1), 217–245 (2007)
2. Bouyssou, D., Marchant, T.: An axiomatic approach to noncompensatory sorting methods in MCDM, II: more than two categories. Eur. J. Oper. Res. **178**(1), 246–276 (2007)
3. Leroy, A., Mousseau, V., Pirlot, M.: Learning the parameters of a multiple criteria sorting method based on a majority rule. In: Brafman, R. (ed.) ADT 2011. LNCS, vol. 6992, pp. 219–233. Springer, Heidelberg (2011)
4. Sobrie, O., Mousseau, V., Pirlot, M.: Learning a majority rule model from large sets of assignment examples. In: Perny, P., Pirlot, M., Tsoukiàs, A. (eds.) ADT 2013. LNCS, vol. 8176, pp. 336–350. Springer, Heidelberg (2013)
5. Jacquet-Lagrèze, E., Siskos, Y.: Assessing a set of additive utility functions for multicriteria decision making: the UTA method. Eur. J. Oper. Res. **10**, 151–164 (1982)
6. Doumpos, M., Zopounidis, C.: Multicriteria Decision Aid Classification Methods. Kluwer Academic Publishers, Dordrecht (2002)
7. Fishburn, P.C.: Noncompensatory preferences. Synth. **33**(1), 393–403 (1976)
8. Bouyssou, D.: Some remarks on the notion of compensation in MCDM. Eur. J. Oper. Res. **26**, 150–160 (1986)
9. Grabisch, M.: The application of fuzzy integrals in multicriteria decision making. Eur. J. Oper. Res. **89**(3), 445–456 (1996)
10. Yu, W.: Aide multicritère à la décision dans le cadre de la problématique du tri: méthodes et applications. Ph.D. thesis, LAMSADE, Université Paris Dauphine, Paris (1992)
11. Roy, B., Bouyssou, D.: Aide multicritère à la décision: méthodes et cas. Economica, Paris (1993)
12. Bouyssou, D., Pirlot, M.: A characterization of concordance relations. Eur. J. Oper. Res. **167**(2), 427–443 (2005)
13. Bouyssou, D., Pirlot, M.: Further results on concordance relations. Eur. J. Oper. Res. **181**, 505–514 (2007)
14. Chateauneuf, A., Jaffray, J.: Derivation of some results on monotone capacities by Möbius inversion. In: Bouchon-Meunier, B., Yager, R.R. (eds.) IPMU 1986. Lecture Notes in Computer Science, vol. 286, pp. 95–102. Springer, Heidelberg (1986)
15. Mousseau, V., Słowiński, R.: Inferring an ELECTRE TRI model from assignment examples. J. Global Optim. **12**(1), 157–174 (1998)
16. Mousseau, V., Figueira, J., Naux, J.P.: Using assignment examples to infer weights for ELECTRE TRI method: some experimental results. Eur. J. Oper. Res. **130**(1), 263–275 (2001)
17. The, A.N., Mousseau, V.: Using assignment examples to infer category limits for the ELECTRE TRI method. J. Multi-criteria Decis. Anal. **11**(1), 29–43 (2002)
18. Dias, L., Mousseau, V., Figueira, J., Clímaco, J.: An aggregation/disaggregation approach to obtain robust conclusions with ELECTRE TRI. Eur. J. Oper. Res. **138**(1), 332–348 (2002)
19. Doumpos, M., Marinakis, Y., Marinaki, M., Zopounidis, C.: An evolutionary approach to construction of outranking models for multicriteria classification: the case of the ELECTRE TRI method. Eur. J. Oper. Res. **199**(2), 496–505 (2009)

20. Cailloux, O., Meyer, P., Mousseau, V.: Eliciting ELECTRE TRI category limits for a group of decision makers. Eur. J. Oper. Res. **223**(1), 133–140 (2012)
21. Zheng, J., Metchebon, S., Mousseau, V., Pirlot, M.: Learning criteria weights of an optimistic Electre Tri sorting rule. Comput. OR **49**, 28–40 (2014)
22. Hüllermeier, E., Tehrani, A.F.: Efficient learning of classifiers based on the 2-additive choquet integral. In: Moewes, C., Nürnberger, A. (eds.) Computational Intelligence in Intelligent Data Analysis. SCI, vol. 445, pp. 17–29. Springer, Heidelberg (2013)
23. Tehrani, A.F., Cheng, W., Dembczynski, K., Hüllermeier, E.: Learning monotone nonlinear models using the choquet integral. Mach. Learn. **89**(1–2), 183–211 (2012)
24. Ersek Uyanık, E., Sobrie, O., Mousseau, V., Pirlot, M.: Listing the families of sufficient coalitions of criteria involved in sorting procedures. In: DA2PL 2014 Workshop From Multiple Criteria Decision Aid to Preference Learning, pp. 60–70, Paris, France (2014)

k-Agent Sufficiency for Multiagent Stochastic Physical Search Problems

Daniel S. Brown[(✉)], Steven Loscalzo, and Nathaniel Gemelli

Air Force Information Directorate, Rome, NY 13441, USA
{daniel.brown.81,steven.loscalzo,nathaniel.gemelli}@us.af.mil

Abstract. In many multi-agent applications, such as patrol, shopping, or mining, a group of agents must use limited resources to successfully accomplish a task possibly available at several distinct sites. We investigate problems where agents must expend resources (e.g. battery power) to both travel between sites and to accomplish the task at a site, and where agents only have probabilistic knowledge about the availability and cost of accomplishing the task at any location. Previous research on Multiagent Stochastic Physical Search (mSPS) has only explored the case when sites are located along a path, and has not investigated the minimal number of agents required for an optimal solution. We extend previous work by exploring physical search problems on both paths and in 2-dimensional Euclidean space. Additionally, we allow the number of agents to be part of the optimization. Often, research into multiagent systems ignores the question of how many agents should actually be used to solve a problem. To investigate this question, we introduce the condition of k-agent sufficiency for a multiagent optimization problem, which means that an optimal solution exists that requires only k agents. We show that mSPS along a path with a single starting location is at most 2-agent sufficient, and quite often 1-agent sufficient. Using an optimal branch-and-bound algorithm, we also show that even in Euclidean space, optimal solutions are often only 2- or 3-agent sufficient on average.

Keywords: Stochastic physical search · Planning under uncertainty · Multiagent optimization · k-Agent sufficiency

1 Introduction

We investigate the problem of multiple agents seeking for a single item that may possibly be obtained at one of several locations. We assume that the availability and actual cost to acquire the item at any site is not fully known beforehand, but that a priori probabilistic cost distributions are known. In particular we examine problems where there is a finite resource that must be expended to both travel and obtain the item of interest. We refer to this class of problems as Multiagent Stochastic Physical Search (mSPS). Examples of this type of problem include battery-powered mining or space exploration robots seeking a precious metal deposit or specific mineral sample, hikers seeking a suitable location to

© Springer International Publishing Switzerland 2015
T. Walsh (Ed.): ADT 2015, LNAI 9346, pp. 171–186, 2015.
DOI: 10.1007/978-3-319-23114-3_11

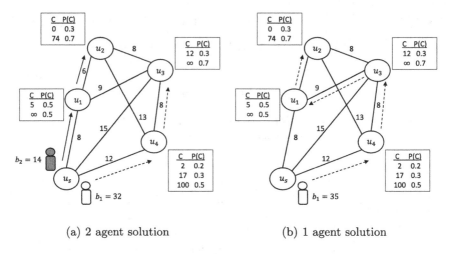

(a) 2 agent solution   (b) 1 agent solution

Fig. 1. (a) Example of a two-agent strategy where agent 1 is allocated a starting budget of $b_1 = 32$ and the path $\pi^1 = \langle u_s, u_4, u_3 \rangle$ and agent 2 is allocated a starting budget of 14 and the path $\pi^2 = \langle u_s, u_1, u_2 \rangle$. Probability distributions over cost are shown next to each site. The joint probability of success of the shown multiagent search strategy is $1 - Pr(\text{failure of agent 1}) \cdot Pr(\text{failure of agent 2}) = 1 - Pr(\text{failure at } u_4) \cdot Pr(\text{failure at } u_3) \cdot Pr(\text{failure at } u_1) \cdot Pr(\text{failure at } u_2) = 1 - 0.5 \cdot 0.7 \cdot 0.7 \cdot 0.5 = 0.8775$. (b) Using a single agent, the same probability of success can be achieved with a strategy that requires less budget.

set up a base camp, or tourists using public transportation to explore several different shopping areas for a desired souvenir. What makes the above problems challenging, is that while actual distance costs may be reliably and accurately estimated using satellite imagery, maps, or taxi fares, the actual cost to accomplish the task (or purchase the item of interest) at a specific location may be unknown until an agent actually visits the site. One of the major challenges is that we assume each agent must use a single budget (e.g. battery power, fuel, or currency) to both travel and obtain the item. This adds extra complexity to the problem because it means that taking a different path to a site can change the probability of success—a longer path will consume more budget, reducing the budget available to obtain the item. Figure 1(a) shows an example problem along with a possible two-agent solution that allocates a total budget of 46 and achieves a joint probability of success of 0.8775. However, Fig. 1(b) shows that an equivalent probability of success can be achieved using only a single agent with a lower total budget of 35.

In many multiagent search problems it is often assumed that the number of agents is fixed and that having multiple searchers is better than a single searcher. However, this may not be the case when searchers start from the same location, when both search and acquisition are costly, and when there is a limit to the total allocatable budget. For example, in many vehicle routing problems, the number of vehicles is part of the problem definition and solutions often assume that all the vehicles will be used [14]. Some vehicle routing problems try to minimize the number of required agents, but these problems do not consider probabilities of success or costs to purchase or acquire an item along a vehicle's route [15]. Work on the multiple traveling salesman (a special case of the mSPS) and its extensions sometimes allows the number of salesman to be variable, but we know of no proven bounds on the maximum number of agents required for different problems [2,11]. In many other multiagent scenarios such as multiagent task allocation, the number of agents is also often assumed to be fixed [5,12]. Previous work on stochastic physical search has either focused on single agent solutions or has assumed that the number of agents is not part of the optimization [1,3,7,9,10].

Our research assumes that there is a given upper bound on the number of agents available to search for the item; however, we do not restrict our solutions to a fixed number of agents. Instead, we algorithmically decide which of the available agents should participate in the search in order to maximize the probability of successfully obtaining the item as well as minimizing the required total budget allocated to the agents. Because our problem is bi-objective we use the standard epsilon constraint method to split the problem into two dual objectives. The *Min-Budget* objective is to minimize the total budget allocated to the agents while guaranteeing a specified minimum probability of success. The *Max-Probability* objective is to maximize the probability of success given an upper bound on the budget that can be allocated. In both of these problems, the agents all start at the same starting location and a solution is an allocation of resources to each agent, along with a search path for each agent.

While stochastic physical search problems capture many real world planning and algorithmic decision problems, very little is known about the solution properties of these problems. Work by Aumann et al. and Hazon et al. has proposed optimal algorithms for the case when sites are located along a path; however, their work never actually computes or analyzes optimal solutions and does not investigate the frequency of optimal solutions that required more than one agent [1,6,7]. Work by Brown et al. and Hudack et al. has examined the Min-Budget and Max-Probability problems on general graphs and 2-dimensional euclidean spaces; however, they only consider the single agent case [3,9]. This paper provides, to the best of our knowledge, the first theoretical and empirical investigation of the solution properties of the multiagent Min-Budget and Max-Probability problems.

We start by examining problems where the item may possibly be obtained from a set of locations on a path. We examine two cases: (1) single-cost and (2) multi-cost. We prove that in both cases, problems are at most 2-agent sufficient, and empirically investigate the frequency of 1-agent sufficient problems. We next investigate solutions to the Min-Budget and Max-Probability problems when locations are in 2-dimensional Euclidean space. We provide a theoretical analysis of when multiple agents are unnecessary and use an exact branch-and-bound algorithm to provide empirical insights into k-agent sufficiency for 2-dimensional problems. We show that in many cases, even when searching in two dimensions, the optimal strategy is to use a very small number of searchers, rarely requiring more than 3 searchers. We conclude by discussing the factors that contribute to a search problem being k-agent sufficient for different values of k along with areas for future research.

2   Problem Definition

A stochastic physical search problem is defined by a graph $G(S^+, E)$ with a set of locations $S^+ = S \cup \{u_s\}$ where $S = \{u_1, \ldots, u_m\}$ is the set of m sites offering an item of interest, u_s is the starting location, and $E \subseteq S^+ \times S^+$ is the set of edges. Each $(i, j) \in E$ has a non-negative cost of travel t_{ij}. For each site $i \in S$ we are given a cost probability mass function $p_i(c)$, which gives the probability that the item will cost c at site u_i. We assume that the actual cost is not revealed until the agent visits the site and that the cost remains fixed thereafter. We further assume that there is a finite number of possible costs in the support of $p_i(c)$, $\forall i \in S$. Finally, we define a set $\mathcal{N}$ of $n^{\max} = |\mathcal{N}|$ agents that are available, but not required, to be used in the search. Each agent n starts at u_s with a starting budget $b_n^*, \forall n \in \mathcal{N}$. We let $b^* = \{b_n^* : n \in N\}$ and let $B^* = \sum_{n \in N} b_n^*$, the total budget allocated to all agents. We assume that once budgets are allocated they are non-transferable, that two or more agents cannot combine their budgets to obtain the item, and that agents cannot share information about sites they have visited with other agents. Following previous work on stochastic physical search problems [7], we assume that success is achieved if any agent is able to purchase the item. We also assume that the item cannot be found at the start site, u_s.

We examine two dual problems (1) Min-Budget: Given a required probability of success p_{succ}^* find the initial budget allocation b^* that satisfies p_{succ}^* and that minimizes B^*. (2) Max-Probability: Given an upper bound on the budget available for allocation of B^* determine the optimal budget allocation b^* so as to maximize the probability of success. A solution to either problem is an allocation of starting budgets $b*$ along with a set of paths Π^* where each individual path $\pi^n \in \Pi^*$ is a sequence of sites in S^+, where π_i^n is the ith site visited along path π^n, and where each path starts at the start site, i.e., $\pi_0^n = u_s, \forall n \in \mathcal{N}$. We assume that success is achieved if any agent arrives at a site where the actual cost is less than or equal to that agent's remaining budget.

2.1 k-Agent Sufficiency

Before we investigate solutions to the Min-Budget and Max-Probability mSPS problems we define the term k-agent sufficiency as it relates to multiagent optimization problems of the kind investigated in this paper.

Definition 1. *A mSPS problem is k-agent sufficient if an optimal solution exists such that $|B^+| = k$ where $B^+ = \{b_n^* \in b^* \mid b_n^* > 0, \forall n \in \mathcal{N}\}$.*

The following result is true for all mSPS problems.

Proposition 1. *If an SPS problem has zero travel costs between all sites, then it is 1-agent sufficient.*

Proof. Assume you have two agents i and j with starting budgets $b_i, b_j > 0$. Since the agents can travel between sites without incurring costs, an equivalent probability of success can be achieved with a single agent, given less starting budget $b' = \max(b_i, b_j) \leq b_i + b_j$. $\qquad\qquad\qquad\qquad\qquad\qquad\square$

Thus, for the remainder of the paper we assume that all travel costs are non-zero.

3 mSPS Along a Path

We first investigate the Min-Budget and Max-Probability problems where the set of locations in S^+ are restricted to be along a path. We note that the following discussion on paths is not purely academic, as many multi-agent coverage algorithms convert their complex environment into a path and many perimeter monitoring and border control tasks could also be represented by sites along a path [4,7,8,13].

To simplify our analysis, we follow the methodology used by Hazon et al. [7], and assume WLOG (without loss of generality) that all locations are along a line such that the travel cost between any two sites u_i and u_j is $t_{ij} = |u_i - u_j|$. We also assume WLOG that the sites are ordered from left to right such that $u_1 \leq u_2 \leq \cdots \leq u_m$. We first examine the case when there is only one possible cost to obtain the item. Despite the simplicity of this problem, we show that the results for k-sufficiency are non-trivial. We then examine the case where there are multiple possible item costs.

Before examining the single and multi-price cases, we note the following.

Proposition 2. *When u_s is the leftmost (rightmost) location, then the problem is 1-agent sufficient and the optimal strategy only moves to the right (left).*

Proof. Any other strategy to cover the same locations would use at least as much budget and achieve no greater probability of success. $\qquad\qquad\qquad\square$

Thus, the most interesting cases, in terms of k-agent sufficiency, are those where the start site u_s is towards the middle of the path.

3.1 Single Price

We first assume that all sites either offer the item for a cost of c_0 or do not offer the item at all (this can simply be modeled as a cost of ∞). All we are given are the a priori probabilities p_i that the item is available for cost c_0 at site i.

We first note the following useful lemma and definition proposed by Aumann et al. [1].

Lemma 1. *Consider a price c_0 and suppose that an agent's optimal strategy starting at point u_s covers the interval $[u_\ell, u_r]$ while the remaining budget is at least c_0. Then WLOG we may assume that the agent's optimal strategy is either $(u_s \rightarrowtail u_r \rightarrowtail u_\ell)$ or $(u_s \rightarrowtail u_\ell \rightarrowtail u_r)$.*

Definition 2. *Agents i and j are said to be* separated *by a strategy if each site in S that is reached by i is not reached by j.*

We now prove the following lemma and theorem which give us our first k-agent sufficiency condition.

Lemma 2. *With a single price, there exists an optimal multiagent strategy where every agent is separated.*

Proof. Assume by contradiction that two agents i and j are not separated in every optimal strategy. Consider the intervals covered by these two agents, $[l_i, r_i]$ and $[l_j, r_j]$, respectively. Let $[L, R] = [l_i, r_i] \cup [l_j, r_j]$ be the full combined coverage area of the two agents. In the case that $[l_i, r_i] \subset [l_j, r_j]$ we can safely remove agent i from the strategy, resulting in a strategy with the same probability of success but lower budget. Otherwise, WLOG we can assume based on Lemma 1 that only i reaches L and only j reaches R. However, now the separated strategy of $u_s \rightarrowtail L$ for agent i and $u_s \rightarrowtail R$ for j guarantees at least the same probability of success with no more budget. This contradicts our assumption.   □

Theorem 1. *For a path with a single possible price, there is always an optimal strategy with fewer than 3 agents. If using two agents is optimal, then only one agent moves left and only one agent moves right in the optimal strategy.*

Proof. This follows as a direct result of Lemmas 1 and 2.   □

The work of Aumann et al. [1] provides an $O(m)$ algorithm for the single agent single price problem. Based on the result of Theorem 1 we can easily adapt the algorithm given by Aumann et al. to obtain an $O(m)$ algorithm for the multiagent single item cost Min-Budget and Max-Probability cases by simply checking each possible single agent coverage region to see if dividing the region between two agents results in lower budget or higher probability of success. Each of these checks can be done in constant time.

3.2   Single Price k-Agent Sufficiency

We now examine when the single price problem is 1-agent sufficient. When there are multiple agents, each one has to carry at least c_0 of budget to enable purchasing when the item is available. Thus the question of 1-agent sufficiency is directly related to the ratio of travel distances and c_0. We note that for increasing values of c_0, there exists a point at which c_0 is so high that it dominates the total travel cost.

Theorem 2. *Suppose that the optimal strategy covers the interval $[u_\ell, u_r]$ while the remaining budget is at least c_0. If $c_0 \geq \max(|u_s - u_\ell|, |u_r - u_s|)$ then the problem is 1-agent sufficient.*

Proof. Assume by contradiction that the optimal solution requires two agents, i and j. Let b_i and b_j be the starting budgets of i and j, where $t_i = b_i - c_0$ and $t_j = b_j - c_0$ are the portions of the budgets allocated for travel. By assumption and by Theorem 1 we have

$$c_0 \geq \max(|u_s - u_\ell|, |u_r - u_s|) = \max(t_i, t_j). \tag{1}$$

For a single searcher to cover both search paths it must have as a minimum travel budget

$$t' = 2\min(t_i, t_j) + \max(t_i, t_j) \tag{2}$$

to enable an out and back trip on the shorter leg, followed by an out trip on the longer leg. Thus the budget b' for a single agent is given by

$$b' = t' + c_0 \tag{3}$$
$$= 2\min(t_i, t_j) + \max(t_i, t_j) + c_0 \tag{4}$$
$$\leq t_i + t_j + \max(t_i, t_j) + c_0 \tag{5}$$
$$\leq t_i + t_j + 2c_0 \tag{6}$$
$$= b_i + b_j \tag{7}$$

Which contradicts our assumption that two agents were required.   □

Figure 2 shows an example of how the number of agents allocated changes for both Min-Budget and Max-Probability as c_0 is increased. To obtain these results we uniform randomly generated 25 sites along a 100 unit long interval and let u_s be the median site. Probabilities of success p_i are randomly chosen between 0 and 0.5 for each site. For the Min-Budget problem, we examined several different values for the required probability of success. We know that the best success probability is achieved by visiting all the sites, giving

$$p_{succ}^{max} = 1 - \prod_{i \in S} 1 - p_i. \tag{8}$$

To vary the solutions we set the required probability of success equal to $\rho \cdot p_{succ}^{max}$ for different values of ρ. The results for Min-Budget are shown in Fig. 2(a). The x-axis shows the cost of the item and the y-axis shows the percentage of 1000 random instances that were 1-agent sufficient. When $\rho = 1$ all the sites must be visited. We see that as expected, when $\rho = 1$, most problems are not 1-agent sufficient, until the cost gets close to $\max(|u_s - u_1|, |u_m - u_s|) \approx 50$, when 1-agent sufficiency is guaranteed by Theorem 2. However, as soon as all of the sites are not required (i.e. $\rho < 1$), the probability of an instance being 1-agent sufficient dramatically increases.

Figure 2(b) shows the results for Max-Probability. Because there is only a single purchase cost we can think of this cost as a fixed start-up cost that is incurred for each agent used. Thus, for low values of c_0 and low starting budget, it is more beneficial to divide and conquer and send one agent left and one agent right. On the other hand, for large values of c_0 the start-up cost to use a second

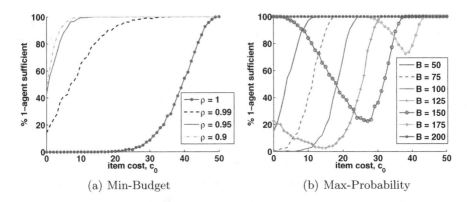

Fig. 2. Percentage of 1000 random 25-site, single-price problems that are 1-agent sufficient. Sites are randomly placed in the interval $[0, 100]$. (a) Min-Budget results for different required probabilities of success $p^*_{succ} = \rho \cdot (1 - \prod_{i \in S} 1 - p_i)$ where p_i is the probability the item is available for the single cost. (b) Max-Probability results for different total budget allotments B.

agent is so high that most optimal solutions only require a single agent. Note that given a starting budget of 100 with $c_0 = 0$ the optimal solution is to always use two agents with each agent coverage one half of the solution space. However, as c_0 increases the two agent solution coverage region decreases because agents cannot reach the farther endpoints and still have enough to purchase the item. Eventually, c_0 is so high that giving two agents c_0 plus budget to travel requires more budget than using a single agent. Given a starting budget of 150 and $c_0 = 0$ one agent has enough to traverse the entire interval. In this case we see that as c_0 increases, eventually two agents can cover a larger region (resulting in higher probability of success) than one agent on its own. However, past a certain point the start-up cost of c_0 starts to dominate the travel costs and single agent solutions become more common.

3.3 Multiple Prices on a Path

We now consider the case where there can be a large number of different realizable costs at the sites. Note that Lemma 1 is not always true for multiple purchase prices. Figure 3 shows a simple example where multiple direction changes are optimal. However, we do have the following result:

Lemma 3. *In the multi-price case, if agent i and agent j both only travel left (right), then one of them is unnecessary.*

Proof. WLOG assume agent i travels past the leftmost (rightmost) point visited by j. Let b_i and b_j be the starting budgets of i and j and let $t_i = |l_i - u_s|$ and $t_j = |l_j - u_s|$ be the distances traveled by each agent, respectively. Consider the

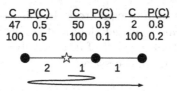

Fig. 3. When there are multiple realizable costs, optimal paths can often include several changes of direction. Shown is the optimal strategy for a single agent solving Max-Probability with starting budget of 51, or equivalently solving Min-Budget with a required probability of success equal to 0.99.

new strategy where only one agent i' travels with budget

$$b_{i'} = t_i + (b_i - t_i + b_j - t_j) \tag{9}$$
$$= b_i + b_j - t_j \tag{10}$$
$$\leq b_i + b_j. \tag{11}$$

This single-agent strategy guarantees no decrease in probability of success with no increase in total budget.   □

We also utilize the following lemma, adapted from Aumann et al. [1].

Lemma 4. *Consider two agents i and j that start at u_s. Then, there is an optimal strategy such that one of the following holds:*

- *j moves only in one direction which is opposite to i's final movement direction. Furthermore, if i's final movement direction is right (left) then j passes the leftmost (rightmost) site that is reached by i.*
- *either i or j is unnecessary.*

Using Lemmas 3 and 4, we can now prove the following theorem, which is the multi-cost analogue of Theorem 1.

Theorem 3. *For the same-start mSPS problem on a path, there is always an optimal strategy with fewer than 3 agents.*

Proof. Assume by contradiction that 3 agents are necessary in every optimal strategy. Denote these agents i, j, and k. Consider agents i and j. WLOG by Lemma 4 assume that i only moves left in the optimal solution, i passes the leftmost site reached by j, and j's final movement direction is right.

Now consider the results of Lemma 4 applied to agents j and k. There are two cases: (1) Agent k only moves left and passes the leftmost site reached by j. Then either $[l_i, r_i] \subset [l_k, r_k]$ or $[l_k, r_k] \subset [l_i, r_i]$. In either case, by Lemma 3 one of the agents is unnecessary. (2) Agent j only moves right and passes the rightmost site reached by agent k, and k's final movement direction is left. If k passes the leftmost point covered by i, then i is unnecessary by Lemma 3. Otherwise, i passes the leftmost site visited by k. Consider two cases: (a) at l_k, i

has remaining budget less than or equal to k. In this case i is not needed and k can travel to l_i. (b) i has more budget than k at l_k. In this case i also has more budget available than k at all sites left of u_s so k only has to travel right. Thus, by Lemma 3 either j or k are unnecessary.   □

3.4   Multi-price k-Agent Sufficiency

We now investigate when the multi-price mSPS problem along a path is 1-agent sufficient. Once again we have the obvious cases that if travel costs are all zero, then the problem is 1-agent sufficient and all the sites are located to one side of the start site. We also have the following multi-price analogue of Theorem 2.

Theorem 4. *Consider a strategy that has two agents with paths π^i and π^j and budgets b_i and b_j, respectively. WLOG let $b_i = t_i + c_i^{pur}$ and $b_j = t_j + c_j^{pur}$ where t_i is the budget needed to travel along π_i and c_i^{pur} is the remainder that is allocated to purchase. If $\max(t_i, t_j) \leq \min(c_i^{pur}, c_j^{pur})$, then the problem is 1-agent sufficient.*

The proof is almost identical to the proof of Theorem 2.

Figure 4 shows an example of how the number of agents allocated changes for Min-Budget and Max-Probability over different cost profiles for multiple costs along a 100 unit path with 10 sites. Unlike the single-price case, the multi-price problem appears to be NP-Hard. Aumann et al. examine the case where all agents have access to a shared budget and show that even this case is NP-Hard [1]. In this paper, we assume that a distinct, non-sharable, initial budget must be allocated to each agent for both the Max-Probability and Min-Budget problems, adding another dimension of complexity to the problem. However, we were able to use a simplified version of the branch-and-bound algorithm described in Sect. 4 to obtain exact results for smaller sized problems.

Similar to the previous empirical results, we see that as the item costs increase, 1-agent sufficient solutions become more common for the Min-Budget case, but as ρ is increased (i.e. more sites are required to be visited) solutions tend towards two agents unless item costs dominate travel costs. For Max-Probability we see that very small or very large starting budgets lead to solutions with fewer agents, but there is always a dip between the low and high budgets where it becomes more beneficial to use two agents. The scaling and location of this dip is determined by the distribution over item costs, with higher item costs penalizing multiagent solutions.

4   mSPS in 2-D Euclidean Space

The results above have all assumed that the sites are located along a simple path. We now assume that sites are located in 2-dimensional Euclidean space, where the cost to travel between sites is the euclidean distance between sites, i.e. $t_{ij} = \|u_i - u_j\|_2$. When solving both the Min-Budget and Max-Probability problems in Euclidean space, we have the following result:

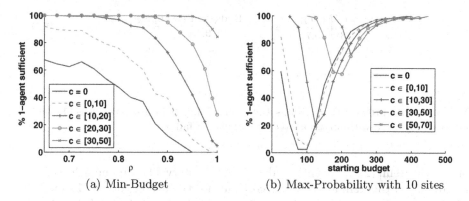

(a) Min-Budget   (b) Max-Probability with 10 sites

Fig. 4. Percentage of 200 random 10-site, single-price problems that are 1-agent sufficient where the item costs are either all zero or uniform randomly chosen from some interval. (a) Min-Budget results over different required probabilities of success $p^*_{succ} = \rho \cdot (1 - \prod_{i=1}^{m} 1 - p_i)$ where p_i is the probability the item is available for the single cost at site i. (b) Max-Probability results over total starting budgets.

Proposition 3. *The optimal solution for both the Min-Budget and Max-Probability Euclidean mSPS problems consists of distinct non-overlapping paths.*

Proof. We prove this by contradiction. Assume in the optimal solution that there is a site $u_k \in S$ that is visited by at least two agents a_1 and a_2. Because agents cannot share budget to purchase an item, the probability of success obtained at that site will remain unchanged if the agent with lower budget does not travel to u_k. WLOG assume agent a_1 is thus chosen not to visit the site and instead goes straight to its next site on its path. By the triangle inequality this path length is less than or equal to the original path length resulting in a strategy with no more budget and at least equivalent probability of success, resulting in the desired contradiction. ☐

Thus, for the Euclidean problem, we can safely exclude overlapping paths from our search space. This separation principle allows us to find optimal solutions to the mSPS problem using an extension of the branch-and-bound formulation proposed by Brown et al. for solving the single-agent SPS problem on general graphs [3]. Any graph more complex than a simple path has been shown to be NP-complete [7]; however, we were able to find optimal solutions for small problems up to about 20 sites in 2-d Euclidean space. This allows us to examine the characteristics of optimal solutions to the mSPS problem in a more general and applicable setting.

We note that for each site $i \in S$ we can use the possible costs at i to form a partition over all possible budgets that may be brought to site i. For example, the cost profile shown in Fig. 5 partitions the budget space into three intervals with the corresponding probabilities of failure if an agent arrives at that site with any budget in that interval. There are three possibilities when the agent arrives at the site: (1) the agent's budget is in the interval $[0, 3)$ and it cannot

C	P(C)
3	0.7
10	0.3

(a)

Budget interval	p_{fail}
$[0, 3)$	1.0
$[3, 10)$	0.3
$[10, \infty)$	0

(b)

Fig. 5. A cost profile (a) partitions the interval $[0, \infty)$ into several possible budget intervals (b), each with an associated probability of failure.

obtain the item, (2) the agent's budget is in the interval $[3, 10)$ and it has enough left to obtain the item at the lower cost but not the higher cost, and so will fail to obtain the item with probability 0.3, and (3) the agent has sufficient budget to obtain the item for any of the possible costs.

Both the Min-Budget and Max-Probability branch-and-bound algorithms need to determine the optimal budget allocation over $n^{\max}$ agents. While the number of all possible budget allocations is infinite, we can ignore most of these intervals and only focus on each budget interval $[c_\ell, c_u)$ induced by the possible costs at each site. Thus rather than branching on individual budget values we branch on possible budget intervals for each agent. This can still result in an exponential number of branches, but does allow exact solutions. We refer the reader to [3] for the full details of the successor function and bounding criteria for the single agent case. To extend the work by Brown et al. to the multiagent case, we simply added a budget interval for each available agent. The successors for each state are found as follows: iterate over all unvisited sites and all available agents; add the site to the agent's path; and update the agent's budget interval, the total budget required for all agents, and the joint probability of failure.

5 k-Agent Sufficiency in 2-Dimensions

We introduce the following definition that allows us to characterize a certain class of 1-agent sufficient mSPS problems for the 2-dimensional Euclidean case

Definition 3. *A mSPS problem is purchase-dominated if*

$$\min\{c : P_i(c) > 0, i \in S\} > \max_{i \in S^+, j \in S^+} t_{ij}, \tag{12}$$

i.e., the minimum purchase cost at any site is greater than the maximum travel cost between any two sites.

Theorem 5. *If a mSPS problem in Euclidean space is purchase-dominated, then it is 1-agent sufficient.*

Proof. WLOG, assume that we have an optimal solution that requires two agents i and j with paths π^i, π^j of corresponding travel costs t_i and t_j. Additionally, each agent may need some additional budget to use for purchasing, c_i^{pur} and

c_j^{pur}. We show that we can achieve the same probability of success using a single agent. To do this assume that a single agent first visits all sites in π^i and then visits all sites (except for the start site) in π^j with a corresponding total path cost equal to $t_i + t_j - t_{u_s, \pi_1^j} + t_{\pi_\omega^i, \pi_1^j}$, where ω is the index to the last element of the path. Additionally, this agent may need some extra budget to allow for purchasing. This agent will need c_i^{pur} to get the probability of failure p_{fail}^i on path π^i. The agent also needs $\min(0, c_j^{pur} - c_i^{pur})$ to get the probability of failure p_{fail}^j on path π^j. Thus the single agent case requires

$$b' = t_i + t_j - t_{u_s, \pi_1^j} + t_{\pi_\omega^i, \pi_1^j} + c_i^{pur} + \min(0, c_j^{pur} - c_i^{pur}) \qquad (13)$$

$$\leq t_i + t_j + t_{\pi_\omega^i, \pi_1^j} + \max(c_i^{pur}, c_j^{pur}) \qquad (14)$$

$$< t_i + t_j + \min(c_i^{pur}, c_j^{pur}) + \max(c_i^{pur}, c_j^{pur}) \qquad (15)$$

$$= t_i + t_j + c_i^{pur} + c_j^{pur} \qquad (16)$$

$$= B^* \qquad (17)$$

Resulting in single agent strategy with no more budget and at least equivalent probability of success as the strategy with two agents.   □

5.1 Results for Clustered Sites

When sites are located in 2-dimensional Euclidean space, there is the potential for many widely separated clusters of sites, which may result in solutions that require more than 2 agents. To examine the effect of clustered sites on the optimal number of agents we ran two experiments, one for Min-Budget and one for Max-Probability. For both experiments we generated data sets consisting of 5 well-separated cluster centers identified in a 100-by-100 region and generated 15 site locations according to varying cluster tendency (ct), or the probability that a site will be near a cluster center. For $ct = 0.0$ all sites were uniformly randomly generated in the region, and for increasing ct it becomes more likely that sites are located in close proximity to the cluster centers until at $ct = 1.0$, there are no uniform-randomly generated sites. The start site is always placed in the center of the region. For the Min-Budget experiment we used $\rho = 0.95$ and generated random item costs in the intervals $[0]$, $[10, 30]$ and $[30, 50]$. For the Max-Probability experiment we used item costs of 0 and explored total budgets of 100, 600, and 1000. We ran 100 replicates for each setting.

The results are shown in Fig. 6. We see that for Min-Budget, the number of agents used grows as the item costs decrease. Additionally, we see that as ct increases, there is a slight increase in the average number of agents used. For Max-Probability there are two trends that are largely insensitive to the clustering tendency: the 100 budget case, and the 1000 budget case. With budget 100, the optimal solutions tend to include two agents to visit sites that are widely separated, but the limited budget tends not to be split any further. With 1000 budget, one agent can typically visit any useful sites. With budget 600 there is a different trend: the number of agents is greatest at $ct = 0.25$ which corresponds

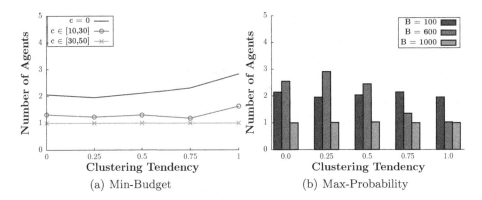

Fig. 6. Average number of agents used in optimal plans for Min-Budget and Max-Probability for 15 sites clustered into 5 clusters in a 100 unit by 100 unit region, with variable clustering tendencies.

to forcing many well-separated sites, so a divide and conquer strategy is used. At $ct = 1.0$, the tight clusters of points make it easier for a single agent to travel within a cluster with very low travel cost.

Our empirical results for these and other settings revealed a tendency towards very few agents in the optimal solution. To obtain a better intuition for this phenomenon, we analyzed increasing numbers of equidistant sites located on a circle centered on the start site.

Theorem 6. *Given an mSPS problem in 2-dimensional Euclidean space with all sites S equidistant from the start site, the problem never requires more than 5 agents.*

Proof. We assume that all sites must be visited in the optimal solution, if all sites are not required, then this can only result in fewer required agents. Additionally, we assume that item costs are all zero, since having positive item costs can never increase the number of agents required in the zero-cost case.

Consider $|S|$ sites equally spaced around a circle of radius r. The case of $|S| = 1$ trivially only requires one agent. Consider two sites as shown in Fig. 7(a). In this case $B^* = 2r$ for two agents, but $B^* = 3r$ for one agent so two agents are necessary. The three site case is shown in Fig. 7(b). In this case using three agents is optimal since the removal of an agent from the solution causes another agent to travel a distance of $\sqrt{2}r + r > 2r$. The cases for 4 and 5 sites are similar—sending an agent along the radius of the circle is cheaper than sending an agent along an edge of the inscribed regular polygon. Figure 7(c) and (d) show the case for $|S| = 6$. This is the break even point where traveling along an edge of the hexagon requires the same budget as traveling along the radius, thus the problem is 1-agent sufficient. For $|S| > 6$ the edges of the inscribed regular $|S|$-gon will be strictly less than r so these cases are all 1-agent sufficient. Even if we relax the assumption that sites are equally spaced, we still only need at most 5 agents.   □

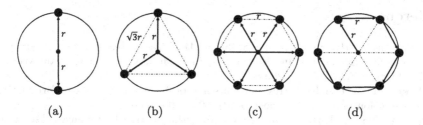

(a)   (b)   (c)   (d)

Fig. 7. Geometric argument that if all sites are equidistant from the start site, then an optimal solution will never require more than 5 agents.

6   Conclusions and Future Work

Often, research into multiagent systems ignores the question of how many agents should actually be used to solve a problem. To investigate this question, we introduced the condition of k-agent sufficiency, as it relates to Multiagent Stochastic Physical Search. We provided the first theoretical and empirical analysis of k-agent sufficiency for mSPS when sites are along a path and in 2-d Euclidean space. We showed that mSPS along a path with a single starting location is always 2-agent sufficient, and quite often 1-agent sufficient. We also showed that even in 2-d Euclidean space with a single starting location, optimal solutions usually require at most 3 agents on average. Our results show strong evidence that optimal solutions to the mSPS problem in 2-d Euclidean space never requires more than 5 agents even if sites located in widely-separated clusters. Using a geometric argument we show why this is true when sites are equidistant from the start site. We conjecture that in general, optimal solutions to mSPS problems in 2-d Euclidean space require at most 5 agents. We hope that these results will inspire other researchers in multiagent planning and optimization to consider cases where multiple agents are not always necessary or even desirable, rather than simply showing that an algorithm or solution method scales to x agents.

We note that there are many assumptions not considered in this paper which may cause the number of agents in an optimal solution to increase. Some of these assumptions include no communication during search, starting all agents from the same initial location, having a limit on the maximum budget per agent, having needs for redundancy or collaboration, and problems where the objective is to minimize time. Future work should examine these extensions to see if there are still k-agent sufficiency conditions. We also hope to leverage the k-agent sufficiency results shown in this paper to develop efficient heuristics and approximation algorithms for solving difficult mSPS problem instances.

Acknowledgments. We would like to thank the Air Office of Scientific Research and the Air Force Research Laboratory's Machine Intelligence for Mission Focused Autonomy Program for funding this research. The views expressed in this article are those of the authors and do not reflect the official policy or position of the USAF, DoD, or the US Government.

References

1. Aumann, Y., Hazon, N., Kraus, S., Sarne, D.: Physical search problems applying economic search models. In: Proceedings of the 23rd AAAI Conference on Artificial Intelligence. AAAI Press (2008)
2. Bektas, T.: The multiple traveling salesman problem: an overview of formulations and solution procedures. Omega **34**(3), 209–219 (2006)
3. Brown, D.S., Hudack, J., Banerjee, B.: Algorithms for stochastic physical search on general graphs. In: Workshops at the Twenty-Ninth AAAI Conference on Artificial Intelligence. AAAI Press (2015)
4. Gabriely, Y., Rimon, E.: Spanning-tree based coverage of continuous areas by a mobile robot. Ann. Math. Artif. Intell. **31**(1–4), 77–98 (2001)
5. Gerkey, B.P., Matarić, M.J.: A formal analysis and taxonomy of task allocation in multi-robot systems. Int. J. Robot. Res. **23**(9), 939–954 (2004)
6. Hazon, N., Aumann, Y., Kraus, S.: Collaborative multi agent physical search with probabilistic knowledge. In: Twenty-First International Joint Conference on Artificial Intelligence (2009)
7. Hazon, N., Aumann, Y., Kraus, S., Sarne, D.: Physical search problems with probabilistic knowledge. Artif. Intell. **196**, 26–52 (2013)
8. Hazon, N., Kaminka, G.A.: Redundancy, efficiency and robustness in multi-robot coverage. In: Proceedings of the 2005 IEEE International Conference on Robotics and Automation, ICRA 2005, pp. 735–741. IEEE (2005)
9. Hudack, J., Gemelli, N., Brown, D., Loscalzo, S., Oh, J.C.: Multiobjective optimization for the stochastic physical search problem. In: Ali, M., Kwon, Y.S., Lee, C.-H., Kim, J., Kim, Y. (eds.) IEA/AIE 2015. LNCS, vol. 9101, pp. 212–221. Springer, Heidelberg (2015)
10. Kang, S., Ouyang, Y.: The traveling purchaser problem with stochastic prices: exact and approximate algorithms. Eur. J. Oper. Res. **209**(3), 265–272 (2011)
11. Sariel-Talay, S., Balch, T.R., Erdogan, N.: Multiple traveling robot problem: a solution based on dynamic task selection and robust execution. IEEE/ASME Trans. Mechatron. **14**(2), 198–206 (2009)
12. Shehory, O., Kraus, S.: Methods for task allocation via agent coalition formation. Artif. Intell. **101**(1), 165–200 (1998)
13. Spires, S.V., Goldsmith, S.Y.: Exhaustive geographic search with mobile robots along space-filling curves. In: Drogoul, A., Fukuda, T., Tambe, M. (eds.) CRW 1998. LNCS, vol. 1456, pp. 1–12. Springer, Heidelberg (1998)
14. Toth, P., Vigo, D.: The Vehicle Routing Problem. Society for Industrial and Applied Mathematics, Philadelphia (2001)
15. Vokřínek, J., Komenda, A., Pěchouček, M.: Agents towards vehicle routing problems. In: Proceedings of the 9th International Conference on Autonomous Agents and Multiagent Systems, pp. 773–780. International Foundation for Autonomous Agents and Multiagent Systems (2010)

Utility and Decision Theory

Geometry on the Utility Space

François Durand[1]($\boxtimes$), Benoît Kloeckner[2], Fabien Mathieu[3],
and Ludovic Noirie[3]

[1] Inria, Paris, France
`francois.durand@inria.fr`
[2] Université Paris-Est, Laboratoire d'Analyse et de Mathématiques Appliquées
(UMR 8050), UPEM, UPEC, CNRS, 94010 Créteil, France
`benoit.kloeckner@u-pec.fr`
[3] Alcatel-Lucent Bell Labs France, Nozay, France
`{fabien.mathieu,ludovic.noirie}@alcatel-lucent.com`

Abstract. We study the geometrical properties of the utility space (the space of expected utilities over a finite set of options), which is commonly used to model the preferences of an agent in a situation of uncertainty. We focus on the case where the model is neutral with respect to the available options, i.e. treats them, *a priori*, as being symmetrical from one another. Specifically, we prove that the only Riemannian metric that respects the geometrical properties and the natural symmetries of the utility space is the round metric. This canonical metric allows to define a uniform probability over the utility space and to naturally generalize the Impartial Culture to a model with expected utilities.

Keywords: Utility theory · Geometry · Impartial culture · Voting

1 Introduction

Motivation. In the traditional literature of Arrovian social choice [2], especially voting theory, the preferences of an agent are often represented by ordinal information only[1]: a strict total order over the available options, or sometimes a binary relation of preference that may not be a strict total order (for example if indifference is allowed). However, it can be interesting for voting systems to consider cardinal preferences, for at least two reasons.

Firstly, some voting systems are not based on ordinal information only, like Approval voting or Range voting.

Secondly, voters can be in a situation of uncertainty, either because the rule of the voting system involves a part of randomness, or because each voter has incomplete information about other voters' preferences and the ballots they will choose. To express preferences in a situation of uncertainty, a classical and elegant model is the one of expected utilities [5,10,12,17]. The utility vector $\overrightarrow{u}$ representing the preferences of an agent is an element of $\mathbb{R}^m$, where m is the

---

[1] This is not always the case: for example, Gibbard [7] considers voters with expected utilities over the candidates.

© Springer International Publishing Switzerland 2015
T. Walsh (Ed.): ADT 2015, LNAI 9346, pp. 189–204, 2015.
DOI: 10.1007/978-3-319-23114-3_12

number of available options or *candidates*; the utility of a lottery over the options is computed as an expected value.

For a broad set of applications in economics, the options under consideration are financial rewards or quantities of one or several economic goods, which leads to an important consequence: there is a natural structure over the space of options. For example, if options are financial rewards, there is a natural order over the possible amounts and it is defined prior to the agents' preferences.

We consider here the opposite scenario where options are symmetrical *a priori*. This symmetry condition is especially relevant in voting theory, by a normative principle of neutrality, but it can apply to other contexts of choice under uncertainty when there is no natural preexistent structure on the space of available options.

The motivation for this paper came from the possible generalizations of the *Impartial Culture* to agents with expected utilities. The Impartial Culture is a classical probabilistic model in voting theory where each agent draws independently her strict total order of preference with a uniform probability over the set of possible orders.

The difficulty is not to define a probability law over utilities such that its projection on ordinal information is the Impartial Culture. Indeed, it is sufficient to define a distribution where voters are independent and all candidates are treated symmetrically. The issue is to choose one in particular: an infinity of distributions meet these conditions and we can wonder whether one of these generalizations has canonical reasons to be chosen rather than the others.

To answer this question, we need to address an important technical point. As we will see, an agent's utility vector is defined up to two constants, and choosing a specific normalization is arbitrary. As a consequence, the utility space is a quotient space of $\mathbb{R}^m$, and *a priori*, there is no canonical way to push a metric from $\mathbb{R}^m$ to this quotient space. Hence, at first sight, it seems that there is no natural definition of a uniform distribution of probability over that space.

More generally, searching a natural generalization of the Impartial Culture to the utility space naturally leads to investigate different topics about the geometry of this quotient space and to get a better understanding of its properties related to algebra, topology and measure theory.

We emphasize that our goal is not to propose a measure that represents real-life preferences. Such approach would follow from observation and experimental studies rather than theoretical work. Instead, we aim at identifying a measure that can play the role of a neutral, reference, measure. This will give a model for uniformness to which real distributions of utilities can be compared. For example, if an observed distribution has higher density than the reference measure in certain regions of the utility space, this could be interpreted as a non-uniform culture for the population under study and give an indication of an overall trend for these regions. With this aim in mind, we will assume a symmetry hypothesis over the different candidates.

Contributions and Plan. The rest of the paper is organized as follows. In Sect. 2, we quickly present Von Neumann–Morgenstern framework and define the utility space. In Sect. 3, we show that the utility space may be seen as a quotient of the dual of the space of pairs of lotteries over the candidates. In Sect. 4, we naturally define an inversion operation, that corresponds to reversing preferences while keeping their intensities, and a summation operation, that is characterized by the fact that it preserves unanimous preferences.

Since the utility space is a manifold, it is a natural wish to endow it with a Riemannian metric. In Sect. 5, we prove that the only Riemannian representation that preserves the natural projective properties and the *a priori* symmetry between the candidates is the round metric. Finally, in Sect. 6, we use this result to give a canonical generalization of the Impartial culture and to suggest the use of Von Mises–Fisher model to represent polarized cultures.

2 Von Neumann–Morgenstern Model

In this section, we define some notations and we quickly recall Von Neumann–Morgenstern framework in order to define the utility space.

Let $m \in \mathbb{N} \setminus \{0\}$. We will consider m mutually exclusive options called *candidates*, each one represented by an index in $\{1, \ldots, m\}$. A *lottery* over the candidates is an m-tuple $(L_1, \ldots, L_m) \in (\mathbb{R}_+)^m$ such that (s.t.) $\sum_{j=1}^m L_j = 1$. The set of lotteries is denoted $\mathcal{L}_m$.

The preferences of an agent over lotteries are represented by a binary relation $\leq$ over $\mathcal{L}_m$.

Von Neumann–Morgenstern theorem states that, provided that relation $\leq$ meets quite natural assumptions[2], it can be represented by a utility vector $\overrightarrow{u} = (u_1, \ldots, u_m) \in \mathbb{R}^m$, in the sense that following the relation $\leq$ is equivalent to maximizing the expected utility. Formally, for any two lotteries L and M:

$$L \leq M \Leftrightarrow \sum_{j=1}^m L_j u_j \leq \sum_{j=1}^m M_j u_j.$$

Mathematical definitions, assumptions and proof of this theorem can be found in [10,12,16], and discussions about the experimental validity of the assumptions are available in [6,12].

For the purposes of this paper, a crucial point is that $\overrightarrow{u}$ is defined up to an additive constant and a positive multiplicative constant. Formally, for any two vectors $\overrightarrow{u}$ and $\overrightarrow{v}$, let us note $\overrightarrow{u} \sim \overrightarrow{v}$ iff $\exists a \in (0, +\infty), \exists b \in \mathbb{R}$ s.t. $\overrightarrow{v} = a\overrightarrow{u} + b\overrightarrow{1}$, where $\overrightarrow{1}$ denotes the vector whose m coordinates are equal to 1. With this notation, if $\overrightarrow{u} \in \mathbb{R}^m$ is a utility vector representing $\leq$, then a vector $\overrightarrow{v} \in \mathbb{R}^m$ is also a utility vector representing $\leq$ iff $\overrightarrow{u} \sim \overrightarrow{v}$.

In order to define the utility space, all vectors representing the same preferences must be identified as only one point. The *utility space* over m candidates,

---

[2] The necessary and sufficient condition is that relation $\leq$ is complete, transitive, archimedean and independent of irrelevant alternatives.

denoted $\mathcal{U}_m$, is defined as the quotient space $\mathbb{R}^m/\sim$. We call *canonical projection* from $\mathbb{R}^m$ to $\mathcal{U}_m$ the function:

$$\tilde{\pi} : \begin{vmatrix} \mathbb{R}^m \to \mathcal{U}_m \\ \overrightarrow{u} \to \tilde{u} = \{\overrightarrow{v} \in \mathbb{R}^m \text{ s.t. } \overrightarrow{v} \sim \overrightarrow{u}\}. \end{vmatrix}$$

For any $\overrightarrow{u} \in \mathbb{R}^m$, we denote without ambiguity $\leq_{\tilde{u}}$ the binary relation over $\mathcal{L}_m$ represented by $\overrightarrow{u}$.

Figure 1 represents the space $\mathbb{R}^3$ of utility vectors for 3 candidates, without projecting to the quotient space. The canonical base of $\mathbb{R}^3$ is denoted $(\overrightarrow{e_1}, \overrightarrow{e_2}, \overrightarrow{e_3})$. Utility vectors $\overrightarrow{u}^{(1)}$ to $\overrightarrow{u}^{(4)}$ represent the same preferences as any other vector of the half-plane $\tilde{u}$, represented in gray. More generally, each non-trivial point $\tilde{u}$ in the quotient utility space corresponds to a half-plane in $\mathbb{R}^m$, delimited by the line $\text{vect}(\overrightarrow{1})$, the linear span of $\overrightarrow{1}$. The only exception is the point of total indifference $\tilde{0}$. In $\mathbb{R}^m$, it does not correspond to a plane but to the line $\text{vect}(\overrightarrow{1})$ itself.

To deal with utilities, conceptually and practically, it would be convenient to have a canonical representative $\overrightarrow{u}$ for each equivalence class $\tilde{u}$. In Fig. 2(a), for each non-indifferent $\tilde{u}$, we choose its representative satisfying $\min(u_i) = -1$ and $\max(u_i) = 1$. The utility space $\mathcal{U}_3$ (except the indifference point) is represented in $\mathbb{R}^3$ by six edges of the unit cube. In Fig. 2(b), we choose the representative

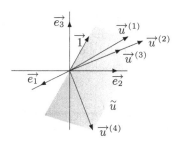

Fig. 1. Space $\mathbb{R}^3$ of utility vectors for 3 candidates (without quotient).

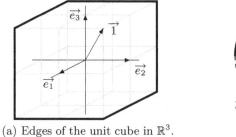 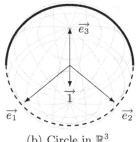

(a) Edges of the unit cube in $\mathbb{R}^3$.　　　(b) Circle in $\mathbb{R}^3$.

Fig. 2. Two representations of $\mathcal{U}_3$.

satisfying $\sum u_i = 0$ and $\sum u_i^2 = 1$. In that case, $\mathcal{U}_3 \setminus \{\tilde{0}\}$ is represented in $\mathbb{R}^3$ by the unit circle in the linear plane that is orthogonal to $\overrightarrow{1}$.

If we choose such a representation, the quotient space $\mathcal{U}_m$ can inherit the Euclidean distance from $\mathbb{R}^m$; for example, we can evaluate distances along the edges of the cube in Fig. 2(a), or along the unit circle in Fig. 2(b). But it is clear that the result will depend on the representation chosen. Hence, it is an interesting question whether one of these two representations, or yet another one, has canonical reasons to be used. But before answering this question, we need to explore in more generality the geometrical properties of the utility space.

3 Duality with the Tangent Hyperplane of Lotteries

In this section, we remark that the utility space is a dual of the space of pairs of lotteries. Not only does it give a different point of view on the utility space (which we think is interesting by itself), but it will also be helpful to prove Theorem 3, which characterizes the summation operator that we will define in Sect. 4.

In the example represented in Fig. 3, we consider $m = 3$ candidates and $\overrightarrow{u} = (\frac{5}{3}; -\frac{1}{3}; -\frac{4}{3})$. The great triangle, or simplex, is the space of lotteries $\mathcal{L}_3$. Hatchings are the agent's indifference lines: she is indifferent between any pair of lotteries on the same line (see [12], Section 6.B). The utility vector $\overrightarrow{u}$ represented here is in the plane of the simplex, but it is not mandatory: indeed, $\overrightarrow{u}$ can be arbitrarily chosen in its equivalence class $\tilde{u}$. Nevertheless, it is a quite natural choice, because the component of $\overrightarrow{u}$ in the direction $\overrightarrow{1}$ (orthogonal to the simplex) has no meaning in terms of preferences.

The utility vector $\overrightarrow{u}$ can be seen as a gradient of preference: at each point, it reveals in what directions the agent can find lotteries she likes better. However, only the direction of $\overrightarrow{u}$ is important, whereas its norm has no specific meaning. As a consequence, the utility space is not exactly a dual space, but rather a quotient of a dual space, as we will see more formally.

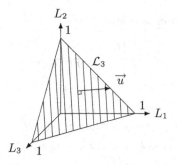

Fig. 3. Space $\mathcal{L}_3$ of lotteries over 3 candidates.

For any two lotteries $L = (L_1, \ldots, L_m)$ and $M = (M_1, \ldots, M_m)$, we call *bipoint* from L to M the vector $\overrightarrow{LM} = (M_1 - L_1, \ldots, M_m - L_m)$. We call *tangent polytope* of $\mathcal{L}_m$ the set $\mathcal{T}$ of bipoints of $\mathcal{L}_m$.

We call *tangent hyperplane* of $\mathcal{L}_m$:

$$\mathcal{H} = \{(\Delta_1, \ldots, \Delta_m) \in \mathbb{R}^m \text{ s.t. } \sum_{j=1}^{m} \Delta_j = 0\}.$$

In Fig. 3, the tangent polytope $\mathcal{T}$ is the set of the bipoints of the great triangle, seen as a part of a vector space (whereas $\mathcal{L}_m$ is a part of an affine space). The tangent hyperplane $\mathcal{H}$ is the whole linear hyperplane containing $\mathcal{T}$.

Let us note $\langle \overrightarrow{u} \mid \overrightarrow{v} \rangle$ the canonical inner product of $\overrightarrow{u}$ and $\overrightarrow{v}$. We call *positive half-hyperplane associated to* $\overrightarrow{u}$ the set $\overrightarrow{u}^+ = \{\overrightarrow{\Delta} \in \mathcal{H} \text{ s.t. } \langle \overrightarrow{u} \mid \overrightarrow{\Delta} \rangle \geq 0\}$. By definition, a lottery M is preferred to a lottery L iff the bipoint $\overrightarrow{LM}$ belongs to this positive half-hyperplane:

$$L \leq_{\tilde{u}} M \Leftrightarrow \langle \overrightarrow{u} \mid \overrightarrow{LM} \rangle \geq 0 \Leftrightarrow \overrightarrow{LM} \in \overrightarrow{u}^+.$$

Let $\mathcal{H}^\star$ be the dual space of $\mathcal{H}$, that is, the set of linear forms on $\mathcal{H}$. For any $\overrightarrow{u} \in \mathbb{R}^m$, we call *linear form associated to* $\overrightarrow{u}$ the following element of $\mathcal{H}^\star$:

$$\langle \overrightarrow{u} \mid : \begin{vmatrix} \mathcal{H} \to \mathbb{R} \\ \overrightarrow{\Delta} \to \langle \overrightarrow{u} \mid \overrightarrow{\Delta} \rangle. \end{vmatrix}$$

We observed that the utility vector can be seen as a gradient, except that only its direction matters, not its norm. Let us formalize this idea. For any $(f, g) \in (\mathcal{H}^\star)^2$, we denote $f \sim g$ iff these two linear forms are positive multiples of each other, that is, iff $\exists a \in (0, +\infty)$ s.t. $g = af$. We denote $\tilde{\pi}(f) = \{g \in \mathcal{H}^\star \text{ s.t. } f \sim g\}$: it is the equivalence class of f, up to positive multiplication.

Proposition 1. *For any $(\overrightarrow{u}, \overrightarrow{v}) \in (\mathbb{R}^m)^2$, we have:*

$$\overrightarrow{u} \sim \overrightarrow{v} \Leftrightarrow \langle \overrightarrow{u} \mid \sim \langle \overrightarrow{v} \mid.$$

The following application is a bijection:

$$\Theta : \begin{vmatrix} \mathcal{U}_m \to \mathcal{H}^\star / \sim \\ \tilde{\pi}(\overrightarrow{u}) \to \tilde{\pi}(\langle \overrightarrow{u} \mid). \end{vmatrix}$$

Proof. $\overrightarrow{u} \sim \overrightarrow{v}$
    $\Leftrightarrow \exists a \in (0, +\infty), \exists b \in \mathbb{R}$ s.t. $\overrightarrow{v} - a\overrightarrow{u} = b\overrightarrow{1}$
    $\Leftrightarrow \exists a \in (0, +\infty)$ s.t. $\overrightarrow{v} - a\overrightarrow{u}$ is orthogonal to $\mathcal{H}$
    $\Leftrightarrow \exists a \in (0, +\infty)$ s.t. $\forall \overrightarrow{\Delta} \in \mathcal{H}, \langle \overrightarrow{v} \mid \overrightarrow{\Delta} \rangle = a \langle \overrightarrow{u} \mid \overrightarrow{\Delta} \rangle$
    $\Leftrightarrow \langle \overrightarrow{u} \mid \sim \langle \overrightarrow{v} \mid.$

The implication $\Rightarrow$ proves that Θ is correctly defined: indeed, if $\tilde{\pi}(\overrightarrow{u}) = \tilde{\pi}(\overrightarrow{v})$, then $\tilde{\pi}(\langle \overrightarrow{u} \mid) = \tilde{\pi}(\langle \overrightarrow{v} \mid)$. The implication $\Leftarrow$ ensures that Θ is injective. Finally, Θ is obviously surjective.

Hence, the utility space can be seen as a quotient of the dual $\mathcal{H}^*$ of the tangent space $\mathcal{H}$ of the lotteries $\mathcal{L}_m$. As noticed before, a utility vector may be seen, up to a positive constant, as a uniform gradient, i.e. as a linear form over $\mathcal{H}$ that reveals, for any point in the space of lotteries, in what directions the agent can find lotteries that she likes better.

4   Inversion and Summation Operators

As a quotient of $\mathbb{R}^m$, the utility space inherits natural operations from $\mathbb{R}^m$: inversion and summation. We will see that both these quotient operators have an intuitive meaning regarding preferences. The summation will also allow to define *lines* in Sect. 5, which will be a key notion for Theorem 5, characterizing the suitable Riemannian metrics.

We define the *inversion* operator of $\mathcal{U}_m$ as:

$$- : \left| \begin{array}{l} \mathcal{U}_m \rightarrow \mathcal{U}_m \\ \tilde{\pi}(\overrightarrow{u}) \rightarrow \tilde{\pi}(-\overrightarrow{u}). \end{array} \right.$$

This operator is correctly defined and it is a bijection; indeed, $\tilde{\pi}(\overrightarrow{u}) = \tilde{\pi}(\overrightarrow{v})$ iff $\tilde{\pi}(-\overrightarrow{u}) = \tilde{\pi}(-\overrightarrow{v})$. Considering this additive inverse amounts to reverting the agent's preferences, without modifying their relative intensities.

Now, we want to push the summation operator from $\mathbb{R}^m$ to the quotient $\mathcal{U}_m$. We use a generic method to push an operator to a quotient space: considering $\tilde{u}$ and $\tilde{v}$ in $\mathcal{U}_m$, their antecedents are taken in $\mathbb{R}^m$ thanks to $\tilde{\pi}^{-1}$, the sum is computed in $\mathbb{R}^m$, then the result is converted back into the quotient space $\mathcal{U}_m$, thanks to $\tilde{\pi}$.

However, the result is not unique. Indeed, let us take arbitrary representatives $\overrightarrow{u} \in \tilde{u}$ and $\overrightarrow{v} \in \tilde{v}$. In order to compute the sum, we can think of any representatives. So, possible sums are $a\overrightarrow{u} + a'\overrightarrow{v} + (b+b')\overrightarrow{1}$, where a and a' are positive and where $b + b'$ is any real number. Converting back to the quotient, we can get for example $\tilde{\pi}(2\overrightarrow{u} + \overrightarrow{v})$ and $\tilde{\pi}(\overrightarrow{u} + 3\overrightarrow{v})$, which are generally distinct. As a consequence, the output is not a point in the utility space $\mathcal{U}_m$, but rather a set of points, i.e. an element of $\mathcal{P}(\mathcal{U}_m)$.

This example shows how we could define the sum of two elements $\tilde{u}$ and $\tilde{v}$. In order to be more general, we will define the sum of any number of elements of $\mathcal{U}_m$. Hence we also take $\mathcal{P}(\mathcal{U}_m)$ as the set of inputs.

We define the *summation* operator as:

$$\sum : \left| \begin{array}{l} \mathcal{P}(\mathcal{U}_m) \rightarrow \mathcal{P}(\mathcal{U}_m) \\ A \quad \rightarrow \left\{ \tilde{\pi} \left(\sum_{i=1}^{n} \overrightarrow{u_i} \right), n \in \mathbb{N}, (\overrightarrow{u_1}, \ldots, \overrightarrow{u_n}) \in \left(\tilde{\pi}^{-1}(A) \right)^n \right\}. \end{array} \right.$$

Example 2. *Let us consider* $\mathcal{U}_4$, *the utility space for 4 candidates. In Fig. 4, for the purpose of visualization, we represent its projection in $\mathcal{H}$, which is permitted by the choice of normalization constants b. Since $\mathcal{H}$ is a 3-dimensional space, let $(\overrightarrow{h_1}, \overrightarrow{h_2}, \overrightarrow{h_3})$ be an orthonormal base of it.*

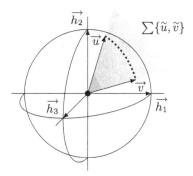

Fig. 4. Sum of two utility vectors in the utility space $\mathcal{U}_4$.

For two non-trivial utility vectors $\tilde{u}$ and $\tilde{v}$, the choice of normalization multipli-cators a allows to choose representatives $\overrightarrow{u}$ and $\overrightarrow{v}$ whose Euclidean norm is 1.

In this representation, the sum $\sum\{\tilde{u}, \tilde{v}\}$ consists of utilities corresponding to vectors $a\overrightarrow{u} + a'\overrightarrow{v}$, where a and a' are nonnegative. Indeed, we took a represen-tation in $\mathcal{H}$, so all normalization constants b vanish. Moreover, a, a' or both can be equal to zero because our definition allows to ignore $\overrightarrow{u}$, $\overrightarrow{v}$ or both. Up to taking representatives of unitary norm for non-trivial utility vectors, let us note that the sum $\sum\{\tilde{u}, \tilde{v}\}$ can be represented by the dotted line and the point $\overrightarrow{0}$ of total indifference.

Geometrically, the sum is the quotient of the convex hull of the inputs. Note that this convex hull is actually a convex cone. We will see below the interpre-tation of the sum in terms of preferences.

Due to our definition of the sum operator, we consider the closed cone: for example, the inputs themselves (e.g. $\tilde{u}$) fit in our definition, and so is the total indifference $\tilde{\pi}(\overrightarrow{0})$. That would not be the case if we took only $\tilde{\pi}(a\overrightarrow{u}+a'\overrightarrow{v}+b\overrightarrow{1})$, where $a > 0$ and $a' > 0$. The purpose of this convention is to have a concise wording for Theorem 3 that follows.

We now prove that, if A is the set of the utility vectors of a population, then $\sum A$ is the set of utility vectors that respect the unanimous preferences of the population.

Theorem 3 (characterization of the sum). *Let $A \in \mathcal{P}(\mathcal{U}_m)$ and $\tilde{v} \in \mathcal{U}_m$.*
The following conditions are equivalent.

1. $\tilde{v} \in \sum A.$
2. $\forall (L, M) \in \mathcal{L}_m^2 : \left[(\forall \tilde{u} \in A, L \leq_{\tilde{u}} M) \Rightarrow L \leq_{\tilde{v}} M \right].$

Proof. First, let us remark that the tangent polytope $\mathcal{T}$ generates the tangent hyperplane $\mathcal{H}$ by positive multiplication. That is:

$$\forall \overrightarrow{\Delta} \in \mathcal{H}, \exists \overrightarrow{LM} \in \mathcal{T}, \exists \lambda \in (0, +\infty) \text{ s.t. } \overrightarrow{\Delta} = \lambda \overrightarrow{LM}.$$

Indeed, $\mathcal{T}$ contains a neighborhood of the origin in vector space $\mathcal{H}$.

Let $\overrightarrow{v} \in \tilde{\pi}^{-1}(\tilde{v})$. We have the following equivalences.

- $\forall (L, M) \in \mathcal{L}_m{}^2, (\forall \tilde{u} \in A, L \leq_{\tilde{u}} M) \Rightarrow L \leq_{\tilde{v}} M$,
- $\forall \overrightarrow{LM} \in \mathcal{T}, \left(\forall \overrightarrow{u} \in \tilde{\pi}^{-1}(A), \langle \overrightarrow{u} \mid \overrightarrow{LM} \rangle \geq 0 \right) \Rightarrow \langle \overrightarrow{v} \mid \overrightarrow{LM} \rangle \geq 0$,
- $\forall \overrightarrow{\Delta} \in \mathcal{H}, \left(\forall \overrightarrow{u} \in \tilde{\pi}^{-1}(A), \langle \overrightarrow{u} \mid \overrightarrow{\Delta} \rangle \geq 0 \right) \Rightarrow \langle \overrightarrow{v} \mid \overrightarrow{\Delta} \rangle \geq 0$ (because $\mathcal{T}$ generates $\mathcal{H}$),
- $\bigcap_{\overrightarrow{u} \in \tilde{\pi}^{-1}(A)} \overrightarrow{u}^+ \subset \overrightarrow{v}^+$,
- $\overrightarrow{v}$ is in the convex cone of $\tilde{\pi}^{-1}(A)$ (because of the duality seen in Sect. 3),
- $\tilde{v} \in \sum A$.

Example 4. *Let us consider a non-indifferent $\tilde{u}$ and let us examine the sum of $\tilde{u}$ and its additive inverse $-\tilde{u}$. By a direct application of the definition, the sum consists of $\tilde{u}$, $-\tilde{u}$ and $\tilde{0}$.*

However, we have just proved that the sum is the subset of utility vectors preserving the unanimous preferences over lotteries. Intuitively, we could think that, since $\tilde{u}$ and $-\tilde{u}$ seem to always disagree, any utility vector $\tilde{v}$ respects the empty set of their common preferences; so, their sum should be the whole space. But this is not a correct intuition.

Indeed, let us examine the example of $\overrightarrow{u} = (1, 0, \ldots, 0)$. About any two lotteries L and M, inverse opinions $\tilde{u}$ and $-\tilde{u}$ agree iff $L_1 = M_1$: in that case, both $\tilde{u}$ and $-\tilde{u}$ are indifferent between L and M. The only points of the utility space meeting this common property are $\tilde{u}$ and $-\tilde{u}$ themselves, as well as the indifference point $\tilde{0}$.

5 Riemannian Representation of the Utility Space

Since the utility space is a manifold, it is a natural desire to endow it with a Riemannian metric. In this section, we prove that there is a limited choice of metrics that are coherent with the natural properties of the space and with the *a priori* symmetry between the candidates.

First, let us note that the indifference point $\tilde{0}$ must be excluded. Indeed, its unique open neighborhood is $\mathcal{U}_m$ in whole, and no distance is consistent with this property[3]. In contrast, $\mathcal{U}_m \setminus \{\tilde{0}\}$ has the same topology as a sphere of dimension $m - 2$, so it can be endowed with a distance.

Now, let us define the round metric. The quotient $\mathbb{R}^m / \text{vect}(\overrightarrow{1})$ is identified to $\mathcal{H}$ and endowed with the inner product inherited from the canonical one of $\mathbb{R}^m$. The utility space $\mathcal{U}_m \setminus \{\tilde{0}\}$ is identified to the unit sphere of $\mathcal{H}$ and endowed with the induced Riemannian structure. We note ξ_0 this Riemannian metric on $\mathcal{U}_m \setminus \{\tilde{0}\}$.

In order to get an intuitive vision of this metric, one can represent any position $\tilde{u}$ by a vector $\overrightarrow{u}$ that verifies $\sum u_i = 0$ and $\sum u_i^2 = 1$. We obtain the

---

[3] Technically, this remark proves that $\mathcal{U}_m$ (with its natural quotient topology) is not a T_1 space [8].

$(m-2)$-dimensional unit sphere of $\mathcal{H}$ and we consider the metric induced by the canonical Euclidean metric of $\mathbb{R}^m$. That is, distances are measured on the surface of the sphere, using the restriction of the canonical inner product on each tangent space. For $m = 3$, such a representation has already been illustrated in Fig. 2(b).

With this representation in mind, we can give a formula to compute distances with ξ_0. We denote J the $m \times m$ matrix whose all coefficients are equal to 1 and $P_{\mathcal{H}}$ the matrix of the orthogonal projection on $\mathcal{H}$:

$$P_{\mathcal{H}} = \text{Id} - \frac{1}{m} J.$$

The canonical Euclidean norm of $\vec{u}$ is denoted $\|\vec{u}\|$. For two non-indifferent utility vectors $\vec{u}$ and $\vec{v}$, the distance between $\tilde{u}$ and $\tilde{v}$ in the sense of metric ξ_0 is:

$$d(\tilde{u}, \tilde{v}) = \arccos \left\langle \frac{P_{\mathcal{H}} \vec{u}}{\|P_{\mathcal{H}} \vec{u}\|} \,\middle|\, \frac{P_{\mathcal{H}} \vec{v}}{\|P_{\mathcal{H}} \vec{v}\|} \right\rangle.$$

If $\vec{u}$ and $\vec{v}$ are already unit vectors of $\mathcal{H}$, i.e. canonical representatives of their equivalence classes $\tilde{u}$ and $\tilde{v}$, then the formula amounts to $d(\tilde{u}, \tilde{v}) = \arccos \langle \vec{u} \mid \vec{v} \rangle$.

A natural property for the distance would be that its geodesics coincide with the unanimity segments defined by the sum. Indeed, imagine that Betty with utilities $\tilde{u}_B$ is succesfully trying to convince Alice with initial utilities $\tilde{u}_A$ to change her mind. Assume that Alice evolves continuously, so that her utility follows a path from $\tilde{u}_A$ to $\tilde{u}_B$. Unless Betty argues in a quite convoluted way, it makes sense to assume on one hand that Alice's path is a shortest path, and on the other hand that whenever Alice and Betty initially agree on the comparison of two lotteries, this agreement continues to hold all along the path. This is precisely what assumption 1a below means: shortest paths preserve agreements that hold at both their endpoints.

We will now prove that for $m \geq 4$, the spherical representation is the only one that is coherent with the natural properties of the space and that respects the *a priori* symmetry between candidates.

Theorem 5 (Riemannian representation of the utility space). *We assume that $m \geq 4$. Let ξ be a Riemannian metric on $\mathcal{U}_m \setminus \{\tilde{0}\}$.*
Conditions 1 and 2 are equivalent.

1. (a) *For any non-antipodal pair of points $\tilde{u}, \tilde{v} \in \mathcal{U}_m \setminus \{\tilde{0}\}$ (i.e. $\tilde{v} \neq -\tilde{u}$), the set $\sum \{\tilde{u}, \tilde{v}\}$ of elements respecting the unanimous preferences of $\tilde{u}$ and $\tilde{v}$ is a segment of a geodesic of ξ; and*
   (b) *for any permutation σ of $\{1, \ldots, m\}$, the action Φ_σ induced on $\mathcal{U}_m \setminus \{\tilde{0}\}$ by*

$$(u_1, \ldots, u_m) \to (u_{\sigma(1)}, \ldots, u_{\sigma(m)})$$

   is an isometry.
2. *$\exists \lambda \in (0, +\infty)$ s.t. $\xi = \lambda \xi_0$.*

Proof. Since the implication $2 \Rightarrow 1$ is obvious, we now prove $1 \Rightarrow 2$. The deep result behind this is a classical theorem of Beltrami, which dates back to the middle of the nineteenth century: see [3] and [4].

The image of a 2-dimensional subspace of $\mathcal{H}$ in $\mathcal{U}_m \setminus \{\tilde{0}\}$ by the canonical projection $\tilde{\pi}$ is called a *line* in $\mathcal{U}_m \setminus \{\tilde{0}\}$. This notion is deeply connected to the summation operator: indeed, the sum of two non antipodal points in $\mathcal{U}_m \setminus \{\tilde{0}\}$ is a segment of the line joining them. Condition 1a precisely means that the geodesics of ξ are the lines of $\mathcal{U}_m \setminus \{\tilde{0}\}$. Beltrami's theorem then states that $\mathcal{U}_m \setminus \{\tilde{0}\}$ has constant curvature. Note that this result is in fact more subtle in dimension 2 (that is, for $m = 4$) than in higher dimensions; see [13], Theorem 1.18 and [14], Theorem 7.2 for proofs.

Since $\mathcal{U}_m \setminus \{\tilde{0}\}$ is a topological sphere, this constant curvature must be positive. Up to multiplying ξ by a constant, we can assume that this constant curvature is 1. As a consequence, there is an isometry $\Psi : \mathcal{S}_{m-2} \to \mathcal{U}_m \setminus \{\tilde{0}\}$, where $\mathcal{S}_{m-2}$ is the unit sphere of $\mathbb{R}^{m-1}$, endowed with its usual round metric. The function Ψ obviously maps geodesic to geodesics, let us deduce the following.

Lemma 6. *There is a linear map* $\Lambda : \mathbb{R}^{m-1} \to \mathcal{H}$ *inducing* Ψ, *that is such that:*

$$\Psi \circ \Pi = \Pi \circ \Lambda,$$

where Π *denotes both projections* $\mathbb{R}^{m-1} \to \mathcal{S}_{m-2}$ *and* $\mathcal{H} \to \mathcal{U}_m \setminus \{\tilde{0}\}$.

Proof. First, Ψ maps any pair of antipodal points of $\mathcal{S}_{m-2}$ to a pair of antipodal points of $\mathcal{U}_m \setminus \{\tilde{0}\}$: indeed in both cases antipodal pairs are characterized by the fact that there are more than one geodesic containing them. It follows that Ψ induces a map Ψ' from the projective space $\mathrm{P}(\mathbb{R}^{m-1})$ (which is the set of lines through the origin in $\mathbb{R}^{m-1}$, identified with the set of pairs of antipodal points of $\mathcal{S}_{m-2}$) to the projective space $\mathrm{P}(\mathcal{H})$ (which is the set of lines through the origin in $\mathcal{H}$, identified with the set of pairs of antipodal points of $\mathcal{U}_m \setminus \{\tilde{0}\}$).

The fact that Ψ sends geodesics of $\mathcal{S}_{m-2}$ to geodesics of $\mathcal{U}_m \setminus \{\tilde{0}\}$ and condition 1a together imply that Ψ' sends projective lines to projective lines.

As is well known, a one-to-one map $\mathbb{R}^n \to \mathbb{R}^n$ which sends lines to lines must be an affine map; a similar result holds in projective geometry, concluding that Ψ' must be a projective map. See e.g. [1] for both results.

That Ψ' is projective exactly means that there is a linear map $\Lambda : \mathbb{R}^{m-1} \to \mathcal{H}$ inducing Ψ', which then induces $\Psi : \mathcal{S}_{m-2} \to \mathcal{U}_m \setminus \{\tilde{0}\}$.

Using Λ to push the canonical inner product of $\mathbb{R}^{m-1}$, we get that there exists an inner product $(\vec{u}, \vec{v}) \to \phi(\vec{u}, \vec{v})$ on $\mathcal{H}$ that induces ξ, in the sense that ξ is the Riemannian metric obtained by identifying $\mathcal{U}_m \setminus \{\tilde{0}\}$ with the unit sphere defined in $\mathcal{H}$ by ϕ and restricting ϕ to it.

The last thing to prove is that ϕ is the inner product coming from the canonical one on $\mathbb{R}^m$. Note that hypothesis 1b is mandatory, since any inner product on $\mathcal{H}$ does induce on $\mathcal{U}_m \setminus \{\tilde{0}\}$ a Riemannian metric satisfying 1a.

Each canonical basis vector $\vec{e_j} = (0, \ldots, 1, \ldots, 0)$ defines a point in $\mathcal{U}_m \setminus \{\tilde{0}\}$ and a half-line $\ell_j = \mathbb{R}_+ e_j$ in $\mathcal{H}$. Condition 1b ensures that these half-lines are permuted by some isometries of $(\mathcal{H}, \phi)$. In particular, there are vectors $\vec{u_j} \in \ell_j$ that have constant pairwise distance (for ϕ).

Lemma 7. *The family* $\vec{u_1}, \ldots, \overrightarrow{u_{m-1}}$ *is, up to multiplication by a scalar, the unique basis of* $\mathcal{H}$ *such that* $\vec{u_j} \in \ell_j$ *and* $\sum_{j<m} \vec{u_j} \in -\ell_m$.

Proof. First, by definition of the $\vec{u_j}$, these vectors form a regular simplex and $\sum_j \vec{u_j} = \vec{0}$. It follows that $\vec{u_1}, \ldots, \overrightarrow{u_{m-1}}$ has the required property and we have left to show uniqueness.

Assume $\vec{v_1}, \ldots, \overrightarrow{v_{m-1}}$ is a basis of $\mathcal{H}$ such that $\vec{v_j} \in \ell_j$ and $\sum_{j<m} \vec{v_j} \in -\ell_m$. Then there are positive scalars $a_1, \ldots, a_m$ such that $\vec{v_j} = a_j \vec{u_j}$ for all $j < m$, and $\sum_{j<m} \vec{v_j} = a_m \sum_{j<m} \vec{u_j}$.

Then $\sum_{j<m} a_j \vec{u_j} = \sum_{j<m} a_m \vec{u_j}$, and since the u_j form a basis, it must hold $a_j = a_m$ for all j.

Now consider the canonical inner product ϕ_0 on $\mathcal{H}$ that comes from the canonical one on $\mathbb{R}^m$. Since permutations of coordinates are isometries, we get that the vectors $\vec{v_j} = \Pi(\vec{e_j})$ (where Π is now the orthogonal projection from $\mathbb{R}^m$ to $\mathcal{H}$) form a regular simplex for ϕ_0, so that $\sum_j \vec{v_j} = \vec{0}$. It follows that $\vec{u_j} = \lambda \vec{v_j}$ for some $\lambda > 0$ and all j. We deduce that the $\vec{u_j}$ form a regular simplex for both ϕ and ϕ_0, which must therefore be multiple from each other. This finishes the proof of Theorem 5.

However, the implication $1 \Rightarrow 2$ in the theorem is not true for $m = 3$. For each non-indifferent utility vector, let us consider its representative verifying $\min(u_i) = 0$ and $\max(u_i) = 1$. This way, $\mathcal{U}_m \setminus \{\tilde{0}\}$ is identified to edges of the unit cube in $\mathbb{R}^3$, as in Fig. 2(a). We use this identification to endow $\mathcal{U}_m \setminus \{\tilde{0}\}$ with the metric induced on these edges by the canonical inner product on $\mathbb{R}^3$. Then conditions 1a and 1b of the theorem are met, but not condition 2.

Another remark is of paramount importance about this theorem. Since we have a canonical representative $\vec{u}$ for each equivalence class $\tilde{u}$ as a unit vector of $\mathcal{H}$, we could be tempted to use it to compare utilities between several agents.

We stress on the fact that this representation cannot be used for interpersonal comparison of utility levels or utility differences.

For example, for two agents, consider the following representatives:

$$\begin{cases} \vec{u} = (0.00, 0.71, -0.71), \\ \vec{v} = (0.57, 0.22, -0.79). \end{cases}$$

The fact that $v_3 < u_3$ does not mean that an agent with preferences $\tilde{v}$ dislikes candidate 3 more than an agent with preferences $\tilde{u}$. Similarly, when changing from candidate 1 to candidate 2, the gain for agent $\tilde{u}$ (+0.71) cannot be compared to the loss for agent $\tilde{v}$ (−0.35).

Theorem 5 conveys no message for interpersonal comparison of utilities. Indeed, utilities belonging to two agents are essentially incomparable in the absence of additional assumptions [9]. Taking canonical representatives on the $(m-2)$-dimensional sphere is only used to compute distances between two points in the utility space.

6 Application: Probability Measures on the Utility Space

Once the space is endowed with a metric, it is endowed with a natural probability measure: the uniform measure in the sense of this metric (this is possible because the space has a finite total measure). We will denote μ_0 this measure, which is thus the normalized Riemannian volume defined by the metric ξ_0.

In practice, to draw vectors according to a uniform probability law over $\mathcal{U}_m \setminus \{\tilde{0}\}$, it is sufficient to use a uniform law on the unit sphere in $\mathcal{H}$. In other words, once one identifies $\mathcal{U}_m \setminus \{\tilde{0}\}$ with the unit sphere in $\mathcal{H}$, then μ_0 is exactly the usual uniform measure.

In the present case, the fact that the round sphere has many symmetries implies additional nice qualities for μ_0, which we sum up in a proposition.

Proposition 8. *Let μ be any probability measure on $\mathcal{U}_m \setminus \{\tilde{0}\}$.*

1. *Assume that for all $\delta > 0$, μ gives the same probability to all the balls in $\mathcal{U}_m \setminus \{\tilde{0}\}$ of radius δ (in the metric ξ_0); then $\mu = \mu_0$.*
2. *If μ is invariant under all isometries for the metric ξ_0, then $\mu = \mu_0$.*

Both characterizations are classical; the first one is (a strengthening of) the definition of the Riemannian volume, and the second one follows from the first and the fact that any two points on the round sphere can be mapped one to the other by an isometry.

In Fig. 5(a), we represent a distribution with 100 agents drawn uniformly and independently on the sphere, with 4 candidates. Like for Fig. 4, we represented only the unit sphere of $\mathcal{H}$.

The solid dark lines draw the *permutohedron*, a geometrical figure representing the ordinal aspect of these preferences. Each *facet* is constituted of all the points who share the same strict order of preference. A utility vector belongs to an *edge* if it has only three distinct utilities: for example, if the agent prefers candidate 1 to 4, 4 to 2 and 3, but is indifferent between candidates 2 and 3. Finally, a point is a *vertex* if it has only two distinct utilities: for example, if the agent prefers candidate 1 to all the others, but is indifferent between the others.

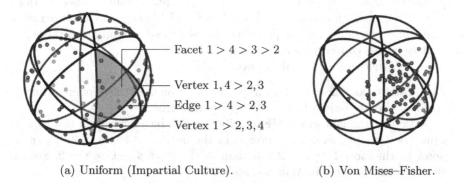

Facet $1 > 4 > 3 > 2$

Vertex $1, 4 > 2, 3$

Edge $1 > 4 > 2, 3$

Vertex $1 > 2, 3, 4$

(a) Uniform (Impartial Culture).   (b) Von Mises–Fisher.

Fig. 5. Two distributions of 100 agents on $\mathcal{U}_4$.

In this distribution, each agent has almost surely a strict order of preference. Each order has the same probability and agents are independent, hence this distribution is a natural generalization of the Impartial Culture when considering expected utilities.

Since the point $\tilde{0}$ is a geometrical singularity, it is difficult to include it naturally in such a measure. If one wants to take it into account, the easiest way is to draw it with a given probability and to use the distribution on $\mathcal{U}_m \setminus \{\tilde{0}\}$ in the other cases. That being said, we have just noticed that all other non-strict orders have a measure equal to 0 ; so for a canonical theoretical model, it can be argued that a natural choice is to attribute a measure 0 to the indifference point also.

Having a distance, hence a uniform measure, allows also to define other measures by their densities with respect to the uniform measure. Here is an example of a law defined by its density. Given a vector $\overrightarrow{u_0}$ in the unit sphere of $\mathcal{H}$ and κ a nonnegative real number, the Von Mises–Fisher (VMF) distribution of pole $\overrightarrow{u_0}$ and concentration κ is defined by the following density with respect to the uniform law on the unit sphere in $\mathcal{H}$:

$$p(\overrightarrow{u}) = C_\kappa e^{\kappa \langle \overrightarrow{u} \mid \overrightarrow{u_0} \rangle},$$

where C_κ is a normalization constant. Given the mean resultant vector of a distribution over the sphere, VMF distribution maximizes the entropy, in the same way that, in the Euclidean space, Gaussian distribution maximizes the entropy among laws with given mean and standard deviation. Hence, without additional information, it is the "natural" distribution that should be used. Figure 5(b) represents such a distribution, with the same conventions as Fig. 5(a). To draw a VMF distribution, we used Ulrich's algorithm revised by Wood [15,18].

Qualitatively, VMF model is similar to Mallows model, which is used for ordinal preferences [11]. In the later, the probability of an order of preference σ is:

$$p(\sigma) = D_\kappa e^{-\kappa \delta(\sigma, \sigma_0)},$$

where σ_0 is the mode of the distribution, $\delta(\sigma, \sigma_0)$ a distance between σ and σ_0 (typically Kendall's tau distance), κ a nonnegative real number (concentration) and D_k a normalization constant. Both VMF and Mallows models describe a culture where the population is polarized, i.e. scattered around a central point, with more or less concentration.

However, there are several differences.

- VMF distribution allows to draw a specific point on the utility sphere, whereas Mallows' chooses only a facet of the permutohedron.
- In particular, the pole of a VMF distribution can be anywhere in this continuum. For example, even if its pole is in the facet $1 > 4 > 3 > 2$, it can be closer to the facet $1 > 4 > \mathbf{2} > \mathbf{3}$ than to the facet $\mathbf{4} > \mathbf{1} > 3 > 2$. Such a nuance is not possible in Mallows model.
- In the neighborhood of the pole, VMF probability decreases like the exponential of the square of the distance (because the inner product is the cosine of

the distance), whereas Mallows probability decreases like the exponential of
the distance (not squared).

– VMF is the maximum entropy distribution, given a spherical mean and dis-
persion (similarly to a Gaussian distribution in a Euclidean space), whereas
Mallows' model is not characterized by such a property of maximum entropy.

The existence of a canonical measure allows to define other probability measures easily in addition to the two we have just described. Such measures can
generate artificial populations of agents for simulation purposes. They can also
be used to fit the data from real-life experiments to a theoretical model, and as
a neutral comparison point for such data.

To elaborate on this last point, let us stress that given a (reasonably regular)
distribution μ on a space such as $\mathcal{U}_m \setminus \{\tilde{0}\}$ there is *a priori* no way to define what
it means for an element $\tilde{u}$ to be more probable than another one $\tilde{v}$. Indeed, both
have probability 0 and what would make sense is to compare the probability of
being close to $\tilde{u}$ to the probability of being close to $\tilde{v}$. But one should compare
neighborhoods of the same size, and one needs a metric to make this comparison.
Alternatively, if one has a reference distribution such as μ_0, then it makes sense
to consider the density $f = \frac{d\mu}{d\mu_0}$, which is a (continuous, say) function on $\mathcal{U}_m \setminus \{\tilde{0}\}$.
Then we can compare $f(\tilde{u})$ and $f(\tilde{v})$ to say whether one of these elements is more
probable than the other according to μ. Note that in the present case, comparing
the probability of δ-neighborhoods for the metric ξ_0 or the density relative to
μ_0 gives the same result in the limit $\delta \to 0$, which is the very definition of the
Riemannian volume.

7   Conclusion

We have studied the geometrical properties of the classical model of expected
utilities, introduced by Von Neumann and Morgenstern, when candidates are
considered symmetrical *a priori*. We have remarked that the utility space may
be seen as a dual of the space of lotteries, that inversion and summation operators
inherited from $\mathbb{R}^m$ have a natural interpretation in terms of preferences and that
the space has a spherical topology when the indifference point is removed.

We have proved that the only Riemannian representation whose geodesics
coincide with the projective lines naturally defined by the summation operator
and which respects the symmetry between candidates is a round sphere.

All these considerations lay on the principle to add as little information as
possible in the system, especially by respecting the *a priori* symmetry between
candidates. This does not imply that the spherical representation of the utility
space $\mathcal{U}_m$ is the most relevant one in order to study a specific situation. Indeed,
as soon as one has additional information (for example, a model that places
candidates in a political spectrum), it is natural to include it in the model.
However, if one wishes, for example, to study a voting system in all generality,
without focusing on its application in a specific field, it looks natural to consider
a utility space with a metric as neutral as possible, like the one defined in this
paper by the spherical representation.

204   F. Durand et al.

Acknowledgments. The work presented in this paper has been partially carried out at LINCS (http://www.lincs.fr).

References

1. Audin, M.: Universitext. Geometry. Springer, Heidelberg (2003)
2. Arrow, K.J.: A difficulty in the concept of social welfare. J. Polit. Econ. **58**(4), 328–346 (1950)
3. Beltrami, E.: Résolution du problème de reporter les points d'une surface sur un plan, de manière que les lignes géodésiques soient représentée par des lignes droites. Annali di Matematica (1866)
4. Beltrami, E.: Essai d'interprétation de la géométrie noneuclidéenne. Trad. par J. Hoüel. Ann. Sci. École Norm. Sup. **6**, pp. 251–288 (1869)
5. Fishburn, P.C.: Utility Theory for Decision Making. Wiley, New York (1970)
6. Fishburn, P.C.: Nonlinear Preference and Utility Theory. Johns Hopkins Series in the Mathematical Sciences. Johns Hopkins University Press, Baltimore (1988)
7. Gibbard, A.: Straightforwardness of game forms with lotteries as outcomes. Econometrica **46**(3), 595–614 (1978)
8. Guénard, F., Lelièvre, G.: Compléments d'analyse. Number, vol. 1 in Compléments d'analyse. E.N.S (1985)
9. Hammond, P.J.: Interpersonal comparisons of utility: why and how they are and should be made. In: Elster, J., Roemer, J.E. (eds.) Interpersonal Comparisons of Well-Being, pp. 200–254. Cambridge University Press, Cambridge (1991)
10. Kreps, D.M.: A Course in Microeconomic Theory. Princeton University Press, Princeton (1990)
11. Mallows, C.L.: Non-null ranking models. i. Biometrika, pp. 114–130 (1957)
12. Mas-Colell, A., Whinston, M.D., Green, J.R.: Microeconomic Theory. Oxford University Press, Oxford (1995)
13. Spivak, M.: A Comprehensive Introduction to Differential Geometry, vol. III, Second edn. Publish or Perish Inc., Wilmington (1979)
14. Spivak, M.: A Comprehensive Introduction to Differential Geometry, vol. IV, Second edn. Publish or Perish Inc., Wilmington (1979)
15. Ulrich, G.: Computer generation of distributions on the m-sphere. Appl. Stat. **33**(2), 158–163 (1984)
16. Von Neumann, J., Morgenstern, O., Kuhn, H.W., Rubinstein, A.: Theory of Games and Economic Behavior. Princeton Classic Editions, Commemorative edn. Princeton University Press, Princeton (2007)
17. Von Neumann, J., Morgenstern, O.: Theory of Games and Economic Behavior. Princeton University Press, Princeton (1944)
18. Wood, A.T.A.: Simulation of the von mises fisher distribution. Commun. Stat. Simul. Comput. **23**(1), 157–164 (1994)

Sequential Extensions of Causal and Evidential Decision Theory

Tom Everitt$^{(\boxtimes)}$, Jan Leike, and Marcus Hutter

Australian National University, Canberra, Australia
{tom.everitt,jan.leike,marcus.hutter}@anu.edu.au

Abstract. Moving beyond the dualistic view in AI where agent and environment are separated incurs new challenges for decision making, as calculation of expected utility is no longer straightforward. The non-dualistic decision theory literature is split between *causal decision theory* and *evidential decision theory*. We extend these decision algorithms to the *sequential* setting where the agent alternates between taking actions and observing their consequences. We find that evidential decision theory has two natural extensions while causal decision theory only has one.

Keywords: Evidential decision theory · Causal decision theory · Planning · Causal graphical models · Dualism · Physicalism

1 Introduction

In artificial-intelligence problems an agent interacts sequentially with an environment by taking actions and receiving percepts [RN10]. This model is *dualistic*: the agent is distinct from the environment. It influences the environment only through its actions, and the environment has no other information about the agent. The dualism assumption is accurate for an algorithm that is playing chess, go, or other (video) games, which explains why it is ubiquitous in AI research. But often it is not true: real-world agents are embedded in (and computed by) the environment [OR12], and then a *physicalistic model*[1] is more appropriate.

This distinction becomes relevant in multi-agent settings with similar agents, where each agent encounters 'echoes' of its own decision making. If the other agents are running the same source code, then the agents' decisions are logically connected. This link can be used for uncoordinated cooperation [LFY+14]. Moreover, a physicalistic model is indispensable for self-reflection. If the agent is required to autonomously verify its integrity, and perform maintenance, repair, or upgrades, then the agent needs to be aware of its own functioning. For this, a reliable and accurate self-modeling is essential. Today, applications of this level of autonomy are mostly restricted to space probes distant from earth or robots navigating lethal situations, but in the future this might also become crucial for sustained self-improvement in generally intelligent agents [Yud08, Bos14, SF14a, RDT+15].

---

[1] Some authors also call this type of model *materialistic* or *naturalistic*.

© Springer International Publishing Switzerland 2015
T. Walsh (Ed.): ADT 2015, LNAI 9346, pp. 205–221, 2015.
DOI: 10.1007/978-3-319-23114-3_13

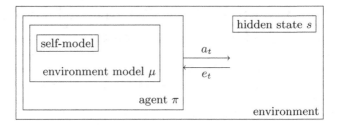

Fig. 1. The physicalistic model. The hidden state s contains information about the agent that is unknown to it. The distribution μ is the agent's (subjective) *environment model*, and π its (deterministic) policy. The agent models itself through the beliefs about (future) actions given by its environment model μ. Interaction with the environment at time step t occurs through an action a_t chosen by the agent and a percept e_t returned by the environment.

In the physicalistic model the agent is embedded inside the *environment*, as depicted in Fig. 1. The environment has a *hidden state* that contains information about the agent that is inaccessible to the agent itself. The agent has an *environment model* that describes the behavior of the environment given the hidden state and includes beliefs about the agent's own future actions (thus modeling itself).

Physicalistic agents may view their actions in two ways: as their selected output, and as consequences of properties of the environment. This leads to significantly more complex problems of inference and decision making, with actions simultaneously being both means to influence the environment and evidence about it. For example, looking at cat pictures online may simultaneously be a *means* of procrastination, and *evidence* of bad air quality in the room.

Dualistic decision making in a known environment is straightforward calculation of expected utilities. This is known as Savage decision theory [Sav72]. For non-dualistic decision making two main approaches are offered by the decision theory literature: *causal decision theory* (CDT) [GH78, Lew81, Sky82, Joy99, Wei12] and *evidential decision theory* (EDT) [Jef83, Bri14, Ahm14]. EDT and CDT both take actions that maximize expected utility, but differ in the way this expectation is computed: EDT uses the action under consideration as evidence about the environment while CDT does not. Section 2 formally introduces these decision algorithms.

Our contribution is to formalize and explore a decision-theoretic setting with a physicalistic reinforcement learning agent interacting *sequentially* with an environment that it is embedded in (Sect. 3). Previous work on non-dualistic decision theories has focused on *one-shot* situations. We find that there are two natural extensions of EDT to the sequential case, depending on whether the agent updates beliefs based on its next action or its entire policy. CDT has only one natural extension. We extend two famous *Newcomblike problems* to the sequential setting to illustrate the differences between our (generalized) decision theories.

Section 4 summarizes our results and outlines future directions. A list of notation can be found on page 15 and the formal details of the examples can be found in the technical report [ELH15].

2   One-Shot Decision Making

In a *one-shot decision problem*, we take one *action* $a \in \mathcal{A}$, receive a *percept* $e \in \mathcal{E}$ (typically called *outcome* in the decision theory literature) and get a *payoff* $u(e)$ according to the *utility function* $u : \mathcal{E} \to [0,1]$. We assume that the set of actions $\mathcal{A}$ and the set of percepts $\mathcal{E}$ are finite. Additionally, the environment contains a *hidden state* $s \in \mathcal{S}$. The hidden state holds information that is inaccessible to the agent at the time of the decision, but may influence the decision and the percept. Formally, the environment is given by a probability distribution P over the hidden state, the action, and the percept that factors according to a causal graph [Pea09].

A *causal graph* over the random variables $x_1, \ldots, x_n$ is a directed acyclic graph with nodes $x_1, \ldots, x_n$. To each node x_i belongs a probability distribution $P(x_i \mid pa_i)$, where pa_i is the set of parents of x_i in the graph. It is natural to identify the causal graph with the factored distribution $P(x_1, \ldots, x_n) = \prod_{i=1}^{n} P(x_i \mid pa_i)$. Given such a causal graph/factored distribution, we define the *do-operator* as

$$P(x_1, \ldots, x_{j-1}, x_{j+1}, \ldots, x_n \mid \mathsf{do}(x_j := b)) = \prod_{\substack{i=1 \\ i \neq j}}^{n} P(x_i \mid pa_i) \qquad (1)$$

where x_j is set to b wherever it occurs in pa_i, $1 \leq i \leq n$. The result is a new probability distribution that can be marginalized and conditioned in the standard way. Intuitively, intervening on node x_j means ignoring all incoming arrows to x_j, as the effects they represent are no longer relevant when we intervene; the factor $P(x_j \mid pa_j)$ representing the ingoing influences to x_j is therefore removed in the right-hand side of (1). Note that the do-operator is only defined for distributions for which a causal graph has been specified. See [Pea09, Chapter 3.4] for details.

2.1   Savage Decision Theory

In the *dualistic* formulation of decision theory, we have a function P that takes an action a and returns a probability distribution P_a over percepts. *Savage decision theory* (SDT) [Sav72, Bri14] takes actions according to

$$\arg\max_{a \in \mathcal{A}} \sum_{e \in \mathcal{E}} P_a(e) u(e). \qquad \text{(SDT)}$$

In the dualistic model it is usually conceptually clear what P_a should be. In the physicalistic model the environment model takes the form of a causal graph over a hidden state s, action a, and percept e, as illustrated in Fig. 2.

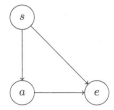

Fig. 2. The causal graph $P(s, a, e) = P(s)P(a \mid s)P(e \mid s, a)$ for one-step decision making. The hidden state s influences both the decision maker's action a and the received percept e.

According to this causal graph, the probability distribution P factors causally into $P(s, a, e) = P(s)P(a \mid s)P(e \mid s, a)$. The hidden state is not independent of the decision maker's action and Savage's model is not directly applicable since we do not have a specification of P_a. How should decisions be made in this context? The literature focuses on two answers to this question: CDT and EDT.

2.2 Causal and Evidential Decision Theory

The literature on causal and evidential decision theory is vast, and we give only a very superficial overview that is intended to bring the reader up to speed on the basics. See [Bri14, Wei12] and references therein for more detailed introductions.

Evidential decision theory (endorsed in [Jef83, Ahm14]) considers the probability of the percept e *conditional on* taking the action a:

$$\arg\max_{a \in \mathcal{A}} \sum_{e \in \mathcal{E}} P(e \mid a)\, u(e) \quad \text{with} \quad P(e \mid a) = \sum_{s \in \mathcal{S}} P(e \mid s, a)P(s \mid a) \qquad \text{(EDT)}$$

Causal decision theory has several formulations [GH78, Lew81, Sky82, Joy99]; we use the one given in [Sky82], with Pearl's calculus of causality [Pea09]. According to CDT, the probability of a percept e is given by the *causal intervention* of performing action a on the causal graph from Fig. 2:

$$\arg\max_{a \in \mathcal{A}} \sum_{e \in \mathcal{E}} P(e \mid \mathsf{do}(a))\, u(e) \quad \text{with} \quad P(e \mid \mathsf{do}(a)) = \sum_{s \in \mathcal{S}} P(e \mid s, a)P(s)$$
$$\text{(CDT)}$$

where $P(e \mid \mathsf{do}(a))$ follows from (1) and marginalization over s.

The difference between CDT and EDT is how the action affects the belief about the hidden state. EDT assigns credence $P(s \mid a)$ to the hidden state s if action a is taken, while CDT assigns credence $P(s)$. A common argument for CDT is that an action under my direct control should not influence my belief about things that are not causally affected by the action. Hence $P(s)$ should be my belief in s, and not $P(s \mid a)$. (By assumption, the action does not *causally* affect the hidden state.) EDT might reply that if action a does not have the same likelihood under all hidden states s, then action a should indeed inform me about the hidden state, regardless of causal connection. The following two classical examples from the decision theory literature describe situations where CDT and EDT disagree. A formal definition of these examples can be found in the technical report [ELH15].

Example 1. (Newcomb's Problem [Noz69]) In Newcomb's Problem there are two boxes: an opaque box that is either empty or contains one million dollars and a transparent box that contains one thousand dollars. The agent can choose between taking only the opaque box ('one-boxing') and taking both boxes ('two-boxing'). The content of the opaque box is determined by a prediction about the agent's action by a very reliable predictor: if the agent is predicted to one-box, the box contains the million, and if the agent is predicted to two-box, the box is empty. In Newcomb's problem EDT prescribes one-boxing because one-boxing is evidence that the box contains a million dollars. In contrast, CDT prescribes two-boxing because two-boxing dominates one-boxing: in either case we are a thousand dollars richer, and our decision cannot causally affect the prediction. Newcomb's problem has been raised as a critique to CDT, but many philosophers insist that two-boxing is in fact the rational choice,[2] even if it means you end up poor.

Note how the decision depends on whether the action influences the belief about the hidden state (the contents of the opaque box) or not.

Newcomb's problem may appear as an unrealistic thought experiment. However, we argue that problems with similar structure are fairly common. The main structural requirement is that $P(s \mid a) \neq P(s)$ for some state or event s that is not causally affected by a. In Newcomb's problem the predictor's ability to guess the action induces an 'information link' between actions and hidden states. If the stakes are high enough, the predictor does not have to be much better than random in order to generate a *Newcomblike decision problem.* Consider for example spouses predicting the faithfulness of their partners, employers predicting the trustworthiness of their employees, or parents predicting their children's intentions. For AIs, the potential for accurate predictions is even greater, as the predictor may have access to the AI's source code. Although rarely perfect, all of these predictions are often substantially better than random.

To counteract the impression that EDT is generally superior to CDT, we also discuss the *toxoplasmosis problem.*

Example 2. (Toxoplasmosis Problem [Alt13])[3] This problem takes place in a world in which there is a certain parasite that causes its hosts to be attracted to cats, in addition to uncomfortable side effects. The agent is handed an adorable

---

[2] In a 2009 survey, 31.4 % of philosophers favored two-boxing, and 21.3 % favored one-boxing (931 responses); see http://philpapers.org/surveys/results.pl. Is that the reason there are so few wealthy philosophers?.

[3] Historically, this problem has been known as the *smoking lesion problem* [Ega07]. We consider the smoking lesion formulation confusing, because today it is universally known that smoking *does* cause lung cancer.

little kitten and is faced with the decision of whether or not to pet it. Petting the kitten feels nice and therefore yields more utility than not petting it. However, people suffering from the parasite are more likely to pet the kitten. Petting the kitten is evidence of having the parasite, so EDT recommends against it. CDT correctly observes that petting the kitten does not *cause* the parasite, and is therefore in favor of petting.

Newcomb's problem and the toxoplasmosis problem cannot be properly formalized in SDT, because SDT requires the percept-probabilities P_a to be specified, but it is not clear what the right choice of P_a would be. However, both CDT and EDT can be recast in the context of SDT by setting P_a to be $P(\cdot \mid \mathsf{do}(a))$ and $P(\cdot \mid a)$ respectively. Thus we could say that the formulation given by Savage needs a specification of the environment that tells us whether to act evidentially, causally, or otherwise.

3   Sequential Decision Making

In this section we extend CDT and EDT to the sequential case. We start by formally specifying the physicalitic model depicted in Fig. 1 in the first subsection, and discuss problems with time consistency in Sect. 3.2, before defining the extensions proper in Sects. 3.3 and 3.4. The final subsection dissects the role of the hidden state.

3.1   The Physicalistic Model

For the remainder of this paper, we assume that the agent interacts sequentially with an environment. At time step t the agent chooses an *action* $a_t \in \mathcal{A}$ and receives a *percept* $e_t \in \mathcal{E}$ which yields a *utility* of $u(e_t) \in \mathbb{R}$; the cycle then repeats for $t + 1$. A *history* is an element of $(\mathcal{A} \times \mathcal{E})^*$. We use $\textit{æ} \in \mathcal{A} \times \mathcal{E}$ to denote one interaction cycle, and $\textit{æ}_{<t}$ to denote a history of length $t - 1$. The percepts between time t and time m are denoted $e_{t:m}$. A *policy* is a function that maps a history $\textit{æ}_{<t}$ to the next action a_t. We only consider deterministic policies.

We assume that the agent is given an environment model μ, but knows neither the hidden state s nor its own future actions. The unknown hidden state may influence both percepts and actions. Actions and percepts in turn influence the entire future. The environment model μ is given by a probability distribution over hidden states and histories that factors as

$$\mu(s, \textit{æ}_{<t}) = \mu(s) \prod_{i=1}^{t-1} \mu(a_i \mid s, \textit{æ}_{<i}) \mu(e_i \mid s, \textit{æ}_{<i} a_i) \tag{2}$$

for any $t \in \mathbb{N}$. While such a factorization is possible for any distribution, we additionally demand that this factorization is *causal* according to the causal graph in Fig. 3. The distribution $\mu(a_t \mid s, \textit{æ}_{<t})$ gives the likelihood of the agent's

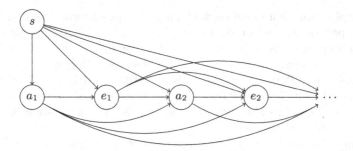

Fig. 3. The (infinite) causal graph for a sequential environment. Each action a_t and each percept e_t is represented by a node in the causal graph. Actions and percepts affect all subsequent actions and percepts: causality follows time. The hidden state s is only ever indirectly (partially) observed.

own actions provided a hidden state $s \in S$ (for example, the prior probability of an infected agent petting the kitten in the toxoplasmosis problem above). For technical reasons, this distribution must always leave some uncertainty about the actions: if the environment model assigned probability zero for an action a', the agent could not deliberate taking action a' since a' could not be conditioned on. Formally, we require $\mu(\cdot \mid s)$ to be *action-positive* for all $s \in S$:

$$\forall æ_{<t}a_t \in (\mathcal{A} \times \mathcal{E})^* \times \mathcal{A}. \ \big(\mu(æ_{<t} \mid s) > 0 \implies \mu(a_t \mid s, æ_{<t}) > 0\big) \qquad (3)$$

The distribution μ is a *model* of the environment, a belief held by the agent, but not the distribution from which the actual history is drawn. The actual history is distributed according to the true environment distribution. Because the environment contains the agent, the agent's algorithm might get modified by it and the actions that the agent actually ends up taking might not be the actions that were planned. In the end, model and reality will disagree: for example, we simultaneously assume the agent's policy π to be deterministic and the environment model to be action positive. Nevertheless, we assume the given environment model is *accurate* in the sense that it faithfully represents the environment in the ways relevant to the agent. In other words, we are interested in problems that arise during planning, not problems that arise due to poor modeling.

3.2 Time Consistency

When planning for the infinite future we need to make sure that utilities do not sum to infinity; typically this is achieved with discounting. Here, we simplify by fixing a finite $m \in \mathbb{N}$ to be the agent's *lifetime*: the agent cares about the sum of the utilities of all percepts $e_1 \dots e_m$ until and including time step m, but does not care what happens after that (presumably the agent is then retired).

In sequential decision theory we need to plan the next $m - t$ actions in time step t. We plan what we would do for all possible future percepts $e_{t:m}$ by choosing a policy $\pi : (\mathcal{A} \times \mathcal{E})^* \to \mathcal{A}$ that specifies which action we take depending on how the history plays out. For example, we take action a_t, and when we subsequently receive the percept e_t, we plan to take action a_{t+1}. Problems arise once we get to the next step and even tough we *did* take action a_t and the percept *did* turn out to be e_t, we change our mind and take a different action $\hat{a}_{t+1}$. This is called *time inconsistency*. Time inconsistency is an artifact of bad planning since the agent incorrectly anticipates her own actions. The choice of discounting can lead to time inconsistency: a sliding fixed-size horizon is time inconsistent, but a fixed finite lifetime is time consistent [LH14].

We achieve time consistency by using a fixed finite lifetime, and by calculating decisions recursively using value functions. A *value function* $V_{\mu,m}$ is a function of type $((\mathcal{A} \times \mathcal{E})^* \cup ((\mathcal{A} \times \mathcal{E})^* \times \mathcal{A})) \to \mathbb{R}$. It gives an estimate of future reward: $V_{\mu,m}^{\pi}(\text{æ}_{<t})$ and $V_{\mu,m}^{\pi}(\text{æ}_{<t}a_t)$ are estimates of how much reward the policy π will obtain in environment μ within lifetime m subsequent to history $\text{æ}_{<t}$ and $\text{æ}_{<t}a_t$ respectively. For any history $\text{æ}_{<t}$, we define $V_{\mu,m}^{\pi}(\text{æ}_{<t}) := V_{\mu,m}^{\pi}(\text{æ}_{<t}\pi(\text{æ}_{<t}))$. We say that a policy π is *optimal and time consistent for the value function* $V_{\mu,m}$ iff $\pi(\text{æ}_{<t}) = \arg\max_a V_{\mu,m}^{\pi}(\text{æ}_{<t}a)$ for all histories $\text{æ}_{<t} \in (\mathcal{A} \times \mathcal{E})^{t-1}$ and all $t \leq m$.

3.3   Sequential Evidential Decision Theory

Evidential decision theory assigns probability $P(e \mid a)$ to action a resulting in percept e (Sect. 2.2). There are two ways to generalize this to the sequential setting, depending on whether we use only the next action or the whole future policy as evidence for the next percept.

Definition 3 (Action-Evidential Decision Theory). *The* action-evidential value of a policy π with lifetime m in environment μ given history $\text{æ}_{<t}a_t$ is

$$V_{\mu,m}^{\text{aev},\pi}(\text{æ}_{<t}a_t) := \sum_{e_t} \mu(e_t \mid \text{æ}_{<t}a_t)\Big(u(e_t) + V_{\mu,m}^{\text{aev},\pi}(\text{æ}_{<t}a_te_t)\Big) \qquad \text{(SAEDT)}$$

and $V_{\mu,m}^{\text{aev},\pi}(\text{æ}_{<t}a_t) := 0$ *for* $t > m$. *Sequential Action-Evidential Decision Theory (SAEDT) prescribes adopting an optimal and time consistent policy* π *for* $V_{\mu,m}^{\text{aev}}$.

It may be argued that (SAEDT) does not take all available (deliberative) information into account. When considering the consequences of an action, future developments of the environment-policy interactions could also be used as evidence. That is, we could condition not only on the next action, but on the future policy as a whole (within the lifetime). In order to define conditional probabilities with respect to (deterministic) policies, we define the following events. For a given policy π, let $\Pi_{t:m}$ be the set of all strings consistent with π between time step t and m:

$$\Pi_{t:m} := \{\text{æ}_{1:\infty} \mid \forall t \leq i \leq m. \ \pi(\text{æ}_{<i}) = a_i\}$$

The likelihood of a next percept e_t provided a history $\text{æ}_{<t}$ and a (future) policy π followed from time step t until lifetime m (denoted $\pi_{t:m}$) is then defined as

$$\mu(e_t \mid \text{æ}_{<t}, \pi_{t:m}) := \mu(e_t \mid \text{æ}_{<t} \cap \Pi_{t:m}). \qquad (4)$$

This is an *atemporal* conditional because we are conditioning on future actions up until the end of the agent's lifetime. The conditional (4) is well-defined because we only take the actions from time step t to m into account; conditioning on policies with infinite lifetime leads to technical problems because such policies typically have μ-measure zero.

Definition 4 (Policy-Evidential Decision Theory). *The* policy-evidential value of a policy π with lifetime m in environment μ given history $\ae_{<t}a_t$ *is*

$$V^{\text{pev},\pi}_{\mu,m}(\ae_{<t}a_t) := \sum_{e_t} \mu(e_t \mid \ae_{<t}a_t, \pi_{t+1:m}) \cdot \left(u(e_t) + V^{\text{pev},\pi}_{\mu,m}(\ae_{<t}a_te_t) \right)$$

(SPEDT)

and $V^{\text{pev},\pi}_{\mu,m}(\ae_{<t}) := 0$ *for* $t > m$. Sequential Policy-Evidential Decision Theory (SPEDT) *prescribes adopting an optimal and time consistent policy* π *for* $V^{\text{pev}}_{\mu,m}$.

For one-step decisions ($m = t + 1$), SAEDT and SPEDT coincide.

To all our embedded agents, past actions constitute evidence about the hidden state. For evidential agents, this principle is extended to future actions. SAEDT and SPEDT differ in how far they extend it. The action-evidential agent only updates his belief on the action about to take place. In that sense, he only updates his belief about the next percept on events taking place *before* this percept. The policy-evidential agent takes the principle much further, using "thought-experiments" of what action he *would take in hypothetical situations*, most of which will never be realized. This is illustrated in the next example.

Example 5 (Sequential Toxoplasmosis). In our sequential variation of the toxoplasmosis problem the agent has some probability of encountering a kitten. Additionally, the agent has the option of seeing a doctor (for a fee) and getting tested for the parasite, which can then be safely removed. In the very beginning, an SPEDT agent updates his belief on the fact that if he encountered a kitten, he would not pet it, which lowers the probability that he has the parasite and makes seeing the doctor unattractive. An SAEDT agent only updates his belief about the parasite when he actually encounters a kitten, and thus prefers seeing the doctor. See Fig. 4 for more details and a graphical illustration.

The observant reader may ask whether SPEDT could be enticed to make some percepts unlikely by choosing improbable actions subsequent to them. For example, could an SPEDT agent decide on a policy of selecting highly improbable actions in case it rained to make histories with rain less likely? The answer is no, as most such policies would not be time consistent. If it does rain, the highly improbable action would usually not be the best one, and so the policy would not be prescribed by Definition 4.

3.4 Sequential Causal Decision Theory

In sequential causal decision theory we ask what would happen if we causally intervened on the node a_t of the next action and fix it to $\pi(\ae_{<t})$ according to the policy π. This is expressed by the notation $\text{do}(a_t := \pi(\ae_{<t}))$, or $\text{do}(\pi(\ae_{<t}))$ for short.

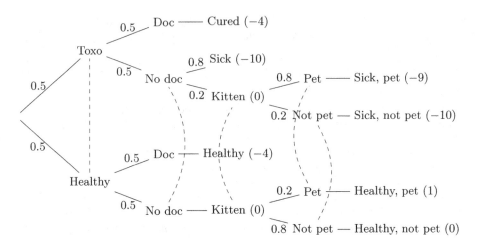

Fig. 4. One formalization of the sequential toxoplasmosis problem. Dashed lines connect states indistinguishable to the agent. The numbers on the edges indicate probabilities of the environment model μ, and the numbers in parenthesis indicate utilities of the associated percepts. In the first step, the environment selects the hidden state that is unknown to the agent. The agent then decides whether to go to the doctor. If he does not go, he may encounter a kitten which he can choose to pet or not. SAEDT and SPEDT will disagree whether going to the doctor is the best option in this scenario. [ELH15] contains the full calculations.

Definition 6 (Sequential Causal Decision Theory). *The* causal value of a policy π with lifetime m in environment μ given history $\text{\ae}_{<t}a_t$ *is*

$$V_{\mu,m}^{\text{cau},\pi}(\text{\ae}_{<t}a_t) := \sum_{e_t \in \mathcal{E}} \mu(e_t \mid \text{\ae}_{<t}, \text{do}(a_t))\Big(u(e_t) + V_{\mu,m}^{\text{cau},\pi}(\text{\ae}_{<t}a_t e_t)\Big) \quad \text{(SCDT)}$$

and $V_{\mu,m}^{\text{cau},\pi}(\text{\ae}_{<t}a_t) := 0$ for $t > m$. Sequential Causal Decision Theory (SCDT) prescribes adopting an optimal and time consistent policy π for $V_{\mu,m}^{\text{cau}}$.

For sequential evidential decision theory we discussed two versions (SAEDT) and (SPEDT), based on next action and future policy respectively. In SCDT we perform the causal intervention $\text{do}(a_t := \pi(\text{\ae}_{<t}))$. We could also consider a policy-causal decision theory by replacing $\mu(e_t \mid \text{\ae}_{<t}, \text{do}(a_t))$ with $\mu(e_t \mid \text{\ae}_{<t}, \text{do}(\pi_{t:m}))$ in Definition 6. The causal intervention $\text{do}(\pi_{t:m})$ of a policy π between time step t and time step m is defined as

$$\mu(e_t \mid \text{\ae}_{<t}, \text{do}(\pi_{t:m})) := \sum_{e_{t+1:m}} \mu(e_{t:m} \mid \text{\ae}_{<t}, \text{do}(a_t := \pi(\text{\ae}_{<t}), \ldots, a_m := \pi(\text{\ae}_{<m}))).$$

$$(5)$$

However, since the interventions are causal, we do not get any extra evidence from the future interventions. Therefore policy-causal decision theory is the same as action-causal decision theory:

Proposition 7 (Policy-Causal = Action-Causal). *For all histories* $æ_{<t} \in (\mathcal{A} \times \mathcal{E})^*$ *and all* $e_t \in \mathcal{E}$, *we have* $\mu(e_t \mid æ_{<t}, \text{do}(\pi_{t:m})) = \mu(e_t \mid æ_{<t}, \text{do}(\pi(æ_{<t})))$.

We defer the proof to the end of this section. The following two examples illustrate the difference between SCDT and SAEDT/SPEDT in sequential settings.

Example 8 (Newcomb with Precommitment). In this variation to Newcomb's problem the agent first has the option to pay \$300,000 to sign a contract that binds the agent to pay \$2000 in case of two-boxing. An SAEDT or SPEDT agent knows that he will one-box anyways and hence has no need for the contract. An SCDT agent knows that she favors two-boxing, but signs the contract only if this occurs before the prediction is made (so it has a chance of causally affecting the prediction). With the contract in place, one-boxing is the dominant action, and thus the SCDT agent is predicted to one-box.

Example 9 (Newcomb with Looking). In this variation to Newcomb's problem the agent may look into the opaque box before making the decision which box to take. An SCDT agent is indifferent towards looking because she will take both boxes anyways. However, an SAEDT or SPEDT agent will avoid looking into the box, because once the content is revealed he two-boxes.

3.5   Expansion Over the Hidden State

The difference between sequential versions of EDT and CDT is how they update their prediction of a next percept e_t (Definitions 3, 4 and 6). The following proposition expands the different beliefs in terms of the hidden state.

Proposition 10. *For all histories* $æ_{<t}a_te_t \in (\mathcal{A} \times \mathcal{E})^*$ *the following holds for the next-percept beliefs of SAEDT, SPEDT and SCDT respectively:*

$$\mu(e_t \mid æ_{<t}a_t) = \sum_{s \in \mathcal{S}} \mu(s \mid æ_{<t}a_t)\mu(e_t \mid s, æ_{<t}a_t) \tag{6}$$

$$\mu(e_t \mid æ_{<t}, \pi_{t:m}) = \sum_{s \in \mathcal{S}} \mu(s \mid æ_{<t}, \pi_{t:m})\mu(e_t \mid s, æ_{<t}, \pi_{t:m}) \tag{7}$$

$$\mu(e_t \mid æ_{<t}, \text{do}(a_t)) = \sum_{s \in \mathcal{S}} \mu(s \mid æ_{<t})\mu(e_t \mid s, æ_{<t}a_t) \tag{8}$$

Proof. For the action-evidential conditional we take the joint distribution with s, and then split off e_t:

$$\mu(e_t \mid æ_{<t}a_t) = \frac{\sum_{s \in \mathcal{S}} \mu(s, æ_{<t}a_te_t)}{\mu(æ_{<t}a_t)} = \frac{\sum_{s \in \mathcal{S}} \mu(s, æ_{<t}a_t)\mu(e_t \mid s, æ_{<t}a_t)}{\mu(æ_{<t}a_t)}$$

$$= \sum_{s \in \mathcal{S}} \mu(s \mid æ_{<t}a_t)\mu(e_t \mid s, æ_{<t}a_t)$$

Similarly for the policy-evidential conditional:

$$\mu(e_t \mid \text{æ}_{<t}, \pi_{t:m}) = \frac{\sum_{s \in \mathcal{S}} \mu(s, \text{æ}_{<t}\pi(\text{æ}_{<t})e_t, \pi_{t+1:m})}{\mu(\text{æ}_{<t}, \pi_{t:m})}$$

$$= \frac{\sum_{s \in \mathcal{S}} \mu(s, \text{æ}_{<t}\pi(\text{æ}_{<t}), \pi_{t+1:m})\mu(e_t \mid s, \text{æ}_{<t}\pi(\text{æ}_{<t}), \pi_{t+1:m})}{\mu(\text{æ}_{<t}, \pi_{t:m})}$$

$$= \frac{\sum_{s \in \mathcal{S}} \mu(s, \text{æ}_{<t}, \pi_{t:m})\mu(e_t \mid s, \text{æ}_{<t}\pi(\text{æ}_{<t}), \pi_{t+1:m})}{\mu(\text{æ}_{<t}, \pi_{t:m})}$$

$$= \sum_{s \in \mathcal{S}} \mu(s \mid \text{æ}_{<t}, \pi_{t:m})\mu(e_t \mid s, \text{æ}_{<t}\pi(\text{æ}_{<t}), \pi_{t+1:m})$$

$$= \sum_{s \in \mathcal{S}} \mu(s \mid \text{æ}_{<t}, \pi_{t:m})\mu(e_t \mid s, \text{æ}_{<t}, \pi_{t:m})$$

For the causal conditional we turn to the rules of the do-operator [Pea09, Theorem 3.4.1]. The first equality below holds by definition. In the denominator of the second equality we can use Rule 3 (deletion of actions) to remove $\text{do}(a_t)$ because the do-operator removes all incoming edges to a_t and makes a_t independent of the history $\text{æ}_{<t}$. In the numerator of the second equality we use the definition of do (1):

$$\mu(e_t \mid \text{æ}_{<t}, \text{do}(a_t)) = \frac{\mu(\text{æ}_{<t}, e_t \mid \text{do}(a_t))}{\mu(\text{æ}_{<t} \mid \text{do}(a_t))}$$

$$= \frac{\sum_{s \in \mathcal{S}} \mu(s, \text{æ}_{<t})\mu(e_t \mid s, \text{æ}_{<t}a_t)}{\mu(\text{æ}_{<t})}$$

$$= \sum_{s \in \mathcal{S}} \mu(s \mid \text{æ}_{<t})\mu(e_t \mid s, \text{æ}_{<t}a_t) \qquad \square$$

Proposition 10 shows that between SCDT and SAEDT, the difference in opinion about e_t only depends on differences in their (acausal) *posterior belief* $\mu(s \mid \ldots)$ about the hidden state. SCDT and SAEDT thus become equivalent in scenarios where there is only one hidden state s^* with $\mu(s^*) = 1$, as this renders $\mu(s^* \mid \text{æ}_{<t}) = \mu(s^* \mid \text{æ}_{<t}a_t) = \mu(s^*) = 1$. SPEDT, on the other hand, may disagree with the other two also after a hidden state has been fixed.

From a problem modeler's perspective, it is also instructive to consider the effect of moving uncertainty between the hidden state and environmental stochasticity. For two different environment models μ and μ', the action and percept probabilities may be identical (i.e., $\mu(a_t \mid \text{æ}_{<t}) = \mu'(a_t \mid \text{æ}_{<t})$ and $\mu(e_t \mid \text{æ}_{<t}a_t) = \mu'(e_t \mid \text{æ}_{<t}a_t)$) even though μ and μ' have non-isomorphic sets of hidden states $\mathcal{S}$ and $\mathcal{S}'$. For example, given any μ, an environment model μ' with a single hidden state s_0, $\mu'(s_0) = 1$, may be constructed from μ by $\mu'(s_0, \text{æ}_{<t}) := \sum_{s \in \mathcal{S}} \mu(s, \text{æ}_{<t})$. The transformation will not affect SAEDT and SPEDT, as the definitions of their value functions only depends on the 'observable' action- and percept-probabilities $\mu(a_t \mid \text{æ}_{<t})$ and $\mu(e_t \mid \text{æ}_{<t}a_t)$ which are preserved between μ and μ'. But the transformation will change SCDT's behavior in any μ where SCDT disagrees with SAEDT, as SCDT and SAEDT are equivalent in μ' that

only has a single hidden state. That SCDT depends on what uncertainty is captured by the hidden state is unsurprising given that the hidden state has a special place in the causal structure of the problem. Ultimately, the modeler must decide what uncertainty to put in the hidden state, and what to attribute to environmental stochasticity. A general principle for how to do this is still an open question [SF14b].

The value functions of SAEDT, SPEDT and SCDT can be rewritten in the following *iterative forms*, where the latter form uses Proposition 10. Numbers above equality signs reference a justifying equation. Let $a_i := \pi(\text{æ}_{<i})$ for $i \geq t$:

$$V_{\mu,m}^{\text{aev},\pi}(\text{æ}_{<t}) = \sum_{k=t}^{m} \sum_{e_{t:k}} u(e_k) \prod_{i=t}^{k} \mu(e_i \mid \text{æ}_{<i}a_i)$$

$$\overset{(6)}{=} \sum_{k=t}^{m} \sum_{e_{t:k}} u(e_k) \prod_{i=t}^{k} \sum_{s \in \mathcal{S}} \mu(s \mid \text{æ}_{<i}a_i)\mu(e_i \mid s, \text{æ}_{<i}a_i)$$

$$V_{\mu,m}^{\text{pev},\pi}(\text{æ}_{<t}) = \sum_{k=t}^{m} \sum_{e_{t:k}} u(e_k) \prod_{i=t}^{k} \mu(e_i \mid \text{æ}_{<i}, \pi_{i:m})$$

$$\overset{(7)}{=} \sum_{k=t}^{m} \sum_{e_{t:k}} u(e_k) \prod_{i=t}^{k} \sum_{s \in \mathcal{S}} \mu(s \mid \text{æ}_{<i}\pi_{i:m})\mu(e_i \mid s, \text{æ}_{<i}, \pi_{i:m})$$

$$V_{\mu,m}^{\text{cau},\pi}(\text{æ}_{<t}) = \sum_{k=t}^{m} \sum_{e_{t:k}} u(e_k) \prod_{i=t}^{k} \mu(e_i \mid \text{æ}_{<i}, \text{do}(a_i))$$

$$\overset{(8)}{=} \sum_{k=t}^{m} \sum_{e_{t:k}} u(e_i) \prod_{i=t}^{k} \sum_{s \in \mathcal{S}} \mu(s \mid \text{æ}_{<i})\mu(e_i \mid s, \text{æ}_{<i}a_i)$$

Proof of Proposition 7. By the definition (5) of $\text{do}(\pi_{t:m})$,

$$\mu(e_t \mid \text{æ}_{<t}, \text{do}(\pi_{t:m})) = \sum_{e_{t+1:m}} \mu(e_{t:m} \mid \text{æ}_{<t}, \text{do}(a_t := \pi(\text{æ}_{<t}), \ldots, a_m := \pi(\text{æ}_{<m})))$$

$$= \sum_{s,e_{t+1:m}} \mu(s \mid \text{æ}_{<t})\mu(e_{t:m} \mid s, \text{æ}_{<t}, \text{do}(\pi(\text{æ}_{<t}), \ldots, \pi(\text{æ}_{<m})))$$

$$\overset{(1)}{=} \sum_{s,e_{t+1:m}} \mu(s \mid \text{æ}_{<t}) \prod_{i=t}^{m} \mu(e_i \mid s, \text{æ}_{<i}\pi(\text{æ}_{<i}))$$

$$= \sum_{s} \mu(s \mid \text{æ}_{<t})\mu(e_t \mid s, \text{æ}_{<t}\pi(\text{æ}_{<t}))$$

$$\overset{(8)}{=} \mu(e_t \mid \text{æ}_{<t}, \text{do}(\pi(\text{æ}_{<t})))$$

The second equality follows from the equivalence $P(\cdot) = \sum_s P(s)P(\cdot \mid s)$ applied to the distribution $\mu(\cdot \mid \text{æ}_{<t}, \text{do}(a_t := \pi(\text{æ}_{<t}), \ldots, a_m := \pi(\text{æ}_{<m})))$, and the third equality by (repeated) application of (1) to $\mu(\text{æ}_{t:m} \mid s, \text{æ}_{<t}) = \prod_{i=t}^{m} \mu(a_i \mid s, \text{æ}_{<i})\mu(e_i \mid s, \text{æ}_{<i}a_i)$. □

4   Discussion

Our paper is a first stab at the problem of how physicalistic agents should make sequential decisions. CDT and EDT provide an existing basis for non-dualistic decision making, which we extended to the sequential setting. There are two natural ways for making sequential evidential decisions: do I update my beliefs about the hidden state based on my next action ('what I do next', SAEDT) or my whole policy ('the kind of agent I am', SPEDT)? By Proposition 7, this distinction does not exist for causal decision theory, because with that theory the agent does not consider its own actions evidence at all. Therefore we have only one version of sequential causal decision theory, SCDT.

Table 1. Decisions made by SAEDT, SPEDT and SCDT in Examples 1, 2, 5, 8, and 9. The latter three examples are sequential. Winning moves are in italics; in Newcomb with looking the winning move is to be indifferent and one-box. Because Savage decision theory is dualistic, these problems cannot be properly formalized in it.

	SAEDT	SPEDT	SCDT
Newcomb	*1-box*	*1-box*	2-box
Newcomb w/ precommit	*not commit, 1-box*	*not commit, 1-box*	commit, 1-box
Newcomb w/ looking	not look, 1-box	not look, 1-box	indifferent, 2-box
Toxoplasmosis	not pet	not pet	*pet*
Seq. Toxoplasmosis	doctor, not pet	not doctor, not pet	*doctor, pet*

To illustrate the differences between the decision theories, we discussed three variants of Newcomb's problem (Examples 1, 8, and 9) and two variants of the toxoplasmosis problem (Examples 2 and 5). We implemented SCDT, SAEDT, and SPEDT; Table 1 shows their behavior on those examples.[4]

So which decision theory is better? The answer to this question depends on which decision you consider to be *correct* (or even *rational*) in each of the problems. We posit that ultimately, what counts is not whether your decision algorithm is theoretically pleasing, but *whether you win*. Winning means getting the most utility. If maximizing utility involves making crazy decisions, then this is what you should do!

In Newcomb's problem, winning means one-boxing, because you end up richer. In the toxoplasmosis problem, winning means petting the kitten, because that yields more utility. (S)CDT performs suboptimally in the Newcomb variations, while the evidential decision theories perform suboptimally in the toxoplasmosis variations. This entails that neither CDT nor EDT are the final answer to the problem of non-dualistic decision making.

Furthermore, neither CDT nor EDT agents are fully physicalistic: they do not model the environment to contain themselves [SF14b]. For example, when

---

[4] Source code available at http://jan.leike.name/.

playing a prisoner's dilemma against your own source code [SF15], your opponent defects if and only if you defect. This *logical* connection between your action and your opponent's is disregarded in the formalization based on causal graphical models that we discuss here because it is not causal.

Timeless decision theory [Yud10] and *updateless decision theory* [SF14b] are recent attempts of more physicalistic decision theories. However, so far both have eluded explicit formalization [SF15]. We conclude that finding a physicalistic decision theory remains an important open problem in artificial intelligence research.

List of Notation

$:=$ defined to be equal

$\mathbb{N}$ the natural numbers, starting with 0

$\mathbb{R}$ the real numbers

$\mathcal{A}$ the (finite) set of possible actions

$\mathcal{E}$ the (finite) set of possible percepts

$\mathcal{S}$ the set of hidden states

u the utility function $u : \mathcal{E} \to [0,1]$

a_t the action in time step t

e_t the percept in time step t

$æ_{<t}$ the first $t-1$ interactions, $a_1 e_1 a_2 e_2 \ldots a_{t-1} e_{t-1}$

$æ_{i:k}$ the interactions between and including time step i and time step k, $a_i e_i a_{i+1} e_{i+1} \ldots a_k e_k$

$æ_{1:\infty}$ a history of infinite length

s a hidden state

π a deterministic policy, i.e., a function $\pi : (\mathcal{A} \times \mathcal{E})^* \to \mathcal{A}$

$\pi_{t:k}$ policy π restricted to the time steps between and including t and k

$V_{\mu,m}^{\text{aev},\pi}$ action-evidential value of policy π in environment μ up to time step m, defined in (SAEDT)

$V_{\mu,m}^{\text{pev},\pi}$ policy-evidential value of policy π in environment μ up to time step m, defined in (SPEDT)

$V_{\mu,m}^{\text{cau},\pi}$ causal value of policy π in environment μ up to time step m, defined in (SCDT)

k, i time steps, natural numbers

t (current) time step

m lifetime of the agent

P_a distribution over percepts induced by action a in (SDT)

P distribution over percepts and actions in one-shot decision making

μ an accurate environment model

Acknowledgments. This work was in part supported by ARC grant DP120100950. It started at a MIRIxCanberra workshop sponsored by the Machine Intelligence Research Institute. Mayank Daswani and Daniel Filan contributed in the early stages of this paper and we thank them for interesting discussions and helpful suggestions. We also thank Nate Soares for useful feedback.

References

[Ahm14] Ahmed, A.: Evidence, Decision and Causality. Cambridge University Press, Cambridge (2014)

[Alt13] Altair, A.: A comparison of decision algorithms on Newcomblike problems. Technical report, Machine Intelligence Research Institute (2013). http://intelligence.org/files/Comparison.pdf

[Bos14] Bostrom, N.: Superintelligence: Paths, Dangers, Strategies. Oxford University Press, Oxford (2014)

[Bri14] Briggs, R.: Normative theories of rational choice: expected utility. In: Zalta, E.N. (ed.) The Stanford Encyclopedia of Philosophy. Fall 2014 edition (2014)

[Ega07] Egan, A.: Some counterexamples to causal decision theory. Philos. Rev. **116**(1), 93–114 (2007)

[ELH15] Everitt, T., Leike, J., Hutter, M.: Sequential extensions of causal and evidential decision theory. Technical report, Australian National University (2015). http://arxiv.org/abs/1506.07359

[GH78] Gibbard, A., Harper, W.L.: Counterfactuals and two kinds of expected utility. In: Harper, W.L., Stalnaker, R., Pearce, G. (eds.) Foundations and Applications of Decision Theory. The University of Western Ontario Series in Philosophy of Science, vol. 15, pp. 125–162. Springer, Heidelberg (1978)

[Jef83] Jeffrey, R.C.: The Logic of Decision, 2nd edn. University of Chicago Press, Chicago (1983)

[Joy99] Joyce, J.M.: The Foundations of Causal Decision Theory. Cambridge University Press, Cambridge (1999)

[Lew81] Lewis, D.: Causal decision theory. Australas. J. Philos. **59**(1), 5–30 (1981)

[LFY+14] LaVictoire, P., Fallenstein, B., Yudkowsky, E., Barasz, M., Christiano, P., Herreshoff, M.: Program equilibrium in the prisoner's dilemma via Lö's theorem. In: AAAI Workshop on Multiagent Interaction without Prior Coordination (2014)

[LH14] Lattimore, T., Hutter, M.: General time consistent discounting. Theor. Comput. Sci. **519**, 140–154 (2014)

[Noz69] Nozick, R.: Newcomb's problem and two principles of choice. In: Rescher, N. (ed.) Essays in Honor of Carl G. Hempel. SL, vol. 24, pp. 114–146. Springer, Heidelberg (1969)

[OR12] Orseau, L., Ring, M.: Space-time embedded intelligence. In: Bach, J., Goertzel, B., Iklé, M. (eds.) AGI 2012. LNCS, vol. 7716, pp. 209–218. Springer, Heidelberg (2012)

[Pea09] Pearl, J.: Causality, 2nd edn. Cambridge University Press, New York (2009)

[RDT+15] Russell, S., Dewey, D., Tegmark, M., Kramar, J., Mallah, R.: Research priorities for robust and beneficial artificial intelligence. Technical report, Future of Life Institute (2015). http://futureoflife.org/static/data/documents/research_priorities.pdf

[RN10] Russell, S.J., Norvig, P.: Artificial Intelligence. A Modern Approach, 3rd edn. Prentice Hall, Saddle River (2010)

[Sav72] Savage, L.J.: The Foundations of Statistics. Dover Publications, New York (1972)

[SF14a] Soares, N., Fallenstein, B.: Aligning superintelligence with human interests: A technical research agenda. Technical report, Machine Intelligence Research Institute (2014). http://intelligence.org/files/TechnicalAgenda. pdf

[SF14b] Soares, N., Fallenstein, B.: Toward idealized decision theory. Technical report, Machine Intelligence Research Institute (2014). http://intelligence. org/files/TowardIdealizedDecisionTheory.pdf

[SF15] Soares, N., Fallenstein, B.: Two attempts to formalize counterpossible reasoning in deterministic settings. In: Bieger, J., Goertzel, B., Potapov, A. (eds.) Artificial General Intelligence. LNCS, vol. 9205. Springer, Heidelberg (2015)

[Sky82] Skyrms, B.: Causal decision theory. J. Philos. **79**(11), 695–711 (1982)

[Wei12] Weirich, P.: Causal decision theory. In: Zalta, E.N. (ed.) The Stanford Encyclopedia of Philosophy. Winter 2012 edition (2012)

[Yud08] Yudkowsky, E.: Artificial intelligence as a positive and negative factor in global risk. In: Bostrom, N., M ćirković, M. (eds.) Global Catastrophic Risks, pp. 308–345. Oxford University Press (2008)

[Yud10] Yudkowsky, E.: Timeless decision theory. Technical report, Machine Intelligence Research Institute (2010). http://intelligence.org/files/TDT.pdf

MOPIC Properties in the Representation of Preferences by a 2-Additive Choquet Integral

Brice Mayag[(✉)]

LAMSADE, University Paris Dauphine, Place du Maréchal de Lattre de Tassigny,
75775 Paris Cedex 16, France
brice.mayag@dauphine.fr

Abstract. In the context of Multiple Criteria Decision aiding (MCDA), we present necessary conditions to obtain a representation of a cardinal information by a Choquet integral w.r.t a 2-additive capacity. A cardinal information is a preferential information provided by a Decision Maker (DM) containing a strict preference, a quaternary and indifference relations. Our work is focused on the representation of a cardinal information by a particular Choquet integral defined by a 2-additive capacity. Used as an aggregation function, it arises as a generalization of the arithmetic mean, taking into account the interaction between two criteria. Then, it is a good compromise between simple models like arithmetic mean and complex models like general Choquet integral. We consider also the set of fictitious alternatives called binary alternatives or binary actions from which the Choquet integral w.r.t a 2-additive capacity can be entirely specified. The proposed MOPIC (MOnotonicity of Preferential Information for Cardinal) conditions can be viewed as an alternative to balanced cyclones which are complex necessary and sufficient conditions, used in the characterization of a 2-additive Choquet integral through a cardinal information.

Keywords: MCDA · Preference modeling · Choquet integral · MAC-BETH

1 Introduction

MultiCriteria Decision Aid (MCDA) aims at representing the preferences of a Decision-Maker (DM) on a set of alternatives (or actions, options) X evaluated over a finite set of attributes or criteria $N = \{1, \ldots, n\}$ $(n > 1)$ often conflicting. An alternative can be identified as an element $x = (x_1, \ldots, x_n)$ of the Cartesian product $X = X_1 \times \cdots \times X_n$, where $X_1, \ldots, X_n$ represent the set of points of view or attributes. The Multi-Attribute Utility Theory (MAUT) is one of the decision models usually used to represent the preferences of the DM.

In practice, MAUT elaborates a preference relation over X by asking to the DM some pairwise comparisons of alternatives on a finite subset X' of X. X' is called a learning data set (or reference set) and has a small size in general. Hence we get a preference relation $\succsim_{X'}$ on X'. The question is then: how to construct

© Springer International Publishing Switzerland 2015
T. Walsh (Ed.): ADT 2015, LNAI 9346, pp. 222–235, 2015.
DOI: 10.1007/978-3-319-23114-3_14

a preference relation $\succsim_X$ on X, so that $\succsim_X$ is an extension of $\succsim_{X'}$? To this end, people usually suppose that $\succsim_X$ is representable by an overall utility function:

$$x \succsim_X y \Leftrightarrow F(U(x)) \geq F(U(y)) \tag{1}$$

where $U(x) = (u_1(x_1), \ldots, u_n(x_n))$, $u_i : X_i \rightarrow \mathbb{R}$ is called a utility function, and $F : \mathbb{R}^n \rightarrow \mathbb{R}$ is an aggregation function. Usually, we consider a family of aggregation functions characterized by a parameter vector θ (e.g., a weight distribution over the criteria). The parameter vector θ can be deduced from the knowledge of $\succsim_{X'}$, that is, we determine the possible values of θ for which (1) is fulfilled over X'.

The aggregation function F we study is the Choquet integral w.r.t. a 2-additive capacity, the latter being the parameter vector. To identify this parameter vector, we assume that the DM can provide cardinal information (preferences given with preference intensity) on a particular set of alternatives X' called the set of binary actions denoted by $\mathcal{B}$. A binary action is a fictitious alternative which takes either the neutral value **0** for all criteria, or the neutral value **0** for all criteria except for one or two criteria for which it takes the satisfactory value **1**. The binary actions are used in many applications through the MACBETH methodology [1,2,4].

In [10], a characterization of the representation of a cardinal information on binary actions by a 2-additive Choquet integral have been proposed. This result is based on some cycles called *cyclone* in a directed graph where multiple edges are allowed between two vertices. The main disadvantage of this axiomatization is the difficulty to implement these conditions in practice, because cyclone are very complex to detect and not easy to understand during the decision process. Therefore, an alternative to these axioms is the use of methods dealing with inconsistencies based on techniques of linear programming, and relaxing constraints when a linear program is infeasible [9].

Our aim is to provide some necessary and simple conditions, called MOPIC properties, which can be tested before the use of the previous algorithm. These conditions, directly related to the monotonicity constraints of a 2-additive capacity, can help the DM to better understand and explain his inconsistent judgments when his preferences are not compatible with MOPIC properties. We also provide some sufficient conditions to obtain a cardinal information representable by a Choquet integral w.r.t. a 2-additive capacity. These sufficient conditions concern a particular case of cardinal information called "quasi-ordinal information".

The next section presents basic concepts we need in the representation of a cardinal information by a 2-additive Choquet integral, while the last section is dedicated to the four MOPIC properties.

2   Basic Concepts

We assume that, given two alternatives x and y the DM is able to assess the *difference of attractiveness* between x and y when he prefers strictly x to y. The difference of attractiveness will be provided under the form of semantic categories

d_s, $s = 1, \ldots, q$ defined so that, if $s < t$, any difference of attractiveness in the class d_s is smaller than any difference of attractiveness in the class d_t. The MACBETH approach [2] uses the following six semantic categories: d_1 = very weak, d_2 = weak, d_3 = moderate, d_4 = strong, d_5 = very strong, d_6 = extreme. For a subset A of N, the notation $z = (x_A, y_{N-A})$ means that z is defined by $z_i = x_i$ if $i \in A$, and $z_i = y_i$ otherwise.

2.1   Choquet Integral w.r.t. a 2-Additive Capacity

The Choquet integral is a well-known aggregation function used in MCDA taking into account interaction phenomena between criteria. We define the Choquet integral w.r.t a 2-additive capacity [6] below.

Definition 1.   *1. A capacity on N is a set function $\mu : 2^N \to [0,1]$ such that:*
  (a) $\mu(\emptyset) = 0$
  (b) $\mu(N) = 1$
  (c) $\forall A, B \in 2^N$, $[A \subseteq B \Rightarrow \mu(A) \leq \mu(B)]$ (monotonicity).
2. The Möbius transform (see [3]) of a capacity μ on N is a function $m : 2^N \to \mathbb{R}$ defined by:

$$m(T) := \sum_{K \subseteq T} (-1)^{|T \setminus K|} \mu(K), \forall T \in 2^N. \tag{2}$$

3. A capacity μ on N is 2-additive if
  • For all subsets T of N such that $|T| > 2$, $m(T) = 0$;
  • There exists a subset B of N such that $|B| = 2$ and $m(B) \neq 0$.

Notation. Our notation for a capacity μ and its Möbius transform m are simplified by using the following shorthand: $\mu_i := \mu(\{i\})$, $\mu_{ij} := \mu(\{i,j\})$, $m_i := m(\{i\})$, $m_{ij} := m(\{i,j\})$, for all $i, j \in N$, $i \neq j$. Whenever we use i and j together, it always means that they are different.

Given $x := (x_1, \ldots, x_n) \in \mathbb{R}_+^n$, the Choquet integral w.r.t. a 2-additive capacity μ (called for short a *2-additive Choquet integral*) can be written as follows (see [7]):

$$C_\mu(x) = \sum_{i=1}^n v_i x_i - \frac{1}{2} \sum_{\{i,j\} \subseteq N} I_{ij} |x_i - x_j| \tag{3}$$

where

1. The index v_i given by

$$v_i := \sum_{K \subseteq N \setminus i} \frac{(n - |K| - 1)! |K|!}{n!} (\mu(K \cup i) - \mu(K)) \tag{4}$$

represents the importance of the criterion i and corresponds to the Shapley value of μ.
2. The index I_{ij} given by

$$I_{ij} := \mu_{ij} - \mu_i - \mu_j \tag{5}$$

represents the interaction between the two criteria i and j.

2.2 Binary Actions and Cardinal Information

We assume that the DM is able to identify for each criterion i two reference levels:

1. A reference level $\mathbf{1}_i$ in X_i which he considers as good and completely satisfying if he could obtain it on criterion i, even if more attractive elements could exist. This special element corresponds to the *satisficing level* in the theory of bounded rationality of Simon [12].
2. A reference level $\mathbf{0}_i$ in X_i which he considers neutral on i. The neutral level is the absence of attractiveness and repulsiveness. The existence of this neutral level has roots in psychology [13], and is used in bipolar models like Cumulative Prospect Theory [14].

We set for convenience $u_i(\mathbf{1}_i) = 1$ and $u_i(\mathbf{0}_i) = 0$. A *binary action or binary alternative* is an element of the set

$$\mathcal{B} = \{\mathbf{0}_N, \ (\mathbf{1}_i, \mathbf{0}_{N-i}), \ (\mathbf{1}_{ij}, \mathbf{0}_{N-ij}), \ i,j \in N, i \neq j\} \subseteq X$$

where

- $\mathbf{0}_N = (\mathbf{1}_\emptyset, \mathbf{0}_N) =: a_0$ is an action considered neutral on all criteria.
- $(\mathbf{1}_i, \mathbf{0}_{N-i}) =: a_i$ is an action considered satisfactory on criterion i and neutral on the other criteria.
- $(\mathbf{1}_{ij}, \mathbf{0}_{N-ij}) =: a_{ij}$ is an action considered satisfactory on criteria i and j and neutral on the other criteria.

Using the Choquet integral, we get the following consequences:

1. For any capacity μ,

$$C_\mu(U((\mathbf{1}_A, \mathbf{0}_{N-A}))) = \mu(A), \ \forall A \subseteq N. \tag{6}$$

2. For any 2-additive capacity,

$$C_\mu(U(a_0)) = 0 \tag{7}$$

$$C_\mu(U(a_i)) = \mu_i = v_i - \frac{1}{2} \sum_{k \in N, \ k \neq i} I_{ik} \tag{8}$$

$$C_\mu(U(a_{ij})) = \mu_{ij} = v_i + v_j - \frac{1}{2} \sum_{k \in N, \ k \notin \{i,j\}} (I_{ik} + I_{jk}) \tag{9}$$

The last two equations come from general relations between the capacity μ and interaction (see [5] for details). Generally the DM is able to compare some alternatives using his knowledge of the problem, his experience, etc. These alternatives form a set of reference alternatives and allow to determine the parameters of a model (weights, utility functions, subjective probabilities,...) in the decision process (see [8] for more details). As shown by the previous Eqs. (7), (8), (9), it should be sufficient to get some preferential information from the DM only on binary actions. To entirely determine the 2-additive capacity, this preferential information is expressed by the following relations:

- $P = \{(x, y) \in \mathcal{B} \times \mathcal{B} : \text{the DM strictly prefers } x \text{ to } y\}$,
- $I = \{(x, y) \in \mathcal{B} \times \mathcal{B} : \text{the DM is indifferent between } x \text{ and } y\}$,
- For the semantic category "d_k", $k \in \{1, ..., q\}$, $P_k = \{(x, y) \in P \text{ such that DM }$ judges the difference of attractiveness between x and y belonging to the class "d_k"\}. If there is no ambiguity, a category d_s will be simply designated by s.

Without loss of generality, we will suppose that all the relations P_k are nonempty (we can always redefine the number q when some P_k are empty). The relation P is irreflexive and asymmetric while I is reflexive and symmetric.

Definition 2. *The cardinal information on $\mathcal{B}$ is the structure $\{P, I, P_1, \ldots, P_q\}$.*

The cardinal information is used also in the MACBETH methodology [2]. Now we will suppose P to be nonempty for any cardinal information $\{P, I, P_1, \ldots, P_q\}$ ("non triviality axiom") and $P = P_1 \cup P_2 \cup \cdots \cup P_q$.

2.3 The Representation of the Cardinal Information by a 2-Additive Choquet Integral

A cardinal information $\{P, I, P_1, \ldots, P_q\}$ is said to be *representable by a 2-additive Choquet integral* if there exists a 2-additive capacity μ such that:

1. $\forall x, y \in \mathcal{B}, \; x \, P \, y \Rightarrow C_\mu(U(x)) > C_\mu(U(y))$,
2. $\forall x, y \in \mathcal{B}, \; x \, I \, y \Rightarrow C_\mu(U(x)) = C_\mu(U(y))$,
3. $\forall x, y, z, w \in \mathcal{B}, \forall s, t \in \{1, \ldots, q\}$ s.t. $s < t$, $\left[\begin{array}{c} (x, y) \in P_t \\ (z, w) \in P_s \end{array} \right\} \Rightarrow C_\mu(U(x)) - $

$$C_\mu(U(y)) > C_\mu(U(z)) - C_\mu(U(w)) \Big]$$

Necessary and sufficient conditions to represent a cardinal information by a 2-additive Choquet integral are given in [10]. These conditions are based on some cycles called *cyclones* in a directed graph (multigraph) where multiple edges are allowed between two vertices. Figure 1 represents an example of a multigraph computed by using this characterization when $N = 1, 2, 3, 4$, $P_1 = \{(a_{23}, a_2)\}$ and $P_2 = \{(a_1, a_{23}); (a_4, a_0)\}$. Because cyclones are very difficult to detect and understand in practice , there is a real need to find some simple necessary or sufficient conditions to represent a cardinal information. The next section provides MOPIC properties which are necessary conditions, simplest than those "cyclones", to get this representation.

Before defining MOPIC properties, we need to introduce the *monotonicity relation M* on $\mathcal{B}$ translating the simple monotonicity conditions, on pairs and singletons, of a 2-additive capacity. For each (x, y) in $\{(a_i, a_0), i \in N\}$ $\cup \{(a_{ij}, a_i), i, j \in N, i \neq j\}$,

$$x \, M \, y \text{ if not } (x \, (P \cup I) \, y). \tag{10}$$

A path of $(P \cup I \cup M)$ from x to y is denoted by $x \, TC \, y$ and $x \, TC_{P_l} \, y$ if this path contains an element of P_l, $l \in \{1, \ldots, q\}$. If there exists a cycle of $(I \cup M)$ containing x and y, we use the notation $x \sim y$. For all $i, j \in N$, the notation $i \vee j$ denotes one of the two criteria i or j.

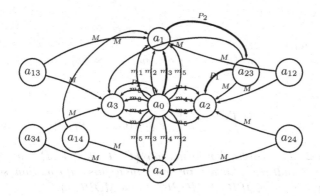

Fig. 1. A Graph of 4 criteria with some "cyclones"

Proposition 1. *Let be μ a 2-additive capacity and $\{P, I, P_1, \ldots, P_q\}$ a cardinal information on $\mathcal{B}$. For all x, y, z, w in $\mathcal{B}$.*

$$\text{If} \begin{cases} x \ P_l \ y \\ z \ TC_{P_h} \ w \\ l < h \end{cases} \text{then } C_\mu(U(z)) - C_\mu(U(w)) > C_\mu(U(x)) - C_\mu(U(y)).$$

Proof. If $z \ TC_{P_h} \ w$ then there exist $z_1, z_2 \in \mathcal{B}$ such that $z \ TC \ z_1 \ P_h \ z_2 \ TC \ w$. hence we have $C_\mu(U(z)) - C_\mu(U(w)) \geq C_\mu(U(z_1)) - C_\mu(U(z_2))$. As $l < h \Rightarrow C_\mu(U(z_1)) - C_\mu(U(z_2)) > C_\mu(U(x)) - C_\mu(U(y)$, we have $C_\mu(U(z)) - C_\mu(U(w)) > C_\mu(U(x)) - C_\mu(U(y))$.

3 MOPIC Properties

3.1 General Definitions

The MOPIC (*MOnotonicity of Preferential Information for the Cardinal case*) properties are defined as follows:

Definition 3 (MOPIC Properties). *For all distinct i, j, k in N and $l, l' \in \{1, \ldots, q\}$, a cardinal information $\{P, I, P_1, \ldots, P_q\}$ on $\mathcal{B}$ fulfills the property:*

*1. **MOPIC-1** if*

$$\left.\begin{array}{l} a_i \ TC_{P_l} \ a_0 \\ a_{ij} \ P_{l'} \ a_j \\ l > l' \end{array}\right\} \Rightarrow \forall k \in N \setminus \{i,j\}, \ [not(a_{ik} \sim a_k) \text{ and } not(a_{jk} \sim a_k)]. \quad (11)$$

*2. **MOPIC-2** if*

$$\left.\begin{array}{l} a_{i \vee j} \ TC_{P_l} \ a_0 \\ a_{ij} \ P_{l'} \ a_k \\ l > l' \end{array}\right\} \Rightarrow [not(a_{ik} \sim a_h) \text{ and } not(a_{jk} \sim a_h)], \ h \in \{i,j\} \setminus i \vee j.$$

$$(12)$$

3. *MOPIC-3 if*

$$\left.\begin{array}{l} a_k \ TC_{P_l} \ a_0 \\ a_{ij} \ P_{l'} \ a_{i\vee j} \\ \quad l > l' \end{array}\right\} \Rightarrow [not(a_{ik} \sim a_h) \ and \ not(a_{jk} \sim a_h)], \ h \in \{i,j\} \setminus i \vee j. \quad (13)$$

4. *MOPIC-4 if*

$$\left.\begin{array}{l} a_k \ TC_{P_l} \ a_{ij} \\ \quad l > l' \end{array}\right\} \Rightarrow [not(a_{ik} \ P_{l'} \ a_{i\vee j})]. \quad (14)$$

Proposition 2. *If a cardinal information $\{P, I, P_1, \ldots, P_q\}$ on $\mathcal{B}$ is representable by a 2-additive Choquet integral, then this information satisfies the properties MOPIC-1, MOPIC-2, MOPIC-3 and MOPIC-4.*

Proof. The assertion has been proved in [10] for MOPIC-1, MOPIC-2 and MOPIC-3. We give only the proof for MOPIC-4 which is a new property.

Let be i, j, k in N and $l, l' \in \{1, \ldots, q\}$ such that $l > l'$. If $a_k \ TC_{P_l} \ a_{ij}$ and $a_{ik} \ P_{l'} \ a_{i\vee j}$ then by using Proposition 1 we have $C_\mu(U(a_k)) - C_\mu(U(a_{ij})) > C_\mu(U(a_{ik})) - C_\mu(U(a_{i\vee j}))$ i.e. $C_\mu(U(a_k)) + C_\mu(U(a_{i\vee j})) > C_\mu(U(a_{ij})) + C_\mu(U(a_{ik}))$. Hence we get $\mu_k + \mu_i + \mu_j > \mu_k + \mu_{i\vee j} > \mu_{ik} + \mu_{ij}$ which contradicts the monotonicity condition $\mu_{ik} + \mu_{ij} \geq \mu_k + \mu_i + \mu_j$.

The four MOPIC properties are inspired from the necessary and sufficient conditions called "MOPI" conditions introduced in the characterization of an ordinal information $\{P, I\}$ by a 2-additive Choquet integral [11]. The MOPI properties are defined as follows:

Definition 4 (MOPI Property). *Let $i, j, k \in N$, i fixed. A Monotonicity of Preferential Information in $\{i, j, k\}$ w.r.t. i is the following property (denoted by $(\{i,j,k\},i)$-MOPI):*

$$\left.\begin{array}{l} a_{ij} \sim a_{i\vee j} \\ a_{ik} \sim a_{i\vee k} \\ i \vee j \neq i \vee k \end{array}\right\} \Rightarrow [not(a_l \ TC_P \ a_0), l \in \{i,j,k\} \setminus \{i \vee k, i \vee j\}] \quad (15)$$

We proved in [10] that MOPIC-1, MOPIC-2 and MOPIC-3 are mutually independent. The following examples show that MOPIC-4 is independent from the other MOPIC properties.

Example 1. *1. $N = \{1,2,3\}$, $P_1 = \{(a_{13}, a_1)\}$ and $P_2 = \{(a_3, a_{12})\}$. MOPIC-4 is not satisfied because we have $a_3 \ P_2 \ a_{12}$ and $a_{13} \ P_1 \ a_1$ (see Fig. 2), while MOPIC-1, MOPIC-2 and MOPIC-3 are satisfied.*

2. $N = \{1,2,3\}$, $I = \{(a_{13}, a_3)\}$, $P_1 = \{(a_{12}, a_1); (a_{13}, a_1)\}$ and $P_2 = \{(a_2, a_0); (a_3, a_{12})\}$. MOPIC-4 is not satisfied (as shown in Fig. 2), MOPIC-2 and MOPIC-3 are satisfied, MOPIC-1 is not satisfied because $a_2 \ P_2 \ a_0$, $a_{12} \ P_1 \ a_1$ and $a_{13} \sim a_3$ (see Fig. 3).

3. $N = \{1,2,3\}$, $I = \{(a_{23}, a_2)\}$, $P_1 = \{(a_{12}, a_3), (a_{12}, a_1)\}$ and $P_2 = \{(a_1, a_0), (a_2, a_{13})\}$. MOPIC-2 is not satisfied because $a_1 \ P_2 \ a_0$, $a_{12} \ P_1 \ a_3$ and $a_{23} \sim a_2$. Because we have $a_2 \ P_2 \ a_{13}$ and $a_{12} \ P_1 \ a_1$ then MOPIC-4 is not satisfied (see Fig. 4).

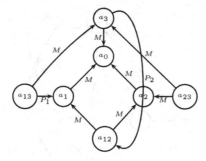

Fig. 2. MOPIC-4 is not satisfied. MOPIC-1, MOPIC-2 and MOPIC-3 are satisfied

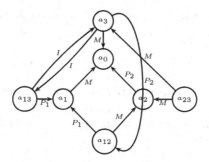

Fig. 3. MOPIC-4 and MOPIC-1 are not satisfied. MOPIC-2 and MOPIC-3 are satisfied

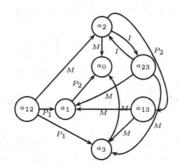

Fig. 4. MOPIC-4 and MOPIC-2 are not satisfied

4. $N = \{1, 2, 3\}$, $I = \{(a_{13}, a_1)\}$, $P_1 = \{(a_{13}, a_2); (a_{12}, a_2)\}$ and $P_2 = \{(a_3, a_0); (a_1, a_{23})\}$. MOPIC-2 is not satisfied because $a_3 \ P_2 \ a_0$, $a_{12} \ P_1 \ a_1$ and $a_{13} \sim a_2$. On the other hand $a_1 \ P_2 \ a_{23}$ and $a_{13} \ P_1 \ a_2$ implies that MOPIC-4 is not satisfied (see Fig. 5).

In order to better explain inconsistencies when it is not possible to represent a cardinal information, we give the following necessary condition based on a particular dominance relation called k-dominance (k like "kardinal").

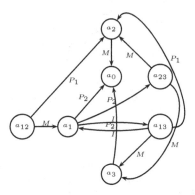

Fig. 5. MOPIC-4 and MOPIC-3 are not satisfied

Definition 5 (*k*-dominance). *Let be* $\{P, I, P_1, \ldots, P_q\}$ *a cardinal informa-tion on* $\mathcal{B}$.

A binary action y *is* ***k-dominated*** *(by a binary action* x) *if there exist* $x_1, \ldots, x_r \in \mathcal{B}$, $x'_1, \ldots, x'_r \in \mathcal{B}$, $l_1, \ldots, l_{r-1} \in \{1, \ldots, q\}$, $h_1, \ldots, h_{r-1} \in \{1, \ldots, q\}$ *and a bijection* g *from* $\{(x_i, x_{i+1})\}_{i=1,\ldots,r-1}$ *to* $\{(x'_j, x'_{j+1})\}_{j=1,\ldots,r-1}$ *such that:*

$$\begin{cases} x = x_1 \; P_{l_1} \; x_2 \; P_{l_2} \ldots P_{l_{r-2}} \; x_{r-1} \; P_{l_{r-1}} \; x_r = y \\ \\ x = x'_1 \; TC_{P_{h_1}} \; x'_2 \; TC_{P_{h_2}} \ldots TC_{P_{h_{r-1}}} \; x'_{r-1} \; TC_{P_{h_r}} \; x'_r = y \end{cases} \quad (16)$$

$$[g((x_i, x_{i+1})) = (x'_j, x'_{j+1})] \Rightarrow \begin{cases} x_i \; P_{l_i} \; x_{i+1} \\ x'_j \; TC_{P_{h_j}} \; x'_{j+1} \\ l_i < h_j \end{cases} \quad (17)$$

The Eq. (16) means that there exist two different paths of $(P \cup I \cup M)$ from x to y.

Example 2. *In Fig. 6, we can see that* y *is* k*-dominated by* x *by considering* $r = 4$, $x_1 = x$, $x_2 = t_2$, $x_3 = t_3$, $x_4 = y$, $x'_1 = x$, $x'_2 = z_1$, $x'_3 = z_2$, $x'_4 = y$ *and the bijection* g *given by:* $g((x_1, x_2)) = (x'_2, x'_3)$, $g((x_2, x_3)) = (x'_1, x'_2)$, $g((x_3, x_4)) = (x'_3, x'_4)$.

The k-dominance appears as a concept related to Proposition 1. It is independent to monotonicity conditions of a 2-additive capacity. Indeed, it is not difficult to check that, no binary action is k-dominated in preferences given in Example 1. The following proposition shows that this non-dominance property is necessary in the representation of a cardinal information.

Proposition 3. *Let be* $\{P, I, P_1, \ldots, P_q\}$ *a cardinal information on* $\mathcal{B}$.
If $\{P, I, P_1, \ldots, P_q\}$ *is representable by a 2-additive Choquet integral, then no binary action in* $\mathcal{B}$ *is* k*-dominated.*

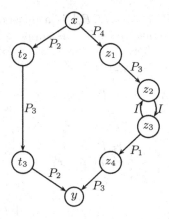

Fig. 6. y is k-dominated by x

Proof. It is a consequence of Definition 5 and Proposition 1.

Even if all the properties presented in this section are necessary but not sufficient, it is important to include them in the algorithm which deals with inconsistencies in the representation of a cardinal information. Their implementation is simple compared to the complex cyclones obtained by the characterization theorem.

In the next section, we give through the Theorem 1, some sufficient conditions to obtain a cardinal information representable by a cardinal information. These conditions are based on the notion of quasi-ordinal information related to the k-dominance.

3.2   Some Sufficient Conditions to Represent a Cardinal Information

Definition 6 (Quasi-ordinal Information). *A cardinal information*
$\{P, I, P_1, \ldots, P_q\}$ *on* $\mathcal{B}$ *is said to be* **quasi-ordinal** *if for all* $x, y, z, w \in \mathcal{B}$
and $l, h \in \{1, \ldots, q\}$, *the following conditions hold:*

$$\left.\begin{array}{l} x\ P_l\ y \\ z\ P_h\ w \\ not(x \sim z) \\ l > h \end{array}\right\} \Rightarrow x\ TC\ z \quad and \quad \left.\begin{array}{l} x\ P_l\ y \\ z\ P_h\ w \\ x \sim z \\ l > h \end{array}\right\} \Rightarrow w\ TC_P\ y \qquad (18)$$

Remark 1. *A quasi-ordinal information is not a necessary condition. To check it, it is sufficient to take* $N = \{1, 2, 3\}$, $I = \emptyset$, $P_1 = \{(a_{12}, a_1)\}$ *and* $P_2 = \{(a_2, a_3)\}$. *Therefore* $\{P, I, P_1, P_2\}$ *is a cardinal information representable by a 2-additive Choquet integral (but it is not quasi-ordinal) if we choose as* μ: $\mu(N) = 1$, $\mu(\emptyset) = \mu_3 = 0$, $\mu_1 = \frac{2}{7}$, $\mu_2 = \frac{3}{7}$ *and* $\mu_{12} = \mu_{13} = \mu_{23} = \frac{4}{7}$.

Proposition 4. *If* $\{P, I, P_1, \ldots, P_q\}$ *is a quasi-ordinal information on* $\mathcal{B}$ *and* $(P \cup I \cup M)$ *contains no strict cycle (a cycle containing an element of* P*), then each binary action of* $\mathcal{B}$ *is not k-dominated.*

Proof. We assume that $\{P, I, P_1, \ldots, P_q\}$ is a quasi-ordinal information and $(P \cup I \cup M)$ contains no strict cycle. Let us suppose the existence of $y \in \mathcal{B}$ k-dominated by $x \in \mathcal{B}$ i.e. there exist $x_1, \ldots, x_r \in \mathcal{B}$, $x'_1, \ldots, x'_r \in \mathcal{B}$ and a bijection g from $\{(x_i, x_{i+1})\}_{i=1,\ldots,r-1}$ to $\{(x'_j, x'_{j+1})\}_{j=1,\ldots,r-1}$ such that:

$$\bullet \quad \begin{cases} x = x_1 \; P_{l_1} \; x_2 \; P_{l_2} \ldots P_{l_{r-2}} \; x_{r-1} \; P_{l_{r-1}} \; x_r = y \\ x = x'_1 \; TC_{P_{h_1}} \; x'_2 \; TC_{P_{h_2}} \ldots TC_{P_{h_{r-1}}} \; x'_{r-1} \; TC_{P_{h_r}} \; x'_r = y \end{cases}$$

$$\bullet \quad [g((x_i, x_{i+1})) = (x'_j, x'_{j+1})] \Rightarrow \begin{cases} x_i \; P_{l_i} \; x_{i+1} \\ x'_j \; TC_{P_{h_j}} \; x'_{j+1} \\ l_i < h_j \end{cases}.$$

If we denote $g((x_1, x_2)) = (x'_{j_1}, x'_{j_2})$ then we get $\begin{cases} x_1 \; P_{l_1} \; x_2 \\ x'_{j_1} \; TC_{P_h} \; x'_{j_2} \\ l_1 < h \\ h \in \{h_1, \ldots, h_r\} \end{cases}.$

$x'_{j_1} \; TC_{P_h} \; x'_{j_2} \Rightarrow$ there exist $z, w \in \mathcal{B}$ such that $x'_{j_1} \; TC \; z \; P_h \; w \; TC x'_{j_2}$. Furthermore $x = x_1$ implies $x \; TC \; z$ by using hypothesis $x = x'_1 \; TC_{P_{h_1}} \; x'_2 \; TC_{P_{h_2}} \ldots TC_{P_{h_{r-1}}} \; x'_{r-1} \; TC_{P_{h_r}} \; x'_r = y$ and $\{x'_{j_1}, x'_{j_2}\} \subseteq \{x'_1, \ldots, x'_r\}$.

If $not(x_1 \sim z)$ then we should have $\begin{cases} x_1 \; P_{l_1} \; x_2 \\ z \; P_h \; w \\ not(x_1 \sim z) \\ l_1 < h \end{cases}$ and $z \; TC \; x_1$, because $\{P, I, P_1, \ldots, P_q\}$ is a quasi-ordinal information on $\mathcal{B}$. This leads to a contradiction with $x = x_1 \; TC \; z$. So we necessarily have $\begin{cases} x_1 \; P_{l_1} \; x_2 \\ z \; P_h \; w \\ x_1 \sim z \\ l_1 < h \end{cases}$

As $\{P, I, P_1, \ldots, P_q\}$ is a quasi-ordinal information on $\mathcal{B}$, we get $x_2 \; TC_P \; w$. To end the proof, let us show that x_2 and w are containing in a strict cycle.

If $g((x_2, x_3)) = (x'_{j_3}, x'_{j_4})$ then there exist $z', w' \in \mathcal{B}$ such that:

$\begin{cases} x_2 \; P_{l_2} \; x_3 \\ x'_{j_3} \; TC_{P_{h'}} \; x'_{j_4} \\ l_2 < h' \\ h' \in \{h_1, \ldots, h_r\} \end{cases}$ and $x'_{j_3} \; TC \; z' \; P_{h'} \; w' \; TC x'_{j_4}$. We also have $w \; TC \; z'$ because $x'_{j_2} \; TC \; x'_{j_3}$.

- If $z' \sim x_2$ then $z' \; TC \; x_2 \; TC_P \; w \; TC \; z'$ is a strict cycle of $(P \cup I \cup M)$. A contradiction with the hypothesis $(P \cup I \cup M)$ contains no strict cycle.
- If $not(z' \sim x_2)$ then $z' \; TC \; x_2$ because $\{P, I, P_1, \ldots, P_q\}$ is a quasi-ordinal information on $\mathcal{B}$. Hence $z' \; TC \; x_2 \; TC_P \; w \; TC \; z'$ is a strict cycle of $(P \cup I \cup M)$. A contradiction with the hypothesis $(P \cup I \cup M)$ contains no strict cycle.

We have shown that if there is a binary action y k-dominated by a binary alternative x, then there exist $x_2, w \in \mathcal{B}$ contained in a strict cycle. A contradiction.

Theorem 1. *Let be* $\{P, I, P_1, \ldots, P_q\}$ *a cardinal information on* $\mathcal{B}$*. If the following three conditions hold:*

1. $(P \cup I \cup M)$ *contains no strict cycle;*
2. *Every subset* K *of* N *such that* $|K| = 3$ *satisfies the MOPI condition;*
3. $\{P, I\ P_1, \ldots, P_q\}$ *is a quasi-ordinal information.*

Then $\{P, I, P_1, \ldots, P_q\}$ *is representable by a 2-additive Choquet integral.*

Proof (Sketch of). The proof is done in the spirit of the proof of the characterization theorem of the representation of an ordinal information $\{P, I\}$ on $\mathcal{B}$ by a 2-additive Choquet integral (see [11]).

As $(P \cup I \cup M)$ contains no strict cycle and every subset K of N such that $|K| = 3$ satisfies the MOPI condition, it is possible to compute a partition $\mathcal{B}_0, \mathcal{B}_1, \ldots, \mathcal{B}_m$ of $\mathcal{B}$ by applying a topological sorting. If we denote by $[x]$ the set of all binary actions belonging to the same set of the partition with x, then we can define the following relation $P_\sim$: $[t]\ P_\sim\ [u] \Leftrightarrow \exists t' \in [x], \exists u' \in [u]$ such that $t'\ (P \cup M)\ u'$.

Let us consider the following two applications:

1. $\mu : \mathcal{B} \to \mathbb{R}$ defined such as for each $i \in \{0, \ldots, m\}$,

$$\forall x \in \mathcal{B}_i,\ \mu(\phi(x)) = \begin{cases} 0 & \text{if } i = 0 \\ (2n)^i & \text{else.} \end{cases}$$

2. $\nu : 2^N \to [0,1]$ defined by:

$$\begin{cases} \nu_\emptyset = 0 \\ \nu_i = \frac{\mu_i}{\alpha}, \ \forall i \in N \\ \nu_{ij} = \frac{\mu_{ij}}{\alpha}, \ \forall i, j \in N \\ \nu(K) = \sum_{\{i,j\} \subseteq K} \nu_{ij} - (|K| - 2) \sum_{i \in K} \nu_i, \ \forall K \subseteq N, |K| > 2. \end{cases}$$

where $\alpha = \sum_{\{i,j\} \subseteq N} \mu_{ij} - (n - 2) \sum_{i \in N} \mu_i.$

As proved in [11], ν is a 2-additive capacity on $\mathcal{B}$ such that:

1. $\forall x, y \in \mathcal{B},\ x\ P\ y \Rightarrow C_\mu(U(x)) > C_\mu(U(y))$,
2. $\forall x, y \in \mathcal{B},\ x\ I\ y \Rightarrow C_\mu(U(x)) = C_\mu(U(y))$.

Let be $x, y, z, w \in \mathcal{B},\ s, t \in \{1, \ldots, q\}$ such that $\begin{cases} (x, y) \in P_t \\ (z, w) \in P_s \ . \\ s < t \end{cases}$

1. *Case 1: we suppose* $not(x \sim z)$

   $\{P, I, P_1, \ldots, P_q\}$ being a quasi-ordinal information, we have $[x]\ P_\sim\ [z]$ because $x\ TC\ z$. Therefore, if $x \in \mathcal{B}_l$, $y \in \mathcal{B}_{l'}$, $z \in \mathcal{B}_h$ and $w \in \mathcal{B}_{h'}$ then $l > l'$, $h > h'$ and $l > h$. If $k_0 = \max\{l', h, h'\}$, we have:

   $$(2n)^l \geq (n+n)(2n)^{k_0} > (2n)^h + (2n)^{h'} + (2n)^{l'} > (2n)^h + (2n)^{h'} - (2n)^{l'},$$

   because $n \geq 2$ and $k_0 \neq h'$.

   Hence $C_\mu(U(x)) - C_\mu(U(y)) > C_\mu(U(z)) - C_\mu(U(w))$.

2. *Case 2: we suppose* $x \sim z$

   $\{P, I, P_1, \ldots, P_q\}$ being a quasi-ordinal information, we have $[w]\ P_\sim\ [y]$ because $w\ TC_P\ y$. Therefore, if $x \in \mathcal{B}_l$, $y \in \mathcal{B}_{l'}$, $z \in \mathcal{B}_h$ and $w \in \mathcal{B}_{h'}$ then $l = h$, $h > h'$ and $h' > l'$.

   We have $C_\mu(U(x)) - C_\mu(U(y)) > C_\mu(U(z)) - C_\mu(U(w))$ because $C_\mu(U(x)) = C_\mu(U(z))$ and $C_\mu(U(w)) > C_\mu(U(y))$.

4   Conclusion

We presented some necessary and simple conditions to represent a cardinal information by a 2-additive Choquet integral. As the characterization of such preferences is not an easy task, these conditions can help the DM during the decision process to manage his inconsistencies. The sufficient conditions we gave can be easily extended to a more general case than binary actions, because the k-dominance and the quasi-ordinal information do not depend on the 2-additive structure of our problem. Therefore we plan in future works to elaborate more MOPIC properties and more interesting sufficient conditions in the representation of a cardinal information.

References

1. Bana e Costa, C.A., Correa, E.C., De Corte, J.-M., Vansnick, J.-C.: Facilitating bid evaluation in public call for tenders: a socio-technical approach. Omega **30**, 227–242 (2002)
2. Bana e Costa, C.A., De Corte, J.-M., Vansnick, J.-C.: Macbeth. Int. J. Inf. Technol. Decis. Making **11**(2), 359–387 (2012)
3. Chateauneuf, A., Jaffray, J.Y.: Some characterizations of lower probabilities and other monotone capacities through the use of Möbius inversion. Math. Soc. Sci. **17**, 263–283 (1989)
4. Clivillé, V., Berrah, L., Mauris, G.: Quantitative expression and aggregation of performance measurements based on the MACBETH multi-criteria method. Int. J. Prod. Econ. **105**, 171–189 (2007)
5. Grabisch, M.: k-order additive discrete fuzzy measures and their representation. Fuzzy Sets Syst. **92**, 167–189 (1997)
6. Grabisch, M., Labreuche, C.: Fuzzy measures and integrals in MCDA. In: Figueira, J., Greco, S., Ehrgott, M. (eds.) Multiple Criteria Decision Analysis: State of the Art Surveys, vol. 78, pp. 565–608. Springer, New York (2005)

7. Grabisch, M., Labreuche, Ch.: A decade of application of the Choquet and Sugeno integrals in multi-criteria decision aid. 4OR **6**, 1–44 (2008)
8. Marchant, T.: Towards a theory of MCDM: stepping away from social choice theory. Math. Soc. Sci. **45**(3), 343–363 (2003)
9. Mayag, B., Grabisch, M., Labreuche, C.: An interactive algorithm to deal with inconsistencies in the representation of cardinal information. In: Hüllermeier, E., Kruse, R., Hoffmann, F. (eds.) IPMU 2010. CCIS, vol. 80, pp. 148–157. Springer, Heidelberg (2010)
10. Mayag, B., Grabisch, M., Labreuche, Ch.: A characterization of the 2-additive Choquet integral through cardinal information. Fuzzy Sets Syst. **184**, 84–105 (2011)
11. Mayag, B., Grabisch, M., Labreuche, Ch.: A representation of preferences by the Choquet integral with respect to a 2-additive capacity. Theory Decis. **71**, 297–324 (2011)
12. Simon, H.: Rational choice and the structure of the environment. Psychol. Rev. **63**(2), 129–138 (1956)
13. Slovic, P., Finucane, M., Peters, E., MacGregor, D.G.: The affect heuristic. In: Gilovitch, T., Griffin, D., Kahneman, D. (eds.) Heuristics and Biases: The Psychology of Intuitive Judgment, pp. 397–420. Cambridge University Press (2002)
14. Tversky, A., Kahneman, D.: Advances in prospect theory: cumulative representation of uncertainty. J. Risk Uncertainty **5**, 297–323 (1992)

Profitable Deviation Strong Equilibria

Laurent Gourvès[(⊠)]

CNRS UMR 7243, Université Paris Dauphine,
Place du Maréchal de Lattre de Tassigny, 75775 Paris Cedex 16, France
laurent.gourves@dauphine.fr

Abstract. This paper deals with states that are immune to group deviations. Group deviations help the players of a strategic game to escape from undesirable states but they compromise the stability of a system. We propose and analyse a solution concept, called profitable deviation strong equilibrium, which is between two well-known equilibria: the strong equilibrium and the super strong equilibrium. The former precludes joint deviations by groups of players who all benefit. The latter is more demanding in the sense that at least one member of a deviating coalition must be better off while the other members cannot be worst off. We study the existence, computation and convergence to a profitable deviation strong equilibrium in three important games in algorithmic game theory: job scheduling, max cut and singleton congestion game.

Keywords: Algorithmic game theory · Equilibrium · Group deviation

1 Introduction

A central question in game theory is to define a solution concept which captures the plausible outcomes of a game. The notion of equilibrium (a state that is stable against a given type of deviation) has a prominent place in this area. Indeed, if some players find profitable to deviate from a given state and if they can do it then this state is an unlikely outcome of the game.

In a world populated with selfish individuals, the *Nash equilibrium* is a credible outcome because no single player can unilaterally deviate and be better off [24]. This notion can be refined by allowing group deviations. In a *strong equilibrium* [4], there is no *improving move*, i.e. no way for a group of players to coordinate their actions such that every member is better off. In a *super strong equilibrium*, no *weak improving move* (a group deviation in which no member is worst off and at least one member is better off) exists. This latter solution concept is attractive because there is room for both selfishness and altruism in the moves that it precludes. Some disinterested players can take part in a deviating coalition if their individual costs remain unchanged.

The more deviations an equilibrium excludes, the more sustainable it is. Moreover, sophisticated moves can help the players to escape from some undesirable

---

Supported by the project ANR-14-CE24-0007-01 CoCoRICo-CoDec.

T. Walsh (Ed.): ADT 2015, LNAI 9346, pp. 236–252, 2015.
DOI: 10.1007/978-3-319-23114-3_15

states (e.g. those having a high social cost). However, as the set of possible deviations grows, the induced set of stable states reduces. Indeed, only very restricted classes of games admit a super strong equilibrium. Therefore, it is challenging to devise a solution concept (equivalently, the moves that it rules out) in which deviations comprising disinterested players (those who are not worst off) are allowed and the solution concept is guaranteed to exist in some non-trivial games. An additional desideratum is that the game admits a potential, i.e. the players naturally converge to a stable state from any initial configuration.

This paper deals with some games which possess a strong equilibrium but not a super strong equilibrium. Our motivation is to propose a solution concept that is between them. Ideally, such a solution concept would meet the aforementioned desiderata. The main difficulty is to find a tradeoff between the guaranteed convergence to an equilibrium and the sophistication of the possible deviations, especially when disinterested players are involved.

This question has already been posed and treated in different ways. Let us mention some of them. We can assume that the players are embedded in a social network [18], e.g. there is a partition of the player set so that two coalitions involved in a weak improving move cannot overlap [12]. Another way is to consider that every player minimizes a function which combines his individual cost and a social cost [6]; this behavior can even push a single player to slightly increase his own cost if it is profitable to some other players. A third way is to consider the Nash equilibria whose vector of individual costs is not Pareto dominated by the cost vector of another state [7,8].

The above concepts do not necessarily refine, as we require, the notion of strong equilibrium. In this paper we propose a different approach. We stress the fact that some games are not stable against weak improving moves because in such moves, it is not mandatory for a coalition member who benefits to concretely change his strategy. So we find important to take this information into account in order to understand which players are legitimate in the coalition.

Take a strong equilibrium σ and consider a coalition of players having a weak improving move which turns σ into a new state σ'. Separate the entire set of players according to 3 criteria: Does a player belong to the coalition? Is a player better off, indifferent or worst off? Does a player concretely move? The situation is depicted in the next table where no member of the coalition can be worst off and $A \cup C$ can not be empty.

		Better off	Indifferent	Worst off
oalition	move	A	B	
	¬move	C		
¬coalition	¬move		D	E

Indifferent players who stick to their strategy are dummy so we can assume that they do not belong to the coalition. If $E = \emptyset$ then nobody is worst off and σ' is clearly more desirable than σ. Now suppose $E \neq \emptyset$. If $B = \emptyset$ then σ cannot be a strong equilibrium. If $A = \emptyset$ then the coalition reduces to $B \cup C$. The deviation from σ to σ' can be immediately followed by the reverse deviation performed by the coalition $B \cup E$ (i.e. from σ' to σ where B moves but E does not move).

The case $A = \emptyset$ is problematic and for the games studied in this paper, counterexamples showing that a game does not necessarily possess a super strong equilibrium fall in this case. Therefore, we propose to impose that for a coalition involved in a weak improving move, the subset of players who move and benefit cannot be empty.

2   Definitions and Contribution

A *strategic game* has a set $N = \{1, \ldots, n\}$ of players. Each player i has a finite set of actions A_i. The *strategy* of a player is a distribution of probabilities over his action set. A strategy is said to be *pure* if one action is chosen with probability 1; otherwise it is said to be *mixed*. In this paper we focus on pure strategies. The word pure is omitted for the sake of readability.

Let $\sigma = (\sigma_1, \ldots, \sigma_n)$ be a *state* of the game where $\sigma_i \in A_i$ is the strategy of player i. The set of all possible states is $\Sigma = A_1 \times A_2 \times \ldots \times A_n$. Let σ_{-i} denote σ without player i's strategy, i.e. $\sigma_{-i} := (\sigma_1, \ldots, \sigma_{i-1}, \sigma_{i+1}, \ldots, \sigma_n)$. The state obtained by the replacement of player i's strategy by s'_i in a state σ is denoted by (σ_{-i}, s'_i).

Depending on the situation, the players either maximize or minimize an individual function. In the first case every player i has a *utility function* $u_i : \Sigma \to \mathbb{R}$. In the second case, every player i has a *cost function* $c_i : \Sigma \to \mathbb{R}$.

For a coalition of players $C \subseteq N$, σ_{-C} denotes σ in which the strategy of every member of C is removed. Similarly σ_C denotes σ restricted to the strategy of the members of C. Given a coalition C and two states σ^1 and σ^2, $(\sigma^2_{-C}, \sigma^1_C)$ is the state in which player i's strategy is σ^1_i if $i \in C$, otherwise it is σ^2_i.

For a state σ, a coalition C has an *improving move* if there is σ'_C such that $u_i(\sigma_{-C}, \sigma'_C) > u_i(\sigma)$ for all $i \in C$ ($c_i(\sigma_{-C}, \sigma'_C) < c_i(\sigma)$ for all $i \in C$, resp.). The improving move is *unilateral* if $|C| = 1$. A coalition C has a *weak improving move* with respect to σ if there is σ'_C such that $u_i(\sigma_{-C}, \sigma'_C) \geq u_i(\sigma)$ for all $i \in C$ and $u_i(\sigma_{-C}, \sigma'_C) > u_i(\sigma)$ for at least one $i \in C$ ($c_i(\sigma_{-C}, \sigma'_C) \leq c_i(\sigma)$ for all $i \in C$ and $c_i(\sigma_{-C}, \sigma'_C) < c_i(\sigma)$ for at least one $i \in C$, resp.).

A state σ is a *Nash equilibrium* (NE) if there is no unilateral improving move, a *strong equilibrium* (SE) if there is no improving move, a *super strong equilibrium*[1] (SSE) if there is no weak improving move.

In this paper we introduce a solution concept which is between the SE and the SSE. A coalition C has a *profitable deviation weak improving move* with respect to σ if there is σ'_C such that $u_i(\sigma_{-C}, \sigma'_C) \geq u_i(\sigma)$ for all $i \in C$ and $u_i(\sigma_{-C}, \sigma'_C) > u_i(\sigma)$ for at least one $i \in C$ such that $\sigma_i \neq \sigma'_i$ ($c_i(\sigma_{-C}, \sigma'_C) \leq c_i(\sigma)$ for all $i \in C$ and $c_i(\sigma_{-C}, \sigma'_C) < c_i(\sigma)$ for at least one $i \in C$ such that $\sigma_i \neq \sigma'_i$, resp.). A state σ is a *profitable deviation strong equilibrium* (PDSE) if there is no profitable deviation weak improving move.

The relation NE $\supseteq$ SE $\supseteq$ PDSE $\supseteq$ SSE clearly holds. See Fig. 1 where for each solution concept, we also provide the deviation that it is immune to.

Throughout the article we mention other solution concepts:

---

[1] Also known as *strictly strong Nash equilibrium* [28].

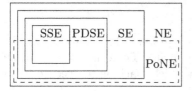 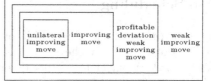

Fig. 1. Hierarchy of solution concepts and deviations.

- A *Pareto optimal Nash equilibrium* (PoNE for short) is a state σ that is a NE and moreover, there is no σ' such that $\forall i \in N$, $c_i(\sigma') \leq c_i(\sigma)$ and $\exists i' \in N$, $c_{i'}(\sigma') < c_{i'}(\sigma)$ [7,8]. Actually, we can see a PoNE as a SSE in which only coalitions of size 1 or $|N|$ are allowed. But by definition, a PoNE is not necessarily a SE.
- Given a partition $\mathcal{P}$ of the player set, a state is a $\mathcal{P}$-SSE (*partition super strong equilibrium*) if no group of players $P \in \mathcal{P}$ has a weak improving move [12].
- A *consideration graph* $\mathcal{G} = (N, F)$ is undirected and unweighted. Its node set is the player set. The maximal cliques of the graph correspond to the possible coalitions. The neighborhood of $C \subseteq N$ is $\mathcal{N}_{\mathcal{G}}(C) := \{j : i \in C \wedge (i, j) \in F\}$. For a state σ, a coalition C has a *considerate weak improving move* if there is σ'_C such that $u_i(\sigma'_C, \sigma_{-C}) \geq u_i(\sigma)$ for all $i \in C \cup \mathcal{N}_{\mathcal{G}}(C)$ and $u_j(\sigma'_C, \sigma_{-C}) > u_j(\sigma)$ for at least one player $j \in C$. A state is a *considerate super strong equilibrium* (CSSE) if there is no considerate weak improving move [18]. Note that CSSE generalize $\mathcal{P}$-SSE when the consideration graph consists of disjoints cliques.

The *dynamics* associated with a given equilibrium and its related type of deviation is defined as every series of states $\sigma_0, \sigma_1, \ldots$ such that σ_{i+1} is obtained by a profitable deviation performed on σ_i. We say that the dynamics converges if, starting from any given state of Σ, the series ends and its last state is an equilibrium. In other words, the game admits a potential [23] which is a highly desirable property. Note that several deviations may be possible for a given state (e.g. best vs better response). Here, we do not make any assumption on which deviation is performed when more than one is possible. Therefore, the property of convergence fails each time a cycle of better moves exists.

We study three types of games which have received a lot of attention in the field of algorithmic game theory: a job scheduling game (Sect. 3), a max cut game (Sect. 4) and monotone singleton congestion games (Sect. 5). These games (or some special cases) are known to admit a SE but they do not necessarily possess a SSE. Thus, we are mainly interested in identifying the conditions which guarantee the existence of a PDSE and the possibility of the players to converge to a PDSE from any initial state. In addition, we try to understand how the PDSE relates with other equilibria. The three games are treated in separate sections. For the sake of redability, we have grouped our results and some elements of previous works. Due to space limitations, some proofs are put in an appendix.

3   Job Scheduling

In the *job scheduling* game [1,9–11,13], every player $i \in N$ controls exactly one job, also denoted by i for the sake of simplicity. There is a set of machines $M = \{1, \ldots, m\}$. A player has to decide on which machine to execute its job so M is the strategy space of everyone.

In the general case, the machines are *unrelated*, meaning that the *weight* of job i on machine j is a positive value denoted by $w_j(i)$. This quantity is known for every pair of indices from $N \times M$. The machines are said to be *identical* if for every $j, j' \in M$ and $i \in N$, it holds that $w_j(i) = w_{j'}(i)$. In the case of identical machines, the weight of job i is denoted by $w(i)$. The machines are said to be *uniformly related* (related for short) if there is a speed s_j for every $j \in M$ and a work $w(i)$ for every job $i \in N$ such that $w_j(i) = \frac{w(i)}{s_j}$. Therefore, identical machines is a special case of related machines, and related machines is a special case of unrelated machines.

The *load* of machine j under the state σ is defined as $L_j(\sigma) = \sum_{\{i:\sigma_i=j\}} w_j(i)$. The cost of player i under the state σ is the load of the machine where its job is placed, i.e. $c_i(\sigma) = L_{\sigma_i}(\sigma)$.

The scheduling game always admits a pure NE [11,13] and also a SE [1]. Indeed, associate a vector $\Lambda(\sigma)$ with every state σ where $\Lambda_i(\sigma)$ is the i-th largest load of a machine under σ (i.e. order the elements of the multi-set $\{L_1(\sigma), \ldots, L_m(\sigma)\}$ in a non increasing way to get $\Lambda(\sigma)$). $\Lambda(\sigma)$ is said to be *lexicographically smaller* than $\Lambda(\sigma')$, denoted by $\Lambda(\sigma) \prec \Lambda(\sigma')$, if there exists k such that $\Lambda_k(\sigma) < \Lambda_k(\sigma')$ and $\Lambda_j(\sigma) = \Lambda_j(\sigma')$ for all $j < k$. Following [1], a state whose corresponding vector is lexicographically minimal must be a SE. However, even if the game is with two (or more) identical machines, a SSE does not necessarily exist [1].

Our contribution is summarized in Table 1.

Table 1. Summary of results for the scheduling game.

	$m = 2$	$m \geq 3$
Identical	NE = PDSE	
Related	PoNE $\neq$ PDSE $\neq$ SE	PDSE not guaranteed
Unrelated	PDSE guaranteed	

Proposition 1. *For the scheduling game with two unrelated machines, a PDSE always exists and the dynamics converges.*

Proof. Suppose σ is the current state and σ' denotes the strategy profile obtained after a profitable deviation weak improving move is performed by a non-empty set of players C. We consider two cases in order to show that $\Lambda(\sigma') \prec \Lambda(\sigma)$.

– There exists $x \in \{1, 2\}$ such that all the players of C are on machine x and $3 - x$ in σ and σ', respectively. Clearly, machine x is more loaded than machine $3 - x$ in σ : $L_x(\sigma) > L_{3-x}(\sigma)$. Since the players in C benefit, we have $L_x(\sigma) >$

$L_{3-x}(\sigma')$. Since some players, with positive weights, have left machine x, we have $L_x(\sigma) > L_x(\sigma')$. Therefore $\Lambda(\sigma') \prec \Lambda(\sigma)$.

- Some players of C move from machine 1 to machine 2 and vice versa. Since the cost of a member of C does not increase with the deviation, we have $L_1(\sigma) \geq L_2(\sigma')$ and $L_2(\sigma) \geq L_1(\sigma')$, and at least one of these inequalities is strict because the deviation is profitable for at least one player. Therefore $\Lambda(\sigma') \prec \Lambda(\sigma)$.

In conclusion, we can start from any strategy profile and let the players perform some deviations until they reach a state that is (locally) lexicographically minimal.   □

Observation 1. *For the scheduling game with two related machines, it holds that:*

(i) a SE or PoNE is not necessarily a PDSE;
(ii) a PDSE is not necessarily a PoNE.

Proposition 2. *For the scheduling game with two identical machines, a NE must be a PDSE.*

Proof. Let σ be a NE of the scheduling game with two identical machines. For the sake of contradiction, suppose σ is not a PDSE. A coalition of players C can perform a profitable deviation weak improving move leading to a new state σ'. If there is a unique machine, say machine 1, such that $\sigma_i = 1$ for all $i \in C$ then σ is not a NE, contradiction. Indeed, every member of C can, alone, deviate and benefit. Henceworth suppose $\bigcup_{i \in C}\{\sigma_i\} = \{1, 2\}$. By hypothesis, one has $c_i(\sigma) \geq c_i(\sigma')$ for all $i \in C$ and this inequality is strict for at least one member of C. In other words, $L_1(\sigma) \geq L_2(\sigma')$ and $L_2(\sigma) \geq L_1(\sigma')$ and one of these inequalities is strict. This gives $L_1(\sigma) + L_2(\sigma) > L_1(\sigma') + L_2(\sigma')$, but it contradicts $L_1(\sigma) + L_2(\sigma) = L_1(\sigma') + L_2(\sigma')$ which holds because the machines are identical.   □

It is noteworthy that a NE can be efficiently computed if the machines are identical or related [13].

Proposition 3. *There exists an instance of the scheduling game with $m \geq 3$ identical machines which does not have any PDSE.*

Proof. The instance consists of $m \geq 3$ machines and $m + 2$ jobs. The $m + 1$ first jobs have weight 1. Job $m + 2$ has weight $\frac{1}{2}$.

By the pigeon hole principle, there is at least one machine which accommodates at least two unit jobs. We consider three cases.

- One machine, say j, accommodates at least three unit jobs. There must be another machine, say j', which does not accommodate any unit job. The cost of a player on j is at least 3, and if this player moves to j', his cost would be either $\frac{3}{2}$ or 1, depending on the presence of job $m + 2$ on j'. Therefore the current strategy profile cannot be a NE.

– Two machines, say j and j', both accommodate two unit jobs. There must be another machine, say j'', which does not accommodate any unit job. As in the previous case, the current strategy profile cannot be a NE.
– One machine accommodates exactly two unit jobs (suppose this is machine 1) while any other machine accommodates exactly one unit job. If job $m+2$ is on machine 1 then the state is not a NE because the half job can move to machine 2 and decrease its cost. Suppose wlog. that job $m+2$ is on machine 2. There exists a profitable deviation in this case. Let 1 and 2 be the two unit jobs on machine 1. Let 3 and 4 be the unit jobs on machines 2 and 3, respectively. The deviation, performed by coalition $\{1, 2, 4, m+2\}$ is the following. Job 4 goes to machine 1, job 2 goes to machine 2, jobs 1 and $m+2$ go to machine 3. Thus the cost of job 2 remains 2, the cost of job 4 remains 1, the cost of job $m+2$ remains $\frac{3}{2}$ and the cost of job 1 drops from 2 to $\frac{3}{2}$. Only the utility of job 3, who does not belong to the coalition, decreases.

In conclusion, there exists a deviation from any given strategy profile. □

4   Max Cut Game

The *max cut* game [3,14,15,17] is defined on a simple undirected graph $G = (V, E)$. Every edge $(i, j) \in E$ has a non negative integer weight $w(i, j)$. A player is associated with each vertex so $N = V$ and we interchangeably mention a player and its corresponding vertex. The strategy of a player is 0 or 1. For a given state σ, the utility of player i is $u_i(\sigma) = \sum_{j \in V:(i,j) \in E \wedge \sigma_i \neq \sigma_j} w(i, j)$. The *cut* induced by a state is the set of edges having one extremity in the 0-part and the other extremity in the 1-part. Thus a player's utility is proportional to the weight of its contribution to the cut.[2]

Every instance of the max cut game admits a SE [14]. In fact, each improving move induces a decrease in the weight of the cut associated with a state. However, it is not the case for weak improving moves.

Observation 2. *The max cut game does not always admit a SSE.*

Proof. Take a complete graph on 3 nodes, each edge having weight 1. If the three players have the same strategy then the corresponding state is not a NE. If there are two players with the same strategy, say players 1 and 2 playing 0 while player 3 plays 1, then players 1 and 2 can collude such that player 2 deviates. The payoff of player 2 remains unchanged (it is 1) but player 1's utility becomes 2 instead of 1. □

The existence of a partition super strong equilibrium ($\mathcal{P}$-SSE) is guaranteed in every instance of the max cut game and for every partition $\mathcal{P}$ of the set of players.

Theorem 1. *For every partition $\mathcal{P}$ of N, the max cut game has a $\mathcal{P}$-SSE and the dynamics converges.*

---

[2] Note that the max cut game can be defined as a congestion game, but not as a singleton congestion game, so the results of Sect. 5 do not apply.

Proof. We are given a partition $\mathcal{P}$ of N. A state σ induces a cut in the original graph G. Let $w_P(\sigma)$ denote the weight of the cut restricted to the graph induced by $P \in \mathcal{P}$. Define $\Lambda(\sigma)$ as a couple whose first coordinate $\Lambda_1(\sigma)$ is equal to $w_N(\sigma)$, i.e. the cut induced by σ. The second coordinate $\Lambda_2(\sigma)$ is equal to $\sum_{P \in \mathcal{P}} w_P(\sigma)$.

As an example, consider a complete graph on three 3 nodes a, b and c. Each edge has weight 1. The partition is $\{\{a, b\}, \{c\}\}$. The state $\sigma = (0, 0, 1)$ (a and b both play 0 while c plays 1) is such that $\Lambda(\sigma) = (2, 0)$. The whole cut has weight 2, corresponding to the first coordinate. For the second coordinate, the weight of the cut within both $\{a, b\}$ and $\{c\}$ is 0.

For two consecutive states σ and σ' of the dynamics, we can show that $\Lambda(\sigma) \prec \Lambda(\sigma')$. The weak improving move which turned σ into σ' has been performed by a coalition $P \in \mathcal{P}$.

One can partition the player set in six subsets P_0^0, P_0^1, P_1^0, P_1^1, $\bar{P}_0$ and $\bar{P}_1$. For $(x, y) \in \{0, 1\}^2$, P_x^y denotes the players of P who play x and y in σ and σ', respectively. For $x \in \{0, 1\}$, $\bar{P}_x$ denotes the players of $N \setminus P$ who play x in both σ and σ'. For two disjoint subsets A and B of N, $w(A, B)$ denotes the total weight of the edges having one extremity in A and the other one in B (clearly $w(A, B) = w(B, A)$).

Suppose there is one player in $P_0^1 \cup P_1^0$ for whom the deviation is profitable. We deduce that $\sum_{i \in P_0^1 \cup P_1^0} u_i(\sigma) < \sum_{i \in P_0^1 \cup P_1^0} u_i(\sigma')$. One can rewrite this inequality as $2w(P_0^1, P_1^0) + w(P_0^1, P_1^1 \cup \bar{P}_1) + w(P_1^0, P_0^0 \cup \bar{P}_0) < 2w(P_0^1, P_1^0) + w(P_0^1, P_0^0 \cup \bar{P}_0) + w(P_1^0, P_1^1 \cup \bar{P}_1) \Leftrightarrow w(P_0^1, P_1^0 \cup P_1^1 \cup \bar{P}_1) + w(P_1^0, P_0^0 \cup \bar{P}_0) < w(P_0^1, P_0^0 \cup P_0^0 \cup \bar{P}_0) + w(P_1^0, P_1^1 \cup \bar{P}_1)$. Add $w(P_0^0, P_1^1) + w(\bar{P}_0, \bar{P}_1)$ on both sides of the inequality to get that $w(P_0^1 \cup P_0^0 \cup \bar{P}_0, P_1^0 \cup P_1^1 \cup \bar{P}_1) < w(P_1^0 \cup P_0^0 \cup \bar{P}_0, P_0^1 \cup P_1^1 \cup \bar{P}_1)$. In other words, the weight of the cut induced by σ (left part) is strictly smaller that the weight of the cut induced by σ' (right part). Thus $\Lambda(\sigma) \prec \Lambda(\sigma')$ holds because the first coordinate of each of these vectors is the weight of the corresponding cuts.

Now suppose the deviation leaves the utility of every member of $P_0^1 \cup P_1^0$ unchanged. At least one player in $P \setminus (P_0^1 \cup P_1^0) = P_0^0 \cup P_1^1$ must benefit from the deviation. This gives us

$$\sum_{i \in P_0^1 \cup P_1^0} u_i(\sigma) = \sum_{i \in P_0^1 \cup P_1^0} u_i(\sigma') \tag{1}$$

$$\sum_{i \in P_0^0 \cup P_1^1} u_i(\sigma) < \sum_{i \in P_0^0 \cup P_1^1} u_i(\sigma') \tag{2}$$

Rewrite Inequality (1) as $2w(P_0^1, P_1^0) + w(P_0^1, P_1^1 \cup \bar{P}_1) + w(P_1^0, P_0^0 \cup \bar{P}_0) = 2w(P_0^1, P_1^0) + w(P_0^1, P_0^0 \cup \bar{P}_0) + w(P_1^0, P_1^1 \cup \bar{P}_1) \Leftrightarrow w(P_0^1, P_0^0 \cup P_1^1 \cup \bar{P}_1) + w(P_1^0, P_0^0 \cup \bar{P}_0) = w(P_1^0, P_0^0 \cup P_0^0 \cup \bar{P}_0) + w(P_0^1, P_1^1 \cup \bar{P}_1)$. Adding $w(P_0^0 \cup \bar{P}_0, P_1^1 \cup \bar{P}_1)$ on both sides leads to

$$w(P_0^1 \cup P_0^0 \cup \bar{P}_0, P_1^0 \cup P_1^1 \cup \bar{P}_1) = w(P_1^0 \cup P_0^0 \cup \bar{P}_0, P_0^1 \cup P_1^1 \cup \bar{P}_1) \tag{3}$$

This inequality means that the deviation leaves the weight of the cut unchanged, i.e. $\Lambda_1(\sigma) = \Lambda_1(\sigma')$. With similar operations, one can rewrite Inequality (2) as

$$w(P_0^0, P_1^0) + w(P_1^1, P_1^0) + w(P_0^0, P_0^1) < w(P_0^0, P_0^1) + w(P_1^1, P_1^0) + w(P_0^0, P_1^1) \tag{4}$$

Adding $w(P_0^1, P_1^0)$ on both sides gives $w(P_0^0 \cup P_0^1, P_1^1 \cup P_1^0) < w(P_0^0 \cup P_1^0, P_1^1 \cup P_0^1) \Leftrightarrow w_P(\sigma) < w_P(\sigma')$. Now observe that for any $P' \in \mathcal{P} \setminus P$, $w_{P'}(\sigma) = w_{P'}(\sigma')$ holds because the players of P' do not participate to the deviation. It follows that $\Lambda_2(\sigma) < \Lambda_2(\sigma')$. In conclusion $\Lambda(\sigma) \prec \Lambda(\sigma')$.   □

Proposition 4. *Every instance of the max cut game has a PDSE and the dynamics converges.*

Proof. Let σ and σ' be two consecutive states of the dynamics. We are going to show that the weight of the cut induced by σ' is larger than the weight of the cut induced by σ. The deviation is performed by a coalition $C \subseteq N$.

We can partition N in four subsets N_0^0, N_0^1, N_1^0 and N_1^1. For $(x, y) \in \{0,1\}^2$, N_x^y denotes the players who play x and y in σ and σ', respectively. For two disjoint subsets A and B of N, $w(A, B)$ denotes the total weight of the edges having one extremity in A and the other extremity in B.

It is known that if the deviation is an improving move then the weight of the cut increases [14] so we focus on profitable deviation weak improving moves. By definition, there must be at least one player $i^* \in N_0^1 \cup N_1^0$ such that $u_{i^*}(\sigma') > u_{i^*}(\sigma)$. In addition, for every player $i \in N_0^1 \cup N_1^0$ we have $u_i(\sigma') \geq u_i(\sigma)$. If we sum these inequalities over $N_0^1 \cup N_1^0$ then we get that $2w(N_0^1, N_0^0) + w(N_0^1, N_0^0) + w(N_1^0, N_1^1) > 2w(N_0^1, N_1^0) + w(N_0^1, N_1^1) + w(N_1^0, N_0^0)$. Add $w(N_0^0, N_1^1) - w(N_0^1, N_1^0)$ on both sides to get that $w(N_0^0, N_1^1) + w(N_0^1, N_1^1) + w(N_0^1, N_0^0) + w(N_1^0, N_1^1) > w(N_0^0, N_1^1) + w(N_0^1, N_1^1) + w(N_0^1, N_1^1) + w(N_1^0, N_0^0)$. In other words, the weight of the cut induced by σ is strictly smaller that the weight of the cut induced by σ'.   □

Proposition 5. *In the max cut game it holds that*

(i) a SE, or PoNE, is not necessarily a PDSE;
(ii) a PDSE must be a PoNE.

Proof. For item (i) consider the instance depicted on Fig. 2.A where solid edges have weight 1 and the dashed edge has weight ε where $1 > \varepsilon > 0$. Suppose players a and b play 0 while players c and d play 1. This state is a SE and also a PoNE but it is not a PDSE since players a and d can switch their strategies such that player d's utility remains 1 whereas player a's utility becomes 2 instead of $1 + \varepsilon$.

For item (ii) consider a state σ that is not a PoNE. If it is because a unilateral deviation is possible, then σ is not a PDSE. Suppose a Pareto improving deviation is possible for the grand coalition, inducing a new state σ'. We have $u_i(\sigma') \geq u_i(\sigma)$ for every $i \in N$. If $u_{i^*}(\sigma') > u_{i^*}(\sigma)$ and $\sigma'_{i^*} \neq \sigma_{i^*}$ hold for some $i^* \in N$ then σ cannot be a PDSE (the members of $C' = \{i \in N : \sigma_i \neq \sigma'_i\}$ can flip their strategy and $i^* \in C'$). The last case is when $\sigma'_i = \sigma_i$ for every i such that $u_i(\sigma') > u_i(\sigma)$. Consider the state σ'' such that $\sigma''_i = 1 - \sigma'_i$, $\forall i \in N$. We have $u_i(\sigma') = u_i(\sigma'')$ for all $i \in N$. Therefore $C'' = \{i \in N : \sigma_i \neq \sigma''_i\}$ is equal to $N \setminus \{i \in N : \sigma_i \neq \sigma'_i\}$. It follows that σ cannot be a PDSE because the members of C'' can flip their strategy and there must be $i^* \in C''$ such that $u_i(\sigma'') > u_i(\sigma)$ and $\sigma''_i \neq \sigma_i$.   □

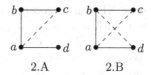

2.A   2.B

Fig. 2. Illustration of Proposition 5.

We can observe that the weight of a cut induced by a PDSE is, in the worst case, 2/3 of the maximum weight cut. Indeed, the price of anarchy for SE is 2/3 [14]. Because a PDSE is a SE, every PDSE induces a cut whose weight is at least 2/3 of the maximum weight of a cut. This ratio is assymptotically tight by considering the instance of Fig. 2.B. Solid edges have weight $M \gg 1$ and dashed edges have weight 1. The state which separates $\{a, b\}$ from $\{c, d\}$ is a PDSE whose induced cut has weight $2(M + 1)$. The state which separates $\{a, c\}$ from $\{b, d\}$ induces a cut of weight $3M$. This example can be extended to unweighted graphs (each edge has weight 1 and there is only one edge between two nodes) by replacing nodes b and a by two sets of nodes $\{b_1, \ldots, b_k\}$ and $\{a_1, \ldots, a_k\}$, respectively. The edge set is $\{(a_i, b_i), (b_i, c), (a_i, d)\}$, for $i = 1..k$, plus $\{(a_1, c), (b_1, d)\}$.

Computing a NE of the max cut game is, in general, a PLS-complete problem [27]. Therefore, it is unlikely to design a polynomial algorithm for computing a stable state. However, we can efficiently compute a NE if the graph is unweighted: start from any initial state and, while the current state is not a NE, perform profitable unilateral deviations. At each step the cut contains a greater number of edges, so the algorithm stops after at most $|E|$ rounds. An interesting question is to settle the complexity of computing a PDSE of the max cut game in unweighted graphs (or more generally computing a SE).

5   Singleton and Monotone Congestion Games

A *congestion game* [25] is a tuple $\langle N, M, (f_j)_{j \in M}, (A_i)_{i \in N} \rangle$ where N is a finite set of n players, M is a finite set of m *resources*, $f_j : \mathbb{N} \to \mathbb{R}_{\geq 0}$ is a *resource function* associated with resource $j \in M$, and A_i is the strategy set of player $i \in N$. In a congestion game, every A_i is a collection of subsets of M, so $A_i \subseteq 2^M$. The argument of f_j is the number of players having resource j in their action (the identity of these players is irrelevant). The *congestion vector* associated with state σ is denoted by $\ell(\sigma)$. It is an m-dimensional vector which contains, for every resource j, the number of players having j in their strategy. In a congestion game, the cost $c_i(\sigma)$ of player i under strategy profile σ is defined as $\sum_{j \in \sigma_i} f_j(\ell_j(\sigma))$.

A congestion game is said to be *singleton* when the strategy space of every player is a collection of single resources, *monotone* when the resource functions are either all non decreasing or all non increasing and *symmetric* when $A_1 = A_2 = \ldots = A_n$. It is known that every congestion game admits a NE [25]; for every singleton congestion game with non decreasing resource functions, a NE must be a SE [20,28] and a NE can be computed in polynomial time [22]; every singleton congestion game with non increasing resource functions admits a SE [26]; a SSE is not guaranteed to exist in a symmetric singleton congestion game with 2 non decreasing resource functions [1,12,28]. A congestion game that is singleton, symmetric and monotone with strictly increasing functions is also known as a *resource selection game*. A resource selection game always admits a considerate equilibrium [18] and a partition equilibrium [2].

Various complexity results related to the recognition or computation of SE and PoNE in congestion games can be found in [19].

The next two subsections deal with monotone singleton congestion games, that is non decreasing resource functions (Sect. 5.1) and non increasing resource functions (Sect. 5.2). Note that for $m \geq 2$ resources with non monotone cost functions, we cannot guarantee the existence of a SE, so the existence of PDSE is excluded in this case.

5.1 Non Decreasing Resource Functions

Known and new results are summarized in Table 2.

Table 2. Singleton congestion games with non decreasing resource functions.

$m = 2$	$m \geq 3$
SSE not guaranteed [12,28]	PDSE not guaranteed
PDSE guaranteed	SE guaranteed [20,28]
– dynamics converges	NE = SE
– PoNE $\neq$ PDSE $\neq$ SE	

Proposition 6. *When $m = 2$, every instance of the singleton congestion game with non decreasing resource functions admits a PDSE and the dynamics converges.*

Proof. Given any strategy profile σ, let $\Lambda(\sigma)$ be a two-dimensional vector whose first and second coordinates are the maximum and the minimum between $f_1(\ell_1(\sigma))$ and $f_2(\ell_2(\sigma))$, respectively. If a profitable deviation weak improving move exists, turning state σ into σ', then $\Lambda(\sigma') \prec \Lambda(\sigma)$ because the resource functions are non increasing (see the proof of Proposition 1 for the details).   □

Next observation states that though the set of NE and the set of SE coincide [20], the set of PDSE can be a strict subset of these sets.

Observation 3. *In symmetric singleton congestion games with two resources having non decreasing resource functions, it holds that:*

(i) a SE is not necessarily a PDSE;
(ii) a PDSE is not necessarily a PoNE.

Proposition 7. *There exists an instance of the symmetric singleton congestion game with $m \geq 3$ resources having non decreasing functions which does not have any PDSE.*

Proof. The instance falls in the case of resource selection. It consists of $m \geq 3$ resources and $m + 2$ agents. For each resource $j \in M$, one has $f_j(x) = x$ for $x = 1..n$. We consider two cases.

- One resource, say j, is played by at least three players. There must be another resource, say j', which is played by at most one player. The cost of a player on j is at least 3, and if this player moves to j', his cost would be at most 2. Therefore the current strategy profile cannot be a NE.
- Two resources, say j and j', are both selected by two players. There must a third resource, say j'', which is played by at most one player. Let a and b be the players on j; let c and d be the players on j'; if there is a player on j'' then denote it by e. The cost of players a, b, c and d is 2. Suppose b moves to j', c moves to j and d moves to j''. The new cost of c is 2, the new cost of b is 1 and the new cost of d is at most 2. Therefore, a profitable deviation weak improving move exists.

In conclusion, there exists a deviation from any given strategy profile.   □

5.2   Non Increasing Resource Functions

The results are summarized in Table 3.

Table 3. Singleton congestion games with non increasing resource functions.

Symmetric	Asymmetric
SSE guaranteed for all m	SSE not guaranteed ($m \geq 2$)
– dynamics converges	PDSE not guaranteed ($m \geq 3$)
iff $m = 2$	PDSE guaranteed for $m = 2$
	– dynamics converges
	– SE = PDSE $\neq$ PoNE

Observation 4. *Asymmetric singleton congestion games with $m = 2$ resources and non increasing resource functions are not guaranteed to possess a SSE.*

Proof. There are 2 resources and 3 players. For $i \in \{1, 2\}$ player i's single strategy is resource i. The strategy of player 0 is $\{1, 2\}$. The resource functions are defined as $f_j(x) = 2/x$ for the two resources. This instance does not have any SSE because if player 0 plays resource r then his deviation to the other resource $3 - r$ is profitable to player $3 - r$ and the cost of player 0 remains unchanged.   □

Observation 5. *Asymmetric singleton congestion games with $m \geq 3$ resources and non increasing resource functions are not guaranteed to possess a PDSE.*

Proof. Let us first describe an instance with 3 resources $\{r_1, r_2, r_3\}$ and 3 players $\{1, 2, 3\}$. $A_i = \{r_i, r_{i+1}\}$ for $i \in \{1, 2, 3\}$ (with the convention $r_4 = r_1$). The resource functions are identical: $f_{r_j}(x) = 2/x$ for $j = 1..3$. If the players chose pairwise distinct resource then the strategy profile is not a NE (therefore it is not a PDSE). Indeed every player has cost 2 whereas one player can deviate and select the same resource as another player, and his cost becomes 1. Suppose from now on that a resource is selected by two players, another resource is selected by one player and the last resource is not selected. By the symmetry of the instance, one can suppose w.l.o.g. that the current strategy profile is (r_1, r_2, r_1). In this case players 2 and 3 can collude and deviate to strategy profile (r_1, r_3, r_3). The cost of player 3 remains 1 whereas the cost of player 2 is 1 instead of 2. Thus a profitable deviation weak improving move always exists.

Next instance consists of $m \geq 4$ resources. Let us first suppose that $m = 2k$ for some $k \geq 2$. The resource set is $\{r_1, \ldots, r_k\} \cup \{r'_1, \ldots, r'_k\}$. There are $k + 1$ players $\{0, 1, \ldots, k\}$ such that $A_i = \{r_i, r'_i\}$ for $i \in \{1, \ldots, k\}$ and A_0 is the entire resource set. The resource functions are defined as $f_j(x) = 2/x$ for all j. Let σ be a strategy profile. There is (at least) one player, say p, whose strategy σ_p is different from the strategy of player 0. Thus the cost of player p is 2 while the cost of player 0 at least 1. If $\sigma_p = r_p$ then $\bar{\sigma}_p$ denotes r'_p, otherwise $\sigma_p = r'_p$ and $\bar{\sigma}_p$ denotes r_p. Players 0 and p can deviate such that they both play $\bar{\sigma}_p$. The new cost of these players is 1 so the deviation is a profitable deviation weak improving move. Therefore σ cannot be a PDSE. Now suppose that $m = 2k + 1$ for some $k \geq 2$. We slightly modify the instance by adding a single resource r_0 which is only available to player 0. With similar arguments, the instance cannot admit a PDSE. $\qquad \square$

Proposition 8. *Every asymmetric singleton congestion game with 2 non increasing resource functions admits a PDSE which can be built in $O(1)$ time. In addition, the dynamics are guaranteed to converge.*

Proof. Denote the two resources by 1 and 2. The set of players can be partitioned in 3 sets N_1, N_2 and N_3. N_1 and N_2 contain the players whose single strategy is 1 and 2, respectively. N_3 contains the remaining agents who can select either 1 or 2. The players of $N_1 \cup N_2$ have no choice so a strategy profile only depends on the strategy of the members of N_3. If $f_1(|N_1 \cup N_3|) \leq f_2(|N_2 \cup N_3|)$ then put the players of N_3 on resource 1, otherwise put them on resource 2. The resulting strategy profile must be a PDSE because the players of N_3 have the lowest possible cost and they are the only players who can deviate.

Given any strategy profile σ, let $\Lambda(\sigma)$ be a two-dimensional vector whose first and second coordinates are $\max_{j \in \{1,2\}} f_j(\ell_j(\sigma))$ and $\min_{j \in \{1,2\}} f_j(\ell_j(\sigma))$, respectively. If a profitable deviation weak improving move is performed, turning state σ into σ', then it is not difficult to see that $\Lambda(\sigma') \prec \Lambda(\sigma)$. $\qquad \square$

Note that the proof of Proposition 8 still holds if we do not impose the game to be singleton because the non-singleton 2-resource game reduces to a singleton game (see [16], proof of Theorem 1).

Proposition 9. *For asymmetric singleton congestion games with 2 non increasing resource functions, the set of SE and the set of PDSE coincide.*

Proof. A PDSE is, by definition, a SE. For the sake of contradiction, take a SE σ which admits a profitable deviation weak improving move but this deviation is not an improving move. Let σ' denote the state after the deviation. The game is such that some players have a single strategy and the others can choose between the two resources. Clearly, the members of the coalition doing the profitable deviation weak improving move belong to the latter. In the profitable deviation weak improving move, let x_{12} (x_{21} resp.) denote the number of players moving from resource 1 to resource 2 (from resource 2 to resource 1 resp.). If $x_{12} = x_{21}$ then the members exchange their costs, so the existence of a player who is better off implies the existence of a player who is worst off, contradiction. Suppose wlog. that $x_{12} > x_{21}$. We consider two cases:

- The deviating agent who is better off moves from resource 1 to resource 2, i.e. $f_2(\ell_2(\sigma')) < f_1(\ell_1(\sigma))$. In this case, σ admits a deviation by $x_{12} - x_{21} > 0$ players, from resource 1 to resource 2, such that all the members are better off.
- The deviating agent who is better off moves from resource 2 to resource 1, i.e. $f_1(\ell_1(\sigma')) < f_2(\ell_2(\sigma))$. Thus $x_{21} \neq 0$ and σ admits a deviation by x_{21} players, from resource 2 to resource 1, such that all the members are better off. Indeed, $\ell_1(\sigma') = \ell_1(\sigma) - x_{12} + x_{21}$ implies $\ell_1(\sigma') < \ell_1(\sigma) + x_{21}$. Since f_1 is non increasing and by the hypothesis, $f_1(\ell_1(\sigma) + x_{21}) \leq f_1(\ell_1(\sigma')) < f_2(\ell_2(\sigma))$.

In any case, σ is not a SE, contradiction. □

Observation 6. *For asymmetric singleton congestion games with 2 non increasing resource functions, a PDSE is not necessarily a PoNE, and vice versa.*

Proposition 10. *Every symmetric singleton congestion game with non increasing resource functions admits a SSE which can be built in $O(m)$ time. In addition, the dynamics are guaranteed to converge only when $m \leq 2$.*

Proof. Let $j^\star = \arg\min_{j \in M} f_j(n)$ and consider $\sigma^\star$, the state where every player plays $j^\star$. Under $\sigma^\star$, the cost of every player is $f_{j^\star}(n)$, and this is the lowest possible cost because the resource functions are non increasing. Therefore no player can decrease its cost with a unilateral or group deviation.

Suppose there are $m = 2$ resources denoted by 1 and 2. Given any strategy profile σ, let $\Lambda(\sigma)$ be a two-dimensional vector whose first and second coordinates are $\max_{j \in \{1,2\}} f_j(\ell_j(\sigma))$ and $\min_{j \in \{1,2\}} f_j(\ell_j(\sigma))$, respectively. If a weak improving move is performed, turning state σ into σ', then it is not difficult to see that $\Lambda(\sigma') \prec \Lambda(\sigma)$.

Suppose there are $m \geq 3$ resources and 3 players. Define $f_j(x)$ as $1/x$ for every $j \in M$. We are going to describe three states σ^1, σ^2 and σ^3 such that a profitable deviation weak improving move can be a transition between σ^i and σ^{i+1} for $i \in \{1, 2, 3\}$ and $\sigma^{3+1} = \sigma^1$. In σ^1 players 1 and 2 are on resource 1 while player 3 is on resource 2. In σ^2 players 1 and 3 are on resource 3 while player 2 is on resource 1. In σ^3 players 2 and 3 are on resource 2 while player 1 is on resource 3. The cost profiles associated with σ^1, σ^2 and σ^3 are $(\frac{1}{2}, \frac{1}{2}, 1)$, $(\frac{1}{2}, 1, \frac{1}{2})$ and $(1, \frac{1}{2}, \frac{1}{2})$, respectively. □

The computation of a PDSE for a singleton congestion game with 2 (non decreasing or non increasing) resources is direct since there are essentially $n + 1$ possible states: x players on the first resource and $n - x$ on the other, with $0 \leq x \leq n$.

6   Conclusion

To conclude, our positive results concerning the existence of a PDSE occur when two strategies are available to each player. For scheduling games and monotone singleton congestion games with 3 or more strategies, there are instances without any PDSE. This observation triggers two questions. Can we characterize the 2-strategy games which admit a PDSE? Can we learn from the counterexamples with 3 (or more) strategies how to refine the notion of PDSE such that a similar solution concept exists when more than 2 strategies are available?

As a future work it would be interesting to extend our results to other 2-strategy games. In the *max sat game* (see for example [5]) there is a set of weighted disjunctive clauses defined over a set of variables. Each variable is controlled by a player who can play true or false. The payoff of a player is the weight of the satisfied clauses where its variable appears. We can also think of *player-specific* singleton congestion games with two resources [22].

References

1. Andelman, N., Feldman, M., Mansour, Y.: Strong price of anarchy. Games Econ. Behav. **65**(2), 289–317 (2009)
2. Anshelevich, E., Caskurlu, B., Hate, A.: Partition equilibrium always exists in resource selection games. Theory Comput. Syst. **53**(1), 73–85 (2013)
3. Anshelevich, E., Postl, J., Wexler, T.: Assignment games with conflicts: price of total anarchy and convergence results via semi-smoothness. CoRR abs/1304.5149 (2013). http://arxiv.org/abs/1304.5149
4. Aumann, R.J.: Acceptable points in general cooperative n-person games. In: Tucker, A.W., Luce, R.D. (eds.) Contribution to the Theory of Games. Annals of Mathematics Studies, 40, vol. IV, pp. 287–324. Princeton University Press (1959)
5. Caragiannis, I., Fanelli, A., Gravin, N.: Short sequences of improvement moves lead to approximate equilibria in constraint satisfaction games. In: Lavi, R. (ed.) SAGT 2014. LNCS, vol. 8768, pp. 49–60. Springer, Heidelberg (2014)
6. Chen, P., de Keijzer, B., Kempe, D., Schäfer, G.: Altruism and its impact on the price of anarchy. ACM Trans. Econ. Comput. **2**(4), 17 (2014). http://doi.acm.org/10.1145/2597893
7. Chien, S., Sinclair, A.: Strong and pareto price of anarchy in congestion games. In: Albers, S., Marchetti-Spaccamela, A., Matias, Y., Nikoletseas, S., Thomas, W. (eds.) ICALP 2009, Part I. LNCS, vol. 5555, pp. 279–291. Springer, Heidelberg (2009)

8. Epstein, L., Kleiman, E.: On the quality and complexity of Pareto equilibria in the job scheduling game. In: Sonenberg, L., Stone, P., Tumer, K., Yolum, P. (eds.) 10th International Conference on Autonomous Agents and Multiagent Systems (AAMAS 2011). IFAAMAS, Taipei, Taiwan, 2–6 May 2011, vol. 1–3, pp. 525–532 (2011). http://portal.acm.org/citation.cfm?id=2031692&CFID=54178199&CFTOKEN=61392764

9. Epstein, L., Kleiman, E., van Stee, R.: Maximizing the minimum load: the cost of selfishness. In: Leonardi, S. (ed.), [21], pp. 232–243 (2009)

10. Epstein, L., van Stee, R.: Maximizing the minimum load for selfish agents. Theor. Comput. Sci. **411**(1), 44–57 (2010)

11. Even-Dar, E., Kesselman, A., Mansour, Y.: Convergence time to Nash equilibrium in load balancing. ACM Trans. Algorithms **3**(3) (2007)

12. Feldman, M., Tennenholtz, M.: Partition equilibrium. In: Mavronicolas, M., Papadopoulou, V.G. (eds.) SAGT 2009. LNCS, vol. 5814, pp. 48–59. Springer, Heidelberg (2009)

13. Fotakis, D.A., Kontogiannis, S.C., Koutsoupias, E., Mavronicolas, M., Spirakis, P.G.: The structure and complexity of Nash equilibria for a selfish routing game. In: Widmayer, P., Triguero, F., Morales, R., Hennessy, M., Eidenbenz, S., Conejo, R. (eds.) ICALP 2002. LNCS, vol. 2380, p. 123. Springer, Heidelberg (2002)

14. Gourvès, L., Monnot, J.: On strong equilibria in the max cut game. In: Leonardi, S. (ed.) WINE 2009. LNCS, vol. 5929, pp. 608–615. Springer, Heidelberg (2009)

15. Gourvès, L., Monnot, J.: The max k-cut game and its strong equilibria. In: Kratochvíl, J., Li, A., Fiala, J., Kolman, P. (eds.) TAMC 2010. LNCS, vol. 6108, pp. 234–246. Springer, Heidelberg (2010)

16. Gourvès, L., Monnot, J., Moretti, S., Thang, N.: Congestion games with capacitated resources. Theory of Computing Systems pp. 1–19 (2014)

17. Hoefer, M.: Cost Sharing and Clustering under Distributed Competition. Ph.D. thesis, Universität Konstanz (2007)

18. Hoefer, M., Penn, M., Polukarov, M., Skopalik, A., Vöcking, B.: Considerate equilibrium. In: Walsh, T. (ed.) IJCAI. pp. 234–239. IJCAI/AAAI (2011)

19. Hoefer, M., Skopalik, A.: On the complexity of Pareto-optimal Nash and strong equilibria. Theory Comput. Syst. **53**(3), 441–453 (2013). doi:10.1007/s00224-012-9433-0

20. Holzman, R., Law-Yone, N.: Strong equilibrium in congestion games. Games and Economic Behavior **21**(1–2), 85–101 (1997)

21. Leonardi, S. (ed.): Internet and Network Economics, 5th International Workshop, WINE 2009, Rome, Italy, December 14–18, 2009. Proceedings, Lecture Notes in Computer Science, vol. 5929. Springer (2009)

22. Milchtaich, I.: Congestion games with player-specific payoff functions. Games and Economic Behavior 13(1), 111–124 (1996), http://www.sciencedirect.com/science/article/pii/S0899825696900275

23. Monderer, D., Shapley, L.S.: Potential games. Games and Economic Behavior 14(1), 124–143 (1996), http://www.sciencedirect.com/science/article/pii/S0899825696900445

24. Nash, J.: Non-cooperative Games. The Annals of Mathematics **54**(2), 286–295 (1951)

25. Rosenthal, R.W.: A class of games possessing pure-strategy Nash equilibria. International Journal of Game Theory **2**, 65–67 (1973)

26. Rozenfeld, O., Tennenholtz, M.: Strong and Correlated Strong Equilibria in Monotone Congestion Games. In: Spirakis, P.G., Mavronicolas, M., Kontogiannis, S.C. (eds.) WINE 2006. LNCS, vol. 4286, pp. 74–86. Springer, Heidelberg (2006)

27. Schäffer, A.A., Yannakakis, M.: Simple local search problems that are hard to solve. SIAM J. Comput. **20**(1), 56–87 (1991). doi:10.1137/0220004
28. Voorneveld, M.: Potential Games and Interactive Decisions with Multiple Criteria. Ph.D. thesis, Tilburg University (1999)

Democratix: A Declarative Approach to Winner Determination

Günther Charwat[1]([⊠]) and Andreas Pfandler[1,2]([⊠])

[1] Institute of Information Systems, TU Wien, Vienna, Austria
{gcharwat,pfandler}@dbai.tuwien.ac.at
[2] School of Economic Disciplines, University of Siegen, Siegen, Germany

Abstract. Computing the winners of an election is an important task in voting and preference aggregation. The declarative nature of answer-set programming (ASP) and the performance of state-of-the-art solvers render ASP very well-suited to tackle this problem. In this work we present a novel, reduction-based approach for a variety of voting rules, ranging from tractable cases to problems harder than NP. The encoded voting rules are put together in the extensible tool Democratix, which handles the computation of winners and is also available as a web application. To learn more about the capabilities and limits of the approach, the encodings are evaluated thoroughly on real-world data as well as on random instances.

1 Introduction

Voting and preference aggregation are central topics in the field of computational social choice. Here one is interested in how opinions (or *preferences*) can be aggregated in order to obtain a collective decision. Application areas range from (political) elections to multi-agent systems. Further applications are network design and ranking algorithms for search engines (see, e.g., [15,22]). Although voting and preference aggregation are vivid and growing research areas, the number of available implementations and tools is still rather limited. In particular, a general, customizable, and freely available system could support and encourage experimental research in this interdisciplinary area.

In this paper we present a novel reduction-based approach for winner determination: Hereby, we express voting rules in the formalism of answer-set programming (ASP) (see, e.g., [21,30]). ASP allows one to model problems declaratively, which not only leads to readable and maintainable code but also results in succinct encodings (compared to imperative languages). These encodings oftentimes closely resemble the mathematical definitions of the respective voting rules, thereby yielding an "executable specification". Furthermore, due to the developments of the last years, sophisticated solvers have become available for ASP (e.g., [19,26]). All encoded voting rules are readily available in our tool *Democratix* that allows the user to automatically obtain the winners of elections and also to specify further voting rules. This makes the tool especially well-suited for experimenting with new voting rules, and allows one to model new rules "hands-on"

© Springer International Publishing Switzerland 2015
T. Walsh (Ed.): ADT 2015, LNAI 9346, pp. 253–269, 2015.
DOI: 10.1007/978-3-319-23114-3_16

together with experts from other fields (similar to [32]). To enable a broader range of users to work with Democratix, the tool is also made available as a tutorial-like web application.

So far, preference aggregation in combination with ASP has hardly ever been explored. One exception is the work of Konczak [25], where the possible/necessary winner problem in the setting of incomplete preferences is solved for several cases which are polynomially decidable. In contrast, here we consider twelve different voting rules over fully specified preferences. For four of these rules it is harder than NP to decide whether a given candidate is among the winners. Furthermore, some work exists on implementations for specific voting rules, including Kemeny winner determination (cf. [1,9,13,14]) and approximation of Dodgson and Young elections [12]. Additionally, some commercial tools (e.g., OpenSTV [31]) are available as well as software that supports some polynomial voting rules (e.g., http://vote.sourceforge.net/). Moreover, there exist user-friendly web-platforms for preference aggregation, such as Whale[3] [6] and the recently developed Pnyx system [8]. In contrast to Democratix, these platforms are designed for end-users rather than to be a research tool and comprise a fixed set of voting rules. Another branch of research in social choice, where reduction-based approaches have been successfully employed, is automated theorem proving. One example is the application of the satisfiability problem (SAT) for finding strategyproof social choice functions [7] and in the area of "ranking sets of objects" [20]. Before that, reductions to SAT and constraint satisfaction problems (CSP) have been applied for proving, e.g., Arrow's theorem [35].

Despite this progress, to the best of our knowledge, there does not exist a uniform system that permits the declarative specification of hard voting rules. Democratix is the first tool that computes, using an exact approach, all winners with respect to declaratively specified voting rules that can be harder than NP. In more detail, our main contributions are the following:

- Democratix comprises novel ASP encodings for a variety of voting rules, ranging from tractable (such as *Plurality*, *Borda* and other scoring rules, *Maximin*, *Copeland$^\alpha$*, *Bucklin*, and *Black*) to intractable (*Kemeny*, *Dodgson*, *Young*, and *Slater*) rules. It features a uniform interface for all voting rules and hence can easily be extended and integrated into other tools. This makes the tool especially suitable for experimenting with voting rules in a declarative way.
- The tool is available as a web application that allows one to evaluate the voting rules on any election with complete strict-orders given in the PrefLib format [28]. Several interactive examples help to make the tool also accessible to non-experts. Furthermore, we think that the web application is useful for demonstrations and teaching, as examples can be executed and modified directly in the browser.
- We evaluate our approach using the instances from PrefLib and a collection of randomly generated elections. Benchmark results show the capabilities and limits of our approach. Results indicate that our approach works very well for all tractable rules. For the hard rules, we obtain a mixed picture. We thus study how the runtime is influenced by different representations of a voting rule by comparing three alternative encodings of Kemeny's rule.

The web application, the Democratix source-code, and the encodings of the voting rules are available at: http://democratix.dbai.tuwien.ac.at/

This work is structured as follows: In Sect. 2 we recall the required basics of voting theory and ASP; followed by Sect. 3, where we present our encodings. An overview of the Democratix system and the web application is given in Sect. 4. In Sect. 5, we provide an experimental evaluation of the tool. Finally, we conclude in Sect. 6 and provide an outlook on further developments.

2   Preliminaries

2.1   Voting Theory

Let C be a finite set of *candidates* with $|C| = m$ and $V = \{1, 2, \ldots, n\}$ a finite set of *voters*. Furthermore, let $\succ$ be a *preference relation*, i.e., a strict total order over C. The top-ranked candidate of $\succ$ is at position 1, the successor at position 2, $\ldots$, and the last-ranked candidate is at position m. The vote of voter $i \in V$ is the preference relation $\succ_i$. A collection of preference relations $\mathcal{P} = (\succ_1, \ldots, \succ_n)$ is called a *preference profile*. A voter i *prefers* candidate c over candidate c' if $c \succ_i c'$. We denote by $\mathsf{prf}(\mathcal{P}, c, c')$ the number of voters in $\mathcal{P}$ that prefer c over c'. An *election* is given by $E = (C, V, \mathcal{P})$. A *voting rule* $\mathcal{F}$ is a mapping from an election E to a non-empty subset of the candidates $W \subseteq C$, i.e., the *winners* of the election. We briefly recall the voting rules discussed in this paper.

Scoring rules. The class of (positional) *scoring rules* can be expressed by scoring vectors $\alpha = (\alpha_1, \ldots, \alpha_m)$, where $\alpha_i \in \mathbb{N}$ for $1 \leq i \leq m$ with $\alpha_1 \geq \alpha_2 \geq \cdots \geq \alpha_m$ and $\alpha_1 > \alpha_m$. To evaluate an election according to a scoring rule, the candidate ranked at position i gains α_i points. The winners of the election are the candidates having maximum score.

The *plurality* rule can easily be expressed via the vector $\alpha = (1, 0, 0, \ldots, 0)$. Similarly, the *veto* rule is expressed by $\alpha = (1, 1, \ldots, 1, 0)$. In another rule, *k-approval*, the candidates at position 1 to k gain one point each. Finally, *Borda's* rule uses the scoring vector $\alpha = (m - 1, m - 2, \ldots, 0)$.

Condorcet. The Condorcet winner is a candidate $c \in C$ such that for all $c' \in C \setminus \{c\}$ the condition $\mathsf{prf}(\mathcal{P}, c, c') > \frac{n}{2}$ holds. Notice that there are elections without Condorcet winner.

Dodgson and Young. Since having a Condorcet winner is a very favorable property there are several voting rules that try to modify an election as little as possible to obtain a Condorcet winner. There are various notions characterizing this minimality of change. One such rule is the voting rule attributed to Dodgson. For this rule, the Dodgson score of a candidate c is defined to be the minimum number of swaps of adjacent candidates in the votes necessary to make c a Condorcet winner. The winners are the candidates with minimum Dodgson score. In contrast to swapping candidates, in the related Young rule votes are removed until a Condorcet winner exists.

Kemeny. Kemeny's rule is based on the distance between votes. For two votes v_1, v_2 and two candidates c_1, c_2 we define disagree(v_1, v_2, c_1, c_2) to be 0 if v_1 and v_2 rank the candidates c_1 and c_2 in the same way, and to be 1 otherwise. The distance between two votes v_1 and v_2 is defined as dist$(v_1, v_2) = \sum_{\{c_1,c_2\} \subseteq C}$ disagree(v_1, v_2, c_1, c_2). The distance between a preference relation $\succ$ and an election $E = (C, V, \mathcal{P} = (\succ_1, \ldots, \succ_n))$ is given by the Kemeny score kemeny$(\succ, E) = \sum_{1 \leq i \leq n}$ dist$(\succ, \succ_i)$. A preference relation $\succ$ with minimum kemeny$(\succ, E)$ is called a Kemeny consensus with respect to E. The winners are the top-ranked candidates in any Kemeny consensus.

Notice that determining the winner according to a scoring rule as well as finding the Condorcet winner can be done in polynomial time. However, deciding whether a candidate is a winner was shown to be Θ_2^P-complete for Dodgson [23], Young [33] as well as Kemeny [24]. Recall that the class Θ_2^P contains all problems that can be decided in polynomial time by a deterministic Turing machine using $\mathcal{O}(\log n)$ calls to an NP-oracle, where n is the input size.

2.2   Answer-Set Programming

We introduce normal logic programs under the answer-set semantics (see [10, 21, 30]), thereby restricting ourselves to the syntax and semantics relevant for this work. For more details on ASP see, e.g., [16, 18].

We fix a countable set $\mathcal{U}$ of *domain elements*, also called *constants*. An *atom* is an expression $p(t_1, \ldots, t_a)$, where p is a *predicate* of arity $a \geq 0$ and each t_i is either a variable or an element from $\mathcal{U}$. We use "_" to denote an anonymous variable. An atom is *ground* if it is free of variables. $B_\mathcal{U}$ denotes the set of all ground atoms over $\mathcal{U}$.

A *normal rule* r with $0 \leq k \leq n$ is of the form

$$h \leftarrow b_1, \ldots, b_k, \ not\, b_{k+1}, \ldots, \ not\, b_n.$$

The *head* of a rule r is the set $H(r) = \{h\}$, containing exactly one element. The *body* of r is $B(r) = B^+(r) \cup B^-(r)$ with $B^+(r) = \{b_1, \ldots, b_k\}$ and $B^-(r) = \{b_{k+1}, \ldots, b_n\}$. Here, $h, b_1, \ldots, b_n$ are atoms, and "*not*" stands for *default negation*. An atom x is a positive literal, while *not* x is a default negated literal. We denote by $b(t_1; \ldots; t_l)$ the sequence of unary atoms $b(t_1), \ldots, b(t_l)$. Extending normal rules we have *integrity constraints* where $H(r) = \emptyset$ and $B(r) \neq \emptyset$.

A rule r is *safe* if each variable in r occurs in $B^+(r)$. A rule r is *ground* if no variable occurs in r. A *fact* is a ground rule with empty body. A program π is a finite set of safe rules. $I \subseteq B_\mathcal{U}$ is an *answer set* of π iff it is a subset-minimal set satisfying the *Gelfond-Lifschitz reduct* $\pi^I = \{H(r) \leftarrow B^+(r) \mid I \cap B^-(r) = \emptyset, r \in Gr(\pi)\}$, where $Gr(\pi)$ is the grounding of π.

Additionally we consider the class of *optimization programs*, i.e., programs containing also *weak constraints*

$$\leftsquigarrow b_1, \ldots, b_k, \ not\, b_{k+1}, \ldots, \ not\, b_n. \quad [w]$$

where all b_j with $1 \leq j \leq n$ are as in rules and the weight w is a positive integer variable occurring in $b_1, \ldots, b_k$ or a constant. Answer sets are minimized w.r.t. the *costs*, i.e., the sum of weights.

In addition to atoms, the body of a rule can contain *aggregates* of the form $x := \text{aggr}_t\{p(t_1, \ldots, t_a) : p_1 : \cdots : p_l\}$ where aggr $\in \{\text{sum}, \text{min}, \text{max}\}$, $p(t_1, \ldots, t_a)$ is an atom, $p_1, \ldots, p_l$ are *conditional atoms* and t is an integer variable occurring in $t_1, \ldots, t_a$. Variable x gets assigned an integer that corresponds to the value of aggr evaluated on the values of t in all grounded instantiations of p in interpretation I, such that $p_1, \ldots, p_l$ are in I. Furthermore, $x := \text{count}\{p(t_1, \ldots, t_a)\}$ counts the number of grounded occurrences of $p(t_1, \ldots, t_a)$ in I. In addition, we allow standard relations and arithmetic expressions. All these extensions are readily supported by modern ASP solvers.

3   Winner Determination with ASP

In this section we present our approach for encoding voting rules in ASP and describe some of the encodings in detail. Special focus is devoted to voting rules that are harder than NP. For Kemeny's rule, we additionally highlight how different design choices can be put into practice in the Democratix system. Additionally, we show how voting rules can be combined.

Encoding Elections.   For the encodings to follow we assume the election to be given as a set of ASP facts. Let $E = (C, V, \mathcal{P})$ be an election with $C = \{c_1, \ldots, c_m\}$ and $V = \{1, \ldots, n\}$. Furthermore, let $prefs(\mathcal{P}) = \{\succ_1, \ldots, \succ_l\}$, $l \leq n$, denote the set of *distinct* preference relations occurring in the profile $\mathcal{P}$, and $vc(\mathcal{P}, \succ)$ denote the number of times preference relation $\succ$ occurs in $\mathcal{P}$. For $1 \leq i \leq l$, let the preference relation $\succ_i \in prefs(\mathcal{P})$ be of the form $c_{i_1} \succ_i c_{i_2} \succ_i \cdots \succ_i c_{i_m}$. The input of the ASP encoding is now given as follows: Each distinct preference relation $\succ_i$ is represented by m facts $\mathsf{p}(i, j, c_{i_j})$, where $1 \leq j \leq m$, and a single fact $\mathsf{vcnt}(i, vc(\mathcal{P}, \succ_i))$. (This representation usually reduces the size of the encoded election as each distinct preference relation is encoded only once.) Additionally, for convenience, three unary facts $\mathsf{vnum}(n)$, $\mathsf{cnum}(m)$, and $\mathsf{pnum}(l)$ are added to the input. Note that it is easy to adapt this representation to handle also incomplete votes as well as non-anonymous voting rules. For voting rules such as Copeland$^\alpha$ and k-approval it is also possible to pass parameters along with the input instance in form of ASP facts.

The output of the ASP solver applied to the encoding of a voting rule together with the encoding of the election is either one or several answer sets containing winner predicates, or `UNSATISFIABLE` (e.g., in case we want to compute the Condorcet winner, but there is none for the given election). For encodings of problems that contain weak constraints, only the minimal answer sets, i.e., the answer sets with minimum cost, are returned by the solver.

Encoding 1: Borda

$$\mathsf{cand}(I) \leftarrow \mathsf{cnum}(M), 1 \leq I \leq M. \tag{1}$$

$$\mathsf{posScore}(P, C, S \cdot VC) \leftarrow \mathsf{p}(P, Pos, C), \mathsf{cnum}(M), S := M - Pos, \mathsf{vcnt}(P, VC). \tag{2}$$

$$\mathsf{score}(C, N) \leftarrow \mathsf{cand}(C), N := \underset{S}{\mathsf{sum}}\{\mathsf{posScore}(\_, C, S)\}. \tag{3}$$

$$\mathsf{maxScore}(M) \leftarrow M := \underset{S}{\mathsf{max}}\{\mathsf{score}(\_, S)\}. \tag{4}$$

$$\mathsf{winner}(C) \leftarrow \mathsf{cand}(C), \mathsf{score}(C, M), \mathsf{maxScore}(M). \tag{5}$$

Scoring rules. The family of scoring rules can be expressed very naturally in ASP. Here, we start with an ASP program for Borda's rule, depicted in Encoding 1. The first rule is used to obtain the cand relation, which contains all elements of $\{1, \ldots, m\}$. Then, we determine for each candidate in every preference relation the score according to his position (rule 2). Observe that this score is multiplied by VC, i.e., the number of occurrences of the preference relation in $\mathcal{P}$. Next, we sum the scores of a candidate over all votes (rule 3). Finally, in rules (4) and (5) the winner(s) are determined.

It is simple to modify Encoding 1 to specify other scoring rules, such as plurality, k-approval, and veto. The plurality rule can be obtained by replacing rule (2) with $\mathsf{posScore}(P, C, VC) \leftarrow \mathsf{p}(P, 1, C), \mathsf{vcnt}(P, VC)$. Similarly, for veto we replace rule (2) with $\mathsf{posScore}(P, C, VC) \leftarrow \mathsf{p}(P, Pos, C), \mathsf{cnum}(M), Pos \neq M,$ $\mathsf{vcnt}(P, VC)$. Finally, for k-approval we replace rule (2) with $\mathsf{posScore}(P, C, VC)$ $\leftarrow \mathsf{p}(P, Pos, C), \mathsf{vcnt}(P, VC), \mathsf{kApp}(K), Pos \leq K.$, where variable K in $\mathsf{kApp}(K)$ is a parameter.

Encoding 2: Condorcet

$$\mathsf{cand}(I) \leftarrow \mathsf{cnum}(M), 1 \leq I \leq M. \tag{1}$$

$$\mathsf{prefer}(P, C_1, C_2) \leftarrow \mathsf{p}(P, Pos_1, C_1), \mathsf{p}(P, Pos_2, C_2), Pos_1 < Pos_2. \tag{2}$$

$$\mathsf{preferCnt}(C_1, C_2, N) \leftarrow \mathsf{cand}(C_1; C_2), C_1 \neq C_2, \tag{3}$$

$$N := \underset{VC}{\mathsf{sum}}\{\mathsf{vcnt}(P, VC) : \mathsf{prefer}(P, C_1, C_2)\}.$$

$$\mathsf{noWin}(C) \leftarrow \mathsf{preferCnt}(C, \_, N), \mathsf{vnum}(V), N \cdot 2 \leq V. \tag{4}$$

$$\mathsf{winner}(C) \leftarrow \mathsf{cand}(C), not\ \mathsf{noWin}(C). \tag{5}$$

$$\mathsf{anyWinner} \leftarrow \mathsf{winner}(\_). \tag{6}$$

$$\leftarrow not\ \mathsf{anyWinner}. \tag{7}$$

Condorcet. Determining whether a given election has a Condorcet winner is a central subtask in several voting rules. As in the previous encoding, in rule (1) of Encoding 2 the unary relation cand is constructed. Rules (2) and (3) are used to compute the value of $\mathsf{prf}(\mathcal{P}, c_i, c_j)$ for any pair of distinct candidates in C. A candidate c cannot be a Condorcet winner if there is some other candidate c' such that $\mathsf{prf}(\mathcal{P}, c, c') \leq \frac{n}{2}$. This search is encoded in rule (4) where $\mathsf{noWin}(C)$ is derived if such a counterexample can be found for candidate C. In case no counterexample exists, we have indeed found the Condorcet winner (rule 5). Rule (6) derives the atom anyWinner if there is some winner. The constraint in rule (7) ensures that no answer set is returned if no Condorcet winner exists.

Notice that only stratified default negation and no weak constraints are used in the previous encodings. Hence the encodings lie in the P fragment of ASP (data-complexity). We remark that for such programs an ASP solver can compute the unique answer set (if it exists) without backtracking. We now turn to harder voting rules, i.e., voting rules for which the problem of winner determination is Θ_2^P-complete. To capture these problems, the following encodings make use of non-stratified default negation and weak constraints.

Kemeny. Kemeny's rule is particularly well-suited for illustrating the *guess, check & optimize* approach of ASP. In the following, we discuss three different encodings for this voting rule. The intuitive approach is to guess a preference relation and compute the Kemeny score. The Kemeny consensuses are then obtained by minimizing over all guessed preference relations. The winners are the top-ranked candidates in a Kemeny consensus.

Encoding 3: Kemeny (direct, dir)

$$\text{dom}(I) \leftarrow \text{cnum}(M), 1 \leq I \leq M. \tag{1}$$

$$\text{wrank}(P, C_2, C_1) \leftarrow \text{p}(P, Pos_1, C_1), \text{p}(P, Pos_2, C_2), Pos_1 < Pos_2. \tag{2}$$

$$\text{wrankC}(C_2, C_1, N) \leftarrow \text{dom}(C_1; C_2), N := \underset{VC}{\text{sum}}\{\text{vcnt}(P, VC) : \text{wrank}(P, C_2, C_1)\}. \tag{3}$$

$$\text{gpref}(Pos, C) \leftarrow \text{dom}(Pos; C), not\ \text{npref}(Pos, C). \tag{4}$$

$$\text{npref}(Pos, C) \leftarrow \text{dom}(Pos; C), not\ \text{gpref}(Pos, C). \tag{5}$$

$$\leftarrow \text{gpref}(Pos, C_1), \text{gpref}(Pos, C_2), C_1 \neq C_2. \tag{6}$$

$$\leftarrow \text{gpref}(Pos_1, C), \text{gpref}(Pos_2, C), Pos_1 \neq Pos_2. \tag{7}$$

$$\text{occupied}(Pos) \leftarrow \text{gpref}(Pos, \_). \tag{8}$$

$$\leftarrow \text{dom}(Pos), not\ \text{occupied}(Pos). \tag{9}$$

$$\text{rank}(C_1, C_2) \leftarrow \text{gpref}(Pos_1, C_1), \text{gpref}(Pos_2, C_2), Pos_1 < Pos_2. \tag{10}$$

$$\rightsquigarrow \text{rank}(C_1, C_2), \text{wrankC}(C_1, C_2, N). \quad [N] \tag{11}$$

$$\text{winner}(C) \leftarrow \text{gpref}(1, C). \tag{12}$$

This first variant is depicted in Encoding 3. In rule (1) the unary relation dom is obtained, which is used to identify candidates and positions in preferences. We then determine for each preference relation the candidates C_2 that are worse-ranked than C_1 (rule 2) and sum up the overall number of voters that do not prefer C_2 over C_1 (rule 3). Note that rules (1–3) can be computed independently of the guess during grounding. In rules (4–9) the preference relation is guessed by assigning to each candidate exactly one position.[1] We obtain the relation rank whenever C_1 is better-ranked than C_2 in our guessed preference relation (rule 10). What remains is to select only preference relations with minimal Kemeny

---

[1] Note that these rules can be simplified by using the so-called *choice rule* (see, e.g., [11]). Currently, this construct is, however, not supported by all ASP solvers.

score. (rule 11). A candidate ranked first in such a Kemeny consensus is a winner (rule 12).

Another approach to model Kemeny's rule proceeds as follows. The preference relation is obtained *implicitly* by guessing the relative order for each pair of candidates within the relation. We call this alternative encoding *Kemeny (alt)*, given in Encoding 4. As we will see in Sect. 5, the representation in ASP can influence the runtime notably. In particular, within Encoding 4 we apply the following optimizations: (a) Rule (4.2) combines rules (3.2–3.3) of Encoding 3. This reduces the size of the grounding, since wrank is not derived explicitly. (b) By the condition $C_1 < C_2$ in rule (4.2) only half of the candidates are compared. (c) The guess in rules (4.3–4.4) directly contains the costs (N resp. $U - N$) for a candidate C_1 being preferred over a candidate C_2. Since the weak constraint in rule (4.9) directly minimizes over these costs, the ASP solver is guided towards guessing first on prefer predicates with low costs. (d) Rules (4.5–4.7) guarantee that the guess forms a valid preference relation. With xpref, the transitive closure over prefer is obtained, and relations containing a cycle are removed. (e) Rule (4.8) is redundant but increases performance: Each candidate is either ranked before or after each other candidate.

Encoding 4: Kemeny (alternative, alt)

$$\text{cand}(I) \leftarrow \text{cnum}(M), 1 \leq I \leq M. \tag{1}$$

$$\text{wrankC}(C_2, C_1, N) \leftarrow \text{cand}(C_1; C_2), C_1 < C_2, N := \text{sum}_{VC}\{\text{vcnt}(P, VC) : \tag{2}$$
$$\text{p}(P, Pos_1, C_1) : \text{p}(P, Pos_2, C_2) : Pos_1 < Pos_2\}.$$

$$\text{prefer}(C_2, C_1, N) \leftarrow \text{wrankC}(C_2, C_1, N), \text{vnum}(U), not\ \text{prefer}(C_1, C_2, U - N). \tag{3}$$

$$\text{prefer}(C_1, C_2, U - N) \leftarrow \text{wrankC}(C_2, C_1, N), \text{vnum}(U), not\ \text{prefer}(C_2, C_1, N). \tag{4}$$

$$\text{xpref}(C_1, C_2) \leftarrow \text{prefer}(C_1, C_2, \_). \tag{5}$$

$$\text{xpref}(C_1, C_3) \leftarrow \text{xpref}(C_1, C_2), \text{xpref}(C_2, C_3). \tag{6}$$

$$\leftarrow \text{xpref}(C, C). \tag{7}$$

$$\leftarrow \text{cand}(C_1; C_2), C_1 \neq C_2, not\ \text{xpref}(C_1, C_2), not\ \text{xpref}(C_2, C_1). \tag{8}$$

$$\rightsquigarrow \text{prefer}(\_, \_, N).\ [N] \tag{9}$$

$$\text{someBetter}(C_2) \leftarrow \text{prefer}(C_1, C_2, \_). \tag{10}$$

$$\text{winner}(C) \leftarrow \text{cand}(C), not\ \text{someBetter}(C). \tag{11}$$

Finally, we show how to encode Kemeny's rule building upon the *weighted majority graph* $G = (V, A, w)$ of an election $E = (C, V, \mathcal{P})$. Intuitively, G is a directed, weighted graph with an arc between two candidates c and c', whenever c is preferred over c' by more voters in E, and the arc is weighted with the majority margin. For $c, c' \in C$, let $\text{mar}(\mathcal{P}, c, c') = \text{prf}(\mathcal{P}, c, c') - \text{prf}(\mathcal{P}, c', c)$. Then, G is constructed as $V = C$, $A = \{(c, c') \mid c, c' \in C, \text{mar}(\mathcal{P}, c, c') \geq 0\}$ and for $(c, c') \in A$, we have $w((c, c')) = \text{mar}(\mathcal{P}, c, c')$. The goal is now to construct an acyclic graph of G by inverting arcs. A Kemeny consensus has minimal total weight over the inverted arcs, and winners are candidates without incoming arc in a Kemeny consensus.

In Encoding 5, rules (2–4) are used to construct the weighted majority graph from the input. Note that this conversion of the input format is directly realized in the encoding. With rules (5–6) we guess whether an arc is inverted. Based on this guess, rules (7–9) remove solutions where the graph is not acyclic, and the weak constraint in rule (10) adds the weight of each inverted arc to the costs. Finally, rules (11–12) select the candidate without incoming arc as winner. Observe that we can obtain also an encoding of Slater's rule from this by setting the weight of all arcs to one. Still, improving performance by using techniques such as described in [17] are left for future work.

Encoding 5: Kemeny (Weighted Majority Graph, MG)

$$\text{cand}(I) \leftarrow \text{cnum}(M), 1 \leq I \leq M. \tag{1}$$

$$\text{mar}(C_1, C_2, D) \leftarrow \text{cand}(C_1; C_2), C_1 < C_2, \text{vnum}(N), \tag{2}$$
$$D = 2 \cdot S - N, S = \underset{VC}{\text{sum}}\{\text{vcnt}(P, VC) :$$
$$\text{p}(P, Pos_1, C_1) : \text{p}(P, Pos_2, C_2) : Pos_1 < Pos_2\}.$$

$$\text{arc}(C_1, C_2, D) \leftarrow \text{mar}(C_1, C_2, D), D \geq 0. \tag{3}$$

$$\text{arc}(C_2, C_1, -D) \leftarrow \text{mar}(C_1, C_2, D), D < 0. \tag{4}$$

$$\text{garc}(C_1, C_2, 0) \leftarrow \text{arc}(C_1, C_2, D), not\ \text{garc}(C_2, C_1, D). \tag{5}$$

$$\text{garc}(C_2, C_1, D) \leftarrow \text{arc}(C_1, C_2, D), not\ \text{garc}(C_1, C_2, 0). \tag{6}$$

$$\text{reach}(C_1, C_2) \leftarrow \text{garc}(C_1, C_2, \_). \tag{7}$$

$$\text{reach}(C_1, C_3) \leftarrow \text{garc}(C_1, C_2, \_), \text{reach}(C_2, C_3). \tag{8}$$

$$\leftarrow \text{reach}(C, C). \tag{9}$$

$$\rightsquigarrow \text{garc}(\_, \_, D).\quad [D] \tag{10}$$

$$\text{incoming}(C_2) \leftarrow \text{garc}(C_1, C_2, \_). \tag{11}$$

$$\text{winner}(C) \leftarrow \text{cand}(C), not\ \text{incoming}(C). \tag{12}$$

Dodgson. For Dodgson's rule, one could guess all $(m!)^n$ possible preference profiles, check whether there exists a Condorcet winner and minimize over the number of swaps. In order to avoid unnecessary guesses we impose the following constraints (see [4, Observation 1]). It is sufficient to shift at most one candidate per vote, i.e., one candidate is swapped successively towards the top.

To allow for a simpler presentation of this encoding, we assume that the input is given in extensive form. In extensive form we do not make use of the vcnt predicate to represent preferences occurring multiple times in profile $\mathcal{P}$. Instead, for a preference relation $\succ_i$ of the form $c_{i_1} \succ_i c_{i_2} \succ_i \cdots \succ_i c_{i_m}$ we introduce $\text{vc}(\mathcal{P}, \succ_i)$ many facts $\text{v}(f(i, x), j, c_{i_j})$ where $1 \leq j \leq m$, $1 \leq x \leq \text{vc}(\mathcal{P}, \succ_i)$, and f is a bijection that assigns to each pair (i, x) a distinct voter in V. Notice that this conversion to extensive form can be easily realized during the preparation of the input or directly in the ASP encoding.

In Encoding 6 we first obtain the voters and the domain (positions and candidates) as in the previous encodings (rules 1–2). In rules (3–4) we guess the shifts in the votes. For a voter V the candidate at position Pos_1 will be shifted to Pos_2. At most one shift per voter (rules 5–6) to a better position (rule 7) is allowed. The preference profile is now recomputed: The candidate C_1 is moved from Pos_1 to Pos_2 (rule 8) and each candidate originally at Pos with $Pos_2 \leq Pos < Pos_1$ is shifted by one position downwards (rule 9). In the newly computed votes nv, the shifted candidates are assigned to their new positions (rule 10) and the remaining positions are filled with the respective candidates of the original vote (rules 11–12). Rules (13–18) encode the computation of the Condorcet winner, similar to Encoding 2. Finally, rule (19) minimizes over the number of swaps. Note that one shift consists of $Pos_1 - Pos_2$ elementary exchanges, i.e., swaps, of adjacent candidates.

Encoding 6: Dodgson

$$\mathsf{voter}(I) \leftarrow \mathsf{vnum}(N), 1 \leq I \leq N. \tag{1}$$

$$\mathsf{dom}(I) \leftarrow \mathsf{cnum}(M), 1 \leq I \leq M. \tag{2}$$

$$\mathsf{shift}(V, Pos_1, Pos_2) \leftarrow \mathsf{voter}(V), \mathsf{dom}(Pos_1; Pos_2), not\ \mathsf{noshift}(V, Pos_1, Pos_2). \tag{3}$$

$$\mathsf{noshift}(V, Pos_1, Pos_2) \leftarrow \mathsf{voter}(V), \mathsf{dom}(Pos_1; Pos_2), not\ \mathsf{shift}(V, Pos_1, Pos_2). \tag{4}$$

$$\leftarrow \mathsf{shift}(V, Pos_1, \_), \mathsf{shift}(V, Pos_1', \_), Pos_1 \neq Pos_1'. \tag{5}$$

$$\leftarrow \mathsf{shift}(V, \_, Pos_2), \mathsf{shift}(V, \_, Pos_2'), Pos_2 \neq Pos_2'. \tag{6}$$

$$\leftarrow \mathsf{shift}(V, Pos_1, Pos_2), Pos_1 \leq Pos_2. \tag{7}$$

$$\mathsf{sv}(V, Pos_2, C_1) \leftarrow \mathsf{shift}(V, Pos_1, Pos_2), \mathsf{v}(V, Pos_1, C_1). \tag{8}$$

$$\mathsf{sv}(V, PShift, C) \leftarrow \mathsf{shift}(V, Pos_1, Pos_2), \mathsf{v}(V, Pos, C), \tag{9}$$
$$Pos_2 \leq Pos, Pos < Pos_1, PShift := Pos + 1.$$

$$\mathsf{nv}(V, PShift, C) \leftarrow \mathsf{sv}(V, PShift, C). \tag{10}$$

$$\mathsf{occupied}(V, Pos) \leftarrow \mathsf{sv}(V, Pos, \_). \tag{11}$$

$$\mathsf{nv}(V, Pos, C_1) \leftarrow \mathsf{v}(V, Pos, C_1), not\ \mathsf{occupied}(V, Pos). \tag{12}$$

$$\mathsf{prefer}(V, C_1, C_2) \leftarrow \mathsf{nv}(V, Pos_1, C_1), \mathsf{nv}(V, Pos_2, C_2), Pos_1 < Pos_2. \tag{13}$$

$$\mathsf{preferCnt}(C_1, C_2, N) \leftarrow \mathsf{dom}(C_1; C_2), C_1 \neq C_2, N := \mathsf{count}\{\mathsf{prefer}(\_, C_1, C_2)\}. \tag{14}$$

$$\mathsf{noWin}(C) \leftarrow \mathsf{preferCnt}(C, \_, N), \mathsf{vnum}(V), N \cdot 2 \leq V. \tag{15}$$

$$\mathsf{winner}(C) \leftarrow \mathsf{dom}(C), not\ \mathsf{noWin}(C). \tag{16}$$

$$\mathsf{anyWinner} \leftarrow \mathsf{winner}(\_). \tag{17}$$

$$\leftarrow not\ \mathsf{anyWinner}. \tag{18}$$

$$\leftsquigarrow \mathsf{shift}(\_, Pos_1, Pos_2).\quad [Pos_1 - Pos_2] \tag{19}$$

Young's rule can be encoded quite similarly to Dodgson's rule. The basic idea is to replace rules (3–12) by a set of rules that guess the votes to be deleted. Now, in rules (13–19) we check whether this gives a Condorcet winner and minimize over the number of votes to be deleted.

Combining Voting Rules. Having a plethora of voting rules at hand it is a natural question to ask how one can *combine* existing voting rules. For instance, Black's rule is a combination of Condorcet voting and Borda's rule. Another example is the recent work of Narodytska *et al.* [29] where the properties of combinations of rules are studied.

Our approach of using ASP encodings of voting rules readily supports the combination of voting rules. Besides using a sequence of ASP solver calls, a much more elegant way is to specify a monolithic encoding. Here one has to make sure that the predicates occurring in the heads of the rules originating from different encodings are made disjoint and that the input relations do not occur in the heads. Notice that the former condition can be ensured by prefixing while the latter condition should hold in most reasonable encodings anyway. Black's rule is very well suited to illustrate these ideas.

Black. Black's rule returns the Condorcet winner if it exists, and otherwise returns the Borda winners. In Encoding 7, the winners are contained in the relations $winner_{Cond}(C)$ and $winner_{Borda}(C)$, respectively. The effort needed for "gluing" the encodings together is minimal. We add the atom computeBorda to the body of each rule that is exclusively used to compute the winners of Borda's rule. In rules (1–2) it is checked whether a Condorcet winner exists. If there is no Condorcet winner, rule (3) fires and enables the computation of the Borda winners (rule 4).

Encoding 7: Black

$$winner_{Cond}(C) \leftarrow \cdots \qquad\qquad (1)$$
$$condorcet \leftarrow winner_{Cond}(\_). \qquad\qquad (2)$$
$$computeBorda \leftarrow not\ condorcet. \qquad\qquad (3)$$
$$winner_{Borda}(C) \leftarrow computeBorda,\ldots \qquad\qquad (4)$$
$$winner(C) \leftarrow winner_{Cond}(C). \qquad\qquad (5)$$
$$winner(C) \leftarrow winner_{Borda}(C). \qquad\qquad (6)$$

Another case where combining voting rules is applicable, is the following: For voting rules that measure the distance to elections with a Condorcet winner (e.g., Dodgson, Young), it might be favorable to first check whether the instance already has a Condorcet winner. Only if there is no Condorcet winner, the grounding for the guess has to be computed. Note that such an encoding follows the same pattern as used for Black's rule.

4   The Democratix System

All implemented voting rules are put together in the tool Democratix. The application handles parsing of input instances in PrefLib format [28] to ASP facts. Internally, the ASP solver Clingo [19] is called with the input instance and the encoding of the voting rule as input. The tool is easily extendible, thereby allowing integration of further (e.g., combined) voting rules and handling of, e.g., incomplete preferences. Additionally, the tool is readily prepared to be used with other ASP solvers such as further solvers from the Potassco family [19], DLV [26] and WASP [2]. Democratix is implemented in Python, licensed as open source under the GNU General Public License (GPLv3), and can be run both on Unix-based and Windows systems.

We also provide easy access to Democratix via a web front-end that is available at http://democratix.dbai.tuwien.ac.at/. A screenshot is depicted in Fig. 1.

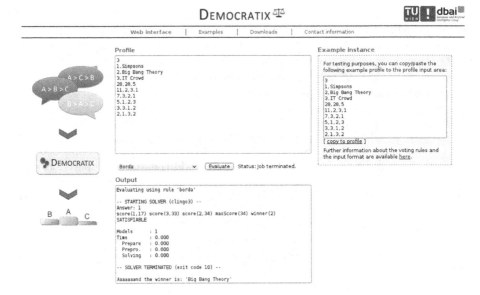

Fig. 1. Screenshot of the web front-end for Democratix

There, instances from PrefLib can be evaluated directly with respect to the voting rules considered in this work. In the example section of the web page, the voting rules are presented and explained in a tutorial-like style. The front-end features interactive evaluation and modification of the provided examples. We believe that the web front-end is therefore particularly well-suited to make voting theory also accessible to non-experts.

Our long-term goal is to constantly extend the Democratix system, e.g., by including further voting rules and providing support for incomplete preferences. We think that our declarative ASP-based approach is the right choice to provide concise, well-readable and maintainable extensions for the system. To this end, we would also like to invite the community to contribute to the system.

5   Evaluation

We evaluate Democratix on basis of the complete strict-order instances provided by PrefLib [28]. This allows us to gain a detailed insight into the runtime behavior of our tool on various kinds of instances. Additionally, to study the capabilities and limits of our general, ASP-based approach, we run the hard voting rules on randomly generated elections (using "PrefLibTools-0.1" under the *impartial culture model*). To highlight how different implementation variants influence runtime, we compare our three encodings for Kemeny's rule in detail. Additionally, we consider an Integer Linear Programming (ILP) formulation for Kemeny's rule that follows ideas presented in [13] and is implemented in the Pnyx system [8].

Table 1. PrefLib instances (*.soc*): number of solved, timeout (TO) and memout (MO) instances. The maximal time (over all) and the average time (over solved) instances is given in seconds (s). In the cells the entries left (right) of "/" refer to instances with (without) Condorcet winner (281/33 instances in total)

Name	Solved	TO	MO	Time (max,s)	Time (avg,s)
Plurality	281/33	0/0	0/0	0.39/0.06	0.05/0.05
Veto	281/33	0/0	0/0	0.58/0.06	0.05/0.05
Borda	281/33	0/0	0/0	0.65/0.06	0.05/0.05
Bucklin	281/33	0/0	0/0	1.11/1.11	0.08/0.23
Maximin	281/33	0/0	0/0	1.64/1.33	0.08/0.24
Black	281/33	0/0	0/0	1.57/1.28	0.08/0.24
Copeland	281/33	0/0	0/0	1.63/1.53	0.09/0.29
Condorcet	281/33	0/0	0/0	1.71/1.23	0.09/0.23
Young (+C)	281/31	0/2	0/0	1.78/TO	0.08/1.42
Dodgson (+C)	281/22	0/11	0/0	1.60/TO	0.08/98.77
Slater (+C)	281/14	0/19	0/0	0.69/TO	0.06/4.80
Kemeny (dir+C)	281/2	0/27	0/4	1.57/TO	0.08/0.05
Kemeny (alt+C)	281/3	0/30	0/0	1.63/TO	0.08/3.83
Kemeny (MG+C)	281/14	0/19	0/0	0.69/TO	0.06/0.49
Kemeny (ILP)	274/27	7/6	0/0	TO/TO	14.87/53.64

We are not aware of any further exact and freely available (ILP) implementations for the hard voting rules considered in this work.

We performed benchmarks on a server with two Intel Xeon E5345 @ 2.33GHz processors and 48 GB RAM running Debian 7.8, kernel 3.2.63–2. Each run was limited to a single core, 16GB of memory and a time limit of 10 min. Results for Democratix were obtained using ASP solver Clingo 4.4.0. The ILP formulation was tested with the open source solver GLPK 4.45. The evaluation of proprietary (but potentially more efficient) ILP solvers is left as future work. Note that we also tested voting rules not presented in detail in Sect. 3.

PrefLib (SOC). Table 1 contains the results for all 314 complete, strict-order instances of the PrefLib library (as of April 20, 2015) [3,27,28,31]. The results show that for all voting rules, where the problem of winner determination is in P, our ASP-based implementation is very fast. For the problems above NP, we tested our combined encodings that implement a Condorcet pre-check (+C). All instances that were not solved are out of the 33 instances that have no Condorcet winner. For Young's rule, the unsolved instances contain a rather high number of voters ($n \in \{532, 578\}$), in particular when considering that there are 2^n possible combinations of removed votes from the preference profile. In general, Dodgson has $(m!)^n$ possibilities of swapping positions of candidates in the votes, which is also observable in our results, as both the number of candi-

Table 2. Random instances without Condorcet winner: maximum size (in terms of the number of voters/candidates) of instances solved within 10 min and 16GB memory

Name	$(m = 5, n =?)$	$(m =?, n = 5)$
Young (+C)	100	1521
Dodgson (+C)	141	73
Slater (+C)	≥ 100000	22
Kemeny (MG+C)	≥ 100000	24
Kemeny (ILP)	≥ 100000	68

dates and voters strongly influences the runtime. In Slater's and Kemeny's rule (MG implementation) we have $\mathcal{O}(2^{m^2})$ potential guesses on the edge direction between candidates. For the direct and alternative implementation of Kemeny's rule there are $m!$ possible preference relations to be guessed. Our results show that the runtime is indeed mainly influenced by m.

Bounds. To give an indication of the tool's applicability in particular settings, we generated instances with a fixed number of candidates $m = 5$ (voters $n = 5$) and increasing n (m) up to $n = 100000$ ($m = 2000$). Due to the applied Condorcet pre-check, all instances having a Condorcet winner were solved by Democratix within the time and memory limits. Table 2 gives the maximum number of voters (candidates) up to which random instances without Condorcet winner could be solved by Democratix. All instances with a bound of $n = 100000$ were solved in less than one second. Note that the reported bounds are dependent on the generated instances, the benchmark server and solver-internal heuristics. Nevertheless, this gives some indication of whether Democratix is suitable for evaluating a given instance. A detailed comparison between the different Kemeny implementations is given below.

Kemeny. We study the performance of our ASP encodings for Kemeny's rule as well as of an ILP formulation of the rule. Our focus lies on the different implementations for Kemeny's rule, therefore we did not apply a Condorcet pre-check here. Figure 2 shows the average runtime over 10 instances for each fixed n and m. Most notably, the implementations based on the majority graph (Kemeny (MG) and Kemeny (ILP)) perform best, and almost equally good on instances up to $m = 20$. However, as m grows, Kemeny (MG) tends to have more outliers (w.r.t. time) compared to Kemeny (ILP). One reason is that there can be millions of solutions for larger instances, which are all computed by Democratix. The ILP-based implementation, however, currently reports only one solution (winner) due to restrictions of GLPK that would require multiple solver invocations. We remark that although restricting Democratix to return a single winner would be possible in principle, this is not favorable as our goal is to compute all winners.

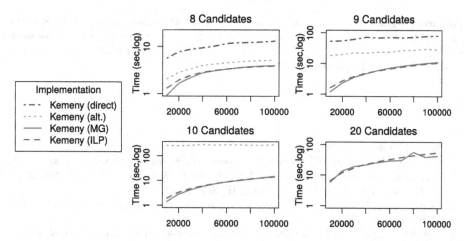

Fig. 2. Kemeny, random instances: average runtime for direct (dash-dotted), alternative (dotted), election graph (solid) and ILP (dashed) implementation with $m \in \{8, 9, 10, 20\}$ and $n \in [10000; 100000]$ on the x-axis

6   Conclusion

In this work we have introduced a reduction-based approach for computing the winners of an election. To this end, we have presented ASP-encodings for a variety of voting rules and explored their runtime behavior. The encodings are integrated in our tool, Democratix, that serves as a uniform and extensible system. The tool is also provided in form of an easy-to-use web application to make it accessible to a broader range of users.

The strengths of our approach clearly lie in the readability of the encodings and the extensibility of the tool. Still, it performs well on instances of reasonable size even for hard voting rules. Furthermore, any progress made in the optimization techniques for ASP will directly improve the performance of Democratix. Conversely, our encodings of voting rules in combination with sufficiently hard preference data could serve as challenging problems for evaluating the performance of ASP solvers. Taken together, we believe that this work is a starting point for an ASP-based tool to support experimental research in the area of voting theory and preference aggregation. In addition, the tool can be used to obtain prototype implementations for new voting rules, which, in a next step, can serve as a baseline when comparing the performance of other approaches.

An important direction for future work is to develop encodings for other types of preferences such as incomplete preferences. Furthermore, we intend to investigate how the structure of elections (e.g., "distance to Condorcet") influences runtime and how our tool compares to further rule-tailored systems. Also, the use of preprocessing techniques (cf. [5, 34]) can be a promising way to further increase the performance. Another interesting step is the study of an ASP-based approach for problems beyond winner determination (e.g., manipulation, bribery and control) and to integrate the resulting encodings into Democratix.

Acknowledgements. The authors are grateful to the anonymous COMSOC-2014 and ADT-2015 referees for their very helpful comments and suggestions. This work was supported by the Austrian Science Fund (FWF): P25518, P25607, Y968; and the German Research Foundation (DFG): ER 738/2–1.

References

1. Ali, A., Meila, M.: Experiments with Kemeny ranking: what works when? Math. Soc. Sci. **64**(1), 28–40 (2012)
2. Alviano, M., Dodaro, C., Faber, W., Leone, N., Ricca, F.: WASP: a native ASP solver based on constraint learning. In: Cabalar, P., Son, T.C. (eds.) LPNMR 2013. LNCS, vol. 8148, pp. 54–66. Springer, Heidelberg (2013)
3. Bennett, J., Lanning, S.: The Netflix prize. In: Proceedings of KDD Cup and Workshop (2007)
4. Betzler, N., Guo, J., Niedermeier, R.: Parameterized computational complexity of Dodgson and Young elections. Inf. Comput. **208**(2), 165–177 (2010)
5. Betzler, N., Bredereck, R., Niedermeier, R.: Theoretical and empirical evaluation of data reduction for exact Kemeny rank aggregation. Auton. Agent. Multi-Agent Syst. **28**(5), 721–748 (2014)
6. Bouveret, S.: Whale³: Which alternative is elected (2013). http://whale3. noiraudes.net/whale3/
7. Brandt, F., Geist, C.: Finding strategyproof social choice functions via SAT solving. In: Proceedings of AAMAS 2014, pp. 1193–1200. IFAAMAS (2014)
8. Brandt, F., Chabin, G., Geist, C.: Pnyx: a powerful and user-friendly tool for preference aggregation. In: Proceedings of AAMAS 2015, pp. 1915–1916. IFAAMAS (2015)
9. Bredereck, R.: Fixed-parameter algorithms for computing Kemeny scores - theory and practice. Pre-diploma thesis, Department of Mathematics and Computer Science, University of Jena (2009)
10. Brewka, G., Eiter, T., Truszczyński, M.: Answer set programming at a glance. Commun. ACM **54**(12), 92–103 (2011)
11. Calimeri, F., Faber, W., Gebser, M., Ianni, G., Kaminski, R., Krennwallner, T., Leone, N., Ricca, F., Schaub, T.: ASP-Core-2: 4th ASP competition official input language format (2013)
12. Caragiannis, I., Covey, J., Feldman, M., Homan, C., Kaklamanis, C., Karanikolas, N., Procaccia, A., Rosenschein, J.: On the approximability of Dodgson and Young elections. Artif. Intell. **187**, 31–51 (2012)
13. Conitzer, V., Davenport, A., Kalagnanam, J.: Improved bounds for computing Kemeny rankings. In: Proceedings of AAAI 2006, pp. 620–626. AAAI Press (2006)
14. Davenport, A., Kalagnanam, J.: A computational study of the Kemeny rule for preference aggregation. In: Proceedings of AAAI 2004, pp. 697–702. AAAI Press (2004)
15. Dwork, C., Kumar, R., Naor, M., Sivakumar, D.: Rank aggregation methods for the web. In: Proceedings of WWW 2001, pp. 613–622. ACM (2001)
16. Eiter, T., Ianni, G., Krennwallner, T.: Answer set programming: a primer. In: Tessaris, S., Franconi, E., Eiter, T., Gutierrez, C., Handschuh, S., Rousset, M.-C., Schmidt, R.A. (eds.) Reasoning Web. LNCS, vol. 5689, pp. 40–110. Springer, Heidelberg (2009)
17. Gebser, M., Janhunen, T., Rintanen, J.: ASP encodings of acyclicity properties. In: Proceedings of KR 2014. AAAI Press (2014)

18. Gebser, M., Kaminski, R., Kaufmann, B., Schaub, T.: Answer Set Solving in Practice. Synthesis Lectures on Artificial Intelligence and Machine Learning. Morgan and Claypool, San Rafael (2012)

19. Gebser, M., Kaufmann, B., Kaminski, R., Ostrowski, M., Schaub, T., Schneider, M.: Potassco: the Potsdam answer set solving collection. AI Comm. **24**(2), 107–124 (2011)

20. Geist, C., Endriss, U.: Automated search for impossibility theorems in social choice theory: ranking sets of objects. J. Artif. Intell. Res. **40**, 143–174 (2011)

21. Gelfond, M., Lifschitz, V.: Classical negation in logic programs and disjunctive databases. New Generat. Comput. **9**(3/4), 365–386 (1991)

22. Ghosh, S., Mundhe, M., Hernandez, K., Sen, S.: Voting for movies: the anatomy of a recommender system. In: Proceedings of Agents 1999, pp. 434–435. ACM (1999)

23. Hemaspaandra, E., Hemaspaandra, L., Rothe, J.: Exact analysis of Dodgson elections: Lewis Carroll's 1876 voting system is complete for parallel access to NP. J. ACM **44**(6), 806–825 (1997)

24. Hemaspaandra, E., Spakowski, H., Vogel, J.: The complexity of Kemeny elections. Theor. Comput. Sci. **349**(3), 382–391 (2005)

25. Konczak, K.: Voting theory in answer set programming. In: Proceedings of WLP 2006. INFSYS Research Report, vol. 1843-06-02, pp. 45–53. Technische Universität Wien (2006)

26. Leone, N., Pfeifer, G., Faber, W., Eiter, T., Gottlob, G., Perri, S., Scarcello, F.: The DLV system for knowledge representation and reasoning. ACM Trans. Comput. Log. **7**(3), 499–562 (2006)

27. Mao, A., Procaccia, A., Chen, Y.: Better human computation through principled voting. In: Proceedings of AAAI 2013, pp. 1142–1148. AAAI Press (2013)

28. Mattei, N., Walsh, T.: PrefLib: a library for preferences. In: Perny, P., Pirlot, M., Tsoukiàs, A. (eds.) ADT 2013. LNCS (LNAI), vol. 8176, pp. 259–270. Springer, Heidelberg (2013)

29. Narodytska, N., Walsh, T., Xia, L.: Combining voting rules together. In: Proceedings of ECAI 2012. FAIA, vol. 242, pp. 612–617. IOS Press (2012)

30. Niemelä, I.: Logic programming with stable model semantics as a constraint programming paradigm. Ann. Math. Artif. Intell. **25**(3–4), 241–273 (1999)

31. O'Neill, J.: www.OpenSTV.org (2013)

32. Ricca, F., Grasso, G., Alviano, M., Manna, M., Lio, V., Iiritano, S., Leone, N.: Team-building with answer set programming in the Gioia-Tauro seaport. Theor. Pract. Log. Prog. **12**(03), 361–381 (2012)

33. Rothe, J., Spakowski, H., Vogel, J.: Exact complexity of the winner problem for Young elections. Theor. Comput. Syst. **36**(4), 375–386 (2003)

34. Simjour, N.: Parameterized enumeration of neighbour strings and Kemeny aggregations. Ph.D. thesis, University of Waterloo (2013)

35. Tang, P., Lin, F.: Computer-aided proofs of Arrow's and other impossibility theorems. Artif. Intell. **173**(11), 1041–1053 (2009)

Multiobjective Optimisation

Approximation Schemes for Multi-objective Optimization with Quadratic Constraints of Fixed CP-Rank

Khaled Elbassioni and Trung Thanh Nguyen[(✉)]

Department of Electrical Engineering and Computer Science,
Masdar Institute of Science and Technology, 54224 Abu Dhabi, UAE
ttnguyen@masdar.ac.ae

Abstract. Motivated by the power allocation problem in AC (alternating current) electrical systems, we study the multi-objective (combinatorial) optimization problem where a constant number of (nonnegative) linear functions are simultaneously optimized over a given feasible set of 0–1 points defined by quadratic constraints. Such a problem is very hard to solve if no specific assumptions are made on the structure of the constraint matrices. We focus on the case when the constraint matrices are *completely positive* and have fixed *cp-rank*. We propose a polynomial-time algorithm which computes an ϵ-*Pareto curve* for the studied multi-objective problem when both the number of objectives and the number of constraints are fixed, for any constant $\epsilon > 0$. This result is then applied to obtain polynomial-time approximation schemes (PTASes) for two NP-hard problems: *multi-criteria power allocation* and *sum-of-ratios optimization*.

1   Introduction

Many real-world decision-making problems can be formulated as multi-criteria (or multi-objective) combinatorial optimization problems which take into consideration optimizing not a single objective function, but two or more functions, simultaneously, subject to a finite set of discrete points defined by constraint functions. Over the past few decades, there has been a significant progress on studying the solution concepts to this kind of optimization, leading to several potential approaches for solving the problem: *budget approach*, *goal programming* and *Pareto curve* (see, e.g., the textbook [18] and the survey [19] for an overview of recent techniques in the area). In this paper we will follow the Pareto curve approach.

An important characteristic of an optimization problem involving more than one objective is that a feasible solution optimizing simultaneously all the objective functions unlikely exists due to the conflict between different objectives. Therefore, in such a problem we are interested in the *tradeoff* between the objectives instead of looking for a single optimal solution. This is captured by a *Pareto curve* - a set consisting of all feasible solutions which are not *dominated* by any others. Here a solution s is said to dominate a solution s' if, for every objective

© Springer International Publishing Switzerland 2015
T. Walsh (Ed.): ADT 2015, LNAI 9346, pp. 273–287, 2015.
DOI: 10.1007/978-3-319-23114-3_17

f_i, it holds that $f_i(s) \geq f_i(s')$, and the strict inequality holds for at least one i. A solution belonging to the Pareto curve is called *Pareto optimal* solution. General speaking, computing the Pareto curve is essentially an intractable task since the number of Pareto optimal solutions is typically exponential in the input size, even with a small number of objectives. In fact, for the bicriteria version of several well-known problems such as spanning tree, shortest path and matching, checking whether or not a feasible solution is Pareto optimal is NP-hard [15]. Hence, a natural goal of research in this setting is, given an instance of a multi-objective optimization problem, to efficiently construct a *good approximation* of the Pareto curve which contains only a *small* number of elements. For this purpose, many algorithms based on heuristic approaches have been proposed in the last decades but the major drawback of this kind of techniques is that they do not provide any guarantee on the quality of the set of solutions found. An alternative way, that is of our interest in this article, is to study theoretically efficient approximation algorithms that guarantee a specific approximation factor.

Although there is a large body of work devoted to multi-criteria combinatorial optimization problems with or without (linear) constraints, the case with quadratic constraints has still not been investigated. The motivation of using quadratic constraints arose from the allocation of power in AC (alternating current) electrical systems where the users' power demand is often characterized by two components *active* power and *reactive* power (see, e.g., [25] for more details). Mathematically, these power demands (and their relevant components such as voltage, current) can be expressed in terms of complex numbers in which, the real and imaginary parts of complex-valued power correspond to the active and reactive power, whereas the magnitudes express as the so-called *apparent* power. In this case, a limitation on the magnitude of apparent power allocated to the users will result in a quadratic constraint. In a bit more detail, let us consider the power allocation problem introduced in [37]. This problem, called *Complex Demand Knapsack Problem*[1], can be seen as a variant of the classical Knapsack problem [27], in which each user (or item) i has a complex demand $d_i = a_i + \mathbf{i}b_i$, rather than a real number often interpreted as the weight or size of the item, and gives a utility (or profit) u_i iff its demand d_i is fully satisfied. The goal is to maximize the sum of utilities of the chosen users such that the magnitude of the sum of satisfied demands should not exceed the capacity. In this setting, the constraint is exactly bounding a sum of two squares of linear functions $\sum_{i=1}^{n} a_i x_i$ and $\sum_{i=1}^{n} b_i x_i$, assuming that there are n users.

Extending the work in [13,37] to the case with more than one objective function is very natural since this circumstance is frequently faced especially in the context of power allocation. Consider, for example, a scenario where a company which manages the power consumption of a group of smart grid users. Each user is required to report her power demand to the company who then, based on the obtained preferences, tries to optimize the energy allocation so that two objectives can be satisfied: the first objective is to maximize the profits (or

---

[1] Actually, this combinatorial optimization problem was first studied by Woeginger [35] under the name 2-*weighted Knapsack*, in terms of inapproximability.

revenue) and the second one is to minimize the cost of carbon dioxide emissions which would be generated by energy consumption of the users.

Our contribution. A main contribution of this paper is a polynomial-time algorithm for computing a minimum set of solutions that approximates the Pareto curve (to within any constant factor $\epsilon > 0$) for a class of multi-objective optimization problems with a fixed number of linear objective functions, and a fixed number of quadratic constraints, assuming that all the constraint matrices have fixed *cp-rank*. Our approach essentially follows the same idea used for designing approximation scheme for multi-objective Knapsack problem [21], but here it requires more technical details in handling the quadratic constraints.

Related work. Although, to the best of our knowledge, the multi-criteria combinatorial optimization with quadratic constraints has not been investigated so far, we can mention here some related work. In the model with/without linear constraints, many multi-objective optimization problems have been studied intensively in the literature, including: shortest path [16, 34], Knapsack [4, 9, 21, 24, 29], spanning tree [22, 26], matching [7], scheduling [14], and TSP [8]. Papadimitrou and Yannakakis [31] propose an efficient tool for designing fully polynomial-time approximation schemes. Their technique, however, is applicable only for the problem whose exact version can be solved in pseudo-polynomial time (for example, multi-objective Knapsack problem).

For the (mono-criterion) complex demand Knapsack problem, Yu and Chau [37] presented a 1/2-approximation algorithm and this was recently improved to a PTAS in [13]. In the first paper, the authors also ruled out the existence of an FPTAS for the problem by using a reduction from Equi-Partition[2] problem. It is noticed that a similar result was already given in an independent work by Woeginger in [35, 36]. In [20], Elbassioni and Nguyen extended the PTAS result to a wider class of binary quadratic programming involving constraint matrices of fixed cp-rank. They also obtained a constant factor approximation algorithm for the case with submodular objective function.

Organization. The rest of the paper is organized as follows. In Sect. 2 we model the multi-objective optimization problem with quadratic constraints and give basic notions on the Pareto curve and its approximate version. Section 3 is devoted to present the proof of our main result. In Sect. 4 we apply the main result to obtain a PTAS for two problems: Multi-criteria Power Allocation and Sum-Of-Ratios Optimization. Finally, we will give conclusion and mention some open questions for future work in Sect. 5.

---

[2] The Equi-Partition is defined as follows: Given $(a_1, a_2, \ldots, a_{2n}) \in \mathbb{Z}^{2n}$ with $\sum_{i=1}^{2n} a_i = 2k$, does exist a subset $S \subset \{1, 2, \ldots, 2n\}$, $|S| = n$, such that $\sum_{i \in S} a_i = \sum_{i \notin S} a_i = k$?.

2   Preliminaries

2.1   Basic Notions and Problem Model

We consider the following binary multi-objective quadratic problem (BMOQ):

$$(\text{BMOQ}) \qquad \{\max f(x), \min g(x)\} \qquad (1)$$
$$\text{subject to} \qquad x \in X \subseteq \{0,1\}^n,$$

where $f(x) = \{f_i(x)\}_{i=1}^p$ and $g(x) = \{g_j(x)\}_{j=1}^q$ are nonempty sets of nonnegative linear functions (i.e., $p, q \geq 1$), and X is a set of 0–1 vectors of dimension n belonging to the intersection of m quadratic constraints:

$$X = \left\{ x \in \{0,1\}^n \mid x^T A_k x \leq C_k; k = 1, \ldots, m \right\}, \qquad (2)$$

where A_k is nonnegative positive semi-definite matrix (w.l.o.g we assume that A_k is symmetric), and C_k is positive number, for all $k \in [m] = \{1, \ldots, m\}$. We write the linear functions f_i and g_j as $f_i(x) = a_i^T x$ and $g_j(x) = b_j^T x$, with $a_i, b_j \in \mathbb{R}_+^n$ for all $i \in [p], j \in [q]$. We assume that p, q, m are constant integers. If all the matrices A_k are diagonal matrices then, by taking $x_\ell^2 = x_\ell, \ell \in [n]$ into account, one can see that our problem becomes the multi-objective multi-dimensional Knapsack problem which is known to be NP-hard but admits a PTAS [21]. Finding approximation algorithms for the BMOQ problem without any specific assumption made on the constraint matrices A_k would be a challenge, even with one objective. It is therefore of great interest to discover special cases for which the problem can be efficiently solved or approximated, taking into account the structure of the constraint matrices. In this paper we restrict our attention to the case where every matrix A_k is *completely positive matrix* of fixed *cp-rank* [10]. This is motivated by the application in the power allocation in AC electric systems.

Definition 1. *A positive semi-definite matrix $A \succeq 0$ is said to be* completely positive *if it can be decomposed as $A = LL^T$, where $L \in \mathbb{R}_+^{n \times r}$ is a nonnegative matrix. The smallest positive number r such that such a matrix L exists is called the* cp-rank *of A, denoted* cp-rank$(A) = r$, *and the corresponding decomposition is called* cp-decomposition. *If no such matrix L exists then* cp-rank$(A) = +\infty$.

Although computing cp-decomposition of a given matrix A is NP-hard [17], it was proved in [20] that the problem is solvable in polynomial time if cp-rank(A) is constant. Hence, for simplicity we will assume that for every completely positive matrix A_k of fixed cp-rank, we know explicitly it's cp-decomposition $A_k = LL^T$. Note that the quadratic constraint involved in the complex demand Knapsack problem (or the 2-weighted Knapsack) is exactly a sum of two squares of linear functions and thus, the cp-rank of the constraint matrix equals 2.

For convenience, we identify a vector $x \in \{0,1\}^n$ with a subset $S \subseteq [n]$, i.e., write $S = \{\ell \in [n] \mid x_\ell = 1\}$. Hence, for a function defined on the power set $2^{[n]}$, $f(x) \equiv f(S)$. In the rest of the paper, we will call $\ell \in [n]$ the ℓ-th item, and for every objective $f_i = a_i^T x = \sum_{\ell=1}^n a_{i\ell} x_\ell$, we call $a_{i\ell}$ the profit of item ℓ with respect to f_i.

For $\rho > 0$, a vector $x \in \{0,1\}^n$ (or the corresponding set $S \subseteq [n]$) is said to be ρ-*approximate solution* for maximization problem Π if x is a feasible solution satisfying $v(x) \geq \rho \cdot \text{OPT}$, where $v(x)$ and OPT are the respective value of the solution x and an optimal solution. A polynomial-time approximation scheme (PTAS) is an algorithm that runs in polynomial in the size of the instance, for every fixed ϵ, and outputs a $(1 - \epsilon)$-optimal solution; a fully polynomial-time approximation scheme (FPTAS) is a PTAS where the running time is polynomial in $\frac{1}{\epsilon}$;

2.2 Preliminaries on Pareto Curve

Let $\mathcal{I}$ be an instance of a multi-objective optimization problem BMOQ. For any pair of feasible solutions $x, y \in X$, we say that x is *weakly dominated* by y if $f_i(y) \geq f_i(x)$ for all $i \in [p]$, and $g_j(y) \leq g_j(x)$ for all $j \in [q]$. If at least one of those inequalities is strict, we said that x is (*strictly*) dominated by y. All the undominated solutions form a Pareto curve of the instance problem $\mathcal{I}$. The following are formal definitions taken from [29]:

Definition 2. *The Pareto curve $\mathcal{P}$ of $\mathcal{I}$ is a subset of X, such that for any solution $x \in \mathcal{P}$ there is no solution $x' \in X$ such that $f_i(x') \geq f_i(x)$ for all $i \in [p]$ and $g_j(x') \leq g_j(x)$ for all $j \in [q]$, with strict inequality for at least one of them.*

Definition 3. *For $\epsilon > 0$, an ϵ-approximate Pareto curve of $\mathcal{I}$, denoted by $\mathcal{P}_\epsilon$, is a set of solutions, such that for all $x \in X$, there exists a solution $x' \in \mathcal{P}_\epsilon$ such that $f_i(x') \geq (1 - \epsilon)f_i(x)$ and $g_j(x') \leq g_j(x)/(1 - \epsilon)$ for all $i \in [p], j \in [q]$.*

3 Approximation Scheme for BMOQ Problem

In this section we will give the proof for the following main result:

Theorem 1. *Given an instance $\mathcal{I}$ of the BMOQ problem, one can construct an ϵ-approximate Pareto curve $\mathcal{P}_\epsilon$ for $\mathcal{I}$ in polynomial time in the size of input, for any fixed $\epsilon > 0$.*

Let $\mathcal{I} = (\{f_i\}_{i=1}^p, \{g_j\}_{j=1}^q, \{A_k, C_k\}_{k=1}^m)$ be an instance of the BMOQ problem, where $f_i(x) = a_i^T x$, $g_j(x) = b_j^T x$ for all $i \in [p], j \in [q]$, and p, q, m are fixed. Let $\epsilon > 0$ be a fixed constant and let $\rho = \epsilon/2$. We will prove that Algorithm 1 below produces a set $\mathcal{P}_\epsilon$, which is an ϵ-approximate Pareto curve to the instance $\mathcal{I}$, in polynomial time in size of the instance. Our algorithm makes use of the PTAS result given in [20] as a black box.

The basic idea behind the algorithm is to try to guess the lower bounds of the ith objective f_i for every $i \in [p-1]$, and the upper bounds of the jth objective g_j, for every $j \in [q]$, and then to maximize over $x \in X$ the objective $f_p(x)$ while requiring the remaining objectives to be satisfied by these bounds. In the first part of the algorithm, the lower and upper bounds are guessed by considering all the possibilities in such a way that the number of alternatives is bounded by a polynomial in the size of input. To do so, for each single objective f_i, we compute a $(1 - \rho)$-approximate solution U_i by using the PTAS in [20], such that

$$f_i(U_i) \geq (1 - \rho) \, \text{OPT}(f_i) \tag{3}$$

where $\text{OPT}(f_i)$ is the value of the optimal solution to the problem of maximizing function f_i over X. This implies that for any feasible solution S:

$$f_i(S) \leq \text{OPT}(f_i) \leq (1 - \rho)^{-1} f_i(U_i).$$

Similarly, we obtain a $(1 - \rho)$-approximate solution V_j to the problem of maximizing function g_j over X such that:

$$g_j(S) \leq (1 - \rho)^{-1} g_j(V_j),$$

for any feasible solution S.

We next define

$$\bar{f}_i \triangleq \left\lceil \log_{(1-\rho)^{-1}} f_i(U_i) \right\rceil \quad \text{and} \quad \bar{g}_j \triangleq \left\lceil \log_{(1-\rho)^{-1}} g_j(V_j) \right\rceil \tag{4}$$

for every $i \in [p], j \in [q]$. For each objective f_i, we consider $\bar{f}_i + 2$ lower bounds

$$\alpha_0^i = 0, \quad \text{and} \quad \alpha_{\mu_i}^i = (1 - \rho)^{1-\mu_i}, \quad \text{for } \mu_i = 1, \ldots, \bar{f}_i + 1 \tag{5}$$

and for each objective g_j, we consider $\bar{g}_j + 2$ upper bounds

$$\beta_{\eta_j}^j = (1 - \rho)^{1-\eta_j}, \quad \text{for } \eta_j = 1, \ldots, \bar{g}_j + 2 \tag{6}$$

From (4) it follows that $\bar{f}_i$ and $\bar{g}_j$ are bounded by a polynomial in the size of input and thus so do μ_i and η_j. Let Γ be a subset of $\mathbb{Z}_+^{p+q-1}$ which contains all the tuples of the form $(\mu_1, \ldots, \mu_{p-1}, \eta_1, \ldots, \eta_q)$:

$$\Gamma \triangleq \{(\mu_1, \ldots, \mu_{p-1}, \eta_1, \ldots, \eta_q) \mid 0 \leq \mu_i \leq \bar{f}_i + 1, 1 \leq \eta_j \leq \bar{g}_j + 2\} \tag{7}$$

It is easy to see that the size of Γ is bounded by $O(\|\mathcal{I}\|^{p+q-1})$ and thus is polynomial in the size of the instance for any constant numbers p and q.

Now, turning to the second part of the algorithm. For each tuple $t = (\mu_1, \ldots, \mu_{p-1}, \eta_1, \ldots, \eta_q)$, we define the corresponding lower bounds and upper bounds $\alpha_{\mu_i}^i, \beta_{\eta_j}^j, i \in [p-1], j \in [q]$ based on (5) and (6). Given these bounds on the objective functions $f_i, g_j, i \in [p-1], j \in [q]$, the goal is to maximize the function f_p over X. For doing so, we define $\phi \triangleq \sum_{k=1}^m (r_k + 1)$, where $r_k = \text{cp-rank}(A_k)$, and define λ as

$$\lambda \triangleq \left\lceil \frac{(\phi + p + q - 1)(1 + \rho)}{\rho} \right\rceil. \tag{8}$$

We extend the idea in [23], i.e., try to guess, for each of objectives f_i, a set $X_i \subset [n]$ consisting of λ items of highest profit that appear in the optimal solution. Furthermore, we denote by $\hat{X}_i$ the set of all items that are not in the optimal solution, provided that X_i is a subset of the optimal solution. Formally, we have

$$\hat{X}_i = \{\ell \in [n] \backslash X_i \mid a_{i\ell} > \min\{a_{i\ell'} \mid \ell' \in X_i\}\}, \tag{9}$$

where $a_{i\ell}$ is the profit of the item ℓ with respect to the objective f_i. Let Σ be the set of all the tuples $(X_1, \ldots, X_p)$. Since X_i has at most λ elements, the number of possibilities of choosing such a subset is n^λ. This implies that the size of Σ should be bounded by $O(n^{\lambda p})$ and thus is polynomial in the size of input for every constant numbers λ and p.

Now for each tuple $T = (X_1, \ldots, X_p) \in \Sigma$ such that $(\cup_{i\in[p]} X_i) \cap (\cup_{i\in[p]} \hat{X}_i) = \varnothing$, we define the following nonlinear (convex) program (CP$[t,T]$):

$$\text{(CP}[t,T]) \quad \max f_p(x) \tag{10}$$

$$\text{subject to} \quad x^T A_k x \leq C_k, \qquad\qquad k \in [m], \tag{11}$$

$$a_i^T x \geq \alpha_{\mu_i}^i, \qquad\qquad i \in [p-1], \tag{12}$$

$$b_j^T x \leq \beta_{\eta_j}^j, \qquad\qquad j \in [q], \tag{13}$$

$$x_k = 1, \qquad\qquad k \in \cup_{i\in[p]} X_i, \tag{14}$$

$$x_k = 0, \qquad\qquad k \in \cup_{i\in[p]} \hat{X}_i, \tag{15}$$

$$x_k \in [0,1], \qquad k \in [n] \setminus \cup_{i\in[p]} (X_i \cup \hat{X}_i). \tag{16}$$

We will prove that there exists a feasible solution to (CP$[t,T]$) which has only a few number of fractional components and thus, by a simple rounding down, we can obtain an integral solution that is a good approximation of the optimum. Note that (CP$[t,T]$) is convex as the matrix A_k is positive semi-definite for every $k \in [m]$ and thus, it can be solved efficiently (see, e.g., [30]). Let x^* be an $(1-\rho)$-optimal (rational) solution to (CP$[t,T]$). Denote $\mathcal{N} = [n] \setminus \cup_{i\in[p]} (X_i \cup \hat{X}_i)$ and $\mathcal{X} = \cup_{i\in[p]} X_i$. For every $k \in [m]$, define

$$t_k \triangleq L_k^T[*; \mathcal{N}] x_{\mathcal{N}}^*, \quad \text{and} \quad t_k' \triangleq 1_{\mathcal{X}}^T A_k[\mathcal{X}; \mathcal{N}] x_{\mathcal{N}}^*$$

and for $i \in [p-1], j \in [q]$, define:

$$\theta_i \triangleq a_i^T x_{\mathcal{N}}^*, \quad \text{and} \quad \theta_j' \triangleq b_j^T x_{\mathcal{N}}^*.$$

where, $v_{\mathcal{X}}$ denotes the restriction of the vector $v \in [0,1]^n$ to the set of components indexed by $\mathcal{X}$, and $A_k[\mathcal{X}; \mathcal{N}]$ denotes the restriction of the matrix A_k to the columns and rows indexed by the sets $\mathcal{X}$ and $\mathcal{N}$, respectively; similarly, $L_k^T[*; \mathcal{N}]$ means the restriction of L_k^T to the set of columns defined by $\mathcal{N}$.

Note that from the feasibility of x^* to (CP$[t,T]$), it follows that

$$(x_{\mathcal{N}}^*)^T L_k[\mathcal{N}; *] L_k^T[*; \mathcal{N}] x_{\mathcal{N}}^* + 2 \cdot 1_{\mathcal{X}}^T A_k[\mathcal{X}; \mathcal{N}] x_{\mathcal{N}}^* \leq C_k - 1_{\mathcal{X}}^T A_k[\mathcal{X}; \mathcal{X}] 1_{\mathcal{X}},$$

implying that

$$t_k^2 + 2t_k' \leq C_k - 1_{\mathcal{X}}^T A_k[\mathcal{X}; \mathcal{X}] 1_{\mathcal{X}}, \tag{17}$$

for all $k \in [m]$. Furthermore, we have:

$$b_j^T x_{\mathcal{N}}^* \leq \beta_{\eta_j}^j - b_j^T x_{\mathcal{X}}^*, \quad \text{or} \quad \theta_j' \leq \beta_{\eta_j}^j - b_j^T x_{\mathcal{X}}^*. \tag{18}$$

Let $\mathcal{Q}(\mathcal{X}) \subseteq [0,1]^{\mathcal{N}}$ be a polytope defined by the following system of linear inequalities:

$$\begin{cases} L_k^T[*; \mathcal{N}]y \leq t_k, \ 1_{\mathcal{X}}^T A_k[\mathcal{X}; \mathcal{N}]y \leq t'_k, \text{ for } k \in [m] \\ a_i^T y \geq \theta_i, \text{ for } i \in [p-1], \\ b_j^T y \leq \theta'_j, \text{ for } j \in [q]. \end{cases}$$

We can find (if exists) a basic feasible solution (BFS) y in this polytope such that $a_p^T y \geq a_p^T x_{\mathcal{N}}^*$. By rounding down this fractional solution y and setting $x_k \in \{0,1\}$ according to the assumption $k \in \cup_{i \in [p]}(X_i \cup \hat{X}_i)$, we obtain an integral solution $\bar{x} \in \{0,1\}^n$.

Fact 1. *The BFS y has at most $\phi + p + q - 1$ fractional components.*

Proof. Since $\mathcal{Q}(\mathcal{X})$ involves $\phi + p + q - 1$ linear inequalities, one can compute in polynomial time (see a standard textbook on linear programming, e.g., [33]) a BFS y of at most $\phi + p + q - 1$ fractional components. ∎

Fact 2. *The rounded solution $\bar{x} \in \{0,1\}^n$ is feasible to the constraints (11) and (13).*

Proof. We need to show that $\bar{x}A_k x \leq C_k$ for every $k \in [m]$. Indeed, we have:

$$\bar{x}^T A_k \bar{x} = \bar{x}_{\mathcal{N}}^T L_k[\mathcal{N}; *]L_k^T[*; \mathcal{N}]\bar{x}_{\mathcal{N}} + 2 \cdot 1_{\mathcal{X}}^T A_k[\mathcal{X}; \mathcal{N}]\bar{x}_{\mathcal{N}} + 1_{\mathcal{X}}^T A_k[\mathcal{X}; \mathcal{X}]1_{\mathcal{X}}$$
$$\leq t_k^2 + 2t'_k + 1_{\mathcal{X}}^T A_k[\mathcal{X}; \mathcal{X}]1_{\mathcal{X}} \leq C_k,$$

where the last inequality follows from (17).
On the other hand,

$$b_j^T \bar{x} = b_j^T \bar{x}_{\mathcal{N}} + b_j^T x_{\mathcal{X}}^* \leq \theta' + b_j^T x_{\mathcal{X}}^* \leq \beta_{\eta_j}^j,$$

where the last inequality follows from (18). ∎

There are totally $n^{\lambda p}$ tuples of the form $(X_1, \ldots, X_p)$. For each such tuple such that the condition in Step 18 is satisfied and $(\text{CP}[t, T])$ is feasible, the algorithm outputs an integral solution $\bar{x}$. All possible integral solutions obtained in this way are collected in the set $\mathcal{P}_\epsilon$. This completes the description of the algorithm. The polynomial running time and the correctness of the algorithm are followed by the Lemmas 1 and 2.

Lemma 1. *For a fixed number of linear objective functions and for any fixed $\epsilon > 0$, the running time of Algorithm 1 is upper bounded by a polynomial in the size of the input.*

Proof. One can see easily that the complexity of the algorithm is dominated by the work finding approximate solution to the single objective maximization problems in steps 3 and 7, and by the main loop for solving convex programs as well as finding BFS to the polytopes defined in Step 16. Note that Step 3 and Step 7 can be done in polynomial time by applying the PTAS result in [20].

Algorithm 1. PARETO($\{a_i, b_j, A_k, C_k\}_{i \in [p], \ j \in [q], \ k \in [m]}, \epsilon$)

Input: Objectives $\{a_i^T x, b_j^T x\}_{i \in [p], \ j \in [q]}$; cp-matrix $A_k \in \mathbb{R}_+^{n \times n}$ of fixed cp-rank(A_k) = r_k, capacity $C_k \in \mathbb{R}_+$, for $k \in [m]$; and accuracy parameter ϵ

Output: $(1 - \epsilon)$-approximate Pareto curve $\mathcal{P}_\epsilon$

1: $\mathcal{P}_\epsilon \leftarrow \varnothing$; $\rho \leftarrow \epsilon/2$

2: **for** $i \in [p]$ **do**

3:     Find a $(1 - \rho)$-approximate solution U_i w.r.t the objective f_i

4:     $\bar{f}_i \leftarrow \left\lceil \log_{(1-\rho)^{-1}} f_i(U_i) \right\rceil$

5:     $\alpha_{j0} \leftarrow 0$; $\alpha_{\mu_i}^i \leftarrow (1 - \rho)^{1-\mu_i}$ for all $\mu_i \in [\bar{f}_i + 1]$

6: **for** $j \in [q]$ **do**

7:     Find a $(1 - \rho)$-approximate solution V_j w.r.t the objective g_j

8:     $\bar{g}_j \leftarrow \left\lceil \log_{(1-\rho)^{-1}} g_j(V_j) \right\rceil$

9:     $\beta_{\eta_j}^j \leftarrow (1 - \rho)^{1-\eta_j}$ for all $\eta_j \in [\bar{g}_j + 2]$

10: $\Gamma = \{(\mu_1, \ldots, \mu_{p-1}, \eta_1, \ldots, \eta_q) | \ 0 \le \mu_i \le \bar{f}_i + 1, 1 \le \eta_j \le \bar{g}_j + 2\}$

11: $\phi \leftarrow \sum_{k=1}^m (r_k + 1)$; $\lambda \leftarrow \left\lceil \dfrac{(\phi + p + q - 1)(1 + \rho)}{\rho} \right\rceil$

12: $\Sigma \leftarrow \{(X_1, \ldots, X_p) | \ X_i \subseteq [n], |X_i| \le \lambda, i \in [p]\}$

13: **for each** $(X_1, \ldots, X_p) \in \Sigma$ **do**

14:     **for** $j \in [p]$ **do**

15:         $\hat{X}_i \leftarrow \{\ell \in [n] \backslash X_i | a_{i\ell} > \min\{a_{i\ell'} | \ell' \in X_i\}\}$

16: **for each** tuple $t = (\mu_1, \ldots, \mu_{p-1}, \eta_1, \ldots, \eta_q) \in \Gamma$ **do**

17:     **for each** tuple $T = (X_1, \ldots, X_p) \in \Sigma$ **do**

18:         **if** $(\cup_{i \in [p]} X_i) \bigcap (\cup_{i \in [p]} \hat{X}_i) = \varnothing$ **then**

19:             Find (if exists) a $(1 - \rho)$-optimal solution x^* to the convex program (CP$[t, T]$)

20:             Define a polytope $\mathcal{Q}(\mathcal{X}) \subseteq [0, 1]^{\mathcal{N}}$

21:             Find an BFS y of $\mathcal{Q}(\mathcal{X})$ such that $a_p^T y \ge a_p^T x_{\mathcal{N}}^*$

22:             $\bar{x} \leftarrow \{(\bar{x}_1, \ldots, \bar{x}_n) | \bar{x}_k = \lfloor y_k \rfloor$ for $k \in \mathcal{N}, \ x_k = 1,$ for $k \in \cup_{i \in [p]} X_i, \ x_k = 0,$ for $k \in \cup_{i \in [p]} \hat{X}_i\}$ ▷ *rounding down solution y*

23:             $\mathcal{P}_\epsilon \leftarrow \mathcal{P}_\epsilon \cup \bar{x}$

24: **return** $\mathcal{P}_\epsilon$

On the other hand, since there are totally $\|\mathcal{I}\|^{p+q-1}$ tuples $(\mu_1, \ldots, \mu_{p-1}, \eta_1, \ldots, \eta_q)$ considered in Loop 16, and for each such a tuple at most $n^{\lambda p}$ tuples $(X_1, \ldots, X_p)$ are examined in Loop 17, the number of convex programs and the number of polytopes need to be solved are bounded by a polynomial in the size of the input as p, q, λ are constant numbers.

Now it remains to show the approximation factor of Algorithm 1.

Lemma 2. *The output $\mathcal{P}_\epsilon$ of Algorithm 1 is exactly a ϵ-approximate Pareto curve for the instance $\mathcal{I}$, for any fixed $\epsilon > 0$.*

Proof. Let S be an arbitrary feasible solution to the instance $\mathcal{I}$ and let $\mathcal{P}_\epsilon$ be the set of solution returned by Algorithm 1, we will to prove that there always exists a solution $S' \in \mathcal{P}_\epsilon$ such that

$$f_i(S') \ge (1 - \epsilon) f_i(S) \quad \text{and} \quad g_j(S') \le g_j(S)/(1 - \epsilon),$$

for every $i \in [p]$ and $j \in [q]$.

Let us define

$$\hat{\mu}_i = \max_{0 \le \mu_i \le \bar{f}_i + 1} \{\mu_i | \, \alpha^i_{\mu_i} \le f_i(S)\}, \quad i = 1, \ldots, p-1, \tag{19}$$

and

$$\hat{\eta}_j = \min_{1 \le \eta_j \le \bar{g}_j + 2} \{\eta_j | \, \beta^j_{\eta_j} \ge g_j(S)\}, \quad j = 1, \ldots, q. \tag{20}$$

By the definition of $\hat{\mu}_i$ and of $\alpha^i_{\hat{\mu}_i}$, we must have that

$$\alpha^i_{\hat{\mu}_i} \le f_i(S) < \alpha^i_{\hat{\mu}_i + 1} = (1-\rho)^{-1}\alpha^i_{\hat{\mu}_i}, \quad i = 1, \ldots, p-1. \tag{21}$$

Similarly, we have:

$$\beta^j_{\hat{\eta}_j} \ge g_j(S) \ge \beta^j_{\hat{\eta}_j - 1} = (1-\rho)\beta^j_{\hat{\eta}_j}, \quad j = 1, \ldots, q. \tag{22}$$

Let $\delta = \min\{\lambda, |S|\}$. For each $i \in [p]$, let X_i be the set that contains δ highest-profit items in S with respect to the objective f_i. From (22) and the fact that $\delta \le \lambda$, it follows that S must be feasible to $(\mathrm{CP}[t,T])$ with respect to $t = (\hat{\mu}_1, \ldots, \hat{\mu}_{p-1}, \hat{\eta}_1, \ldots, \hat{\eta}_q)$ and $T = (X_1, \ldots, X_p)$. Let y be a BFS to the polytope $\mathcal{Q}(\mathcal{X})$ and let $\bar{x}$ be the integral solution obtained by rounding down in Step 22. Define $S' = \{\ell | \, \bar{x}_\ell = 1\}$, we have

$$g_j(S') = b_j^T \bar{x} \le \beta^j_{\hat{\eta}_j} \le g_j(S)/(1-\rho) \le g_j(S)/(1-\epsilon), \tag{23}$$

for all $j \in [q]$.

It remains to show that $f_i(S') \ge (1-\epsilon)f_i(S)$ for all $i \in [p]$. Let x^* be the $(1-\rho)$-optimal solution to $(\mathrm{CP}[t,T])$, then for every $i \in [p-1]$:

$$a_i^T x^* \ge \alpha^i_{\hat{\mu}_i} \ge (1-\rho)f_i(S),$$

where the first inequality is due to the feasibility of x^* to $(\mathrm{CP}[t,T])$ and the last inequality follows from (21). For $i = p$, we have $a_i^T x^* \ge (1-\rho)\mathsf{opt} \ge (1-\rho)f_i(S)$, where opt is the value of an optimal solution to $(\mathrm{CP}[t,T])$. We consider two cases below.

Case I: $\delta = |S|$, this means S has at most λ items. In this case the rounded solution S' will contain (and thus, weakly dominates) S.

Case II: S has more than λ items. Let $a_{i\ell}$ be the smallest value among the λ items of highest value in S with respect to objective f_i, then $a_i^T x^* \ge \lambda a_{i\ell}$. On the other hand, since $\bar{x}$ is obtained by rounding down y, and the fact that y has at most $\phi + p + q - 1$ fractional components, it follows

$$\begin{aligned}
f_i(S') = a_i^T \bar{x} = a_i^T \bar{x}_{\mathcal{N}} + a_i^T \mathbf{1}_{\mathcal{X}} &\ge a_i^T y - a_{i\ell}(\phi + p + q - 1) + a_i^T \mathbf{1}_{\mathcal{X}} \\
&\ge a_i^T x^* - a_{i\ell}(\phi + p + q - 1) \\
&\ge a_i^T x^* - a_i^T x^* \frac{\phi + p + q - 1}{\lambda} \\
&= a_i^T x^* \left(1 - \frac{\phi + p + q - 1}{\lambda}\right)
\end{aligned}$$

By the definition of λ as in (8), it follows that:

$$\frac{\phi + p + q - 1}{\lambda} \leq \frac{\rho}{1 + \rho},$$

and thus,

$$f_i(S') \geq a_i^T x^* \left(1 - \frac{\rho}{1 + \rho}\right) = \frac{1}{1 + \rho} a_i^T x^*.$$

Note that for every $i \in [p]$, $a_i^T x^* \geq (1 - \rho)f_i(S)$. Hence, it follows

$$f_i(S') \geq \frac{1 - \rho}{1 + \rho} f_i(S) > (1 - 2\rho)f_i(S) = (1 - \epsilon)f_i(S). \tag{24}$$

From (23) and (24), the proof is completed.

4   Applications

This section is devoted to present two main applications of the result obtained in the previous section for the multi-objective optimization problem with quadratic constraints. These applications include *Multi-criteria Power Allocation Problem* and *Sum-of-ratios Optimization*. We first model these problems under the complex power setting and then explain how the PTAS can be achieved.

Multi-criteria Power Allocation (MPA): Yu and Chau [37] studied the complex demand Knapsack problem, a variant of traditional Knapsack problem involving a single quadratic constraint, to address a new power allocation problem raising in AC electrical systems. Their model captures a situation in power systems in which the power is expressed as a complex number rather than a real number. Here we consider a more general scenario with multiple objectives and multiple constraints, where the power are allocated to users in a number of different periods. In this model, time is assume to be uniformly divided into discrete units ranging from 1 to T, and we refer to each integer t in the range $\{1, T\}$ as a timeslot (or period). There are a set of n users, each user i is specified by a tuple $(I_i = [r_i, s_i), d_i, p_i, c_i)$, where r_i and s_i are the arrival-time and the departure-time, respectively; $d_i = a_i + \mathbf{i}b_i$ is the complex demand of user i during the interval I_i; p_i and c_i are respective profit and cost (of carbon dioxide emissions) provided by the user i if her demand is fulfilled. We assume that there is a limit on the magnitude of total power supply, say C_t, at every timeslot $t \in \{1, \ldots, T\}$. The goal is to find a subset of users to be allocated power such that the total profit is maximized and the total cost is minimized while keeping the total power consumed at each timeslot t not exceed the capacity C_t. Formally, we can model the Multi-criteria Power Allocation problem (MPA) as follows:

$$(\text{MPA}) \quad \max \sum_{i=1}^{n} p_i x_i, \ \min \sum_{i=1}^{n} c_i x_i, \tag{25}$$

$$\text{subject to} \quad \| \sum_{i:t\in[r_i,s_i)} d_i x_i \| \leq C_t, \quad t \in [T], \tag{26}$$

$$x_i \in \{0,1\}, \tag{27}$$

The constraints (26) can be rewritten in the following form

$$(\sum_{i:t\in[r_i,s_i)} a_i x_i)^2 + (\sum_{i:t\in[r_i,s_i)} b_i x_i)^2 \leq C_t, \quad t \in [T] \tag{28}$$

When the power demand d_i is a real number for every user $i \in [n]$, the problem MPA with a single objective has been studied extensively in the literature under the names *resource allocation* [5,11], *bandwidth allocation* [28], *unsplittable flow problem* [2,3,12], *temporal Knapsack problem* [6]. The best known approximation result for the problem is a polynomial time $(2 + \epsilon)$-approximation algorithm [1]. When the number of timeslots T is constant, the problem is special case of multi-dimensional Knapsack problem which is known to have a PTAS [23].

Obviously, BMOQ includes MPA as a special case by setting $p = q = 1$ and $a_1^T = (p_1, \ldots, p_n)$, $b_1^T = (c_1, \ldots, c_n)$. Consequently, we have following result:

Corollary 1. *Let $\mathcal{I}$ be an instance of the problem MPA and assume that T is fixed. One can compute in polynomial time in the size of the instance a set $\mathcal{P}_\epsilon$ to $\mathcal{I}$ for any fixed $\epsilon > 0$.*

Sum-of-Ratios Optimization (SOR): We consider the binary quadratically constrained programming with a rational objective function.

$$(\text{SOR}) \quad \max \ w(x) = \sum_{i=1}^{p} \frac{a_{i1}x_1 + \cdots + a_{in}x_n}{b_{i1}x_1 + \cdots + b_{in}x_n}$$

$$\text{subject to} \quad x \in X \subseteq \{0,1\}^n,$$

where $a_i = (a_{i1}, \ldots, a_{in}), b_i = (b_{i1}, \ldots, b_{in}) \in \mathbb{R}_+^n$ for all $i \in [p]$. We assume that $a_i^T x > 0, b_i^T x > 0$ for all $x \in X$ and for all $i \in [p]$. This problem belongs to the class of fractional programming which has important applications in many areas such as transportation, retail management, online advertising (see the survey paper by [32] and the references therein, for more applications). Mittal and Schulz [29] presented a FPTAS for the SOR problem under the condition that every mono-criteria linear objective optimization can be solved in pseudo-polynomial time. This immediately implies an FPTAS for the multi-objective Knapsack problem, where X is defined by only one linear constraint. When X involves quadratic constraints, their method is not applicable.

Theorem 2. *Suppose that p is fixed and the feasible set X is defined as in (2), the problem (SOR) admits a PTAS.*

Proof. Let $x^* \in X$ be an optimal solution to an instance problem (SOR) and and let $\epsilon > 0$ be any fixed constant. We will prove that one can compute in polynomial time in the input size a solution y^* such that $w(y^*) \geq (1 - 2\epsilon)w(x^*)$. To do that, let $a_i = (a_{i1}, \ldots, a_{in}), b_i = (b_{i1}, \ldots, b_{in})$ for all $i \in [p]$, and apply the Theorem 1 to the instance $\mathcal{I} = (\{a_i, b_i\}_{i \in [p]}, \{A_k, C_k\}_{k \in [m]})$, where m, p are constant and A_k is a cp-matrix of fixed cp-rank. As a consequence, we obtain the approximation $\mathcal{P}_\epsilon$ of the Pareto curve of $\mathcal{I}$, which consists of a polynomial number of elements. Then, there must be a solution $y \in \mathcal{P}_\epsilon$ such that $a_i^T y \geq (1 - \epsilon)a_i^T x^*$ and $b_i^T y \leq b_i^T x^*/(1 - \epsilon)$, for all $i \in [p]$. Hence,

$$w(y) = \sum_{i=1}^{p} \frac{a_i^T y}{b_i^T y} \geq (1 - \epsilon)^2 \sum_{i=1}^{p} \frac{a_i^T x^*}{b_i^T x^*} = (1 - \epsilon)^2 w(x^*).$$

On the other hand, let $y^* = \text{argmax}\{w(x)| \ x \in \mathcal{P}_\epsilon\}$ and note that such a solution can be found efficiently due to the polynomial size of $\mathcal{P}_\epsilon$. Finally, we have $w(y^*) \geq w(y) \geq (1 - \epsilon)^2 w(x^*) \geq (1 - 2\epsilon)w(x^*)$.

5 Conclusion

In this paper we have presented a polynomial-time approximation scheme for computing the ϵ-Pareto curve for a class of multi-criteria combinatorial optimization problems involving a constant number of quadratic constraints, under the assumption that the constraint matrices have fixed cp-rank; and this is the best result we can achieve, unless $P = NP$. Finding approximation algorithms for the case with many (non constant) number of constraints would be an interesting specific task for future work, even with a single linear objective function. On the other hand, computing non-trivial inapproximability bounds, parametrized by the cp-rank of the constraints matrices is also another interesting open problem.

Acknowledgments. We would like to thank Gerhard Woeginger for helpful discussions, especially for pointing us the papers [35,36]. We thank the ADT-15 reviewers for their helpful comments, suggestions and insights that have helped us improve our manuscript. This work was supported by the MI-MIT Flagship project 13CAMA1.

References

1. Anagnostopoulos, A., Grandoni, F., Leonardi, S., Wiese, A.: A mazing $2+\epsilon$ approximation for unsplittable flow on a path. In: SODA, pp. 26–41 (2014)
2. Bansal, A., Chakrabarti, N., Epstein, A., Schieber, B.: A quasi-ptas for unsplittable flow on line graphs. In: STOC, pp. 721–729 (2006)
3. Bansal, N., Friggstad, Z., Khandekar, R., Salavatipour, M.R.: A logarithmic approximation for unsplittable flow on line graphs. In: SODA, pp. 702–709 (2009)
4. Bansal, N., Korula, N., Nagarajan, V., Srinivasan, A.: Solving packing integer programs via randomized rounding with alterations. Theory of Computing **8**(1), 533–565 (2012)

5. Bar-Noy, A., Bar-Yehuda, R., Freund, A., Naor, J., Schieber, B.: A unified approach to approximating resource allocation and scheduling. In STOC, pp. 735–744 (2000)
6. Bartlett, M., Frisch, A.M., Hamadi, Y., Miguel, I., Tarim, S.A., Unsworth, C.: The temporal knapsack problem and its solution. In: Barták, R., Milano, M. (eds.) CPAIOR 2005. LNCS, vol. 3524, pp. 34–48. Springer, Heidelberg (2005)
7. Bazgan, C., Gourvès, L., Monnot, J.: Approximation with a fixed number of solutions of some multiobjective maximization problems. J. Discrete Algorithms **22**, 19–29 (2013)
8. Bazgan, C., Gourvès, L., Monnot, J., Pascual, F.: Single approximation for the biobjective max TSP. Theor. Comput. Sci. **478**, 41–50 (2013)
9. Bazgan, C., Hugot, H., Vanderpooten, D.: Solving efficiently the 0–1 multiobjective knapsack problem. Comput. & OR **36**(1), 260–279 (2009)
10. Berman, A., Shaked-Monderer, N.: Completely Positive Matrices. World Scientific, Singapore (2003)
11. Calinescu, G., Chakrabarti, A., Karloff, H., Rabani, Y.: Improved approximation algorithms for resource allocation. In: Cook, W.J., Schulz, A.S. (eds.) IPCO 2002. LNCS, vol. 2337, pp. 401–414. Springer, Heidelberg (2002)
12. Chakrabarti, A., Chekuri, C., Gupta, A., Kumar, A.: Approximation algorithms for the unsplittable flow problem. Algorithmica **47**(1), 53–78 (2007)
13. Chau, C., Elbassioni, K., Khonji, M.: Truthful mechanisms for combinatorial ac electric power allocation. In: AAMAS, pp. 1005–1012 (2014)
14. Cheng, T., Janiak, A., Kovalyov, M.: Bicriterion single machine scheduling with resource dependent processing times. SIAM J. Optim. **8**(2), 617–630 (1998)
15. Diakonikolas, I.: Approximation of Multiobjective Optimization Problems. PhD thesis, Deptartment of Computer Science, Columbia University, May 2011
16. Diakonikolas, I., Yannakakis, M.: Small approximate pareto sets for biobjective shortest paths and other problems. SIAM J. Comput. **39**(4), 1340–1371 (2009)
17. Dickinson, P., Gijben, L.: On the computational complexity of membership problems for the completely positive cone and its dual. Comput. Optim. Appl. **57**(2), 403–415 (2014)
18. Ehrgott, M.: Multicriteria Optimization. Springer, Heidelberg (2005)
19. Ehrgott, M., Gandibleux, X.: Multiple Criteria Optimization: State of the Art Annotated Bibliographical Surveys. Kluwer, Boston (2002)
20. Elbassioni, K., Nguyen,T.T.: Approximation schemes for binary quadratic programming problems with low cp-rank decompositions. CoRR (2014). abs/1411.5050
21. Erlebach, T., Kellerer, H., Pferschy, U.: Approximating multiobjective knapsack problems. Manage. Sci. **48**(12), 1603–1612 (2002)
22. Escoffier, B., Gourvès, L., Monnot, J.: Fair solutions for some multiagent optimization problems. Auton. Agent. Multi-Agent Syst. **26**(2), 184–201 (2013)
23. Frieze, A., Clarke, M.: Approximation algorithms for the m-dimensional 0–1 knapsack problem: worst-case and probabilistic analyses. Eur. J. Oper. Res. **15**, 100–109 (1984)
24. Gandibleux, X., Freville, A.: Tabu search based procedure for solving the 0–1 multiobjective knapsack problem: the two objective case. J. Heuristics **6**, 361–383 (2000)
25. Grainger, J., Stevenson, W.: Power System Analysis. McGraw-Hill, New York (1994)
26. Hong, S.-P., Chung, S.-J., Park, B.H.: A fully polynomial bicriteria approximation scheme for the constrained spanning tree problem. Oper. Res. Lett. **32**(3), 233–239 (2004)

27. Kellerer, H., Pferschy, U., Pisinger, D.: Knapsack Problems. Springer, Heidelberg (2004)
28. Leonardi, S., Marchetti-Spaccamela, A., Vitaletti, A.: Approximation algorithms for bandwidth and storage allocation problems under real time constraints. In: FSTTCS, pp. 409–420 (2000)
29. Mittal, S., Schulz, A.: A general framework for designing approximation schemes for combinatorial optimization problems with many objectives combined into one. Oper. Res. **61**(2), 386–397 (2013)
30. Nemirovski, A., Todd, M.: Interior-point methods for optimization. Acta Numerica **17**, 191–234 (2008)
31. Papadimitriou, C., Yannakakis, M.: On the approximability of trade-offs and optimal access of web sources. In: FOCS, pp. 86–92 (2000)
32. Schaible, S., Shi, J.: Fractional programming: the sum-of-ratios case. Optim. Methods Softw. **18**, 219–229 (2003)
33. Schrijver, A.: Theory of Linear and Integer Programming. Wiley, New York (1986)
34. Tsaggouris, G., Zaroliagis, C.D.: Multiobjective optimization: improved FPTAS for shortest paths and non-linear objectives with applications. Theory Comput. Syst. **45**(1), 162–186 (2009)
35. Woeginger, G.J.: When does a dynamic programming formulation guarantee the existence of an FPTAS?. In: SODA, pp. 820–829 (1999)
36. Woeginger, G.J.: When does a dynamic programming formulation guarantee the existence of a fully polynomial time approximation scheme (FPTAS)? INFORMS J. Comput. **12**(1), 57–74 (2000)
37. Yu, L., Chau, C.: Complex-demand knapsack problems and incentives in ac power systems. In: AAMAS, pp. 973–980 (2013)

Choquet Integral Versus Weighted Sum in Multicriteria Decision Contexts

Thibaut Lust$^{(\boxtimes)}$

Sorbonne Universités, UPMC Université, Paris 06, CNRS, LIP6, UMR,
7606, F-75005 Paris, France
thibaut.lust@lip6.fr

Abstract. In this paper, we address the problem of comparing the performances of two popular aggregation operators, the weighted sum and the Choquet integral, for selecting the best alternative among a set of alternatives, all evaluated according to different criteria. While the weighted sum is simple to use and very popular, the Choquet integral is still hard to use in practice but leads theoretically to better results in terms of concordance with the preferences of a decision maker. However, given the efforts needed to set the parameters of the Choquet integral, it is important to measure, for a given decision problem, if it is really worth defining the Choquet integral or if a simple weighted sum could have been used to determine the best alternative. We will compute the probability that a recommendation to a decision maker could only been obtained with the Choquet integral and not with a weighted sum. When the number of criteria increases, the results show that this probability tends to one. However, a high value of probability can only be attained for particular data sets.

Keywords: Choquet integral · Weighted sum · Multicriteria decision making · Multiobjective optimization

1 Introduction

When a decision maker is confronted to a set of alternatives presenting each one meaningful advantages and disadvantages, the temptation to use a weighted sum to select the best alternative is high. However, it is well-known that this easy to use and simple aggregation method presents the disadvantage that, in some cases, alternatives cannot be elicited even if they correspond to the preferences of the decision maker.

For example, if we consider a set of four alternatives $\{y^1, \ldots, y^4\}$ where each alternative y^i is evaluated with three criteria to maximize: $y^1 = (18, 10, 10)$, $y^2 = (10, 18, 10)$, $y^3 = (10, 10, 18)$ and $y^4 = (14, 12, 11)$, the alternative y^4 could never be selected with a weighted sum (it is impossible to have at the same time $14\lambda_1 + 12\lambda_2 + 11\lambda_3 \geq 18\lambda_1 + 10\lambda_2 + 10\lambda_3$, $14\lambda_1 + 12\lambda_2 + 11\lambda_3 \geq 10\lambda_1 + 18\lambda_2 + 10\lambda_3$ and $14\lambda_1 + 12\lambda_2 + 11\lambda_3 \geq 10\lambda_1 + 10\lambda_2 + 18\lambda_3$, with $\lambda_1, \lambda_2, \lambda_3 \geq 0$ and $\lambda \neq 0$). However, the alternative y^4 is the most balanced alternative among the different

© Springer International Publishing Switzerland 2015
T. Walsh (Ed.): ADT 2015, LNAI 9346, pp. 288–304, 2015.
DOI: 10.1007/978-3-319-23114-3_18

criteria and is a good candidate for a decision maker who prefers well-balanced alternatives.

Other models could be used to handle this problem [1,2] like outranking methods (ranking of the alternatives based on pairwise comparisons) [3], additive value function models [4] or more evolved aggregation functions like the weighted minimum, weighted maximum, ordered weighted average operators (OWA) [5], weighted ordered weighted averaging operator (WOWA) [6], Choquet integral [7], etc.

In this paper, we will focus on the Choquet integral.

The Choquet integral is a powerful tool in multicriteria decision making and decision under uncertainty [8–10]. A Choquet integral can be seen as an integral on a non-additive measure (or capacity or fuzzy measure). It presents extremely wide expressive capabilities and can model many specific aggregation operators, including, but not limited to, the weighted sum, the minimum, the maximum, all the statistic quantiles, OWA, WOWA, etc. The Choquet integral can also be used in the additive value function model instead of the weighted sum [11]. In this case, the performance vector is replaced by marginal utility values.

However, this high expressiveness capability has a price: while the definition of a simple weighted sum operator with m criteria requires $m - 1$ parameters, the definition of the Choquet integral with m criteria requires setting of $2^m - 2$ values, which can be a problem even for low values of m.

Given the effort needed to set the parameters of the Choquet integral comparing to the weighted sum, we will measure in this paper the interest of using the Choquet integral instead of the weighted sum. More precisely, for different alternatives evaluated with m criteria, we will evaluate the probability that an alternative that optimizes a defined Choquet integral could not be obtained with a simple weighted sum. This is particularly important in the general context where the alternatives to compare are not explicitly given but are obtained from a multiobjective optimization problem. In different works [12–16], the authors define a Choquet integral and then search for an optimal solution according to the defined Choquet integral. Two difficulties are thus introduced: the elicitation of the Choquet integral and the optimization of the Choquet integral for the particular multiobjective problem studied. Given this complexity, it is worth studying the real strength of the Choquet integral and to see, if a simpler method (the weighted sum) could have been used to obtain the same optimal solution.

To our knowledge, only one group of authors have performed experiments to assess the powerfulness of the Choquet integral. In [17], Meyer and Pirlot compare the ability of related models to represent rankings of alternatives. They compare different aggregators, including the weighted sum, the Choquet integral and additive value functions. To do so, they randomly generate alternatives and define a ranking of the alternatives. Then they check if the models can represent the ranking. They show that the Choquet integral model can represent significantly more orders than the weighted sum and that the difference becomes quite large when the number of criteria is high. If their work can appear similar to our work, there are two important differences:

- They consider rankings of alternatives while we only check the ability of an aggregator to reach one optimal alternative.
- Given a set alternatives we will not pick up randomly a best alternative, as they do, but we will generate randomly a Choquet integral (with a uniform law), and check if the alternative optimizing the Choquet integral could not have been obtained with a weighted sum too. Therefore, a probability will be defined, according to the set of possible Choquet integrals, and not according to the set of possible alternatives. That gives a more general way to measure the strength of the Choquet integral since results defined independently from the set of alternatives considered could be given.

The paper is organized as follows: we first introduce the main notions of this paper: the multicriteria context and the two aggregation operators used (weighted sum and Choquet integral). In Sects. 3 and 4, we present the main contributions of the paper: we expose how we have compared the weighted sum and the Choquet integral operators: the comparison is based on the probability to reach an alternative optimal for a Choquet integral but not for a weighted sum. In the results section (Sect. 5), lower and upper bounds are computed, according to the number of criteria considered, and independently from the problem studied. We will see that this probability tends to one according to the number of criteria considered. We expose then some experimental results on randomly generated data sets. We will see that some conditions have to be respected to attain high probability values.

2   Definitions

We consider a general model with a set $\mathcal{Y}$ of n alternatives $\{y^1, \ldots, y^n\}$ evaluated with a set $\mathcal{M}$ of m criteria $\{1, \ldots, m\}$. The performance vector associated to an alternative y^i is denoted $(y_1^i, \ldots, y_m^i)$. We will consider w.l.o.g that the performance values are in the $[0, 1]$ interval and that the criteria have to be maximized.

The representation of an alternative in the criteria space is called a point and these two notions will be considered as equivalent in the rest of the paper.

We first recall the notion of Pareto dominance.

Definition 1. *The Pareto dominance relation (P-dominance for short) is defined, for all* $y^1, y^2 \in \mathbb{R}^m$, *by:*

$$y^1 \succ_P y^2 \iff [\forall k \in \mathcal{M}, y_k^1 \geq y_k^2 \text{ and } y^1 \neq y^2]$$

We will only work with sets $\mathcal{Y}$ of Pareto non-dominated alternatives, that is $\forall y^1 \in \mathcal{Y}, \nexists y^2 \in \mathcal{Y} \mid y^2 \succ_P y^1$.

2.1   Weighted Sum

The most popular aggregation operator is the weighted sum (WS), where positive importance weights $\lambda_i (i = 1, \ldots, m)$ are allocated to the criteria.

Definition 2. *Given a vector* $y \in \mathbb{R}^m$ *and a weight set* $\lambda \in \mathbb{R}^m$ *(with* $\lambda_i > 0$ *and* $\sum_{i=1}^m \lambda_i = 1$*), the WS* $f_\lambda^{ws}(y)$ *of* y *is equal to:*

$$f_\lambda^{ws}(y) = \sum_{i=1}^m \lambda_i y_i$$

In a set of $\mathcal{Y}$ of Pareto non-dominated alternatives, the alternatives that optimize a WS are called WS-optimal alternatives or supported Pareto-optimal alternatives [18] (SP alternatives). Note that there could exist alternatives that do not optimize a WS, and they are generally called non-supported Pareto-optimal alternatives (N-SP alternatives).

2.2   Choquet Integral

The Choquet integral has been introduced by Choquet [7] in 1953 and has been intensively studied, especially in the field of multicriteria decision analysis, by several authors (see [9,10,19] for a brief review). Lately, the Choquet integral has also been used in the AI field, for classification problems [20,21], constraint programming [22] or state space search [23].

We first define the notion of capacity, on which the Choquet integral is based.

Definition 3. *A capacity is a set function* $v\colon 2^{\mathcal{M}} \to [0,1]$ *such that:*

- $v(\emptyset) = 0$, $v(\mathcal{M}) = 1$ *(boundary conditions)*
- $\forall \mathcal{A}, \mathcal{B} \in 2^{\mathcal{M}}$ *such that* $\mathcal{A} \subseteq \mathcal{B}, v(\mathcal{A}) \leq v(\mathcal{B})$ *(monotonicity conditions)*

Therefore, for each subset of criteria $\mathcal{A} \subseteq \mathcal{M}$, $v(\mathcal{A})$ represents the importance of the subset $\mathcal{A}$.

Definition 4. *A capacity is said additive if for each subset* $\mathcal{A}, \mathcal{B} \subseteq \mathcal{M}, v(\mathcal{A} \cup \mathcal{B}) = v(\mathcal{A}) + v(\mathcal{B})$.

Definition 5. *The Choquet integral of a vector* $y \in \mathbb{R}^m$ *with respect to a capacity* v *is defined by:*

$$f_v^C(y) = \sum_{i=1}^m \left(v(Y_i^\uparrow) - v(Y_{i+1}^\uparrow) \right) y_i^\uparrow$$

$$= \sum_{i=1}^m (y_i^\uparrow - y_{i-1}^\uparrow) v(Y_i^\uparrow)$$

where $y^\uparrow = (y_1^\uparrow, \ldots, y_m^\uparrow)$ *is a permutation of the components of* y *such that* $0 = y_0^\uparrow \leq y_1^\uparrow \leq \cdots \leq y_m^\uparrow$ *and* $Y_i^\uparrow = \{j \in \mathcal{M}, y_j \geq y_i^\uparrow\} = \{i^\uparrow, (i+1)^\uparrow, \ldots, m^\uparrow\}$ *for* $i \leq m$ *and* $Y_{(m+1)}^\uparrow = \emptyset$.

The Choquet integral is a versatile aggregation operator, as it can express preferences to a wider set of solutions than a weighted sum, through the use of a non-additive capacity. For example, the Choquet integral can attain N-SP alternatives, while it is impossible with the weighted sum [24].

Example 1. Let us consider the four alternatives exposed in the introduction and the following capacity: $v(\{1\}) = v(\{2\}) = v(\{3\}) = 0.2$ and $v(\{1,2\}) = v(\{1,3\}) = v(\{2,3\}) = 0.4$. We obtain $f_v^C(y^1) = 10 + (10 - 10) * v(\{1,2\}) + (18 - 10) * v(\{1\}) = 11.6$, $f_v^C(y^2) = 10 + (10 - 10) * v(\{2,3\}) + (18 - 10) * v(\{2\}) = 11.6$, $f_v^C(y^3) = 10 + (10 - 10) * v(\{1,3\}) + (18 - 10) * v(\{3\}) = 11.6$, $f_v^C(y^4) = 11 + (12 - 11) * v(\{1,2\}) + (14 - 12) * v(\{1\}) = 11.8$. For this capacity, y^4 is thus the best alternative.

In this example, we see that the alternative y^4 can optimize a Choquet integral, but cannot optimize a WS. An alternative presenting this property will be called an *exclusive* Choquet optimal (*C*-optimal) alternative.

In the following, we will define more precisely the notion of exclusive *C*-optimal alternative.

3   Exclusive *C*-optimal Alternatives

We first define the notion of WS-optimal set.

Definition 6. *Given a set $\mathcal{Y}$ of alternatives, the WS-optimal set, called $\mathcal{Y}_{ws}$, is the set containing an optimal alternative, for each possible WS, that is $\forall \lambda \in \mathcal{L}, \exists y^j \in \mathcal{Y}_{ws} \mid f_\lambda^{ws}(y^j) \geq f_\lambda^{ws}(y^i) \; \forall y^i \in \mathcal{Y}$, where $\mathcal{L}$ represents the set of possible weights defined over m criteria.*

Similarly, we can define the notion of Choquet-optimal set (*C*-optimal set).

Definition 7. *Given a set $\mathcal{Y}$ of alternatives, the C-optimal set, called $\mathcal{Y}_C$, is the set containing an optimal alternative, for each possible Choquet integral, that is $\forall v \in \mathcal{V}, \exists y^j \in \mathcal{Y}_C \mid f_v^C(y^j) \geq f_v^C(y^i) \; \forall y^i \in \mathcal{Y}$, where $\mathcal{V}$ represents the set of possible capacities defined over m criteria.*

The *C*-optimal set contains thus all potential *C*-optimal alternatives. A characterization of the *C*-optimal set has been proposed in [25]. We briefly recall it here.

Let σ be a permutation on $\mathcal{M}$. Let $\mathcal{O}_\sigma$ be the subset of alternatives $y \in \mathbb{R}^m$ such that $y \in \mathcal{O}_\sigma \iff y_{\sigma_1} \geq y_{\sigma_2} \geq \cdots \geq y_{\sigma_m}$.

Let $a_{\mathcal{O}_\sigma}$ be the following application:

$$a_{\mathcal{O}_\sigma} : \mathbb{R}^m \to \mathbb{R}^m, (a_{\mathcal{O}_\sigma}(y))_{\sigma_i} = (\min(y_{\sigma_1}, \ldots, y_{\sigma_i})), \forall i \in \mathcal{M}$$

For example, if $m = 3$, for the permutation $(2,3,1)$, we have:

$$a_{\mathcal{O}_\sigma}(y) = \big(\min(y_2, y_3, y_1), \min(y_2), \min(y_2, y_3) \big)$$

We denote by $\mathcal{A}_{\mathcal{O}_\sigma}(\mathcal{Y})$ the set containing the alternatives obtained by applying the application $a_{\mathcal{O}_\sigma}(y)$ to all the alternatives $y \in \mathcal{Y}$. As $(a_{\mathcal{O}_\sigma}(y))_{\sigma_1} \geq (a_{\mathcal{O}_\sigma}(y))_{\sigma_2} \geq \cdots \geq (a_{\mathcal{O}_\sigma}(y))_{\sigma_m}$, we have $\mathcal{A}_{\mathcal{O}_\sigma}(\mathcal{Y}) \subseteq \mathcal{O}_\sigma$.

In the following, we will denote $\mathcal{O}_\sigma$ as simply $\mathcal{O}$ for the sake of simplicity, and we will consider, w.l.o.g., that the permutation σ is equal to $(1, 2, \ldots, m)$, that is $y \in \mathcal{O} \Leftrightarrow y_1 \geq y_2 \geq \cdots \geq y_m$.

In [25], Lust and Rolland show the following results in order to characterize the *C*-optimal set $\mathcal{Y}_C$ of a set $\mathcal{Y}$:

Theorem 1.
$$\mathcal{Y}_C \cap \mathcal{O}_\sigma = \mathcal{Y} \cap \mathcal{Y}_{ws}(\mathcal{A}_{\mathcal{O}_\sigma}(\mathcal{Y}))$$

where $\mathcal{Y}_{ws}(\mathcal{A}_{\mathcal{O}_\sigma}(\mathcal{Y}))$ designs the set of WS-optimal alternatives of the set $\mathcal{A}_{\mathcal{O}_\sigma}(\mathcal{Y})$.

This theorem characterizes the alternatives that can be C-optimal in a set $\mathcal{Y}$ of points as being, in each subspace of the criteria space $\mathcal{Y}$ where $y_{\sigma_1} \geq y_{\sigma_2} \geq \ldots \geq y_{\sigma_m}$, the WS-optimal points in $\mathcal{A}_{\mathcal{O}_\sigma}(\mathcal{Y})$.

Proof: see [25].

Property 1. $\mathcal{Y}_{ws} \subseteq \mathcal{Y}_C$

Proof. If the capacity v is additive, the Choquet integral of a vector $y \in \mathbb{R}^m$ is equal to $\sum_{i=1}^m \left(v(Y_i^\uparrow) - v(Y_{i+1}^\uparrow) \right) y_i^\uparrow = \sum_{i=1}^m v(\{i\}) y^i$. Therefore, the WS is a particular Choquet integral for which the capacity is additive. All WS-optimal alternatives are thus also C-optimal alternatives.

Example 2. For the four alternatives of the introduction, we have $\mathcal{Y}_{ws} = \{y^1, y^2, y^3\}$ and $\mathcal{Y}_C = \{y^1, y^2, y^3, y^4\}$.

Definition 8. *Given a set $\mathcal{Y}$ of alternatives, the exclusive C-optimal set $\mathcal{Y}_{eC}$ is equal to $\{\mathcal{Y}_C \setminus \mathcal{Y}_{ws}\}$.*

The set $\mathcal{Y}_{eC}$ is thus composed of the alternatives that optimize a Choquet integral, without optimizing a WS.

Definition 9. *Given a set $\mathcal{Y}$ of alternatives, let us consider the exclusive C-optimal set $\mathcal{Y}_{eC}$. For all alternatives $y^i \in \mathcal{Y}_{eC}$, let $\mathcal{V}^i$ the set of capacities for which the alternative $y^i \in \mathcal{Y}_{eC}$ is C-optimal in $\mathcal{Y}$. Let $\mathcal{V}_e = \bigcup \mathcal{V}^i$ the union of these sets. The set $\mathcal{V}_e$ is called the exclusive capacity set. All exclusive C-optimal alternatives are optimal for capacities $v \in \mathcal{V}_e$ (it does not exist an exclusive C-optimal alternative for a capacity $v \notin \mathcal{V}_e$).*

We can now define the probability p to get an exclusive C-optimal alternative when a Choquet integral is randomly generated (that is a capacity v is randomly generated, with a uniform law).

Definition 10. *Let $v \in \mathcal{V}$ a capacity randomly generated with a uniform law. Let $F(v) = 1$ if $v \in \mathcal{V}^e$ and 0 otherwise and y the best alternative for the Choquet integral f_v^C. The probability p_e that y is exclusive C-optimal is equal to:*

$$p_e = \int_{\mathcal{V}} F(v) dv$$

Example 3. Let $m = 2$ and $\mathcal{Y} = \{(1, 0), (0, 1), (0.2, 0.7)\}$. The point $(0.2, 0.7)$ cannot be WS-optimal as $(0.2 + 0.7 < 1)$. But it can be C-optimal if the capacity v respects the following conditions[1]: $0.2 + 0.5 v_2 \geq v_2$ and $0.2 + 0.5 v_2 \geq v_1$,

---

[1] In the following, a capacity associated to a set will be simply written $v_{e_1, \ldots, e_n}$ where $e_1, \ldots, e_n$ denotes the elements belonging to the set.

that is v such that $v_1 \leq 0.2 + 0.5v_2$ and $v_2 \leq 0.4$. Therefore we have $\mathcal{Y}_{ws} = \{(1,0),(0,1)\}, \mathcal{Y}_C = \mathcal{Y}$ and $\mathcal{Y}_{eC} = \{(0.2,0.7)\}$. If we generate randomly a capacity v, the probability p_e to get the point $(0.2, 0.7)$ optimal for the defined Choquet integral is thus equal to $\int_0^{0.4} \int_0^{0.2+0.5v_2} 1 dv1 \, dv2 = 0.12$. We see also through this example that the probability takes a different value compared to the probability to get an exclusive C-optimal alternative if the alternative is randomly selected (which is equal to $1/3$) , as done in the work of Meyer and Pirlot [17].

4   Maximal Proportion of Exclusive C-optimal Alternatives

The value of the probability p_e is problem-dependent, but bounds, according to the number of criteria (and independent from the problem studied), can however be generated. The minimal value of p_e is zero, whatever the number the criteria, since it is always possible to generate sets composed of only WS-optimal alternatives. The maximal value of p_e is more difficult to compute, but also more interesting: given a problem with m criteria, what is the maximal value of p_e, that is the maximal chance to reach an exclusive C-optimal alternative with a Choquet integral?

In the following, we will construct artificial sets of alternatives, in order to estimate the maximal value of p_e. The WS-optimal set of the artificial sets will always have the same form and composed of only m points, in order to favor the Choquet integral. We will use m I^j points $(j \in \mathcal{M})$, such that $I_i^j = 1$ if $i = j$ and 0 otherwise $(i \in \mathcal{M})$. Such a WS-optimal set will be called $\mathcal{Y}_{ws}^I$. For example, for $m = 3$, $\mathcal{Y}_{ws}^I = \{(1,0,0),(0,1,0),(0,0,1)\}$.

We first introduce some new definitions.

Definition 11. *Let $\mathcal{Y}$ a set of alternatives. Let us consider two additional alternatives y^1 and y^2 and $\mathcal{V}_1$ and $\mathcal{V}_2$ the set of capacities for which y^1 and y^2 are C-optimal in $\mathcal{Y}$. We say that y^1 covers y^2 if $\mathcal{V}_2 \subseteq \mathcal{V}_1$.*

Definition 12. *Let $\mathcal{Y}$ a set of alternatives with $\mathcal{Y}_{ws} = \mathcal{Y}_{ws}^I$, $\mathcal{Y}_{eC}$ the exclusive C-optimal set and $\mathcal{V}_e$ the exclusive capacity set. The set $\mathcal{V}_e$ is called a maximal exclusive capacity set (denoted $\mathcal{V}_e^*$) if any other possible exclusive C-optimal alternatives (not necessarily in $\mathcal{Y}$) are covered by an alternative of $\mathcal{Y}_{eC}$. The set $\mathcal{Y}_{eC}$ associated with $\mathcal{V}_e^*$ is called a maximal cover set and is denoted $\mathcal{Y}_{eC}^*$.*

For a problem with m criteria, if we can generate $\mathcal{V}_e^*$, we can then compute the maximal value of p_e as follows:

Definition 13. *Let $v \in \mathcal{V}$ a capacity randomly generated with a uniform law. Let $G(v) = 1$ if $v \in \mathcal{V}_e^*$ and 0 otherwise and y the best alternative for the Choquet integral f_v^C. The maximal probability p_e^* that y is exclusive C-optimal is equal to:*

$$p_e^* = \int_{\mathcal{V}} G(v) dv$$

For a problem with m criteria, we have thus to generate $\mathcal{V}_e^*$ to obtain p_e^*.

We will start our study with alternatives evaluated with only two criteria and determine, in this case, the maximal value that p_e can take. We will then generalize for any number of criteria.

4.1 Two-Criteria Problems

We will try to generate $\mathcal{Y}_{eC}^*$ and $\mathcal{V}_e^*$, in order to express p_e^*. We have thus to generate a set $\mathcal{Y}$ of alternatives such that any other possible exclusive C-optimal alternatives will be covered by an alternative of $\mathcal{Y}_{eC}^*$. We first detail the results obtained with the simple case where $m = 2$.

Property 2. Let $m = 2$ and $\mathcal{Y}$ a set of alternatives with $\mathcal{Y}_{ws} = \mathcal{Y}_{ws}^I = \{(1,0),(0,1)\}$. The maximal exclusive capacity set $\mathcal{V}_e$ is equal to $\{(v_1,v_2) \mid v_1 < \frac{1}{2}$ and $v_2 < \frac{1}{2}\}$ and the maximal cover set $\mathcal{Y}_{eC}^*$ is equal to $\simeq (\frac{1}{2},\frac{1}{2})$, where the notation $\simeq (\frac{1}{2},\frac{1}{2})$ means that we have a point close to $(\frac{1}{2},\frac{1}{2})$ (but the sum of its component is less than 1).

Proof. With two criteria, we have $\mathcal{V} = \{v_1, v_2\}$ with $v_1 \in [0,1]$ and $v_2 \in [0,1]$. We need at least three points to have one exclusive C-optimal alternative. Let us consider the two points of $\mathcal{Y}_{ws}^I$ $((1,0)$ and $(0,1))$ and the point (a,b) such that $a \geq b$ and $a + b < 1$. We have $f_v^C(1,0) = v_1$, $f_v^C(0,1) = v_2$ and $f_v^C(a,b) = b + (a-b)v_1$. To have (a,b) C-optimal we need to have $b + (a-b)v_1 \geq v_1$ and $b + (a-b)v_1 \geq v_2$, that is $v_1 \leq \frac{b}{1-a+b}$ and $v_2 \leq b + (a-b)v_1$.

To maximize $f_v^C(a,b)$, $a + b \to 1$, and thus $b \to 1 - a$. We obtain:

$$\begin{cases} v_1 < \frac{1-a}{1-a+1-a} = \frac{1}{2} \\ v_2 < (1-a) + (2a-1)v_1 \end{cases}$$

Let

$$\begin{cases} G(v) = 1 \text{ if } v_1 < \frac{1}{2} \text{ and } v_2 < (1-a) + (2a-1)v_1 \\ \quad\quad = 0 \text{ otherwise.} \end{cases}$$

We get:

$$\int_{\mathcal{V}} G(v)dv = \int_0^{\frac{1}{2}} \int_0^{(1-a)+(2a-1)v_1} 1 \, dv_2 \, dv_1 = \frac{3}{8} - \frac{1}{4}a$$

As $a \in [\frac{1}{2}, 1]$, the maximal value is obtained when $a = \frac{1}{2}$ and $b \to \frac{1}{2}$.

We see that it is enough to consider only one additional point $(a,b) \to (\frac{1}{2},\frac{1}{2})$. Indeed, if we consider another points, we will still have $v_1 < \frac{1}{2}$ and $v_2 < (1-a) + (2a-1)v_1$ that is $v_2 < \frac{1}{2}$ when $v_1 \to \frac{1}{2}$. Therefore the maximal value of $\int_{\mathcal{V}} G(v)dv$ will still be bounded by $\frac{1}{4}$ $(\frac{1}{2} * \frac{1}{2})$; the value of p_e^* is thus equal to $\frac{1}{4}$.

Consequently, for two-criteria sets, $\mathcal{V}_e^* = \{(v_1,v_2) \mid v_1 < \frac{1}{2}$ and $v_2 < \frac{1}{2}\}$, $\mathcal{Y}_{eC}^* = \{\simeq (\frac{1}{2},\frac{1}{2})\}$ and $p_e^* = \frac{1}{4}$.

4.2 m-criteria Problems

For three criteria, we have followed the same reasoning as for two-criteria problems, by adding points (a, b, c) such that $a + b + c < 1$, in an initial set composed of the three points of $\mathcal{Y}_{ws}^{I}$ $((1,0,0)$, $(0,1,0)$ and $(0,0,1))$. We have obtained $\mathcal{V}_e^* = \{(v_1, v_2, v_3, v_{12}, v_{13}, v_{23}) \mid (v_1 < \frac{1}{3}, v_2 < \frac{1}{3}, v_3 < \frac{1}{3})$ or $(v_1 < \frac{1}{2}v_{12}, v_2 < \frac{1}{2}v_{12}, v_3 < \frac{1}{2}v_{12})$ or $(v_1 < \frac{1}{2}v_{13}, v_2 < \frac{1}{2}v_{13}, v_3 < \frac{1}{2}v_{13})$ or $(v_1 < \frac{1}{2}v_{23}, v_2 < \frac{1}{2}v_{23}, v_3 < \frac{1}{2}v_{23})\}$, $\mathcal{Y}_{eC}^* = \{\simeq (\frac{1}{3}, \frac{1}{3}, \frac{1}{3}), \simeq (\frac{1}{2}, \frac{1}{2}, 0), \simeq (\frac{1}{2}, 0, \frac{1}{2}), \simeq (0, \frac{1}{2}, \frac{1}{2})\}$.

By solving analytically different sextuples integrals defined on $\mathcal{V}_e^*$, we have computed p_e^* and obtained a value equal to $\frac{20323}{6^6} = 0.4356$.

Generally, for m criteria, we have the following property:

Property 3. Let $\mathcal{Y}$ a set of alternatives with $\mathcal{Y}_{ws} = \mathcal{Y}_{ws}^{I}$. The maximal cover set $\mathcal{Y}_{eC}^*$ for a number m of criteria is composed of the following points: $\simeq (\frac{1}{m}, \frac{1}{m}, \ldots, \frac{1}{m})$, all permutations of $\simeq (0, \frac{1}{m-1}, \frac{1}{m-1}, \ldots, \frac{1}{m-1})$, all permutations of $\simeq (0, 0, \frac{1}{m-2}, \frac{1}{m-2}, \ldots, \frac{1}{m-2})$, $\ldots$, and all permutations of $\simeq (0, \ldots, 0, \frac{1}{2}, \frac{1}{2})$.

Proof. We have to show that any additional exclusive C-optimal point in $[0, 1]^m$ will be covered by a point of $\mathcal{Y}_{eC}^*$. In Property 3, we have that $\mathcal{Y}$ is composed of $\mathcal{Y}_{ws}^{I}$ and $\mathcal{Y}_{eC}^*$. Let us consider an additional alternative z in $\mathcal{O}$, that is $z_1 \geq z_2 \geq \cdots \geq z_m$. As $\mathcal{Y}_{ws}$ is composed of the I^j points, to be an exclusive C-optimal point, z has to fulfill the following constraint: $\sum_{i=1}^{m} z_i < 1$ (otherwise that point would be a WS-optimal point). Moreover, to be C-optimal, z has to be WS-optimal in $\mathcal{A}_\mathcal{O}(\mathcal{Y})$, where $\mathcal{A}_\mathcal{O}(\mathcal{Y})$ is obtained by applying the application $(a_\mathcal{O}(y))_i = (\min(y_1, \ldots, y_i))$, $\forall i \in \mathcal{M}$, to all $y \in \mathcal{Y}$ (see Theorem 1). $\mathcal{A}_\mathcal{O}(\mathcal{Y})$ is thus equal to $\{\simeq (\frac{1}{m}, \frac{1}{m}, \ldots, \frac{1}{m}), \simeq (\frac{1}{m-1}, \ldots, \frac{1}{m-1}, 0), \simeq (\frac{1}{m-2}, \ldots, \frac{1}{m-2}, 0, 0), \ldots, \simeq (\frac{1}{2}, \frac{1}{2}, 0, \ldots, 0), (1, 0, \ldots, 0)\}$ (the point $(0, \ldots, 0)$ belongs also to this set but it can be removed since this point is P-dominated in $\mathcal{A}_\mathcal{O}(\mathcal{Y})$). We can remark that the vertices of the polyhedron defined by the constraint inequalities $(z_1 \geq z_2 \geq \cdots \geq z_m)$ and $(\sum_{i=1}^{m} z_i \leq 1)$ are the points $\{(\frac{1}{m}, \frac{1}{m}, \ldots, \frac{1}{m}), (\frac{1}{m-1}, \ldots, \frac{1}{m-1}, 0), (\frac{1}{m-2}, \ldots, \frac{1}{m-2}, 0, 0), \ldots, (\frac{1}{2}, \frac{1}{2}, 0, \ldots, 0), (1, 0, \ldots, 0)\}$. Therefore, if the points of $\mathcal{A}_\mathcal{O}(\mathcal{Y})$ are sufficiently close to the vertices of the polyhedron, it will not possible to find a point z in $\mathcal{A}_\mathcal{O}(\mathcal{Y})$ which is WS-optimal [26]. Any additional points that do not optimize a WS will not be C-optimal and thus covered by a point of $\mathcal{Y}_{eC}^*$. Therefore $\mathcal{Y}_{eC}^*$ is a maximal cover set.

5   Results

5.1   Computation of $p_e^*(m)$

In this section, we estimate the probability p_e^* that some alternative might be exclusively Choquet-optimal, as the number of criteria varies. That is, we estimate $p_e^*(m)$ for varying m. We have already analytically determined this probability for the cases of 2 and 3 criteria ($p_e^*(2) = 0.25$ and $p_e^*(3) = 0.4356$); but for larger number of criteria we turn to numerical estimation, because analytically solving the integrals is exponential in the number of criteria. To do so, roughly speaking, we randomly generate candidate Choquet integrals with a uniform distribution, and count the number of times that some alternative of the maximal cover set $\mathcal{Y}^*_{eC}$ (see Property 3) is superior to some alternative of the initial set of alternatives $(\mathcal{Y}^I_{ws})$. Dividing that by the number of samples gives our numerical estimate.

However, because of the monotonicity constraints, it is not trivial to generate randomly (with a uniform distribution) capacities. One way to deal with this problem is to generate $(2^m - 2)$ random values between 0 and 1, and to check if the values respect the monotonicity constraints of a capacity. If it works for three or four criteria, it is quickly unusable since the monotonicity constraints become harder to satisfy when the number of criteria increases.

Recently, Combarro *et al.* [27] have proposed a way to generate randomly capacities. They established that the random generation of capacities involves the generation of random linear extensions of capacities (that is linear extensions of the monotonocity constraints). Once a linear extension is generated, it is enough to compute a point that respects the linear extension, which can be easily done (see [28]). A method to generate linear extensions in a random way appears in [29]. However, this procedure implies the generation of graphs (lattice of ideals) whose the number of vertices increases according to the sequence of numbers defined by Dedekind [30]. This procedure can only be applied to generate capacities until $m = 5$ [27].

So, for greater number of criteria $(m > 5)$, we exploit a heuristic method. We have used the Markov chain Monte Carlo (MCMC) method introduced in [31]. The method works with iterative modifications of a starting admissible linear extension. It has been shown that this procedure evolves in limit to a uniform linear extension [32], no matter the initial linear extension.

The results obtained are shown in Fig. 1. We vary the number of criteria m from 2 to 8. For $m \geq 4$ we have used random generations, with 1000000 trials.

We see that if $p^*_e(m)$ was quite low for $m = 2$ and $m = 3$, the value of $p^*_e(m)$ increases rapidly with m. We have $p^*_e(4) = 0.659$, $p^*_e(5) = 0.868$, $p^*_e(6) = 0.977$, $p^*_e(7) = 0.997$ and for $p = 8$, we are really close to 1 $(p^*_e(8) = 0.999)$.

These results testify of the strength of the Choquet integral and its ability to attain alternatives that are not possible to reach with the WS, especially when the number of criteria increases.

However these results have been obtained for the "best possible data set" for the Choquet integral, composed of well-located exclusive C-optimal points and presenting few WS-optimal points (the m points I^j).

We study in the next section how the probability p_e to get exclusive Choquet optimal solutions evolves when the number of WS-optimal points increases.

5.2 Influence of the Number of WS-optimal Points

We will now increase the number of WS-optimal points (or supported points) and see how p_e evolves. We have generated sets of alternatives composed of n_{SP} supported points and n_{NSP} non-supported points. We first generate $\mathcal{Y}^I_{ws}$

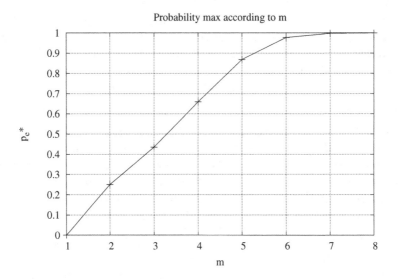

Fig. 1. Evolution of p_e^* according to the number of criteria.

(all the m points of $\mathcal{Y}_{ws}^I$ are supported). We add $(n_{SP} - m)$ points by generating randomly different points y in $[0,1]^m$. We add the following constraint: $\sum_{i=1}^m y_i^2 < 1$ (otherwise all the points will be closed to the point $(1,\dots,1)$). We check with linear programming that the point optimizes a WS. If not, we try another point, until $n_{SP} - m$ points have been generated. Then, we generate n_{NSP} non-supported points. A point will be added in the set if it does not optimize a WS (also checked with linear programming) and if it is a non-dominated point (and if the point dominates another non-supported points, these points are removed from the set). The procedure is repeated until n_{NSP} non-supported points are produced. An example of a set obtained, for two criteria, and for $n_{SP} = 22$ and $n_{NSP} = 20$ is represented in Fig. 2.

The results for $m = 2$ to $m = 6$ are given in Figs. 3, 4, 5, 6 and 7. We vary the number of n_{SP} points and for each value of n_{SP}, we also vary the number of n_{NSP} points. For each combination (n_{SP}, n_{NSP}), 100 different sets are randomly produced and we average the probability p_e over these sets.

For $m = 2$ and $m = 3$, adding SP alternatives decreases p_e, but if enough N-SP alternatives are added (about 400), the decreasing remains reasonable (from 0.25 to 0.2 for $m = 2$ and from 0.42 to 0.3 for $m = 3$). Quite surprisingly, the decreasing of p_e is higher for $m = 4$, $m = 5$ and $m = 6$. Adding only few more SP alternatives reduces p_e of about 50 %. The results are quite impressive for $m = 6$: for 6 SP alternatives and 400 N-SP alternatives, we have p_e equal to 0.91. If we add only one SP alternatives, p_e drops to 0.43.

This phenomenon can be explained by the fact that the Choquet integral is attracted by particular points, that is the points that composed the set $\mathcal{Y}_{eC}^*$. When the number of criteria m increases, it could be more difficult to have

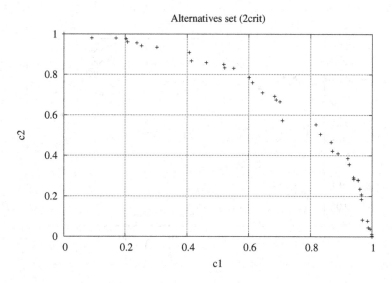

Fig. 2. Example of alternatives set obtained for $n_{SP} = 22$ (supported points) and $n_{NSP} = 20$ (non-supported points) (2 criteria).

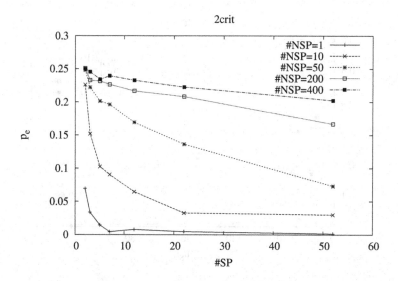

Fig. 3. 2 criteria: p_e according to n_{SP} (supported points) and n_{NSP} (non-supported points).

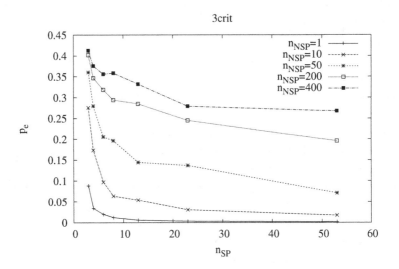

Fig. 4. 3 criteria: p_e according to n_{SP} (supported points) and n_{NSP} (non-supported points).

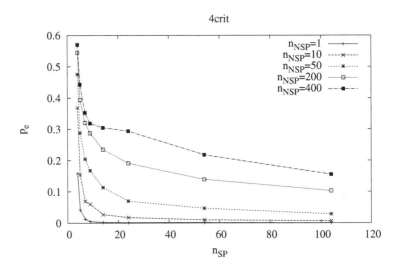

Fig. 5. 4 criteria: p_e according to n_{SP} (supported points) and n_{NSP} (non-supported points).

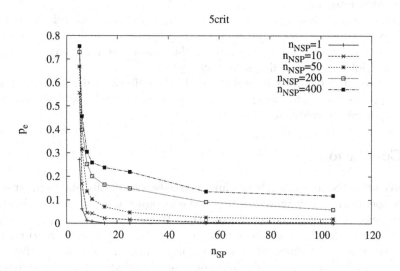

Fig. 6. 5 criteria: p_e according to n_{SP} (supported points) and n_{NSP} (non-supported points).

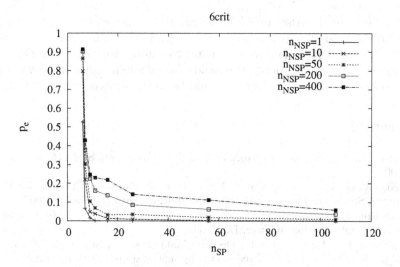

Fig. 7. 6 criteria: p_e according to n_{SP} (supported points) and n_{NSP} (non-supported points).

alternatives close to these points and the WS becomes a tougher opponent for the Choquet integral.

Similar results, not reported here, have been obtained for other constraint functions to produce sets of alternatives ($\sum_{i=1}^{m} y_i^\alpha < 1$ with $1 \leq \alpha \leq 4$) and for alternatives representing solutions of the multiobjective knapsack problem or the multiobjective traveling salesman problem.

Therefore, the Choquet integral should be carefully used if the set of alternatives present supported alternatives, at least four criteria, and not too many non-supported alternatives.

6   Conclusion

We have proposed in this work a comparison of two successful and popular aggregation operators: the weighted sum and the Choquet integral. If it is clear than the Choquet integral is more powerful than the weighted sum, given the hard work that the elicitation of its parameters asks, it is still relevant to compare both operators and to measure the probability that the best alternative determined with a Choquet integral could not been obtained with a simple weighted sum. The results show that the maximal value that this probability can take is close to one, when the number of criteria is higher than four. However, to reach this high probability, particular data sets have been constructed, in favor of the Choquet integral. When the number of WS-optimal alternatives increases in a set, the probability decreases rapidly, and especially if the number of criteria is higher than four. This work opens many new perspectives: different operators could be compared using this framework, like WOWA and the Choquet integral, for example. Also, we could restrict the capacities of the Choquet integral to certain families (convex capacities, k-additive capacities, etc.) and to examine which families of capacities enable to mainly reach exclusive C-optimal alternatives. Finally, extending this study in a combinatorial space where the alternatives are not explicitly given would be worth studying.

References

1. Torra, V.: Aggregation operators and models. Fuzzy Sets Syst. **156**(3), 407–410 (2005)
2. Beliakov, G., Pradera, A., Calvo, T.: Aggregation Functions: A Guide for Practitioners. Studies in Fuzziness and Soft Computing. Springer, Berlin (2007)
3. Bouyssou, D.: Outranking methods. In: Floudas, C., Pardalos, P. (eds.) Encyclopedia of Optimization. Kluwer, Dordrecht (2001)
4. Fishburn, P.: Utility Theory for Decision Making. Wiley, New York (1970)
5. Yager, R.: On ordered weighted averaging aggregation operators in multicriteria decision making. IEEE Trans. Syst. Man Cybern. **18**, 183–190 (1998)
6. Torra, V.: The weighted OWA operator. Int. J. Intell. Syst. **12**, 153–166 (1997)
7. Choquet, G.: Theory of capacities. Ann. de l'Institut Fourier **5**, 131–295 (1953)
8. Chateauneuf, A.: Modeling attitudes towards uncertainty and risk through the use of choquet integral. Ann. Oper. Res. **52**(1), 1–20 (1994)

9. Grabisch, M., Marichal, J.C., Mesiar, R., Pap, E.: Aggregation Functions. Encyclopedia of Mathematics and its Applications. Cambridge University Press, Cambridge (2009)

10. Grabisch, M., Labreuche, C.: A decade of application of the Choquet and Sugeno integrals in multi-criteria decision aid. Ann. OR **175**(1), 247–286 (2010)

11. Grabisch, M., Labreuche, C.: Fuzzy measures and integrals in mcda. In: Multiple Criteria Decision Analysis: State of the Art Surveys. Volume 78 of International Series in Operations Research and Management Science, pp. 563–604. Springer, New York (2005)

12. Galand, L., Perny, P., Spanjaard, O.: Choquet-based optimisation in multiobjective shortest path and spanning tree problems. Eur. J. Oper. Res. **204**(2), 303–315 (2010)

13. Galand, L., Perny, P., Spanjaard, O.: A branch and bound algorithm for Choquet optimization in multicriteria problems. Proc. Lect. Notes Econo. Math. Syst. **634**, 355–365 (2011)

14. Fouchal, H., Gandibleux, X., Lehuédé, F.: Preferred solutions computed with a label setting algorithm based on Choquet integral for multi-objective shortest paths. In, : IEEE Symposium on Computational Intelligence in Multicriteria Decision-Making (MDCM). Paris, France, IEEE, pp. 143–150 (2011)

15. Galand, L., Lesca, J., Perny, P.: Dominance rules for the Choquet integral in multiobjective dynamic programming. In: IJCAI 2013, Proceedings of the 23rd International Joint Conference on Artificial Intelligence, Beijing, China, August 3–9, 2013 (2013)

16. Branke, J., Corrente, S., Greco, S., Slowinski, R., Zielniewicz, P.: Using Choquet integral as preference model in interactive evolutionary multiobjective optimization. Technical report, Warwick Business School, Warwick, Uninted Kingdom (2014)

17. Meyer, P., Pirlot, M.: On the expressiveness of the additive value function and the Choquet integral models. In: DA2PL 2012 : from Multiple Criteria Decision Aid to Preference Learning, pp. 48–56 (2012)

18. Ehrgott, M.: Multicriteria Optimization, 2nd edn. Springer, Berlin (2005)

19. Grabisch, M.: The application of fuzzy integrals in multicriteria decision making. Eur. J. Oper. Res. **89**, 445–456 (1996)

20. Tehrani, A., Cheng, W., Hüllermeier, E.: Choquistic regression: generalizing logistic regression using the Choquet integral. In: Proceedings of the EUSFLAT2011, 7th International Conference of the European Society for Fuzzy Logic and Technology, pp. 868–875 (2011)

21. Valls, A., Medina, C., Moreno, A., Puig, D.: Selection of gabor filters with Choquet integral for texture analysis in mammogram images. In: Gibert, K., Botti, V.J., Bolao, R.R. (eds.) CCIA. Volume 256 of Frontiers in Artificial Intelligence and Applications, pp. 67–76. IOS Press (2013)

22. Le Huédé, F., Grabisch, M., Labreuche, C., Savéant, P.: Integration and propagation of a multi-criteria decision making model in constraint programming. J. Heuristics **12**(4–5), 329–346 (2006)

23. Galand, L., Perny, P.: Search for Choquet-optimal paths under uncertainty. In: UAI07, vol. 2007, pp. 125–132 (2007)

24. Lust, T., Rolland, A.: Choquet optimal set in biobjective combinatorial optimization. Comput. Oper. Res. **40**(10), 2260–2269 (2013)

25. Lust, T., Rolland, A.: On the computation of choquet optimal solutions in multicriteria decision contexts. In: Ramanna, S., Lingras, P., Sombattheera, C., Krishna, A. (eds.) MIWAI 2013. LNCS, vol. 8271, pp. 131–142. Springer, Heidelberg (2013)

26. Boyd, S., Vandenberghe, L.: Convex Optimization. Cambridge University Press, New York (2004)
27. Combarro, E.F., DíAz, I., Miranda, P.: On random generation of fuzzy measures. Fuzzy Sets Syst. **228**, 64–77 (2013)
28. Rubinstein, R., Kroese, D.: Simulation and the Monte Carlo Method, 2nd edn. Wiley, Chichester (2008)
29. Loof, K.D., Meyer, H.D., Baets, B.D.: Exploiting the lattice of ideals representation of a poset. Fundam. Inf. **71**(2–3), 309–321 (2006)
30. Dedekind, R.: Ueber zerlegungen von zahlen durch ihre grossten gemeinsamen teiler. Fest. Hoch. Braunschweig u. ges. Werke **II**, 103–148 (1897)
31. Karzanov, A., Khachiyan, L.: On the conductance of order Markov chains. Order **8**(1), 7–15 (1991)
32. Bubley, R., Dyer, M.: Faster random generation of linear extensions. Discrete Math. **201**, 81–88 (1999)

Exact Methods for Computing All Lorenz Optimal Solutions to Biobjective Problems

Lucie Galand[1] and Thibaut Lust[2]([✉])

[1] PSL, Université Paris-Dauphine LAMSADE, CNRS UMR 7243,
75775 Paris Cedex 16, France
lucie.galand@dauphine.fr
[2] Sorbonne Universités, UPMC Université Paris 06, CNRS, LIP6,
UMR 7606, 75005 Paris, France
thibaut.lust@lip6.fr

Abstract. This paper deals with biobjective combinatorial optimization problems where both objectives are required to be well-balanced. Lorenz dominance is a refinement of the Pareto dominance that has been proposed in economics to measure the inequalities in income distributions. We consider in this work the problem of computing the Lorenz optimal solutions to combinatorial optimization problems where solutions are evaluated by a two-component vector. This setting can encompass fair optimization or robust optimization. The computation of Lorenz optimal solutions in biobjective combinatorial optimization is however challenging (it has been shown intractable and NP-hard on certain problems). Nevertheless, to our knowledge, very few works address this problem. We propose thus in this work new methods to generate Lorenz optimal solutions. More precisely, we consider the adaptation of the well-known two-phase method proposed in biobjective optimization for computing Pareto optimal solutions to the direct computing of Lorenz optimal solutions. We show that some properties of the Lorenz dominance can provide a more efficient variant of the two-phase method. The results of the new method are compared to state-of-the-art methods on various biobjective combinatorial optimization problems and we show that the new method is more efficient in a majority of cases.

Keywords: Multiobjective combinatorial optimization · Fairness · Lorenz dominance · Two-phase method

1 Introduction

In many decision problems, a decision (or a solution) has to be evaluated with respect to several dimensions. In multicriteria decision making the dimensions reflect several aspects to take into account (one aspect per criterion). In multiagent decision making they reflect the point of view of several agents, and they can reflect several scenarios that can occur in robust decision making. We consider in this paper a general framework where a solution is evaluated with respect to

© Springer International Publishing Switzerland 2015
T. Walsh (Ed.): ADT 2015, LNAI 9346, pp. 305–321, 2015.
DOI: 10.1007/978-3-319-23114-3_19

a component vector, which could be a vector of criteria, a vector of scenarios or a vector of agents' utilities. Since there is generally not a solution that optimizes all the components, one has to determine compromise solutions. In this setting, the concept of *Pareto dominance* enables to focus on the subset of solutions to a decision problem for which one cannot make a component better off without worsening another component. However the number of *Pareto-optimal* (P-optimal) solutions to a decision problem can be very large, which could make a final choice of one (or a few) solution(s) among the P-optimal ones difficult for a decision maker. The notion of Lorenz dominance has been proposed in economics to measure the inequalities in income distributions. It refines the Pareto dominance by selecting only the better distributed solutions. Furthermore, it has been used for characterizing equitable solutions in multicriteria optimization [1,2] and robust solutions in decision under uncertainty [3]. It has also been studied within the framework of convex-cone theory [4] in multiobjective programming [5]. The *Lorenz-optimal* (L-optimal) solutions can be determined with a two-stage procedure that first generates all the P-optimal solutions and second selects only the L-optimal ones among them. But the efficiency of the two-stage procedure depends on the efficiency of the procedure that generates the P-optimal solutions. Besides the number of L-optimal solutions can be very small compared to the number of P-optimal solutions, which would make the two-stage procedure quite inadequate. In the last decade, some procedures have been proposed to deal with the direct determination of the L-optimal solutions in combinatorial optimization (see e.g. the works of Perny *et al.* [3], Baatar and Wiecek [5], Moghaddam *et al.* [6], and Endriss [7]), which is generally a difficult problem (NP-complete and intractable [3,7]). Nevertheless, the amount of works related to Lorenz optimization is quite small compared to the amount of works related to Pareto optimization in combinatorial optimization. The aim of this work is therefore to study the direct determination of L-optimal solutions in combinatorial optimization. More precisely, we propose in this paper to adapt one of the most famous method proposed in biobjective optimization, namely the *two-phase method* [8], to Lorenz optimization. The two-phase method is a generic approach that enables to determine the P-optimal solutions by computing first the subset of P-optimal solutions that optimize a weighted sum, and second the other P-optimal solutions. It has been widely applied on various problems of biobjective combinatorial optimization (see e.g. the works of Visée *et al.* [9], Ehrgott and Skriver [10], Przybylski *et al.* [11], and Raith and Ehrgott [12]). In this paper, we propose two variants of the adaptation of the two-phase method to Lorenz optimization and we study the efficiency of these procedures on two biobjective combinatorial optimization problems: the biobjective shortest path problem and the biobjective set covering problem.

In Sect. 2, we formally define the Lorenz dominance and we present the problem of Lorenz optimization in multi-objective combinatorial optimization. Section 3 is devoted to some characterizations of L-optimal solutions. In Sect. 4, we present a straight adaptation of the two-phase method to Lorenz optimization, and a variant that uses the characterization results of Sect. 3 to improve its efficiency. We present in Sect. 5 some numerical experiments, and discuss the efficiency of the adapted two-phase method compared to some state-of-the-art methods. We conclude in Sect. 6.

2   Multi-objective Combinatorial Optimization

2.1   Notations and Definitions

A multi-objective combinatorial optimization (MOCO) problem can be formulated as follows:

$$\text{``min''}_x \ f(x) = Cx = (f_1(x), f_2(x), \ldots, f_p(x))^T$$

$$\text{s.t. } Ax \leq b$$

$$x \in \{0,1\}^n$$

where $A \in \mathbb{R}^{m \times n}$, $b \in \mathbb{R}^m$, $C \in \mathbb{R}^{p \times n}$ and the quotation marks means that we want to minimize a vector and not a single scalar value. A feasible solution x is a binary vector of n variables that satisfies the m constraints of the problem. Each solution x is evaluated by p objective functions f_k, $k = 1, \ldots, p$ such that $f_k(x)$ is the value of solution x for objective k. The feasible set in decision space is given by $\mathcal{X} = \{x \in \{0,1\}^n : Ax \leq b\}$. One compares the solutions according to their evaluation in $\mathbb{R}^p$, called the *objective space*. The feasible set in the objective space, that is the evaluation of the feasible set, is given by $\mathcal{Y} = f(\mathcal{X}) = \{f(x) : x \in \mathcal{X}\} \subset \mathbb{R}^p$. An element of the set $\mathcal{Y}$ is called a *cost-vector* or a *point*. W.l.o.g. we consider through the paper that the p objective functions have to be minimized.

Definition 1. *The* Pareto dominance *relation (P-dominance for short) is defined for all* $y^1, y^2 \in \mathbb{R}^p$ *by:* $y^1 \succ_P y^2 \iff [\forall k \in \{1, \ldots, p\}, y_k^1 \leq y_k^2 \text{ and } y^1 \neq y^2]$

Within a feasible set $\mathcal{X}$, any element x^1 is said to be *P-dominated* when $f(x^2) \succ_P f(x^1)$ for some x^2 in $\mathcal{X}$, and *P-efficient* (or P-optimal) when there is no x^2 in $\mathcal{X}$ such that $f(x^2) \succ_P f(x^1)$. The *P-efficient set* denoted by $\mathcal{X}_P$ contains all the P-efficient solutions. The image $f(x)$ in the objective space of a P-efficient solution x is called a *Pareto-non-dominated* point. The image $\mathcal{Y}_P = f(\mathcal{X}_P)$ of the P-efficient set $\mathcal{X}_P$ in $\mathcal{Y}$, is called the *Pareto front*.

The Lorenz dominance is based on the construction of particular vectors, called *generalized Lorenz vectors*, that are obtained as follows:

Definition 2. *For all* $y \in \mathbb{R}^p$, *the* generalized Lorenz vector *of* y *is the vector* $L(y)$ *defined by:* $L(y) = (y_{(1)}, y_{(1)} + y_{(2)}, \ldots, y_{(1)} + y_{(2)} + \ldots + y_{(p)})$, *where* $y_{(1)} \geq y_{(2)} \geq \ldots \geq y_{(p)}$ *represent the components of* y *sorted in non-increasing order. The* k^{th} *component of* $L(y)$ *is* $L_k(y) = \sum_{i=1}^{k} y_{(i)}$.

Definition 3. *The* Lorenz dominance *relation (L-dominance for short) is defined for all* $y^1, y^2 \in \mathbb{R}^p$ *by:* $y^1 \succ_L y^2 \iff [L(y^1) \succ_P L(y^2)]$

The space in which the generalized Lorenz vectors of a solution x are represented is called the *Lorenz space*. Within a feasible set $\mathcal{X}$, any element x^1 is said to be *L-dominated* when $f(x^2) \succ_L f(x^1)$ for some x^2 in $\mathcal{X}$, and *L-efficient* (or L-optimal) when there is no x^2 in $\mathcal{X}$ such that $f(x^2) \succ_L f(x^1)$. The *L-efficient set* denoted by $\mathcal{X}_L$ contains all the L-efficient solutions. The image $f(x)$ in the objective space or the image $L(f(x))$ in the Lorenz space of a L-efficient solution x is called a *L-non-dominated* point. The image $\mathcal{Y}_L = f(\mathcal{X}_L)$ of the L-efficient set in $\mathcal{Y}$ is called the *Lorenz front*. The generalized Lorenz vectors of the Lorenz front are given by $L(\mathcal{Y}_L)$. The Lorenz dominance is closely related to the *Transfer Principle* [13], which means that for some cost-vector $y \in \mathbb{R}^p$ with $y_i > y_j$, slightly improving y_j to the detriment of y_i while preserving the mean of the costs would produce a better distribution of the costs, and consequently a more balanced solution. This principle enables to compare vectors with the same mean. The generalized Lorenz extension considered here enables to compare vectors with different means thanks to the *P-monotonicity* axiom [14], which means that if a cost-vector y^1 P-dominates another cost-vector y^2 then y^1 L-dominates y^2. Consequently L-optimal solutions are a subset of P-optimal solutions.

Following Definition 3, finding the L-efficient solutions to a MOCO problem boils down to finding the P-efficient solutions to the same MOCO problem where the costs are given by the generalized Lorenz vectors:

$$\text{"}\min_{x \in \mathcal{X}}\text{"} \quad L(f(x))$$

where $L(f(x)) = (f_{(1)}(x), f_{(1)}(x) + f_{(2)}(x), \ldots, f_{(1)}(x) + f_{(2)}(x) + \ldots + f_{(p)}(x))$ with $f_{(1)}(x) \geq f_{(2)}(x) \geq \ldots \geq f_{(p)}(x)$ represent the components of $f(x)$ sorted in non-increasing order. In the special case where $p = 2$, the two objective functions to be minimized are: $L_1(f(x)) = \max(f_1(x), f_2(x))$ and $L_2(f(x)) = f_1(x) + f_2(x)$. We thus look for solutions that establish a good compromise between the value of the worst performance and the sum of the costs.

Example 1. *Let us consider the point $y = (6, 3)$. All the points L-dominated by y are in the hatched area called "Lorenz worse" in the biobjective space of Fig. 1 (left part). The points that L-dominate y are in the hatched area called "Lorenz better" in the same figure. To illustrate the L-dominance, the symmetric point of y, the point $(3, 6)$, is also represented in the figure by a circle. The points in the not hatched area are incomparable to y with L-dominance. The generalized Lorenz vector of the point $(6, 3)$, that is the point $(6, 9)$, is represented in the Lorenz space (right part of the figure).*

2.2 Algorithmic Issues

Intractability. As there is not generally a unique optimal solution when multiple objectives are involved, the number of optimal solutions turns out to be a crucial point to evaluate the hardness of the problem. It leads us to the notion of *intractability* [15]. In our setting, a Lorenz optimization problem is intractable if

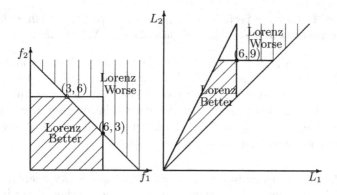

Fig. 1. Representation of the Lorenz dominance.

the number of L-efficient solutions is exponential in the size of the instance. The biobjective shortest path problem, the biobjective spanning tree problem and the biagent Markov decision process problem have been proved intractable when looking for L-efficient solutions [3,16]. As Ehrgott did in a Pareto optimization setting [15], one can show that even the unconstrained problem is intractable in a Lorenz optimization setting. The multi-objective unconstrained (MOUC) problem is defined as follows:

$$ \text{``} \min_{x_i \in \{0,1\}^n} \text{''} \sum_{i=1}^{n} c_k^i x_i \quad k = 1 \dots p $$

Proposition 1. *Problem MOUC is intractable for Lorenz optimization.*

Proof. For $p = 2$, by setting $c_1^i = 2^{(i-1)}$ and $c_2^i = -2^i$, we obtain $\mathcal{Y} = \{(0,0), (1,-2), (2,-4), \dots, (2^n - 1, -2^{n+1} + 2)\}$. If we represent the generalized Lorenz vectors of all $y \in \mathcal{Y}$, we obtain: $L(\mathcal{Y}) = \{(0,0), (1,-1), (2,-2), \dots, (2^n - 1, -2^n + 1)\}$. All the generalized Lorenz vectors have the same sum $L_1 + L_2$ and a distinct value on the first dimension L_1 (and consequently on the second dimension L_2 as well). Thus all the generalized Lorenz vectors are P-non-dominated in the Lorenz space and then we have $\mathcal{Y} = \mathcal{Y}_L$. Furthermore, by construction, each feasible solution has a distinct image in the objective space, i.e. $|\mathcal{Y}| = |\mathcal{X}|$. As the number of feasible solution is $|\mathcal{X}| = 2^n$, we have thus $|\mathcal{X}_L| = |\mathcal{Y}_L| = 2^n$. □

NP-completeness. The complexity of a Lorenz optimization problem is defined by the complexity of its decision version: given a vector $v = (v_1, \dots, v_p)$, does there exist a solution to the Lorenz optimization problem that L-dominates v? The decision version of the biobjective shortest path problem and the biobjective spanning tree problem has been proved NP-complete [3], and the decision version of the multi-agent allocation problem has been proved NP-complete for many languages [7]. Besides, one can easily show that the decision version of problem MOUC, which is obviously in NP, is also NP-complete.

Proposition 2. *Given a vector* $v = (v_1, \ldots, v_p)$, *deciding whether there exists a solution to problem MOUC that L-dominates* v *is NP-complete.*

Proof. We use a reduction from the partition problem. Consider an instance of this problem with a finite set $A = \{a^1, \ldots, a^n\}$ and a size $s(a^i) \in \mathbb{N}$ for each element a^i in A. We construct a biobjective MOUC instance by setting $c_1^i = s(a^i)$ and $c_2^i = -2s(a^i)$, and we ask if there exists a solution that L-dominates the vector $(\sum_i s(a^i)/2 + \varepsilon, -\sum_i s(a^i) + \varepsilon)$ for some small $\varepsilon > 0$. Answering this question amounts to solving the partition problem. □

Other Difficulties. In addition to the previous complexity results, the determination of Lorenz optimal solutions can encounter another algorithmic issue. It has indeed been shown for some Lorenz optimization problems that one cannot resort directly to an approach based on the Lorenz optimality of partial solutions, like dynamic programming or greedy procedures [16,17].

2.3   State-of-the-Art

To our knowledge, only a few works address the problem of Lorenz optimization for MOCO problems. We now briefly list the methods proposed in the literature.

Ranking Method. The ranking method has been proposed by Perny *et al.* [3] in a robust optimization setting. This method works simply by computing the solutions in nondecreasing order of their sum using a k-best algorithm (that is an algorithm that generates the k best solutions to an optimization problem). This enumeration can be stopped when all the L-efficient solutions have been generated. Note that the sum of the costs of a L-dominated solution can be lower than the sum of the costs of a L-efficient solution, thus the ranking method actually computes a superset of the set $\mathcal{X}_L$. As one cannot know in advance the number k of solutions to be enumerated, one has to define a valid stopping criterion that ensures that all L-efficient solutions have been generated. This method can be used with any number of objectives, but its efficiency strongly relies on the efficiency of the k-best algorithm.

ε-Constraint based Method. This method, proposed by Baatar and Wiecek [5], is based on the classic ε-constraint procedure for Pareto optimization [18]. It generates L-efficient solutions in the nondecreasing order of their sum. In addition, it uses Euclidean norm optimization to ensure to determine an L-efficient solution for a given sum of the costs (and not a L-dominated one). Each solution is computed by solving two mathematical programs with appropriate constraints and objective functions. This method works with any number of objectives, but solving such mathematical programs can be inefficient in practice. Besides it generates only one L-efficient solution per Lorenz vector which implies that it computes the set $L(\mathcal{Y}_L)$ but not the Lorenz front $\mathcal{Y}_L$.

Dynamic Programming based method. One cannot use directly a dynamic programming procedure to generate the L-non-dominated points to a MOCO problem (see Sect. 2.2). However, since dynamic programming can be used with P-dominance, and since L-optimal solutions are also P-optimal, Perny *et al.* have proposed to adapt a multi-objective dynamic programming based procedure to Lorenz optimization by using a valid additional dominance rule [3].

3 Supported Solutions

3.1 Definitions

In multi-objective optimization, there exists an important classification of the P-efficient solutions: *supported* P-efficient solutions (SP solutions) and *non-supported* P-efficient solutions (NSP solutions). The images of the SP solutions in the objective space are located on the convex hull of the Pareto front and the images of the NSP solutions are located inside the convex hull of the Pareto front. More precisely, we can characterize these solutions as follows [15]:

Supported P-efficient solutions: a solution x is supported P-efficient iff there exists a vector λ ($\lambda_k > 0, \forall k \in \{1, \ldots, p\}$) such that x is an optimal solution to the weighted sum single-objective problem: $\min_{x \in \mathcal{X}} \sum_{k=1}^{p} \lambda_k f_k(x)$. The set of SP solutions is denoted by $\mathcal{X}_{SP}$ and the set of supported P-non-dominated points, called SP points, by $\mathcal{Y}_{SP}(= f(\mathcal{X}_{SP}))$. Note that in biobjective optimization, the set $\mathcal{X}_{SP}$ can be easily computed with a *dichotomic search* [19,20] which gives the different weighting vectors that allow to generate all the SP points.

Non-supported P-efficient solutions: P-efficient solutions that are not supported. The set of NSP solutions is denoted by $\mathcal{X}_{NSP}$ and the set of non-supported P-non-dominated points, called NSP points, by $\mathcal{Y}_{NSP}(= f(\mathcal{X}_{NSP}))$.

One can easily transpose these notions to Lorenz dominance by applying the definitions of SP and NSP solutions in the Lorenz space. In that respect, we define *supported L-efficient solutions* (SL solutions) as follows: a solution x is supported L-efficient iff there exists a vector λ ($\lambda_k > 0, \forall k \in \{1, \ldots, p\}$) such that x is an optimal solution to the weighted sum single-objective problem defined on the generalized Lorenz vector of $f(x)$: $\min_{x \in \mathcal{X}} \sum_{k=1}^{p} \lambda_k L_k(f(x))$, where $f_{(1)}(x) \geq f_{(2)}(x) \geq \ldots \geq f_{(p)}(x)$. Note that $\sum_{k=1}^{p} \lambda_k L_k(f(x)) = (\sum_{k=1}^{p} \lambda_k) f_{(1)}(x) + (\lambda_2 + \ldots \lambda_p) f_{(2)}(x) + \ldots + \lambda_p f_{(p)}(x)$. Let w be a weight vector defined by $w_k = \sum_{i=k}^{p} \lambda_i$, then $\sum_{k=1}^{p} \lambda_k L_k(f(x)) = w_1 f_{(1)}(x) + w_2 f_{(2)}(x) + \ldots + w_p f_{(p)}(x)$. Such an aggregation function is well-known in fair optimization, it corresponds to a particular family of Ordered Weighted Averages (OWA), where the weights are strictly decreasing. The OWA function has been introduced by Yager [21]:

Definition 4. *Given a vector $y \in \mathbb{R}^p$ and a weighting vector $w \in [0,1]^p$, the ordered weighted average (OWA) of y with respect to w is defined by:* $f^{owa}(y,w) = \sum_{k=1}^{p} w_k y_{(k)}$ *where* $y_{(1)} \geq \ldots \geq y_{(p)}$.

It has been shown that any solution minimizing an OWA function endowed with strictly decreasing and strictly positive weights is L-efficient [22]. One can then define the SL solutions as follows:

Definition 5. *The SL solutions are the solutions that minimize an OWA function for some strictly decreasing and strictly positive weighting vector.*

However, in general there exist L-efficient solutions that do not optimize any OWA functions. We call these solutions *non-supported L-efficient* solutions (NSL solutions). The set of SL (resp. NSL) solutions is denoted by $\mathcal{X}_{SL}$ (resp. $\mathcal{X}_{NSL}$) and the set of supported (resp. non-supported) L-non-dominated points, called SL (resp. NSL) points, by $\mathcal{Y}_{SL}$ (resp. $\mathcal{Y}_{NSL}$). Unfortunately, there are no trivial relations between the sets $\mathcal{X}_{SP}$ and $\mathcal{X}_{SL}$, or between the sets $\mathcal{Y}_{SP}$ and $\mathcal{Y}_{SL}$, as illustrated in the following example.

Example 2. *Let us consider the following 4 solutions: x^1 with $f(x^1) = (6, 18)$, x^2 with $f(x^2) = (9, 16)$, x^3 with $f(x^3) = (12, 14)$ and x^4 with $f(x^4) = (20, 2)$. The generalized Lorenz vectors of the 4 solutions are all P-non-dominated: $L(f(x^1)) = (18, 24)$, $L(f(x^2)) = (16, 25)$, $L(f(x^3)) = (14, 26)$ and $L(f(x^4)) = (20, 22)$. It means that the 4 solutions are L-efficient. In this simple example, we have a SP solution that is not SL (x^1), a solution that is neither SP nor SL (x^2), a SL solution that is not SP (x^3) and a solution that is both SP and SL (x^4).*

3.2 Biobjective Case

Even though there is no trivial relation between the supported Pareto and Lorenz efficient sets, we show in this section that, in the case of two objectives, there are interesting properties on the location of the SL solutions with respect to the location of some SP solutions in the objective space. Before presenting the properties, let us introduce some notation for the biobjective case. We denote by $O^i \subset \mathbb{R}^2_{\geq}$ the space in the positive orthant of the objective space where all the points y are such that $y_i \leq y_j$ for $j \neq i$. The *bisector* is the line $y_1 = y_2$ in the objective space. Any point in O^1 (resp. O^2) is said to be *above* (resp. *below*) the bisector. Let us also denote by x^* a L-efficient solution that minimizes the sum of the costs (it is an optimal solution to leximin $(f_1(x) + f_2(x), \max(f_1(x), f_2(x)))$, where an optimal solution to leximin (y_1, y_2) is an optimal solution w.r.t. y_1 that minimizes y_2 among all the optimal solutions w.r.t. y_1), and y^* its image in the objective space. In this section, we suppose, w.l.o.g., that $y_2^* > y_1^*$, i.e. $y^* \in O^1$. We will also use the notion of "betweenness" in the sequel: we say that a point y is *between* two points y^i and y^j ($i \neq j$) in the objective space when $y_1^i \leq y_1^j \Rightarrow y_1^i \leq y_1 \leq y_1^j$. Furthermore, two SP points y^1 and y^2 are said to be *adjacent* when there is no other solution x^3 in $\mathcal{X}_{SP}$ such that $f(x^3)$ is between y^1 and y^2.

Property 1. *All SP solutions x such that $f(x) \in O^1$ and $f_2(x) < y_2^*$ are SL.*

Proof. Let x^1 be a solution in $\mathcal{X}_{SP}$ such that $x^1 \in O^1$ and $f_2(x^1) < y_2^*$. As x^1 is in $\mathcal{X}_{SP}$, $\exists \lambda$ s.t. $\lambda_1 f_1(x^1) + \lambda_2 f_2(x^1) \leq \lambda_1 f_1(x) + \lambda_2 f_2(x), \forall x \in \mathcal{X}$. Moreover, as $f_2(x^1) < y_2^*$, $\lambda_2 > \lambda_1$ (otherwise x^* would be better than x^1 for the weighted sum). This implies that for any y in O^1, $\lambda_1 y_1 + \lambda_2 y_2 = f^{owa}(y, \lambda)$, and therefore for any solution $x \in \mathcal{X}$ s.t. $f(x) \in O^1$, $f^{owa}(f(x^1), \lambda) \leq f^{owa}(f(x), \lambda)$ (1). Let us now consider the case where the image of a solution $x \in \mathcal{X}$ is not in O^1, which means that $f_1(x) > f_2(x)$. As $\lambda_2 > \lambda_1$, we have $\lambda_2(f_1(x) - f_2(x)) > \lambda_1(f_1(x) - f_2(x))$, that is $f^{owa}(f(x), \lambda) = \lambda_2 f_1(x) + \lambda_1 f_2(x) > \lambda_1 f_1(x) + \lambda_2 f_2(x)$. As x^1 is in $\mathcal{X}_{SP}$ and in O^1, we have thus $f^{owa}(f(x^1), \lambda) = \lambda_2 f_2(x^1) + \lambda_1 f_1(x^1) < f^{owa}(f(x), \lambda)$ (2). From (1) and (2), we have that solution x^1 is in $\mathcal{X}_{LS}$.   □

Note that all solutions x such that $f(x)$ is in O^1 and $f_2(x) > y_2^*$ are L-dominated by y^* since $L(y^*) = (y_2^*, y_1^* + y_2^*) \succ_P (f_2(x), f_1(x) + f_2(x)) = L(f(x))$.

The next property enables us to locate in the objective space the SL points that are NSP according to the bisector. We say that two points are located on *opposite sides of the bisector* if one is above the bisector and the other one is below.

Property 2. *The image of a solution that is SL but not SP is between the two adjacent SP points located on opposite sides of the bisector.*

Proof. Let x^1 and x^2 be two P-efficient solutions such that their images $y^1 = f(x^1)$ and $y^2 = f(x^2)$ are two adjacent SP points such that $y_1^1 \leq y_1^2$ (and thus $y_2^1 \geq y_2^2$), and y^1 and y^2 are both either in O^1, either in O^2. Let us assume for example that they are in O^1. Let x^3 be a NSP solution whose image $y^3 = f(x^3)$ is between y^1 and y^2 in the objective space. As points y^1, y^2 and thus y^3 are in O^1, we have $f^{owa}(y^i, \lambda) = \lambda_1 y_2^i + \lambda_2 y_1^i$ for any $i = 1, 2$ or 3. Since x^1 and x^2 are SP and x^3 is NSP, there is always a weight λ such that $f^{owa}(y^1, \lambda)$ or $f^{owa}(y^2, \lambda)$ is less than $f^{owa}(y^3, \lambda)$. Thus solution x^3 cannot be L-efficient. Obviously, we come to the same conclusions when y^1 and y^2 are in O^2.   □

Property 3. *Let y^2 be an SP point (image of a solution) such that y^2 is in set O^2. Then all solutions x such that $f_1(x) > y_1^2$ are not L-efficient.*

Proof. It is sufficient to show that all solutions $x \in \mathcal{X}_P$ such that $f_1(x) > y_1^2$ are L-dominated by a solution x^2 the image of which is y^2. Let us consider one of these solutions x. As x is P-efficient, $f_2(x) < y_2^2$, which implies that $f(x) \in O^2$, then $L_1(f(x)) = f_1(x) > y_1^2 = L_1(y^2)$. To be L-efficient x must therefore satisfy $L_2(f(x)) < L_2(y^2)$, that is $f_1(x) + f_2(x) > y_1^2 + y_2^2$ (1). As solution x^2 is SP, there exists a weighting vector λ such that $\lambda_1 y_1^2 + \lambda_2 y_2^2 \leq \lambda_1 f_1(x) + \lambda_2 f_2(x)$ (2) and $\lambda_1 y_1^2 + \lambda_2 y_2^2 \leq \lambda_1 f_1(x^*) + \lambda_2 f_2(x^*)$ (3). From (1) and (2) and since $f_1(x) > y_1^2$, one obtains $\lambda_1 > \lambda_2$. And from (3) and the fact that solution x^* minimizes the sum of the costs, we obtain $\lambda_2 \geq \lambda_1$, which leads to a contradiction. Therefore, we have $L_2(f(x)) \geq L_2(y^2)$. Thus, solution x^2 L-dominates solution x.   □

The image in the objective space of all L-efficient solutions in O^2 are thus between y^* and any SP point in O^2. Let x^{max} be an SP solution such that $f(x^{max}) = y^{max}$ is in O^2 and minimizes the cost of the first criterion among all the P-non-dominated points in O^2. Then, from the previous properties, we can deduce that all L-non-dominated points are between y^* and y^{max}. We show in the next section how we can use these properties to compute all the L-non-dominated points of $\mathcal{Y}_L$, without generating the entire set $\mathcal{Y}_P$.

4   New Methods

4.1   Straight Adaptation of the Two-Phase Method

The two-phase method has been developed by Ulungu and Teghem [8] to find the P-efficient solutions to MOCO problems with two objectives. We describe only at a high level how this method can be adapted in order to generate all the L-non-dominated points of a biobjective optimization problem. The adaptation closely follows the original method. As the name of the method suggests it, the method works in two distinct phases:

Phase 1: generation of all the SL solutions ($\mathcal{X}_{SL}$). This consists in applying the first phase of the original two-phase method for Pareto optimization in the Lorenz space instead of the objective space. This amounts to optimizing OWA functions with different weights until all the SL solutions have been detected. The weight sets w used in the different OWA functions are defined by $w_k = \sum_{i=k}^{p} \lambda_i$ ($k \in \{1, \ldots, p\}$) from the weight sets λ defined in the Lorenz space and computed by the dichotomic search (see Sect. 3.1).

Phase 2: generation of $\mathcal{X}_{NSL}$. The SL points generated at Phase 1 are used to reduce the search space, since for biobjective problems, NSL solutions are always located, in the Lorenz space, in the interior of the right triangle defined by two adjacent SL points $L(y^1)$ and $L(y^2)$ and the point $(L_1(y^2), L_2(y^1))$ (when $L_1(y^1) < L_1(y^2)$). The exploration of the triangles can be performed with a branch and bound algorithm or with a k-best algorithm.

Even if this straight adaptation of the two-phase method is theoretically interesting, the main drawback is in the first phase: the OWA function that has to be optimized is non-linear and therefore even generating only the SL solutions could be computationally expensive. We propose in the next section a new method where the optimization of OWA functions is avoided.

4.2   Supported Pareto-Efficient Solutions Based Method

From Properties 1 and 3, we have that all L-non-dominated points are between the points y^* and y^{max} (see Sect. 3.2). We use this property to define a new two-phase method based on the computation of some SP solutions. This new method is called *method SP* in the sequel.

Phase 1. The first phase of method SP consists in generating all the SP points located between y^* and y^{max}. From Property 1, we know indeed that all these solutions, except perhaps solution x^{max}, are SL. In order to do so, one first finds a solution x^1 that optimizes L_2. Three cases can then occur:

1. $f_1(x^1) = f_2(x^1)$: in this case, $f(x^1)$ is the only L-non-dominated point of the problem since it optimizes both L_2 and L_1: the method SP stops and returns solution x^1.

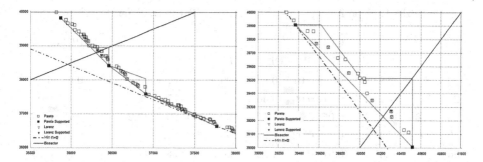

Fig. 2. Search zonee: in the objective space (left) and in the Lorenz space (right).

2. $f_1(x^1) < f_2(x^1)$ (i.e. $f(x^1) \in O^1$): in this case, we perform a dichotomic search between the points $f(x^1)$ and $f(x^2)$ where x^2 optimizes $\min_{x \in \mathcal{X}}(f_2(x))$.

Note that we only need to compute the point y^{max} in O^2, and consequently the search is mainly performed in O^1.

3. $f_1(x^1) > f_2(x^1)$ (i.e. $f(x^1) \in O^2$): analogous to case 2.

For the cases 2 and 3, the SP point y^{max} is also stored. Note that at the end of the first phase, we have generated a subset of $\mathcal{Y}_{SL}$, that is the set of points that are also in $\mathcal{Y}_{SP}$. However, from Properties 2 and 3, we know that the remaining SL points are between the two adjacent SP points of Property 2 (in fact one is y^{max}), which are now known.

Phase 2. Let Y be the set of the points generated at Phase 1. To each point y of Y is associated an area in the objective space containing all the points that are not L-dominated by y (called the *L-non-dominance zone* of y). The intersection of all the non-dominance zones of the points in Y defines the search zone to be explored at Phase 2. Two illustrations of search zone are given in Fig. 2 in the objective space. The black squares represent the SP points generated in Phase 1. The search zone is represented by the triangles and polygon drawn between two adjacent SP points in the figures. Suppose that the set $Y = \{y^1, y^2, \ldots, y^t\}$ is ordered w.r.t. L_2, that is for any $i = 1, \ldots, t-1$ we have $L_2(y^i) \leq L_2(y^{i+1})$. By Property 1, the $t-1$ points $y^1, \ldots, y^{t-1}$ are all either above the bisector, or below. Suppose, w.l.o.g that they are below the bisector (as in left part of Fig. 2). The point y^t is thus the only point in Y that could be above the bisector. Besides, note that y^t is also the only point in Y that may be L-dominated. Since two points y^i and y^{i+1} ($i < t-2$) in Y are adjacent, any L-non-dominated point between these two points in the objective space relies inside the triangle defined by the three points y^i, y^{i+1} and $p^i = (L_1(y^i), L_2(y^{i+1}) - L_1(y^i))$. Let Z^i denote such a triangle. When the points y^{t-1} and y^t are on opposite sides of the bisector, any L-non-dominated point between these two points in the objective space relies inside a zone that could be a triangle or a particular polygon, as the polygon in Fig. 2 (right part). Let Z^{t-1} denote this zone (triangle or polygon). When the points y^{t-1} and y^t are on the same side of the bisector, the zone Z^{t-1} is

Table 1. Type A

Size			CPU(s)		
$l \times v$	#P	#L	Rkg	SP	DP
20×20	90.8	2.9	0	0.01	0.08
20×50	101.1	2.6	0.01	0.06	0.46
20×100	91.4	2.5	0.05	0.46	1.17
30×20	171.4	3.2	0	0.02	0.29
30×50	161.4	3.1	0.01	0.14	1.23
30×100	147.7	2.3	0.09	0.73	5.12
50×20	349.8	4.7	0	0.04	1.19
50×50	307.5	2.6	0.04	0.44	6.75
50×100	276.7	3.1	0.18	1.45	22.6

Table 2. Type B

Size			CPU(s)		
$l \times v$	#P	#L	Rkg	SP	DP
20×20	11.4	4.9	1.28	0.01	0
20×50	17.1	9.7	/	0.1	0.04
20×100	20	13.2	/	0.73	0.15
30×20	20.9	11.7	/	0.02	0.01
30×50	28.7	17.5	/	0.29	0.13
30×100	30.3	20.7	/	1.63	0.38
50×20	42.1	21.5	/	0.14	0.06
50×50	60.4	36.55	/	1.23	0.34
50×100	55.7	35.7	/	4.65	1.21

simply a triangle defined as the other triangles Z^i ($i < t - 2$). The search zone in the objective space to be explored is therefore defined by the $t - 2$ triangles Z^i ($i < t - 2$) and the polygon Z^{t-1}.

The exploration of each of the zones Z^i ($i < t$) consists in enumerating solutions with respect to the weighted sum of their costs with the weighting vector w^i defined such that points y^i and y^{i+1} have the same weighted sum. The enumeration is performed with a k-best algorithm (for an unknown k) and can be stopped as soon as the value of the weighted sum of a point is greater than the following upper bound: $U^i = \max_{y^j, y^{j+1} \in Y^i} w_1^i p_1^j + w_2^i p_2^j = \max_{y^j, y^{j+1} \in Y^i} (w_2^i - w_1^i) L_1(y^j) + w_1^i L_2(y^{j+1})$, where y^j and y^{j+1} are consecutive in the ordered set Y^i of the already detected points in Z^i ordered w.r.t. L_2. In the case where the bisector crosses the zone Z^{t-1} to be explored, one can easily show that a valid upper bound for Z^{t-1} is: $U^{t-1} = \max \{ \max_{y^j, y^{j+1} \in Y_1} (w_2^{t-1} - w_1^{t-1}) L_1(y^j) + w_1^{t-1} L_2(y^{j+1}), L_1(y^k) \}$, where y^k is the point that minimizes L_1 in Y^{t-1}. Once all the L-non-dominance zones Z^i have been explored, the method SP can stop, all the L-non-dominated points have been detected.

5   Experimental Results

We have applied the method SP to the biobjective shortest path problem (BOSPP) and to the biobjective set covering problem (BOSCP). The experiments have been run on an Intel Xeon CPU E5-2430 at 2.20 GHz for BOSPP, and on an Intel Core i7-3820 CPU at 3.60 GHz for BOSCP.

Biobjective Shortest Path Problem. Given a digraph $G = (V, E)$, where V is a set of vertices and $E \subseteq V \times V$ is a set of arcs, and two nodes s and t in V, one looks for the L-non-dominated points, images of feasible shortest paths from s to t in G. The value of each arc e in E is given by a vector of two costs: (c_1^e, c_2^e).

It is assumed that all the costs c_i^e are non-negative. In our experiments, we solve BOSPP on *layered digraphs* with randomly generated costs. A layered digraph is a graph in which the set V is partitioned into l subsets, called *layers*, $L_1, L_2, \ldots, L_l$, such that all arcs of E are between consecutive layers. To this *layered* digraph we add a vertex s and $|L_1|$ arcs from s to every vertex of L_1, and a vertex t and $|L_k|$ arcs from every vertex of L_l to t. Therefore, any path from s to t in such a graph is made of $l + 1$ arcs exactly. In our instances, all the layers have the same size and there is an arc (v, v') from any vertex v of a layer L_i to any vertex v' of the layer L_{i+1}. For each instance size, 4 different kinds of objectives, A, B, C and UN, are defined. The costs of the instances of type A, B and C are drawn uniformly at random from $[1, 100]$, and the costs of type B are positively correlated and those of type C are negatively correlated (see the instances of type A, B and C proposed by Bazgan *et al.* [23]). In addition, we consider a new type UN where the first cost is randomly generated, and the second cost is also randomly generated, but with a normal distribution. We use the following normal distribution: the mean is 50 and the variance is 20. We compare the running times of the method SP to the running times of two other methods proposed by Perny *et al.* [3] for Lorenz optimization for the multi-objective shortest path problem: the ranking method, called Rkg, and an extension of dynamic programming to Lorenz optimization, called DP (see Sect. 2.3). For method Rkg and for the Phase 2 of method SP, a k-best algorithm is needed. We have used a modified version of Eppstein's algorithm [24] proposed by Jiménez and Marzal [25]. The results obtained are summarized in Tables 1, 2, 3 and 4. At each row of the tables is given the size of the graph $l \times v$ where l is the number of layers and v the number of vertices per layer, the average number of P-non-dominated points (#P) and L-non-dominated points (#L), and the average CPU running times in seconds of the three tested methods on 20 instances of the same type and the same size. Note that the number of P-non-dominated points is computed by applying a multi-objective dynamic programming algorithm (see e.g. the algorithms of Stewart and White [26], and Mandow and Pérez-de-la-Cruz [27]). The running time of this algorithm is not indicated in the tables, but it is always greater than the running time of method DP. The symbol '/' in the tables means that the average running time is more than 15 min. The results show that the efficiency of method SP compared to the two other methods depends on the number of P-non-dominated points (#P) and on the proportion of L-non-dominated points. For instances A, the costs are rather balanced and #L is small compared to #P. Method SP is quite efficient on these instances but method Rkg is a bit faster. It means that the ranking method has to enumerate only a small number of feasible solutions. This comes from the fact that the costs are well-balanced. Method DP is clearly less efficient on these instances. The instances B are also balanced, but #P is significantly smaller than for instances A. This means that a lot of feasible solutions are P-dominated, and as the costs of all the feasible solutions are quite close, the method Rkg is particularly inefficient on these instances, it can only solve very small instances, whereas the method SP is quite efficient. The exploration of several zones instead of one is thus much more appropriate on these instances. Note that the method DP is very efficient when

Table 3. Type C

Size			CPU(s)		
$l \times v$	#P	#L	Rkg	SP	DP
20 × 10	1053.8	3.5	0.01	0.02	2.23
20 × 20	1806.5	3.75	0.03	0.01	23
20 × 50	/	3114	120.84	18.65	/
30 × 10	2005.1	4.6	0.05	0.01	11.52
30 × 20	3930.5	5	0.2	0.05	165.75
30 × 30	/	42.6	6.64	0.26	/
50 × 10	5140.9	3.6	0.24	0.02	215.5
50 × 20	/	7.15	1.46	0.1	/
50 × 30	/	5119.7	/	137.16	/

Table 4. Type UN

Size			CPU(s)		
$l \times v$	#P	#L	Rkg	SP	DP
20 × 20	99.8	8.6	0.47	0.02	0.1
20 × 50	109.3	6.1	0.08	0.09	0.59
20 × 100	106.8	6.2	0.1	0.53	1.91
30 × 20	171.1	15.2	110.7	0.04	0.18
30 × 50	184.9	10.4	8.27	0.19	1.41
30 × 100	172.1	8.7	2.24	0.94	6.18
50 × 20	376.4	28	/	0.09	1.43
50 × 50	362.2	20.2	/	0.83	8.92
50 × 100	346.2	13.2	108.1	2.23	28.6

#P and #L are small (even if the proportion of L-non-dominated points is high). Instances C and UN are unbalanced, and we can observe in Tables 3 and 4 that the method SP is clearly the most efficient on these instances. It comes from the fact that this method explores the objective space by taking into account the form of the convex hull of the images of the solutions, contrary to method Rkg. When the convex hull of the images of the solutions is not symmetric with respect to the bisector, the method SP reveals thus to be much more suitable. Besides, one can observe that the method DP is penalized by an important number of P-non-dominated points. It is particularly inefficient on instances C for which the number of P-non-dominated points is large.

Biobjective Set Covering Problem. We have a set of m rows (or items), and each row can be covered by a subset of n columns (or sets), each column j has two costs c_l^j ($l = 1, 2$). A feasible solution to the BOSCP is a subset of columns, among the n columns ($j = 1, \ldots, n$) such that all the rows are covered by at least one column. Our aim is to find the L-non-dominated points. We have used all the instances generated by Gandibleux et al. [28], from the size 100×10 (100 columns, 10 rows) to the size 1000×200 (1000 columns, 200 rows).

Table 5. Type A

#	#P	#L	CPU(s) Rkg	SP	Pareto
11	39	1	**0.019**	0.15	8.64
41	107	1	**0.026**	0.074	18.01
42	208	4	**0.072**	0.085	35.83
43	46	3	**0.15**	0.30	12.80
61	257	6	**0.76**	0.81	83.66
62	98	2	**0.27**	0.99	58.10
81	424	4	**0.20**	0.57	148.66
82	132	3	**0.33**	0.83	116.51
101	157	1	**0.59**	1.25	375.84
102	83	1	**0.37**	1.04	104.76
201	274	2	**9.20**	27.15	6850

Table 6. Type B

#	#P	#L	CPU(s) Rkg	SP	Pareto
11	43	3	**0.017**	0.047	6.26
41	108	2	**0.049**	0.051	16.63
42	276	2	0.24	**0.15**	52.04
43	28	1	**0.023**	0.18	7.51
61	338	2	**0.17**	0.38	114.01
62	99	1	**0.082**	0.58	60.20
81	354	4	**0.25**	2.09	130.26
82	88	2	**0.06**	0.26	38.16
101	141	5	**1.52**	2.20	225.50
102	86	1	**0.29**	1.62	211.48
201	282	6	**19.61**	22.07	4278

Table 7. Type UN-A

#	#L	CPU(s) Rkg	SP
11	2	**0.016**	0.15
41	3	**0.062**	0.13
42	5	0.26	**0.23**
43	3	**0.17**	0.71
61	5	2.39	**0.40**
62	3	1.24	**0.83**
81	1	**0.067**	0.27
82	3	3.95	**1.30**
101	3	7.70	**6.35**
102	4	2.06	3.53
201	5	209.07	**9.40**

Table 8. Type UN-B

#	#L	CPU(s) Rkg	SP
11	1	**0.014**	0.036
41	3	**0.067**	0.18
42	3	**0.059**	0.11
43	6	1.20	**0.63**
61	4	**0.34**	0.70
62	3	**0.73**	0.84
81	4	20.71	**0.41**
82	3	**1.40**	2.41
101	2	**0.61**	1.79
102	2	**0.87**	2.59
201	10	1085.22	**87.7**

For each size instance, two different kinds of objectives A, B are defined. In the case of instances of type A, the costs of each objective are randomly generated with a uniform distribution. For the type B, the costs of the first objective is also randomly generated with a uniform distribution and the ones of the second objective are made dependent in the following way: $c_2^j = c_1^{n-j+1}, \forall j = 1, \ldots, n$. As done with BOSPP, we also consider the new type UN with the following normal distribution: the mean is equal to the mean value of the first cost and the variance is equal to half the mean. Two types of instance are considered: type UN-A (first cost corresponds to the first cost of type A instance) and type UN-B (first cost corresponds to the first cost of type B instance). We compare the running times of the method SP with the running times of the ranking method Rkg. We also give, for the type A and type B instances, the number of P-non-dominated points ($\#P$) and the CPU time needed for generating these points. The results are from the method of Florios and Mavrotas [29], based on the ε-constraint method (on an Intel core 2 quad CPU at 3.00 GHz). In both methods, a k-best algorithm is necessary to enumerate the k-best solutions. Contrary to the shortest path problem, to our knowledge, no k-best algorithm has been previously developed. We have thus used the commercial CPLEX solver and implemented a procedure based on incumbent callback with solution injection (we inject in the search tree the current best solutions generated to get the next best solution), in order to enumerate the k-best solutions. The results are given in Tables 5, 6, 7 and 8. For each type of instance, we report the name of the instance (same name as used by Gandibleux et al.), $\#P$ (when available), $\#L$, and the CPU times in seconds of the tested methods. From Tables 5 and 6, we see that the $\#L$ represent only a small part of $\#P$: there are only between 1 and 6 L-non-dominated points. The running time of both methods are small and quite lower than the running time of the Pareto generation. For these instances, it is thus particularly interesting to apply a method dedicated to the generation of L-non-dominated points. However, the running time of the method Rkg is almost always slightly lower than the new method SP.

From Tables 7 and 8, the running times of both methods Rkg and SP are comparable for most of the instances, except for the last instance 201 (with 1000 columns and 200 rows). For the type UN-A (resp. UN-B), Rkg needs 209.07s (resp. 1085.22s) while SP only needs 9.40s (resp. 87.7s). We see thus that, as soon as the two objectives are unbalanced, the CPU time needed by method Rkg can be very high compared to method SP.

6   Conclusion

We have proposed new properties and new generic methods to generate Lorenz-optimal solutions to biobjective combinatorial optimization problems. The method has been evaluated experimentally on the biobjective shortest path problem and the biobjective set covering problem and showed good results compared to state-of-the-art methods, especially for "unbalanced instances". This work is dedicated to the efficient adaptation of the classic two-phase method to Lorenz optimization. Future work could be to efficiently adapt other classic methods proposed for Pareto optimization to Lorenz optimization. Besides, studying the location of the optimal points in the objective space is also a good starting point for developing efficient methods to generate the L-efficient solutions to MOCO problems, where the number of objectives is not limited to 2.

References

1. Kostreva, M.M., Ogryczak, W.: Linear optimization with multiple equitable criteria. RAIRO - Operations Research 33 (7 1999) 275–297
2. Kostreva, M., Ogryczak, W., Wierzbicki, A.: Equitable aggregations and multiple criteria analysis. Eur. J. Oper. Res. **158**(2), 362–377 (2004)
3. Perny, P., Spanjaard, O., Storme, L.X.: A decision-theoretic approach to robust optimization in multivalued graphs. Annals OR **147**(1), 317–341 (2006)
4. Yu, P.: Cone convexity, cone extreme points, and nondominated solutions in decision problems with multiobjectives. J. Optim. Theory Appl. **14**(3), 319–377 (1974)
5. Baatar, D., Wiecek, M.: Advancing equitability in multiobjective programming. Computational &. Applied Mathematics **52**(1–2), 225–234 (2006)
6. Moghaddam, A., Yalaoui, F., Amodeo, L.: Lorenz versus Pareto dominance in a single machine scheduling problem with rejection. In: Takahashi, R., Deb, K., Wanner, E., Greco, S. (eds.) Evolutionary Multi-Criterion Optimization. Lecture Notes in Computer Science, vol. 6576, pp. 520–534. Springer, Berlin Heidelberg (2011)
7. Endriss, U.: Reduction of economic inequality in combinatorial domains. In: AAMAS. (2013) 175–182
8. Ulungu, E., Teghem, J.: The two-phases method: An efficient procedure to solve biobjective combinatorial optimization problems. Foundation of Computing and Decision Science **20**, 149–156 (1995)
9. Visée, M., Teghem, J., Pirlot, M., Ulungu, E.: Two-phases method and branch and bound procedures to solve the bi-objective knapsack problem. J. Global Optim. **12**, 139–155 (1998)

10. Ehrgott, M., Skriver, A.: Solving biobjective combinatorial max-ordering problems by ranking methods and a two-phases approach. Eur. J. Oper. Res. **147**(3), 657–664 (2003)
11. Przybylski, A., Gandibleux, X., Ehrgott, M.: Two-phase algorithms for the biobjective assignement problem. Eur. J. Oper. Res. **185**(2), 509–533 (2008)
12. Raith, A., Ehrgott, M.: A two-phase algorithm for the biobjective integer minimum cost flow problem. Computers &. Oper. Res. **36**(6), 1945–1954 (2009)
13. Hardy, G., Littlewood, J., Pólya, G.: Inequalities. Cambridge University Press, Cambridge Mathematical Library (1952)
14. Shorrocks, A.F.: Ranking income distributions. Economica **50**(197), 3–17 (1983)
15. Ehrgott, M.: Multicriteria Optimization, 2nd edn. Springer, Berlin (2005)
16. Perny, P., Weng, P., Goldsmith, J., Hanna, J.: Approximation of Lorenz-Optimal Solutions in Multiobjective Markov Decision Processes. In: Conference on Uncertainty in Artificial Intelligence. (2013)
17. Perny, P., Spanjaard, O.: An Axiomatic Approach to Robustness in Search Problems with Multiple Scenarios. In: Proceedings of the 19th conference on Uncertainty in Artificial Intelligence. (2003) 469–476
18. Laumanns, M., Thiele, L., Zitzler, E.: An adaptive scheme to generate the Pareto front based on the epsilon-constraint method. In Branke, J., Deb, K., Miettinen, K., Steuer, R., eds.: Practical Approaches to Multi-Objective Optimization. Number 04461 in Dagstuhl Seminar Proceedings (2005)
19. Cohon, J.: Multiobjective Programming and Planning. Academic Press, New York (1978)
20. Aneja, Y., Nair, K.: Bicriteria transportation problem. Manage. Sci. **25**, 73–78 (1979)
21. Yager, R.: On ordered weighted averaging aggregation operators in multicriteria decision making. In: IEEE Trans. Systems, Man and Cybern. Volume 18. (1998) 183–190
22. Ogryczak, W.: Inequality measures and equitable approaches to location problems. Eur. J. Oper. Res. **122**(2), 374–391 (2000)
23. Bazgan, C., Hugot, H., Vanderpooten, D.: Solving efficiently the 0–1 multiobjective knapsack problem. Computers &. Oper. Res. **36**(1), 260–279 (2009)
24. Eppstein, D.: Finding the k shortest paths. SIAM J. Computing **28**(2), 652–673 (1998)
25. Jiménez, V.M., Marzal, A.: A lazy version of Eppstein's k shortest paths algorithm. In: Proceedings of the 2Nd International Conference on Experimental and Efficient Algorithms. WEA'03, Berlin, Heidelberg, Springer-Verlag (2003) 179–191
26. Stewart, B.S., White III, C.C.: Multiobjective A*. J. ACM **38**(4), 775–814 (1991). October
27. Mandow, L., Pérez-De-la Cruz, J.L.: A new approach to multiobjective A* search. In: Proceedings of the 19th International Joint Conference on Artificial Intelligence. IJCAI'05, San Francisco, CA, USA, Morgan Kaufmann Publishers Inc. (2005) 218–223
28. Gandibleux, X., Vancoppenolle, D., Tuyttens, D.: A first making use of GRASP for solving MOCO problems. In: 14th International Conference in Multiple Criteria Decision-Making, Charlottesville (1998)
29. Florios, K., Mavrotas, G.: Generation of the exact Pareto set in multi-objective traveling salesman and set covering problems. Appl. Math. Comput. **237**, 1–19 (2014)

On Possibly Optimal Tradeoffs in Multicriteria Spanning Tree Problems

Nawal Benabbou and Patrice Perny[✉]

CNRS, LIP6 UMR 7606, Sorbonne Universités, UPMC Univ Paris 06,
4 Place Jussieu, 75005 Paris, France
{nawal.benabbou,patrice.perny}@lip6.fr

Abstract. In this paper, we propose an interactive approach to determine a compromise solution in the multicriteria spanning tree problem. We assume that the Decision Maker's preferences over spanning trees can be represented by a weighted sum of criteria but that weights are imprecisely known. In the first part of the paper, we propose a generalization of Prim's algorithm to determine the set of possibly optimal tradeoffs. In the second part, we propose an incremental weight elicitation method to reduce the set of feasible weights so as to identify a necessary optimal tradeoff. Numerical tests are given to demonstrate the practical feasibility of the approach.

Keywords: Multicriteria optimization · Spanning tree problem · Criteria weight elicitation · Minimax regret · Possibly optimal solutions

1 Introduction

The multicriteria spanning tree (MCST) problem is an unexpectedly difficult problem investigated in the field of multicriteria optimization. It appears naturally in various situations, for example, when a set of clients must be connected through a communication or transportation network. In such a case one may look for a minimum spanning tree, minimum referring simultaneously to different aspects such as prices, construction time or distance. In the single objective case, the minimum spanning tree problem is known as an easy problem that can be solved in polynomial time using standard greedy algorithms due to Kruskal [14] and Prim [17]. The usual generalization of this problem in the setting of multicriteria optimization concerns graph instances whose edges are valued by cost vectors in $\mathbb{R}^n$, n being the number of criteria (cost functions) to be minimized. The MCST problem consists in determining all Pareto-optimal tradeoffs (cost vectors), and for each of them, one associated spanning tree.

Unfortunately, as soon as $n \geq 2$, the problem becomes intractable: the number of Pareto-optimal cost vectors associated to spanning trees is in the worst case exponential in the number nodes. There exists indeed instances of graphs in which all spanning trees lead to a distinct Pareto-optimal cost vector [12]. Since a complete graph with v vertices includes v^{v-2} distinct spanning trees [7], it is clearly not possible to list them all in polynomial time. No efficient algorithm is known to solve standard instances of the MCST problem except

© Springer International Publishing Switzerland 2015
T. Walsh (Ed.): ADT 2015, LNAI 9346, pp. 322–337, 2015.
DOI: 10.1007/978-3-319-23114-3_20

perhaps in the bi-objective case where some specificities can be exploited to propose practically efficient procedures [22] (see also [20] for a survey). This suggests resorting to more discriminating models than Pareto-dominance, so as to better discriminate between feasible spanning trees. In this paper, assuming that Decision Maker's preferences can be represented by a weighted sum of criteria, we study the interactive elicitation of criteria weights, so as to progressively enrich the initial Pareto-dominance relation and to allow a fast determination of the most attractive feasible tradeoffs for the Decision Maker. Although the weighted sum is a quite simple model, it remains indeed difficult to define a priori the weights of criteria, especially when criterion values are expressed on different scales such as time, distances or money. This explains the interest for weight elicitation methods in the context of multiobjective optimization.

Incremental elicitation methods are now standardly used in various decision problems, for example to elicit von Neumann-Morgenstern utilities [4,8,23], multiattribute utilities [3,5,18], and individual preferences in voting [13,15,16]. They allow a fast identification of an optimal or near-optimal solution without requiring a full elicitation of the decision model. Incremental elicitation is often seen as a problem of robust decision making under preference uncertainty. Some preference information is lacking, uncertain or not known precisely, and we want to find a solution that is likely to remain attractive after any further specification of preference information.

An even more challenging issue is to implement preference elicitation mechanisms on combinatorial optimization problems. In such problems, the set of feasible solutions is not given explicitly and we have to elicit preferences over elementary components of solutions so as to efficiently determine the possibly optimal solutions or, if possible, identify a necessary optimal solution. Several recent studies investigate this line, let us mention, for example, the elicitation strategies proposed for soft constraint satisfaction problems with missing preferences [11], the incremental elicitation of preferences over policies in Markov Decision Processes [19,24], the incremental elicitation of agent preferences in stable matching problems [9], and the incremental elicitation of criteria weights in multobjective state-space search [1,2]. To go further in this direction, we address here the weight elicitation problem in multicriteria spanning tree problems. Instead of trying to enumerate all Pareto-optimal tradeoffs corresponding to spanning trees, we want to design an interactive search procedure collecting preference information so as to progressively narrow the set of possibly optimal spanning trees until a recommendation can be made with some guarantee. To this end, we propose to interweave an incremental weight elictation procedure and a multiobjective greedy search algorithm.

The paper is organized as follows: Sect. 2 recalls some background and notations concerning the MCST problem. Then assuming the set of admissible criteria weights is characterized by a convex polyhedron, we propose an exact algorithm to determine the set of possibly optimal spanning trees in a MCST problem in Sect. 3. In Sect. 4, a new interactive approach combining an incremental weight elicitation procedure and a greedy search is proposed for the fast determination of a spanning tree achieving an optimal tradeoff. Finally numerical tests are presented in Sect. 5 to illustrate the efficiency of the proposed approach.

2 Background and Notations

We consider a connected graph $G = (V, E)$ where each edge $e \in E$ is valued by a cost vector $x_e \in \mathbb{R}_+^n$ giving the cost of e with respect to different criteria. The set of criteria is denoted $N = \{1, \ldots, n\}$, and every criterion is assumed to be additive over the edges. Thus, to any set of edges $E' \subseteq E$ is associated a cost vector $x_{E'} = \sum_{e \in E'} x_e$. A *spanning tree* of G is a connected subgraph of G which contains no cycle while including every node of G. For the sake of simplicity, since a spanning tree T of G is completely characterized by its set of edges, then T will indifferently denote the tree or its set of edges. The set of cost vectors associated with the spanning trees of G is denoted X_G and represents the image of all solutions in the space of criteria. Throughout the paper, we will use the following dominance relations:

- *Weak Pareto dominance:* $y \precsim x \Leftrightarrow \forall i \in N, y_i \leq x_i$
- *Pareto dominance:* $y \prec x \Leftrightarrow [\forall i \in N, y_i \leq x_i]$ and $[\exists k \in N, y_k < x_k]$

The MCST problem consists in determining the set $\mathrm{ND}(X_G) = \{x \in X_G : \forall y \in X_G, \mathrm{not}(y \prec x)\}$ of non-dominated vectors in X_G, also known as the Pareto set. In this paper, we assume that the DM's preferences can be represented by a linear function $f_\omega(x) = \sum_{i \in N} \omega_i x_i$ measuring the overall cost of spanning trees. Initially, parameter ω is not known. Instead we consider a set Ω, named uncertainty set, containing all admissible normalized weighting vectors. The set Ω is initially defined as the simplex $\Omega_0 = \{\omega \in \mathrm{int}(\mathbb{R}_+^n) : \sum_{i=1}^n \omega_i = 1\}$ where int represents the interior of the cone. Whenever some preference statements of type "x is at least as good as y" are expressed by the DM, they induce linear constraints of type $\sum_{i \in N} \omega_i(x_i - y_i) \leq 0$ over the set Ω of admissible weights. Hence, throughout the paper, we assume without loss of generality that Ω is a convex polyhedron. Given the uncertainty set Ω, we first consider the problem of determining the set $\mathrm{PO}_\Omega(X_G)$ of possibly f_ω-optimal cost vectors in X_G, i.e. the set of cost vectors $x \in X_G$ that minimize f_ω for some $\omega \in \Omega$. Formally:

$$\mathrm{PO}_\Omega(X_G) = \bigcup_{\omega \in \Omega} \arg \min_{x \in X_G} f_\omega(x)$$

Since ω has strictly positive components, $\mathrm{PO}_{\Omega_0}(X_G)$ is included in $\mathrm{ND}(X_G)$. Moreover, $\mathrm{PO}_{\Omega'}(X_G) \subseteq \mathrm{PO}_\Omega(X_G)$ for any $\Omega' \subseteq \Omega$. Therefore, any new preference statement reduces the set of possibly optimal spanning trees.

3 Determination of Possibly Optimal Spanning Trees

In this section, we propose a multiobjective extension of Prim's algorithm [17] enabling to compute the set $\mathrm{PO}_\Omega(X_G)$ for a given convex polyhedron Ω. Recall that Prim's algorithm is a greedy search. Starting from an initial node v_0, we first select an adjacent edge of minimum cost. Then, Prim's algorithm iteratively selects a min-cost edge in the cocycle $C(V')$, where V' is the set of nodes covered so far. Recall that the cocycle of a set V' is the set of all edges adjacent to V', i.e., $C(V') = \{(v_1, v_2) \in E : v_1 \in V', v_2 \in V \backslash V'\}$. Similarly, we propose here a greedy search that determines, at any step, the set of possibly optimal edges in the current cocycle $C(V')$ and, for each edge, the associated set of weights. The determination of possibly optimal elements can be performed by a polynomial algorithm named Ω-Filter (see Algorithm 1), enabling to computing possibly optimal vectors on explicit sets [2]. Note that in Algorithm 1 the test appearing in line 3 can easily be performed using a LP-solver. It is indeed sufficient to solve a LP with $n + 1$ variables and $m + q$ constraints, where m is the size of the input set and q is the number of available preference statements.

Algorithm 1. Ω-Filter(Y)

    Input: Ω; $Y = \{y^1, \ldots, y^m\}$;
    Output: Y_m: the set $PO_\Omega(Y)$
1 $Y_0 \leftarrow Y$
2 **for** $i = 1 \ldots m$ **do**
3   **if** $\max\limits_{\omega \in \Omega} \min\limits_{y \in Y_{i-1}} [f_\omega(y) - f_\omega(y^i)] < 0$ **then**
4 | $Y_i \leftarrow Y_{i-1} \backslash \{y^i\}$
5   **else**
6 | $Y_i \leftarrow Y_{i-1}$
7   **end**
8 **end**
9 **return** Y_m

Now, we propose Algorithm 2 to compute the set of possibly optimal tradeoffs. The result is obtained by recursive calls to Ω-Prim, from the initial call Ω-Prim$(\{v_0\}, \emptyset, \Omega)$ where $v_0 \in V$ is an arbitrary starting node. In this recursive procedure, the set POE (line 5) represents the set of possibly optimal edges in the cocycle $C(V')$ and SELECT is a procedure that returns one edge per vector. The set SOL consists of ordered pairs (E', Ω'), where E' is a spanning tree of G and Ω' is the associated set of weights. Note that the depth of the search tree associated to recursive calls is $|V| - 1$ since one edge is added to E' at every call.

Let us first prove that the output set SOL obtained by the call to Ω-Prim with the input $(\{v_0\}, \emptyset, \Omega)$ includes at least one ordered pair for every possibly optimal tradeoff.

Theorem 1. *For all possibly optimal tradeoff* $x \in PO_\Omega(X_G)$, *there exists an ordered pair* (E', Ω') *in the output set SOL such that* $x_{E'} = x$.

Proof. Let x be a possibly optimal tradeoff. By definition, there exists $\omega \in \Omega$ such that $x \in \arg\min_{x' \in X_G} f_\omega(x')$. Therefore, there exists an execution of Prim's

---

Algorithm 2. Ω-Prim(V', E', Ω')

---

1 **if** $V' = V$ **then**
2 $\quad$ SOL $\leftarrow \{(E', \Omega')\}$;
3 **else**
4 $\quad$ SOL $\leftarrow \emptyset$
5 $\quad$ POE $\leftarrow$ SELECT(Ω-Filter(ND($\{x_{e'} : e' \in C(V')\}$)))
6 $\quad$ **for** $e \in POE$ **do**
7 $\quad\quad$ $E'' \leftarrow E' \cup \{e\}$
8 $\quad\quad$ $V'' \leftarrow V' \cup \{v_e\}$, v_e being the endpoint of e such that $v_e \notin V'$
9 $\quad\quad$ $\Omega'' \leftarrow \{\omega \in \Omega' : \forall e' \in$ POE, $f_\omega(x_e) \leq f_\omega(x_{e'})\}$
10 $\quad\quad$ SOL $\leftarrow$ SOL $\cup$ Ω-Prim(V'', E'', Ω'')
11 $\quad$ **end**
12 **end**
13 **return** SOL

---

algorithm over the graph $G = (V, E)$ endowed with the scalar valuation $f_\omega(x_e)$, $e \in E$, that leads to tradeoff x. Let $x^i, i \in \{1, \ldots, |V| - 1\}$, be the sequence of cost vectors that would be selected by Prim's algorithm to obtain x from the initial node v_0. We want to prove that Algorithm 2 returns the tradeoff x.

At the first recursive call, the tuple $(\{v_0\}, \emptyset, \Omega)$ is considered. According to line 5, one edge per possibly optimal cost vector in the cocycle $C(\{v_0\})$ is selected to grow the tree (which is empty at that point). Note that there exists an edge of cost x^1 in the cocycle $C(\{v_0\})$ since x^1 would be selected by Prim's algorithm from the initial node v_0. Moreover, since $\omega \in \Omega$ and x^1 minimizes f_ω over $\{x_e : e \in C(\{v_0\})\}$ by definition, we know that x^1 is possibly optimal in the cocycle $C(\{v_0\})$. Therefore, there exists an edge e^1 of cost x^1 that is selected to grow the tree and, according to line 10, Ω-Prim is then called on the input (V'', E'', Ω'') where $E'' = \emptyset \cup \{e_1\} = \{e_1\}$ (line 7). Note that we necessarily have $\omega \in \Omega''$ according to line 9. Therefore, we can similarly prove that, during the latter call to Ω-Prim, an edge e^2 of cost x^2 is selected to grow the tree $\{e^1\}$. Hence, by iterating, we obtain a sequence of calls to Ω-Prim leading to the construction of a tree $T = \{e^1, \ldots, e^{|V|-1}\}$ such that $x_{e^i} = x^i$ for all $i \in \{1, \ldots, |V| - 1\}$, and so we have $x_T = \sum_{i=1}^{|V|-1} x^i = x$. Finally, tree T and the associated set of weights are returned at the end of the last call (according to line 2). $\square$

Conversely, we prove now that all ordered pairs returned by Ω-Prim$(\{v_0\}, \emptyset, \Omega)$ correspond to possibly optimal cost vectors in graph G.

Theorem 2. *For all ordered pairs (E', Ω') in the output set SOL, the spanning tree E' is such that $x_{E'} \in PO_\Omega(X_G)$.*

Proof. Let (E', Ω') be in the output set SOL. We want to prove that $x_{E'} \in PO_\Omega(X_G)$. Let Ω-Prim(V_i, E_i, Ω_i), $i \in \{0, \ldots, |V| - 1\}$, be the sequence of calls leading to (E', Ω'). Note that we have $(V_0, E_0, \Omega_0) = (\{v_0\}, \emptyset, \Omega)$ and

$(V_{|V|-1}, E_{|V|-1}, \Omega_{|V|-1}) = (V', E', \Omega')$. For all $i \in \{1, \ldots, |V| - 1\}$, let e_i denote the edge of E' such that $E_i = E_{i-1} \cup \{e_i\}$ (line 7). Then, we have $\Omega_i = \{\omega \in \Omega_{i-1} : e_i \in \arg\min_{e \in C(V_{i-1})} f_\omega(x_e)\}$ (line 9). Therefore, for all $i \in \{0, \ldots, |V| - 2\}$, we have $\Omega_i \supseteq \Omega_{i+1}$, and so $\Omega_i \supseteq \Omega_{|V|-1}$. Moreover, since Ω-Prim has been called with the input (V_i, E_i, Ω_i), we know that e_i is possibly optimal in $C(V_{i-1})$ given the uncertainty set Ω_{i-1}. Therefore, $\Omega_i \neq \emptyset$ for all $i \in \{1, \ldots, |V| - 1\}$, and in particular $\Omega_{|V|-1} \neq \emptyset$. Let $\omega \in \Omega_{|V|-1} \subseteq \Omega_i$ for all $i \in \{1, \ldots, |V| - 1\}$. Then, for all $i \in \{1, \ldots, |V| - 1\}$, edge e_i minimizes the function f_ω over $C(V_{i-1})$. Therefore, considering the graph G endowed with the valuation $f_\omega(x_e)$ for all edges $e \in E$, there exists an execution of Prim's algorithm that yealds the spanning tree E'. Hence $x_{E'} \in \arg\min_{x \in X_G} f_\omega(x)$. □

The two previous results show that the set of tradeoffs present in the ordered pairs returned by Ω-Prim$(\{v_0\}, \emptyset, \Omega)$ corresponds exactly to the set of $PO_\Omega(X_G)$. Moreover, the first components of these ordered pairs provide one spanning tree for every possibly optimal tradeoff. Finally, the sets Ω' appearing in the ordered pairs define the optimality regions of the returned spanning trees, and by contruction, their union covers the entire uncertainty set Ω.

For the sake of illustration, we present the result of Ω-Prim$(\{v_0\}, \emptyset, \Omega)$ on a randomly generated instance of bi-criteria MCST problem with 10 nodes and cost vectors in $[0, 1000]^2$.

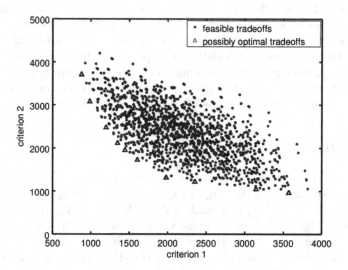

Fig. 1. X_G and $PO_\Omega(X_G)$ for $\Omega = [0, 1]$

Here, for any tradeoff $x \in X_G \subset \mathbb{R}^2$, $f_\omega(x) = \omega x_1 + (1 - \omega)x_2$, $\omega \in [0, 1]$. The elements of X_G are represented by points in Fig. 1 and the output of Ω-Prim$(\{v_0\}, \emptyset, \Omega)$ for $\Omega = [0, 1]$ is represented by triangles. We obtained 10 distinct elements in $PO_\Omega(X_G)$. Then, Fig. 2 represents the output of Ω-Prim$(\{v_0\}, \emptyset, \Omega)$ for $\Omega = [0.5, 0.7]$. We can see that the set of possibly optimal

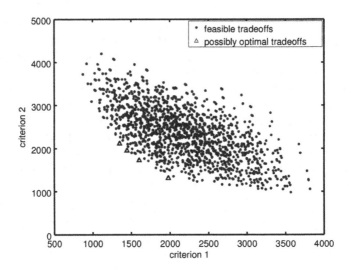

Fig. 2. X_G and $\mathrm{PO}_\Omega(X_G)$ for $\Omega = [0.5, 0.7]$

tradeoffs is reduced to four elements. To obtain a single possibly optimal tradeoff, further restrictions of Ω are needed.

Let us recall that, when Ω is the entire simplex, there exist complete bi-valued graphs G of any size such that $\mathrm{PO}_\Omega(X_G) = X_G$, X_G being the set of all feasible tradeoffs obtained from all spanning trees of G [12]. For such instances, the output set of Algorithm 2 is exponential in the number of vertices and Algorithm 2 has exponential running times. This shows the potential interest of reducing Ω during the search, using preference elicitation methods, so as to progressively narrow the set of possibly optimal tradeoffs.

4   Incremental Weight Elicitation

For any given uncertainty set Ω, Algorithm 2 introduced in the previous section enables the determination of $\mathrm{PO}_\Omega(X_G)$, the set of possibly optimal tradeoffs for functions $f_\omega(x) = \sum_{i \in N} \omega_i x_i$ with $\omega \in \Omega$. Whenever Algorithm 2 returns a single tradeoff (the unique optimal cost vector), the associated spanning tree is a necessary optimal solution to the problem, even if other trees having the same cost vector may exist. Necessary means here that this is the optimal tradeoff for all functions $f_\omega, \omega \in \Omega$. Alternatively, when the set $\mathrm{PO}_\Omega(X_G)$ includes several tradeoffs, one may be interested in collecting additional preference statements so as to reduce Ω and therefore $\mathrm{PO}_\Omega(X_G)$. This suggests to interleave calls to Algorithm 2 and preference queries restricting Ω so as to repeatedly compute $\mathrm{PO}_\Omega(X_G)$ for a nested sequence of Ω until obtaining a singleton. Such a procedure would not be efficient because some parts of the job would be made several times. To implement this idea more efficiently, we propose in this section a more integrated approach interweaving a preference elicitation procedure with the greedy search presented in Sect. 3, so as to allow a faster determination of an optimal or near-optimal spanning tree.

We consider here a standard incremental elicitation procedure, where preference queries are asked one at a time to be as informative as possible. The strategy for the selection of the most appropriate query is based on the Minimax Regret criterion, as proposed in [4,23]. According to this decision criterion, the most promising cost vector is characterized by the following definitions of regrets, for all $x, y \in X_G$:

Definition 1. *Pairwise Max Regret:* $\text{PMR}(x, y, \Omega) = \max_{\omega \in \Omega} \{f_\omega(x) - f_\omega(y)\}$

Max Regret: $\text{MR}(x, X_G, \Omega) = \max_{y \in X_G} \text{PMR}(x, y, \Omega)$

Minimax Regret: $\text{MMR}(X_G, \Omega) = \min_{x \in X_G} \text{MR}(x, X_G, \Omega)$

$\text{MR}(x, X_G, \Omega)$ is the worst-case regret of choosing tradeoff x instead of any y in X_G. According to the minimax regret criterion, the optimal cost vectors are those minimizing MR, i.e. those achieving the MMR value. Choosing a MR-optimal vector allows one to guarantee that the worst-case loss is minimized. Given the uncertainty set Ω, the worst-case loss measured by MMR might still be too large for certifying the quality of the solution. Therefore, the Minimax Regret criterion can be used to select the most effective queries to reduce the MMR-value (the answers will further restrict the set Ω). Ideally, we would like to obtain MMR $= 0$, which corresponds to the identification of a necessarily optimal tradeoff. However, to reduce the elicitation burden, it is more efficient to use an admissibility threshold $\lambda > 0$ representing the maximum admissible gap to optimality and to stop asking queries as soon as MMR drops below λ.

Hence, the *Current Solution Strategy* [4,23] consists in generating the following preference query: the DM is asked to compare two potentially good cost vectors in X_G: an MR-optimal vector $x^* \in \arg\min_{x \in X_G} \text{MR}(x, X_G, \Omega)$ and another vector y^* in $\arg\max_{y \in X_G} \text{PMR}(x^*, y, \Omega)$. The set Ω is then reduced by inserting the linear constraint induced by the answer ($f_\omega(x^*) - f_\omega(y^*) \geq 0$ or $f_\omega(x^*) - f_\omega(y^*) \leq 0$), so as to keep consistency with DM's preferences. The CSS is based on the repeated computation of $\text{PMR}(x, y, \Omega)$ for many pairs (x, y) of feasible solutions (all in the worst-case), which may induce prohibitive computation times in our context due to the size of X_G. Therefore, instead of computing X_G and then applying this elicitation scheme, we propose to integrate the CSS to Algorithm 2. Within Algorithm 2, we suggest collecting preference information to discriminate between the edges of the current cocycle and to reduce the set Ω accordingly. We implement this idea by computing minimax regrets on cocycles and asking preference queries according to the CSS. More precisely, at step i (selection of ith edge), preference queries are generated until the MMR drops under a threshold λ_i, where λ_i is a fraction of the admissibility threshold λ such that $\sum_{i \in N} \lambda_i = \lambda$. More precisely, we propose the following algorithm:

Note that the number of iterations of the while loop (line 6) is bounded above by $|C(V')| \leq n^2$, n being the number of criteria, which guarantees the

---

Algorithm 3

---

1 $V' \leftarrow \{v_0\}$
2 $E' \leftarrow \emptyset$
3 **for** $i = 1 \ldots |V| - 1$ **do**
4 $\quad$ ND $\leftarrow \{e \in C(V') : \forall e' \in C(V'), \text{not}(x_{e'} \prec x_e)\}$
5 $\quad X \leftarrow \{x_e : e \in \text{ND}\}$
6 $\quad$ **while** $MMR(X, \Omega) > \lambda_i$ **do**
7 $\quad\quad$ Ask one preference query to the DM according to the CSS
8 $\quad\quad$ Update Ω by inserting the linear constraint associated to the answer
9 $\quad$ **end**
10 $\quad$ Select $e \in \text{ND}$ such that $x_e \in \arg\min_{x \in X} \text{MR}(x, X, \Omega)$
11 $\quad E' \leftarrow E' \cup \{e\}$
12 $\quad V' \leftarrow V' \cup \{v_e\}$, where v_e is the endpoint of e that is not in V'
13 **end**
14 **return** E'

---

termination of the algorithm after a finite number of steps. Hence the following theorem shows that Algorithm 3 returns a spanning tree T with a MR value smaller or equal to λ.

Theorem 3. *Let Ω_f be the final set Ω when Algorithm 3 stops. The returned spanning tree T is such that $MR(x_T, X_G, \Omega_f) \leq \lambda$.*

Proof. Let T be the spanning tree returned by Algorithm 3 and Ω_f be the final set Ω when Algorithm 3 stops. We want to prove that $\text{MR}(x_T, X_G, \Omega_f) \leq \lambda$. This amounts to proving that, for any spanning tree T_0 of graph G, we have $\text{PMR}(x_T, x_{T_0}, \Omega_f) \leq \lambda$.

Let e_i, $i \in \{1, \ldots, |V| - 1\}$ denote the ith edge inserted in T and let Ω_i denote the uncertainty set Ω at the end of ith iteration step. Let j be the first iteration step such that $e_j \notin T_0$. Since T_0 is a spanning tree of G, then there exists a chain c in T_0 linking the two endpoints of e_j. Since edge e_j links one node in V_{j-1} to one node in $V \backslash V_{j-1}$, then there exists an edge e'_j in the chain c that links one node in V_{j-1} to one node in $V \backslash V_{j-1}$. Therefore $e'_j \in C(V_{j-1})$. Two cases may occur: either $e'_j \in \text{ND}_j$ or $e'_j \notin \text{ND}_j$ where ND_j is the set non-dominated edges in the cocycle $C(V_{j-1})$ (line 4). Let us prove that, in both cases, we have $\text{PMR}(x_{e_j}, x_{e'_j}, \Omega_f) \leq \lambda_j$. Assume first that $e'_j \in \text{ND}_j$. In that case, since edge e_j has been selected to build the spanning tree T, then we know that $x_{e_j} \in \arg\min_{x \in X_j} \text{MR}(x, X_j, \mathcal{P}_j)$ (line 10) where $X_j = \{x_e : e \in \text{ND}_j\}$. Moreover, since $\text{MMR}(X_j, \Omega_j) \leq \lambda_j$ at the end of the while loop (line 6), then we necessarily have $\text{PMR}(x_{e_j}, x_{e'_j}, \Omega_j) \leq \lambda_j$. Since $\Omega_j \subseteq \Omega_f$ by construction (line 8), then we have $\text{PMR}(x_{e_j}, x_{e'_j}, \Omega_f) \leq \lambda_j$. Assume now that $e'_j \notin \text{ND}_j$. In that case, there exists $e \in \text{ND}_j$ such that $x_e \prec x_{e'_j}$ by construction of ND_j. Since the function f_ω is increasing with the Pareto-dominance, then $f_\omega(x_e) \leq f_\omega(x_{e'_j})$ for any set of weights ω. Therefore $\text{PMR}(x_{e_j}, x_e, \Omega_f) \geq \text{PMR}(x_{e_j}, x_{e'_j}, \Omega_f)$. Then, similarly to the first case, we can prove that we have $\text{PMR}(x_{e_j}, x_e, \Omega_f) \leq \lambda_j$ since $x_e \in \text{ND}_j$, and so we have $\text{PMR}(x_{e_j}, x_{e'_j}, \Omega_f) \leq \lambda_j$.

Now, consider the spanning tree T_1 obtained from T_0 by replacing edge e'_j by edge e_j. Let e_k be the first edge inserted in T that is not in T_1. Then, similarly to T_0, we can prove that we have $\text{PMR}(x_{e_k}, x_{e'_k}, \Omega_f) \leq \lambda_k$. Therefore, by iterating, we have $\text{PMR}(x_{e_l}, x_{e'_l}, \Omega_f) \leq \lambda_l$ for all $l \in \{1, \dots, |V| - 1\}$ such that $e_l \notin T_0$. Moreover, for all $l \in \{1, \dots, |V| - 1\}$ such that $e_l \in T_0$, we have $\text{PMR}(x_{e_l}, x_{e_l}, \Omega_f) = 0 \leq \lambda_l$. Consider $\pi : T \to T_0$ the one-to-one correspondence such that $\pi(e_l) = e'_l$ if $e_l \notin T_0$ and $\pi(e_l) = e_l$ otherwise. Hence we have $\text{PMR}(x_{e_l}, x_{\pi(e_l)}, \Omega_f) \leq \lambda_l$ for all $l \in L = \{1, \dots, |V| - 1\}$. Finally:

$$
\begin{aligned}
\text{PMR}(x_T, x_{T_0}, \Omega_f) &= \max_{\omega \in \Omega_f} \{ f_\omega(x_T) - f_\omega(x_{T_0}) \} \\
&= \max_{\omega \in \Omega_f} \{ f_\omega(\sum_{l \in L} x_{e_l}) - f_\omega(\sum_{l \in L} x_{\pi(e_l)}) \} \\
&= \max_{\omega \in \Omega_f} \sum_{l \in L} [f_\omega(x_{e_l}) - f_\omega(x_{\pi(e_l)})] \text{ by linearity of } f_\omega \\
&\leq \sum_{l \in L} \max_{\omega \in \Omega_f} [f_\omega(x_{e_l}) - f_\omega(x_{\pi(e_l)})] \\
&= \sum_{l \in L} \text{PMR}(x_{e_l}, x_{\pi(e_l)}, \Omega_f) \\
&\leq \sum_{l \in L} \lambda_l \\
&= \lambda
\end{aligned}
$$

Hence $\text{PMR}(x_T, x_{T_0}, \Omega_f) \leq \lambda$ which establishes the result. $\qquad\square$

The following example presents an execution of Algorithm 3 on a small instance of $G = (V, E)$.

Example 1. *Consider the randomly generated instance of $G = (V, E)$ with four nodes and three criteria, given in Fig. 3, and assume that the DM's preferences are represented by a weighted sum with the hidden weight $\omega = (0.2, 0.5, 0.3)$. Initially, Ω is set to the simplex represented by the triangle ABC in the space (ω_1, ω_2), ω_3 being implicitly defined by $1 - \omega_1 - \omega_2$ (see Fig. 4). The list of selected edges E' is initialized to the empty set.*

We start with node $v_0 = 0$ and $\lambda = 0$. At the first iteration step, the cocycle consists of the edges $(0, 1)$, $(0, 2)$ and $(0, 3)$. The non-dominated edges are $(0, 2)$ and $(0, 3)$ with cost vectors $(1, 4, 8)$ and $(9, 1, 2)$ respectively. At this step, the procedure asks the DM to compare these two vectors, and the DM declares that she prefers $(9, 1, 2)$ to $(1, 4, 8)$ (since $f_\omega((9, 1, 2)) = 2.9$ whereas $f_\omega((1, 4, 8)) = 4.6$). Then, the algorithm updates the uncertainty set Ω by inserting the linear constraint represented by the (EH) line in Fig. 4, the admissible area being at the left of this line. Finally, the edge $(0, 3)$ is inserted in E'.

At the second iteration step, the cocycle contains the edges $(0, 1)$, $(0, 2)$, $(3, 1)$ and $(3, 2)$. The non-dominated edges are $(0, 2)$ and $(3, 1)$ with cost vectors $(1, 4, 8)$ and $(4, 3, 6)$ respectively. No question is needed here since the minimax regret over $X = \{(1, 4, 8), (4, 3, 6)\}$ does not exceed $\lambda_2 = 0$. The edge $(3, 1)$ with cost $(4, 3, 6)$ is inserted in E'.

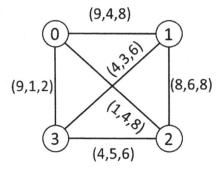

Fig. 3. An instance with 3 criteria.

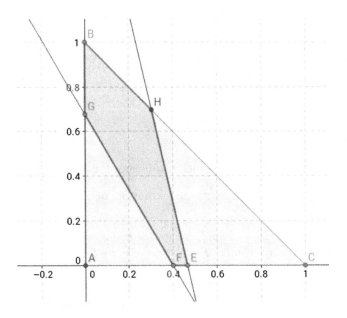

Fig. 4. Ω seen in the space (ω_1, ω_2).

At the third iteration step, the cocycle contains $(0, 2)$, $(1, 2)$ *and* $(3, 2)$. *The non-dominated edges are* $(0, 2)$ *and* $(3, 2)$ *with cost vectors* $(1, 4, 8)$ *and* $(4, 5, 6)$ *respectively. The procedure asks the DM to compare these two vectors, and the DM declares that she prefers* $(1, 4, 8)$ *to* $(4, 5, 6)$ *(since* $f_\omega((4, 5, 6)) = 5.1$ *whereas* $f_\omega((1, 4, 8)) = 4.6$). *Then, the algorithm updates the uncertainty set* Ω *by inserting the linear constraint represented by the* (FG) *line in Fig. 4, the admissible area being at the right of this line. Finally, the edge* $(0, 2)$ *is inserted in* E'.

Then the algorithm stops with the optimal tree $E' = \{(0, 3), (3, 1), (0, 2)\}$. *It can easily be checked on Fig. 4 that the final* Ω *set defined by the polygon* $EFGBH$ *contains the point* $(0.2, 0.5)$ *corresponding to the hidden vector* $\omega = (0.2, 0.5, 0.3)$. *No weighting vector in the polygon can induce a regret greater than* λ.

In Example 1, the algorithms terminates with the tradeoff $(14, 8, 16)$ associated to the tree $\{(0, 3), (3, 1), (0, 2)\}$. This solution is on the boundary of the convex hull of the set of feasible tradeoffs since it minimizes the weighted sum of criteria for $\omega = (0.2, 0.5, 0.3)$. This is always the case when $\lambda = 0$. When $\lambda > 0$ Algorithm 3 may produce an unsupported tradeoff, i.e. a Pareto-optimal tradeoff belonging to the interior of the convex hull of the feasible set. This is due to the use of the minimax regret, as shown in the following:

Example 2. *Let G be a graph with 3 vertices $\{a, b, c\}$ and 3 arcs $(a, b), (a, c), (b, c)$ of cost $(1, 5), (5, 1), (2, 2)$ respectively. We have 3 spanning trees leading to the 3 following tradeoffs: $(3, 7), (7, 3), (6, 6)$. If Ω is the entire simplex, we see that the optimal solution according ot the MMR criterion is $(6, 6)$ with a max regret equal to 3 while $(3, 7)$ and $(7, 3)$ have a max regret equal to 4. Note that $(6, 6)$ is an unsupported solution. It can indeed be easily checked that there exists no positive weighting vector such that $(6, 6)$ is simultaneously better than $(3, 7)$ and $(7, 3)$ with respect to a weighted sum (to be minimized). Nevertheless, vector $(6, 6)$ is a good compromise solution that may be well justified considering the weight uncertainty.*

It is worth noting that the use of linear aggregation functions with imprecise weights may lead to recommend unsupported Pareto-optimal tradeoffs (see Example 2). This enhances the recommendation possibilities offered by the weighted sum. Moreover, when a supported solution is wanted, the elicitation process can be continued until a necessary winner is identified (as in Example 1).

5   Numerical Tests

We performed numerical experiments of Algorithms 2 and 3 on random instances of the MCST problem. The numerical tests were performed on a Intel Core i7-4770 CPU with 15 GB de RAM.

For Algorithm 2, we considered graphs with 10 nodes and edge densities ranging from 50 % to 100 %. Recall that a complete graph with 10 nodes includes 10^8 spanning trees. In our experiments, the number of criteria n varies from 2 to 5 and the costs are drawn within $\{1, 1000\}$. Moreover, in order to study the impact of the size of the uncertainty set Ω, we ran Algorithm 2 with different sets Ω obtained as follows: we first consider that Ω is the entire simplex. Then, a second series of tests is performed after inserting a preference statement and the associated linear constraint to restrict Ω. Finally, a third series of tests is performed after adding a second preference statement. We will denote q the number of preference statements used to restrict the set Ω. Linear optimizations required by Ω-Filter are performed using the Gurobi library of Java. The results are reported in Table 1 resulting from an average over 30 runs.

We observe that the computation times decrease with the size of Ω but increase with the number of criteria. The algorithm runs in less than 12 min for all instances including complete graphs with 5 criteria. Notice that, when Ω is the entire simplex, our algorithm computes quite efficiently the set of all supported solutions, i.e.

those Pareto-optimal solutions that belong to the boundary of the convex hull of the feasible tradeoffs. They are often considered as a useful basis to further exploration of the Pareto set in the so-called two-phase methods (see e.g. [10]).

Table 1. Computation times (in seconds) for Algorithm 2 with $|N| = 10$.

n	Density	$q = 0$	$q = 1$	$q = 2$
2	50 %	0.15	0.12	0.09
3	50 %	0.89	0.78	0.75
4	50 %	10.70	9.82	8.03
5	50 %	17.79	17.40	15.73
2	75 %	0.42	0.36	0.27
3	75 %	8.60	5.41	3.16
4	75 %	17.40	16.23	14.41
5	75 %	267.93	219.25	168.83
2	100 %	0.54	0.46	0.28
3	100 %	12.95	10.35	9.34
4	100 %	91.48	75.45	69.55
5	100 %	668.21	634.26	621.46

In a second series of experiments, we have simulated preference elicitation sessions using Algorithm 3. Simulated DMs answer to queries according to a linear scalarizing function f_ω where ω is randomly chosen in $\Omega_0 = \{int(\mathbb{R}^n_+) : \sum_{i \in N} \omega_i = 1\}$. The use of preference information significantly improves computation times compared to Algorithm 2, which makes it possible to solve larger instances. Hence, to test Algorithm 3, we consider graphs with a number of nodes ranging from 25 to 100, a number of criteria n ranging from 2 to 8 and an edge density of 50 %.

To study the impact of threshold λ on the number of queries and the computation time, we ran tests with two distinct values: $\lambda = 0.05$ and $\lambda = 0.1$. The values λ_i required at every step of the algorithm are set to $\lambda_i = \lambda/(|V| - 1)$ for all $i \in \{1, \ldots, |V| - 1\}$. Here also we used the Gurobi-solver to compute PMR values at every step. Moreover, we used standard pruning rules for min aggregators to compute the MMR more efficiently [6]. We observed empirically that, due to these pruning rules, the number of PMR computations performed is linear instead of quadratic in the size of the input set. Results have been obtained by averaging over 30 runs and are summarized in Table 2.

We observe that our interactive procedure is quite efficient considering the highly combinatorial nature of the spanning tree problem and the number of criteria under consideration (recall that the state of the art literature on the computation of Pareto-optimal spanning trees considers no more than two or three criteria). The relative efficiency of our procedure is due to the possibility of collecting some preference information during the search. It considerably speeds

Table 2. Performance of Algorithm 3 (times in seconds, queries).

		$\lambda = 0.05$		$\lambda = 0.1$	
n	$|V|$	Time	Queries	Time	Queries
2	25	0.9	3.8	1.1	3.4
2	50	2.0	3.6	2.9	2.8
2	75	5.4	4.4	7.0	3.4
2	100	8.3	4.2	10.7	2.8
4	25	4.6	16.2	7.2	11.8
4	50	23.5	17.0	42.6	12.6
4	75	75.8	19.8	116.5	12.2
4	100	127.1	20.4	227.7	12.1
6	25	12.9	29.2	13.6	15.4
6	50	81.1	32.0	123.2	22.8
6	75	256.8	30.8	498.1	25.0
6	100	750.3	33.0	1182.4	25.4
8	25	32.2	49.2	40.3	34.6
8	50	234.6	62.0	370.8	35.2
8	75	731.0	65.6	1105.8	35.8
8	100	2518.9	71.4	3389.0	40.8

up the determination of a relevant Pareto-optimal tradeoff. For example, we need only twelve preference queries on average to determine the optimal tradeoff and a corresponding spanning tree on instances with 100 nodes and 4 criteria (with $\lambda = 0.1$).

We also observe that as the value of λ increases (from 0.5 to 0.1), the requirement on the performance guarantee is weakened and consequently the number of queries is reduced. However, reducing the number of queries tends to keep a larger uncertainty set during the execution which impacts negatively on computation times, as can be observed in the table.

6   Conclusion

We have proposed and experimented a procedure (Algorithm 2) to determine possibly optimal spanning trees when preferences are assumed to be representable by a linear aggregation of criteria but weights are imprecisely known. This procedure which is a generalization of Prim's algorithm uses cocycles of partial trees to decompose the set of admissible weights into regions characterizing all possibly optimal tradeoffs. It can be used in particular to determine the set of supported Pareto-optimal tradeoffs but also some subset of it when some preference information is available. Note that Algorithm 2 yields all possibly optimal tradeoffs. Another line could be to focus only on extreme points of the convex hull of feasible tradeoffs. A recent work on the multiobjective spanning tree problem establishes that there are only polynomially many of these extreme points [21][1]. This result could probably be used to design new efficient weight elicitation procedures.

In the second part of the paper, Algorithm 2 has been sophisticated to obtain an interactive greedy search procedure for the MCST problem (Algorithm 3). This approach interweaves incremental elicitation and search and enables a fast determination of an optimal solution or near-optimal (with performance guarantee). This combination of elicitation and combinatorial optimization algorithms could be compared to the approach proposed in [2] for multiobjective state space search. There is however a significant difference since the state space search proposed in [2] relies on dynamic programming algorithms whereas the procedure proposed here relies on a greedy approach. An interesting extension of this work would be to relax the assumption of linearity for f_ω. The use of a non-linear scalarizing function may indeed offer the possibility to determine new possibly optimal tradeoffs in the Pareto set. This seems to be a challenging issue because our multiobjective greedy search is no longer valid for non-linear aggregation functions.

Acknowledgments. This work is part of the ELICIT project supported by the French National Research Agency through the Idex Sorbonne Universités under grant ANR-11-IDEX-0004-02.

References

1. Benabbou, N., Perny, P.: Combining preference elicitation and search in multiobjective state-space graphs. In: Proceedings of IJCAI 2015 (2015)
2. Benabbou, N., Perny, P.: Incremental weight elicitation for multiobjective state space search. In: Proceedings of AAAI 2015, pp. 1093–1099 (2015)
3. Benabbou, N., Perny, P., Viappiani, P.: Incremental elicitation of choquet capacities for multicriteria decision making. In: Proceedings of ECAI 2014, pp. 87–92 (2014)
4. Boutilier, C., Patrascu, R., Poupart, P., Schuurmans, D.: Constraint-based optimization and utility elicitation using the minimax decision criterion. Artif. Intell. **170**(8–9), 686–713 (2006)
5. Braziunas, D., Boutilier, C.: Minimax regret based elicitation of generalized additive utilities. In: Proceedings of UAI 2007, pp. 25–32 (2007)
6. Braziunas, D.: Decision-theoretic elicitation of generalized additive utilities. Ph.D. thesis, University of Toronto (2011)
7. Cayley, A.: A theorem on trees. Q. J. Math. **23**, 376–378 (1889)
8. Chajewska, U., Koller, D., Parr, R.: Making rational decisions using adaptive utility elicitation. In: Proceedings of AAAI 2000, pp. 363–369 (2000)
9. Drummond, J., Boutilier, C.: Preference elicitation and interview minimization in stable matchings. In: Proceedings of AAAI 2014, pp. 645–653 (2014)
10. Ehrgott, M.: Multicriteria Optimization. Springer, Heidelberg (2006)

---

[1] We wish to thank an anonymous reviewer for pointing out this reference.

11. Gelain, M., Pini, M.S., Rossi, F., Venable, K.B., Walsh, T.: Elicitation strategies for soft constraint problems with missing preferences: properties, algorithms and experimental studies. Artif. Intell. J. **174**(3–4), 270–294 (2010)
12. Hamacher, H.W., Ruhe, G.: On spanning tree problems with multiple objectives. Ann. Oper. Res. **52**, 209–230 (1994)
13. Kalech, M., Kraus, S., Kaminka, G.A.: Practical voting rules with partial information. Auton. Agent. Multi-Agent Syst. **22**(1), 151–182 (2010)
14. Kruskal, J.B.: On the shortest spanning subtree of a graph and the traveling salesman problem. Proc. Am. Math. Soc. **7**, 48–50 (1956)
15. Lu, T., Boutilier, C.: Robust approximation and incremental elicitation in voting protocols. In: Proceedings of IJCAI 2011, pp. 287–293 (2011)
16. Naamani-Dery, L., Kalech, M., Rokach, L., Shapira, B.: Reaching a joint decision with minimal elicitation of voter preferences. Inf. Sci. **278**, 466–487 (2014)
17. Prim, R.C.: Shortest connection networks and some generalizations. Bell Syst. Tech. J. **36**, 1389–1401 (1957)
18. White III, C.C., Sage, A.P., Dozono, S.: A model of multiattribute decisionmaking and trade-off weight determination under uncertainty. IEEE Trans. Syst. Man Cybern. **14**(2), 223–229 (1984)
19. Regan, K., Boutilier, C.: Eliciting additive reward functions for markov decision processes. In: Proceedings of IJCAI 2011, pp. 2159–2164 (2011)
20. Ruzika, S., Hamacher, H.W.: A survey on multiple objective minimum spanning tree problems. In: Lerner, J., Wagner, D., Zweig, K.A. (eds.) Algorithmics of Large and Complex Networks, pp. 104–116. Springer, Heidelberg (2009)
21. Seipp, F.: On adjacency, cardinality, and partial dominance in discrete multiple objective optimization. Ph.D. thesis, Technischen Universitat Kaiserslautern (2013)
22. Sourd, F., Spanjaard, O.: A multiobjective branch-and-bound framework: application to the biobjective spanning tree problem. INFORMS J. Comput. **20**(3), 472–484 (2008)
23. Wang, T., Boutilier, C.: Incremental utility elicitation with the minimax regret decision criterion. In: Proceedings of IJCAI 2003, pp. 309–316 (2003)
24. Weng, P., Zanuttini, B.: Interactive value iteration for Markov decision processes with unknown rewards. In: Proceedings of IJCAI 2013, pp. 2415–2421 (2013)

Argumentation

Verification in Attack-Incomplete Argumentation Frameworks

Dorothea Baumeister, Daniel Neugebauer[✉], and Jörg Rothe

Heinrich-Heine-Universität Düsseldorf, 40225 Düsseldorf, Germany
{baumeister,neugebauer,rothe}@cs.uni-duesseldorf.de

Abstract. We tackle the problem of expressing incomplete knowledge about the attack relation in abstract argumentation frameworks. In applications, incomplete argumentation frameworks may arise as intermediate states in an elicitation process, when merging different beliefs about an argumentation framework's state, or in cases where the complete information cannot be fully obtained. To this end, we employ a model introduced by Cayrol et al. [10] and analyze the question of whether certain justification criteria are possibly (or necessarily) fulfilled, i.e., whether they are fulfilled in some (or in every) completion of the incomplete argumentation framework. We formally extend the definition of existing criteria to these incomplete argumentation frameworks and provide characterization and complexity results for variants of the verification problem.

1 Introduction

Argumentation frameworks are used to model discussions and deliberations among agents, be it human beings or software agents. The aim is to find sets of arguments that can be considered "justified" by satisfying certain properties. In a pathbreaking paper, Dung [17] introduced a formal model to describe argumentation frameworks and their semantics, which abstracts from the content of arguments and regards their interaction only. More background on abstract argumentation in artificial intelligence can be found in the book by Rahwan and Simari [30].

We revisit a generalized model for abstract argumentation frameworks originally proposed by Cayrol et al. [10] who extend the classical model to an attack-incomplete setting. In *attack-incomplete argumentation frameworks*, all arguments are known, but the set of all possible attacks between them is partitioned into attacks that are either known to definitely exist, or known to definitely never exist, or currently unknown to exist but that may potentially arise in the future. We study central properties and semantics of argumentation frameworks, such as conflict-freeness, admissibility, stability, preferredness, completeness, and groundedness [17], which we extend to the attack-incomplete setting by asking whether they are *possibly* or *necessarily* fulfilled. As our technical contribution, we provide characterization and complexity results for variants of the standard verification problem in attack-incomplete argumentation frameworks.

© Springer International Publishing Switzerland 2015
T. Walsh (Ed.): ADT 2015, LNAI 9346, pp. 341–358, 2015.
DOI: 10.1007/978-3-319-23114-3_21

Related Work and Motivation: Our work is motivated by the "Online Participation" project, an interdisciplinary graduate college of HHU Düsseldorf and
other institutions[1] in which researchers from economics, communication theory,
political sciences, social sciences, law, and computer science are participating.
A central goal in this project is to build an internet platform that can be used
for online discussions and deliberations. While these—as mentioned above—can
be modeled abstractly by argumentation frameworks, a major drawback of the
classical model due to Dung [17] is that it assumes complete knowledge of the
arguments and the attack relation, that is, the *process* of arguing is assumed
to have been completed already. However, such complete information is rarely
available in practical applications; rather, one would like to model such an online
discussion *dynamically*, evolving over time.

First ideas regarding dynamic changes in argumentation frameworks applying the theory of belief revision are due to Cayrol et al. [11], who also survey
the literature on the dynamics of abstract argumentation frameworks [12]. They
limit themselves to the addition or deletion of one argument, together with a
respective change in the attack relation. Their work focuses on a classification of
how and why those changes can alter the set of extensions of the given argumentation framework. Boella et al. [6] define general principles for the abstraction
of arguments and attacks for the grounded semantics mainly. Liao et al. [26]
investigate the question of how one can efficiently compute the status of an
argument (i.e., whether it is accepted, rejected, or undecided) upon changing
the arguments and attacks. Coste-Marquis et al. [14] study how belief revision
postulates can be applied to argumentation systems.

Also, the concept of incomplete knowledge in abstract argumentation has
recently received some attention. In *probabilistic argumentation frameworks* (see,
for example, the work of Li et al. [25], Rienstra [31], Fazzinga et al. [19,20],
Hunter [22], and Doder and Woltran [16]), arguments and/or attacks have an
associated probability, which represents an agent's degree of belief that the argument or attack is in force, or their reluctance to disregard the argument or attack.
This can be considered as a quantified model of uncertainty that allows to derive
the probability of certain criteria to hold. Baumeister et al. [5] study a model of
argument-incomplete argumentation frameworks.

Cayrol et al. [10] propose argumentation frameworks with an additional
"ignorance relation" among arguments that contains the attacks for which there
is uncertainty. We adopt their extended framework model, but take a different perspective: In their work [10], new semantics for attack-incomplete argumentation frameworks are defined, which puts a lot of focus on the incomplete
framework itself, rather than on its completions. Opposed to that, we analyze
whether standard semantics apply in some (or all) completions of an incomplete
framework. This is a natural question arising when dealing with incomplete

---

[1] Besides four faculties of HHU Düsseldorf and the Fachhochschule für öffentliche Verwaltung NRW, the practice partners of this project include registered societies, limited
liability companies, and the municipal councils of Köln, Bonn, and Münster, among
others. We refer to the website http://www.fortschrittskolleg.de for more details.

knowledge and has already been considered for similar notions of uncertainty in various areas. In the related field of computational social choice (see, e.g., the book chapter by Brandt et al. [9]), and especially so in voting, classical complete-information settings have been extended to allow for incomplete information as well. The book chapters by Boutilier and Rosenschein [7] and Baumeister and Rothe [1] survey the known results on incomplete information and communication in voting, in particular covering the concepts of *possible* and *necessary* *winners* in elections that have been introduced by Konczak and Lang [23] and studied in terms of computational complexity both for the original problems (see, e.g., [23,34]) and for a number of variants, such as possible winners when new alternatives are added [13], when there is uncertainty about which voting rule is used [2], and when there is uncertainty about the voters' weights in weighted elections [3].[2] The notions of possible and necessary winners have also been transferred to other fields where information may be incomplete, including fair division [8], algorithmic game theory [24], and judgment aggregation [4].

In Sect. 2, we describe the classical model of abstract argumentation frameworks, and we provide the needed notions from complexity theory. In Sect. 3, we introduce attack-incomplete argumentation frameworks and in Sect. 4 we present our results. In Sect. 5, we give our conclusions and state some open questions.

2   Preliminaries

In this section, we introduce the classical argumentation framework model and the notation used in this paper and provide some basic notions of complexity theory. Our models are based on the seminal work of Dung [17] who introduced an abstract model for argumentation frameworks; while using his notions and concepts, we adopt some notation from the book chapter by Dunne and Wooldridge [18].

An *argumentation framework* is a pair $AF = \langle \mathscr{A}, \mathscr{R} \rangle$ that contains a set $\mathscr{A}$ of n *arguments* and a binary *attack relation* $\mathscr{R} \subseteq \mathscr{A} \times \mathscr{A}$ of up to n^2 pairs of arguments. We say that a *attacks* b if $(a, b) \in \mathscr{R}$. Given an argumentation framework $AF = \langle \mathscr{A}, \mathscr{R} \rangle$, the set of *attackers* of a set B of arguments is $\{a \in \mathscr{A} \mid \exists b \in B : (a, b) \in \mathscr{R}\}$. We say that a set D of arguments *defends* a set B of arguments if for each attacker a of B, there is an argument $d \in D$ with $(d, a) \in \mathscr{R}$. Accordingly, D does not defend B if there is an attacker of B that is not attacked by any $d \in D$.

Every argumentation framework can be illustrated as a directed graph $G = (V, E)$ by identifying $V = \mathscr{A}$ and $E = \mathscr{R}$ (see Example 1 and Fig. 1).

Example 1. A very basic argumentation framework is $AF = \langle \mathscr{A}, \mathscr{R} \rangle$ with the argument set $\mathscr{A} = \{a, b, c, d\}$ and the attacks $\mathscr{R} = \{(a, b), (a, c), (a, d), (b, d), (c, c), (d, a), (d, b)\}$ (see Fig. 1 for its graph representation).

---

[2] Other models of incomplete-information settings in voting include dynamic social choice with evolving preferences [29] and online manipulation in sequential elections [21].

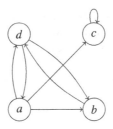

Fig. 1. Graph representation of the argumentation framework in Example 1

We now formally define the properties of sets of arguments in argumentation frameworks that were introduced in Dung's initial work and that are central to this paper:

1. The most basic property is conflict-freeness, which simply forbids attacks within a subset of the arguments. Formally, a subset $S \subseteq \mathscr{A}$ is *conflict-free* if there are no arguments a and b in S such that $(a, b) \in \mathscr{R}$.
2. An argument $a \in \mathscr{A}$ is *acceptable with respect to* $S \subseteq \mathscr{A}$ if S defends a, i.e., if for all $b \in \mathscr{A}$ with $(b, a) \in \mathscr{R}$, there is at least one $c \in S$ with $(c, b) \in \mathscr{R}$.
3. Further, a conflict-free set S of arguments is called *admissible* if every argument $a \in S$ is acceptable with respect to S.

Dung defines several *semantics* based on these properties in his original work, namely the *preferred, stable, complete,* and *grounded semantics*. Subsequent papers on argumentation frameworks proposed a variety of further semantics, but we will only be concerned with the semantics mentioned above.[3]

1. A set $S \subseteq \mathscr{A}$ is *preferred* if S is a maximal (with respect to set inclusion) admissible set.
2. A conflict-free set $S \subseteq \mathscr{A}$ is *stable* if it attacks all other arguments, i.e., if for every argument $b \in \mathscr{A} \setminus S$, there exists an $a \in S$ with $(a, b) \in \mathscr{R}$.
3. The *complete* semantics is defined via the *characteristic function* of an argumentation framework AF, which is $F_{AF} : 2^{\mathscr{A}} \to 2^{\mathscr{A}}$ with

$$F_{AF}(S) = \{a \in \mathscr{A} \mid a \text{ is acceptable with respect to } S\}.$$

The characteristic function is monotonic with respect to set inclusion and there is always an $i \in \mathbb{N}$ for which the i-fold composition of F_{AF} has a fixed point. A set $S \subseteq \mathscr{A}$ is *complete* if it is a fixed point of F_{AF}, or equivalently, if every $a \in \mathscr{A}$ that is acceptable with respect to S is contained in S.
4. The (unique) *grounded set* of an argumentation framework AF is the least (with respect to set inclusion) fixed point of F_{AF}, i.e., the complete set obtained when starting with the empty set.

---

[3] In addition to these semantics we also use conflict-freeness and admissibility as criteria. While these are generally not considered to be semantics, we will not always explicitly distinguish between semantics and basic properties for the sake of conciseness.

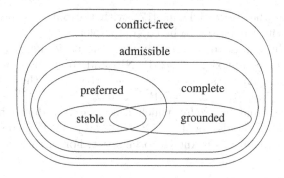

Fig. 2. Relations among the various criteria and semantics for sets of arguments

Dung investigates how these properties are correlated, and provides many results for this that we can make use of. Figure 2 displays all relations among the various criteria and semantics that we use. If an area labeled with criterion **s** is fully included in an area labeled with criterion **s'**, this indicates that in all argumentation frameworks all sets of arguments that fulfill **s** also fulfill **s'**. The converse is not necessarily true, i.e., all displayed set inclusions are strict. Further, none of the areas are disjoint, so one and the same set of arguments might fulfill all criteria/semantics simultaneously.

Given an argumentation framework AF and a semantics **s**, a set S of arguments that fulfills the conditions imposed by **s** in AF is also called an **s** *extension of* AF. If it is clear from the context, we omit stating explicitly the argumentation framework that the subset is an extension of.

Dunne and Wooldridge [18] give an overview of a number of decision problems, each defined for various semantics. We will focus on only one of them, namely, the verification problem.

---

s-VERIFICATION

Given: An argumentation framework $\langle \mathscr{A}, \mathscr{R} \rangle$ and a subset $S \subseteq \mathscr{A}$.

Question: Is S an **s** extension?

Here, the letter **s** is a placeholder for a specific semantics. For better readability, we will sometimes use CF as a shorthand for *conflict-freeness*, AD for *admissibility*, PR for *preferredness*, ST for *stability*, CP for *completeness*, and GR for *groundedness*.

Other previously considered decision problems are, for example, s-EXISTENCE, s-CREDULOUS-ACCEPTANCE, and s-SKEPTICAL-ACCEPTANCE (for a definition, see, for example, [18]). Many of these problems are hard to decide: They are complete for the complexity classes NP, coNP, or even Π_2^p. By contrast, s-VERIFICATION is easy for most semantics **s** studied here, which follows immediately from the work of Dung [17], with the only exception being PR-VERIFICATION, which is known to be coNP-complete [15].

We assume the reader to be familiar with the basic notions of complexity theory, such as the complexity classes P, NP, and coNP mentioned above and with the notions of hardness and completeness (based on the polynomial-time many-one reducibility, $\leq_m^p$). $\Sigma_2^p = \text{NP}^{\text{NP}}$ and $\Pi_2^p = \text{coNP}^{\text{NP}}$ are the second level of the polynomial hierarchy, which has been introduced by Meyer and Stockmeyer [27,33]. It holds that $\text{P} \subseteq \text{NP} \subseteq \Sigma_2^p \cup \Pi_2^p$ and $\text{P} \subseteq \text{coNP} \subseteq \Sigma_2^p \cup \Pi_2^p$, and none of these inclusions is known to be strict. For further details, see, e.g., [28,32].

3   Attack-Incomplete Argumentation Frameworks

We will now consider argumentation frameworks with incomplete knowledge about the attack relation, where a set of n arguments is fixed and only a subset of all n^2 possible attacks is known to either definitely exist or to definitely not exist—the state of the remaining attacks is currently unknown. We call this an *attack-incomplete argumentation framework*.

3.1   Model and Formal Definitions

An extension of standard argumentation frameworks to attack-incomplete argumentation frameworks was proposed by Cayrol et al. [10], which allows to distinguish between definite attacks, impossible attacks, and possible attacks. We apply their extended model using a slightly different notation.

Definition 1. *An* attack-incomplete argumentation framework *is a triple* $\langle \mathscr{A}, \mathscr{R}^+, \mathscr{R}^- \rangle$, *where* $\mathscr{A}$ *is a nonempty set of arguments and* $\mathscr{R}^+$ *and* $\mathscr{R}^-$ *are disjoint subsets of* $\mathscr{A} \times \mathscr{A}$. $\mathscr{R}^+$ *denotes the set of all ordered pairs of arguments between which an attack is known to definitely exist, while* $\mathscr{R}^-$ *denotes the set of all ordered pairs of arguments between which an attack is known to never exist. The set of possible attacks* $(\mathscr{A} \times \mathscr{A}) \smallsetminus (\mathscr{R}^+ \cup \mathscr{R}^-)$, *which is implicitly given through* $\mathscr{R}^+$ *and* $\mathscr{R}^-$, *is denoted as* $\mathscr{R}^?$.

Let $AF = \langle \mathscr{A}, \mathscr{R}^+, \mathscr{R}^- \rangle$ be a given attack-incomplete argumentation framework. An argumentation framework $AF^* = \langle \mathscr{A}, \mathscr{R}^* \rangle$ with $\mathscr{R}^+ \subseteq \mathscr{R}^* \subseteq \mathscr{R}^+ \cup \mathscr{R}^?$ is called a *completion of AF*. Every attack-incomplete argumentation framework obviously has $2^{\|\mathscr{R}^?\|}$ different completions. In particular, we call the completion that discards all possible attacks $(\mathscr{R}^* = \mathscr{R}^+)$ the *minimal completion of AF*, and the completion that includes all possible attacks $(\mathscr{R}^* = \mathscr{R}^+ \cup \mathscr{R}^?)$ is called the *maximal completion of AF*.

We now extend the notions for classical argumentation frameworks that we described in Sect. 2 to attack-incomplete argumentation frameworks, distinguishing between properties holding either *possibly* or *necessarily*. Generally, a property holds *possibly* for an attack-incomplete argumentation framework AF if

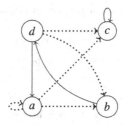

Fig. 3. An attack-incomplete argumentation framework

there *exists* a completion AF^* of AF for which the property holds, and a property holds *necessarily* if it holds *for all* completions of AF.[4] Note that if a property holds necessarily, it also holds possibly; even if $\mathscr{R}^? = \emptyset$, there exists $2^0 = 1$ completion, which happens to be both minimal and maximal in this case.

Example 2. Figure 3 gives the graph representation of an attack-incomplete argumentation framework $AF = \langle \mathscr{A}, \mathscr{R}^+, \mathscr{R}^- \rangle$ with the set $\mathscr{A} = \{a, b, c, d\}$ of arguments, where definite attacks $\mathscr{R}^+ = \{(b,d), (c,c), (d,a)\}$ are drawn as solid arcs, possible attacks $\mathscr{R}^? = \{(a,a), (a,b), (a,c), (d,b), (d,c)\}$ as dotted arcs, and attacks that are known to never exist (i.e., $\mathscr{R}^- = \{(a,d), (b,a), (b,b), (b,c), (c,a), (c,b), (c,d), (d,d)\}$) are not displayed.

For this example, it holds that the sets $\emptyset$, $\{b\}$, and $\{d\}$ are the only necessarily conflict-free extensions of AF, while the sets $\emptyset$, $\{a\}$, $\{b\}$, $\{d\}$, and $\{a,b\}$ are the only possibly conflict-free extensions of AF.

Deciding whether a given property holds possibly (respectively, necessarily) adds an existential (respectively, universal) quantifier over an exponential space to the standard problem, potentially making it intractable or increasing its level of intractability. However, some problems remain easy to solve. This is obvious, for example, for the possible and necessary attack between two given arguments: An argument $a \in \mathscr{A}$ is possibly attacked (respectively, necessarily attacked) by an argument $b \in \mathscr{A}$ if and only if $(b,a) \notin \mathscr{R}^-$ (respectively, $(b,a) \in \mathscr{R}^+$), which can clearly be verified in polynomial time. In Sect. 4, we will present our results on the complexity of deciding whether a set of arguments is a possible or necessary s-extension for all considered semantics s.

3.2 Comparison with the Model of Cayrol et al. [10]

Although we use the notion of attack-incomplete argumentation framework due to Cayrol et al. [10] (called *Partial Argumentation Framework (PAF)* in their

---

[4] Unlike the concepts of *credulous* and *skeptical* acceptance in the related literature, which denote membership of arguments in, respectively, *some* and *all extensions* of a specific argumentation framework, our notions of properties holding *possibly* and *necessarily* describe criteria holding in, respectively, *some* and *all argumentation frameworks* (i.e., *completions*), and are therefore settled one level of abstraction higher.

Table 1. Comparison of properties of sets of arguments

Property from [10]		Our property
S is R-conflict-free	$\Longleftrightarrow$	S is possibly conflict-free
S is RI-conflict-free	$\Longleftrightarrow$	S is necessarily conflict-free
a is R-acceptable w.r.t. S	$\Longleftrightarrow\!\!\!\!/$	a is possibly acceptable w.r.t. S
a is RI-acceptable w.r.t. S	$\Longleftrightarrow$	a is necessarily acceptable w.r.t. S
S is R-admissible	$\Longleftrightarrow\!\!\!\!/$	S is possibly admissible
S is RI-admissible	$\Longleftrightarrow$	S is necessarily admissible
S is R-preferred	$\Longleftrightarrow\!\!\!\!/$	S is possibly preferred
S is RI-preferred	$\Longleftrightarrow\!\!\!\!/$	S is possibly preferred

work), we do not take the same perspective on properties and semantics: While they define new semantics for the PAFs *themselves*, we will analyze whether the conditions of standard semantics are fulfilled in some or all *completions* of it. This avoids the strange case where an incomplete framework satisfies some property, despite none of its completions satisfying this property; for example, in the model by Cayrol et al. it may be the case that a set S of arguments is the only RI-preferred extension[5] of an attack-incomplete argumentation framework, even though it is not a preferred extension for any of the framework's completions.

While the formal conditions imposed by both approaches coincide in some cases, they are generally different. Table 1 gives an overview of all criteria and semantics introduced by Cayrol et al.[6] and their counterparts in our model, and indicates whether or not they are equivalent. A formal proof of why equivalence does or does not hold in each individual case is omitted due to space constraints.

4   Possible and Necessary Verification

The problem s-VERIFICATION for standard argumentation frameworks naturally yields two problems for attack-incomplete argumentation frameworks, s-ATT-INC-POSSIBLE-VERIFICATION and s-ATT-INC-NECESSARY-VERIFICATION, for each semantics s.

---

s-ATT-INC-POSSIBLE-VERIFICATION (s-ATTINCPV)

Given:	An attack-incomplete argumentation framework $AF = \langle \mathscr{A}, \mathscr{R}^+, \mathscr{R}^- \rangle$ and a set $S \subseteq \mathscr{A}$.
Question:	Is there a completion AF^* of AF such that S is an s extension in AF^*?

---

[5] A set of arguments is RI-*preferred* if it is maximal among all necessarily admissible sets, where $R \cong \mathscr{R}^+$ and $I \cong \mathscr{R}^?$ in our notation.

[6] For formal definitions of these criteria, see their work [10].

	s-ATT-INC-NECESSARY-VERIFICATION (s-ATTINCNV)
Given:	An attack-incomplete argumentation framework $AF = \langle \mathscr{A}, \mathscr{R}^+, \mathscr{R}^- \rangle$ and a set $S \subseteq \mathscr{A}$.
Question:	For all completions AF^* of AF, is S an s extension in AF^*?

As already mentioned, the original problem can be solved efficiently for the admissible, stable, complete, and grounded semantics. We prove that both new problems can still be solved efficiently for these semantics, even though the number of completions is exponential in the number of possible attacks. We define best-case and worst-case completions for the different semantics, a given attack-incomplete argumentation framework AF, and a given set S of arguments. Intuitively, a best-case completion includes all attacks that are beneficial for S with respect to the considered semantics, whereas a worst-case completion includes those attacks that harm the conditions imposed by the semantics. We prove that these completions are critical completions for the respective decision problem, i.e., the answer to the VERIFICATION variant corresponding to the attack-incomplete framework is the same as that to VERIFICATION for the respective completion.

4.1 Verifying Conflict-Freeness, Admissibility, and Stability

For conflict-freeness, admissibility, or stability of a set S of arguments, all attacks against elements of S are never beneficial and possibly harmful, and all attacks against arguments outside of S are never harmful and possibly beneficial. Thus the simple and straightforward "optimistic" and "pessimistic" completions, defined as follows, can serve as critical completions for these three criteria.

Definition 2. *Let $AF = \langle \mathscr{A}, \mathscr{R}^+, \mathscr{R}^- \rangle$ be an attack-incomplete argumentation framework and let $S \subseteq \mathscr{A}$. The optimistic completion of AF for S is $AF_S^{\text{opt}} = \langle \mathscr{A}, \mathscr{R}_S^{\text{opt}} \rangle$ with $\mathscr{R}_S^{\text{opt}} = \mathscr{R}^+ \cup \{(a, b) \in \mathscr{R}^? \mid b \notin S\}$. The pessimistic completion of AF for S is $AF_S^{\text{pes}} = \langle \mathscr{A}, \mathscr{R}_S^{\text{pes}} \rangle$ with $\mathscr{R}_S^{\text{pes}} = \mathscr{R}^+ \cup \{(a, b) \in \mathscr{R}^? \mid b \in S\}$.*

Example 3. Figure 4 displays the optimistic and the pessimistic completion for $S = \{a, b\}$ in the argumentation framework from Example 2: Possible attacks that are added to the set of definite attacks in the respective completion are drawn as boldfaced arcs, possible attacks that are not added to the set of definite attacks in the respective completion are omitted in Fig. 4(b) and (c), and the arguments in S are displayed by boldfaced circles.

Lemma 1. *Let $AF = \langle \mathscr{A}, \mathscr{R}^+, \mathscr{R}^- \rangle$ be an attack-incomplete argumentation framework, let $S \subseteq \mathscr{A}$, and let AF_S^{opt} be the optimistic completion of AF for S.*

1. S is possibly conflict-free in AF if and only if S is a conflict-free extension of AF_S^{opt}.

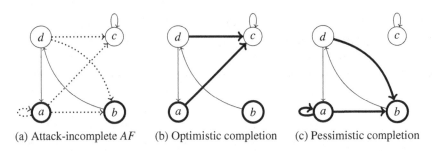

(a) Attack-incomplete *AF*   (b) Optimistic completion   (c) Pessimistic completion

Fig. 4. Optimistic and pessimistic completions for $S = \{a, b\}$

2. $a \in S$ is possibly acceptable with respect to S in AF if and only if a is acceptable with respect to S in AF_S^{opt}.
3. S is possibly admissible in AF if and only if S is an admissible extension of AF_S^{opt}.
4. S is possibly stable in AF if and only if S is a stable extension of AF_S^{opt}.

Proof. The converse is trivial in all cases: If S fulfills a given criterion in AF_S^{opt}, this immediately yields that S possibly fulfills the criterion in AF. We now prove the other direction of the equivalence individually for each criterion:

1. If a set S of arguments is not conflict-free in AF_S^{opt}, then there must be an attack between elements of S in $\mathscr{R}_S^{\text{opt}}$, which must be already in $\mathscr{R}^+$ due to how $\mathscr{R}_S^{\text{opt}}$ is constructed, and which therefore exists in every completion of AF. Thus S is not a possibly conflict-free set in AF.
2. If there is some $a \in S$ that is not acceptable with respect to S in AF_S^{opt}, then it is attacked by some $b \in \mathscr{A}$ in $\mathscr{R}_S^{\text{opt}}$ and there is no attack from an element of S against b in $\mathscr{R}_S^{\text{opt}}$. By construction, $\mathscr{R}_S^{\text{opt}}$ does not contain any possible attacks (members of $\mathscr{R}^?$) that attack elements of S, and it contains all possible attacks that can defend S. Therefore, all attacks in $\mathscr{R}_S^{\text{opt}}$ against elements of S are already in $\mathscr{R}^+$, so the undefended attack from b against a is in every completion of AF. Since a cannot be acceptable with respect to S in any completion of AF, a is not possibly acceptable with respect to S in AF.
3. Assume that S is not an admissible extension in AF_S^{opt}, i.e., S is not conflict-free in AF_S^{opt} or there is some $a \in S$ that is not acceptable with respect to S in AF_S^{opt}. In either case, the previous results imply that S is not conflict-free in any completion of AF or a is not acceptable with respect to S in any completion of AF. Thus S is not a possibly admissible extension in AF.
4. If a set S of arguments is not stable in AF_S^{opt}, S is necessarily not conflict-free in AF or there is an $a \in \mathscr{A} \setminus S$ that is not attacked by S in AF_S^{opt}, and therefore—by construction of AF_S^{opt}—a cannot be attacked by S in any completion of AF. In both cases, there is no completion of AF for which S is stable, so S is not a possibly stable extension of AF.

This completes the proof.   □

An analogous result holds for the pessimistic completion and the same properties holding necessarily. The proof of Lemma 2 is omitted due to space constraints.

Lemma 2. *Let $AF = \langle \mathscr{A}, \mathscr{R}^+, \mathscr{R}^- \rangle$ be an attack-incomplete argumentation framework, $S \subseteq \mathscr{A}$, and let AF_S^{pes} be the pessimistic completion of AF for S.*

1. *S is necessarily conflict-free in AF if and only if S is a conflict-free extension of AF_S^{pes}.*
2. *$a \in S$ is necessarily acceptable with respect to S in AF if and only if a is acceptable with respect to S in AF_S^{pes}.*
3. *S is necessarily admissible in AF if and only if S is an admissible extension of AF_S^{pes}.*
4. *S is necessarily stable in AF if and only if S is a stable extension of AF_S^{pes}.*

Note that in the second part of Lemmas 1 and 2 it is required that $a \in S$; the properties do not hold for the general case where $a \in \mathscr{A}$.

Finally, we can conclude that **s-ATTINCPV** and **s-ATTINCNV** are in P for conflict-freeness, admissibility, and stability.

Theorem 1. *For $\mathbf{s} \in \{\mathrm{CF}, \mathrm{AD}, \mathrm{ST}\}$, both* **s-ATTINCPV** *and* **s-ATTINCNV** *are in* P.

Proof. The optimistic and pessimistic completions can obviously be constructed in polynomial time. As already mentioned, the problem **s-VERIFICATION** can be solved in polynomial time for a given completion. Lemmas 1 and 2 then provide that the answer to, respectively, **s-ATTINCPV** and **s-ATTINCNV** is the same as that to **s-VERIFICATION** for the respective completion.   □

4.2   Verifying Groundedness and Completeness

Recall that, for a given argumentation framework AF, the set of complete extensions is the set of fixed points of the characteristic function F_{AF}, and the (unique) grounded extension is the fixed point of the characteristic function F_{AF} when starting with the empty set. For the complete and the grounded semantics, a critical completion of an attack-incomplete argumentation framework for a given set S of arguments can be constructed by choosing attacks in a way that makes it most likely (respectively, most unlikely) for S to be a fixed point of F_{AF}. We call a completion in which S is most likely to be a fixed point of F_{AF} a "fixed completion," and a completion in which it is most unlikely to be a fixed point an "unfixed completion."

Definition 3. *Let $AF = \langle \mathscr{A}, \mathscr{R}^+, \mathscr{R}^- \rangle$ be an attack-incomplete argumentation framework and $S \subseteq \mathscr{A}$. The fixed completion AF_S^{fix} of AF is the completion that is obtained by the following algorithm. The algorithm defines a finite sequence $(AF_i)_{i \geq 0}$ of attack-incomplete argumentation frameworks, with the fixed completion being the minimal completion of the sequence's last element.*

1. *Include definite attacks: Let $AF_0 = AF$.*
2. *Include external conflicts: Let $AF_1 = \langle \mathscr{A}, \mathscr{R}_1^+, \mathscr{R}^- \rangle$ with $\mathscr{R}_1^+ = \mathscr{R}^+ \cup \{(a,b) \in \mathscr{R}^? \mid a \notin S$ and $b \notin S\}$.*
3. *Include defending attacks: Let $T = \{t \in \mathscr{A} \setminus S \mid \exists s \in S : (t,s) \in \mathscr{R}_1^+\}$ (i.e., each argument in T necessarily attacks S) and let $AF_2 = \langle \mathscr{A}, \mathscr{R}_2^+, \mathscr{R}^- \rangle$ with $\mathscr{R}_2^+ = \mathscr{R}_1^+ \cup \{(a,b) \in \mathscr{R}_1^? \mid a \in S$ and $b \in T\}$.*
4. *Avoid arguments outside of S to be acceptable with respect to S: For the current i (initially, $i = 2$), let $AF_i^{\min}$ be the minimal completion of AF_i and $T_i = F_{AF_i^{\min}}(S) \setminus S$ (i.e., T_i is the set of arguments that are not in S, but that are acceptable with respect to S in the current minimal completion). Let $AF_{i+1} = \langle \mathscr{A}, \mathscr{R}_{i+1}^+, \mathscr{R}^- \rangle$ with $\mathscr{R}_{i+1}^+ = \mathscr{R}_i^+ \cup \{(a,b) \in \mathscr{R}_i^? \mid a \in S$ and $b \in T_i\}$, and set $i \leftarrow i + 1$.*
5. *Repeat Step 4 until no more attacks are added.*
6. *The fixed completion of AF is $AF_S^{\mathrm{fix}} = \langle \mathscr{A}, \mathscr{R}_S^{\mathrm{fix}} \rangle$ with $\mathscr{R}_S^{\mathrm{fix}} = \mathscr{R}_i^+$.*

Definition 4. *Let $AF = \langle \mathscr{A}, \mathscr{R}^+, \mathscr{R}^- \rangle$ be an attack-incomplete argumentation framework and $S \subseteq \mathscr{A}$. The unfixed completion AF_S^{unf} of AF is the completion that is obtained by the following algorithm. The algorithm defines a finite sequence $(AF_i)_{i \geq 0}$ of attack-incomplete argumentation frameworks, with the unfixed completion being the minimal completion of the sequence's last element.*

1. *Include definite attacks: Let $AF_0 = AF$.*
2. *Include attacks against S: Let $AF_1 = \langle \mathscr{A}, \mathscr{R}_1^+, \mathscr{R}_1^- \rangle$ with $\mathscr{R}_1^+ = \mathscr{R}^+ \cup \{(a,b) \in \mathscr{R}^? \mid b \in S\}$ and $\mathscr{R}_1^- = \mathscr{R}^-$.*
3. *Exclude external conflicts: Let $AF_2 = \langle \mathscr{A}, \mathscr{R}_2^+, \mathscr{R}_2^- \rangle$ with $\mathscr{R}_2^+ = \mathscr{R}_1^+$ and $\mathscr{R}_2^- = \mathscr{R}_1^- \cup \{(a,b) \in \mathscr{R}_1^? \mid a \notin S$ and $b \notin S\}$.*
4. *Exclude defending attacks: Let $T = \{t \in \mathscr{A} \setminus S \mid \exists s \in S : (t,s) \in \mathscr{R}_2^+\}$ (i.e., each argument in T necessarily attacks S) and let $AF_3 = \langle \mathscr{A}, \mathscr{R}_3^+, \mathscr{R}_3^- \rangle$ with $\mathscr{R}_3^+ = \mathscr{R}_2^+$ and $\mathscr{R}_3^- = \mathscr{R}_2^- \cup \{(a,b) \in \mathscr{R}_2^? \mid a \in S$ and $b \in T\}$.*
5. *Try to let arguments outside of S be acceptable with respect to S: Let $T = \mathscr{A} \setminus S = \{t_1, \ldots, t_k\}$. For the current i (initially, $i = 3$) and for each $t_j \in T$, do:*
   (a) *For $S' = S \cup \{t_j\}$, let $AF_{i,S'}^{\mathrm{opt}}$ be the optimistic completion of AF_i for S' and let $AF_i^{\min}$ be the minimal completion of AF_i.*
   (b) *If t_j is acceptable with respect to S in $AF_{i,S'}^{\mathrm{opt}}$, but not acceptable with respect to S in $AF_i^{\min}$, let $AF_{i+1} = \langle \mathscr{A}, \mathscr{R}_{i+1}^+, \mathscr{R}_{i+1}^- \rangle$ with $\mathscr{R}_{i+1}^+ = \mathscr{R}_i^+ \cup \{(a,b) \in \mathscr{R}_i^? \mid a \in S$ and $(b, t_j) \in \mathscr{R}_i^+\}$ and $\mathscr{R}_{i+1}^- = \mathscr{R}_i^-$, and set $i \leftarrow i+1$. (To accept an argument t_j that is not currently accepted by S but possibly accepted by S, include all possible attacks by S against t_j's attackers.)*
6. *Repeat Step 5 until no more attacks are added.*
7. *The unfixed completion of AF is $AF_S^{\mathrm{unf}} = \langle \mathscr{A}, \mathscr{R}_S^{\mathrm{unf}} \rangle$ with $\mathscr{R}_S^{\mathrm{unf}} = \mathscr{R}_i^+$.*

Lemma 3. *For an attack-incomplete argumentation framework $AF = \langle \mathscr{A}, \mathscr{R}^+, \mathscr{R}^- \rangle$ and a set $S \subseteq \mathscr{A}$ of arguments, the fixed completion AF_S^{fix} and the unfixed completion AF_S^{unf} can be constructed in polynomial time.*

Proof. All individual steps in both constructions can obviously be carried out in time polynomial in the number of arguments. It remains to prove that the loops in, respectively, Step 4 and Step 5 run at most a polynomial number of times. For the fixed completion, in each execution of a loop there is either (at least) one possible attack that is added to $\mathscr{R}_{i+1}^+$, or no action is taken in which case the loop terminates. Therefore, the number of times a loop is executed is bounded by the number of possible attacks in the attack-incomplete argumentation framework AF, which is at most n^2, where n is the number of arguments. For the unfixed completion, the only difference is the sub-loop in Step 5, which however has a predefined number of iterations that is bounded by the number n of arguments. Therefore, the total number of loop iterations in the construction of the unfixed completion is bounded by n^3. This completes the proof. $\square$

Now we prove that the fixed completion indeed is a critical completion for the complete and the grounded semantics.

Lemma 4. *Let $AF = \langle \mathscr{A}, \mathscr{R}^+, \mathscr{R}^- \rangle$ be an attack-incomplete argumentation framework, $S \subseteq \mathscr{A}$, and let AF_S^{fix} be the fixed completion of AF for S. (1) S is a possibly complete extension of AF if and only if S is a complete extension of AF_S^{fix}. (2) S is a possibly grounded extension of AF if and only if S is the grounded extension of AF_S^{fix}.*

Proof. Again, the converse is trivial in both cases. Further, if S is not an admissible extension in AF_S^{fix}, then S is not admissible in any completion of AF, due to the same arguments that we used for the optimistic completion and, therefore, neither possibly complete nor possibly grounded in AF. So, we may assume that S is admissible in AF_S^{fix}.

(1) Assume that S is not a complete extension of AF_S^{fix}, i.e., S is not a fixed point of $F_{AF_S^{\mathrm{fix}}}$. We will show that this implies that S is not possibly complete in AF. Since S is not a fixed point of $F_{AF_S^{\mathrm{fix}}}$, there is an argument $b \notin S$ which is acceptable with respect to S in AF_S^{fix}.

We prove that, then, there must be some $c \notin S$ for which all attackers of c are attacked by S in AF^* ($c = b$ may or may not be the case) by individually covering all cases in which attacks are added to $\mathscr{R}_S^{\mathrm{fix}}$:

All attacks from $\mathscr{R}^?$ between arguments outside of S, which are added to $\mathscr{R}_S^{\mathrm{fix}}$ in Step 2, cannot make an argument $b \notin S$ acceptable with respect to S: If S did not attack all attackers of an argument before, it cannot do so after *more* attackers are added.

All attacks that are added in Step 3 are crucial for S to be admissible, and must therefore also be included in $\mathscr{R}^*$. In a case where multiple arguments in S attack a single attacker of S, it would be sufficient to include one of these defending attacks, but including all of them does not make a difference, since the criterion of being acceptable with respect to S does not distinguish between different elements of S.

All attacks that are added in Step 4 are attacks by S against arguments that are currently acceptable with respect to S. Since all possible attacks among arguments outside of S were already included in Step 2, the only way to destroy acceptability of these arguments is by S directly attacking them. Therefore, none of the attacks added in Step 4 can be omitted without making the respective argument acceptable with respect to S (again, it is not necessary to distinguish between multiple attacks by different arguments in S against the same argument). It is possible for a given $b \notin S$ to be acceptable with respect to S in AF_S^{fix} and not in AF^*, but this happens only if S attacks an attacker (or several attackers) of b in AF_S^{fix} that would otherwise be acceptable with respect to S, and which therefore must be acceptable with respect to S in AF^*. In either case, if an argument outside of S is acceptable with respect to S in AF_S^{fix}, then some argument outside of S must be acceptable with respect to S in each completion AF^* of AF in which S is admissible. Therefore, if S is not a complete extension of AF_S^{fix}, it is not a complete extension of any completion AF^* of AF, and therefore not a possibly complete extension of AF.

(2) Let AF^* be an arbitrary completion of AF and assume that S is its grounded extension. We prove that, then, S is also the grounded extension of AF_S^{fix}. Let $A_i = F_{AF^*}^i(\emptyset)$ and $B_i = F_{AF_S^{\text{fix}}}^i(\emptyset)$, where F^i is the i-fold composition of the respective characteristic function F. Since S is grounded in AF^*, it is complete in AF_S^{fix} due to our previous result, and it holds that $A_i \subseteq S$ for all $i \geq 0$ and there exists a $j \geq 0$ such that for all $i \geq j$, it holds that $A_i = S$. We will prove that $A_i \subseteq B_i \subseteq S$ for all $i \geq 0$. Combined, these statements show that there exists some j such that $B_i = S$ for all $i \geq j$, which is equivalent to S being the grounded extension of AF_S^{fix}.

First, we prove that $A_i \subseteq B_i$ for all $i \geq 0$. For $i = 0$, we have $A_i = B_i = \emptyset$. For $i = 1$, A_i (respectively, B_i) is the set of all unattacked arguments in AF^* (respectively, in AF_S^{fix}). We know that $A_1 \subseteq S$. Since the fixed completion does not include any possible attacks against elements of S, all $a \in S$ that are unattacked in AF^* are unattacked in AF_S^{fix}, too, which proves $A_1 \subseteq B_1$. If we now have $A_k \subseteq B_k$ for some $k \geq 1$, this implies $A_{k+1} \subseteq B_{k+1}$: Assume that this were not true, i.e., that $A_k \subseteq B_k$, but there is an argument $a \in A_{k+1}$ with $a \notin B_{k+1}$. Then, a is acceptable with respect to A_k in AF^*, but not acceptable with respect to B_k in AF_S^{fix}. We know that—since $A_{k+1} \subseteq S$—no possible attacks against A_{k+1} (and in particular, against a) are included in AF_S^{fix} and all possible defending attacks by arguments in A_{k+1} against arguments outside of S are included in AF_S^{fix}. Further, no element of S attacks a in AF_S^{fix}, since $a \in S$ and S is complete in AF_S^{fix}. Therefore, a is acceptable with respect to A_k in AF_S^{fix}; otherwise it could not be acceptable with respect to A_k in AF^*. Now, the only way for a to not be acceptable with respect to B_k in AF_S^{fix} is if there were some $b \in B_k \setminus A_k$ that necessarily attacks a. Then there would have to be a defending attack by an argument $d \in A_k$ against b in AF^*, since a is acceptable with respect to A_k in AF^*. This implies that $b \notin S$, since S is conflict-free in AF^*. Finally, since (d, b) is a possible (or even a necessary) defending attack by an element of S against $b \notin S$, $(d, b) \in \mathscr{R}_S^{\text{fix}}$ holds by construction of the fixed completion, which contradicts that B_k is admissible in AF_S^{fix}. Therefore, a must be acceptable with respect to B_k in AF_S^{fix}, which proves that $A_{k+1} \subseteq B_{k+1}$.

Now we prove that $B_i \subseteq S$ for all $i \geq 0$: Assume that $B_i \not\subseteq S$ for some $i \geq 0$. Then it also holds that $G_S^{\text{fix}} \not\subseteq S$ for the grounded extension G_S^{fix} of AF_S^{fix}. It further holds that $S \subset G_S^{\text{fix}}$, since there exists a $j \geq 0$ such that $S \subseteq B_i$ for all $i \geq j$, as established before. However, this contradicts the fact that S is complete in AF_S^{fix}, since the grounded extension G_S^{fix} of AF_S^{fix} is its least complete extension with respect to set inclusion and the complete set S cannot be a strict subset of G_S^{fix}. This completes the proof. $\qquad\square$

Analogously, the unfixed completion is a critical completion for the complete and the grounded semantics. Due to limitation of space, the proof of Lemma 5 is omitted.

Lemma 5. *Let* $AF = \langle \mathscr{A}, \mathscr{R}^+, \mathscr{R}^- \rangle$ *be an attack-incomplete argumentation framework,* $S \subseteq \mathscr{A}$, *and let* AF_S^{unf} *be the unfixed completion of AF for S. (1) S is a necessarily complete extension of AF if and only if S is a complete extension of* AF_S^{unf}. *(2) S is a necessarily grounded extension of AF if and only if S is the grounded extension of* AF_S^{unf}.

Finally, our results allow us to establish that s-ATTINCPV and s-ATTINCNV are in P for the complete and grounded semantics.

Theorem 2. *For* $\mathbf{s} \in \{\text{CP}, \text{GR}\}$, *both* s-ATTINCPV *and* s-ATTINCNV *are in P.*

Proof. Lemma 3 provides polynomial-time constructability for the fixed and unfixed completion. Given a completion, s-VERIFICATION can be solved in polynomial time, and Lemmas 4 and 5 imply that the answer to, respectively, s-ATTINCPV and s-ATTINCNV is the same as that to s-VERIFICATION for the respective completion. $\qquad\square$

4.3 Verifying Preferredness

As mentioned above, the VERIFICATION problem for the preferred semantics is coNP-complete. For PR-ATTINCPV and PR-ATTINCNV, we have the following results.

Theorem 3. *The problem* PR-ATTINCPV *is in* Σ_2^p *and* coNP-*hard, and* PR-ATTINCNV *is* coNP-*complete.*

Proof. In PR-ATTINCPV one has to check whether, given an attack-incomplete argumentation framework $AF = \langle \mathscr{A}, \mathscr{R}^+, \mathscr{R}^- \rangle$ and a set $S \subseteq \mathcal{A}$, there is a completion $AF^* = \langle \mathcal{A}, \mathcal{R}^* \rangle$ such that S is preferred in AF^*. To check whether S is preferred in AF^*, one has to check whether for all sets $S' \subseteq \mathcal{A}$ with $S \subset S'$ it holds that S is an admissible extension and S' is not an admissible extension. Thus this problem is in Σ_2^p.

To see that PR-ATTINCNV is in coNP, consider the complementary problem. Here one has to check whether there is a completion AF^* of the given attack-incomplete AF such that the given set S is *not* preferred. To see this, it is enough to check whether there is a strict superset of S that is admissible or whether S

Table 2. Overview of complexity results both in the standard model (s-Verification) and in the attack-incomplete model of this paper (s-AttIncPV and s-AttIncNV)

s	Verification	AttIncPV		AttIncNV	
CF	in P	in P	[10]	in P	[10]
AD	in P	in P	(Theorem 1)	in P	[10]
ST	in P	in P	(Theorem 1)	in P	(Theorem 1)
CP	in P	in P	(Theorem 2)	in P	(Theorem 2)
GR	in P	in P	(Theorem 2)	in P	(Theorem 2)
PR	coNP-complete	coNP-hard, in Σ_2^p	(Theorem 3)	coNP-complete	(Theorem 3)

itself is not admissible. Since admissibility can be checked in polynomial time, the complement of PR-AttIncNV is in NP and hence PR-AttIncNV is in coNP.

On the other hand, coNP-hardness for both problems follows by a direct reduction from the original PR-Verification problem, which is coNP-complete [15]. For a given instance $(\langle \mathscr{A}, \mathscr{R} \rangle, S)$ of PR-Verification, the constructed instance of both pr-AttIncPV and pr-AttIncNV is $(\langle \mathscr{A}, \mathscr{R}, (\mathscr{A} \times \mathscr{A}) \setminus \mathscr{R} \rangle, S)$. The only completion of $\langle \mathscr{A}, \mathscr{R}, (\mathscr{A} \times \mathscr{A}) \setminus \mathscr{R} \rangle$ is $\langle \mathscr{A}, \mathscr{R} \rangle$. Now, it is easy to see that $(\langle \mathscr{A}, \mathscr{R} \rangle, S) \in$ PR-Verification if and only if $(\langle \mathscr{A}, \mathscr{R}, (\mathscr{A} \times \mathscr{A}) \setminus \mathscr{R} \rangle, S) \in pr\text{-AttIncPV}$, which in turn is equivalent to $(\langle \mathscr{A}, \mathscr{R}, (\mathscr{A} \times \mathscr{A}) \setminus \mathscr{R} \rangle, S) \in pr\text{-AttIncNV}$. □

5 Conclusions and Open Questions

We have investigated argumentation frameworks in a setting where we don't have full knowledge of the attacks. We adapted the s-Verification decision problems with respect to notions of possibility and necessity to fit the model of Cayrol et al. [10], and we analyzed their complexity for the fundamental semantics admissibility, stability, completeness, groundedness, and preferredness. This may be useful to predict those sets of arguments that will be "good" solutions once all attacks are known eventually.

Table 2 summarizes our results, and also gives the previously known results for argumentation frameworks without uncertainty that are due to Dimopoulos and Torres [15], Dung [17], and Dunne and Wooldridge [18], as well as the results for incomplete argumentation frameworks provided by Cayrol et al. [10]. We have shown positive results (characterizations) for all considered semantics except preferredness, for which the exact complexity in the case of possible verification remains open. As a task for future work, we propose to generalize other decision problems like s-Credulous-Acceptance, s-Skeptical-Acceptance, s-Existence, and s-Nonemptiness to fit the model of attack-incompleteness and analyze their complexity. Additionally, one could have a closer look at other semantics like semi-stable, ideal, or prudent semantics (see [18] for the definition of these decision problems and semantics).

Acknowledgments. We thank the anonymous reviewers for their helpful comments. This work was supported in part by an NRW grant for gender-sensitive universities and the project "Online Participation," both funded by the NRW Ministry for Innovation, Science, and Research.

References

1. Baumeister, D., Rothe, J.: Preference aggregation by voting (chapter 4). In: Rothe, J. (ed.) Economics and Computation: An Introduction to Algorithmic Game Theory, Computational Social Choice, and Fair Division, pp. 197–325. Springer, Heidelberg (2015). doi:10.1007/978-3-662-47904-9_4
2. Baumeister, D., Roos, M., Rothe, J.: Computational complexity of two variants of the possible winner problem. In: Proceedings of AAMAS 2011, pp. 853–860. IFAAMAS (2011)
3. Baumeister, D., Roos, M., Rothe, J., Schend, L., Xia, L.: The possible winner problem with uncertain weights. In: Proceedings of ECAI 2012, pp. 133–138. IOS Press (2012)
4. Baumeister, D., Erdélyi, G., Erdélyi, O., Rothe, J.: Complexity of manipulation and bribery in judgment aggregation for uniform premise-based quota rules. Math. Soc. Sci. **76**, 19–30 (2015)
5. Baumeister, D., Rothe, J., Schadrack, H.: Verification in argument-incomplete argumentation frameworks. In: Walsh, T. (ed.) ADT 2015. LNCS (LNAI), vol. 9346, pp. xx–yy. Springer, Heidelberg (2015)
6. Boella, G., Kaci, S., van der Torre, L.: Dynamics in with single extensions: abstraction principles and the grounded extension. In: Sossai, C., Chemello, G. (eds.) ECSQARU 2009. LNCS, vol. 5590, pp. 107–118. Springer, Heidelberg (2009)
7. Boutilier, C., Rosenschein, J.: Incomplete information and communication in voting (chapter 10). In: Brandt, F., Conitzer, V., Endriss, U., Lang, J., Procaccia, A. (eds.) Handbook of Computational Social Choice. Cambridge University Press, Cambridge (2015)
8. Bouveret, S., Endriss, U., Lang, J.: Fair division under ordinal preferences: computing envy-free allocations of indivisible goods. In: Proceedings of ECAI 2010, pp. 387–392. IOS Press (2010)
9. Brandt, F., Conitzer, V., Endriss, U.: Computational social choice. In: Weiß, G. (ed.) Multiagent Systems, 2nd edn, pp. 213–283. MIT Press, Cambridge (2013)
10. Cayrol, C., Devred, C., Lagasquie-Schiex, M.C.: Handling ignorance in argumentation: semantics of partial argumentation frameworks. In: Mellouli, K. (ed.) ECSQARU 2007. LNCS (LNAI), vol. 4724, pp. 259–270. Springer, Heidelberg (2007)
11. Cayrol, C., de Saint-Cyr, F., Lagasquie-Schiex, M.: Revision of an argumentation system. In: Proceedings of KR 2008, pp. 124–134. AAAI Press (2008)
12. Cayrol, C., de Saint-Cyr, F., Lagasquie-Schiex, M.: Change in abstract argumentation frameworks: adding an argument. J. Artif. Intell. Res. **38**, 49–84 (2010)
13. Chevaleyre, Y., Lang, J., Maudet, N., Monnot, J., Xia, L.: New candidates welcome! Possible winners with respect to the addition of new candidates. Math. Soc. Sci. **64**(1), 74–88 (2012)
14. Coste-Marquis, S., Konieczny, S., Mailly, J., Marquis, P.: On the revision of argumentation systems: minimal change of arguments statuses. In: Proceedings of KR 2014. AAAI Press (2014)

15. Dimopoulos, Y., Torres, A.: Graph theoretical structures in logic programs and default theories. Theoret. Comput. Sci. **170**(1), 209–244 (1996)
16. Doder, D., Woltran, S.: Probabilistic argumentation frameworks – a logical approach. In: Straccia, U., Calì, A. (eds.) SUM 2014. LNCS, vol. 8720, pp. 134–147. Springer, Heidelberg (2014)
17. Dung, P.: On the acceptability of arguments and its fundamental role in non-monotonic reasoning, logic programming and n-person games. Artif. Intell. **77**(2), 321–357 (1995)
18. Dunne, P., Wooldridge, M.: Complexity of abstract argumentation (chapter 5). In: Rahwan, I., Simari, G. (eds.) Argumentation in Artificial Intelligence, pp. 85–104. Springer, New York (2009)
19. Fazzinga, B., Flesca, S., Parisi, F.: On the complexity of probabilistic abstract argumentation. In: Proceedings of IJCAI 2013, pp. 898–904. AAAI Press/IJCAI (2013)
20. Fazzinga, B., Flesca, S., Parisi, F.: Efficiently estimating the probability of extensions in abstract argumentation. In: Liu, W., Subrahmanian, V.S., Wijsen, J. (eds.) SUM 2013. LNCS, vol. 8078, pp. 106–119. Springer, Heidelberg (2013)
21. Hemaspaandra, E., Hemaspaandra, L., Rothe, J.: The complexity of online manipulation of sequential elections. J. Comput. Syst. Sci. **80**(4), 697–710 (2014)
22. Hunter, A.: Probabilistic qualification of attack in abstract argumentation. Int. J. Approximate Reasoning **55**(2), 607–638 (2014)
23. Konczak, K., Lang, J.: Voting procedures with incomplete preferences. In: Proceedings of Multidisciplinary IJCAI 2005 Workshop on Advances in Preference Handling, pp. 124–129 (2005)
24. Lang, J., Rey, A., Rothe, J., Schadrack, H., Schend, L.: Representing and solving hedonic games with ordinal preferences and thresholds. In: Proceedings of AAMAS 2015. IFAAMAS (2015)
25. Li, H., Oren, N., Norman, T.J.: Probabilistic argumentation frameworks. In: Modgil, S., Oren, N., Toni, F. (eds.) TAFA 2011. LNCS, vol. 7132, pp. 1–16. Springer, Heidelberg (2012)
26. Liao, B., Jin, L., Koons, R.: Dynamics of argumentation systems: a division-based method. Artif. Intell. **175**(11), 1790–1814 (2011)
27. Meyer, A., Stockmeyer, L.: The equivalence problem for regular expressions with squaring requires exponential space. In: Proceedings of FOCS 1972, pp. 125–129. IEEE Press (1972)
28. Papadimitriou, C.: Computational Complexity, 2nd edn. Addison-Wesley, New York (1995). Reprinted with corrections
29. Parkes, D., Procaccia, A.: Dynamic social choice with evolving preferences. In: Proceedings of AAAI 2013, pp. 767–773. AAAI Press (2013)
30. Rahwan, I., Simari, G. (eds.): Argumentation in Artificial Intelligence. Springer, New York (2009)
31. Rienstra, T.: Towards a probabilistic Dung-style argumentation system. In: Proceedings of AT 2012, pp. 138–152 (2012)
32. Rothe, J.: Complexity Theory and Cryptology: An Introduction to Cryptocomplexity. EATCS Texts in Theoretical Computer Science. Springer, Heidelberg (2005)
33. Stockmeyer, L.: The polynomial-time hierarchy. Theoret. Comput. Sci. **3**(1), 1–22 (1976)
34. Xia, L., Conitzer, V.: Determining possible and necessary winners given partial orders. J. Artif. Intell. Res. **41**, 25–67 (2011)

Verification in Argument-Incomplete Argumentation Frameworks

Dorothea Baumeister, Jörg Rothe, and Hilmar Schadrack[✉]

Heinrich-Heine-Universität Düsseldorf, 40225 Düsseldorf, Germany
{baumeister,rothe,schadrack}@cs.uni-duesseldorf.de

Abstract. Incomplete knowledge in argumentation frameworks may occur during the single steps of an elicitation process, when merging different beliefs about the current state of an argumentation framework, or when it is simply not possible to obtain complete information. The semantics of argumentation frameworks with such incomplete knowledge have previously been modeled in terms of an inco mplete attack relation among the given arguments by Cayrol et al. [12] or when adding an argument that interacts with already present arguments [14]. We propose a more general model of argument-incomplete argumentation frameworks with a variable set of arguments, and we study the related verification problems for various semantics in terms of their computational complexity.

1 Introduction

A discussion between human beings is a form of communicating opinions and thoughts about a given subject. These opinions and their interactions are often highly complex and thus hard to formalize mathematically. The goal of abstract argumentation is to model discussions between (human or software) agents by abstracting from the actual content of arguments and from the reasons of why they attack each other, and instead to consider a given set of arguments along with an attack relation on it and to find certain subsets of the arguments that fulfill certain justification criteria. In 1995, Dung [18] introduced a formal model to describe discussions abstractly. His model uses a graph structure where the nodes represent arguments, and the attacks between arguments are modeled through directed edges. He also introduced various semantics, i.e., criteria that can express different kinds and levels of justification for certain subsets of the arguments. His highly influential model has been used by many researchers, who developed additional ideas of how to extend it so as to make it an elegant, rich, and attractive model for abstract group argumentation. We refer the reader to the book by Rahwan and Simari [30] for more background on abstract argumentation in artificial intelligence.

In this paper, we develop a new model for argumentation frameworks on the basis of Dung's work [18], namely, *argument-incomplete argumentation frameworks*. Our goal is to extend the standard model by allowing uncertainty over the set of arguments. In our model, we have a set of arguments that already

T. Walsh (Ed.): ADT 2015, LNAI 9346, pp. 359–376, 2015.
DOI: 10.1007/978-3-319-23114-3_22

are known to exist, and another set of arguments that in addition contains those arguments which might become relevant in a later state of the discussion. We study properties and semantics, namely conflict-freeness, admissibility, preferredness, stability, completeness, and groundedness, suitably adapted to the argument-incomplete setting, where we distinguish between properties holding possibly and necessarily. Besides the formal model, the main contribution of this work are complexity results of appropriately extended variants of the verification problem [19], asking whether or not a given set of arguments will fulfill a previously specified property either possibly or necessarily.

Related Work and Motivation: In real-world discussions we cannot assume to know all arguments or attacks in advance, or how important they are for the discussion, or to fully capture the dynamics of a discussion. Therefore, we are trying to take an early step to model situations in which complete information is not available by allowing the set of arguments to be uncertain. This can happen, for example, in a well developed discussion in which many, or even all, possible arguments are known already but where certain external limitations can change, which may have an impact on the validity or importance of the arguments. It may be safe to assume that some of the arguments are always valid, but which of the uncertain arguments are valid may depend on the circumstances. For example, if the citizens of a town discuss whether a public swimming pool, an opera house, or a library are to be built, it will have an impact on some of the arguments if it suddenly turns out that the budget deficit is higher than expected—some arguments may then be invalid or less important, while others remain valid and crucial. It would be interesting to know which sets of arguments (possibly also containing uncertain arguments) are justified for different limitations.

As another example, consider the case of different knowledge bases of agents. All the agents share the same set of possible arguments, but they disagree on their importance. Hence, every agent has her own "belief stage" resulting in different individual views on the argumentation framework. Such belief-staged argumentation frameworks can be modeled by an argument-incomplete argumentation framework, and so can the aggregated opinion of the agents, obtained by agreeing on some arguments to be important, leaving the others as uncertain.

Modeling discussions via argument-incomplete argumentation frameworks may help to answer the question of whether it is possible to make early decisions about which sets of arguments will fulfill certain criteria possibly (i.e., in at least one way regarding currently uncertain arguments that may arise—or turn out to be important—in the future) or necessarily (i.e., in any way regarding currently uncertain arguments that may arise—or turn out to be important—in the future).

The need for a model that is capable of capturing these ideas is also motivated by the interdisciplinary graduate school "Online Participation"[1] run by Heinrich-Heine-Universität Düsseldorf in cooperation with Fachhochschule für öffentliche Verwaltung and with a number of practice partners and municipal councils. Researchers from the social sciences, political science, communication science,

---

[1] We refer to the website http://www.fortschrittskolleg.de for further details.

and computer science are involved in creating a broad knowledge base about how to discuss in online platforms, as well as in designing a software tool for performing this in practice. This work aims to provide a solid theoretical foundation.

Incomplete argumentation frameworks have been introduced by Cayrol et al. [12], who define so-called "partial argumentation frameworks" by distinguishing the attacks into those that are definitely part of the argumentation framework, those that are definitely not part of the argumentation framework, and those that are not certain—but possible—to occur. They further define a *completion* of such a partial argumentation framework as a standard argumentation framework that contains all the arguments of the partial argumentation framework, at least the attacks definitely contained, and maybe also some of the possible attacks. We use a similar idea in our model of argument-incomplete argumentation frameworks.

The model of Cayrol et al. [13,14] to study changes in the argument set has a different goal than our model. They introduce change operations that allow for the addition or deletion of one attack, or one argument together with a set of attacks regarding this argument. Their work focuses on a classification of how such changes can possibly alter the outcome.

Other approaches regarding changes in the argument set are due to Boella et al. [7], who define general principles for the abstraction of arguments and attacks, mainly for the grounded semantics. They address the question of which arguments or attacks can be removed such that the extensions remain unchanged.

In so-called "probabilistic argumentation frameworks" (introduced by Li et al. [26]; for more information, see the work of Doder and Woltran [17]), every argument and attack has an associated probability that yields the likelihood of that argument or attack to be part of an induced argumentation framework. This can be seen as an intermediate state between complete knowledge and incomplete information. Li et al. [26] show that computing the probability of a set of arguments being justified regarding a semantics can be intractable. Therefore, the authors approximate it by means of a Monte-Carlo simulation. Fazzinga et al. [20] show that this computation indeed is hard for, e.g., the complete, grounded, and preferred semantics, but is easy for stability and admissibility. They further discuss approximation algorithms [21].

Baumeister et al. [6] describe a model of *attack*-incomplete argumentation frameworks.

The idea of extending models of complete information to allow for incomplete information is not new; it has been applied, e.g., in the field of computational social choice, especially in voting theory (see, e.g., the book chapters by Boutilier and Rosenschein [8] and Baumeister and Rothe [1]). Konczak and Lang [23] introduced the notions of *possible* and *necessary winners* in elections, which then were studied in different variants (see, e.g., Konczak and Lang [23], Xia and Conitzer [33], Lang et al. [24], and Chevaleyre et al. [15]) and for different settings (see, e.g., the work of Baumeister et al. [2,3]). Other fields in which the notions of possibility and necessity are used include judgment aggregation [4], fair division [5,9–11], and algorithmic game theory [25].

This paper is structured as follows. In Sect. 2, we introduce the known model of abstract argumentation frameworks, and we provide the needed notions from complexity theory. In Sect. 3, we describe our model for attack-incomplete argumentation frameworks and present our results. Section 4, finally, gives our conclusions and comments on some open questions.

2   Preliminaries

In this section, we introduce formal definitions of the central notions related to (classical) argumentation frameworks. The basic ideas are due to Dung [18], and we will be using some notation from the book chapter by Dunne and Wooldridge [19].

An *argumentation framework* AF is a pair $\langle \mathscr{A}, \mathscr{R} \rangle$, where $\mathscr{A}$ denotes a set of arguments, and $\mathscr{R} \subseteq \mathscr{A} \times \mathscr{A}$ the attack relation. For every pair $(a, b) \in \mathscr{R}$ we say a *attacks* b, and for simplicity we often write $a \to b$. If $a \to b$ and $b \to a$, we simply write $a \leftrightarrow b$. Every argumentation framework $AF = \langle \mathscr{A}, \mathscr{R} \rangle$ can be seen as a directed Graph $G_{AF} = (V, E)$ by using the arguments as vertices and the attacks as edges, i.e., $V = \mathscr{A}$ and $E = \mathscr{R}$.

Before going into further detail of the abstract argumentation scheme by Dung, we will present an easy example, which will be used again later on.

Example 1. Assume we have seven arguments, $\{a, b, c, d, e, f, g\}$, and nine attacks: $a \to b, a \to c, b \to d, c \to d, e \to d, e \leftrightarrow f, e \to g$, and $g \to g$. Then the appropriate argumentation framework is

$$AF = \langle \mathscr{A}, \mathscr{R} \rangle = \langle \{a, b, c, d, e, f, g, \},$$
$$\{(a, b), (a, c), (b, d), (c, d), (e, d), (e, f), (e, g), (f, e), (g, g)\} \rangle.$$

The graph representation G_{AF} of this argumentation framework is shown in Fig. 1.

We now define properties in argumentation frameworks, mainly for sets of arguments. All of them were introduced by Dung [18]. We start with the three most basic properties: conflict-freeness, acceptability, and admissibility. Let $AF = \langle \mathscr{A}, \mathscr{R} \rangle$ be an argumentation framework.

- A set $S \subseteq \mathscr{A}$ is called *conflict-free* if there are no arguments $a, b \in S$ such that $a \to b$.
- An argument $a \in \mathscr{A}$ is called *acceptable with respect to* $S \subseteq \mathscr{A}$ if for all arguments $b \in \mathscr{A}$ with $b \to a$, we have at least one argument $c \in S$ such that $c \to b$.
- A conflict-free set $S \subseteq \mathscr{A}$ is called *admissible* if every argument $a \in S$ is acceptable with respect to S.

More advanced properties are preferredness, stability, completeness, and groundedness, and Dung calls them *semantics* in his work [18].

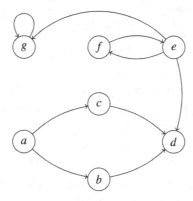

Fig. 1. The graph $G_{AF} = (\mathscr{A}, \mathscr{R})$ for the argumentation framework $AF = \langle \mathscr{A}, \mathscr{R} \rangle$ from Example 1

- A set $S \subseteq \mathscr{A}$ is called *preferred* if S is a maximal (with respect to set inclusion) admissible set.
- A set $S \subseteq \mathscr{A}$ is called *stable* if S is conflict-free and for all arguments $b \in \mathscr{A} \setminus S$, there is at least on argument $a \in S$ with $a \to b$.
- A set $S \subseteq \mathscr{A}$ is called *complete* if S is admissible and contains all arguments $a \in \mathscr{A}$ that are acceptable with respect to S.
- A set $S \subseteq \mathscr{A}$ is called *grounded* if S is the least (with respect to set inclusion) fixed point of the characteristic function of the argumentation framework. The *characteristic function* F_{AF} of the argumentation framework AF is a function $F_{AF} : 2^{\mathscr{A}} \to 2^{\mathscr{A}}$ defined by

$$F_{AF}(S) = \{a \in \mathscr{A} \mid a \text{ is acceptable with respect to } S\}.$$

The characteristic function is monotonic with respect to set inclusion, and there always is exactly one least fixed point. Hence, there always is exactly one grounded set for a given argumentation framework. Additionally, starting with an arbitrary subset of the arguments, there always is an $i \in \mathbb{N}$ such that the i-fold composition of the characteristic function has a fixed point. All those fixed points are exactly the complete sets of the given argumentation framework.

Dung [18] shows how those above defined properties are related to each other. In particular, he proved that there always is a preferred set, and that every admissible set is a subset of a preferred set, every stable set is preferred, and every preferred set is complete. As already mentioned, there is exactly one grounded set in a given argumentation framework, and it is obviously complete. It is easy to see that a stable or a preferred set can be the grounded set, but there are argumentation frameworks in which the grounded set is neither preferred nor stable. Finally, every complete set is admissible and every admissible set is conflict-free, due to their definitions. Figure 2 summarizes these results.

In this work, we will focus on the six properties for sets of arguments introduced above. In the literature, conflict-freeness and admissibility are not called

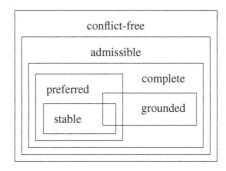

Fig. 2. A summary over how the different properties for sets of arguments used in this paper are correlated

semantics, as they are often considered to be basic conditions. However, for the sake of simplicity, we will call all these properties semantics.

Dung [18] also introduced the term "extension". For a given argumentation framework AF and a semantics **s**, a subset S of the arguments is called **s** *extension of AF* if S fulfills the conditions of semantics **s**.

We will now illustrate the above semantics in an example.

Example 2. Consider again the argumentation framework from Example 1, illustrated in Fig. 1. The only stable extension in this argumentation framework is $\{a, e\}$. Besides this, there is another preferred extension, namely $\{a, d, f\}$. There are no more preferred extensions. The unique grounded extension is $\{a\}$, but it is neither preferred nor stable. Besides those three extensions, there are no further sets that are complete. The only other admissible sets are $\emptyset$, $\{e\}$, $\{f\}$, and $\{a, f\}$. g is not part of any extension, as the self-attack always yields a conflict, and b or c is never part of any admissible set, because argument a which attacks them is never attacked itself.

We will now briefly mention the notions from complexity theory that we use in this paper. We assume the reader to be familiar with the basic notions, like the complexity classes P, NP, and coNP, as well as hardness, completeness, polynomial-time many-one-reducibility, $\leq_m^p$, and the notion of (oracle) Turing machines. The complexity class DP was introduced by Papadimitriou and Yannakakis [29] as the class of the differences of two NP problems. DP is also the second level of the boolean hierarchy over NP. $\Sigma_2^p = \text{NP}^{\text{NP}}$ contains those problems that are solvable by a nondeterministic oracle Turing machine with access to an NP oracle, and was introduced, together with $\Pi_2^p = \text{coNP}^{\text{NP}}$, by Meyer and Stockmeyer [27,31] as the second level of the polynomial hierarchy. It is known that $P \subseteq NP \subseteq DP \subseteq \Sigma_2^p$, but it is not known whether any of these inclusion is strict. For further details, see [28].

Dunne and Wooldridge [19] give an overview over decision problems defined for argumentation frameworks. Among others, they investigate VERIFICATION, CREDULOUS-ACCEPTANCE, SKEPTICAL-ACCEPTANCE, EXISTENCE, and NON-EMPTINESS for several semantics. Many of those decision problems are hard to decide, as they are complete for NP, coNP, DP, or even Π_2^p. We will focus on VERIFICATION (while we are going to change their notation slightly), which is easy to decide for all semantics studied here, except for preferredness for which it is known to be coNP-complete [16]. All easiness results follow immediately from the work of Dung [18].

---

s-VERIFICATION

Given: An argumentation framework $\langle \mathscr{A}, \mathscr{R} \rangle$ and a subset $S \subseteq \mathscr{A}$.
Question: Is S an **s** extension of AF?

The boldfaced letter **s** is a placeholder for any of the six semantics defined above. For better readability, we will use CF for *conflict-freeness*, AD for *admissibility*, PR for *preferredness*, ST for *stability*, CP for *completeness*, and GR for *groundedness*.

3   Argument-Incomplete Argumentation Frameworks

We now turn to our model extending classical argumentation frameworks: argument-incomplete argumentation frameworks. Cayrol et al. [12] introduced a model of incomplete argumentation framework in which the exact attacks are unknown. Cayrol et al. [14] discuss a model that allows for one of the following four so-called *change operations*: The addition (1) or deletion (2) of one attack, the addition of one argument together with at least one attack regarding this argument (3), and the deletion of one argument together with all corresponding attacks (4). Their goal is to find different classifications for how the set of *all extensions* (of a given semantics and argumentation framework) alter after applying *one* change. In contrast, we want to *verify* those *sets of arguments* (of a given semantics and argumentation framework) that can *become once* or *remain always* extensions for *arbitrarily many* changes in the argument set. After describing our model, we will study the computational complexity of the verification problem for various semantics in argument-incomplete argumentation frameworks.

3.1   Model

In our setting we do not know exactly which arguments will be part of the final discussion, but we know a set of arguments that are important already, and have a vague idea of which arguments may be important in the future. Formally, our model is defined as follows.

Definition 1. *An* argument-incomplete argumentation framework *is a triple* $\langle \mathscr{A}', \mathscr{A}, \mathscr{R} \rangle$, *where* A' *and* A *with* $A' \subseteq A$ *are sets of arguments, and* $\mathcal{R} \subseteq A \times A$ *is an attack relation.*

The arguments in $\mathscr{A}'$ are those arguments that definitely are already part of the discussion. $\mathscr{A}$ contains, additionally to the arguments in $\mathscr{A}'$, also those arguments that could possibly join the discussion in the future.

Definition 2. *Let $IAF = \langle \mathscr{A}', \mathscr{A}, \mathscr{R} \rangle$ be an argument-incomplete argumentation framework. For a set $\mathscr{A}^*$ of arguments with $\mathscr{A}' \subseteq \mathscr{A}^* \subseteq \mathscr{A}$, define the restriction of $\mathscr{R}$ to $\mathscr{A}^*$ by*

$$\mathscr{R}|_{\mathscr{A}^*} = \{(a,b) \in \mathscr{R} \mid a,b \in \mathscr{A}^*\}.$$

$AF^* = \langle \mathscr{A}^*, \mathscr{R}|_{\mathscr{A}^*} \rangle$ *is called a* completion *of IAF.*

Why is it plausible to assume that the number of possible arguments is finite? An answer to this question is that in real-world applications it is relatively safe to assume that the number of participants that are part of an discussion is limited, and that no individual has an infinite number of ideas to propose as arguments. In such scenarios, the total number of arguments must be finite. Also, why is it plausible to assume that all arguments are known in advance? Because in real-world applications arguments that do not have enough support by the participants are not important for the discussion. As soon as a new argument is introduced, however, we can ask how this argument is related to the existing arguments, and thus learn step by step new arguments and attacks once they come to life and then maybe become significant.

We now extend the definitions for classical argumentation frameworks to argument-incomplete ones, distinguishing between properties holding either *possibly* or *necessarily*.

Definition 3. *Let $IAF = \langle \mathscr{A}', \mathscr{A}, \mathscr{R} \rangle$ be an argument-incomplete argumentation framework. For a property* **s**, *call a subset $S \subseteq \mathscr{A}$ of arguments*

- possibly **s** *in IAF if there is a completion $AF^* = \langle \mathscr{A}^*, \mathscr{R}|_{\mathscr{A}^*} \rangle$ of IAF such that $S|_{\mathscr{A}^*} = S \cap \mathscr{A}^*$ is* **s** *in AF^* and*
- necessarily **s** *in IAF if for all completions $AF^* = \langle \mathscr{A}^*, \mathscr{R}|_{\mathscr{A}^*} \rangle$ of IAF, $S|_{\mathscr{A}^*} = S \cap \mathscr{A}^*$ is* **s** *in AF^*.*

We call a set S a *possibly* (respectively *necessarily*) **s** *extension of IAF* if S is possibly (respectively necessarily) **s** in IAF.

Remark 1. The following concluding remarks hold for all argument-incomplete argumentation frameworks.

- The possible and necessary semantics inherit the correlations of the properties from Dung's model, i.e., for example, possible stability implies possible preferredness.
- There always is a possibly preferred extension and a possibly grounded extension.
- A possibly grounded extension is not unique, but there is at most one necessarily grounded extension.
- Let S be a possibly **s** extension of $\langle \mathscr{A}', \mathscr{A}, \mathscr{R} \rangle$, $\langle \mathscr{A}^*, \mathscr{R}|_{\mathscr{A}^*} \rangle$ a completion such that S is an **s** extension of $\langle \mathscr{A}^*, \mathscr{R}|_{\mathscr{A}^*} \rangle$, and $a \in \mathscr{A} \setminus \mathscr{A}^*$. Then $S \cup \{a\}$ is a possibly **s** extension of $\langle \mathscr{A}', \mathscr{A}, \mathscr{R} \rangle$ as well. We call those possibly **s** extensions *trivial supersets (of S).*

Example 3. Consider again the argumentation framework from Example 1 but assume that some arguments, namely b, e, and g, are not certain yet. Then we have the argument-incomplete argumentation framework

$$IAF = \langle \mathscr{A}', \mathscr{A}, \mathscr{R} \rangle = \langle \{a, c, d, f\}, \{a, b, c, d, e, f, g\},$$
$$\{(a, b), (a, c), (b, d), (c, d), (e, d), (e, f), (e, g), (f, e), (g, g)\} \rangle.$$

Figure 3 illustrates this argument-incomplete argumentation framework. Solid vertices represent members of $\mathscr{A}'$, while dotted vertices stand for elements of $\mathscr{A} \setminus \mathscr{A}'$. Dashed arcs symbolize the incoming and outgoing attacks of the elements $\mathscr{A} \setminus \mathscr{A}'$. Note that the attacks drawn as black arcs will always be in any completion, while those attacks (x, y) drawn as dashed arcs are part of a completion if and only if x and y both are arguments in that completion.

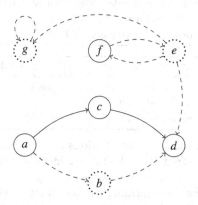

Fig. 3. The representation of the argument-incomplete argumentation framework $IAF = \langle \mathscr{A}', \mathscr{A}, \mathscr{R} \rangle$ from Example 3

First, let us consider possibly **s** extensions of IAF. Obviously, all **s** extensions of the original argumentation framework AF are possibly **s** extensions of IAF. Hence, $\{a, e\}$ is possibly stable (and also possibly preferred, possibly complete, and possibly admissible). However, $\{a, d, f\}$ is also possibly stable, as it is stable in the completion $\langle \mathscr{A}', \mathscr{R}|_{\mathscr{A}'} \rangle$, and there are no other possibly stable nor possibly preferred sets, except for trivial supersets. $\{a\}$ is the grounded extension of AF, and it remains to be a possibly grounded extension of IAF, but additionally $\{a, d, f\}$ is possibly grounded, as it is the unique grounded extension of $\langle \mathscr{A}', \mathscr{R}|_{\mathscr{A}'} \rangle$. It is easy to see that there are no more possibly grounded extensions, and that the only possibly complete sets are $\{a\}$, $\{a, e\}$, and $\{a, d, f\}$ (both except for trivial supersets). Besides the admissible sets of AF, there is one more possibly admissible set of IAF (except for trivial supersets), namely $\{a, d\}$, which is admissible in, e.g., the completion $\langle \mathscr{A}', \mathscr{R}|_{\mathscr{A}'} \rangle$.

To find all necessarily **s** extensions of IAF, it is sufficient to check whether the possibly **s** extensions mentioned above, except for trivial supersets, are **s** extensions of all completions. Hence, there is no necessarily stable extension, as $\{a, e\} \cap \mathscr{A}'$ is not stable in $\langle \mathscr{A}', \mathscr{R}|_{\mathscr{A}'}\rangle$ and $\{a, d, f\}$ is not stable in $\langle \mathscr{A}, \mathscr{R}\rangle$, and the same completions also prevent $\{a\}$ and $\{a, d, f\}$ from being necessarily grounded. The only necessarily preferred extension is $\{a, d, f\}$, because $\{a, e\} \cap \mathscr{A}'$ is not preferred in $\langle \mathscr{A}', \mathscr{R}|_{\mathscr{A}'}\rangle$. $\{a\}$ and $\{a, e\}$ are not necessarily complete because of the completion $\langle \mathscr{A}', \mathscr{R}|_{\mathscr{A}'}\rangle$. Lastly, all possibly admissible sets except for $\{a, d\}$ are also necessarily admissible.

3.2   Possible and Necessary Verification

Using s-VERIFICATION as a starting point, we define two decision problems for argument-incomplete argumentation frameworks.

---

s-ARG-INC-POSSIBLE-VERIFICATION (s-ARGINCPV)

Given: An argument-incomplete argumentation framework $IAF = \langle \mathscr{A}', \mathscr{A}, \mathscr{R}\rangle$ and a set $S \subseteq \mathscr{A}$.

Question: Is S a possibly s extension of IAF?

---

s-ARG-INC-NECESSARY-VERIFICATION (s-ARGINCNV)

Given: An argument-incomplete argumentation framework $IAF = \langle \mathscr{A}', \mathscr{A}, \mathscr{R}\rangle$ and a set $S \subseteq \mathscr{A}$.

Question: Is S a necessarily s extension of IAF?

Note that we do not make any restrictions on the choice of the set S in the problem instance, but restrict it to the arguments that occur in the completion when asking whether it is an extension. This captures the setting where all elements in $S \cap \mathcal{A}'$ must be contained in our final restriction of S, whereas this is not decided yet for the elements in $S \cap (\mathcal{A} \setminus \mathcal{A}')$. Furthermore, it is not a restriction that only elements from $\mathcal{A}'$ are sure to be in our final set, since other elements that should definitely be in the final set may be added to our argumentation framework in advance. Hence, this is a reasonable choice in argument-incomplete argumentation frameworks.

We will start the discussion of argument-incomplete argumentation frameworks with an easy result for conflict-freeness.

Proposition 1. CF-ARGINCPV *and* CF-ARGINCNV *both are in P.*

Proof. First, it holds that $(\langle \mathscr{A}', \mathscr{A}, \mathscr{R}\rangle, S) \in$ CF-ARGINCPV if and only if $S|_{\mathscr{A}'}$ is conflict-free in $\langle \mathscr{A}', \mathscr{R}|_{\mathscr{A}'}\rangle$, because any conflict in $\langle \mathscr{A}', \mathscr{R}|_{\mathscr{A}'}\rangle$ can also be found in any other completion. Second, it holds that $(\langle \mathscr{A}', \mathscr{A}, \mathscr{R}\rangle, S) \in$ CF-ARGINCNV if and only if S is conflict-free in $\langle \mathscr{A}, \mathscr{R}\rangle$, because if there is no conflict in $\langle \mathscr{A}, \mathscr{R}\rangle$, there is no other completion that can possibly have conflicts.   □

Now, we will turn to the other semantics. By definition of the decision problems, it is obvious that for $\mathbf{s} \in \{ad, st, cp, gr\}$, s-ARGINCPV is in NP and s-ARGINCNV is in coNP, as those four properties are easy to check. However, no polynomial-time algorithm is known to check for preferredness.

A trivial upper bound for PR-ArgIncPV can be obtained by the following—perhaps somewhat naive—approach: Check all supersets of the given set S as to whether they are admissible, and output "yes" if and only if the answer is always "no." Each of these checks is possible in polynomial time, and hence PR-ArgIncPV is in Σ_2^p. However, the problem PR-AttIncNV also is in coNP. As in the corresponding problem PR-ArgIncNV for attack-incomplete argumentation frameworks, the complement of PR-ArgIncNV is in NP. It is possible to check in polynomial time whether, given a completion $\langle \mathscr{A}^*, \mathscr{R}|_{\mathscr{A}^*}\rangle$ of IAF and a set $S^* \subseteq \mathscr{A}^* : S|_{A^*} \subset S^*$, either S^* is admissible or S is not admissible. All these trivial upper bounds are summarized in the following lemma.

Lemma 1. *1.* PR-ArgIncPV *is in* Σ_2^p.
2. For $s \in \{$ AD, ST, CP, GR$\}$, s-ArgIncPV *is in NP.*
3. For $s \in \{$ AD, ST, CP, GR, PR$\}$, s-ArgIncNV *is in coNP.*

We will now turn to showing lower bounds of these problems. We start with a straightforward reduction from PR-Verification to show coNP-hardness of the problems PR-ArgIncPV and PR-ArgIncNV.

Proposition 2. PR-ArgIncPV *is* coNP-*hard and the problem* PR-ArgIncNV *is* coNP-*complete.*

Proof. We show coNP-hardness by a reduction from the coNP-complete problem PR-Verification. Let $(\langle \mathscr{A}, \mathscr{R}\rangle, S)$ be a given instance of PR-Verification, and construct from it $(\langle \mathscr{A}, \mathscr{A}, \mathscr{R}\rangle, S)$, considered as an instance of both PR-ArgIncPV and PR-ArgIncNV. In the argument-incomplete argumentation framework, there are no arguments that can possibly join the discussion. Hence, the only completion in both cases is the argumentation framework $\langle \mathscr{A}, \mathscr{R}\rangle$. Now, it is easy to see that

$$(\langle \mathscr{A}, \mathscr{R}\rangle, S) \in \text{PR-Verification}$$
$$\iff (\langle \mathscr{A}, \mathscr{A}, \mathscr{R}\rangle, S) \in \text{PR-ArgIncPV}$$
$$\iff (\langle \mathscr{A}, \mathscr{A}, \mathscr{R}\rangle, S) \in \text{PR-ArgIncNV}.$$

This completes the proof.   □

Theorem 1. AD-ArgIncPV *is NP-complete.*

Proof. As already mentioned, we only need to show NP-hardness. To this end, we reduce from the following NP-complete problem (see the book by Garey and Johnson [22]):

Exact-Cover-By-3-Sets (X3C)	
Given:	A set $B = \{b_1, \ldots, b_{3k}\}$ and a family $\mathscr{S}$ of subsets of B, with $\|S_j\| = 3$ for all $S_j \in \mathscr{S}$.
Question:	Does there exist a subfamily $\mathcal{S}' \subseteq \mathscr{S}$ of size k that exactly covers B, i.e., $\bigcup_{S_j \in \mathcal{S}'} S_j = B$?

Given an instance $(B, \mathscr{S}) = (\{b_1, \ldots, b_{3k}\}, \{S_1, \ldots, S_m\})$ of X3C, we construct an instance $(\langle \mathscr{A}', \mathscr{A}, \mathscr{R} \rangle, S)$ of AD-ARGINCPV as follows:[2]

$$\mathscr{A}' = \{x\} \cup B,$$
$$\mathscr{A} = \{x\} \cup B \cup \mathscr{S},$$
$$\mathscr{R} = \{(b_i, x) \mid b_i \in B\} \cup$$
$$\{(S_j, b_{j_1}), (S_j, b_{j_2}), (S_j, b_{j_3}) \mid S_j = \{b_{j_1}, b_{j_2}, b_{j_3}\} \in \mathscr{S}\} \cup$$
$$\{(S_i, S_j), (S_j, S_i) \mid S_i, S_j \in \mathscr{S} \text{ and } S_i \cap S_j \neq \emptyset\},$$
$$S = \{x\} \cup \mathscr{S}.$$

In particular, $\mathscr{A}$ contains one argument b_i for every element $b_i \in B$, $1 \leq i \leq 3k$, one argument S_j for every set S_j in $\mathscr{S}$, $1 \leq j \leq m$, and one additional argument x. All arguments corresponding to elements of B attack x, and each argument S_j attacks the three arguments corresponding to those elements of B that belong to S_j in $\mathscr{S}$. Additionally, there are attacks between S_i and S_j if the corresponding sets in $\mathscr{S}$ are not disjoint. Finally, $\mathscr{A}'$ and S act as opponents: x belongs to both, but the arguments corresponding to elements in B belong to $\mathscr{A}'$ only, whereas the arguments corresponding to the sets in $\mathscr{S}$ belong to S only. See Fig. 4 for two examples of this construction: Fig. 4a shows a yes-instance of AD-ARGINCPV created from a yes-instance of X3C, and Fig. 4b shows a no-instance of AD-ARGINCPV created from a no-instance of X3C.

We claim that $(B, \mathscr{S}) \in$ X3C if and only if $(\langle \mathscr{A}', \mathscr{A}, \mathscr{R} \rangle, S) \in$ AD-ARGINCPV.

($\Longrightarrow$) Clearly, if $(B, \mathscr{S})$ is a yes-instance of X3C, we can add exactly those arguments S_i to $\mathscr{A}'$ that correspond to an exact cover of B. Let $\mathscr{A}^*$ be the argument set of this completion. In $\mathscr{A}^*$, every b_i, $1 \leq i \leq 3k$, is attacked by exactly one argument S_j, $1 \leq j \leq m$, as of the exact cover. Hence, $x \in S|_{\mathscr{A}^*}$ is defended against every attack. Additionally, the arguments S_j in $\mathscr{A}^*$ have no attacks between them, because the corresponding sets are pairwise disjoint, which implies that no new attacks on the elements of $S|_{\mathscr{A}^*}$ are introduced. But this means that $S|_{\mathscr{A}^*}$ is admissible in $\langle \mathscr{A}^*, \mathscr{R}|_{\mathscr{A}^*} \rangle$.

($\Longleftarrow$) If there is a completion with the argument set $\mathscr{A}^*$, this completion must defend x against every b_i, $1 \leq i \leq 3k$. This means that there must exist a cover of the elements of B by the sets of $\mathscr{S}$. But because the arguments S_j attack each other whenever they are not disjoint, this cover must be exact; otherwise, the set $S|_{\mathscr{A}^*}$ would not be conflict-free. Hence, there exists an exact cover of B. $\square$

We now try to tighten the bounds of s-ARGINCPV for each $\mathbf{s} \in \{$ST, CP, GR, PR$\}$. The first step is proving NP-hardness in all four cases. By Lemma 1, this gives NP-completeness for $\mathbf{s} \in \{$ST, CP, GR$\}$. Later on, we will use this results for $\mathbf{s} =$ PR, as well as the result for coNP-hardness, to show DP-hardness.

Theorem 2. *For* $\mathbf{s} \in \{$ST, CP, GR$\}$, *s-ARGINCPV is NP-complete, and PR-ARGINCPV is NP-hard.*

---

[2] We slightly abuse notation and use the same identifiers for both instances; it will always be clear from the context, though, which instance an element belongs to.

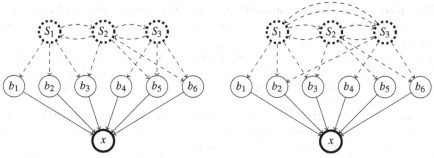

(a) $\mathscr{S} = \{\{b_1,b_2,b_3\},\{b_3,b_5,b_6\},\{b_4,b_5,b_6\}\}$. $(B,\mathscr{S})$ is a yes-instance of X3C that yields a yes-instance of AD-ARGINCPV.

(b) $\mathscr{S} = \{\{b_1,b_2,b_3\},\{b_3,b_5,b_6\},\{b_2,b_4,b_6\}\}$. $(B,\mathscr{S})$ is a no-instance of X3C that yields a no-instance of AD-ARGINCPV.

Fig. 4. Two examples of the reduction from X3C to AD-ARGINCPV. Both X3C instances have $B = \{b_1,\ldots,b_6\}$. All the arguments belong to $\mathscr{A}$, $\mathscr{A}'$ contains the solid arguments only, and the thick arguments are part of S.

Proof. Again, membership of the three former problems in NP is clear. It remains to show hardness for all four problems. We do this by showing that the reduction used in Theorem 1 also works for those four problems. To this end, we will prove that

$$(\langle \mathscr{A}', \mathscr{A}, \mathscr{R}\rangle, S) \in \text{AD-ARGINCPV}$$
$$\iff (\langle \mathscr{A}', \mathscr{A}, \mathscr{R}\rangle, S) \in \text{ST-ARGINCPV}$$
$$\iff (\langle \mathscr{A}', \mathscr{A}, \mathscr{R}\rangle, S) \in \text{PR-ARGINCPV}$$
$$\iff (\langle \mathscr{A}', \mathscr{A}, \mathscr{R}\rangle, S) \in \text{GR-ARGINCPV}$$
$$\iff (\langle \mathscr{A}', \mathscr{A}, \mathscr{R}\rangle, S) \in \text{CP-ARGINCPV}$$

holds for the instance $(\langle \mathscr{A}', \mathscr{A}, \mathscr{R}\rangle, S)$ constructed in the proof of Theorem 1.

$((\langle \mathscr{A}', \mathscr{A}, \mathscr{R}\rangle, S) \in$ AD-ARGINCPV implies $(\langle \mathscr{A}', \mathscr{A}, \mathscr{R}\rangle, S) \in$ ST-ARGINCPV): If $S|_{\mathscr{A}^*}$ is admissible for a completion $\langle \mathscr{A}^*, \mathscr{R}|_{\mathscr{A}^*}\rangle$, it in particular is conflict-free. We know from the reduction that $\langle \mathscr{A}^*, \mathscr{R}|_{\mathscr{A}^*}\rangle$ only contains arguments S_j that do not attack each other, and all these arguments belong to $S|_{\mathscr{A}^*}$. Hence, the only arguments outside of $S|_{\mathscr{A}^*}$ are the b_i's. But all of them are attacked, as explained in the proof of Theorem 1. Therefore, $S|_{\mathscr{A}^*}$ is a stable extension of $\langle \mathscr{A}^*, \mathscr{R}|_{\mathscr{A}^*}\rangle$.

$((\langle \mathscr{A}', \mathscr{A}, \mathscr{R}\rangle, S) \in$ PR-ARGINCPV implies $(\langle \mathscr{A}', \mathscr{A}, \mathscr{R}\rangle, S) \in$ GR-ARGINCPV): If $S|_{\mathscr{A}^*}$ is preferred for a completion $\langle \mathscr{A}^*, \mathscr{R}|_{\mathscr{A}^*}\rangle$, it is admissible, and thus the only arguments that are not attacked by any other argument are those S_j that correspond to an exact cover. This means for the characteristic function of this completion $\langle \mathscr{A}^*, \mathscr{R}|_{\mathscr{A}^*}\rangle$ that the output of the first step is the set that contains exactly those S_j. In the second step, we add argument x, because all those S_j defend x against all attacks from the arguments b_i. No new arguments are added in step three.

Therefore, this set is the grounded extension of the argumentation framework $\langle \mathscr{A}^*, \mathscr{R}|_{\mathscr{A}^*} \rangle$. But this set is exactly the set $S|_{A^*}$. Hence, $S|_{A^*}$ is the grounded extension of $\langle \mathscr{A}^*, \mathscr{R}|_{\mathscr{A}^*} \rangle$.

It is easy to see the three remaining implications needed to prove these five statements equivalent: Every stable set is preferred, every grounded set is complete, and every complete set is admissible. This completes the proof. □

We now strengthen the NP-hardness lower bound for PR-ARGINCPV given in Theorem 2 to DP-hardness. The following lemma due to Wagner [32] gives a sufficient condition for proving hardness for DP.

Lemma 2 (Wagner [32]). *Let A be some NP-hard problem, and let B be any set. If there exists a polynomial-time computable function f such that, for any two instances z_1 and z_2 of A for which $z_2 \in A$ implies $z_1 \in A$, we have*

$$(z_1 \in A \text{ and } z_2 \notin A) \iff f(z_1, z_2) \in B,$$

then B is DP-hard.

Theorem 3. PR-ARGINCPV *is DP-hard.*

Proof. We will use Wagner's lemma to show DP-hardness: Let PR-ARGINCPV be the set B from Wagner's lemma, and let X3C be the NP-complete problem A in that lemma. Let z_1 and z_2 be two instances of X3C such that $z_2 \in$ X3C implies $z_1 \in$ X3C. We construct an instance $(\langle \mathscr{A}', \mathscr{A}, \mathscr{R} \rangle, S)$ of PR-ARGINCPV as follows:

- Construct an instance $(\langle \mathscr{A}'_1, \mathscr{A}_1, \mathscr{R}_1 \rangle, S_1)$ from the X3C instance z_1 exactly as in the proof of Theorem 1.
- The construction of an instance $(\langle \mathscr{A}'_2, \mathscr{A}_2, \mathscr{R}_2 \rangle, S_2)$ from the X3C instance z_2, however, is obtained as the composition of two reductions: Since PR-VERIFICATION is coNP-complete and X3C is NP-complete, there exists a reduction f such that $z_2 \notin$ X3C if and only if $f(z_2) \in$ PR-VERIFICATION. Now, letting g be the reduction from Proposition 2, we have $z_2 \notin$ X3C if and only if $g(f(z_2)) \in$ PR-ARGINCPV.
- Given two instances of PR-ARGINCPV, $(\langle \mathscr{A}'_1, \mathscr{A}_1, \mathscr{R}_1 \rangle, S_1)$ and $(\langle \mathscr{A}'_2, \mathscr{A}_2, \mathscr{R}_2 \rangle, S_2)$, let $(\langle \mathscr{A}', \mathscr{A}, \mathscr{R} \rangle, S) = (\langle \mathscr{A}'_1 \cup \mathscr{A}'_2, \mathscr{A}_1 \cup \mathscr{A}_2, \mathscr{R}_1 \cup \mathscr{R}_2 \rangle, S_1 \cup S_2)$ if $\mathscr{A}_1 \cap \mathscr{A}_2 = \emptyset$ (otherwise, simply rename the elements in one instance). Hence, this new instance consists of two disconnected argument-incomplete argumentation frameworks.

This completes the reduction. We claim that ($z_1 \in$ X3C and $z_2 \notin$ X3C) if and only if $(\langle \mathscr{A}', \mathscr{A}, \mathscr{R} \rangle, S) \in$ PR-ARGINCPV.

($\Longrightarrow$) If $z_1 \in$ X3C and $z_2 \notin$ X3C, then $(\langle \mathscr{A}'_1, \mathscr{A}_1, \mathscr{R}_1 \rangle, S_1)$ and $(\langle \mathscr{A}'_2, \mathscr{A}_2, \mathscr{R}_2 \rangle, S_2)$ both are yes-instances of PR-ARGINCPV. Thus we must have a completion for the first and a completion for the second argument-incomplete argumentation framework such that S_1 restricted to the arguments in this first completion and S_2 restricted to the arguments in the second completion are preferred in their respective completion. But then, using the same completions for each part of $\langle \mathscr{A}', \mathscr{A}, \mathscr{R} \rangle$, we have that S restricted to those arguments must be preferred in this argumentation framework. This is true because no new attacks are introduced in $\langle \mathscr{A}', \mathscr{A}, \mathscr{R} \rangle$ and, therefore, neither are any new conflicts added nor do the elements of S have to be defended by any other arguments than before. Hence, $(\langle \mathscr{A}', \mathscr{A}, \mathscr{R} \rangle, S)$ is a yes-instance of PR-ARGINCPV.

($\Longleftarrow$) Conversely, assume that $(\langle \mathscr{A}', \mathscr{A}, \mathscr{R} \rangle, S)$ is a yes-instance of PR-ARGINCPV, and assume further that $(\langle \mathscr{A}_i', \mathscr{A}_i, \mathscr{R}_i \rangle, S_i)$ is a no-instance of PR-ARGINCPV for some $i \in \{1, 2\}$. Then there is no completion $\langle \mathscr{A}_i^*, \mathscr{R}_i|_{\mathscr{A}_i^*} \rangle$ of $\langle \mathscr{A}_i', \mathscr{A}_i, \mathscr{R}_i \rangle$ such that $S_i|_{\mathscr{A}_i^*}$ is preferred in it. That means that for every completion $\langle \mathscr{A}_i^*, \mathscr{R}_i|_{\mathscr{A}_i^*} \rangle$, $S_i|_{\mathscr{A}_i^*}$ either is not conflict-free, or is not admissible, or that there exists a superset of $S_i|_{\mathscr{A}_i^*}$ in $\langle \mathscr{A}_i^*, \mathscr{R}_i|_{\mathscr{A}_i^*} \rangle$ that is admissible. We consider these cases separately:

1. If $S_i|_{\mathscr{A}_i^*}$ is not conflict-free in $\langle \mathscr{A}_i^*, \mathscr{R}_i|_{\mathscr{A}_i^*} \rangle$, this conflict also exists in $S|_{\mathscr{A}^*}$ for any completion $\langle \mathscr{A}^*, \mathscr{R}|_{\mathscr{A}^*} \rangle$ of $\langle \mathscr{A}', \mathscr{A}, \mathscr{R} \rangle$ with $\mathscr{A}^* \cap \mathscr{A}_i = \mathscr{A}_i^*$.
2. If $S_i|_{\mathscr{A}_i^*}$ is not admissible in $\langle \mathscr{A}_i^*, \mathscr{R}_i|_{\mathscr{A}_i^*} \rangle$, there must be an undefended attack. However, by the same argument as above, this attack is still undefended in any completion $\langle \mathscr{A}^*, \mathscr{R}|_{\mathscr{A}^*} \rangle$ of $\langle \mathscr{A}', \mathscr{A}, \mathscr{R} \rangle$ with $\mathscr{A}^* \cap \mathscr{A}_i = \mathscr{A}_i^*$.
3. If there is a superset of $S_i|_{\mathscr{A}_i^*}$ preventing it from being preferred in $\langle \mathscr{A}_i^*, \mathscr{R}_i|_{\mathscr{A}_i^*} \rangle$, this superset translates into a superset of $S|_{\mathscr{A}^*}$ for any completion $\langle \mathscr{A}^*, \mathscr{R}|_{\mathscr{A}^*} \rangle$ of $\langle \mathscr{A}', \mathscr{A}, \mathscr{R} \rangle$ with $\mathscr{A}^* \cap \mathscr{A}_i = \mathscr{A}_i^*$, thus also preventing $S|_{\mathscr{A}^*}$ from being preferred in $\langle \mathscr{A}^*, \mathscr{R}|_{\mathscr{A}^*} \rangle$.

Hence, none of these cases can happen, because $(\langle \mathscr{A}', \mathscr{A}, \mathscr{R} \rangle, S)$ is a yes-instance of PR-ARGINCPV. But this means that $S_i|_{\mathscr{A}_i^*}$ is a preferred extension of a completion $\langle \mathscr{A}_i^*, \mathscr{R}_i|_{\mathscr{A}_i^*} \rangle$ of $\langle \mathscr{A}_i', \mathscr{A}_i, \mathscr{R}_i \rangle$, a contradiction. $\square$

4 Conclusions and Open Questions

We have analyzed a setting for argumentation frameworks in which only a subset of all arguments is currently known to be part of the discussion. To this end, we have introduced a formal model for argument-incomplete argumentation frameworks including adaptions of the criteria of conflict-freeness, admissibility, preferredness, stability, completeness, and groundedness. These adaptions were defined by means of the notions of possibility and necessity. On this basis, we adapted the decision problem s-VERIFICATION and defined two variants, namely s-ARGINCPV and s-sARGINCNV, that fit our model.

Table 1 summarizes already known results for the s-VERIFICATION problem due to Dung [18], Dunne and Wooldridge [19], and Dimopoulos and Torres [16], as well as our results. In contrast to the results of s-VERIFICATION, s-ARGINCPV is hard to decide in all cases, except for the trivial property conflict-freeness. Besides the straightforward results for conflict-freeness and preferredness, the exact complexity of s-ARGINCNV remains open, as well as that of PR-ARGINCPV.

As a future task, we propose to investigate other decision problems, e.g., CREDULOUS-ACCEPTANCE, SKEPTICAL-ACCEPTANCE, EXISTENCE, and NON-EMPTINESS, adapt their notion to fit our model, and analyze their complexity.

Table 1. Overview of complexity results both in the standard model (s-Verification) and in the argument-incomplete model of this paper (s-ArgIncPV and s-ArgIncNV)

s	Verification	ArgIncPV	ArgIncNV
CF	in P	in P (Lemma 1)	in P (Lemma 1)
AD	in P	NP-complete (Theorem 1)	in coNP (Lemma 1)
ST	in P	NP-complete (Theorem 2)	in coNP (Lemma 1)
CP	in P	NP-complete (Theorem 2)	in coNP (Lemma 1)
GR	in P	NP-complete (Theorem 2)	in coNP (Lemma 1)
PR	coNP-complete	DP-hard (Theorem 3), in Σ_2^p (Lemma 1)	coNP-complete (Lemma 2)

On the other hand, it would be interesting to generalize other semantics (e.g., the ideal, semi-stable, or prudent semantics [19]) in the context of argument-incomplete argumentation frameworks.

Acknowledgments. This work was supported in part by an NRW grant for gender-sensitive universities and the project "Online Participation," both funded by the NRW Ministry for Innovation, Science, and Research.

References

1. Baumeister, D., Rothe, J.: Preference aggregation by voting (chapter 4). In: Rothe, J. (ed.) Economics and Computation: An Introduction to Algorithmic Game Theory, Computational Social Choice, and Fair Division, pp. 197–325. Springer (2015). doi:10.1007/978-3-662-47904-9_4
2. Baumeister, D., Roos, M., Rothe, J.: Computational complexity of two variants of the possible winner problem. In: Proceedings of AAMAS 2011, pp. 853–860. IFAAMAS (2011)
3. Baumeister, D., Roos, M., Rothe, J., Schend, L., Xia, L.: The possible winner problem with uncertain weights. In: Proceedings of ECAI 2012, pp. 133–138. IOS Press (2012)
4. Baumeister, D., Erdélyi, G., Erdélyi, O., Rothe, J.: Complexity of manipulation and bribery in judgment aggregation for uniform premise-based quota rules. Math. Soc. Sci. **76**, 19–30 (2015)
5. Baumeister, D., Bouveret, S., Lang, J., Nguyen, T., Nguyen, N., Rothe, J.: Scoring rules for the allocation of indivisible goods. In: Proceedings of ECAI 2014, pp. 75–80. IOS Press (2014)
6. Baumeister, D., Neugebauer, D., Rothe, J.: Verification in attack-incomplete argumentation frameworks. In: Walsh, T. (ed.) ADT 2015. LNCS (LNAI), vol. 9346, pp. xx–yy. Springer, Heidelberg (2015)
7. Boella, G., Kaci, S., van der Torre, L.: Dynamics in argumentation with single extensions: abstraction principles and the grounded extension. In: Sossai, C., Chemello, G. (eds.) ECSQARU 2009. LNCS, vol. 5590, pp. 107–118. Springer, Heidelberg (2009)

8. Boutilier, C., Rosenschein, J.: Incomplete information and communication in voting. In: Brandt, F., Conitzer, V., Endriss, U., Lang, J., Procaccia, A. (eds.) Handbook of Computational Social Choice. Cambridge University Press, Cambridge (2015). (Chapter 10)

9. Bouveret, S., Lang, J.: Efficiency and envy-freeness in fair division of indivisible goods: logical representation and complexity. J. Artif. Intell. Res. **32**, 525–564 (2008)

10. Bouveret, S., Endriss, U., Lang, J.: Fair division under ordinal preferences: computing envy-free allocations of indivisible goods. In: Proceedings of ECAI 2010, pp. 387–392. IOS Press (2010)

11. Brams, S., Edelman, P., Fishburn, P.: Fair division of indivisible items. Theor. Decis. **55**(2), 147–180 (2003)

12. Cayrol, C., Devred, C., Lagasquie-Schiex, M.C.: Handling ignorance in argumentation: semantics of partial argumentation frameworks. In: Mellouli, K. (ed.) ECSQARU 2007. LNCS (LNAI), vol. 4724, pp. 259–270. Springer, Heidelberg (2007)

13. Cayrol, C., de Saint-Cyr, F., Lagasquie-Schiex, M.: Revision of an argumentation system. In: Proceedings of KR 2008, pp. 124–134. AAAI Press (2008)

14. Cayrol, C., de Saint-Cyr, F., Lagasquie-Schiex, M.: Change in abstract argumentation frameworks: adding an argument. J. Artif. Intell. Res. **38**, 49–84 (2010)

15. Chevaleyre, Y., Lang, J., Maudet, N., Monnot, J., Xia, L.: New candidates welcome! Possible winners with respect to the addition of new candidates. Math. Soc. Sci. **64**(1), 74–88 (2012)

16. Dimopoulos, Y., Torres, A.: Graph theoretical structures in logic programs and default theories. Theoret. Comput. Sci. **170**(1), 209–244 (1996)

17. Doder, D., Woltran, S.: Probabilistic argumentation frameworks – a logical approach. In: Straccia, U., Calì, A. (eds.) SUM 2014. LNCS, vol. 8720, pp. 134–147. Springer, Heidelberg (2014)

18. Dung, P.: On the acceptability of arguments and its fundamental role in non-monotonic reasoning, logic programming and n-person games. Artif. Intell. **77**(2), 321–357 (1995)

19. Dunne, P., Wooldridge, M.: Complexity of abstract argumentation. In: Rahwan, I., Simari, G. (eds.) Argumentation in Artificial Intelligence, pp. 85–104. Springer, New York (2009). (Chapter 5)

20. Fazzinga, B., Flesca, S., Parisi, F.: On the complexity of probabilistic abstract argumentation. In: Proceedings of IJCAI 2013, pp. 898–904. AAAI Press/IJCAI (2013)

21. Fazzinga, B., Flesca, S., Parisi, F.: Efficiently estimating the probability of extensions in abstract argumentation. In: Liu, W., Subrahmanian, V.S., Wijsen, J. (eds.) SUM 2013. LNCS, vol. 8078, pp. 106–119. Springer, Heidelberg (2013)

22. Garey, M., Johnson, D.: Computers and Intractability: A Guide to the Theory of NP-Completeness. W. H. Freeman and Company, New York (1979)

23. Konczak, K., Lang, J.: Voting procedures with incomplete preferences. In: Proceedings of Multidisciplinary IJCAI 2005 Workshop on Advances in Preference Handling, pp. 124–129 (2005)

24. Lang, J., Pini, M., Rossi, F., Salvagnin, D., Venable, K., Walsh, T.: Winner determination in voting trees with incomplete preferences and weighted votes. J. Auton. Agents Multi-Agent Syst. **25**(1), 130–157 (2012)

25. Lang, J., Rey, A., Rothe, J., Schadrack, H., Schend, L.: Representing and solving hedonic games with ordinal preferences and thresholds. In: Proceedings of AAMAS 2015. IFAAMAS (2015)

26. Li, H., Oren, N., Norman, T.J.: Probabilistic argumentation frameworks. In: Modgil, S., Oren, N., Toni, F. (eds.) TAFA 2011. LNCS, vol. 7132, pp. 1–16. Springer, Heidelberg (2012)

27. Meyer, A., Stockmeyer, L.: The equivalence problem for regular expressions with squaring requires exponential space. In: Proceedings of FOCS 1972, pp. 125–129. IEEE Press (1972)

28. Papadimitriou, C.: Computational Complexity, 2nd edn. Addison-Wesley, New York (1995). Reprinted with corrections

29. Papadimitriou, C., Yannakakis, M.: The complexity of facets (and some facets of complexity). J. Comput. Syst. Sci. **28**(2), 244–259 (1984)

30. Rahwan, I., Simari, G. (eds.): Argumentation in Artificial Intelligence. Springer, New York (2009)

31. Stockmeyer, L.: The polynomial-time hierarchy. Theoret. Comput. Sci. **3**(1), 1–22 (1976)

32. Wagner, K.: More complicated questions about maxima and minima, and some closures of NP. Theoret. Comput. Sci. **51**(1–2), 53–80 (1987)

33. Xia, L., Conitzer, V.: Determining possible and necessary winners given partial orders. J. Artif. Intell. Res. **41**, 25–67 (2011)

Bribery and Control

Complexity of Optimal Lobbying in Threshold Aggregation

Ilan Nehama[(✉)]

The Hebrew University of Jerusalem, Jerusalem, Israel
ilan.nehama@mail.huji.ac.il

Abstract. This work studies the computational complexity of Optimal Lobbying under Threshold Aggregation. Optimal Lobbying is the problem a lobbyist or a campaign manager faces in a voting scenario of a multi-issue referendum when trying to influence the result. The Lobby is faced with a profile that specifies for each voter and each issue whether the voter approves or rejects the issue, and seeks to find the smallest set of voters it can influence to change their vote, for a desired outcome to be obtained. This problem also describes problems arising in other scenarios of aggregation, such as principal-agents incentives scheme in a complex combinatorial problem, and bribery in Truth-Functional Judgement Aggregation. We study cases when the issues are aggregated by a threshold aggregator, that is, an anonymous monotone function, and the desired outcomes set is upward-closed. We analyze this problem with regard to two parameters: the minimal number of supporters needed to pass an issue, and the size of the maximal minterm of the desired set. For these parameters we separate tractable cases from untractable cases and in that generalize the *NP-complete* result of Christian et al. [8]. We show that for the extreme values of the parameters, the problem is solvable in polynomial time, and provide algorithms. On the other hand, we prove the problem is not solvable in polynomial time for the non-extremal values, which are common values for the parameters.

Keywords: Optimal Lobbying · Threshold function · Time complexity

1 Introduction

This paper studies the problem of Optimal Lobbying in *Multi-Issue Elections*. In multi-issue elections n voters are voting on m issues; Each voter i declares

---

Extended abstract of this work was presented at AAMAS 2013 [25].

This work was presented in the 5th Israeli Game Theory Conference, Dagtuhl seminar on "Judgment Aggregation for Artificial Intelligence", and MOVE-Federmann joint Workshop on "Game Theory and its Applications" in Barcelona. The author would like to thank the participants in these workshops, as well as to Yehuda (John) Levy and the anonymous referees for their comments that helped improve this paper.

This research was supported by ISF grants 230/10 and 1435/14 administered by the Israeli Academy of Sciences and by the Google Inter-university Center for Electronic Markets and Auctions.

T. Walsh (Ed.): ADT 2015, LNAI 9346, pp. 379–395, 2015.
DOI: 10.1007/978-3-319-23114-3_23

his Boolean position on each issue j: x_i^j; And the outcome on each issue j, o^j, is decided by aggregating the votes on the issue using some *aggregation function* φ: $o^j = \varphi(x_1^j, \ldots, x_n^j)$. The *Optimal Lobbying* problem formalizes the challenge that faces an outside entity (*the Lobby*) that desires to affect the outcome of the vote and can do so by changing the votes of some of the voters, but at a cost. Formally, given the *profile* (x_i^j) stating for each voter his vote for each of the issues, the Lobby's goal is to find the minimal set K of voters such that changing the votes $(x_i^1 \cdots x_i^m)$ of voters $i \in K$ results in some desired outcome, where the *desired outcomes set* is captured by its indicator function $\psi(o^1, \ldots, o^m)$.

This model captures many voting scenarios, e.g., voting on a series of clauses of a bill in the parliament or elections using one ballot for several positions and decisions. The budget constraint on the number of voters ($|K|$) captures that the Lobby may need to compensate voters for the change or the need to invest time and money in personalized advertising. As we discuss in Sect. 2.1, OPTIMAL LOBBYING also models other problems in scenarios of aggregating complex opinions.

Clearly the difficulty of lobbying depends on both the aggregation function φ and on the desired outcomes function ψ. A natural example is where aggregation is done by simple majority vote and the desired outcome is defined by unanimity (i.e., the Lobby wants to achieve a majority on *all* issues). This scenario was studied by Christian et al. [8] who showed that the problem is *NP*-hard. This was generalized by Bredereck et al. [6] who showed this problem is *NP*-hard even under some (extreme) input constraints.[1] On the other hand, it is easy to verify that for some aggregation rules and desired outcomes sets the problem is easy. For example, if we use unanimity for aggregation and the desired outcomes set is also defined by unanimity, then it is easy to find (by a greedy algorithm) the minimal set of voters to influence. The problem is also easy when issues are aggregated using majority and the Lobby wants at least one issue to pass (i.e., the Lobby wants to achieve a majority on *any* issue). In many real-life situations one finds non-majority issue-aggregation functions (e.g., when approval of two-thirds of the voters is needed to rule against the status quo) or desired outcomes sets consisting of more than one outcome (e.g., when there is a trade-off in the eyes of the Lobby between several issues or issues combinations), hence there is a place to extend the study of OPTIMAL LOBBYING to these cases as well.

In this paper, we mainly study the computational complexity of OPTIMAL LOBBYING for the following natural families of aggregation functions and of desired outcomes sets: The aggregation function is an anonymous and monotone function, that is, an issue passes if at least t voters approve it, for a predefined *threshold* t;[2] And the desired outcomes set is an upward-closed set, i.e., if x is a desired outcome, and all the issues that pass in x, also pass in y, then y is a desired outcome too. In the full version we show a more general analysis for

---

[1] When each voter approves at most three issues and the budget is $\left(\lceil \frac{n+1}{2} \rceil - 1\right)$, one less than the required majority threshold.

[2] In the Judgement Aggregation literature, such voting method is also called *Uniform Quota Rule* [10].

cases in which the assumptions do not hold. We present a systematic study of which combinations of an aggregation method and a desired outcomes set allow efficient lobbying, and which give rise to a computationally hard problem. This study generalizes all previously known computational complexity results in this setting. It turns out that the complexity of the lobbying problem hinges mainly on two parameters:

1. t - The *Threshold of Aggregation* – the minimal number of votes that is needed to pass an issue.
2. z - The *Maximal minterm size* of the desired outcomes set – that is, the maximal size of a minimal inclusion-wise desired issues set. Equivalently, the minterms are the terms in the minimal DNF form of the function ψ.[3]

We essentially show that the problem is tractable if and only if either of these two parameters is *bounded by a constant*. When both of them are at least *polynomial* in the input size (i.e., $(mn)^c$ for some $0 < c < 1$) then the problem becomes *NP*-hard. This is true for both the decision and search variants of the OPTIMAL LOBBYING problem.

We show two dichotomy theorems. One for the *unanimity case* in which the issues are aggregated using the unanimity function (that is, an issue passes if and only if all voters support it), and the second for the *non-unanimity* case. It is especially interesting to note the sharp threshold phenomenon between the unanimity case and the *almost-unanimity*. For instance, if the Lobby allows at most one issue to fail, if the issues are aggregated using unanimity we show the problem to be solvable in polynomial time, but if the issues are aggregated using almost-unanimity (even in the extreme case in which at most one voter is allowed to vote against without causing the issue to be rejected) we show the problem to be *NP-complete*.

Throughout this paper we use asymptotic notions for boundaries on the parameters t and z. E.g., we say they are *bounded* when there is a constant independent of n and m bounding them from above, *super-constant* when there is no such constant, and we say they are *polynomial* when there is a polynomial in n and m bounding them (either from above or below). In cases where the notion is not clear enough, we add a formal definition of the bounds.

Theorem 1 (For the full formal statement see Theorems 5 and 11). *Let the aggregation function be unanimity and the desired outcomes be the outcomes in which at least z issues pass.*

- *If z or $(m - z)$ is bounded,* OPTIMAL LOBBYING *can be solved in polynomial time.*
- *If both z and $(m - z)$ are polynomial (i.e., $z \in [m^\epsilon, m - m^\epsilon]$ for some $\epsilon > 0$), then* OPTIMAL LOBBYING *is NP-complete.*

---

[3] For example, the minterms of the set represented by $\psi = \left(x^1 \wedge x^2\right) \vee \left(x^2 \wedge x^3 \wedge x^4\right)$ are 1100 and 0111, and the minterms of the set of majorities for $2k - 1$ issues are all the issues sets of size k.

– *There exists a constant $\alpha^\star > 1$ s.t. if both z and $(m - z)$ are poly-logarithmic of degree $\alpha^\star$ ($z \in \left[(\log m)^{\alpha^\star}, m - (\log m)^{\alpha^\star}\right]$), then* OPTIMAL LOBBYING *is not in P, assuming ETH.*[4]

Theorem 2 (For the full formal statement see Theorems 3, 4 and 6).
Let the aggregation function be the threshold function with threshold $t < n$ and let z be the maximal size of a minimal desired outcome (inclusion-wise).

– *If t or z is bounded,* OPTIMAL LOBBYING *can be solved in polynomial time.*
– *If both t and z are polynomial (i.e., $t \geqslant n^\epsilon$ and $z \geqslant m^\epsilon$ for some $\epsilon > 0$), then* OPTIMAL LOBBYING *is NP-complete.*
– *There exists a constant $\alpha^\star > 1$ s.t. if both t and z are poly-logarithmic of degree $\alpha^\star$ (i.e., $t \geqslant (\log n)^{\alpha^\star}$ and $z \geqslant (\log m)^{\alpha^\star}$), then* OPTIMAL LOBBYING *is not in P, assuming ETH (see Footnote 4).*

2   The Optimal Lobbying Problem

The problem OPTIMAL LOBBYING models a society of n voters – $[n] = \{1, 2, \ldots, n\}$ – that decides on m Boolean issues using a voting method, and a lobbyist that desires to influence the decision to be in a set desired by it. It consists of a profile, a voting method, and a desired outcomes set.

Profile: The profile defines the vote of each of the voters on each of the issues. We model it by a Boolean matrix $X \in \{0, 1\}^{n \times m}$, where $\mathbf{n}$ is the number of *voters* and $\mathbf{m}$ is the number of *issues*. That is, an entry X_i^j denotes the vote of the $i^{\underline{th}}$ voter for the $j^{\underline{th}}$ issue and we consider 1 as an acceptance vote and 0 as a rejection vote. Throughout this paper, we use superscript notation when indexing issues and subscript notation when indexing voters.

Voting Method: The voting method used to aggregate the votes for a specific issue into an aggregated accept/reject opinion. It is defined by a function $\varphi \colon \{0, 1\}^n \to \{0, 1\}$ (applied on each of the issues – the columns of the matrix X).

Desired Outcomes: We model the Lobby as having a *dichotomous preferences*, that is, each outcome is either desired or undesired. We model the preference using a Boolean function $\psi \colon \{0, 1\}^m \to \{0, 1\}$ returning for each outcome (a vector of length m), whether it is desired. We use propositional formulas over $x^1, \ldots, x^m$ to describe ψ, e.g., the desired outcomes set defined by $\psi = x^1 \vee \left(x^2 \wedge x^4\right)$ is all the outcomes in which either the first issue passes or both the second and forth issues pass.

Formally, the problem the Lobby is facing, while knowing the voting method φ and the desired outcomes set ψ, is modeled by the following optimization problem:

---

[4] Exponential Time Hypothesis (ETH) [21]: 3-SAT cannot be solved in time less than $2^{\delta n}$ for some $\delta > 0$.

Problem (OL(φ, ψ) – Optimal Lobbying with voting method φ and desired outcomes set ψ).

INSTANCE: A VOTING PROFILE $X \in \{0,1\}^{n \times m}$.

TASK: FIND A COALITION OF VOTERS C OF MINIMAL SIZE AND A VOTING PROFILE $Y \in \{0,1\}^{n \times m}$ THAT DIFFERS FROM X ONLY FOR THE ROWS (VOTERS) IN C S.T. $\psi\left(\varphi\left(Y^1\right), \ldots, \varphi\left(Y^m\right)\right) = 1$ (Y^j BEING THE $j^{\underline{th}}$ COLUMN OF Y).

We define the corresponding decision problem to be:

Problem (OLD(φ, ψ) – Optimal Lobbying Decision Problem).

INSTANCE: A VOTING PROFILE $X \in \{0,1\}^{n \times m}$.
         A BUDGET $k \leqslant n$.

QUESTION: CAN THE LOBBY CHANGE THE VOTES OF AT MOST k VOTERS TO GET A VOTING PROFILE $Y \in \{0,1\}^{n \times m}$ S.T. $\psi\left(\varphi\left(Y^1\right), \ldots, \varphi\left(Y^m\right)\right) = 1$?

φ and ψ are not part of the input but are parameters of the problem, assuming oracle access to them. This way we circumvent the question of representation compactness of these parameters. For instance, in the characterization theorems and in the proofs, we refer to the DNF representation of ψ and in particular to the maximal minterm in it, while not assuming that ψ is given in DNF form, that this form is compact, or that it is easy to compute the maximal minterm. The common cases of Optimal Lobbying usually consists of functions that are polynomially computable and representable, so results of similar flavor to the ones we show are generated naturally from this paper. In the full version we discuss the reasons we think this is the right way to model and analyze the problem.

In this paper, we study the computational complexity of OLD(φ, ψ) when ψ is an upward-closed desired outcomes set and φ is an anonymous monotone function. A family of sets is upward-closed if for any two sets of issues $A \subseteq B$, if A is in the family then so is B. Note that upward-closedness of the desired outcomes set is equivalent to monotonicity of its indicator function. We note that an anonymous monotone function φ can be equivalently defined by the threshold t defined by: φ returns 1 if and only if at least t out of the n voters approved the issue.[5] Two special cases are the *unanimity* functions, Unan_n, which are the threshold aggregation functions with threshold that is equal to the number of voters ($t = n$), and the *majority* functions, Maj_n, which are the threshold aggregation functions with threshold that is equal to a majority $\left(t = \left\lceil \frac{n}{2} \right\rceil\right)$.

---

[5] As a referee commented, t might also be a subjective threshold of the Lobby and not the objective threshold of the voting method. E.g., in a scenario in which there are several competing lobbyists, the Lobby will want to put together a non-minimal coalition of voters so that it is tougher for the competing lobbyist to break this coalition. In such case, t would be the subjective threshold that includes also this safety margins.

2.1 Motivation

We find the motivation for analyzing $OLD(\varphi, \psi)$ in several fields.

Optimal Lobbying: As shown in the introduction, this is a generalization of the OPTIMAL LOBBYING problem defined by Christian et al. [8]. This problem models a variety of situations in which the opinions of several voters are aggregated and we wish to analyze the complexity of the problem a lobbyist or a campaign manager faces. For the ease of the story, we have in mind a human Lobby, but it is reasonable to imagine in a complex voting procedure that the lobbying task is delegated to a computerized agent needing to find the best way to get a desired result while the budget is a cost of negotiating with the different agents/voters.

Incentivizing in a Complex Combinatorial Project: The OPTIMAL LOBBYING problem can also be interpreted as the problem of a principal to incentivize the minimal number of workers in a complex project. A principal is interested in the success of a meta-project that is composed of m independent projects. The success of each of the projects depends on the effort exerted by a group of workers; i.e., there is a known technology function φ that, given the set of workers who exerted effort, returns whether the project succeeds. In addition, there is a production function ψ that, given the projects that succeeded, returns whether the meta-project succeeds.

For each of the workers, the principal knows in which projects he will exert effort "naturally" (for example, the projects that are close to him, are easy for him, or in which his effort is monitored). The principal can choose to incentivize a worker to exert effort (in some or all projects) in a costly way (e.g., offer a monetary payment to the worker or use a device to monitor his effort). Therefore, the principal faces a trade-off between the success of the meta-project and the cost of incentivizing the workers.

Finding the minimal set of workers that will cause the meta-project to succeed by exerting effort is equivalent to $OL(\varphi, \psi)$. The problem of finding an incentives scheme in simple projects was presented and studied by (among others) Holmstrom [20] and Babaioff et al. [1] so this work is a generalization of these works. In this day and age, when huge complex projects can be run on the network in a distributed manner, e.g., on Mechanical Turk, the coordination is done by an agent that seeks to find the best incentivization scheme in order to maximize the probability of success.

Bribery: Bribery problems in Judgement Aggregation framework deal with the problem of finding the best group of voters whom one should cause to change their vote towards a preferred outcome, given an aggregation method and a restriction on the votes, e.g., preference aggregation.[6] The time complexity of bribery has been studied extensively in the framework of voting (e.g., [11,13,14]), which is a variant of preference aggregation. OPTIMAL LOBBYING can be defined

---

[6] For introduction to the field of Judgement Aggregation, one can read [22–24].

as Bribery for Truth-Functional Agendas[7] [23] where the preference of the briber is on one of the conclusions issues. Optimal Lobbying and Preference aggregation are the only two agenda families for which the bribery problem was defined (Baumeister et al. [5] define a problem of bribery for judgement aggregation, but in their definition the agenda is part of the input and they don't show which agendas are hard to bribe). We see OPTIMAL LOBBYING as a step toward studying bribery in the more general framework of Judgement Aggregation.

Computational Complexity Theory: The definition of OPTIMAL LOBBYING seems to be a minor tweak of classic *NP-complete* problems like HITTING SET (Problem 3), SET COVER [18, Problem SP5], and SET MULTI-COVER [27, p. 112], and vice versa. Yet, we did not find an embedding between OPTIMAL LOBBYING and neither of them (the complexity of $\text{OLD}\left(\text{MAJ}_n, \bigwedge_{j=1}^{m} x^j\right)$ is not derived trivially, as far as we found, from the complexity of these problems). We think that the tweak of *majority constraints* (the constraints being to cover a majority of the items), albeit it looks small, changes the problem dramatically. To the best of our knowledge, the literature of computational complexity did not deal with such majoritized versions of the classic problems. Hence, we think that these results might be of independent interest, and that such majoritized variants of classic problems should be explored further, studying the impact of this change on the complexity. We hope it will add a new trait of interesting problems and contribute to the study of complexity theory.

Why Analyze Computational Complexity of Social Choice Problems? There is a long strand of works analyzing the computational hardness of problems in Social Choice, specifically of possible attacks (manipulation, bribery, control, etc.), and this work joins this strand. Yet, we thought there is place to detail the value we find in such works, both as a practical barrier for a possible attacker (the Lobby in our case) and as results showing what cannot be proved (barrier to the theoretician).

The literature on computation hardness as a **barrier for manipulation in elections** started in the late 80s and early 90s by Bartholdi, Tovey and Trick [3,4] and Bartholdi and Orlin [2]. They defined the property of a voting rule being *computationally resistant* as the *NP*-hardness of the problem a manipulator is facing. The intuition behind this definition is that a manipulator, when faced with this *NP*-hard problem, will prefer not to manipulate but to submit his true vote. This line of thought was continued for other forms of attacks (e.g., bribery and group manipulation) and voting scenarios (e.g., multi-winner elections); For a survey on these works, see [15,16]. This intuition is also supported by experiments like the ones done by Harrison et al. [19]. They showed that, in

---

[7] In a *truth-functional agenda*, in addition to the unconstrained issues (the premises), there are *conclusion* issues. Each conclusion j is characterized by a Boolean function α_j over the premises and a vote is legitimate if the vote on a conclusion issue is consistent with applying the function α_j on the vote on the premise issues. I.e., $\{x \in \{0,1\}^m \mid x^j = \alpha_j(\text{premises}) \text{ for every conclusion issue } j\}$.

an actual survey, participants might give up strategic voting and answer survey questions truthfully while the questions are not incentive compatible, due to computational complexity to figure out the optimal strategy. In that they partially answer the common critique on this approach that it relies on *NP*-hardness as a measure of computational difficulty while it is a worst-case and asymptotic notion.

From the **researchers or designers point of view**, we would like to have a "nice" characterization of the profiles and of the desired outcomes sets that are vulnerable to bribery. While there is no clear formal definition of "nice" characterizations or of "useful" ones, it seems that a necessary condition for such a characterization should be that it can be transformed to a polynomial (or an almost polynomial) algorithm. Indeed, most of the characterizations in the literature satisfy this property. In that sense, proving that there is no polynomial (or sub exponential) algorithm for the bribery problem shows that there is no hope to find a nice characterization as well.

3   Results

We analyze the computational complexity of $OLD(\varphi, \psi)$ for a threshold function φ and a monotone function ψ. We find that the two parameters that characterize (nearly-fully) the complexity of the problem are t (the threshold of φ) and z (the size of the maximal minterm of ψ). For all the cases in which we prove the decision problem is solvable in polynomial time, we provide also direct polynomial algorithms for the search problem as well, showing no discrepancy between the two.

For the non-unanimity case $t < n$, we characterize the complexity of the problem as a function of t and z. In this case, the problem is tractable if either the threshold of the issue-aggregation function is bounded by a constant (which does not grow with the input size), or if the size of all minterms of the desired outcomes set are bounded by a constant. On the other hand, we show that when these parameters are polynomially large, the problem is *NP-complete*, and under mild computation assumptions it is not tractable even for poly-logarithmic values, and by that we get as specific cases the results shown by Christian et al. [8] and Bredereck et al. [7, (Partial Lobbying)]. There remains a small gap between the two ranges we deal with, when the values are super-constant (i.e., not bounded from above by a constant) or poly-logarithmic with a small degree, and in that sense this is only an almost-full characterization.

On the other hand, for the cases involving issue-aggregation using the unanimity function $(t = n)$, we show that the complexity cannot be characterized solely by the size of the maximal minterm. We show that the problem is tractable when the desired outcomes set is simple (can be described using a polynomial number of minterms). While we conjecture the number of minterms characterize the intractability results as well, we characterize the complexity of $OLD(Unan, \psi)$ only for desired outcomes sets defined by a threshold, that is, the Lobby is indifferent between the issues but has a quota of desired issues to pass.

In this case we show that, unlike the non-unanimity case, the behaviour is symmetric in the sense that the complexity is characterized by $z' = \min(z, m - z)$ - the minimum between the number of issues needed to pass in order to guarantee success of the Lobby, and the number of issues needed to fail in order to guarantee failure of the Lobby. Similar to the non-unanimity case, when this parameter z' is bounded by a constant, the problem is tractable, and when it is super-constant, we show that the problem is intractable. Specifically, we show that when it is polynomially large, the problem is *NP-complete*, and under mild computation assumptions it is not tractable even for poly-logarithmic values. In this case too, the intermediate small range between constant and poly-logarithmic remains open.

Tractability Results

When the parameters are very small or large, we find polynomial algorithms to find the optimization problem. Due to space constraints we omit the algorithms and only describe the main idea behind them. The full description can be found in the full version of the paper.

The first algorithm is a simple brute-force enumeration of all "cheap" coalitions and when the threshold is small enough, we get a polynomial algorithm.

Theorem 3. OL (φ, ψ) *can be solved in time* $O(m \cdot tn^t)$. *In particular, when the threshold* t *is bounded by a constant not dependent on* m *and* n, *then* OL (φ, ψ) *can be solved in polynomial time.*

Our second algorithm uses the anonymity of threshold functions, and solves the problem by iterating over all "representative coalitions."[8]

Theorem 4. *If* ψ *can be written as the disjunction of* r *minterms, each of which is a conjunction of at most* z *variables, then* OL (φ, ψ) *can be solved in time* $O\left(r \cdot n^{2^z}\right)$. *In particular, when all minterms of* ψ *are bounded by a constant not dependent on* m *and* n, *then* OL (φ, ψ) *can be solved in polynomial time.*

The third algorithm solves the case in which the issue-aggregation function φ is the unanimity function ($t = n$). In utilizes the fact that the only way for the Lobby to get a desired result is to choose a minterm $\vee_{j \in M} x^j$ and convince all the voters that reject an issue in M to change their vote.

Theorem 5. *If* ψ *can be written as the disjunction of* r *minterms, each of which is a conjunction of at most* z *variables, then* OL $(Unan, \psi)$ *can be solved in time* $O(r \cdot zn)$. *In particular, when* ψ *can be written as the disjunction of* poly (m) *minterms, then* OL $(Unan, \psi)$ *can be solved in polynomial time.*

Intractability Results

We prove separately and using different reductions intractability results for the cases where "the threshold is smaller than n (non-unanimity)" and the cases where "$t = n$ (unanimity)."

---

[8] Note that bounding all the minterms is equivalent to bounding the maximal minterm.

Non-Unanimity Issue-Aggregation $t < n$. For the case $t < n$ we prove the following intractability result.

Theorem 6. *Let φ be a threshold function with threshold $t < n$, and ψ a monotone function with maximal minterm of size z.*

1. *If t is polynomial $(\exists \epsilon > 0 \quad t \geqslant n^{\epsilon})$, and z is polynomial $(\exists \epsilon > 0 \quad z \geqslant m^{\epsilon})$, then $\mathrm{OLD}\,(\varphi, \psi)$ is NP-complete.*
2. *For any problem $A \in NP$ and any $\beta \in (0, 1)$, there exists $\alpha > 1$ s.t. if $\mathrm{OLD}\,(\varphi, \psi)$ is solvable in polynomial time and if t and z are poly-logarithmic of degree α $(t \geqslant (\log n)^{\alpha}, z \geqslant (\log m)^{\alpha})$, then A is solvable in time $2^{O(n^{\beta})}$.*

As a corollary we get the following,

Corollary 7

1. *(Christian et al. [8]) $\mathrm{OLD}\,(\varphi, \psi)$ is NP-complete for φ the simple majority function and $\psi = \wedge_{j=1}^{m} x^{j}$, i.e., requiring all issues to pass.*
2. *Assuming ETH, there exists $\alpha^{\star} > 1$ s.t. if t and z are poly-logarithmic of degree $\alpha^{\star}$ $(t \geqslant (\log n)^{\alpha^{\star}}, z \geqslant (\log m)^{\alpha^{\star}})$, then $\mathrm{OLD}\,(\varphi, \psi)$ is not in P.*
3. *Assuming $NP \nsubseteq SUBEXP$,[9] if t and z are super-poly-logarithmic (there is no $\alpha > 1$ s.t. $t \geqslant (\log n)^{\alpha}$ and $z \geqslant (\log m)^{\alpha}$ for all large enough n and m), then $\mathrm{OLD}\,(\varphi, \psi)$ is not in P.*

We prove Theorem 6 by constructing the three reductions described below. The first two (Lemmas 8 and 9) are reductions from HITTING SET and VERTEX COVER, respectively, to $\mathrm{OLD}\big(\varphi, \wedge_{j=1}^{m} x^{j}\big)$ and the third reduction (Lemma 10) finishes the proof by reducing $\mathrm{OLD}\big(\varphi, \wedge_{j=1}^{m} x^{j}\big)$ to the general case $\mathrm{OLD}(\varphi, \psi)$. The first two reductions are different and aim at different ranges of the threshold t. Although any reduction from HITTING SET is also a reduction from VERTEX COVER, we prefer constructing a reduction from the former to get a stronger result (wider range of the parameters).

Problem (Hitting Set).

INSTANCE: COLLECTION $C = \big\{C^{j}\big\}$ OF $|C|$ SUBSETS OF A UNIVERSE S.
A POSITIVE INTEGER $k \leqslant |S|$.

QUESTION: IS THERE A HITTING SET $H \subseteq S$ OF SIZE k, I.E., A SET $H \subseteq S$
S.T. $\forall j \; C^{j} \cap H \neq \emptyset$?

REFERENCE: THE PROBLEM IS *NP-complete* – [18, PROBLEM SP8].

Lemma 8. *Given a family of threshold functions φ over n voters with threshold t that satisfies "t and $(n - t)$ diverge to infinity," there exists a reduction*

---

[9] $SUBEXP = \cap_{\epsilon > 0} DTIME\left(2^{s^{\epsilon}}\right)$ is the class of problems solvable in sub-exponential time. The assumption $NP \nsubseteq SUBEXP$ means assuming that there exists at least one NP problem that cannot be solved in time less than $2^{n^{\Omega(1)}}$.

that runs in time linear in the output size;[10] *and given an instance of* HITTING SET – $(S, C = \{C^j\}, k)$, *it produces an instance of* OLD $(\varphi, \wedge_{j=1}^m x^j)$ – $(X \in \{0,1\}^{n \times m}, k')$ – *s.t. the following is satisfied:*

- $m = n + |C|$, • $k' = k$, • $\min(t, n - t) = \max\left(|S|, \max_j |C^j|\right)$, • *and*

(S, C, k) *is satisfiable iff* (X, k') *is satisfiable.*

- *In addition the reduction's output satisfies that each issue is supported by either $t - 1$ or $t - k$ supporters.*

Proof Sketch of Lemma 8. Given an instance of HITTING SET – $(S, C = \{C^j\}, k)$, we construct the instance of OLD$(\varphi, \wedge_{j=1}^m x^j)$ – (X, k) by using the same threshold k and defining the profile X to be:

Voters \ Issues	$|C|$ issues	$m - |C|$ issues
$|S|$ voters	A	0
$n - |S|$ voters	In the jth column there are $t - 1 - (|S| - |C^j|)$ ones	- In each column there are $t - k$ ones - No all zeroes line
Total (out of n voters)	$t - 1$	$t - k$

when entries of the matrix $A \in \{0,1\}^{|S| \times |C|}$, $A_{i,j}$, are 1 if $i \notin C^j$ and 0 otherwise.

Due to the way we defined the bottom-right sub-profile, the Lobby can convince at most k voters in order to pass all issues, if and only if it can do so by convincing k voters from the top $|S|$ voters. Due to the definition of the top-left sub-profile, it can achieve that only by finding a hitting set of size k.   □

The reduction from VERTEX COVER is similar to the above but differs in the gadgets used for embedding of the incidence matrix in the profile. The full reductions can be found in the full version.

Problem (Vertex Cover).

INSTANCE: AN UNDIRECTED GRAPH $G = (V, E)$.
A POSITIVE INTEGER $k \leqslant |V|$.
QUESTION: IS THERE A COVER $C \subseteq V$ OF SIZE k IN G, I.E., A SET $C \subseteq V$
S.T. FOR EACH EDGE $e = \{u, v\}$, u OR v BELONGS TO C?
REFERENCE: THE PROBLEM IS *NP-complete* – [18, PROBLEM GT1].

Lemma 9. *Given a family of threshold functions φ over n voters with threshold t that satisfies "$t \leqslant n - 1$ and $\frac{t}{n}$ converges to one," there exists a reduction*

---

[10] In cases in which the output size is polynomial in the input size, which are the ones in which we apply this reduction, we get that the running time is polynomial in the regular sense, i.e., in the input size.

that runs in time linear in the output size; and given an instance of VERTEX COVER – $(G = (V, E), k)$, *it produces an instance of* OLD $\left(\varphi, \wedge_{j=1}^{m} x^j\right)$ – $\left(X \in \{0, 1\}^{n \times m}, k'\right)$ – *s.t. the following is satisfied:*

- $\frac{n}{n-t} = |V|^2$, • $m = |E|$, • $k' = k$, • *and* (G, k) *is satisfiable iff* (X, k') *is satisfiable.*

The third reduction is from OLD $\left(\varphi, \wedge_{j=1}^{m} x^j\right)$ to the general case OLD(φ, ψ).

Lemma 10. *Given a family of threshold functions φ over n voters with threshold t and a family of monotone functions ψ over m variables (issues) with a minterm of size z that diverges to infinity with m, there exists a reduction that runs in time linear in the output size; and given an instance of* OLD $\left(\varphi, \wedge_{j=1}^{m'} x^j\right)$ – $\left(Y \in \{0, 1\}^{n' \times m'}, k'\right)$, *it produces an instance of* OLD (φ, ψ) $\left(X \in \{0, 1\}^{n \times m}, k\right)$ *s.t. the following is satisfied:*

- $n = n'$, • $z = m'$ *(and m is set according to z)*, • $k = k'$, • *and* (Y, k') *is satisfiable if and only if (X, k) is satisfiable.*

In addition, each issue in X either corresponds to one of the issues in Y or is not approved by any of the voters. That is, if the maximal minterm is $\wedge_{j=1}^{z} x^j$, then $\forall j \leqslant z$ $X^j = Y^j$ and $\forall j > z$ $X^j = \overline{0}$ (the all zeroes vector).

Proof Sketch of Lemma 10. W.l.o.g., we assume that $\wedge_{j=1}^{z} x^j$ is a minterm of ψ. Given an instance of OLD $\left(\varphi, \wedge_{j=1}^{m'} x^j\right)$ – $\left(Y \in \{0, 1\}^{n' \times m'}, k'\right)$, we construct the instance of OLD(φ, ψ) – $\left(X \in \{0, 1\}^{n \times m}, k'\right)$ by defining the threshold $k' = k$ and defining the profile X to be:

Voters \ Issues	first z issues	$(m - z)$ issues
n voters	Y	0

for $n = n'$, $z = m'$, and $k = k'$.

Due to the way we define the profile, the only way for the lobby to satisfy ψ is by satisfying the first z issues and this can be done only by solving the original problem $\left(Y \in \{0, 1\}^{n' \times m'}, k'\right)$. $\qquad\square$

We prove the intractability theorem using the above reductions.

Proof Sketch of Theorem 6. Combining the reductions of Lemmas 8 and 10, we get a reduction from HITTING SET to OLD(φ, ψ). This reduction satisfies the desired properties:

- If $z \geqslant m^\epsilon$ and $t, (n - t) \geqslant n^\epsilon$ (for some $\epsilon > 0$), we get a polynomial reduction from the *NP-complete* problem HITTING SET to OLD(φ, ψ), and by that proving the *NP*-hardness of OLD(φ, ψ).
- Similarly, if $z \geqslant (\log m)^\alpha$ and $t, (n - t) \geqslant (\log n)^\alpha$ (for some $\alpha > 1$), we get a reduction from HITTING SET with a blowup smaller than $2^{x^{1/\alpha}}$. Hence, since HITTING SET is *NP-complete*, we get that for any problem A $\in$ *NP* if $z \geqslant$

$(\log m)^\alpha$ and $t, (n - t) \geqslant (\log n)^\alpha$, there is a reduction from A to $\mathrm{OLD}(\varphi, \psi)$ with a blowup smaller than $2^{x^{C/\alpha}}$, for a constant C that depends on A. Given that, for any $\beta \in (0, 1)$, there exists a large enough $\alpha > 1$ s.t. if $\mathrm{OLD}(\varphi, \psi)$ is solvable in polynomial time, $z \geqslant (\log m)^\alpha$, and $t, (n - t) \geqslant (\log n)^\alpha$, then A is solvable in time $2^{O(n^\beta)}$.

We prove the intractability results for the complementary domain $t \in [n - n^\epsilon, n - 1]$, using the reduction from VERTEX COVER (Lemma 9) in a similar way.

Unanimity Issue-Aggregation $t = n$. For the case $t = n$, we prove the following intractability result.

Theorem 11. *Let ψ be a threshold function with threshold z, i.e., the Lobby would like at least z issues to pass.*

1. *If both z and $(m - z)$ are polynomial ($\exists \epsilon > 0 \quad z \in [m^\epsilon, m - m^\epsilon]$), then $\mathrm{OLD}\,(Unan, \psi)$ is NP-complete (under Turing reductions; See Footnote 11).*
2. *For any problem A $\in$ NP and any $\beta \in (0, 1)$, there exists $\alpha > 1$ s.t. if $\mathrm{OLD}\,(Unan, \psi)$ is solvable in polynomial time and if z and $(m - z)$ are poly-logarithmic of degree α ($z \in [(\log m)^\alpha, m - (\log m)^\alpha]$), then A is solvable in time $2^{O(n^\beta)}$.*
3. *Assuming ETH, there exists $\alpha^* > 1$ s.t. if z and $(m - z)$ are poly-logarithmic of degree α^* ($z \in \left[(\log m)^{\alpha^*}, m - (\log m)^{\alpha^*}\right]$), then $\mathrm{OLD}\,(Unan, \psi)$ is not in P.*
4. *Assuming NP $\nsubseteq$ SUBEXP, if z and $(m - z)$ are super-poly-logarithmic (there is no $\alpha > 1$ s.t. $z \geqslant (\log m)^\alpha$ and $(m - z) \geqslant (\log m)^\alpha$ for all large enough m), then $\mathrm{OLD}\,(Unan, \psi)$ is not in P.*

We prove this theorem by constructing a reduction from the following problem:

Problem (EBNCD – Exact Balanced Node Cardinality Decision problem).

INSTANCE: A BIPARTITE GRAPH $G = (L, R, E)$.
           A POSITIVE INTEGER $k \leqslant \min(|L|, |R|)$.
QUESTION: DOES THERE EXIST A BICLIQUE OF SIZE (k, k) IN G, I.E., TWO SETS $A \subseteq L$ AND $B \subseteq R$, BOTH OF SIZE k, S.T. $\forall l \in A, r \in B : (l, r) \in E$?
REFERENCE: THE PROBLEM IS *NP-complete* (UNDER TURING REDUCTIONS) – [9].[11]

---

[11] The reduction presented by Dawande et al. [9] is a Turing reduction from an *NP-complete* problem. Hence, they show that EBNCD is an *NP-complete* problem in a weaker sense, and the same weakness is shared by our result regarding $\mathrm{OLD}(Unan, \psi)$. Notice that nevertheless, this weaker notion still proves that a polynomial algorithm to $\mathrm{OLD}(Unan, \psi)$ implies that *NP=P*.

Lemma 12. *Given a family of threshold functions ψ over m issues with threshold z that satisfies "$(m - z)$ and z diverges to infinity," there exists a reduction that runs in time linear in the output size; and given an instance of EBNCD – $(G = (L, R, E), k)$, it produces an instance of OLD $(Unan, \psi)$ – $\left(X \in \{0,1\}^{n \times m}, \; k'\right)$ – s.t. the following is satisfied:*

- *$n = |L|$,* • *$\min(z, m - z) = \max(k, |R| - k)$,* • *and $k' = n - k$,* • *and (G, k) is satisfiable if (and only if) (X, k') is satisfiable.*

Proof Sketch of Lemma 12. Given an instance of EBNCD – $(G = (L, R, E), k)$, represented by an incidence matrix $A \in \{0,1\}^{|L| \times |R|}$ and a function ψ defined by a threshold k, we construct the instance of OLD$(Unan, \psi)$ – (X, k') by defining the threshold $k' = |L| - k$ and defining the profile X to be:

Voters \ Issues	$|R|$ issues	$z - k$ issues	$m - |R| - (z - k)$ issues
$|L|$ voters	A	1	strictly more than k' zeroes in each column

Due to the way we define the right sub-profile, the only way for the lobby to satisfy ψ is by satisfying at least k of the left issues. This can be done only by finding a (k, k)-biclique in G; that is, k issues, all supported by k voters; and convincing the voters not corresponding to L-vertices of the clique to support the issues corresponding to the R-vertices of the clique. □

4 Related Work

The OPTIMAL LOBBYING problem was first addressed by Christian et al. [8] who (essentially) showed that OLD$\left(\text{MAJ}_n, \bigwedge_{j=1}^{m} x^j\right)$ is *NP-complete* and $W[2] - complete$ with respect to the budget k.[12]

This problem was studied further by Bredereck et al. [7] in terms of its parameterized computational complexity with regard to other parameters. In addition they presented two generalizations of this problem: RESTRICTED LOBBYING, in which the input includes (in addition to the profile and the budget) a parameter o' of the number of issue-votes an influenced voter is allowed to change; and PARTIAL LOBBYING, in which the desired outcomes are the outcomes in which at least a certain number of issues (r) pass instead of all issues (in our terms this is OLD$\left(\text{MAJ}_n, \bigvee_{M \in \binom{[m]}{r}} \bigwedge_{j \in M} x^j\right)$, when r is part of the input and $\binom{[m]}{r}$ denotes the set of all the subsets of size r of $\{1, 2, \ldots, m\}$). Compared to their model, we deal with these desired outcomes sets but we analyze the problem when r is exogenous as well as for more general issue-aggregation functions, and we analyze fully which are the hard values of r.

---

[12] For background on the theory of parameterized complexity and in particular for the definition of $W[2]$, see Downey and Fellows [12], Niedermeier [26], and Flum and Grohe [17].

Another extension of the model of Christian et al. [8] was presented by Baumeister et al. [5]. They define several variants of related problems – Bribery and Manipulation of Premise Based Agendas. The question of bribery they describe is equivalent to the OPTIMAL LOBBYING. They prove hardness of the problem when $\varphi = \mathrm{Maj}_n$ and ψ is part of the input and consisting of a single minterm, i.e., a unique desired outcome for a partial list of the issues. This result is obtained as a corollary of the work by Christian et al. [8] that showed hardness for a specific ψ, $\psi = \wedge_{j=1}^m x^j$, and it can be also be proved as a corollary of more general results that we prove here. The contribution of our work over this result is that in our work the function ψ is not part of the input but exogenous to the problem and hence we characterize the hard ψ-slices of the problem (the functions ψ for which the problem is tractable and those for which it is not).

5   Conclusions

In this paper we defined the OPTIMAL LOBBYING decision and optimization problems (in their general forms) and characterized their computational complexity when (a) all issues are Boolean, (b) all issues are aggregated using the same monotone anonymous function φ, and (c) the desired outcomes set ψ is upward-closed.[13]

There are still some gaps in our characterization that we intend to fill. First, our characterization of the case of aggregating using unanimity characterizes mostly the case of desired outcomes set defined by a threshold. We conjecture that the real parameter that controls the complexity is the number of minterms (when the number of minterms is large, ψ is "close enough" to a threshold function).

Conjecture. *Let the aggregation function be unanimity and let the desired outcomes set be described using r minterms.*

- *If r is polynomial in m, OPTIMAL LOBBYING can be solved in polynomial time (We already proved this part in Theorem 5).*
- *If r is exponential in m ($r \geqslant 2^{m^\alpha}$ for some $\alpha > 0$), then OPTIMAL LOBBYING is NP-complete.*

A smaller gap in our characterization is the gap between bounded values, for which we proved tractability, to poly-logarithmic values, for which we proved intractability. We conjecture that the tractability results can be extended by smarter algorithms.

In a subsequent work, we extend this paper to studying the *parameterized* computational complexity of OPTIMAL LOBBYING. We describe the parameters that we think capture the complexity of an instance and by finer analysis of the reductions, analyze the parameterized complexity with regard to them. A natural extension of both works would be extension of the analysis to other issue-aggregating functions (φ) and other desired outcomes families (ψ), both monotone and non-monotone.

---

[13] In the full version we show that these constraints can be relaxed to get characterization results for a larger set of OPTIMAL LOBBYING problems.

References

1. Babaioff, M., Feldman, M., Nisan, N., Winter, E.: Combinatorial agency. J. Econ. Theor. **147**, 999–1034 (2012)
2. Bartholdi, J.J., Orlin, J.B.: Single transferable vote resists strategic voting. Soc. Choice Welf. **8**, 341–354 (1991)
3. Bartholdi, J.J., Tovey, C.A., Trick, M.A.: The computational difficulty of manipulating an election. Soc. Choice Welf. **6**, 227–241 (1989)
4. Bartholdi, J.J., Tovey, C.A., Trick, M.A.: How hard is it to control an election? Math. Comput. Model. **16**, 27–40 (1992)
5. Baumeister, D., Erdélyi, G., Rothe, J.: How hard is it to bribe the judges? a study of the complexity of bribery in judgment aggregation. In: Brafman, R. (ed.) ADT 2011. LNCS, vol. 6992, pp. 1–15. Springer, Heidelberg (2011)
6. Bredereck, R., Chen, J., Hartung, S., Kratsch, S., Niedermeier, R., Suchý, O., Woeginger, G.J.: A multivariate complexity analysis of lobbying in multiple referenda. J. Artif. Intell. Res. **50**(1), 409–446 (2014)
7. Bredereck, R., Chen, J., Hartung, S., Niedermeier, R., Suchỳ, O., Kratsch, S.: A multivariate complexity analysis of lobbying in multiple referenda. In: Proceedings of the 26th Conference on Artificial Intelligence (AAAI 2012) (2012)
8. Christian, R., Fellows, M., Rosamond, F., Slinko, A.: On complexity of lobbying in multiple referenda. Rev. Econ. Des. **11**, 217–224 (2007)
9. Dawande, M., Keskinocak, P., Swaminathan, J.M., Tayur, S.: On bipartite and multipartite clique problems. J. Algorithms **41**, 388–403 (2001)
10. Dietrich, F., List, C.: Judgment aggregation by quota rules. J. Theor. Polit. **19**, 391–424 (2007)
11. Dorn, B., Schlotter, I.: Multivariate complexity analysis of swap bribery. Algorithmica **64**, 126–151 (2012)
12. Downey, R.G., Fellows, M.R.: Parameterized Complexity. Springer-Verlag New York, Inc., New York (1999)
13. Elkind, E., Faliszewski, P., Slinko, A.: Swap bribery. In: Mavronicolas, M., Papadopoulou, V.G. (eds.) SAGT 2009. LNCS, vol. 5814, pp. 299–310. Springer, Heidelberg (2009)
14. Faliszewski, P., Hemaspaandra, E., Hemaspaandra, L.A.: How hard is bribery in elections? J. Artif. Intell. Res. **35**, 485–532 (2009)
15. Faliszewski, P., Hemaspaandra, E., Hemaspaandra, L.A.: Using complexity to protect elections. Commun. ACM **53**(11), 74–82 (2010)
16. Faliszewski, P., Procaccia, A.D.: AI's war on manipulation: Are we winning? AI Mag. **31**(4), 53–64 (2010)
17. Flum, J., Grohe, M.: Parameterized Complexity Theory. Texts in Theoretical Computer Science. Springer, Heidelberg (2006)
18. Garey, M.R., Johnson, D.S.: Computers and Intractability: A Guide to the Theory of NP-Completeness. Series of Books in the Mathematical Sciences. W.H. Freeman, New York (1979)
19. Harrison, G.W., McDaniel, T.: Voting games and computational complexity. Oxf. Econ. Pap. **60**(3), 546–565 (2008)
20. Holmstrom, B.: Moral hazard in teams. Bell J. Econ. **13**, 324–340 (1982)
21. Impagliazzo, R., Paturi, R.: On the complexity of k-SAT. J. Comput. Syst. Sci. **62**(2), 367–375 (2001)
22. List, C.: Judgment aggregation: a short introduction. In: Gabbay, D.M., Thagard, P., Woods, J., Mäki, U. (eds.) Philosophy of Economics. Elsevier, Amsterdam (2012)

23. List, C., Polak, B.: Introduction to judgment aggregation. J. Econ. Theor. **145**, 441–466 (2010)
24. List, C., Puppe, C.: Judgement aggregation: a survey. In: Pattanaik, P., Anand, P., Puppe, C. (eds.) The Handbook of Rational and Social Choice. Oxford University Press, Oxford (2009)
25. Nehama, I.: Complexity of optimal lobbying in threshold aggregation. In: Proceedings of the 2013 International Conference on Autonomous Agents and Multi-agent Systems, AAMAS 2013, pp. 1197–1198 (2013)
26. Niedermeier, R.: Invitation to Fixed Parameter Algorithms. Oxford Lecture Series in Mathematics and Its Applications. Oxford University Press, Oxford (2006)
27. Vazirani, V.V.: Approximation Algorithms. Springer, Heidelberg (2001)

More Natural Models of Electoral Control by Partition

Gábor Erdélyi[1], Edith Hemaspaandra[2], and Lane A. Hemaspaandra[3]([⊠])

[1] School of Economic Disciplines, University of Siegen, 57076 Siegen, Germany
[2] Department of Computer Science, Rochester Institute of Technology,
Rochester, NY 14623, USA
[3] Department of Computer Science, University of Rochester,
Rochester, NY 14627, USA
lane@cs.rochester.edu

Abstract. "Control" studies attempts to set the outcome of elections through the addition, deletion, or partition of voters or candidates. The set of benchmark control types was largely set in the 1992 paper by Bartholdi, Tovey, and Trick that introduced control, and there now is a large literature studying how many of the benchmark types various election systems are vulnerable to, i.e., have polynomial-time attack algorithms for.

However, although the longstanding benchmark models of addition and deletion model relatively well the real-world settings that inspire them, the longstanding benchmark models of partition model settings that are arguably quite distant from those they seek to capture.

In this paper, we introduce—and for some important cases analyze the complexity of—new partition models that seek to better capture many real-world partition settings. In particular, in many partition settings one wants the two parts of the partition to be of (almost) equal size, or is partitioning into more than two parts, or has groups of actors who must be placed in the same part of the partition. Our hope is that having these new partition types will allow studies of control attacks to include such models that more realistically capture many settings.

1 Introduction, Motivation, and Discussion

Introduction. Elections are an important framework for decision-making, both in human settings and in multiagent systems settings as varied as recommender systems [15], rank aggregation and web-spam filtering [6], similarity search and classification [10], and planning [7].

Given elections' importance, it is natural that people and other agents should attempt manipulative attacks on such systems, and that computational social choice researchers should investigate the computational complexity of conducting such attacks. Such studies have focused primarily on three broad streams of manipulative attacks, known as manipulation, bribery, and control.

Among the three attack streams, control is by far the most troubled as to naturalness. This paper's focus is on control, which studies whether by changing the structure of an election—adding or deleting or partitioning voters or

© Springer International Publishing Switzerland 2015
T. Walsh (Ed.): ADT 2015, LNAI 9346, pp. 396–413, 2015.
DOI: 10.1007/978-3-319-23114-3_24

candidates—an actor can ensure that a given candidate wins. A set of 11 benchmark attacks—"control by adding voters," "control by deleting candidates," etc.—are often studied to seek to understand which types of attacks a given election system computationally resists. The 11 benchmark attacks already were present in the seminal Bartholdi, Tovey, and Trick [1] paper on control, at least as refined in subsequent papers that made the partition tie-breaking options explicit [18] and addressed the one asymmetry in the Bartholdi, Tovey, and Trick paper's definitions [12].

With these benchmark types in hand, many election systems have been evaluated as to how many of these types of attacks they computationally resist, with the goal of identifying natural election systems that resist as many as possible of the 11 benchmark types—or the 22 (or 21, due to a collapse of types recently noticed by Hemaspaandra, Hemaspaandra, and Menton [16]) such types, if one also looks at the "destructive" cases where the goal is instead to prevent a given candidate from winning.

For example, among the papers that do excellent, detailed analyses tallying how many resistances to control attacks are possessed by such broadly-control-resistant election systems as Bucklin, fallback, ranked pairs, Schulze, sincere-strategy preference-based approval, and normalized range voting are [8,9,22–24].

However, although the benchmark control models regarding addition/deletion of candidates/voters are relatively natural, the benchmark models of partitioning have always been far less so. For many natural, real-world (i.e., human or electronic-agent) settings, the longstanding benchmark partition models simply don't come close to capturing the settings that are routinely used to motivate them. And so, despite the fact that an enormous amount of effort—in most of the above papers and very many others—has been focused on the standard partition models over the more than two decades since they were created, we feel that it is valuable to revisit the issue of how to best frame models of partition. Perhaps such a revisitation should have occurred ten or fifteen years ago, but the best we can at this point do is to approach the issue now, and in this paper we do so.

In truth, the appropriateness of a model of partition will depend heavily on the setting one is trying to model, and there indeed are some settings that are well-modeled by the benchmark partition types. Still, there are numerous settings that are not well-modeled by them, and so in this paper we present, and give some initial results regarding, new types of partitioning that we feel are worth studying as capturing naturally important notions of partitioning. In particular, we define three new general flavors of partitioning.

The Classic Version of Control by Partition. Before discussing the three new flavors, we should briefly describe control by partition. In the classic version of control by partition of voters, one is given the votes of each voter (most typically as a linear ordering, e.g., "Bob>Carol>Ted>Alice") and a distinguished candidate one is interested in, and one asks whether there is a way of partitioning the voters into V_1 and V_2 such that if one has subelections among the candidates by V_1 and V_2, and then a final election by all the voters with the candidates being the winners of those subelections, the distinguished candidate is the winner. An example of this would be if on a faculty hiring committee the chair partitioned the committee's members into two subgroups, charging one with evaluating and selecting which candidate(s) would be best at teaching and charging the other with evaluating and selecting which candidate(s) would be best at research (or, if one is cynical, perhaps the charge might be to evaluate grant funding potential), followed by a vote of the full committee on all candidates selected in that first round. In the classic version of candidate partitioning (known as "runoff partition of candidates"), the input is the same, but the question is whether there is a partition of the candidates into C_1 and C_2 such that if all the voters have subelections regarding C_1 and regarding C_2, and then a final election by all voters over just the winners of those subelections, the distinguished candidate is the winner. An example of this would be if in an academic department's faculty hiring process the chair split the candidate pool into two groups (using some explanation that might be a pretext, such as that he or she was distinguishing those who were more likely to excel in research and those who were more likely to excel in teaching), and had the entire faculty vote separately on each group, and then the entire faculty would conduct a vote between the winners of these two initial elections.

Our Variants of Control by Partition. The three variants we suggest and study, in order to create models that are often closer to the real-world settings that partition seeks to capture, are equipartition, multipartition, and partition by groups. In equipartition, the two parts of partitions must be of the same size (or within one if the things being partitioned are odd in cardinality). The motivation for this is that in real-world settings, such as apportioning people or agents into groups, it is very common to want the groups to be of essentially equal size. In our above voter-partition example, the committee members might for example feel that the work load was unfair if the subcommittees weren't essentially equal in size. Even when breaking candidates into two groups, it is often natural to keep the playing field somewhat level by expecting the groups to be of essentially equal sizes. In our above candidate-partition example, if the two groups were different in size, that might be considered unfair to those in the larger group. One can think of having balanced group sizes as a *fairness* condition, and to the best of our knowledge our equipartition model is the first and only step toward fairness of partition size ever taken with regard to the Bartholdi, Tovey, and Trick model in the 23 years since it was introduced. Indeed, the classic models of partition have *no* constraint on the sizes of the parts of the partition; although it might in reality outrage people were this to happen, in the classic candidate-partition model it is completely legal to partition the candidates into c and $C - \{c\}$, so that candidate c gets a free pass into the final election, and it is completely legal in the classic voter-partition model to divide a size 100,000,000 voter set into one group having three voters and another group having all the rest of the voters. One can even imagine settings where that lopsided partitioning is natural, and arguably not unfair to those in the huge group. But in many settings, size-balanced partitioning is the natural and indeed compelling expectation. The tricky thing regarding discussing what is natural and what model is best to capture a given real-world setting is that these issues are highly contextual—the richness of the real world means that no single model will capture all cases. Our goal in this paper, thus, is not to give a new model that we claim will capture all cases, but rather to give new models that in many settings are natural and attractive, and whose addition as models significantly broadens the class of cases one has good models for.

If one is dividing a voter set up into groups, it may well be the case that having just two groups is simply not appropriate. In our above voter-partition example, perhaps the department has a longstanding tradition of first vetting candidates by *three* committees, one each for research, teaching, and service. Then three would be the number of subcommittees the committee would naturally be divided into. Similarly, in our above candidate-partitioning example, it is easy to imagine the chair framing the division of candidates into four parts, with each candidate being assigned to whichever of the school's four institution-wide strategic hiring themes—for example, "accessibility and inclusion," "advanced design and manufacturing," "digital media and imaging science," and "global resilience"—the chair claims he or she fits best with. Thus our second new model for the study of partition-control is *multipartition*, that is, partitioning not into two parts but into k parts.

Our third new model is partition by groups. In partition by groups, each actor has a color, and all actors having the same color must be put in the same part of the partition as each other. For example, regarding candidate partitioning, if in a department the chair is splitting an area-diverse group of hiring candidates into two groups for a culling vote, it might be natural to want candidates from the same subfield to be put in the same culling vote as each other, and so each subarea would have a color. Control by groups is so natural that one might wish to study it not just for partition cases but also for addition/deletion of voters/candidates, and so we provide a result for adding and deleting voters by groups in this paper's section on groups. We mention that group-based (in a somewhat different framing) control by addition of voters has been previously introduced by Bulteau et al. [4], and in this paper's section on groups we discuss that interesting work and its relationship to the present work.

In this paper, we look at these three new models for partitioning, but due to space limitations and for simplicity, we don't seek to prove results about simultaneous combinations of them. However, certainly some (human or electronic) real-world settings will draw on combinations and so that might in the future be a potential area for study.

Although in this paper the new models themselves are very important, we also, for each of them, provide one or more results as to the complexity of control within that model, especially for the most important election system, plurality. In some cases, our versions of control leave the classic control problem's complexity unchanged (although for P cases it often takes far more complicated algorithms to handle the new cases while remaining in P), in some cases our versions increase the complexity from P to NP-complete, and we have even built a system where our version lowers the complexity from NP-complete to P.

For example, regarding equipartition, we show that for plurality, approval, and Condorcet voting, every existing P result can be reestablished even for equipartition. However, we show that for weakCondorcet elections, equipartitioning turns the P classic case into NP-completeness. We prove that control by groups often jumps P classic cases up to NP-completeness. And regarding multipartition and plurality, we build a P algorithm for the most important case.

Meaning of Increasing or Lowering Complexity. Given that our results focus on what happens to the complexity of problems in our variants, it is natural to wonder: Is it good if, for a given problem, our equipartition variant is harder than the classic partition version of the problem? Is it good if our version keeps the complexity the same? Is it good if our version lowers the complexity? The simple answer is that there is no simple answer, and indeed the questions are themselves simplistic. If one views complexity as trying to "shield" elections from undesirable attacks, then having high levels of complexity for a control problem is a good thing. If one is an attacker such as a campaign strategist—or if one is using an election-attack problem to model a real-world problem that one wants to solve (as for example the election problem "bribery" can be used to model resource-allocation problems)—then having polynomial-time algorithms is a good thing. What computational social choice theorists can do, however, is make clear what the control complexity levels are for the most important control types and the most important election systems. Having that knowledge in hand will allow social choice theorists, election-system choosers, campaign strategists, and others to all openly see the lay of the land.

The remainder of the paper is organized as follows. A brief definitions section comes next. In the three sections after that, we cover each of our three new models, giving formal definitions of each as well as results on their behavior. The conclusions and open problems section ends the paper proper.

2   Definitions

An election system is a mapping that, given the candidates and the votes, outputs a subset of the candidates, who are said to be the winners under that election system. We will often use the symbol $\mathscr{E}$ to denote an election system. For approval elections, voters give a 0 or 1 to each candidate, and the candidate(s) having the largest number of 1 votes is the winner(s). For the other election systems that we will study, votes are linear orderings, e.g., "Alice>Bob>Carol." In plurality elections, whichever candidate(s) is in the top spot on the most votes is the winner(s). In Condorcet (resp., weakCondorcet) elections, each candidate who is preferred to each of other candidate d in strictly more than half (resp., greater than or equal to half) the votes is a winner.

For conciseness, we sometimes use bracket notation (borrowed from linguistics) for *independent* choices, e.g., "The $\begin{bmatrix} \text{ball} \\ \text{book} \\ \text{car} \end{bmatrix}$ is $\begin{bmatrix} \text{red} \\ \text{heavy} \end{bmatrix}$" is shorthand for the natural six claims obtained by making each possible choice.

3   Equipartition

Let us now define our equipartition notion and the classic partition problems. (In all our problem definitions, we include "represented via preference lists over C." But in fact the voters are always represented by whatever vote type is that of the election system; throughout this paper that is always linear orders, except that for approval voting and the system of Theorem 3 the votes are instead 0/1-vectors.) Recall that partitioning a set into two (or more) parts means that every element of the set must appear in exactly one of the parts.

$\mathscr{E}$-CCREPC (Control by Runoff Equipartition of Candidates)

Given: A set C of candidates, a collection V of voters represented via preference lists over C, and a distinguished candidate $p \in C$.

Quest.: Is there a partition of C into C_1 and C_2 such that $|\,\|C_1\| - \|C_2\|\,| \leq 1$ and p is the sole winner of the two-stage election where the winners of subelection (C_1, V) that survive the tie-handling rule compete against the winners of subelection (C_2, V) that survive the tie-handling rule? Each subelection (in both stages) is conducted using election system $\mathscr{E}$.

$\mathscr{E}$-CCEPV (Control by Equipartition of Voters)

Given: A set C of candidates, a collection V of voters represented via preference lists over C, and a distinguished candidate $p \in C$.

Quest.: Is there a partition of V into V_1 and V_2 such that $|\,\|V_1\| - \|V_2\|\,| \leq 1$ and p is the sole winner of the two-stage election where the winners of election (C, V_1) that survive the tie-handling rule compete against the winners of (C, V_2) that survive the tie-handling rule? Each subelection (in both stages) is conducted using election system $\mathscr{E}$.

The classic cases, $\mathscr{E}$-CCRPC and $\mathscr{E}$-CCPV, are defined identically, except without the clause forcing the partition parts to be equal in size (or off-by-one if the set being partitioned is of odd cardinality).

For all of the above problems there is the issue of whether, if there are multiple winners of a subelection, all of those winners move on to the final election (called ties-promote, notated TP) or none of them move on to the final election (called ties-eliminate, notated TE; in this model, to move on to the final election one must be the unique winner of a subelection). Thus each problem will always appear with a TE or TP to specify the tie-handling approach that controls the first-round elections, e.g., $\mathscr{E}$-CCPV-TE or $\mathscr{E}$-CCPV-TP.

The literature also contains a "bye" version of partitioning candidates, in which any number of candidates can be assigned to skip ("bye") the first round and the rest compete to get into the second round. Since equipartition is not natural for that "bye" version (because the number of candidates skipping a first round is usually driven by such things as excesses relative to powers of two in the number of candidates), we have not defined here either the classic or an "equi" version of "bye" partition; our full version of this paper, however, will note that many of our results hold for the "equi" version of "bye" partition, and more importantly will give, and prove results about, a model in which one specifies as part of the input how many candidates get a "bye," since that model provides the natural partition-size-sensitive variant of the "bye" candidate-partition problem.

Our definition of "equi" makes the two partition parts be equal in size if the total size is even and be off-by-at-most-one if the total size is odd. In terms of fairness this is as satisfying as one could wish and is as balanced as one can possibly get. And if one were to instead allow the two parts just to be "close" to balanced, then one would have to discuss just how close remains "fair" and/or "close": Larger fixed-constant differences? Proportional differences (e.g., no part contains more than 75 %)? Should the amount of closeness be a fixed part of the problem, or should the problem be parameterized to take the required closeness as being part of the input? We think that these are doors that should be opened. Certainly, one can easily imagine cases where merely ensuring that each of the parts of a partition has at least 1/4 of the overall size would be considered acceptable balancing. However, we leave the study of such models for the future, and hope to include that in a future full version. In this conference version, we focus just on the case of the parts being as closely balanced as possible. This is certainly the fairest notion of balance, and—as long as one keeps in mind that it indeed is simply a model and admittedly a very demanding one—this is a good starting point for understanding what effects balancing may or may not have.

As a final comment about model details, all of the partition problems discussed above are in what is called the unique-winner model, i.e., the goal is to make a given candidate the one and only overall winner. That is the model of the seminal control paper of Bartholdi, Tovey, and Trick [1] and all the immediately subsequent control papers, and probably even today is the most common model when studying control (although we ourselves prefer the alternate model known as the nonunique-winner model or the co-winner model, where one merely needs to make a given candidate become *a* winner). And so throughout all parts of this conference version, we focus solely on the unique-winner model, since doing so creates apples-to-apples contrasts for those cases where our models change the existing complexity from those found in that seminal paper and various other papers. (We have checked the vast majority of our results in both models, and the full version of this paper will cover both models. This is not an issue one can safely take for granted, since there are examples in the literature where the complexity or behavior of the two models differs sharply, e.g., [13,16]).

Let us turn to our results, which give a sense of what holds when one partitions while required to have the parts be "equi." We show that for many important systems, including approval, Condorcet, and plurality, partition problems remain in P even for equipartitioning, albeit typically with substantially more difficult algorithms and new tricks relative to what the non-"equi" cases required. However, in contrast to its sibling, the Condorcet case, we prove that for weak-Condorcet an increase from P to NP-completeness occurs. We also construct an (admittedly artificial) election system having its winner problem in P, such that going from partition to equipartition lowers the control complexity from NP-completeness to P.

For approval, Condorcet, and plurality, each P case of $CC\begin{bmatrix}RPC\\PV\end{bmatrix}$-$\begin{bmatrix}TE\\TP\end{bmatrix}$ [1, 18] remains in P for equipartition.

Theorem 1. *Each of the problems* plurality-CCEPV-TE *and* $\begin{bmatrix}\text{Condorcet}\\\text{approval}\end{bmatrix}$-CCREPC-$\begin{bmatrix}TE\\TP\end{bmatrix}$ *belongs to* P.

We now give a proof of plurality-CCEPV-TE $\in$ P. The proof of this has two interesting issues that do not occur in the standard partition case.

First, even for those inputs where p is at the top of no more than $\lceil \|V\|/2 \rceil$ of the votes, one cannot assume within the proof that if candidate p can be made to win, it can be made to win with some partition that puts all votes with p at their top in the same partition part. Here is an example showing that that assumption, which works in the general case, fails here. If we have 3 votes for b each having a in second place, 5 votes for p, and 6 votes for a, we can make p a sole overall winner by letting V_1 consist of 4 votes for p and 3 votes for a. Then p is the unique winner in V_1, and V_2 consists of 3 votes for a, 3 votes for b, and 1 vote for p, so no candidate from the second election makes it through to the runoff, and p is the overall unique winner. However, if we put all 5 votes for p in V_1, then since there are 7 votes in V_1, at least 4 votes for a are in V_2 and a is the unique winner of the second subelection, and so also of the overall election.

The second twist is that we need to find a "safe" way to legally distribute in an "equi" overall fashion certain "remaining" votes; and our proof does this by pushing them all to one side (violating "equi," typically), and then correcting this in a way that is guaranteed to succeed if success is possible.

Proof (of the Plurality Part of Theorem 1). We now prove the claim that plurality-CCEPV-TE belongs to P. The example given as the "first" interesting issue just above shows that we have to be very careful about what assumptions we make in our proof. However, we indeed can show that our control problem is in P. First of all, note that p can be made a unique winner by EPV-TE if and only if there exists an equipartition (V_1, V_2) such that p is the unique winner of (C, V_1) and

1. (C, V_2) has a unique winner, call it c, and p defeats c, or
2. (C, V_2) has p as a winner, or
3. (C, V_2) has more than one winner.

These three conditions can be checked as follows in polynomial time. ($score_V(p)$ here is the plurality score of p in V, i.e., the number of voters in V having p as their top choice).

1. For every (c, k_p, k_c) such that $c \in C - \{p\}$, p defeats c, $k_p \leq score_V(p)$, and $k_c \leq score_V(c)$, do:
   Put k_p votes for p in V_1 and the remaining votes for p in V_2 and put k_c votes for c in V_2 and the remaining votes for c in V_1. We will now check whether (V_1, V_2) can be extended to a desired equipartition. If p is not the unique winner in V_1 or c is not the unique winner in V_2, then this is not possible and we move on to the next loop iteration. Otherwise, for each $d \in C - \{p, c\}$ put as many votes for d as possible into V_1 while keeping p the unique winner in V_1 (i.e., $\min(score_V(d), k_p - 1)$ votes). If $\|V_1\| < \lfloor \|V\|/2 \rfloor$, then there are not enough votes in V_1 and we move on to the next loop iteration. Otherwise, move votes for candidates in $C - \{p, c\}$ from V_1 to V_2 if this is possible while keeping c the unique winner in V_2 until (V_1, V_2) becomes an equipartition. If this is possible, we have found a successful equipartition. If this is not possible, we move on to the next loop iteration.

2. This is similar to the previous case.
   For every k_p such that $0 < k_p \leq score_V(p)$, do:
   Put k_p votes for p in V_1 and the remaining votes for p in V_2. We now will check whether (V_1, V_2) can be extended to a desired equipartition. For each $d \in C - \{p\}$, put as many votes for d as possible into V_1 while keeping p the unique winner in V_1 (i.e., $\min(score_V(d), k_p - 1)$ votes). If $\|V_1\| < \lfloor \|V\|/2 \rfloor$, there are not enough votes in V_1 and we move on to the next loop iteration. Otherwise, move votes for candidates in $C - \{p\}$ from V_1 to V_2 if this is possible while keeping p a winner in V_2 until (V_1, V_2) becomes an equipartition. If this is possible, we have found a successful equipartition. If this is not possible, we move on to the next loop iteration.

3. The case that p is one of the winners of (C, V_2) has been handled in the previous case, so it suffices to handle the case where (C, V_2) has at least two winners in $C - \{p\}$.
   For every (c, c', k_p, k_c) such that $c, c' \in C - \{p\}$, $c \neq c'$, $k_p \leq score_V(p)$, and $k_c \leq \min(score_V(c), score_V(c'))$, do:
   Put k_p votes for p in V_1 and the remaining votes for p in V_2, put k_c votes for c in V_2 and the remaining votes for c in V_1, and put k_c votes for c' in V_2 and the remaining votes for c' in V_1. We now will check whether (V_1, V_2) can be extended to a desired equipartition. If p is not the unique winner in V_1 or c and c' are not winners in V_2, then this is not possible and we move on to the next loop iteration. Otherwise, for each $d \in C - \{p, c, c'\}$, put as many votes for d as possible into V_1 while keeping p the unique winner in V_1 (i.e., $\min(score_V(d), k_p - 1)$ votes). If $\|V_1\| < \lfloor \|V\|/2 \rfloor$, there are not enough votes in V_1 and we move on to the next loop iteration. Otherwise, move votes for candidates in $C - \{p, c, c'\}$ from V_1 to V_2 if this is possible while keeping c and c' winners in V_2 until (V_1, V_2) becomes an equipartition. If this is possible, we have found a successful equipartition. If this is not possible, we move on to the next loop iteration. □

We have also shown that the NP-complete cases still hold for our systems of interest.[1]

Theorem 2. *For approval, Condorcet, and plurality, each* NP*-complete case of* $\text{CC}\begin{bmatrix}\text{RPC}\\\text{PV}\end{bmatrix}$-$\begin{bmatrix}\text{TE}\\\text{TP}\end{bmatrix}$ *—these can be found in Table 1 of Hemaspaandra, Hemaspaandra and Rothe [18] and are variously due to that paper and Bartholdi, Tovey, and Trick [1]—remains* NP*-complete for the one among* $\text{CC}\begin{bmatrix}\text{REPC}\\\text{EPV}\end{bmatrix}$-$\begin{bmatrix}\text{TE}\\\text{TP}\end{bmatrix}$ *that is its equipartition analogue.*

Now, it might be natural to wonder: Is there a meta-theorem showing that NP-completeness always inherits from partition cases to equipartition cases? If so Theorem 2 would become a freebie consequence of the meta-theorem. However, although we do not have any example of a natural election system where the "equi" case drops the complexity from NP-completeness to P, we have constructed an election system displaying precisely that behavior. And so NP-completeness does not always inherit from the standard case to the "equi" case.

Theorem 3. *There exists an election system* $\mathcal{E}$*, whose winner problem is in* P*, such that* $\mathcal{E}$*-CCPV-TP is* NP*-complete, yet* $\mathcal{E}$*-CCEPV-TP belongs to* P*.*

Proof. We define $\mathcal{E}$ as follows. The votes here are length $\|C\|$ 0/1 vectors (i.e., approval ballots). When speaking of candidates we'll use 0/1/2/3 as shorthands for "0"/"1"/"2"/"3." On input (C, V):

- If $\|C\| \leq 4$ and $(C \cap \{0, 1, 2, 3\} = \{0, 2\}$ or $C \cap \{0, 1, 2, 3\} = \{1, 3\})$, then the winners are the approval winners of $(C - \{0, 1, 2, 3\}, V)$.
- If $\|C\| \leq 4$ and $C \cap \{0, 1, 2, 3\} \neq \{0, 2\}$ and $C \cap \{0, 1, 2, 3\} \neq \{1, 3\}$, there are no winners.
- If $\|C\| > 4$ and $\{0, 1, 2, 3\} \subseteq C$, then $\|V\| \mod 4$ is a winner and if $(C - \{0, 1, 2, 3\}, V)$ has a unique approval winner, then that candidate is also a winner. There are no other winners.
- If $\|C\| > 4$ and $\{0, 1, 2, 3\} \not\subseteq C$, there are no winners.

We first show that $\mathcal{E}$-CCEPV-TP is in P. This is easy. If $\|C\| \leq 4$, there are no winners in the runoff, since 0, 1, 2, and 3 do not participate in the runoff. If $\|C\| > 4$ and $\{0, 1, 2, 3\} \not\subseteq C$, there are no candidates in the runoff. If $\|C\| > 4$ and $\{0, 1, 2, 3\} \subseteq C$, there are at most four candidates in the runoff. The candidates in $\{0, 1, 2, 3\}$ that participate in the runoff are exactly $\|V_1\| \mod 4$ and $\|V_2\| \mod 4$ for partition (V_1, V_2). But if (V_1, V_2) is an equipartition, it is

---

[1] This paper contains some NP-completeness results, the first of which is Theorem 2. NP-completeness is a worst-case theory, and so for our paper's NP-hard cases, seeking results for other notions of hardness would be interesting. See [25,26] for successes of and [19] for limitations of heuristic approaches to election (and other) problems. However, the majority of the present paper's results are about showing that, even for partition-control variants that might seem likely to increase control complexity, polynomial-time control algorithms do exist.

never the case that $\{\|V_1\| \mod 4, \|V_2\| \mod 4\} = \{0, 2\}$ or $\{\|V_1\| \mod 4, \|V_2\| \mod 4\} = \{1, 3\}$ and so there are no winners in the runoff.

To show that $\mathscr{E}$-CCPV-TP is NP-complete, we reduce from approval-CCPV-TE, which is NP-complete [18]. Let (C, V) be an election and let p be the preferred candidate. Assume that $C \cap \{0, 1, 2, 3\} = \emptyset$. Let $\widehat{C} = C \cup \{0, 1, 2, 3\}$ and let $\widehat{V}$ consist of the voters in V (extended to $\widehat{C}$ by not approving of candidates in $\{0, 1, 2, 3\}$) plus two additional voters that don't approve of any candidate if $\|V\|$ is even and one additional voter that doesn't approve of any candidate if $\|V\|$ is odd. We claim that p can be made the unique approval winner in (C, V) by PV-TE if and only if p can be made the unique $\mathscr{E}$ winner in $(\widehat{C}, \widehat{V})$ by PV-TP.

First suppose that (V_1, V_2) is a partition of V that makes p the unique approval winner by PV-TE. If $\|V\|$ is odd, add the one additional voter that doesn't approve of any candidate to V_1 or V_2 in such a way that $\|V_1\| \mod 4 \neq \|V_2\| \mod 4$. If $\|V\|$ is even and $\|V_1\| \mod 4 = \|V_2\| \mod 4$, add the two additional voters to V_1. If $\|V\|$ is even and $\|V_1\| \mod 4 \neq \|V_2\| \mod 4$, add one additional voter to V_1 and one additional voter to V_2. In all cases, we now have a partition $(\widehat{V_1}, \widehat{V_2})$ of $\widehat{V}$ with the same unique approval winners as before (when restricting the candidates to C) and such that $\{\|V_1\| \mod 4, \|V_2\| \mod 4\} = \{0, 2\}$ or $\{\|V_1\| \mod 4, \|V_2\| \mod 4\} = \{1, 3\}$. This immediately implies that this partition makes p the unique winner in $\mathscr{E}$-CCPV-TP.

For the converse, suppose that $(\widehat{V_1}, \widehat{V_2})$ is a partition of $\widehat{V}$ that makes p the unique $\mathscr{E}$ winner by PV-TP. Then $(\widehat{V_1}, \widehat{V_2})$ makes p the unique approval winner in $(C, \widehat{V})$ by PV-TE. Now simply delete the additional voters that don't approve of any candidate to obtain partition (V_1, V_2) of V that makes p the unique approval winner in (C, V) by PV-TE.   □

In contrast to its close relative, Condorcet elections, weakCondorcet elections increase complexity from the partition case to the equipartition case for the RPC-TP case. (This complexity increase is not precluded by the general fact that subcases of problems cannot be harder than the original problem. Although each equipartition of a set is indeed a partition of that set, we are not dealing here with a subcase of a problem, but rather with a control problem whose allowed internal actions are a subset of those of a different control problem, and so there is no automatic prohibition on the complexity increasing.) We do not yet have a complexity classification for weakCondorcet-CCREPC-TE, and consider that an interesting open problem.

Theorem 4. weakCondorcet-CCRPC-$\begin{bmatrix} \text{TE} \\ \text{TP} \end{bmatrix}$ *are in* P, *but weakCondorcet-CCREPC-TP is* NP-*complete.*

We include a proof of the TP cases of the above theorem. Briefly, what is behind the change in complexity here is that we can make one of the subelections represent a vertex cover and we use equipartition to limit the size of that vertex cover. The reason the same approach does not work also for Condorcet elections is that we crucially need that we can have multiple winners.

Proof (of the TP Parts of Theorem 4). To show that weakCondorcet-CCRPC-TP is in P, it suffices to note that for all $p \in C$, if p can be made the unique weakCondorcet winner in (C, V) by RPC-TP, then this is established by candidate partition $(\{p\}, C - \{p\})$. To see this, suppose p is not the unique weak-Condorcet winner in partition $(\{p\}, C - \{p\})$. Then there is a candidate $c \neq p$ such that c is a weakCondorcet winner in $(C - \{p\}, V)$ and c ties-or-defeats p in their head-to-head contest. But then c is a weakCondorcet winner in (C, V) and so in any (C', V) with $C' \subseteq C$ and $c \in C'$. It follows that c is always a winner by RPC-TP, and so p will never be the unique winner.

However, $(\{p\}, C - \{p\})$ is clearly not an equipartition. We will now show that weakCondorcet-CCREPC-TP is NP-complete. We will prove this by a reduction from Cubic Vertex Cover: Given a graph $G = (V, E)$ that is cubic, i.e., where every vertex has degree three, and a positive integer $k \leq \|V\|$, we ask whether G has a vertex cover of size k, i.e., a set of vertices $V' \subseteq V$ of size k such that every edge in E is incident with at least one vertex in V'. Let $\|V\| = n$. Since G is cubic, $\|E\| = 3n/2$.

Using McGarvey's construction [21], we construct, in polynomial time, an election $(C, \widehat{V})$ with the following properties:

- $C = \{p\} \cup V \cup E \cup D$, where $D = \{d_1, \ldots, d_{n/2+2k-1}\}$.
- The set of voters, $\widehat{V}$, is such that we have the following head-to-head contest results:
  - for every $e \in E$, e defeats p,
  - for every $c \in V \cup D$, p defeats c,
  - for every $e = \{v, v'\} \in E$, v defeats e and v' defeats e,
  - all other head-to-head contests are ties.

Suppose V' is a vertex cover of size k of G. Then p can be made the unique weakCondorcet winner by REPC, using partition $(\{p\} \cup D \cup V - V', E \cup V')$. Note that $\|\{p\} \cup D \cup V - V'\| = 1 + n/2 + 2k - 1 + n - k = 3n/2 + k = \|E \cup V'\|$. p is the Condorcet winner in $(\{p\} \cup D \cup V - V', \widehat{V})$, and thus participates in the runoff. Since V' is a vertex cover, for every candidate $e \in E$, there is a candidate $v \in V'$ such that v defeats e in their head-to-head contest. So no candidate in E makes it to the runoff. So p is the Condorcet winner in the runoff, and thus certainly the unique weakCondorcet winner.

For the converse, suppose p can be made the unique weakCondorcet winner by REPC-TP. Let (C_1, C_2) be an equipartition of C with $p \in C_1$ that witnesses this. Then p is a weakCondorcet winner in $(C_1, \widehat{V})$. This implies that $E \subseteq C_2$. Since p is a weakCondorcet winner in the runoff, no candidate from E participates in the runoff. So for every $e \in E$, there is a $c \in C_2$ such that c defeats e in their head-to-head contest. The only candidates that defeat $e = \{v, v'\}$ are v and v'. It follows that $C_2 \cap V$ is a vertex cover of G. Since (C_1, C_2) is an equipartition, $\|C_2 \cap V\| \leq k$. So G has a vertex cover of size k. $\square$

4   Multipartition

In many settings two simply is not the number of parts into which one's voter set must be divided. For example, the Dean may wish to have three study sections each passing forward a best choice on some issue. Multipartition, which we'll define here just for PV but it can just as well be defined for RPC, generalizes the 2-partition PV problem used in the seminal Bartholdi, Tovey, and Trick paper to each k-partitioning. (In the study of judgment aggregation, controlling elections by breaking so-called judges into k groups—in a complicated, quite different framework in which the problem as part of its input already had separated the so-called premises into k groups—has been interestingly studied by Baumeister et al. [2].) For each integer $k \geq 2$, define the following problem.

$\mathscr{E}$-CCP$_k$V (CONTROL BY k-PARTITION OF VOTERS)

Given: A set C of candidates, a collection V of voters represented via preference lists over C, and a distinguished candidate $p \in C$.

Quest.: Is there a partition of V into k parts, $V_1, V_2, \ldots, V_k$, such that p is the sole winner of the two-stage election where the winners of each of the k elections (C, V_i) that survive the tie-handling rule compete against each other in a final election? Each subelection (in both stages) is conducted using election system $\mathscr{E}$.

Plurality-CCP$_2$V-TE is in P [18]. As Theorem 5 states, we in fact have that P membership still holds for each k-partition version.

Theorem 5. *For each $k \geq 2$, plurality-CCP$_k$ V-TE $\in$ P.*

The proof is omitted due to space, but is based on extensively employing Lenstra's [20] powerful method for handling integer linear programming feasibility problems when the number of constraints is bounded.

It would be interesting to study multipartition for other election systems, and also to study multipartition varied to allow the number of partitions to itself not be fixed but rather to be specified as part of the input.

5   Voter Control by Groups

In voter partition by groups, each vote has a color (i.e., a label), and all votes with the same label must be placed into the same partition part. We also define group voter-control problems for deleting voters (where all votes of a given color must be jointly deleted or kept) and for adding voters (where in the pool of potential additional voters each one has a color, and each color group must be added or not added as a block). As discussed in the introduction, these models capture cases where groups cannot be separated. One example might be due to living at the same address in a redistricting problem, and another might be a departmental study group process in which each of the department's area subfaculties must be placed within the same study group. We below define just the voter cases for control by groups, but one could completely analogously define candidate control by groups.

$\mathscr{E}$-CCPVG (CONTROL BY PARTITION OF VOTER GROUPS)

Given: A set C of candidates, a collection V of voters represented via preference lists over C, a partition of V into any number of groups $G_1, \ldots, G_k$, and a distinguished candidate $p \in C$.

Quest.: Is there a partition of V into V_1 and V_2 such that for each i either $G_i \subseteq V_1$ or $G_i \subseteq V_2$ holds and p is the sole winner of the two-stage election where the winners of election (C, V_1) that survive the tie-handling rule compete against the winners of (C, V_2) that survive the tie-handling rule? Each subelection (in both stages) is conducted using election system $\mathscr{E}$.

$\mathscr{E}$-CCDVG (CONTROL BY DELETING VOTER GROUPS)

Given: A set C of candidates, a collection V of voters represented via preference lists over C, a partition of V into any number of groups $G_1, \ldots, G_k$, a nonnegative integer ℓ, and a distinguished candidate $p \in C$.

Quest.: Is there a set $S \subseteq V$, $\|S\| \leq \ell$, such that p is the sole winner of the $\mathscr{E}$ election over C with the vote collection set being V with S (multiset) removed, and for each i either $G_i \subseteq S$ or $G_i \cap S = \emptyset$?

$\mathscr{E}$-CCAVG (CONTROL BY ADDING VOTER GROUPS)

Given: A set C of candidates, a collection V of voters represented via preference lists over C, a collection W of potential additional voters represented via preference lists over C, a partition of W into any number of groups $G_1, \ldots, G_k$, a nonnegative integer ℓ, and a distinguished candidate $p \in C$.

Quest.: Is there a collection $S \subseteq W$, $\|S\| \leq \ell$, such that p is the sole winner of the $\mathscr{E}$ election over C with the vote set being V (multiset) unioned with S, and for each i either $G_i \subseteq S$ or $G_i \cap S = \emptyset$?

Before stating results for this model, let us quickly discuss whether these notions are in overlap with models in the literature. After all, votes are coming and going in blocks, and so one might wonder if this is related to for example the notion of weighted control introduced by Faliszewski, Hemaspaandra, and Hemaspaandra [11]. Briefly, the notions are different to their core in that a weighted vote puts a lot of weight *on that vote*, but in contrast, a vote group may consist of votes that have vastly different preferences from each other. It is true that if one took the Faliszewski, Hemaspaandra, and Hemaspaandra [11] notion of weighted control, and restricted the weights to being input in unary, and for the adding/deleting voters cases shifted from that paper's model of counting as one's limit the number votes and instead adopted the model (mentioned but not adopted in that paper) of limiting by the total weight of votes added/deleted, then *those* weighted control problems would each indeed many-one polynomial-time reduce to our analogous voter control by groups problem; but that seems to be as far as the connection goes between the two papers.

A closer connection is to the work of Bulteau et al. [4], who define and study a very general notion of "combinatorial voter control" for addition of voters. (Their paper is not concerned with partition problems, the main focus of the present paper.) Loosely put, for each voter they have a group of voters who in some sense follow that voter, so that if one takes an action on a voter, the group of the voter follows also. Note that this is a very flexible, general scheme, and for example does not require that the follower function breaks the voters into equivalence classes, as does our coloring scheme. Of course, when proving NP-hardness results, such flexibility *weakens* the results. So in their paper (which is in the nonunique-winner model) they define and study a number of restricted models of the follower function. However, even the most restrictive models of follower functions that they study are incomparable to our model (even when they add nice symmetry-like properties, they focus those on voters with the same preferences, and so are not focused on what we are focused on, which is, in effect, coloring voters, i.e., breaking voters into equivalence classes in whatever arbitrary way is specified by the coloring), and so our NP-hardness result for plurality-CCAVG is incomparable with their NP-hardness results for their model of combinatorial control by adding voters. See also the more recent work in the line of [4] done in [3,5].

Turning to our results for our model, unlike our earlier two models, the addition of groups very broadly increases P complexity levels to NP-completeness. For the three cases covered by the following theorem, the analogous result without groups is well known to be in P [1,18]. And the NP-completeness claims of Theorem 6 are each proved by building an appropriate reduction from an NP-complete problem, in particular X3C (i.e., Exact Cover by 3-Sets, see [14]). Plurality-CCPVG-TP is also NP-complete, but we do not state that in the theorem since this follows immediately from the known result (see [18]) that plurality-CCPV-TP is NP-complete.

Theorem 6. *Each of* plurality-CC$\begin{bmatrix} \text{PVG-TE} \\ \text{AVG} \\ \text{DVG} \end{bmatrix}$ *is* NP-*complete.*

We provide here the proof of the PVG-TE part of Theorem 6.

Proof (of the PVG-TE Part of Theorem 6). We reduce from X3C. We are given a set $B = \{b_1, \ldots, b_{3m}\}$, $m > 1$, and a collection $\mathscr{S} = \{S_1, \ldots, S_n\}$ of subsets $S_i = \{b_{i,1}, b_{i,2}, b_{i,3}\} \subseteq B$ with $\|S_i\| = 3$ for each i, $1 \leq i \leq n$.

We assume $n > m + 1$. We may safely make this assumption, as X3C still remains NP-complete under this restriction. $n < m$ is automatically a no instance and the two cases $n = m$ and $n = m + 1$ are solvable in polynomial time. Thus the problem is still NP-complete under the restriction $n > m + 1$.

Define the election (C, V), where $C = \{p, c, d\} \cup B$ is the set of candidates, p is the distinguished candidate, and V consists of the following $n + 3$ groups of voters. As a shorthand, when specifying votes we will sometimes include a set of candidates, when resolving those as any linear ordering among those voters (e.g., lexicographic) will be fine for the vote's role in the proof. For example, $p > S > b$, where $S = \{z, a, w\}$, may be taken as a shorthand for $p > a > w > z > b$.

- For each i, $1 \leq i \leq n$, there is a group, G_i, with seven voters of the form:

  - $p > C - \{p\}$, $\bullet b_{i,1} > C - \{b_{i,1}, p\} > p$, $\bullet d > C - \{d, p\} > p$, and
  - $p > C - \{p\}$, $\bullet b_{i,2} > C - \{b_{i,2}, p\} > p$, $\bullet d > C - \{d, p\} > p$.
  - $\bullet b_{i,3} > C - \{b_{i,3}, p\} > p$,

- There is a group G_B consisting of the following voters:
  - Let $\ell_j = \|\{S_i \in \mathscr{S} \mid b_j \in S_i\}\|$ for all j, $1 \leq j \leq 3m$. For each j, $1 \leq j \leq 3m$, there are $2n - \ell_j$ voters of the form $b_j > C - \{b_j, p\} > p$.
  - There are $2m$ voters of the form $p > C - \{p\}$.
- There is a group G_c containing $2(n + m) + 1$ voters of the form $c > C - \{c, p\} > p$.
- There is a group G_d containing $2n + 1$ voters of the form $d > C - \{d, p\} > p$.

In this election, each $b_j \in B$ has a score of $2n$, candidate c has a score of $2(n+m)+1$, candidate d has a score of $4n+1$, and the distinguished candidate p has a score of $2(n+m)$. We claim that $\mathscr{S}$ contains an exact cover of B if and only if p can be made the unique winner of the election by control by partition of voter groups in the TE model.

Suppose $\mathscr{S}$ contains an exact cover $\mathscr{S}'$ of B. Partition the set of voters as follows. Let V_2 contain the m groups corresponding to $\mathscr{S}'$ and the groups G_c and G_d. Let $V_1 = V - V_2$. Candidate p is the unique winner of subelection (C, V_1), since p has a score of $2n$, each $b_j \in B$ has a score of $2n - 1$, candidate c has a score of 0, and candidate d has a score of $2n - 2m$. There is no unique winner in subelection (C, V_2), since candidates c and d tie for first place, eliminating each other. Thus only candidate p moves to the final round of the election, and p is the unique winner of the final round.

For the converse direction, suppose p can be made the unique winner of the election by partition of voter groups in the TE model. Since p participates in the final round and we are in the TE model, p is the unique winner of at least one of the subelections, without loss of generality say of (C, V_1). Then voter groups G_c and G_d have to be in subelection (C, V_2). Both candidate c and candidate d would beat p in the final round, and so should not participate in the final round. Since candidates c and d have a higher score than any other candidate and c accumulates all of her $2(n+m)+1$ points in group G_c, they have to eliminate each other in subelection (C, V_2). To this end, exactly m of the G_i groups have to be in subelection (C, V_2). All other G_i groups are in subelection (C, V_1). Right now, regardless of which of the G_i groups are in subelection (C, V_1), candidates p and d are tied in subelection (C, V_1), since for every group G_i, $score_{G_i}(p) = score_{G_i}(d)$. Since we assume that p is the unique winner of subelection (C, V_1), the only remaining voter group G_B has to be in subelection (C, V_1). According to this, candidate p has a score of $2n$ and each candidate in B can have at most $2n - 1$ points in subelection (C, V_1). However, this is only possible if the m G_i groups in subelection (C, V_2) correspond to an exact cover of B, since otherwise there would exist a candidate in B with a score of $2n$ in subelection (C, V_1). $\qquad\Box$

6   Conclusions and Open Directions

We introduced three models of partition control—equipartition, multipartition, and partition by groups—that seek to for many cases more closely model (both human and electronic-agent) real-world situations than the twenty-year-old standard benchmark set and, for the case of equipartition, to incorporate a "fairness"-focused balance condition.

We obtained a number of results on the complexity of our new models with respect to important election systems, especially plurality, the most prevalent of election systems. We established many natural examples where the variants are of the same complexity as their analogous standard benchmark model, and also established many natural examples where the variants' complexity increases relative to the analogous standard benchmark model.

This conference version of the paper focused on the unique-winner model and so-called constructive control, but the full version will cover the nonunique-winner model (in which our results broadly still hold) and so-called destructive control.

Open directions include studying combinations of our new models, seeking additional models to better capture real-world settings, investigating how well various models capture real-world settings, extending the present study to partial-information models, seeking typical-case hardness results, pursuing the additional multipartition studies mentioned at the end of the multipartition section, and resolving the complexity of weakCondorcet-CCREPC-TE to see whether it provides an additional natural example of an increasing complexity level (see Theorem 4 and the comments preceding it). Perhaps most important will be to explore notions of nearly balanced partitions, and we plan to do that in the near future.

Acknowledgments. Supported by COST Action IC1205 and grants DFG-ER-738/{1-1,2-1} and NSF-CCF-{0915792,1101452,1101479}. We thank the anonymous referees.

References

1. Bartholdi III, J., Tovey, C., Trick, M.: How hard is it to control an election? Math. Comput. Model. **16**(8/9), 27–40 (1992)
2. Baumeister, D., Erdélyi, G., Erdélyi, O.J., Rothe, J.: Computational aspects of manipulation and control in judgment aggregation. In: Perny, P., Pirlot, M., Tsoukiàs, A. (eds.) ADT 2013. LNCS, vol. 8176, pp. 71–85. Springer, Heidelberg (2013)
3. Bredereck, R., Faliszewski, P., Niedermeier, R., Talmon, N.: Large-scale election campaigns: combinatorial shift bribery. In: Proceedings of the 14th International Conference on Autonomous Agents and Multiagent Systems, pp. 67–75. International Foundation for Autonomous Agents and Multiagent Systems, May 2015
4. Bulteau, L., Chen, J., Faliszewski, P., Niedermeier, R., Talmon, N.: Combinatorial voter control in elections. Theor. Comput. Sci. **589**, 99–120 (2015)
5. Chen, J., Faliszewski, P., Niedermeier, R., Talmon, N.: Elections with few voters: candidate control can be easy. In: Proceedings of the 29th AAAI Conference on Artificial Intelligence, pp. 2045–2051. AAAI Press, January 2015
6. Dwork, C., Kumar, R., Naor, M., Sivakumar, D.: Rank aggregation methods for the web. In: Proceedings of the 10th International World Wide Web Conference, pp. 613–622. ACM Press, March 2001
7. Ephrati, E., Rosenschein, J.: A heuristic technique for multi-agent planning. Ann. Math. Artif. Intell. **20**(1–4), 13–67 (1997)

8. Erdélyi, G., Fellows, M., Rothe, J., Schend, L.: Control complexity in Bucklin and fallback voting: A theoretical analysis. J. Comput. Syst. Sci. **81**(4), 632–660 (2015)
9. Erdélyi, G., Nowak, M., Rothe, J.: Sincere-strategy preference-based approval voting fully resists constructive control and broadly resists destructive control. Math. Logic Q. **55**(4), 425–443 (2009)
10. Fagin, R., Kumar, R., Sivakumar, D.: Efficient similarity search and classification via rank aggregation. In: Proceedings of the 2003 ACM SIGMOD International Conference on Management of Data, pp. 301–312. ACM Press, June 2003
11. Faliszewski, P., Hemaspaandra, E., Hemaspaandra, L.: Weighted electoral control. J. Artif. Intell. Res. **52**, 507–542 (2015)
12. Faliszewski, P., Hemaspaandra, E., Hemaspaandra, L., Rothe, J.: Llull and Copeland voting computationally resist bribery and constructive control. J. Artif. Intell. Res. **35**, 275–341 (2009)
13. Faliszewski, P., Hemaspaandra, E., Schnoor, H.: Copeland voting: ties matter. In: Proceedings of the 7th International Conference on Autonomous Agents and Multiagent Systems, pp. 983–990. International Foundation for Autonomous Agents and Multiagent Systems, May 2008
14. Garey, M., Johnson, D.: Computers and Intractability: A Guide to the Theory of NP-Completeness. W. H Freeman and Company, New York (1979)
15. Ghosh, S., Mundhe, M., Hernandez, K., Sen, S.: Voting for movies: the anatomy of recommender systems. In: Proceedings of the 3rd Annual Conference on Autonomous Agents, pp. 434–435. ACM Press, May 1999
16. Hemaspaandra, E., Hemaspaandra, L., Menton, C.: Search versus decision for election manipulation problems. Technical Report arXiv:1202.6641 [cs.GT], Computing Research Repository, arXiv.org/corr/, Febuary 2012. Accessed March 2012. Conference Version available as [17]
17. Hemaspaandra, E., Hemaspaandra, L., Menton, C.: Search versus decision for election manipulation problems. In: Proceedings of the 30th Annual Symposium on Theoretical Aspects of Computer Science, pp. 377–388. Leibniz International Proceedings in Informatics (LIPIcs), Febuary/March 2013
18. Hemaspaandra, E., Hemaspaandra, L., Rothe, J.: Anyone but him: the complexity of precluding an alternative. Artif. Intell. **171**(5–6), 255–285 (2007)
19. Hemaspaandra, L., Williams, R.: An atypical survey of typical-case heuristic algorithms. SIGACT News **43**(4), 71–89 (2012)
20. Lenstra Jr., H.: Integer programming with a fixed number of variables. Math. Oper. Res. **8**(4), 538–548 (1983)
21. McGarvey, D.: A theorem on the construction of voting paradoxes. Econometrica **21**(4), 608–610 (1953)
22. Menton, C.: Normalized range voting broadly resists control. Theory Comput. Syst. **53**(4), 507–531 (2013)
23. Menton, C., Singh, P.: Control complexity of Schulze voting. In: Proceedings of the 23rd International Joint Conference on Artificial Intelligence, pp. 286–292. AAAI Press, August 2013
24. Parkes, D., Xia, L.: A complexity-of-strategic-behavior comparison between Schulze's rule and ranked pairs. In: Proceedings of the 26th AAAI Conference on Artificial Intelligence, pp. 1429–1435. AAAI Press, August 2012
25. Rothe, J., Schend, L.: Challenges to complexity shields that are supposed to protect elections against manipulation and control: a survey. Ann. Math. Artif. Intell. **68**(1–3), 161–193 (2013)
26. Walsh, T.: Where are the hard manipulation problems? J. Artif. Intell. Res. **42**, 1–29 (2011)

Elections with Few Candidates: Prices, Weights, and Covering Problems

Robert Bredereck[1], Piotr Faliszewski[2], Rolf Niedermeier[1], Piotr Skowron[3], and Nimrod Talmon[1(✉)]

[1] Institut für Softwaretechnik und Theoretische Informatik,
TU Berlin, Berlin, Germany
{robert.bredereck,rolf.niedermeier}@tu-berlin.de,
nimrodtalmon77@gmail.com
[2] AGH University of Science and Technology, Krakow, Poland
faliszew@agh.edu.pl
[3] University of Warsaw, Warsaw, Poland
p.skowron@mimuw.edu.pl

Abstract. We show that a number of election-related problems with prices (such as, for example, bribery) are fixed-parameter tractable (in FPT) when parameterized by the number of candidates. For bribery, this resolves a nearly 10-year old family of open problems. Our results follow by a general technique that formulates voting problems as covering problems and extends the classic approach of using integer linear programming and the algorithm of Lenstra [19]. In this context, our central result is that WEIGHTED SET MULTICOVER parameterized by the universe size is fixed-parameter tractable. Our approach is also applicable to weighted electoral control for Approval voting. We improve previously known XP-memberships to FPT-memberships. Our preliminary experiments on real-world-based data show the practical usefulness of our approach for instances with few candidates.

1 Introduction

We resolve the computational complexity status of a number of election problems parameterized by the number of candidates, for the case where voters are unweighted but have prices. These include, for example, various bribery problems [10,12] and priced control problems [21] that were known to be in XP since nearly 10 years ago, but were neither known to be fixed-parameter tractable (in FPT), nor to be W[1]-hard. We develop a general technique for showing their membership in FPT, which also applies to weighted voter control for Approval voting, improving results of Faliszewski et al. [14]. We test the running times of our algorithms empirically.

Robert Bredereck–Supported by DFG project PAWS (NI 369/10).
Piotr Faliszewski–Supported by a DFG Mercator fellowship within project PAWS (NI 369/10) while staying at TU Berlin, and by AGH University grant 11.11.230.124 afterward.
Nimrod Talmon–Supported by DFG Research Training Group MDS (GRK 1408).

T. Walsh (Ed.): ADT 2015, LNAI 9346, pp. 414–431, 2015.
DOI: 10.1007/978-3-319-23114-3_25

Algorithmic problems that model the manipulation of elections include, among others, strategic voting problems [1,6] (where we are given an election with honest voters and we ask whether a group of manipulators can cast votes to ensure their preferred candidate's victory), election control problems [2,17] (where we are given an election and ask if we can ensure a given candidate's victory by adding/deleting candidates/voters), or bribery [10,12,24] and campaign management problems [3,5,8,23] (where we want to ensure a given candidate's victory by changing some of the votes, but where each vote change comes at a price and we are bound by a budget). We focus on the case where we have a few candidates but (possibly) many voters. As pointed out by Conitzer et al. [6], this is a very natural setting and it models many real-life scenarios such as political elections or elections among company stockholders.

The complexity of manipulating elections with few candidates is, by now, very well understood. On the one hand, if the elections are weighted (as is the case for the elections held by company stockholders), then our problems are typically NP-hard even if the number of candidates is a small fixed constant [6,12,14]; these results typically follow by reductions from the well-known NP-hard problem PARTITION. One particular example where we do not have NP-hardness is control by adding/deleting voters under the Approval and k-Approval voting rules. Faliszewski et al. [14] have shown that these problems are in XP, that is, that they can be solved in polynomial time if the number of candidates is assumed to be a constant. On the other hand, if the elections are unweighted (as is the case for political elections) and no prices are involved, then we typically get FPT results. These results are often obtained by expressing the respective problems as integer linear programs (ILPs) and by applying the famous algorithm of Lenstra [19] (Lenstra's algorithm solves ILPs in FPT time with respect to the number of integer variables). For example, for control by adding voters we can have a program with a separate integer variable for each possible preference, counting how many voters with each preference we add [13] (the constraints ensure that we do not add more voters with a given preference than are available and that the desired candidate becomes a winner). Since the number of different preferences is a function depending only on the number of candidates, we can solve such an ILP using Lenstra's algorithm in FPT time. Typically, this approach does not work for weighted elections as weights give voters a form of "identity." In the control example, it no longer suffices to specify how many voters to add; we need to know exactly which ones to add (the trick in showing XP-membership for weighted voter control under Approval is to see that for each possible voter preference, we add only the heaviest voters with this preference [14]).

The main missing piece in our understanding of the complexity of manipulating elections with few candidates regards those unweighted-election problems where each voter has some sort of price (for example, as in the bribery problems). In this paper we almost completely fill this gap by showing a general approach for proving FPT membership for a class of bribery-like problems parameterized by the number of candidates, for unweighted elections[1] (as a side effect, we also get FPT membership for weighted control under the Approval and k-Approval rules). The main idea of our solution is to use mixed integer linear programming (MILP) formulations of our problems, divided into two parts. The first part is the same as in the standard ILP solutions for the non-priced variants of our problems, modeling how many voters with each possible preferences are "affected" ("bought" or "convinced" for bribery and campaign managements, "added" or "deleted" for control problems). The second part, and a contribution of this paper, uses non-integer variables to compute the cost of the solution from the first part. The critical insight of our approach is to use the fact that we "affect" the voters in the order of their increasing prices to force all our variables to be integer-valued in the optimal solutions. This way we can compute the cost of each of our solutions using rational-valued variables and solve our MILPs using Lenstra's algorithm (it maintains its FPT running time for MILPs parameterized by the number of integer variables).

Unfortunately, while Lenstra's algorithm is a very powerful tool for proving FPT membership, it might be too slow in practice. Further, as pointed out by Bredereck et al. [4], each time an FPT result is achieved through an application of Lenstra's result, it is natural to ask whether one can derive the same result through a direct, combinatorial algorithm. Coming up with such a direct algorithm seems very difficult. Thus, instead, for our problems we show a combinatorial algorithm obtaining solutions arbitrarily close to the optimal ones in FPT time (formally, we show an FPT approximation scheme). Nonetheless, in practice, one would probably not use Lenstra's algorithm for solving MILPs, but instead, one would use an off-the-shelf optimized heuristic. We provide a preliminary empirical comparison of the running times of the MILP-based algorithm (using an off-the-shelf MILP solver instead of Lenstra's algorithm) and an ILP-based algorithm that reduces our problems directly to integer linear programming (basically without "exploiting" the parameter number of candidates). Our results suggest that FPT algorithms based on solving MILPs can be very efficient in practice.

Our results can be applied to a large class of voting rules and to many election problems. Thus, to better illustrate technical details, we focus on the simplest setting possible. Specifically, we present most of our techniques through a family of classic covering problems. The motivation is threefold: (a) this focus allows us to present our results most clearly, (b) our variants of the covering problem apply directly to a number of election problems for the Approval rule, and (c) various covering problems appear in the analysis of various voting problems, thus our results should translate (more or less directly) to those problems as well. Due to lack of space, we omit some proof details.

---

[1] One problem for which our technique does not apply is SWAP BRIBERY [10]; even though Dorn and Schlotter [8] claim that it is in FPT when parameterized by the number of candidates, their proof applies only to a restricted setting. The complexity of SWAP BRIBERY parameterized by the number of candidates remains open.

2 Preliminaries

We model an election as a pair $E = (C, V)$, where $C = \{c_1, \ldots, c_m\}$ is the set of candidates and $V = (v_1, \ldots, v_n)$ is a collection of voters. Each voter is represented through his or her preferences. For the case of Approval voting, each voter's preferences take the form of a set of candidates approved by this voter. The candidate(s) receiving the most approvals are the winner(s), that is, we assume the nonunique-winner model (if several candidates have the same number of approvals then we view each of them as winning). We write $\mathrm{score}_E(c_i)$ to denote the number of voters approving c_i in election E. We refer to elections that use Approval voting and represent voter preferences in this way as approval elections. In a weighted election, in addition to their preferences, voters also have integer weights. A voter v with weight $\omega(v)$ counts as $\omega(v)$ copies of an unweighted voter. A parameterized problem is a problem with a certain feature of the input distinguished as the parameter. For example, for our election problems, the parameter will always be the number m of candidates in the election. A problem is fixed-parameter tractable (in FPT) if there exists an algorithm that, given an instance I with parameter k, can compute the answer to this problem in time $f(k) \cdot |I|^{O(1)}$, where f is a computable function and $|I|$ is the length of the encoding of I. A parameterized problem is in XP if there exists an algorithm that, given an instance I with parameter k, can compute the answer to this problem in time $|I|^{f(k)}$, where f is some computable function. In other words, XP is the class of those problems that can be solved in polynomial time under the assumption that the parameter is a constant. In contrast, problems which are NP-hard even for constant values of the parameter are said to be Para-NP-hard with respect to the parameter. For further information, we point the readers to textbooks on parameterized complexity theory [9,15,22].

3 Covering and Voting: Technique Showcase

In this section we present our main results and techniques. We start by showing a relation between election problems for the Approval rule and several covering problems. Then we present a technique for obtaining FPT results for these problems, and finally we evaluate our algorithms empirically.

3.1 From Approval Voting to Covering Problems

We are interested in the following three problems.

Definition 1 (Bartholdi et al. [2], Faliszewski et al. [12,21]). In each of the problems Approval-\$BRIBERY (priced bribery), Approval-\$CCAV (priced control by adding voters), and Approval-\$CCDV (priced control by deleting voters), we are given an approval election $E = (C, V)$ with $C = \{p, c_1, \ldots, c_m\}$ and $V = (v_1, \ldots, v_n)$, and an integer budget B. In each of the problems the goal is to decide whether it is possible to ensure that p is a winner, at a cost of at most B. The problems differ in the allowed actions and possibly in some additional parts of the input:

1. In Approval-$BRIBERY, for each voter v_i, $1 \leq i \leq n$, we are given a nonnegative integer price π_i; for this price we can change the voter's approval set in any way we choose.
2. In Approval-$CCAV (CCAV stands for "Constructive Control by Adding Voters") we are given a collection $W = (w_1, \ldots, w_{n'})$ of additional voters. For each additional voter w_i, $1 \leq i \leq n'$, we also have a nonnegative integer price π_i for adding w_i to the original election.
3. In Approval-$CCDV (CCDV stands for "Constructive Control by Deleting Voters"), we have a nonnegative integer price π_i for removing each voter v_i from the election.

In the weighted variants of these problems (which we denote by putting "WEIGHTED" after "Approval"), the input elections (and all the voters) are weighted; in particular, each voter v has an integer weight $\omega(v)$.

The unpriced variants of these problems (denoted by omitting the dollar sign from their names) are defined identically, except that all prices have the same unit value.

The above problems are, in essence, equivalent to certain covering problems, thus in this paper we focus on the complexity of these covering problems. As many of these covering problems consider multisets, the following notation will be useful. If A is a multiset and x is some element, then we write $A(x)$ to denote the number of times x occurs in A (that is, $A(x)$ is x's multiplicity in A). If x is not a member of A, then $A(x) = 0$.

Definition 2. In the WEIGHTED MULTISET MULTICOVER (WMM) problem we are given a multiset $\mathcal{S} = \{S_1, \ldots, S_n\}$ of multisets over the universe $U = \{x_1, \ldots, x_m\}$, integer weights $w_1, \ldots, w_n$ for the multisets[2], integer covering requirements $r_1, \ldots, r_m$ for the elements of the universe, and an integer budget B. We ask whether there is a subfamily $\mathcal{S}' \subseteq \mathcal{S}$ of multisets from $\mathcal{S}$ such that: (a) for each $x_i \in U$ it holds that $\sum_{S_j \in \mathcal{S}'} S_j(x_i) \geq r_i$ (that is, each element x_i is covered by at least the required number of times), and (b) $\sum_{S_j \in \mathcal{S}'} w_j \leq B$ (the budget is not exceeded).

Briefly put, the relation between WMM and various election problems (as those defined above) is that the universe corresponds to the candidates in the election, the multisets correspond to the voters, and the covering requirements depend on particular actions that we are allowed to perform.

Example 1. Consider an instance of Approval-$CCDV with election $E = (C, V)$, where $C = \{p, c_1, \ldots, c_m\}$ and $V = (v_1, \ldots, v_n)$, with prices $\pi_1, \ldots, \pi_n$ for voters not to participate in the election, and with budget B. We can express this instance as an instance of WEIGHTED MULTISET MULTICOVER as follows.

---

[2] There is a name clash between the literature on covering problems and that on elections. In the former, "weights" refer to what voting literature would call "prices." Weights of the voters are modeled as multiplicities of the elements in the multisets. We kept the naming conventions from respective parts of the literature to make our results more accessible to researchers from both communities.

For each voter v_i not approving p, we form a multiset S_i with weight π_i that includes exactly the candidates approved by v_i, each with multiplicity exactly one. For each candidate c_i, $1 \leq i \leq m$, we set its covering requirement to be $\max(\text{score}_E(c_i) - \text{score}_E(p), 0)$. It is easy to see that there is a way to ensure p's victory by deleting voters of total cost at most B if and only if it is possible to solve the presented instance of WEIGHTED MULTISET MULTICOVER with budget B.

It is clear that we do not need the full flexibility of WMM in Example 1; it suffices to use WEIGHTED SET MULTICOVER where each input multiset has multiplicities from the set $\{0, 1\}$ (in other words, the family $\mathcal{S}$ contains sets without multiplicities, but the union operation takes multiplicities into account). This is quite important since unrestricted WMM is NP-hard even for the case of a single-element universe, by a polynomial-time reduction from PARTITION.

Proposition 1. WMM *is* NP-*hard even for the case of a single element in the universe.*

Another variant of WMM is MULTISET MULTICOVER, where we assume each set to have unit weight. By generalizing the proof for Proposition 1, we show that this problem is NP-hard already for two-element universes.

Proposition 2. MULTISET MULTICOVER *is* NP-*hard even for universes of size two.*

From the viewpoint of voting theory, it is more interesting to consider an even more restricted variant of MULTISET MULTICOVER, where for each multiset S_i in the input instance there is a number t_i such that for each element x we have $S_i(x) \in \{0, t_i\}$ (in other words, elements within a single multiset have the same multiplicity). We refer to this variant of the problem as UNIFORM MULTISET MULTICOVER. Using an argument similar to that used in Example 1, it is easy to show that UNIFORM MULTISET MULTICOVER is, in essence, equivalent to Approval-WEIGHTED-CCDV.

In Example 1 we have considered Approval-\$CCDV because, among our problems, it is the most straightforward one to model via a covering problem. Nonetheless, constructions with similar flavor are possible both for Approval-\$CCAC (by, in effect, counting how many times each candidate is not approved) and for Approval-\$BRIBERY (by slightly more complicated tricks). Formally, we have the following result.

Proposition 3. *If, parameterized by the universe size,* WEIGHTED SET MULTICOVER *and* UNIFORM MULTISET MULTICOVER *are in* FPT*, then, parameterized by the number of candidates, each of Approval-*\$CCAV*, Approval-*\$CCDV*, Approval-*\$BRIBERY*,* APPROVAL-WEIGHTED-CCAV*, and* APPROVAL-WEIGHTED-CCDV *textitis in* FPT*.*

3.2   The Main Theoretical Results

Our main theoretical results are FPT algorithms for WEIGHTED SET MULTICOVER and UNIFORM MULTISET MULTICOVER parameterized by the universe size.[3]

The mixed integer linear program that we will construct has two main parts. Part 1 simply specifies what it means to solve the problem at hand, without worrying about the budget. Part 2 uses the fact that we pick the sets that implement the solution expressed in Part 1 in the increasing order of weights, to compute the total cost of the solution through rational-valued variables. The key observation is that there is a solution with the optimal budget B where all rational variables are integer.

Theorem 1. WEIGHTED SET MULTICOVER *is in* FPT *when parameterized by the universe size.*

Proof. Consider an instance of WEIGHTED SET MULTICOVER with universe $U = \{x_1, \ldots, x_m\}$, family $\mathcal{S} = \{S_1, \ldots, S_n\}$ of subsets, weights $w_1, \ldots, w_n$ for the sets, covering requirements $r_1, \ldots, r_m$ for the elements, and budget B. Our algorithm proceeds by solving an appropriate mixed integer linear program.

First, we form a family $U_1, \ldots, U_{2^m}$ of all subsets of U. For each i, $1 \leq i \leq 2^m$, let $\mathcal{S}(U_i) := \{S_j \in \mathcal{S} \mid S_j = U_i\}$. For each i and j, $1 \leq i \leq 2^m$ and $1 \leq j \leq |\mathcal{S}(U_i)|$, we write $\mathcal{S}(U_i, j)$ to denote the set from $\mathcal{S}(U_i)$ with the jth lowest weight (breaking ties in a fixed arbitrary way). Similarly, we write $w(U_i, j)$ to mean the weight of $\mathcal{S}(U_i, j)$ and we define $w(U_i, 0) = 0$ (in other words, we group the sets from $\mathcal{S}$ based on their content and sort them with respect to their weights). Given this setup, we form our mixed integer linear program.

We have 2^m integer variables z_i, $1 \leq i \leq 2^m$. Intuitively, these variables describe how many sets we take from each type. We also have $2^m n$ regular (rational) variables $y_{i,j}$, $1 \leq i \leq 2^m$, $0 \leq j \leq n-1$, which are used to model the total weight of the selected sets. We introduce the following constraints:

Part 1 Constraints. For each i, $1 \leq i \leq 2^m$, we have constraints $z_i \geq 0$ and $z_i \leq |\mathcal{S}(U_i)|$. For each element x_ℓ of the universe, we also have constraint $\sum_{U_i : x_\ell \in U_i} z_i \geq r_\ell$. These constraints ensure that the variables z_i describe a possible solution for the problem (disregarding the budget).

Part 2 Constraints. For each i and j, $1 \leq i \leq 2^m$, $0 \leq j \leq n-1$, we have constraints: $y_{i,j} \geq 0$ and $y_{i,j} \geq z_i - j$. The intended meaning of variable $y_{i,j}$ is as follows. If the solution described by variables $z_1, \ldots, z_{2^m}$ includes fewer than j sets from $\mathcal{S}(U_i)$, then $y_{i,j} = 0$; otherwise, $y_{i,j}$ says that after we added the j lowest-weight sets from family $\mathcal{S}(U_i)$, we still need to add $y_{i,j}$ more sets from this family (however, note that these variables are not required to be integers, thus the following constraint is designed in such a way that inaccurate—too large—values of these variables do not affect correctness).

---

[3] Remarkably, under reasonable complexity-theoretic assumptions, Dom et al. [7] have shown that no polynomial-size kernels exist for SET COVER (which is a special case of WEIGHTED SET MULTICOVER and UNIFORM MULTISET MULTICOVER), parameterized by the universe size and the solution size.

Our final constraint uses variables $y_{i,j}$ to express the requirement that the solution has cost at most B:

$$\sum_{i=1}^{2^m} \sum_{j=0}^{n-1} y_{i,j}(w(U_i, j+1) - w(U_i, j)) \leq B.$$

To understand this constraint, let us first replace each $y_{i,j}$ with $\max(0, z_i - j)$. Now, note that for each fixed value of i, the "internal sum" over j gives the weight of the cheapest z_i sets from $\mathcal{S}(U_i)$ (specifically, we first take z_i times $w(U_i, 1)$ because we know that each set costs at least this much, then to that we add $z_i - 1$ times $w(U_i, 2) - w(U_i, 1)$, because short of the first set from $U(S_i)$, each set costs at least $w(U_i, 2)$, and so on). To see that the constraint is correct, note that, for each $y_{i,j}$, we have that $y_{i,j} \geq \max(0, z_i - j)$ and the smaller the values $y_{i,j}$, the smaller the sum computed in this constraint.

Finally, to solve this mixed integer linear program we invoke Lenstra's famous result in its variant for mixed integer programming (see [19, Section 5]).   □

Using the same technique we can show that UNIFORM MULTISET MULTICOVER is in FPT (Part 2 of our program is slightly different in this case to account for the fact that now we pick sets with particular content in the order of decreasing multiplicities).

Theorem 2. UNIFORM MULTISET MULTICOVER *is in* FPT *when parameterized by the universe size.*

Unfortunately, it is impossible to apply our approach to the more general MULTISET MULTICOVER problem; by Proposition 2, the problem is already NP-hard for two-element universes. It is, however, possible to obtain a certain form of an FPT approximation scheme.[4]

Definition 3. Let ϵ be a real number, $\epsilon > 0$. We say that algorithm $\mathcal{A}$ is an ϵ-*almost-cover algorithm* for MULTISET MULTICOVER if, given an input instance I with universe $U = \{x_1, \ldots, x_m\}$ and covering requirements $r_1, \ldots, r_m$, it outputs a solution that covers each element x_i with multiplicity r_i' such that $\sum_i \max(0, r_i - r_i') < \epsilon \sum_i r_i$.

In other words, on the average an ϵ-almost-cover algorithm can miss each element of the universe by an ϵ-fraction of its covering requirement. For the case where we really need to cover all the elements perfectly, we might first run an ϵ-almost-cover algorithm and then complement its solution, for example, in some greedy way. Indeed, the remaining instance might be much easier to solve.

The key idea regarding computing an ϵ-almost-cover is that it suffices to replace each input multiset by several submultisets, each with a particular "precision level," so that multiplicities of the elements in each submultiset are of a similar order of magnitude.

---

[4] While this result does not, as of yet, have direct application to voting, we believe it is quite interesting in itself.

Theorem 3. *For every rational $\epsilon > 0$, there is an* FPT *ϵ-almost-cover algorithm for* MULTISET MULTICOVER *parameterized by the universe size.*

Proof. Throughout this proof we describe our ϵ-almost-cover algorithm for MULTISET MULTICOVER. We consider an instance I of MULTISET MULTICOVER with a family $\mathcal{S} = \{S_1, \ldots, S_n\}$ of multisets over the universe $U = \{x_1, \ldots, x_m\}$, where the covering requirements for the elements of the universe are $r_1, \ldots, r_m$. We associate each set S from the family $\mathcal{S}$ with the vector $v_S = \langle S(x_1), S(x_2), \ldots, S(x_m)\rangle$ of element multiplicities.

Let $\epsilon > 0$ be the desired approximation ratio. We fix $Z = \lceil\frac{4m}{\epsilon}\rceil$ and $Y = Z + \lceil\frac{4Zm^3}{\epsilon}\rceil$. Notice that $\frac{m}{Z} \leq \frac{\epsilon}{4}$ and $\frac{Zm^3}{Y-Z} \leq \frac{\epsilon}{4}$. Let $X = \left(\frac{2Y^m}{\epsilon} + 1\right)^m$ and let $V_1, \ldots, V_X$ be a sequence of all m-dimensional vectors whose entries come from the $\left(\frac{2Y^m}{\epsilon} + 1\right)$-element set $\{0, \frac{\epsilon}{2}, \epsilon, \frac{3\epsilon}{2}, 2\epsilon, \ldots, Y^m\}$. For each j, $1 \leq j \leq X$, we write $V_j = \langle V_j(x_1), V_j(x_2), \ldots, V_j(x_m)\rangle$. Intuitively, these vectors describe some subset of "shapes" of all possible multisets—interpreted as vectors of multiplicities—over our m-element universe. For each number β, we write βV_i to mean the vector $\langle\lfloor\beta V_{i,1}\rfloor, \lfloor\beta V_{i,2}\rfloor, \ldots, \lfloor\beta V_{i,m}\rfloor\rangle$.

Intuitively, vectors of the form βV_i are approximations of those multisets for which the positive multiplicities of the elements do not differ too much (formally, for those multisets for which the positive multiplicities differ by at most a factor of $\frac{2Y^m}{\epsilon}$). Indeed, for each such set S, we can find a value β and a vector V_j such that for each element x_i it holds that $S(x_i) \geq \beta V_j(x_i) \geq \left(1 - \frac{\epsilon}{2}\right)S(x_i)$. However, this way we cannot easily approximate those sets for which multiplicities differ by a large factor. For example, consider a set S represented through the vector $\langle 0, \ldots, 0, 1, Q\rangle$, where $Q \gg \frac{2Y^m}{\epsilon}$. For each value β and each vector V_j, the vector βV_j will be inaccurate with respect to the multiplicity of element x_{m-1} or inaccurate with respect to the multiplicity of element x_m (or inaccurate with respect to both these multiplicities).

The main step of our algorithm is to modify the instance I so that we replace each set S from the family $\mathcal{S}$ with a sequence of vectors of the form βV_j that altogether add to at most the set S (each such sequence can contain multiple vectors of different "shapes" V_j and of different scaling factors β). The goal is to obtain an instance that on the one hand consists of "nicely-structured" sets (vectors) only, and on the other hand has the following property: If in the initial instance I there exist K sets that cover elements $x_1, \ldots x_m$ with multiplicities $r_1, \ldots, r_m$, then in the new instance there exist K sets that cover elements $x_1, \ldots, x_m$ with multiplicities $r'_1, \ldots, r'_m$, such that $\sum_i \max(0, r_i - r'_i) < \epsilon \sum_i r_i$. We refer to this as the *almost-cover approximation property*.

The procedure for replacing a given set S is presented as Algorithm 1. This algorithm calls the `Emit` function with arguments (β, V) for each vector βV that it wants to output (V is always one of the vectors $V_1, \ldots, V_X$). The emitted sets replace the set S from the input. Below we show that if we apply Algorithm 1 to each set from $\mathcal{S}$, then the resulting instance I' has our almost-cover approximation property.

Let us consider how Algorithm 1 proceeds on a given set S. For the sake of clarity, let us assume there is no rounding performed by Algorithm 1 in function Round_And_Emit (the loop in line 29). We will go back to this issue later.

The algorithm considers the elements of the universe—indexed by variable i throughout the algorithm—in the order given by the vector "sorted" (formed in line 3 of Algorithm 1). Let $\prec$ be the order in which Algorithm 1 considers the elements (so $x_{i'} \prec x_{i''}$ means that $x_{i'}$ is considered before $x_{i''}$), and let $x_1', \ldots, x_m'$ be the elements from the universe renamed so that $x_1' \prec x_2' \prec \cdots \prec x_m'$. Let r be the number of sets that Algorithm 1 emits on our input set S and let these sets be $S_1, S_2, \ldots, S_r$. (This is depicted on Fig. 1, where for the sake of the example we take $m = 6$ and $r = 3$).

Consider the situation where the algorithm emits the k'th set, S_k, and let i_k be the value of variable i right before the call to Round_And_Emit that caused S_k to be emitted. Note that each element x from the universe such that $x_{i_k} \prec x$ has the same multiplicity in S_k as element x_{i_k} (line 19 of Algorithm 1). Let $t_k = \sum_j S_k(x_j')$ be the sum of the multiplicities of the elements from S_k. We make the following observations:

Observation 1: $S_k(x_{i_k}') = Z \cdot S_k(x_{i_k-1}')$.

Observation 2: It holds that $S_{k+1}(x_{i_k}') \geq Y \cdot S_k(x_{i_k-1}') - Z \cdot S_k(x_{i_k-1}') = (Y - Z)S_k(x_{i_k-1}') = \frac{(Y-Z)}{Z}S_k(x_{i_k}')$.

Observation 3: We have that $S_{k+1}(x_{i_k}') \geq \frac{(Y-Z)}{Z}S_k(x_{i_k}') \geq \frac{(Y-Z)}{Zm}t_k$. Further, we have that $S_{k+1}(x_{i_{k+1}}') \geq S_{k+1}(x_{i_k}') \geq \frac{(Y-Z)}{Zm^2}\sum_{j\leq k} t_j$.

Observation 4: For $i < i_k$ it holds that $\sum_{q\leq k} S_q(x_i') = S(x_i')$.

Now let us consider some solution for instance I that consists of K sets, $\mathcal{S}^{\mathrm{opt}} = \{S_1^{\mathrm{opt}}, S_2^{\mathrm{opt}}, \ldots, S_K^{\mathrm{opt}}\} \subseteq \mathcal{S}$. These sets, altogether, cover all the elements from the universe with required multiplicities, that is, it holds that for each i we have $\sum_{S\in\mathcal{S}^{\mathrm{opt}}} S(x_i) \geq r_i$. For each set $S \in \mathcal{S}^{\mathrm{opt}}$ and for each element x_i from the universe, we pick an arbitrary number $y_{S,i}$ so that altogether the following conditions hold:

1. For every set $S \in \mathcal{S}^{\mathrm{opt}}$ and every x_i, $y_{S,i} \leq S(x_i)$.
2. For every x_i, $\sum_{S\in\mathcal{S}^{\mathrm{opt}}} y_{S,i} = r_i$.

Intuitively, for a given set S, the values $y_{S,1}, y_{S,2}, \ldots, y_{S,m}$ describe the multiplicities of the elements from S that are *actually used* to covers the elements. Based on these numbers, we will show how to replace each set from $\mathcal{S}^{\mathrm{opt}}$ with one of the sets emitted for it, so that the resulting family of sets has the almost-cover approximation property.

Consider a set $S \in \mathcal{S}^{\mathrm{opt}}$ for which Algorithm 1 emits r sets, $S_1, S_2, \ldots, S_r$. As in the discussion of Algorithm 1, let $x_1', \ldots, x_m'$ be the elements from the universe in which Algorithm 1 considers them (when emitting sets for S). We write $y_{S,i}'$ to mean the value $y_{S,j}$ such that $x_j = x_i'$. Let $\mathcal{R} = \{S_1, S_2, \ldots, S_r\}$, let $i_{\max} = \mathrm{argmax}_i y_{S,i}'$, and let S_{repl} be the set from $\mathcal{R}$ defined in the following way:

Algorithm 1. The transformation algorithm used in the proof of Theorem 3—the algorithm replaces a given set S with a sequence of vectors of the form βV_j.

```
1  Main(S):
2      multip ← ⟨(1, S(x₁)), (2, S(x₂)), ..., (m, S(xₘ))⟩;
3      sorted ← sort(multip) ;          // sort in ascending order of multiplicities
4      i ← 0 ;
       //  sorted[i].first refers to the i'th item's number
       //  sorted[i].second refers to its multiplicity
5      while sorted[i].second = 0 do
6        |  i ← i + 1 ;
7      Main_Rec(i, sorted) ;
8
9  Main_Rec(i, multip):
10     V ← ⟨0, 0, ..., 0⟩ (vector of m zeros). ;
11     β ← multip[i].second ;
12     V[multip[i].first] ← 1 ;
13     i ← i + 1 ;
14     while i ≤ m do
15         if multip[i].second < Y · multip[i−1].second then
16             V[multip[i].first] ← multip[i].second / β ;
17             i ← i + 1 ;
18         else
19             for j ← i to m do
20               |  V[multip[j].first] ← Z · multip[i−1].second / β ;
21             Round_And_Emit(β, V) ;
22             for j ← 1 to m do
23               |  multip[i].second ← multip[i].second − βV[multip[i].first] ;
24             Main_Rec(i, multip) ;
25             return
26     Round_And_Emit(β, V);
27
28 Round_And_Emit(β, V):
29     for ℓ ← 1 to m do
30       |  V[ℓ] ← ⌊2V[ℓ]/ε⌋ / (ε/2);
31     Emit(β, V);
```

---

1. If for every set $S_k \in \mathcal{R}$ we have $S_k(x'_{i_{\max}}) < y'_{S,i_{\max}}$, then S_{repl} is the set $S_k \in \mathcal{R}$ with the greatest value $S_k(x'_{i_{\max}})$ (the set that covers element $x'_{i_{\max}}$ with the greatest multiplicity). This is the case denoted as "Case (c)" in Fig. 2.
2. Otherwise S_{repl} is the set $S_k \in \mathcal{R}$ that has the lowest value $S_k(x'_{i_{\max}})$, yet no-lower than $y'_{S,i_{\max}}$. This is the case denoted as either "Case (a)" or "Case (b)" in Fig. 2.

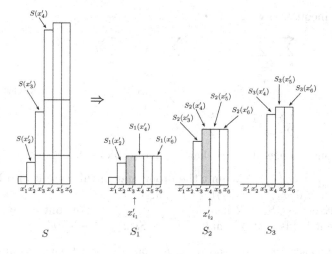

Fig. 1. An example for Algorithm 1: The algorithm replaces S with sets S_1, S_2, and S_3.

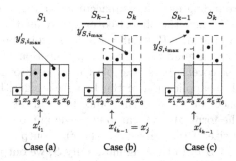

Fig. 2. The cases in the proof of Theorem 3. The bullets represent values $y'_{S,1}, \ldots, y'_{S,m}$.

We now show that S_{repl} is a good candidate for replacing S, that is, that $\sum_i \max(0, y'_{S,i} - S_{\mathrm{repl}}(x'_i)) < \epsilon \sum_i y'_{S,i}$. To this end, we consider the three cases depicted in Fig. 2:

Case (a). It holds that $y'_{S,i_{\max}} < S_1(x'_{i_{\max}})$ (that is, S_1 already covers the most demanding element of the universe to the same extent as S does). This means that we have $\sum_\ell \max(0, y'_{S,\ell} - S_1(x'_\ell)) = 0$. By the criterion for choosing set S_{repl}, we have that $S_{\mathrm{repl}} = S_1$.

Case (b). There exist sets $S_{k-1}, S_k \in \mathcal{R}$ such that $S_k(x'_{i_{\max}}) \geq y'_{S,i_{\max}} > S_{k-1}(x'_{i_{\max}})$ (and thus, $S_{\mathrm{repl}} = S_k$). Let $x'_j = x'_{i_{k-1}}$ (recall from the discussion of Algorithm 1 that i_{k-1} is the index of the universe element which caused emitting S_{k-1}). Let us consider two subcases:

(i) $y'_{S,i_{\max}} \leq S_k(x'_j)$: We first note that for each $i \geq j$ it holds that $y'_{S,i} \leq S_k(x'_i)$. Further, for each $i < j$, we have $y'_{S,i} \leq \sum_{\ell \leq k-1} S_\ell(x'_i)$ (this follows from Observation 4 and the fact that $y'_{S,i} \leq \bar{S}(x'_i)$). Based on this inequality, we get:

$$
\sum_{i<j} y'_{S,i} \leq \sum_{i<j} \sum_{\ell \leq k-1} S_\ell(x'_i) \leq \sum_{\ell \leq k-2} t_\ell + \sum_{i<j} S_{k-1}(x'_i)
$$

$$
\leq \frac{Zm^2}{(Y-Z)} S_{k-1}(x'_j) + \frac{m}{Z} S_{k-1}(x'_j) \qquad \text{(Observations 3 and 1)}
$$

$$
\leq \frac{\epsilon}{2} S_{k-1}(x'_j) \leq \frac{\epsilon}{2} y_{S,i_{\max}}.
$$

In consequence, it holds that $\sum_\ell \max(0, y'_{S,\ell} - S_k(x'_\ell)) < \frac{\epsilon}{2} \sum_\ell y'_{S,\ell}$.

(ii) $y'_{S,i_{\max}} > S_k(x'_j)$: We omit the proof that in this case it also holds that $\sum_\ell \max(0, y'_{S,\ell} - S_k(x'_\ell)) \leq \frac{\epsilon}{2} \sum_\ell y'_{S,\ell}$.

Case (c). Every set $S_k \in \mathcal{R}$ has $S_k(x'_{i_{\max}}) < y'_{S,i_{\max}}$. We omit the proof that in this case it holds that $\sum_\ell \max(0, y_{S,\ell} - S_k(x'_\ell)) \leq \frac{\epsilon}{2} \sum_\ell y'_{S,\ell}$.

The above case analysis almost shows that we indeed have the almost-cover approximation property. It remains to consider the issue of rounding (Line 29 of Algorithm 1). This rounding introduces inaccuracy that is bounded by factor $\frac{\epsilon}{2}$ and thus, indeed, we do have the almost-cover approximation property.

Now, given the new instance I', it suffices to find a solution for I' that satisfies the desired approximation guarantee (that is, a collection $\mathcal{S}'$ of at most K sets that form an ϵ-almost-cover). It is possible to do so through a mixed integer linear program (and an application of the Lenstra's algorithm [19]). We omit the details due to space (we mention that since in I' all the sets are represented through vectors of the form βV_j, we can bound the number of integer variables by a function of the size of the universe). The final output of our algorithm is as follows: For each set S from the original family $\mathcal{S}$, we output S if $\mathcal{S}'$ contains at least one set emitted for S. □

For the case of WEIGHTED SET MULTICOVER we show a more standard variant of an FPT approximation scheme.

Definition 4. Let ε, $0 < \varepsilon < 1$, be a real number. A $(1+\varepsilon)$-approximation algorithm for WEIGHTED SET MULTICOVER is an algorithm that, given an instance of the problem, outputs a solution satisfying all covering requirements, but whose weight is at most $1 + \varepsilon$ times the weight of the optimal one.

As opposed to all the previously presented algorithms (including the one from Theorem 3), the next algorithm does not rely on solving (M)ILP instances. The main idea is to use a refined variant of brute-force search which considers for each type of sets only a set of promising numbers of occurrences in the solution instead of considering every possible number of occurrences.

Theorem 4. *For each ε, $0 < \varepsilon < 1$, there is a $(1+\varepsilon)$-approximation algorithm for WEIGHTED SET MULTICOVER that runs in time $O(\lceil 2^m/\varepsilon + 1 \rceil^{2^m} n^2 m)$.*

We conclude this section by translating the results from the world of covering problems to the world of approval elections. We obtain the following corollary.

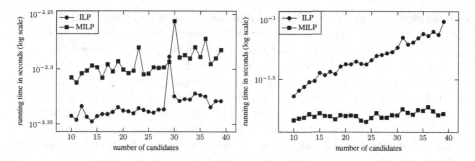

Fig. 3. Running time depending on the number of candidates. Left: plain test series, right: duplicated voters.

Corollary 1. APPROVAL-\$CCAV, APPROVAL-\$CCDV, APPROVAL-\$BRIBERY, *as well as* APPROVAL-WEIGHTED-CCAV *and* APPROVAL-WEIGHTED-CCDV, *are in* FPT, *when parameterized by the number of candidates.*

Contrarily, it is either shown explicitly by Faliszewski et al. [12], or follows trivially, that these problems with both prices and weights are NP-hard already for two candidates (that is, Para-NP-hard with respect to the number of candidates).

3.3 Preliminary Empirical Evaluation

In this section we evaluate our algorithms empirically. Specifically, we focus on the MILP-based algorithm from subsection 3.2, as applied to Approval-\$CCDV, and on the standard ILP algorithm for this problem (see below). In both cases, instead of using the very slow algorithm of Lenstra (with running time being roughly $(2^m)^{(2^m)}$ [16,18,19]), we chose an off-the-shelf solver (CPLEX). The main purpose of the experiments is to explore whether our (M)ILP-based FPT algorithms are practical to use. Thus, we focus on evaluating the running time. (We point the reader, for example, to the work of Erdélyi et al. [11] for an example of a much more detailed experimental analysis of control problems in elections for several voting rules).

Test Data. We use Preflib [20] as a well-known source for real-world elections. Since Preflib contains only few elections with approval preferences (provided through linear orderings with ties containing exactly two groups of tied candidates each), we used elections with strict linear-order preferences and for each voter we uniformly at random chose how many of the top candidates this voter approves of.

Test Series. We focus on Approval-\$CCDV since, among our problems, it requires the least amount of information to be added to the elections to obtain full instances. Specifically, we only need to choose the preferred candidate p and the prices for deleting the voters. We performed two test series, *duplicated voters* and *plain*. In the duplicated voters test series we interpret the Preflib elections as random samples of larger elections and, for each election from PrefLib, duplicate each voter between 1 and 500 times (the multiplication factor was chosen uniformly at random for each voter separately). In the plain test series, the set of voters remains unchanged. For each voter (after the duplication) we set the price for its deletion uniformly at random to be an integer between 1 and 500. Finally, we uniformly at random select one candidate to be the preferred one and create for each Preflib election and each $m \in \{10, 12, \ldots, 40\}$, ten instances with m candidates (by first creating the full instance and removing all but m randomly chosen candidates). All in all, we obtained more than 8000 test instances for each test series.

We stress that our focus is on the running times of our algorithms, and not—for example—on modeling how frequent control might be in real-life settings. The purpose of the experiments is to be a proof-of-concept of the algorithms suggested herein.

Algorithms. We tested two algorithms, both of which transfer the Approval-$CCDV instance into a WEIGHTED SET MULTICOVER instance and apply CPLEX to solve a (mixed) integer program. The first algorithm (referred to as ILP) uses a straightforward integer linear programming formulation with one binary variable for each set (representing presence in the solution), constraints ensuring that each element is appropriately covered by the selected sets, and an objective function minimizing the total costs. The second algorithm (referred to as MILP) uses the mixed integer linear programming formulation from Theorem 1. We did not consider brute-force or approximation approaches since both (M)ILP-based algorithms turned out to be extremely fast for the Preflib data set (and always return optimal solutions).

Results. Surprisingly, both algorithms solved all instances very fast (using at most a few seconds). For the plain test series the running time slightly increases as the number of candidates increases, for both algorithms. A possible explanation is that the program description as well as the number of variables[5] also slightly increases. The ILP is faster by roughly a constant factor which might be caused by its much simpler formulation and the usage of binary variables instead of integer ones. For the duplicated voters test series, the situation changes: the running time increases only slightly with the increase of the number of candidates for the MILP, but it increases significantly for the ILP. A possible explanation is that the ILP has one variable for each voter whereas the MILP has one variable for each class of duplicated voters. See Fig. 3 for an illustration.

4   Generalizations

We now consider the ordinal model of elections, where each voter's preferences are represented as an order, ranking the candidates from the most preferred one

---

[5] By removing candidates during instance generation one also removes voters only approving removed candidates.

to the least preferred one. For example, for $C = \{c_1, c_2, c_3\}$, vote $c_1 \succ c_3 \succ c_2$ means that the voter likes c_1 best, then c_3, and then c_2.

There are many different voting rules for the ordinal election model. Here we concentrate only on scoring rules. A scoring rule for m candidates is a nondecreasing vector $\alpha = (\alpha_1, \ldots, \alpha_m)$ of integers. Each voter gives α_1 points to his or her most preferred candidate, α_2 points to the second most preferred candidate, and so on. Examples of scoring rules include the Plurality rule, defined through vectors of the form $(1, 0, \ldots, 0)$, k-Approval, defined through vectors with k ones followed by $m - k$ zeroes, and Borda count, defined through vectors of the form $(m - 1, m - 2, \ldots, 0)$.

For each voting rule $\mathcal{R}$ in the ordinal model, it is straightforward to define $\mathcal{R}$-$\$$CCAV, $\mathcal{R}$-$\$$CCDV, and $\mathcal{R}$-$\$$BRIBERY. Using our MILP technique, we obtain the following result.

Theorem 5. *For every voting rule $\mathcal{R}$ for which winner determination can be expressed through a set of integer linear inequalities over variables that indicate how many voters with each given preference order are in the election, $\mathcal{R}$-$\$$CCAV, $\mathcal{R}$-$\$$CCDV, and $\mathcal{R}$-$\$$BRIBERY are in FPT when parameterized by the number of candidates.*

For a description of what we mean by "expressing the winner determination problem through integer linear inequalities," we point to the discussions by Dorn and Schlotter [8] or by Faliszewski et al. [14]. For example, the result applies to all scoring rules.

We also partially resolve an open problem posed by Bredereck et al. [5] regarding SHIFT BRIBERY. In this problem we are given an election and a preferred candidate p, and the goal is to ensure p's victory by shifting p forward in some of the votes (the cost of each shift depends on the number of positions by which we shift p). Under the sortable prices assumption, voters with the same preference orders can be sorted so that if voter v' precedes voter v'', then we know that shifting p by each given number of positions i in the vote of v' costs at most as much as doing the same in the vote of v''. Using this assumption, we obtain the following result (all-or-nothing prices are a special case of sortable prices where we always shift p to the top of a given vote or we leave the vote unchanged).

Theorem 6. *For Borda (and for Maximin and Copeland voting rules), SHIFT BRIBERY for sortable price functions and for all-or-nothing price functions is in FPT when parameterized by the number of candidates.*

Bredereck et al. [5] gave an FPT approximation scheme for the the problems from Theorem 6; we use part of their algorithm and apply our MILP technique.

5   Conclusions

We have studied election control and bribery for the case of few voters with prices. We also considered weighted Approval elections with few voters. By developing a very general proof technique, in these settings we have improved known XP-membership results to FPT-membership results. We have also tested our algorithms empirically and found them to be extremely fast. Our empirical results provide some evidence for the correlation between running time and number of candidates, as given by the FPT-classification, at least for Preflib-based Approval-\$CCDV test instances.

Our paper leads to several possible directions for future work. First, the experiments we have presented are only preliminary and they should be extended in a number of ways. Second, it would be very interesting to further explore the relevance of FPT approximation algorithms to other voting scenarios.

References

1. Bartholdi III, J.J., Tovey, C.A., Trick, M.A.: The computational difficulty of manipulating an election. Soc. Choice Welf. **6**(3), 227–241 (1989)
2. Bartholdi III, J.J., Tovey, C.A., Trick, M.A.: How hard is it to control an election. Math. Comput. Model. **16**(8–9), 27–40 (1992)
3. Baumeister, D., Faliszewski, P., Lang, J., Rothe, J.: Campaigns for lazy voters: truncated ballots. In: Proceedings of the 11th International Conference on Autonomous Agents and Multiagent Systems (AAMAS 2012), pp. 577–584, June 2012
4. Bredereck, R., Chen, J., Faliszewski, P., Guo, J., Niedermeier, R., Woeginger, G.J.: Parameterized algorithmics for computational social choice: nine research challenges. Tsinghua Sci. Technol. **19**(4), 358–373 (2014a)
5. Bredereck, R., Chen, J., Faliszewski, P., Nichterlein, A., Niedermeier, R.: Prices matter for the parameterized complexity of shift bribery. In: Proceedings of The Twenty-Eighth AAAI Conference on Artificial Intelligence (AAAI 2014), pp. 1398–1404 (2014b)
6. Conitzer, V., Sandholm, T., Lang, J.: When are elections with few candidates hard to manipulate? J. ACM **54**(3), 1–33 (2007)
7. Dom, M., Lokshtanov, D., Saurabh, S.: Kernelization lower bounds through colors and IDs. ACM Trans. Algorithm. **11**(2), 13:1–13:20 (2014)
8. Dorn, B., Schlotter, I.: Multivariate complexity analysis of swap bribery. Algorithmica **64**(1), 126–151 (2012)
9. Downey, R.G., Fellows, M.R.: Fundamentals of Parameterized Complexity. Springer, London (2013)
10. Elkind, E., Faliszewski, P., Slinko, A.: Swap bribery. In: Mavronicolas, M., Papadopoulou, V.G. (eds.) SAGT 2009. LNCS, vol. 5814, pp. 299–310. Springer, Heidelberg (2009)
11. Erdélyi, G., Fellows, M., Rothe, J., Schend, L.: Control complexity in Bucklin and fallback voting: an experimental analysis. J. Comput. Syst. Sci. **81**(4), 661–670 (2015)
12. Faliszewski, P., Hemaspaandra, E., Hemaspaandra, L.A.: How hard is bribery in elections? J. Artif. Intell. Res. **35**, 485–532 (2009)
13. Faliszewski, P., Hemaspaandra, E., Hemaspaandra, L.A.: Multimode control attacks on elections. J. Artif. Intell. Res. **40**, 305–351 (2011)
14. Faliszewski, P., Hemaspaandra, E., Hemaspaandra, L.A.: Weighted electoral control. In: Proceedings of the 12th International Conference on Autonomous Agents and Multiagent Systems (AAMAS 2013), pp. 367–374 (2013)

15. Flum, J., Grohe, M.: Parameterized Complexity Theory. Springer, Heidelberg (2006)
16. Fredman, M.L., Tarjan, R.E.: Fibonacci heaps and their uses in improved network optimization algorithms. J. ACM **34**(3), 596–615 (1987)
17. Hemaspaandra, E., Hemaspaandra, L.A., Rothe, J.: Anyone but him: the complexity of precluding an alternative. Artif. Intell. **171**(5–6), 255–285 (2007)
18. Kannan, R.: Minkowski's convex body theorem and integer programming. Math. Oper. Res. **12**(3), 415–440 (1987)
19. Lenstra Jr., H.W.: Integer programming with a fixed number of variables. Math. Oper. Res. **8**(4), 538–548 (1983)
20. Mattei, N., Walsh, T.: PREFLIB: a library for preferences HTTP://WWW.PREFLIB.ORG. In: Perny, P., Pirlot, M., Tsoukiàs, A. (eds.) ADT 2013. LNCS, vol. 8176, pp. 259–270. Springer, Heidelberg (2013)
21. Miasko, T., Faliszewski, P.: The complexity of priced control in elections. Manuscript (2014). http://home.agh.edu.pl/faliszew/priced.pdf
22. Niedermeier, R.: Invitation to Fixed-Parameter Algorithms. Oxford University Press, Oxford (2006)
23. Schlotter, I., Faliszewski, P., Elkind, E.: Campaign management under approval-driven voting rules. In: Proceedings of the Twenty-Fifth Conference on Artificial Intelligence (AAAI 2011), pp. 726–731, August 2011
24. Xia, L.: Computing the margin of victory for various voting rules. In: Proceedings of the 13th ACM Conference on Electronic Commerce (EC 2012), pp. 982–999, June 2012

Complexity of Bribery and Control for Uniform Premise-Based Quota Rules Under Various Preference Types

Dorothea Baumeister, Jörg Rothe, and Ann-Kathrin Selker[(✉)]

Institut für Informatik, Heinrich-Heine-Universität Düsseldorf,
40225 Düsseldorf, Germany
{baumeister,rothe,selker}@cs.uni-duesseldorf.de

Abstract. Manipulation of judgment aggregation procedures has first been studied by List [14] and Dietrich and List [8], and Endriss et al. [9] were the first to study it from a computational perspective. Baumeister et al. [2,3,6] introduced the concepts of bribery and control in judgment aggregation and studied their algorithmic and complexity-theoretic properties. However, their results are restricted to Hamming-distance-respecting preferences and their results on bribery apply to the premise-based procedure only. We extend these results to more general preference notions, including closeness-respecting and top-respecting preferences that are due to Dietrich and List and have been applied to manipulation in judgment aggregation by Baumeister et al. [4,5]. In addition, our results apply to uniform premise-based quota rules that generalize the premise-based procedure.

Keywords: Bribery · Control · Judgment aggregation · Computational complexity

1   Introduction

Judgment aggregation refers to methods of collective decision making where the judgments of a number of judges are aggregated so as to arrive at a collective judgment set. Endriss et al. [9] were the first to study manipulation in judgment aggregation from a computational point of view. We study the complexity of problems related to bribery and control in judgment aggregation, notions that were introduced and applied to voting problems in computational social choice by Bartholdi et al. [1] (see also the work of Hemaspaandra et al. [13]) for control and by Faliszewski et al. [10,11] for bribery (see also the book chapter by Faliszewski and Rothe [12] for many more references). These notions have been transferred to (a computational study of) judgment aggregation by Baumeister et al. [2,3, 5,6]. However, their results apply to Hamming-distance-respecting preferences only, and in the case of bribery they have only investigated the premise-based procedure. The main contribution of this paper is to extend their study for three types of control (control by adding, by deleting, and by replacing judges)

© Springer International Publishing Switzerland 2015
T. Walsh (Ed.): ADT 2015, LNAI 9346, pp. 432–448, 2015.
DOI: 10.1007/978-3-319-23114-3_26

and for bribery and microbribery to more general preference notions, including closeness-respecting and top-respecting preferences. We also extend the study of bribery to uniform premise-based quota rules, which generalize the premise-based procedure.

Closeness-respecting and top-respecting preferences have been introduced by Dietrich and List [8] and have been applied to manipulation in judgment aggregation by them and by Baumeister et al. [4,5]. Intuitively, for top-respecting preferences all we know is that the attacker prefers her desired set to any other judgment set, while in closeness-respecting preferences we also know that judgment sets with additional agreements are preferred.

In Sect. 2, we provide the needed notions from judgment aggregation. We study the complexity of control problems in Sect. 3 and that of bribery problems in Sect. 4. In Sect. 5, we summarize our results and propose some open questions for future work.

2   Definitions and Notations

Throughout this paper, we will utilize the judgment aggregation framework due to Endriss et al. [9]. Let $\mathscr{L}_{PS}$ be the set of all propositional formulas that can be built from a set of propositional variables, PS, using the common boolean connectives, i.e., *disjunction* ($\vee$), *conjunction* ($\wedge$), *implication* ($\rightarrow$), and *equivalence* ($\leftrightarrow$) as well as the constants 1 (*true*) and 0 (*false*). We use $\overline{\alpha}$ to refer to the complement of α, that is, $\overline{\alpha} = \neg\alpha$ if α is not negated, and $\overline{\alpha} = \beta$ if $\alpha = \neg\beta$. A set $\Phi \subseteq \mathscr{L}_{PS}$ is said to be *closed under complementation* if $\overline{\alpha} \in \Phi$ for all $\alpha \in \Phi$, and to be *closed under propositional variables* if $PS \subseteq \Phi$. We call a finite nonempty set $\Phi \subseteq \mathscr{L}_{PS}$ without doubly negated formulas that is closed under complementation an *agenda*, and a subset $J \subseteq \Phi$ a *judgment set for* Φ. J is an *individual judgment set* if it is the set of propositions accepted by some judge. Furthermore, J is called *complete* if $\alpha \in J$ or $\overline{\alpha} \in J$ for all $\alpha \in \Phi$, and J is said to be *consistent* if there exists an assignment such that all formulas in J are satisfied. Let $\mathscr{J}(\Phi)$ be the set of all complete and consistent judgment sets of an agenda Φ and let $N = \{1, \ldots, n\}$ be the set of judges. We call $\mathbf{J} = (J_1, \ldots, J_n) \in \mathscr{J}(\Phi)^n$ the *profile* of the judges' individual judgment sets. A resolute[1] *(judgment aggregation) procedure* for an agenda Φ and a set of judges N of size n is a function $F : \mathscr{J}(\Phi)^n \rightarrow 2^{\Phi}$, where 2^{Φ} denotes the power set of Φ. That means that a procedure maps a profile to a *collective judgment set* or *(collective) outcome*.

Let $\|S\|$ be the cardinality of the set S and let $\models$ denote the satisfaction relation. Dietrich and List [7] introduced the class of premise-based quota rules. We will consider only a special case, the uniform premise-based quota rules.

---

[1] There are also irresolute judgment aggregation procedures (i.e., procedures that may output more than one collective judgment set), such as the distance-based procedures introduced by Pigozzi [17] and Miller and Osherson [15], which we won't consider here, though.

Definition 1 (Uniform Premise-based Quota Rule). *Let the agenda Φ be closed under propositional variables. Subdivide Φ into the two disjoint subsets Φ_p (the set of* premises*) containing exactly all literals, and Φ_c (the set of con-clusions), both closed under complementation. Furthermore, subdivide Φ_p into two disjoint subsets, Φ_1 and Φ_2, satisfying that $\varphi \in \Phi_1$ if and only if $\overline{\varphi} \in \Phi_2$. Assign to each literal $\varphi \in \Phi_1$ a rational quota q, $0 \leq q < 1$, and to each literal $\overline{\varphi} \in \Phi_2$ the associated quota $q' = 1 - q$. The* uniform premise-based quota rule with quota q *(denoted by $UPQR_q$) is the procedure mapping each profile $\mathbf{J} = (J_1, \ldots, J_n)$ of individual judgment sets for Φ to the collective outcome $UPQR_q(\mathbf{J}) = \triangle \cup \{\psi \in \Phi_c \mid \triangle \models \psi\}$, where $\triangle = \{\varphi \in \Phi_1 \mid \|\{i \mid \varphi \in J_i\}\| > nq\} \cup \{\varphi \in \Phi_2 \mid \|\{i \mid \varphi \in J_i\}\| \geq nq'\}$.*

Throughout the paper, we will assume that all literals in Φ_1 are not negated. Since Φ is closed under propositional variables and Φ_p contains exactly all literals, the outcomes of $UPQR_q$ are complete and consistent. The threshold for a literal $\varphi \in \Phi_1$ to be accepted is $\lfloor nq + 1 \rfloor$, i.e., φ is contained in the collective outcome if and only if it is contained in at least $\lfloor nq + 1 \rfloor$ individual judgment sets, whereas literals $\overline{\varphi} \in \Phi_2$ need at least $\lceil nq' \rceil$ affirmations. It is possible to determine in polynomial time whether a given formula is an element of the collective outcome of a uniform premise-based quota rule. The special case of $UPQR_{1/2}$ for an odd number of judges is also known as the premise-based procedure (PBP).

We will study judgment aggregation problems where some external agent tries to influence a judgment aggregation process in order to obtain a better outcome. In order to compare two outcomes, we will use various notions of pref-erence types induced by an external agent's desired set. These notions have been introduced by Dietrich and List [8] and have later been refined by Baumeister et al. [5]. Formally, this desired set is a subset of a complete and consistent judgment set.

Let Φ be an agenda, $X, Y \in \mathscr{J}(\Phi)$, and let $\succsim$ be a weak order over $\mathscr{J}(\Phi)$, i.e., a transitive and total binary relation over complete and consistent judgment sets. We say that X *is weakly preferred to* Y whenever $X \succsim Y$, and we say that X *is preferred to* Y, denoted by $X \succ Y$, whenever $X \succsim Y$ and $Y \nsucceq X$. Furthermore, we define $X \sim Y$ by $X \succsim Y$ and $Y \succsim X$.

Definition 2. *Let Φ be an agenda, let U be the set of all weak orders over $\mathscr{J}(\Phi)$, and let J be a possibly incomplete judgment set. Define*

1. *the set $U_J \subseteq U$ of* unrestricted J-induced (weak) preferences *by*

$$U_J = \{\succsim \ \in U \mid \text{for all} X, Y \in \mathscr{J}(\Phi), X \sim Y \text{ whenever} X \cap J = Y \cap J\};$$

2. *the set $TR_J \subseteq U_J$ of* top-respecting J-induced (weak) preferences *by*

$$TR_J = \left\{\succsim \ \in U_J \ \middle| \ \begin{matrix} \text{for all} X, Y \in \mathscr{J}(\Phi), X \succ Y \\ \text{whenever} X \cap J = J \text{ and} Y \cap J \neq J \end{matrix}\right\};$$

3. *the set $CR_J \subseteq U_J$ of* closeness-respecting J-induced (weak) preferences *by*

$$CR_J = \{\succsim \ \in U_J \mid \text{for all} X, Y \in \mathscr{J}(\Phi), \text{if} Y \cap J \subseteq X \cap J \text{then} X \succsim Y\}.$$

Definition 3. *Let Φ be an agenda, let X and Y be complete and consistent judgment sets for Φ, let J be an external agent's desired set, and let $T_J \in \{U_J, TR_J, CR_J\}$ be a type of J-induced (weak) preferences. We say that*

1. *the external agent necessarily/possibly weakly prefers X to Y for type T_J if $X \succsim Y$ for all/some $\succsim \, \in T_J$.*
2. *the external agent necessarily/possibly prefers X to Y for type T_J if $X \succ Y$ for all/some $\succsim \, \in T_J$.*

Let J be the desired set of the external agent. In the case of closeness-respecting preferences, the external agent necessarily prefers a new outcome Y to the actual outcome X if and only if she achieves a new agreement with J while preserving the existing agreements. On the other hand, she possibly prefers Y to X if and only if she achieves a new agreement with J regardless of new differences.

Example 4. Let $\Phi = \{a,\ b,\ c,\ a \wedge b,\ \neg a \vee c,\ \neg a,\ \neg b,\ \neg c,\ \neg(a \wedge b),\ \neg(\neg a \vee c)\}$ be an agenda and let $\mathbf{J} = (J_1, J_2, J_3)$ be a profile. Table 1 shows the individual judgment sets of the three judges as well as the collective outcome $UPQR_{1/2}(\mathbf{J})$ and the external agent's incomplete desired set J. Here a 1 indicates that the formula is contained in the judgment set, whereas a 0 means that the formula's complement is in the set. Assume the external agent changes the profile to some (not further specified) profile $\mathbf{J}'$ with $UPQR_{1/2}(\mathbf{J}') = \{\neg a, \neg b, c, \neg(a \wedge b), \neg a \vee c\}$ and consider closeness-respecting preferences. Since it holds that $\{\neg(a \wedge b), \neg a \vee c\} = J \cap UPQR_{1/2}(\mathbf{J}') \supset J \cap UPQR_{1/2}(\mathbf{J}) = \{\neg a \vee c\}$, the external agent necessarily prefers $UPQR_{1/2}(\mathbf{J}')$ to $UPQR_{1/2}(\mathbf{J})$.

Table 1. Example for closeness-respecting preferences

Judgment set	a	b	c	$a \wedge b$	$\neg a \vee c$
J_1	1	1	0	1	0
J_2	1	0	1	0	1
J_3	0	1	1	0	1
$UPQR_{1/2}$	1	1	1	1	1
J			0	0	1

We assume that the reader is familiar with the complexity classes P and NP as well as with the concept of polynomial-time many-one reducibility (denoted by $\leq_m^p$; see, for example, the textbooks by Papadimitriou [16] and Rothe [18]).

We will use the following three NP-complete decision problems in our reductions. Given a propositional formula φ in conjunctive normal form (CNF) so that neither setting all variables to true nor setting all variables to false will satisfy the formula, the problem RESTRICTED-SAT asks whether there is a satisfying assignment for φ. The problem DOMINATING-SET asks, given a graph

$G = (V, E)$ and a positive integer k, if G has a dominating set of size at most k, i.e., a subset $V' \subseteq V$ where $\|V'\| \leq k$ such that every vertex $v \in V$ belongs to the closed neighborhood of some $v' \in V'$. Finally, given a set X and a collection C containing 3-element subsets of X, the problem EXACT-COVER-BY-3-SETS (X3C) asks if there is an exact cover for X, i.e., a subcollection $C' \subseteq C$ such that each element of X is a member of exactly one element of C'.

3   Control

In this section, we study the complexity of control problems related to the types of preferences defined in the previous section. These types of control in judgment aggregation have been introduced by Baumeister et al. [2,3], but their results are restricted to Hamming-distance-respecting preferences only. Hamming-distance-respecting preferences induce a weak order over all complete and consistent judgment sets for a given agenda, by counting the number of positive formulas on which two judgment sets differ.

3.1   Preliminaries

We now formally define the relevant control problems for the uniform premise-based quota rule with quota q and for some given preference type T, starting with (possible and necessary) control by adding and by deleting judges.

$UPQR_q\text{-}T\text{-}$POSSIBLE-CONTROL-BY-ADDING-JUDGES

Given: An agenda Φ, two profiles $\mathbf{J} \in \mathscr{J}(\Phi)^n$ and $\mathbf{K} \in \mathscr{J}(\Phi)^m$, a desired set J, and a positive integer k.

Question: Is there a subprofile $\mathbf{K}' \subseteq \mathbf{K}$ of size at most k such that for the new profile $\mathbf{J}' = \mathbf{J} \cup \mathbf{K}'$, it holds that $UPQR_q(\mathbf{J}') \succ UPQR_q(\mathbf{J})$ for some $\succsim\, \in T_J$?

$UPQR_q\text{-}T\text{-}$POSSIBLE-CONTROL-BY-DELETING-JUDGES

Given: An agenda Φ, a profile $\mathbf{J} \in \mathscr{J}(\Phi)^n$, a desired set J, and a positive integer k.

Question: Is there a subprofile $\mathbf{J}' \subseteq \mathbf{J}$ of size at most k such that $UPQR_q(\mathbf{J} \smallsetminus \mathbf{J}') \succ UPQR_q(\mathbf{J})$ for some $\succsim\, \in T_J$?

The next control type, control by replacing judges, combines the previous two types. To motivate this control type, Baumeister et al. [2,3] provide real-world examples taken from the regulations on implementing powers in the Council of the European Union or the European Commission.

Concerning the problems $UPQR_q\text{-}T\text{-}$NECESSARY-CONTROL-BY-$\mathscr{C}$ for any one of these control types $\mathscr{C}$, the respective condition must hold for *all* $\succsim$ in T_J,

---
$UPQR_q$-T-POSSIBLE-CONTROL-BY-REPLACING-JUDGES

Given: An agenda Φ, two profiles $\mathbf{J} \in \mathscr{J}(\Phi)^n$ and $\mathbf{K} \in \mathscr{J}(\Phi)^m$, a desired set J, and a positive integer k.

Question: Are there subprofiles $\mathbf{J}' \subseteq \mathbf{J}$ and $\mathbf{K}' \subseteq \mathbf{K}$ of size $\|\mathbf{J}'\| = \|\mathbf{K}'\| \leq k$ such that for the new profile $\mathbf{S} = (\mathbf{J} \smallsetminus \mathbf{J}') \cup \mathbf{K}'$, it holds that $UPQR_q(\mathbf{S}) \succ UPQR_q(\mathbf{J})$ for some $\succsim\, \in T_J$?

---

whereas in $UPQR_q$-EXACT-CONTROL-BY-$\mathscr{C}$ we ask whether the desired set J is contained in the collective outcome after the external agent (called the chair) has exerted control of type $\mathscr{C}$.

A complete desired set J is a special case of an incomplete one. That means that every NP-hardness result for problems with a complete J automatically shows NP-hardness for the problems with incomplete J. It is easy to see that all decision problems in this section are in NP.

Definition 5. *Let Φ be an agenda and let $\mathscr{C}$ be a given control type. A resolute judgment aggregation procedure F is necessarily/possibly immune to control by $\mathscr{C}$ for induced preferences of type $T \in \{U, TR, CR\}$ if for all profiles $\mathbf{J}$ and for each desired set J, the chair necessarily/possibly weakly prefers the outcome $F(\mathbf{J})$ to the outcome $F(\mathbf{J}')$ for type T_J, where $\mathbf{J}'$ denotes the new profile after exerting control of type $\mathscr{C}$.*

3.2 Results for Control

For uniform premise-based quota rules, Baumeister et al. [3] show NP-completeness of exact control by adding and by deleting judges for complete desired sets and for the quota $q = 1/2$, and NP-completeness of exact control by replacing judges for any quota. The proof of the latter result can be modified so as to use a complete desired set.

Our first result gives a link between the exact control problem of a given type and the corresponding possible and necessary control problem with respect to various preference types induced by the chair's desired set.

Proposition 6. *Let $\mathscr{C}$ be a control type and let q, $0 \leq q < 1$, be a rational quota.*

1. *$UPQR_q$-EXACT-CONTROL-BY-$\mathscr{C}$ $\leq^P_m$ $UPQR_q$-T-POSSIBLE-CONTROL-BY-$\mathscr{C}$ for each preference type $T \in \{U, TR, CR\}$.*
2. *$UPQR_q$-EXACT-CONTROL-BY-$\mathscr{C}$ $\leq^P_m$ $UPQR_q$-T-NECESSARY-CONTROL-BY-$\mathscr{C}$ for each preference type $T \in \{TR, CR\}$.*

The simple proof is an adaption of the proof of the corresponding reductions between certain manipulation problems due to Baumeister et al. [5, Thm. 7]. In the construction we use the conjunction of all formulas in the desired set of the EXACT-CONTROL-BY-$\mathscr{C}$ instance as the single element of the desired set in the corresponding preference-based instance. Since the latter set is incomplete, the case of complete desired sets has to be considered separately.

Assuming a complete desired set inducing top-respecting preferences, the chair necessarily prefers only her desired set to any other possible outcome. Thus, in this case, NP-completeness of $UPQR_q$-TR-NECESSARY-CONTROL-BY-$\mathscr{C}$ follows from NP-completeness of the exact control problem of type $\mathscr{C}$ for a complete desired set.

We now consider closeness-respecting preferences.

Theorem 7. $UPQR_{1/2}$-CR-POSSIBLE-CONTROL-BY-ADDING-JUDGES   and $UPQR_{1/2}$-CR-NECESSARY-CONTROL-BY-ADDING-JUDGES *both are* NP-*complete, even for a complete desired set.*

Table 2. Construction for the proof of Theorem 7

Judgment set	α_0	α_1	$\cdots$	α_{3m}	β	$\varphi \vee \beta$
J_1	1	1	$\cdots$	1	0	1
$J_2, \ldots, J_m$	0	1	$\cdots$	1	0	0
J_{m+1}	0	0	$\cdots$	0	0	0
$UPQR_{1/2}$	0	1	$\cdots$	1	0	0
J	0	1	$\cdots$	1	1	1

Proof. The proof works by a reduction from X3C and uses a construction similar to the one employed by Baumeister et al. [3]. Let (X, C) be an X3C instance, where $X = \{x_1, \ldots, x_{3m}\}$ and $C = \{C_1, \ldots, C_n\}$. For the first part of the theorem, let the agenda Φ contain the literals $\alpha_0, \alpha_1, \ldots, \alpha_{3m}, \beta$, the formula $\varphi \vee \beta$ with $\varphi = \alpha_0 \wedge \cdots \wedge \alpha_{3m}$, and the corresponding negations. The profile $\mathbf{J} = (J_1, \ldots, J_{m+1})$, the collective judgment set $UPQR_{1/2}(\mathbf{J})$, and the desired set J can be seen in Table 2.

Let $\mathbf{K} = (K_1, \ldots, K_n)$ be the profile containing the individual judgment sets to be added, where $K_i = \{\neg\beta, \alpha_0, \alpha_j, \neg\alpha_l \mid x_j \in C_i, x_l \notin C_i, 1 \leq j, l \leq 3m\}$. The chair is allowed to add m judgment sets from $\mathbf{K}$.

Since no judge accepts β, the additional agreement of the new outcome with J can only occur for the formula $\varphi \vee \beta$. To add α_0, the chair has to add at least m judges for a total of $2m + 1$ judges. But then every α_i, $1 \leq i \leq 3m$, needs exactly one additional affirmation. Therefore, there is a successful control if and only if there is an exact cover of the given X3C instance. This shows that $UPQR_{1/2}$-CR-POSSIBLE-CONTROL-BY-ADDING-JUDGES is NP-complete.

Concerning the proof of the second part, let the agenda Φ' contain only α_0, $\alpha_1, \ldots, \alpha_{3m}$ and the corresponding negations. Let $\mathbf{J}'$ and $\mathbf{K}'$ be the corresponding profiles restricted to Φ' and let $J' = \{\alpha_0, \alpha_1, \ldots, \alpha_{3m}\}$ be the chair's desired set. Since the chair has to preserve the initial agreements with J, by a similar argumentation as above, there is a successful control if and only if there is an exact cover for the given X3C instance. Thus $UPQR_{1/2}$-CR-NECESSARY-CONTROL-BY-ADDING-JUDGES is NP-complete.   □

Theorem 8. $UPQR_{1/2}$-CR-POSSIBLE-CONTROL-BY-DELETING-JUDGES   and $UPQR_{1/2}$-CR-NECESSARY-CONTROL-BY-DELETING-JUDGES both are NP-complete, even for a complete desired set.

Proof. We adapt a construction used by Baumeister et al. [3]. Let (X, C) be an X3C instance, where $X = \{x_1, \ldots, x_{3m}\}$ and $C = \{C_1, \ldots, C_n\}$. If there exists an element of X that is not contained in any element of C, we construct an arbitrary no-instance for the respective control problem.

For the first part, let Φ be the agenda containing the literals $\alpha_0, \alpha_1, \ldots, \alpha_{3m}$, β, γ, the formula $\varphi \vee \beta$ where $\varphi = \alpha_0 \wedge \cdots \wedge \alpha_{3m} \wedge \neg \gamma$, and all corresponding negations. Let $\mathbf{T} = \mathbf{T_1} \cup \mathbf{T_2}$ be a profile where $\mathbf{T_1} = (J_1, \ldots, J_{n+m})$ and $\mathbf{T_2} = (L_1, \ldots, L_n)$ for a total of $2n + m$ judges. We denote by d_k the number of sets C_i that contain x_k. For each i, $1 \leq i \leq n + m$, J_i is the union of the set $\{\neg\beta, \alpha_j, \neg\alpha_l \mid m + d_j \geq i, m + d_l < i, 1 \leq j, l \leq 3m\}$ with $\{\alpha_0\}$ if $i \leq n + 1$ (and with $\{\neg\alpha_0\}$ otherwise), and with $\{\gamma\}$ if $i \leq m$ (and with $\{\neg\gamma\}$ otherwise), and with the corresponding conclusion $\{\varphi \vee \beta\}$ (respectively, with $\{\neg(\varphi \vee \beta)\}$). Furthermore, for $1 \leq i \leq n$, define

$$L_i = \{\neg\beta, \gamma, \neg\alpha_0, \alpha_j, \neg\alpha_l, \neg(\varphi \vee \beta) \mid x_j \notin C_i, x_l \in C_i, 1 \leq j, l \leq 3m\}.$$

Since β has no affirmation, γ and every a_k, $1 \leq k \leq 3m$, each have $n + m$ affirmations, and since α_0 has $n + 1$ affirmations, it follows that

$$UPQR_{1/2}(\mathbf{T}) = \{\neg\alpha_0, \alpha_1, \ldots, \alpha_{3m}, \neg\beta, \gamma, \neg(\varphi \vee \beta)\}.$$

Let the chair's desired set be $J = \{\neg\alpha_0, \alpha_1, \ldots, \alpha_{3m}, \beta, \gamma, \varphi \vee \beta\}$. He is able to delete m individual judgment sets from the profile $\mathbf{T}$.

Since no judge accepts β, it will never be in the collective outcome. Therefore, the new agreement of the desired set with the new outcome has to occur in the conclusion. To include α_0, the chair has to delete m judges to lower the acception threshold to $n + 1$. These judges' individual judgment sets have to be deleted from $\mathbf{T_2}$ so that γ loses m affirmations and is not contained in the collective outcome anymore. If some x_i is not contained in one of the sets C_j that match the individual judgment sets of the deleted judges, the corresponding α_i loses too many affirmations and is therefore rejected in the new collective outcome. The control action is successful (i.e., $\varphi \vee \beta$ is contained in the new collective outcome) if and only if the sets C_j corresponding to the deleted individual judgment sets form an exact cover of X. This shows that $UPQR_{1/2}$-CR-POSSIBLE-CONTROL-BY-DELETING-JUDGES is NP-complete.

To prove the second part, we create a new agenda Φ' from Φ by removing β, $\varphi \vee \beta$, and the corresponding negations, and by adding the formula $\psi = (\neg\alpha_0 \wedge \gamma) \vee (\alpha_0 \wedge \neg\gamma)$ and its negation. Let $\mathbf{T'} = \mathbf{T_1'} \cup \mathbf{T_2'}$ be the resulting profile that is obtained by restricting $\mathbf{T_1}$ and $\mathbf{T_2}$ to Φ' and by adding the corresponding conclusions to all J_i and L_j. Then it holds that

$$UPQR_{1/2}(\mathbf{T'}) = \{\neg\alpha_0, \alpha_1, \ldots, \alpha_{3m}, \gamma, \psi\}.$$

Let $J' = \{\alpha_0, \alpha_1, \ldots, \alpha_{3m}, \neg\gamma, \psi\}$ and let the chair be able to delete m judgment sets. To preserve the agreement on the conclusion, the chair has to change the collective outcome in regard to α_0 as well as γ. Following the argumentation above, the chair has to delete exactly m judgment sets, can only delete judgment sets from $\mathbf{T_2}$ and therefore only preserves the agreements concerning the α_i if and only if the sets C_j corresponding to the deleted individual judgment sets form an exact cover of X. Thus $UPQR_{1/2}$-CR-Necessary-Control-by-Deleting-Judges is NP-complete. $\qquad\square$

We now turn to control by replacing judges.

Theorem 9. $UPQR_q$-CR-Possible-Control-by-Replacing-Judges *and* $UPQR_q$-CR-Necessary-Control-by-Replacing-Judges *both are NP-complete for each rational quota q, $0 \leq q < 1$, even for a complete desired set.*

Proof. The proof works by a reduction from the problem Dominating-Set. Let (G, k) with $G = (V, E)$ and $V = \{v_1, \ldots, v_n\}$ be a Dominating-Set instance. The neighbors of vertex v_i (including v_i itself) will be denoted by $v_i^1, v_i^2, \ldots, v_i^{j_i}$ for some j_i.

For the first part of the theorem (i.e., for showing NP-completeness of $UPQR_q$-CR-Possible-Control-by-Replacing-Judges), first assume that the quota q is lower than $1/2$. We construct an instance of the control problem as follows. The agenda Φ contains the literals $v_1, \ldots, v_n$, β, γ, the formula $\psi \vee \beta$, where $\psi = \varphi_1 \wedge \cdots \wedge \varphi_n \wedge \gamma$ and $\varphi_i = v_i^1 \vee \cdots \vee v_i^{j_i}$, and all corresponding negations. The profile $\mathbf{J} = (J_1, \ldots, J_m)$ with $m = 2k + 1$ judges, the outcome, and the chair's desired set J can be seen in Table 3(a).

The chair can choose at most k judgment sets from the profile $\mathbf{K} = (K_1, \ldots, K_n)$ with $K_i = \{\neg\beta, \neg\gamma, v_i, \neg v_j, \neg(\psi \vee \beta) \mid 1 \leq j \leq n, i \neq j\}$ to replace judgment sets in $\mathbf{J}$. The formula β will never be contained in the outcome because no judge accepts it. In order to achieve the desired additional agreement between the new outcome and J, the chair has to get the conclusion and therefore ψ accepted. Each v_i needs exactly one additional affirmation to be contained in the new outcome. Note that only judgment sets in the third block can be replaced (or else γ would lose an affirmation, would not be contained in the collective outcome anymore, and thus ψ cannot be evaluated to true). Since $\psi \vee \beta$ is contained in the new outcome if and only if the accepted v_i form a dominating set, and since only k judgment sets can be replaced, the control action is successful under closeness-respecting preferences if and only if G has a dominating set of size k.

In the case of a quota q greater than or equal to $1/2$, the agenda changes slightly. Instead of the formula $\psi \vee \beta$ and its negation the new agenda Φ' contains the formula $\psi' \vee \neg\beta$ with $\psi' = \varphi_1' \wedge \cdots \wedge \varphi_n' \wedge \neg\gamma$ and $\varphi_i' = \neg v_i^1 \vee \cdots \vee \neg v_i^{j_i}$, and its negation, $\neg(\psi' \vee \neg\beta)$. The profile $\mathbf{J}' = (J_1', \ldots J_m')$ with $m = 2k + 1$ judges, the outcome, and the chair's desired set J' can be seen in Table 3(b).

Let $\mathbf{K}' = (K_1', \ldots, K_n')$ be a profile, where

$$K_i' = \{\beta, \gamma, \neg v_i, v_j, \neg(\psi' \vee \neg\beta) \mid 1 \leq j \leq n, i \neq j\}$$

Table 3. Construction for the first part of the proof of Theorem 9

(a) Rational quota q with $0 \leq q < 1/2$

Judgment set	$v_1 \cdots v_n$	β	γ	$\psi \vee \beta$
$J_1, \ldots, J_{\lfloor m \cdot q \rfloor}$	$1 \cdots 1$	0	1	1
$J_{\lfloor m \cdot q \rfloor + 1}$	$0 \cdots 0$	0	1	0
$J_{\lfloor m \cdot q \rfloor + 2}, \ldots, J_m$	$0 \cdots 0$	0	0	0
$UPQR_q$	$0 \cdots 0$	0	1	0
J	$0 \cdots 0$	1	1	1

(b) Rational quota q with $1/2 \leq q < 1$

Judgment set	$v_1 \cdots v_n$	β	γ	$\psi' \vee \neg\beta$
$J'_1, \ldots, J'_{\lceil m \cdot (1-q) \rceil - 1}$	$0 \cdots 0$	1	0	1
$J'_{\lceil m \cdot (1-q) \rceil}$	$1 \cdots 1$	1	1	0
$J'_{\lceil m \cdot (1-q) \rceil + 1}, \ldots, J'_m$	$1 \cdots 1$	1	1	0
$UPQR_q$	$1 \cdots 1$	1	0	0
J'	$1 \cdots 1$	0	0	1

for $1 \leq i \leq n$. Again, the chair is able to replace k judgment sets from $\mathbf{J}'$ with k judgment sets from $\mathbf{K}'$. A formula needs at least $\lceil m(1-q) \rceil$ rejections in order to not be accepted. Since every judge accepts β, its negation will never be contained in the collective outcome. Thus the chair has to get ψ' accepted so as to achieve the desired additional agreement of the new outcome with J'. The argumentation then follows the first case: Since ψ' is true if and only if the rejected v_i form a dominating set and since the k replaceable judgment sets must be from the third block, the control action is successful under closeness-respecting preferences if and only if G has a dominating set of size k.

We prove the second part of the theorem (i.e., NP-completeness of $UPQR_q$-CR-NECESSARY-CONTROL-BY-REPLACING-JUDGES) in a similar way. Unlike in the first part of the proof, the chair now has to necessarily prefer the new outcome to the actual one. That means that all existing agreements have to be preserved. Remove β from the former agenda Φ (respectively, Φ') and replace all appearances of ψ (respectively, ψ') with the formula $\Psi = \psi \vee (\neg v_1 \wedge \cdots \wedge \neg v_n)$ (respectively, $\Psi' = \psi' \vee (v_1 \wedge \cdots \wedge v_n)$). All required changes in the profiles $\mathbf{J}^*$ (respectively, $\mathbf{J}'^*$), the outcomes, and the desired sets J^* (respectively, J'^*) can be seen in Table 4(a) (respectively, in Table 4(b)).

To obtain the profiles $\mathbf{K}^*$ (respectively, $\mathbf{K}'^*$) of judgment sets to choose from, the premises of the judgment sets in $\mathbf{K}$ (respectively, $\mathbf{K}'$) restricted to the corresponding new agenda remain unchanged and the new conclusion is evaluated accordingly. As above the chair is allowed to replace k judgment sets. The chair has to change some premise different from γ in order to achieve a new

442   D. Baumeister et al.

Table 4. Construction for the second part of the proof of Theorem 9

(a) Rational quota q with $0 \leq q < 1/2$

Judgment set	v_1	$\cdots$	v_n	γ	Ψ
$J_1^*,\ldots,J_{\lfloor m\cdot q\rfloor}^*$	1	$\cdots$	1	1	1
$J_{\lfloor m\cdot q\rfloor+1}^*$	0	$\cdots$	0	1	1
$J_{\lfloor m\cdot q\rfloor+2}^*,\ldots,J_m^*$	0	$\cdots$	0	0	1
$UPQR_q$	0	$\cdots$	0	1	1
J^*	1	$\cdots$	1	1	1

(b) Rational quota q with $1/2 \leq q < 1$

Judgment set	v_1	$\cdots$	v_n	γ	Ψ'
$J_1'^*,\ldots,J_{\lceil m\cdot(1-q)\rceil-1}'^*$	0	$\cdots$	0	0	1
$J_{\lceil m\cdot(1-q)\rceil}'^*$	1	$\cdots$	1	0	1
$J_{\lceil m\cdot(1-q)\rceil+1}'^*,\ldots,J_m'^*$	1	$\cdots$	1	1	1
$UPQR_q$	1	$\cdots$	1	0	1
J'^*	0	$\cdots$	0	0	1

agreement. But after this action the second part of Ψ (respectively, Ψ') is not satisfied anymore. In order to preserve the agreement of the outcome with her desired set regarding the conclusion, the chair has to replace the judgment sets from the third block with the judgment sets from $\mathbf{K}^*$ (respectively, $\mathbf{K}'^*$) that correspond to the vertices in a dominating set of G. It follows that the control action is successful if and only if G has a dominating set of size k. □

Finally, we turn to unrestricted and top-respecting preferences.

Proposition 10. *Let $\mathscr{C}$ be one of the control types* ADDING-JUDGES, DELETING-JUDGES, *and* REPLACING-JUDGES, *let* $T \in \{U, TR\}$ *be a preference type, and let the desired set be complete. For each rational quota q, $0 \leq q < 1$, $UPQR_q$-T-POSSIBLE-CONTROL-BY-$\mathscr{C}$ is in* P.

Proof. In the case of unrestricted preferences, the chair possibly prefers every new outcome to the actual outcome. Since her desired set is complete, she only has to check if she can change a premise so as to change the collective judgment set. This is possible in polynomial time for every $\mathscr{C}$.

In the case of top-induced preferences, the chair possibly prefers every new outcome to the actual outcome as long as the latter is not identical to her desired set. Therefore, it also suffices to change some premise if possible. □

Proposition 11. *Let $\mathscr{C}$ be a control type. For each rational quota q, $0 \leq q < 1$, $UPQR_q$-U-* NECESSARY-CONTROL-BY-$\mathscr{C}$ *is possibly immune.*

Proof. In the case of unrestricted preferences, the collective judgment set is always possibly preferred to every other judgment set that can occur as a new outcome after the control action. □

4 Bribery

In this section, we study the complexity of bribery problems related to the types of preferences defined in Sect. 2. Bribery in judgment aggregation has been introduced by Baumeister et al. [5,6]; however, their results are restricted to Hamming-distance-respecting preferences and to the premise-based procedure only.

4.1   Preliminaries

We now formally define the relevant bribery problems for the uniform premise-based quota rule with quota q and for some given preference type T.

$UPQR_q$-Possible-Bribery

Given: An agenda Φ, a profile $\mathbf{J} \in \mathscr{J}(\Phi)^n$, a desired set J, and a positive integer k (the budget).

Question: Is there a new profile $\mathbf{J}' \in \mathscr{J}(\Phi)^n$ with at most k changed individual judgment sets so that $UPQR_q(\mathbf{J}') \succ UPQR_q(\mathbf{J})$ for some $\succsim \in T_J$?

Concerning the analogous problem $UPQR_q$-T-Necessary-Bribery, the condition $UPQR_q(\mathbf{J}') \succ UPQR_q(\mathbf{J})$ is required to hold for *all* $\succsim \in T_J$. In the two corresponding microbribery problems (also introduced by Baumeister et al. [5,6]), the briber is allowed to change k premises instead of k whole judgment sets.

Given an agenda Φ, a profile $\mathbf{J} \in \mathscr{J}(\Phi)^n$, a desired set J, and a positive integer k, in the exact variant of the bribery (respectively, microbribery) problem we ask whether the briber can change up to k individual judgment sets (respectively, premises) such that $J \subseteq UPQR_q(\mathbf{J}')$, where $\mathbf{J}'$ denotes the modified profile. We denote these problems by $UPQR_q$-Exact-Bribery and $UPQR_q$-Exact-Microbribery. Again, it is easy to see that all decision problems in this section are in NP.

Definition 12. *Let Φ be an agenda and let $\mathscr{B}$ be a given bribery type. A resolute judgment aggregation procedure F is* necessarily/possibly immune *to $\mathscr{B}$ for induced preferences of type $T \in \{U, TR, CR\}$ if for all profiles $\mathbf{J}$ and for each desired set J, the briber necessarily/possibly weakly prefers the outcome $F(\mathbf{J})$ to the outcome $F(\mathbf{J}')$ for type T_J, where $\mathbf{J}'$ denotes the new profile after bribery of type $\mathscr{B}$ has been exerted.*

4.2   Results for Bribery

We now present our results for bribery in judgment aggregation.

Theorem 13. *For each rational quota q, $0 \leq q < 1$,*

1. *$UPQR_q$-Exact-Bribery $\leq_m^p UPQR_q$-T-Possible-Bribery for each preference type $T \in \{U, TR, CR\}$;*
2. *$UPQR_q$-Exact-Microbribery $\leq_m^p UPQR_q$-T-Possible-Microbribery for each preference type $T \in \{U, TR, CR\}$.*

The simple proof (an adaption of the proof of Proposition 6) uses an incomplete desired set, so we again have to consider the case of complete desired sets separately.

Baumeister et al. [5] show NP-completeness of exact bribery (respectively, microbribery) with an incomplete desired set for the premise-based procedure (PBP), which—recall—is a special case of $UPQR_{1/2}$ for an odd number of judges. Their proofs can be modified so as to work for every rational quota q with $0 \leq q < 1$ and for every number of judges. For a complete desired set and PBP, Baumeister et al. [5] prove that the exact bribery problem remains NP-complete and provide a P algorithm that solves the exact microbribery problem. The P algorithm for exact microbribery can also easily be adapted to work for every rational quota q with $0 \leq q < 1$ and for every number of judges. Since under top-respecting preferences the briber necessarily prefers only her desired set to any other possible outcome and assuming that the briber's desired set is complete, we thus have that $UPQR_{1/2}$-TR-NECESSARY-BRIBERY is NP-complete, but for each rational quota q, $0 \leq q < 1$, $UPQR_q$-TR-NECESSARY-MICROBRIBERY is in P.

Next we consider closeness-respecting preferences for bribery problems.

Theorem 14. *For each rational quota q, $0 \leq q < 1$, $UPQR_q$-CR-NECESSARY-BRIBERY and $UPQR_q$-CR-POSSIBLE-BRIBERY both are NP-complete, even for a complete desired set.*

Proof. The proof works by a reduction from the problem RESTRICTED-SAT (defined in Sect. 2) and adapts an idea of Endriss et al. [9]. We first show that $UPQR_q$-CR-NECESSARY-BRIBERY is NP-complete. Let φ be a RESTRICTED-SAT instance. For a quota q equal to or greater than $1/2$, the agenda Φ contains the variables of φ (i.e., $\alpha_1, \ldots, \alpha_m$), a literal β, the formula $\psi \vee \beta$ with $\psi = \varphi \vee (\neg \alpha_1 \wedge \cdots \wedge \neg \alpha_m)$ and all corresponding negations. Let m be the briber's budget, let $n = 2m + 1$ be the number of judges, and let $J = \{\alpha_1, \ldots, \alpha_m, \beta, \psi \vee \beta\}$ be the briber's desired set. The profile $\mathbf{J}$ is shown in Table 5(b).

Even by changing m judgments β will never be in the collective judgment set. Therefore, at least one α_i has to be set to 1 to obtain the required additional agreement with J. This is possible because every α_i can be included in the new outcome by changing exactly m judgment sets in the second block of judges. Since the agreement with $\psi \vee \beta$ has to be preserved, the bribery is successful under closeness-respecting preferences if and only if φ has a satisfying assignment.

In the case of $0 \leq q < 1/2$, the agenda has to be slightly changed. The formula $\psi \vee \beta$ and its negation are replaced by the formula $\psi \vee \neg \beta$ and its negation. The corresponding profile $\mathbf{J}' = (J'_1, \ldots, J'_n)$ is shown in Table 5(a). Since it is impossible for the briber to reject β and since all agreements of the collective outcome with J' have to be preserved, the bribery is successful under closeness-respecting preferences if and only if φ has a satisfying assignment.

We now turn to the second part of the theorem (i.e., to NP-completeness of $UPQR_q$-CR-POSSIBLE-BRIBERY). This can be shown in a similar way. Change the agenda described above by replacing the formula $\psi \vee \beta$ with $\varphi \vee \beta$ in the first case (respectively, $\psi \vee \neg \beta$ with $\varphi \vee \neg \beta$ in the second case) including all corresponding negations. Let $\mathbf{J}^*$ (respectively, $\mathbf{J}'^*$) be the profile concerning the

Table 5. Construction for the proof of Theorem 14

(a) Rational quota q with $0 \leq q < 1/2$

Judgment set	$\alpha_1 \cdots \alpha_m$ β	$\psi \vee \neg\beta$
$J'_1,\ldots,J'_n$	$0 \cdots 0$ 1	1
$UPQR_q$	$0 \cdots 0$ 1	1
J'	$1 \cdots 1$ 0	1

(b) Rational quota q with $1/2 \leq q < 1$

Judgment set	$\alpha_1 \cdots \alpha_m$ β	$\psi \vee \beta$
$J_1,\ldots,J_{\lfloor n\cdot q\rfloor-(m-1)}$	$1 \cdots 1$ 0	0
$J_{\lfloor n\cdot q\rfloor-(m-1)+1},\ldots,J_n$	$0 \cdots 0$ 0	1
$UPQR_q$	$0 \cdots 0$ 0	1
J	$1 \cdots 1$ 1	1

corresponding new agenda with the premises of the individual judgment sets as seen in the corresponding part of Table 5 and the conclusions evaluated accordingly. Note that the collective outcomes only differ in the conclusion, which is rejected in both cases. Further, let $J^* = \{\neg\alpha_1,\ldots,\neg\alpha_m,\beta,\varphi\vee\beta\}$ (respectively, $J'^* = \{\neg\alpha_1,\ldots,\neg\alpha_m,\neg\beta,\varphi\vee\neg\beta\}$) be the briber's desired set. Since the only additional agreement the briber can achieve is the conclusion, similar arguments as above complete the proof. □

We now handle the case of microbribery for closeness-respecting preferences.

Theorem 15. *For each rational quota q, $0 \leq q < 1$, $UPQR_q$-CR-NECESSARY-MICROBRIBERY and $UPQR_q$-CR-POSSIBLE-MICROBRIBERY both are NP-complete, even for a complete desired set.*

Proof. This proof works similarly to the proof of Theorem 14. For proving the first part (NP-completeness of $UPQR_q$-CR-NECESSARY-MICROBRIBERY), the only change is the number of judges in the different blocks of judges: Judges $1,\ldots,\lfloor n\cdot q\rfloor$ form the first block, while the second block consists of judges $\lfloor n\cdot q\rfloor + 1,\ldots,n$. Then a similar argumentation as in the proof of Theorem 14 applies. Note that the briber is only allowed to change k premises instead of k whole individual judgment sets.

For the proof of the second part (i.e., for showing NP-completeness of $UPQR_q$-CR-POSSIBLE-MICROBRIBERY), we use the agendas from the corresponding parts of the proof of Theorem 14. In the first case, judges $1,\ldots,\lfloor n\cdot q\rfloor$ accept all premises but β and reject the conclusion, judges $\lfloor n\cdot q\rfloor + 1,\ldots,n$ reject all formulas, the collective outcome contains all negated formulas, and the briber accepts only β and $\varphi\vee\beta$ and rejects the remaining propositions. In the second case, each appearance of β or $\neg\beta$ in the first case is replaced with its

Table 6. Overview of results for $UPQR_{1/2}$-T-POSSIBLE/NECESSARY-CONTROL-BY-$\mathscr{C}$ for $\mathscr{C} \in \{$ADDING-JUDGES, DELETING-JUDGES, REPLACING-JUDGES$\}$ and $T \in \{U, TR, CR\}$

		U	TR	CR
Incomplete DS	POSSIBLE	NP-complete	NP-complete	NP-complete
	NECESSARY	possibly immune	NP-complete	NP-complete
Complete DS	POSSIBLE	P	P	NP-complete
	NECESSARY	possibly immune	NP-complete	NP-complete

Table 7. Overview of results for $UPQR_q$-T-POSSIBLE/NECESSARY-BRIBERY/ MICRO-BRIBERY for $T \in \{U, TR, CR\}$

		U	TR	CR
Incomplete DS	POSSIBLE	NP-complete	NP-complete	NP-complete
	NECESSARY	possibly immune	NP-complete	NP-complete
Complete DS	POSSIBLE	P	P	NP-complete
	NECESSARY	possibly immune	NP-complete^2/P	NP-complete

complement. Once again, the bribery action is successful if and only if φ has a satisfying assignment.   □

The following propositions can be proven in the same way as Propositions 10 and 11.

Proposition 16. *Let* $T \in \{U, TR\}$ *be a preference type, and let the briber's desired set be complete. For each rational quota* q, $0 \leq q < 1$, $UPQR_q$-T-POSSIBLE-BRIBERY *and* $UPQR_q$-T-POSSIBLE-MICROBRIBERY *both are in* P.

Proposition 17. *For each rational quota* q, $0 \leq q < 1$, $UPQR_q$-U-NECESSARY-BRIBERY *and* $UPQR_q$-U-NECESSARY-MICROBRIBERY *both are possibly immune.*

5   Conclusions and Future Work

We have studied bribery and microbribery as well as three types of control in judgment aggregation. While these problems were introduced in previous work by Baumeister et al. [3,5,6], they have been studied only for Hamming-distance-respecting preferences so far. Our contribution is to extend this study to the case of more general preference notions, including closeness-respecting and top-respecting preferences that are due to Dietrich and List [8] and have been applied to manipulation in judgment aggregation by Baumeister et al. [4,5]. Further-more, our results for bribery and microbribery apply to uniform premise-based quota rules that generalize the premise-based procedure. An overview of our complexity results is given in Table 6 for control and in Table 7 for bribery and microbribery. Here, DS stands for "desired set." For control by replacing judges, NECESSARY-CONTROL-BY-ADDING/DELETING-JUDGES for unrestricted preferences, and POSSIBLE-CONTROL-BY-ADDING/DELETING-JUDGES for a complete desired set inducing unrestricted and top-respecting preferences, the results are shown for a general rational quota q, $0 \leq q < 1$. The results for bribery and microbribery are identical, except for the NECESSARY-BRIBERY/MICROBRIBERY problem where we have a complete desired set inducing top-respecting preferences. The entry NP-complete/P here means that this problem is NP-complete for bribery[2] but in P for microbribery.

Regarding Hamming-distance-respecting preferences, note that Baumeister et al. [2,6] have already studied the complexity of bribery and microbribery with *incomplete* desired sets and for the *premise-based procedure* only. However, their proofs can easily be adapted to also apply to complete desired sets and to uniform premise-based quota rules. Similarly, some of the results of Baumeister et al. [2,3] for exact control and for control problems under Hamming-distance-respecting preferences apply to incomplete desired sets only, but the proofs only have to be slightly adapted to work for the case of complete desired sets, too.

Regarding future work, we propose to study the complexity of these problems for different families of judgment aggregation procedures, to study other preferences for the attacker (e.g., by using other distance measures), and to study the complexity of control by bundling judges introduced by Baumeister et al. [4].

Acknowledgments. We thank the anonymous reviewers for their helpful comments. This work was supported in part by a grant for gender-sensitive universities from the NRW Ministry for Innovation, Science, and Research and by DFG grant RO 1202/15-1.

References

1. Bartholdi III, J., Tovey, C., Trick, M.: How hard is it to control an election? Math. Comput. Model. **16**(8/9), 27–40 (1992)
2. Baumeister, D., Erdélyi, G., Erdélyi, O., Rothe, J.: Bribery and control in judgment aggregation. In: Brandt, F., Faliszewski, P. (eds.) Proceedings of the 4th International Workshop on Computational Social Choice, pp. 37–48. AGH University of Science and Technology, Kraków, Poland (2012)
3. Baumeister, D., Erdélyi, G., Erdélyi, O., Rothe, J.: Control in judgment aggregation. In: Proceedings of the 6th European Starting AI Researcher Symposium, pp. 23–34. IOS Press, August 2012
4. Baumeister, D., Erdélyi, G., Erdélyi, O., Rothe, J.: Computational aspects of manipulation and control in judgment aggregation. In: Perny, P., Pirlot, M., Tsoukiàs, A. (eds.) ADT 2013. LNCS, vol. 8176, pp. 71–85. Springer, Heidelberg (2013)
5. Baumeister, D., Erdélyi, G., Erdélyi, O., Rothe, J.: Complexity of manipulation and bribery in judgment aggregation for uniform premise-based quota rules. Math. Soc. Sci. **76**, 19–30 (2015)

---

[2] This result is shown for the quota $q = 1/2$ only.

6. Baumeister, D., Erdélyi, G., Rothe, J.: How hard is it to bribe the judges? a study of the complexity of bribery in judgment aggregation. In: Brafman, R. (ed.) ADT 2011. LNCS, vol. 6992, pp. 1–15. Springer, Heidelberg (2011)

7. Dietrich, F., List, C.: Judgment aggregation by quota rules: majority voting generalized. J. Theor. Politics **19**(4), 391–424 (2007)

8. Dietrich, F., List, C.: Strategy-proof judgment aggregation. Econ. Philos. **23**(3), 269–300 (2007)

9. Endriss, U., Grandi, U., Porello, D.: Complexity of judgment aggregation. J. Artif. Intell. Res. **45**, 481–514 (2012)

10. Faliszewski, P., Hemaspaandra, E., Hemaspaandra, L.: How hard is bribery in elections? J. Artif. Intell. Res. **35**, 485–532 (2009)

11. Faliszewski, P., Hemaspaandra, E., Hemaspaandra, L., Rothe, J.: Llull and Copeland voting computationally resist bribery and constructive control. J. Artif. Intell. Res. **35**, 275–341 (2009)

12. Faliszewski, P., Rothe, J.: Control and bribery in voting. In: Brandt, F., Conitzer, V., Endriss, U., Lang, J., Procaccia, A. (eds.) Handbook of Computational Social Choice, chapter 7. Cambridge University Press (2015, to appear)

13. Hemaspaandra, E., Hemaspaandra, L., Rothe, J.: Anyone but him: the complexity of precluding an alternative. Artif. Intell. **171**(5–6), 255–285 (2007)

14. List, C.: The discursive dilemma and public reason. Ethics **116**(2), 362–402 (2006)

15. Miller, M., Osherson, D.: Methods for distance-based judgment aggregation. Soc. Choice Welf. **32**(4), 575–601 (2009)

16. Papadimitriou, C.: Computational Complexity, 2nd edn. Addison-Wesley, New York (1995)

17. Pigozzi, G.: Belief merging and the discursive dilemma: an argument-based account of paradoxes of judgment. Synthese **152**(2), 285–298 (2006)

18. Rothe, J.: Complexity Theory and Cryptology: An Introduction to Cryptocomplexity. EATCS Texts in Theoretical Computer Science. Springer, Heidelberg (2005)

Social Choice

Beyond Plurality: Truth-Bias
in Binary Scoring Rules

Svetlana Obraztsova[1], Omer Lev[2]([✉]), Evangelos Markakis[3],
Zinovi Rabinovich[4], and Jeffrey S. Rosenschein[2]

[1] Tel Aviv University, Tel Aviv, Israel
svetlana.obraztsova@gmail.com
[2] Hebrew University of Jerusalem, Jerusalem, Israel
{omerl,jeff}@cs.huji.ac.il
[3] Athens University of Economics and Business, Athens, Greece
markakis@gmail.com
[4] Mobileye Vision Technologies Ltd., Jerusalem, Israel
zr@zinovi.net

Abstract. It is well known that standard game-theoretic approaches to voting mechanisms lead to a multitude of Nash Equilibria (NE), many of which are counter-intuitive. We focus on truth-biased voters, a model recently proposed to avoid such issues. The model introduces an incentive for voters to be truthful when their vote is not pivotal. This is a powerful refinement, and recent simulations reveal that the surviving equilibria tend to have desirable properties.

However, truth-bias has been studied only within the context of plurality, which is an extreme example of k-approval rules with $k = 1$. We undertake an equilibrium analysis of the complete range of k-approval. Our analysis begins with the veto rule, the other extreme point of k-approval, where each ballot approves all candidates but one. We identify several crucial properties of pure NE for truth-biased veto. These properties show a clear distinction from the setting of truth-biased plurality. We proceed by establishing that deciding on the existence of NE in truth-biased veto is an NP-hard problem. We also characterise a tight (in a certain sense) subclass of instances for which the existence of a NE can be decided in poly-time. Finally, we study analogous questions for general k-approval rules.

1 Introduction

Voting mechanisms are processes by which preferences can be aggregated, and collective decisions can be made, in various multi-agent contexts. Under most voting rules, potentially beneficial strategic behavior is essentially inherent, as the Gibbard-Satterthwaite theorem famously states [7,16]. Hence, under mild assumptions, voters may have incentives to misreport their preferences. Given this negative result, a natural approach, initiated by Farquharson [5], is to undertake a game-theoretic analysis of voting, viewing voters as strategic agents, and examining the set of Nash equilibria of the underlying game.

© Springer International Publishing Switzerland 2015
T. Walsh (Ed.): ADT 2015, LNAI 9346, pp. 451–468, 2015.
DOI: 10.1007/978-3-319-23114-3_27

However, most voting games contain an enormous amount of Nash equilibria, with even small games reaching hundreds of thousands of equilibria. Furthermore, many of the equilibria are votes which will not occur in the real world (e.g., for most voting rules, if all voters rank the same candidate last, the case where all voters vote for this least favorite option is a Nash equilibrium). Therefore, there has been very little analysis regarding the structure of the different equilibria—such an analysis would not be informative about the voting procedure and its quality. Moreover, without an understanding of strategic effects on voting, the ability to compare voting rules and choose an appropriate one for each setting is very limited.

In the past few years, several ideas have been raised regarding sensible limitations on the structure of games or equilibria, in order to provide a better game-theoretic analysis of voting scenarios. One of the most popular ideas, raised both in the social choice literature [3] and in the computer-science literature [11,17] is *truth-bias*. Truth bias means that in scenarios in which the voter has no way to manipulate via an insincere vote, in order to improve the result, the voter prefers to stick to its actual preferences and vote truthfully. Such a behavior indeed eliminates many nonsensical equilibria, and generally reduces the number of equilibria in voting games [17].

Contribution: While truth-bias has been analyzed and explored for plurality, it has yet to be extended to other voting rules, and this paper advances our understanding of the effect of truth-bias on other rules. It is not clear *a priori* that the same structural properties that were identified for plurality under truth bias will hold for different rules (in fact, as we show, they generally do not). Consequently, we embark on handling the voting rules most closely related to plurality—the veto rule and the more general family of k-approval voting. These rules, together with plurality, are those that contain only two possible values that voters can give to candidates (approve or disapprove, veto or not veto). We focus on analyzing the pure Nash equilibria under truth bias and we examine and identify a variety of their key properties. One of the crucial features in these schemes is the performance of winners and runner-up candidates at non-truthful equilibria, i.e., their score compared against the score of the winner under truthful voting. These scores can vary in different ways, resulting in different equilibrium configurations in each rule. Such comparisons with the existing work on plurality [14,17] are summarized in Table 2 of Sect. 4.

We begin by focusing on the veto voting rule, where each voter chooses a single candidate from whom to withhold a point. We first characterize its truth-biased equilibria. We then further our results to describe an algorithm that—using max-flow considerations—is able to discern, under certain conditions, whether there exists an equilibrium or not.[1] Moreover, we are able to show that our result is tight to an extent, as removing even one of the identified conditions results in an NP-complete problem.

---

[1] Unlike regular, non-truth-biased voting games, with truth-bias there are scenarios where there is no Nash equilibrium at all. This has also been shown for plurality.

Following that, we consider the k-approval rule, where we examine the same set of questions; we demonstrate that the landscape is more complex for k-approval. Our results move us towards a better understanding of the effects of truth bias on each of these voting rules. As we show, we obtain quite different properties under each rule, and we elaborate more on this in Sect. 5.

1.1 Related Work

There have been many modeling approaches towards eliminating the multitude of Nash equilibria in voting games. Some are based on introducing uncertainty, either regarding the support of each candidate [13], or about the reliability of counting procedures [12]. Other research suggests changing the temporal structure of the game; for example, Xia and Conitzer [18] and Desmedt and Elkind [2] consider the case where agents vote publicly and one-at-a-time, and study subgame-perfect equilibria of these extensive-form games. A different approach is the notion of lazy voting [2], where the utility function is changed so that non-pivotal voters have a slight preference to abstain.

Another way to refine the equilibria set is to stick to the basic game theoretic models, but study NEs that are reachable by iterative voting procedures. The iterative voting model was introduced by Meir et al. [11] and later expanded by Lev and Rosenschein [10] and Branzei et al. [1]; this work followed the research into iterative and dynamic mechanisms, chiefly summarised by Laffont [8]. Interestingly, Branzei et al. [1] show that under plurality, the reachable equilibria of this process are of relatively "good" quality.

We focus on a different model than the approaches above for refining the set of equilibria in a voting game, that of truth bias. The notion of adding a truth bias has been introduced into multiple game theoretic models before, albeit for specialised cases. E.g. by Dutta and Sen [4] for a mechanism design model, and by Laslier and Weibull [9] and by Dutta and Laslier [3] for limited voting scenarios. A more robust model was suggested by Thompson et al. [17], which introduced the general framework, and contained various empirical results for the plurality rule in truth-biased games. The theoretical side of that work was enhanced by Obraztsova et al. [14]. More recent work has also attempted to relate this line of work to iterative voting [15], but this again is solely with respect to plurality.

2 Definitions and Notation

We consider a set of m candidates $C = \{c_1, \ldots, c_m\}$ and a set of n voters $V = \{1, \ldots, n\}$. Each voter i has a *preference order* (i.e., a ranking) over C, which we denote by a_i. For notational convenience in comparing candidates, we will often use $\succ_i$ instead of a_i. When $c_k \succ_i c_j$ for some $c_k, c_j \in C$, we say that voter i prefers c_k to c_j.

At an election, each voter submits a preference order b_i, which does not necessarily coincide with a_i. We refer to b_i as the vote or ballot of voter i. The vector of submitted ballots $\mathbf{b} = (b_1, \ldots, b_n)$ is called a *preference profile*. At a profile $\mathbf{b}$, voter i has voted truthfully if $b_i = a_i$. Any other vote from i will be referred to as a non-truthful vote. Similarly the vector $\mathbf{a} = (a_1, \ldots, a_n)$ is the *truthful preference profile*, whereas any other profile is a non-truthful one.

A *voting rule* $\mathcal{F}$ is a mapping that, given a preference profile $\mathbf{b}$ over C, outputs a candidate $c \in C$, the election's winner; we write $c = \mathcal{F}(\mathbf{b})$. In this paper we will consider the veto rule, as well as the more general family of k-approval rules. This is a family of voting rules that let each voter divide candidates into two groups: the first one is the approved group, of size k (the top k candidates ranked by the voter), and each candidate within the group gets a point from this voter. The rest of the candidates compose the second group, and receive no points. The ranking within each group in the submitted ballot does not matter. The candidate with the most points is the winner of the election, and we resolve ties using lexicographic tie-breaking. The veto rule, in which all candidates but one are given a point is an extreme member of this family. The other extreme is plurality, for which many of the issues dealt with in this paper have been resolved by Obraztsova et al. [14]. We denote by $sc(c, \mathbf{b})$ the score of candidate $c \in C$ in a voting profile $\mathbf{b}$.

In this work, we view elections as a non-cooperative game, in which a utility function u_i is associated with every voter i, that is consistent with its true preference order. That is, we require that $u_i(c_k) \neq u_i(c_j)$ for every $i \in V$, $c_j, c_k \in C$, and also that $u_i(c_k) > u_i(c_j)$, if and only if $c_k \succ_i c_j$. We let $p_i(a_i, \mathbf{b}, \mathcal{F})$ denote the utility of voter i, when a_i is its true preference ranking, $\mathbf{b}$ is the submitted profile by all voters, and $\mathcal{F}$ is the voting rule under consideration, and the utility function of voter i can be constructed from its p_i function. In the common model of election, this means $p_i(a_i, \mathbf{b}, \mathcal{F}) = u_i(\mathcal{F}(\mathbf{b}))$. A Nash equilibrium in these games is a profile $\mathbf{b}^{NE}$, where no voter has an incentive to unilaterally deviate, i.e., for every i and for every vote b'_i, we have $p_i(a_i, \mathbf{b}^{NE}, \mathcal{F}) \geq p_i(a_i, (b'_i, \mathbf{b}^{NE}_{-i}), \mathcal{F})$, where $\mathbf{b}^{NE}_{-i}$ is the vector $\mathbf{b}^{NE}$ without player i's vote.

However, such a model is known to result in multiple equilibria, including nonsensical ones. Assume, for example, that all voters have the same preferences, which coincide with the tie-breaking order; then the profile where all of them veto their favourite candidate is an equilibrium. We can construct many other undesirable equilibria. Hence, we instead focus on the more promising *truth-biased model* [17]. In this model, we suppose that voters have a slight preference for voting truthfully when they cannot unilaterally affect the outcome of the election. This bias is captured by inserting a small extra payoff when the voter votes truthfully. This extra gain is small enough so that voters may still prefer to be non-truthful in cases where they *can* affect the outcome. If $\mathbf{a}$ is the real profile, $\mathbf{b}$ is the submitted one, and ϵ a very small value, the payoff function of voter i is given by:

$$p_i(a_i, \mathbf{b}, \mathcal{F}) = \begin{cases} u_i(\mathcal{F}(\mathbf{b})), & \text{if } a_i \neq b_i, \\ u_i(\mathcal{F}(\mathbf{b})) + \epsilon, & \text{if } a_i = b_i. \end{cases}$$

As already described in Sect. 1.1, this model has recently gained popularity, since it achieves a significant refinement of the set of Nash equilibria, and it has been analyzed in previous work under the plurality voting rule.

Now, the following two equilibrium-related problem classes are of interest. The first deals with determining the existence of equilibria in such a voting game, whereas the second asks about the existence of equilibria with a given candidate as a winner.

Definition 1 ($\exists NE$). *An instance of the $\exists NE$ problem is determined by a preference profile* **a**, *and will be denoted by* $\exists NE(\mathbf{a})$. *The profile* **a** *indicates the true preferences of the voters. Given* **a**, $\exists NE(\mathbf{a})$ *is a "yes" instance* $\iff$ *the corresponding election game, with truth-biased voters, admits at least one Nash equilibrium.*

Definition 2 (*WinNE*). *An instance of the WinNE problem is determined by a preference profile* **a**, *and a candidate* $w \in C$, *denoted by* $WinNE(w, \mathbf{a})$. *It is a "yes" instance* $\iff$ *the corresponding election game, with truth-biased voters, admits at least one Nash equilibrium with* w *as the winner.*

3   Truth-Bias Under the Veto Voting Rule

In this section we provide an analysis of pure Nash equilibria, under the assumption that voters are truth-biased and use the veto voting rule. Our main focus is on how non-truthful equilibrium profiles $\mathbf{b} \neq \mathbf{a}$ can arise. Whether the truthful profile itself is a Nash equilibrium can be decided easily. Some proofs are omitted due to space limitations.

3.1   Properties of Nash Equilibria Under Truth-Bias

We begin by defining a class of candidates, which will become useful further on:

Definition 3. *In a profile* **b**, *where the winner is* $\mathcal{F}(\mathbf{b})$, *a runner-up candidate is a candidate* $c \in C$, *for which one of the following conditions hold:*

- $sc(c, \mathbf{b}) = sc(\mathcal{F}(\mathbf{b}), \mathbf{b})$, *and* $\mathcal{F}(\mathbf{b}) \succ c$ *in the tie-breaking rule,*
- $sc(c, \mathbf{b}) = sc(\mathcal{F}(\mathbf{b}), \mathbf{b}) - 1$, *and* $c \succ \mathcal{F}(\mathbf{b})$ *in the tie-breaking rule.*

Essentially, a runner-up candidate is a candidate that could become a winner by gaining one extra point. We will denote the set of runner-up candidates that satisfy the first (respectively, second) condition of the definition above by $\mathbf{R_1}$ (respectively, $\mathbf{R_2}$). Similarly to the analysis of the truth-biased plurality rule in [14], we define here a notion of a *threshold candidate*. Our definition, however, is different, and tailored to our veto rule analysis. Intuitively, the threshold candidate is the candidate that would become a winner if the current winner, $\mathcal{F}(\mathbf{b})$, lost a point, as can be seen from the definition below.

Definition 4. *Given a voting profile* **b**, *a threshold candidate* c *is a runner-up candidate for which one of the following holds:*

– c is the maximal element of $\mathbf{R}_1$ w.r.t. the tie-breaking order, if $\mathbf{R}_1 \neq \emptyset$,
– c is the maximal element of $\mathbf{R}_2$ w.r.t. the tie-breaking order, if $\mathbf{R}_1 = \emptyset$.

The next important lemma considers the score of a winner at an equilibrium.

Lemma 1. Let $\mathbf{b}^{NE} \neq \mathbf{a}$ be a non-truthful Nash equilibrium, with $w = \mathcal{F}(\mathbf{b}^{NE})$. The score of the winner, w, in $\mathbf{b}^{NE}$, is the same as its score at the truthful profile, i.e., $\mathrm{sc}(w, \mathbf{a}) = \mathrm{sc}(w, \mathbf{b}^{NE})$.

Proof. Suppose $\mathrm{sc}(w, \mathbf{b}^{NE}) > \mathrm{sc}(w, \mathbf{a})$. This means that there is a voter $i \in V$, that gives w a point which it would not give under the truthful profile. That is, it is giving a point to its least-favorite candidate. Such a voter can certainly gain by switching back to its truthful vote. In that case, either a new winner emerges, which would be above w in the preference ranking of i, or w remains the winner, but i gets a higher utility by ϵ, due to voting truthfully.

Now suppose $\mathrm{sc}(w, \mathbf{b}^{NE}) < \mathrm{sc}(w, \mathbf{a})$, i.e., a voter $i \in V$ is vetoing w in $\mathbf{b}^{NE}$, but not in the truthful $\mathbf{a}$. Yet, returning to its truthful vote, a_i, w will still remain the winner, and this will increase player i's utility by ϵ, due to the truth-bias. $\square$

In fact, we can further show that not only the winner's score does not change at a non-truthful equilibrium, but the set of voters which support the winner are the same as in the truthful profile. Hence, we obtain the following:

Corollary 1. Let $\mathbf{b}^{NE} \neq \mathbf{a}$ be a non-truthful Nash equilibrium, with $w = \mathcal{F}(\mathbf{b}^{NE})$. The set of voters that veto w in $\mathbf{a}$ is the same set that vetoes w in $\mathbf{b}^{NE}$.

The next properties that we identify are simple to prove but crucial in understanding what equilibria look like under the veto rule.

Lemma 2. For any non-truthful equilibrium profile $\mathbf{b}^{NE} \neq \mathbf{a}$, there always exists a threshold candidate in $\mathbf{b}^{NE}$.

Proof. It suffices to show that there always exist runner-up candidates; hence, there is a threshold runner-up as well. Let $\mathbf{b}^{NE} \neq \mathbf{a}$ be an equilibrium with $w = \mathcal{F}(\mathbf{b}^{NE})$. Suppose we have a non-truthful equilibrium and that there is no runner-up candidate. Consider a voter i that voted non-truthfully. By Corollary 1, the non-truthful voters in $\mathbf{b}^{NE}$ do not veto w (and they do not veto w in $\mathbf{a}$ either). Hence i has vetoed some other candidate. By switching back to its truthful vote, the outcome is not going to change, since there is no runner-up candidate and since w is not going to lose any points. Hence i is better off by ϵ to vote truthfully, a contradiction. Thus there are always runner-up candidates at a non-truthful equilibrium. $\square$

Observation 1. All voters that do not veto the winner or the runner-ups in an equilibrium profile prefer the winner over the threshold candidate (otherwise, they could just veto the winner and make the threshold candidate win).

Example 1. There are cases where the threshold candidate in an equilibrium may have fewer points than in the truthful state (note that this is not true for plurality, as shown in Lemma 2 of Obraztsova et al. [14]); we show it here with an example of 4 candidates. Suppose the tie-breaking rule is $c \succ b \succ d \succ w$, and the truthful profile is:

- 3 voters with ranking: $w \succ b \succ c \succ d$ • 2 voters with ranking: $w \succ d \succ c \succ b$
- 1 voter with ranking: $w \succ b \succ d \succ c$ • 1 voter with ranking: $b \succ c \succ d \succ w$

Then c is the winner of the truthful profile. In turn, in the following profile, w is the winner and the threshold candidate is c, which has fewer points than in the truthful state. Notice that the profile is an equilibrium, and just one voter has moved from the first group to the 3rd one:

- 2 voters with: $w \succ b \succ c \succ d$ • 2 voters with: $w \succ d \succ c \succ b$
- 2 voter with: $w \succ b \succ d \succ c$ • 1 voter with: $b \succ c \succ d \succ w$

Finally, to facilitate our discussion in the next subsections, we define the concept of "voting against" a candidate, and a simple companion lemma.

Definition 5. *We will say that a voter j* **votes against** *candidate c_i in a profile* **b**, *if $b_j \neq a_j$ and c_i is vetoed in b_j.*

Lemma 3. *In every non-truthful NE, all non-truthful voters vote against some runner-up candidate in the NE (not necessarily the same one).*

3.2 Existence of Nash Equilibria

Having identified the properties above, we are now ready to prove our first set of results regarding the complexity of the problems $WinNE$ and $\exists NE$, as defined in Sect. 2. We start with the following negative result.

Theorem 1. *Under the veto rule and with truth-biased voters the problem $WinNE$ is NP-complete.*

Proof. While membership in NP is trivial, completeness requires several steps. We will construct a reduction from exact-cover by 3-sets (X3C).

Definition 6. *Exact cover by 3 sets (X3C) is a problem in which we have a set of $3m$ elements $U = \{u_1, \ldots, u_{3m}\}$ and a set of sets $S = \{S_1, \ldots, S_n\}$ such that for $1 \leq i \leq n$: $S_i \subseteq U$, $|S_i| = 3$. We wish to know if there is a set $T \subseteq S$ such that $|T| = m$ and $\cup_{S \in T} S = U$ (NP-completeness shown in [6])*

Taking an X3C instance, we construct an instance of our problem. Our candidates will be the members of S and U, to which we add two new candidates w and t. To construct our voters, we introduce some markings to aid us: S_i's elements are $\{u_{i_1}, u_{i_2}, u_{i_3}\}$, and we denote by $\mathcal{S}$ the members of S ordered as usual — $S_1 \succ S_2 \succ \ldots \succ S_n$; similarly we use $\mathcal{U}$ for the ordering of U. $\bar{\mathcal{S}}$ marks the opposite direction — $S_n \succ S_{n-1} \succ \ldots \succ S_1$, and ditto for $\bar{\mathcal{U}}$. Our tie-breaking rule is $w \succ t \succ \mathcal{S} \succ \mathcal{U}$. We now describe the set of voters, which consists of the two blocks of voters described in Table 1, along with 3 more blocks described below:

- Block 3: For every $u_i \in U$, we have:
  - m votes of the form: $\mathcal{U} \setminus \{u_i\} \succ \mathcal{S} \succ w \succ t \succ u_i$;
  - $n - 2m - 1$ votes of the form: $\bar{\mathcal{S}} \succ \bar{\mathcal{U}} \setminus \{u_i\} \succ w \succ t \succ u_i$.
- Block 4: For every $S_i \in S$, we have:
  - m votes of the form: $\mathcal{S} \setminus \{S_i\} \succ \mathcal{U} \succ w \succ t \succ S_i$;
  - $n - 2m - 1$ votes of the form: $\bar{\mathcal{U}} \succ \bar{\mathcal{S}} \setminus \{S_i\} \succ w \succ t \succ S_i$.
- Block 5: $n - m$ votes of the form: $t \succ S \succ \mathcal{U} \succ w$.

Table 1. NP-Completeness proof profiles.

Block 1		
$\mathcal{S} \setminus \{S_1\}, \mathcal{U},$		$... \mathcal{S} \setminus \{S_n\}$
$\mathcal{U},$	$\bar{\mathcal{S}} \setminus \{S_2\}, ... \mathcal{U}$	
$w,$	$w,$	$... w$
$S_1,$	$S_2,$	$... S_n$
$t,$	t	$... t$

Block 2		
$... \mathcal{U} \setminus \{u_{i_1}\}, \mathcal{U} \setminus \{u_{i_2}\}, \mathcal{U} \setminus \{u_{i_3}\}, ...$		
$... \mathcal{S} \setminus \{S_i\},$	$\mathcal{S} \setminus \{S_i\},$	$\mathcal{S} \setminus \{S_i\}, ...$
$... w,$	$w,$	$w,$ ...
$... t,$	$t,$	$t,$ ...
$... u_{i_1},$	$u_{i_2},$	$u_{i_3},$ ...
$... S_i,$	$S_i,$	$S_i,$ ...

In the truthful profile, w is not the winner (u_1 is). We claim that there is an equilibrium in which w is the winner if and only if there is a solution to the X3C problem.

Lemma 4. *Given the constructed truthful profile, if a NE profile $\mathbf{b}^{NE}$ exists with w as a winner, t is the threshold candidate in $\mathbf{b}^{NE}$.*

Proof. Consider an equilibrium $\mathbf{b}^{NE}$, and let us assume the opposite: t is not a threshold candidate. Consider the possibility that S_i is the threshold candidate of $\mathbf{b}^{NE}$. Let us take a closer look at a voter from Block 4 of the form $\mathcal{S} \setminus \{S_j\} \succ \mathcal{U} \succ w \succ t \succ S_j$, for some $j \neq i$ so that S_j is not a runner-up. In the equilibrium $\mathbf{b}^{NE}$, this voter can veto only either S_j or t. Furthermore, since t beats all candidates except w in tie breaking, if t is not the threshold candidate then it cannot be a runner-up candidate at all. As a result, the inspected voter cannot veto t, according to Lemma 3. Thus, we have found a voter that votes truthfully in the NE profile $\mathbf{b}^{NE}$ and prefers S_i to w, contradicting Observation 1.

Consider now the case where all S_i's are runner-up candidates. In this case, each S_i has gained two more points relatively to their respective scores in the truthful profile $\mathbf{a}$. Only voters in Block 2 are capable of deviating to accomplish this (the reasoning in the paragraph above is why Block 4 votes will not deviate). In particular, for each S_i there are at least two such voters that deviate to veto u_{i_l} and u_{i_k} for some $l \neq k \in \{1, 2, 3\}$. As a result there are at least 2 additional voters (in comparison to the truthful profile) that veto candidates from $\mathcal{U}$. The Pigeonhole Principle dictates that there is at least one u_{i_l} that has received two more veto votes in comparison to the truthful profile. In addition, notice that there can be no voter in the NE profile $\mathbf{b}^{NE}$ that deviates in favor of u_{i_l}. As a result, u_{i_l} cannot be a runner-up candidate, yet has voters that deviate to veto it, in contradiction to Lemma 3. Hence, it is impossible for all S_i's to be runner-up candidates simultaneously.

Similarly, we will obtain a contradiction for the case where we assume that u_i is the threshold candidate, and, therefore, none of the S_i's are runner-ups. □

We conclude that in a NE profile $\mathbf{b}^{NE}$, t must be the threshold candidate and it will have $(n - m)$ points. Let us now proceed with the remainder of the proof of Theorem 1.

If there is $T = \{S'_1, \ldots, S'_m\} \subseteq S$ which is a solution to the X3C problem, we have an equilibrium in which w is the winner: the voters from Block 1 whose penultimate candidate is $S'_i \in T$ will veto S'_i. The voters in Block 2 who veto $S'_i \in T$ instead veto their penultimate candidates $u_{i_{1/2/3}}$. In such a situation all candidates are vetoed by $n - m$ voters (apart from those in $S \setminus T$, which are vetoed by $n - m + 2$ voters), and therefore w is the winner. All voters are vetoing runner-ups which they prefer less than w or t. Hence, changing their vote will make the candidate they currently veto the winner, and as they would rather have w win, they do not change their vote. Furthermore, all voters from Blocks 1 through 4 that do not veto a runner-up candidate, can only deviate so that t becomes a winner. Since they prefer w to t, none of them will actually have an incentive to deviate. Finally, none of the voters in Block 5 can change the election outcome and will remain truthful.

Now, assume that there is no solution to the X3C problem. At least m voters from Block 1 will veto the S_i's (the only candidates less-preferred than w). However, in order for them not to revert to their truthful vote, those S_i's need to be runner-up candidates, so all votes in Block 2 who would truthfully veto those S_i's, need to veto their respective u_i's instead. In addition, those u_i's need to be runner-ups as well (or those Block 2 votes will revert to the truthful vote), and as they are ranked below S in the tie-breaking rule, they need to have $m - n$ vetoes in order to be runner-ups. This means that each u_i is vetoed only once in Block 2. So we have m (or more) S_i's containing exactly one copy of each u_i; i.e., we found an exact cover of U, contradicting the assumption that X3C has no solution. $\square$

The above proof also implies Corollary 2. Furthermore, its variant can also be used to prove a more general theorem (due to lack of space, the proof is not presented here).

Theorem 2. *Consider the veto rule and truth-biased voters. Then the problem* $\exists NE$ *is NP-complete.*

It is possible to further expand upon the result of Theorem 1. To this end, we identify two conditions that help us characterize the set of computationally hard instances. In particular, consider an instance $WinNE(w, \mathbf{a})$, determined by a candidate $w \in C$ and a truthful profile $\mathbf{a}$. We consider the following conditions:

C1: Let $t \in C$ be the candidate right below w in the tie-breaking order (i.e., the tie-breaking order is in the form $\cdots \succ w \succ t \succ \cdots$). Then $sc(t, \mathbf{a}) \geq sc(w, \mathbf{a})$.
C2: Let t be as in C1. Then, for every voter i that does not veto w in the truthful profile $\mathbf{a}$, it holds that $w \succ_i t$.

Corollary 2. *Under the veto voting rule, the problem* $WinNE$ *is NP-complete, even for the family of instances that satisfy condition C2 and do not satisfy condition C1.*

In fact, together with Theorem 3, the picture becomes more clear regarding hardness results: violating either one of the conditions C1, C2 makes the problem $WinNE$ hard.

Theorem 3. *Under the veto voting rule, the problem $WinNE$ is NP-complete, even for the family of instances where C1 holds but C2 does not.*

Proof. As with Theorem 1, we will construct a reduction from X3C. We will use the same notation, where S is the set of sets and U is the set of elements in an instance of X3C, and convert the members of these sets into distinct candidates. However, unlike the previous proof, in addition to the candidates from S and U, we will introduce four special candidates w, t, p_1 and p_2.

Based on the candidate set defined above, we will construct a set of voters and their truthful preference profile **a**, so that a solution to $WinNE(w, \mathbf{a})$ would entail a solution to the X3C instance.

We will order the candidates to form the following tie-breaking preference order: $w \succ t \succ p_1 \succ p_2 \succ S \succ U$, where candidates from S and U appear in their natural lexicographic order.

We now construct a set of voters, grouped into five distinct blocks, according to their truthful preference profile. In each block we only explicitly describe the order of a few least-preferred candidates. All candidates that are not explicitly mentioned in a profile, appear in an arbitrary order, and are marked by $\ldots$.

- **Block 1:** A set of n voters, one for each candidate in S, with preference profile of the form $\cdots \succ t \succ w \succ S_i \succ p_1$;
- **Block 2:** A set of $n - m$ voters with preference profile of the form $\cdots \succ t \succ w \succ p_2$ and one additional voter with profile of the form $\cdots \succ w \succ t \succ p_2$;
- **Block 3:** For each $\{u_{i_1}, u_{i_2}, u_{i_3}\} = S_i \in S$ a set of $n - m + 2$ voters. Three with profiles of the form $\cdots \succ w \succ t \succ u_{i_j} \succ S_i$, where $j \in \{1, 2, 3\}$, and all others of the form $\cdots \succ w \succ t \succ S_i$;
- **Block 4:** For each $u_k \in U$ a set of $n - m - 1$ voters with profiles of the form $\cdots \succ w \succ t \succ u_k$;
- **Block 5:** A set of $n - m - 1$ voters with profiles of the form $\cdots \succ w \succ t$
- **Block 6:** A set of $n - m$ voters with profiles of the form $\cdots \succ t \succ w$.

Let us now show why the existence of an equilibrium profile that solves $WinNE(w, \mathbf{a})$, where **a** is as described above, entails a solution to the X3C instance. To this end, consider the voters' behaviour in a NE profile **b**, where w is the winner.

According to Lemma 1, $sc(w, \mathbf{b}) = sc(w, \mathbf{a})$, yet by our construction $sc(t, \mathbf{a}) = sc(w, \mathbf{a}) + 1$. Hence, for w to become a winner in **b**, t has to receive at least one additional veto and will also be the threshold candidate (if t is not a runner-up, it means several are no longer vetoing it, but as they are vetoing some non-winner instead, they can revert to being truthful and gain ϵ utility. Due to its loss to w in tie-breaking, it means t is the threshold candidate). According to Lemma 3 and the truth bias assumption, none of those who vetoed t in **a** would switch to veto another candidate. In fact, there is only one voter that needs to deviate from its truthful profile and veto t.

Consider the voters of Block 2. $(n - m)$ of them prefer t to w, and would not veto the former. Yet, if they lie in **b**, they have to veto a runner up. As a result, none of them can deviate from their truthful profile in equilibrium. On the other hand, the last voter of the block can (and should) deviate, and veto t.

Similarly, consider the score of p_1 in the truthful profile **a** and compare it to that of $\mathrm{sc}(w, \mathbf{a})$. For w to be the winner of the equilibrium profile **b**, m voters from Block 1 need to deviate in the equilibrium and stop vetoing p_1. These newly vetoed candidates have to be less preferred than w by the deviating voters. For voters of Block 1, this means vetoing a candidate from the set S. As a result, there are m candidates $S_i \in S$ that are being vetoed by the voters from Block 1 in the equilibrium profile **b**.

These *chosen* S_i's, however, need to be runner-up candidates. To achieve that, exactly 3 candidates that veto S_i's in Block 3 must deviate in the equilibrium profile **b**. These can only be the voters with preference profiles of the form $\cdots \succ w \succ t \succ u_{i_j} \succ S_i$, where $j \in \{1, 2, 3\}$.

Since no voter in Block 4 can deviate, those voters from Block 3 that deviate to veto u_{i_j}s can only do so consistently with Lemma 3, if the total number of times that u_{i_j} is being vetoed is equal to $(n - m)$. This can happen only if each $u_{i_j} \in U$ is vetoed exactly once by voters from Block 3.

As a result, the sub-set S_i's that are vetoed by voters in Block 1 constitutes a solution to the given X3C instance. The opposite direction, that is, constructing a NE profile given a solution to the X3C instance, is trivial. □

The results of this subsection show that there are critical properties of the truthful profile **a** that make the existence of an equilibrium with a given winner a hard problem. However, as we show in the next subsection, combining these properties, namely condition $C1$ and $C2$, creates a polynomial-time decidable sub-class.

3.3 A Polynomially Solvable Subclass

In the previous subsection, we demonstrated two conditions on a candidate and the truthful profile that, if violated, make $WinNE$, under truth-biased voters, NP-hard. In this subsection, we complete our treatment on this categorization of profiles by considering the subset where both aforementioned conditions hold. In fact, we provide a constructive proof, via a reduction to a max-flow problem, showing that in this sub-class of truthful profiles, $WinNE(w, \mathbf{a})$ can be decided in polynomial time.

Theorem 4. *Consider a candidate $w \in C$ and a truthful profile* **a** *for which conditions $C1$ and $C2$ hold. Then $WinNE(w, \mathbf{a})$, i.e., the existence of a NE profile* $\mathbf{b}^{NE}$*, where $\mathcal{F}(\mathbf{b}^{NE}) = w$, is decidable in polynomial time.*

The statement of the above theorem is tight, given Theorems 1 and 3. Namely, should either one of the conditions $C1$ or $C2$ be violated, determining the existence of a NE with w as the winner becomes NP-hard. The conditions C1 and C2 ensure that we can focus on a particular threshold candidate (namely candidate t) for constructing a Nash equilibrium profile. While $C1$ ensures some manipulation will be necessary (and the preference order ensures it will be a threshold), it is $C2$ that ensures that t can be a valid threshold candidate, since without it, as Observation 1 noted, this is not possible.

Proof. Consider an instance of the problem specified by a potential winner $w \in C$ and the real profile $\mathbf{a}$. Let t also be the candidate right next to w, as specified by conditions C1 and C2. The proof is based on a polynomial reduction to the max-flow problem in a graph. We will construct a graph (and later correct the flow) in such a way that the set of flow-saturated edges will indicate the feasibility of obtaining a Nash equilibrium. Furthermore, positive flow at certain nodes in the graph will indicate a switch in the voters' equilibrium ballots from their truthful profile.

Given a truthful voting profile $\mathbf{a}$, we will construct the graph as follows. Vertices will be associated with each candidate and each voter; we also add a *source* and a *sink* node. The set of graph vertices will therefore be $\{source, sink\} \cup C \cup V$.

The set of edges, E, in the graph will consist of three subsets.

- **Potential deviators.** Edges that link voters and potentially vetoed candidates.

For a voter i where the last candidate in its real preference order is some $r \in C$, i.e., the preference order is in the form $\ldots \succ w \succ c_1 \succ \ldots \succ c_l \succ r$, add the following directed edges with *unit* flow capacity: $(r, v_i), (v_i, c_1), \ldots, (v_i, c_l)$.

The resulting palm-leaf sub-structure is depicted in Fig. 1. It essentially captures the ability of the voter to change its veto in a manner that will benefit w without deteriorating the voter's utility (note, of course, that there are multiple such sub-graphs in the graph, and vertices may be part of several such structures).

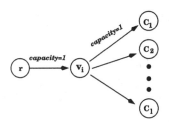

Fig. 1. Polytime special veto subclass. Palm sub-structure for Theorem 4.

- **Sustainable deviations.** Edges from the *source* node. These edges and capacities reflect the number of additional points a candidate may absorb until it becomes a runner-up candidate w.r.t. w. Hence (recall that t is the candidate next to w in the tie-breaking order, as specified by conditions C1 and C2):

For each candidate c so that $t \succ c$ in tie-breaking and $\mathrm{sc}(w, \mathbf{a}) - \mathrm{sc}(c, \mathbf{a}) > 0$, a directed edge $(source, c)$ is added with capacity $\mathrm{sc}(w, \mathbf{a}) - \mathrm{sc}(c, \mathbf{a})$.

For each candidate c so that $c \succ w$ in tie-breaking and $sc(w, \mathbf{a}) - sc(c, \mathbf{a}) > 1$, a directed edge $(source, c)$ is added with capacity $sc(w, \mathbf{a}) - sc(c, \mathbf{a}) - 1$.

• **Necessary deviations.** Edges to the *sink* node. These edges and capacities reflect the number of additional veto votes a candidate needs to sustain to make its score less than that of w. Otherwise, w would not be able to become the winner.

For each candidate c so that $t \succ c$ in tie-breaking and $sc(w, \mathbf{a}) - sc(c, \mathbf{a}) < 0$, a directed edge $(c, sink)$ is added with capacity $sc(c, \mathbf{a}) - sc(w, \mathbf{a})$.

For each candidate c so that $c \succ w$ in tie-breaking and $sc(w, \mathbf{a}) - sc(c, \mathbf{a}) < 1$, a directed edge $(c, sink)$ is added with capacity $1 - (sc(w, \mathbf{a}) - sc(c, \mathbf{a}))$.

Given Corollary 1, the non-truthful votes at an equilibrium profile come from voters that were not vetoing w in $\mathbf{a}$, and they now lie by vetoing some candidate other than their truthful vetoed candidate, which is less-preferred than w. From the construction of the graph, it is easy to see that if the maximal flow through the above graph is **less** than the sum of all incoming capacities to the *sink* node, then there can be no equilibrium profile that makes w a winner. To see this, observe that only candidate vertices connect directly to the sink, and these are precisely candidates that have higher scores than w. Total capacity of all these edges equals the number of voters that necessarily have to change their vote. Furthermore, the flow has to go through voter vertices, connected to candidates that are less-preferred to w (and hence indicate a switch from the truthful vote to a non-truthful one). Finally, if all edges to the sink are saturated in a maximum flow, we will show that a Nash equilibrium profile $\mathbf{b}^{NE}$, with $\mathcal{F}(\mathbf{b}^{NE}) = w$, can be recovered from the flow. In what follows we will demonstrate this formally.

Let $f : E \rightarrow \mathcal{R}$ be a maximal acyclic, integer flow through the constructed graph. Such a flow can be obtained in time polynomial in the number of voters and candidates. Furthermore, all edges from a candidate node to a voter node that have positive flow on them will be saturated (as their capacity is 1).

We will now modify the flow, while maintaining its total capacity, to maximize the flow through the *source* outgoing edges, and minimize the flow through voter nodes. Since we will later associate a flow through a voter node with the voter deviating from the truthful vote, minimizing the flow through voter nodes will reflect and ensure that the voting profile recovered from it will be truth biased (i.e., no unnecessary lying takes place, otherwise some voter would have an incentive to switch back to the truthful vote).

Let $D = \{c|\ \exists e = (source, c) \in E\}$ be the set of all nodes directly connected to the source. Notice that D is a subset of candidate nodes. Let $q \in D$ be a node for which: (i) there is a voter v so that $(v, q) \in E$; (ii) $f((v, q)) > 0$; and (iii) the edge $(source, q)$ is not saturated. We will repeat the following flow modification until no such q exists.

Consider a flow path to q through voter nodes. In particular, let $\pi = (source = n_0, n_1, \ldots, n_l = q)$ be an acyclic path from the source to q, so that $e_k = (n_{k-1}, n_k) \in E$ for all $k \in [l]$ and $f(e_k) > 0$. Notice that since f is an integer flow and all edges between candidate nodes and voter nodes have unit capacity, all the edges of the path have a unit flow apart from the initial edge from the source to n_1. We will modify the flow f and construct an augmented flow $\widehat{f}$ by canceling the flow through π, and replacing it with an additional unit flow from the *source* to q. More formally, let $\widehat{f} = f$. We then set $\widehat{f}(e_k) = 0$ for all $k \in [1 : l - 1]$, $\widehat{f}((source, n_1)) = f((source, n_1)) - 1$ and $\widehat{f}((source, q)) = f((source, q)) + 1$. We then repeat the modification procedure, if necessary, for $\widehat{f}$. Notice that the flow modification procedure does not change the total flow from the *source* to the *sink* node.

Assume now that the flow is such that for all nodes $q \in D$, either the edge $(source, q)$ is saturated, or q has no positive incoming flow from the voter nodes. Then for every voter $v_i \in V$, if there is an edge (v_i, c_j) for some $c_j \in C$, so that $f(v_i, c_j) > 0$, i.e., saturated, we let v_i change its vote to veto c_j. Otherwise, v_i votes truthfully. Let $\mathbf{b}^{NE}$ be the resulting strategy profile. It is easy to see that $\mathbf{b}^{NE}$ is indeed an equilibrium.

Notice that the equilibrium profile, $\mathbf{b}^{NE}$, was constructed in poly-time. Recall, the steps consisted of: (a) constructing the graph, which takes time polynomial in the number of candidates, m, and voters, n; (b) finding a maximal acyclic integer flow (poly-time algorithms exist, and any one is suitable); (c) a set of flow modifications. Finding the path π necessary for the flow modification takes polynomial time in m and n, e.g., by following the flow f back through the saturated edges. Furthermore, the number of repetitions of the flow modification process is polynomial in the number of candidates and voters as well. This is because the flow through any candidate node from voter nodes is bounded from above by the number of voters, and is reduced by one in every modification. As a result, the running time of the whole algorithm is polynomial.   □

Note that we cannot have an analogous separation for the problem $\exists NE$, since the conditions C1 and C2 depend on the winner under consideration. Hence, we can clearly have a polynomial time algorithm for $\exists NE$, if C1 and C2 hold for every $w \in C$ (running the algorithm for $WinNE(w, \mathbf{a})$ for every w), but we cannot conclude anything if these conditions do not hold across all candidates.

4   Truth-Bias Under k-approval

The veto rule and the plurality rule are two extreme points of the general family of k-approval voting. For k-approval with $k \geq 1$, voters "approve" of the first k candidates in their submitted ballot, hence each such candidate receives one point from that voter. In this section, we briefly show that we cannot hope to have analogous results as in Sect. 3. Clearly the NP-hardness results continue to hold, since veto is included in k-approval, but the rest of the properties that we identified in the previous section do not hold for the more general class of k-approval rules.

For the analysis, we will adopt the general terminology of Sect. 3. In more detail, we will use the same definition of a runner-up candidate (as well as the definitions for the sets $\mathbf{R}_1$, $\mathbf{R}_2$, and for threshold candidates), as the ones used for the veto case. Additionally, we will denote the set of approved candidates in a profile a_i (respectively, b_i) by $\mathcal{A}_i$ (respectively, $\mathcal{B}_i$). Finally we will use the terms "votes in favor" and "votes against" in the sense of the following definition.

Definition 7. *Let* **a** *be the truthful profile, and let* **b** *be the submitted profile. A voter* i **votes in favor** *of a candidate* c_j, *if* $c_j \notin \mathcal{A}_i$ *and* $c_j \in \mathcal{B}_i$. *Similarly,* i **votes against** c_j, *if* $c_j \in \mathcal{A}_i$ *and* $c_j \notin \mathcal{B}_i$.

Lemma 5. *Given a Nash equilibrium profile* $\mathbf{b}^{NE} \neq \mathbf{a}$, *for every non-truthful voter, exactly one, but never both, of the following conditions hold: (a) the voter votes in favor of the winner, (b) the voter votes against some* $r \in R_1 \cup R_2$.

The lemma above already shows some differentiation from what holds for veto. Now, note that for veto, it is always case (b) from Lemma 5 that holds, and never case (a). For plurality, on the contrary, it was established in [14] that it is case (a) that holds, and never case (b). For k-approval, with arbitrary k, it can be either of the two cases.

Next, we establish that a threshold candidate always exists as in the previous section.

Proposition 1. *For every equilibrium* $\mathbf{b}^{NE} \neq \mathbf{a}$, *a threshold candidate always exists.*

Unlike the case of veto or plurality, in the k-approval case, it is possible that in a non-truthful equilibrium, neither the winner nor the threshold candidate will maintain the same score as in the truthful profile. This is demonstrated in the following examples.

Example 2. Consider the following two profiles using 2-approval, with the tie-breaking order given by the sequence $a \succ b \succ c \succ d \succ e$. The truthful profile (of 7 voters) is:

- $a \succ b \succ c \succ d \succ e$
- $e \succ d \succ a \succ c \succ b$
- $e \succ c \succ a \succ b \succ d$
- 2 voters with preference $d \succ b \succ a \succ c \succ e$
- 2 voters with preference $a \succ d \succ b \succ c \succ e$

The equilibrium profile changes the last but one voter (i.e., one out of the two identical voters with preference $a \succ d \succ b \succ c \succ e$), as well as the other one of that type and the last voter to $a \succ e \succ b \succ c \succ d$ and $e \succ a \succ c \succ b \succ d$

In this example, the score of the winning candidate (more specifically, candidate a) in the equilibrium profile is higher than in the truthful profile. On the other hand, the score of the threshold candidate in the equilibrium (in this example d) decreases in the equilibrium compared to the truthful profile score.

Example 3. The score of the threshold candidate does not necessarily decrease at an equilibrium. A further example demonstrates that its score can also become larger in an equilibrium than it is in the truthful profile. Using again 2-approval, and the tie-breaking order $d \succ a \succ b \succ c$, this example consists of the following true preferences:

$$a \succ b \succ c \succ d \qquad a \succ c \succ b \succ d$$
$$c \succ d \succ a \succ b \qquad d \succ b \succ a \succ c$$
$$a \succ d \succ b \succ c$$

A NE can be constructed by making only one change — the last voter changes to $a \succ b \succ d \succ c$. The threshold candidate score in the equilibrium (here of candidate b) increases.

Finally, regarding the winner's score in a Nash equilibrium, we prove that it cannot fluctuate and go up or down as the score of the threshold candidate, but instead it is bounded, according to the following proposition.

Proposition 2. *Let* $w = \mathcal{F}(\mathbf{b}^{NE})$ *for a NE* $\mathbf{b}^{NE} \neq \mathbf{a}$. *Then* $\mathrm{sc}(w, \mathbf{b}^{NE}) \geq \mathrm{sc}(w, \mathbf{a})$.

Our results along with a comparison to the plurality rule are summarised in Table 2. One of the main conclusions drawn from Table 2 is the following: the analysis of the plurality rule by Obraztsova et al. [14] established that the score of the threshold candidate in a non-truthful equilibrium remains the same as in the truthful profile. This property was exploited in that work for characterising the set of equilibria. Regarding the veto rule, the score of the threshold candidate may change, but the score of the winner remains unchanged among equilibria. This is a property that we exploited in Sect. 3 in our positive result. In contrast, when it comes to the general family of k-approval, the examples in this section show that the score of both the winner and the threshold candidate may change. The threshold candidate in particular is far less constrained, and can either lose or gain points in the equilibrium compared to the truthful profile. Hence, equilibrium profiles are not as well-structured in this case, and therefore we cannot hope to use the ideas from Sect. 3 or from the plurality analysis of Obraztsova et al. [14] to come up with more positive algorithmic results. Our analysis suggests that one may have to look at special cases of k-approval to obtain structural properties that can yield characterizations, or can be exploited algorithmically.

Table 2. Summary of our complexity results and other properties.

Conditions	Veto			Plurality	k-approval
	$\neg C1$ and $C2$	$C1$ and $\neg C2$	$C1$ and $C2$		
$WinNE(w, \mathbf{a})$	NP-hard		P	NP-hard	NP-hard
Winner score may grow in NE	No			Yes	Yes
Winner score may drop in NE	No			No	No
Runner-up score may grow in NE	Yes			No	Yes
Runner-up score may drop in NE	Yes			No	Yes

5   Discussion and Future Work

In this paper we initiated the investigation of truth-biased voters in voting systems that do not use plurality. We focused on the spectrum of voting rules ranging from plurality to veto, encompassing all scoring rules that allow only two values. In these cases, we have seen that plurality, veto, and k-approval have their own distinct properties. For example, both plurality and k-approval share the property that the score of an equilibrium winner might be higher than the truthful score of that candidate (which cannot happen in veto). However, in k-approval, unlike plurality, equilibria exist where runner-up candidates also change their scores.

While the problems we studied are NP-complete for both veto and k-approval, we showed a tight subset of cases for veto where there is a polynomial time algorithm for knowing (and finding, as the algorithm is constructive) if there is a Nash equilibrium with a winner of our choice.

There are several further research areas to pursue. First, we can combine this research with other voting approaches. For example, while there has been research on *iterative voting* with truth-bias, it has only focused on plurality.

A different approach remains strictly within the framework of truth-bias, and tries to further enhance our understanding of truth-biased voters, and expand the research of it to more voting rules (most interesting, to non-scoring rules, such as maximin), allowing us to further understand the effects of truth-bias. However, this approach presents unique challenges, as while truth-bias in binary scoring effectively eliminates the most egregious of nonsensical equilibria, expanding it to other voting rules requires a wider net, in a sense, to eliminate such equilibria.

Acknowledgements. This research was supported in part by Israel Science Foundation grant #1227/12, Israel Ministry of Science and Technology grant #3-6797, and by Microsoft Research through its PhD Scholarship Programme. It has also been supported by the EU (European Social Fund) and Greek national funds through the Operational Program "Education and Lifelong Learning" of the National Strategic Reference Framework (NSRF) - Research Funding Program: THALES. The work of S. Obraztsova was partially supported by ERC grant #337122 under the EU FP7/2007–2013 and RFFI grant 14-01-00156-a.

References

1. Branzei, S., Caragiannis, I., Morgenstern, J., Procaccia, A.D.: How bad is selfish voting? In: AAAI (2013)
2. Desmedt, Y., Elkind, E.: Equilibria of plurality voting with abstentions. In: ACM EC, pp. 347–356 (2010)
3. Dutta, B., Laslier, J.F.: Costless honesty in voting. In: 10th International Meeting of the Society for Social Choice and Welfare, Moscow (2010)
4. Dutta, B., Sen, A.: Nash implementation with partially honest individuals. Games Econ. Behav. **74**(1), 154–169 (2012)
5. Farquharson, R.: Theory of Voting. Yale University Press, New Haven (1969)
6. Garey, M.R., Johnson, D.S.: Computers and intractability: a guide to the theory of NP-completeness. Series of Books in the Mathematical Sciences. W. H. Freeman, New York (1979)
7. Gibbard, A.: Manipulation of voting schemes. Econometrica **41**(4), 587–602 (1973)

8. Laffont, J.J.: Incentives and the allocation of public goods. In: Handbook of Public Economics, vol. 2, chap. 10, pp. 537–569. Elsevier (1987)
9. Laslier, J.F., Weibull, J.W.: A strategy-proof condorcet jury theorem. Scand. J. Econ. (2012)
10. Lev, O., Rosenschein, J.S.: Convergence of iterative voting. In: AAMAS, pp. 611–618 (2012)
11. Meir, R., Polukarov, M., Rosenschein, J.S., Jennings, N.R.: Convergence to equilibria of plurality voting. In: AAAI, pp. 823–828 (2010)
12. Messner, M., Polborn, M.K.: Miscounts, duverger's law and duverger's hypothesis. Technical Report 380, Innocenzo Gasparini Institute for Economic Research, Bocconi University (2011)
13. Myerson, R.B., Weber, R.J.: A theory of voting equilibria. Am. Polit. Sci. Rev. **87**(1), 102–114 (1993)
14. Obraztsova, S., Markakis, E., Thompson, D.R.M.: Plurality voting with truth-biased agents. In: Vöcking, B. (ed.) SAGT 2013. LNCS, vol. 8146, pp. 26–37. Springer, Heidelberg (2013)
15. Rabinovich, Z., Obraztsova, S., Lev, O., Markakis, E., Rosenschein, J.S.: Analysis of equilibria in iterative voting schemes. In: Proceedings of the 29th AAAI Conference on Artificial Intelligence (AAAI), pp. 1007–1013. Austin, Texas January 2015
16. Satterthwaite, M.A.: Strategy-proofness and arrow's conditions: existence and correspondence theorems for voting procedures and social welfare functions. J. Econ. Theor. **10**(2), 187–217 (1975)
17. Thompson, D.R.M., Lev, O., Leyton-Brown, K., Rosenschein, J.S.: Empirical aspects of plurality election equilibria. In: AAMAS (2013)
18. Xia, L., Conitzer, V.: Stackelberg voting games: computational aspects and paradoxes. In: AAAI, pp. 805–810 (2010)

Winner Determination and Manipulation in Minisum and Minimax Committee Elections

Dorothea Baumeister[1], Sophie Dennisen[2], and Lisa Rey[1(✉)]

[1] Institut für Informatik, Heinrich-Heine-Universität Düsseldorf,
40225 Düsseldorf, Germany
lrey@cs.uni-duesseldorf.de
[2] Institut für Informatik, Technische Universität Clausthal,
38678 Clausthal-Zellerfeld, Germany

Abstract. In a committee election, a set of candidates has to be determined as winner of the election. Baumeister and Dennisen [2] proposed to extend the minisum and minimax approach, initially defined for approval votes, to other forms of votes. They define minisum and minimax committee election rules for trichotomous votes, incomplete linear orders and complete linear orders, by choosing a winning committee that minimizes the dissatisfaction of the voters. Minisum election rules minimize the voter dissatisfaction by choosing a winning committee with minimum sum of the disagreement values for all individual votes, whereas in a minimax winning committee the maximum disagreement value for an individual vote is minimized. In this paper, we investigate the computational complexity of winner determination in these voting rules. We show that winner determination is possible in polynomial time for all minisum rules we consider, whereas it is NP-complete for three of the minimax rules. Furthermore, we study different forms of manipulation for these committee election rules.

Keywords: Computational social choice · Committee elections · Winner determination · Manipulation · Complexity

1 Introduction

There are diverse situations where the preferences of different people have to be aggregated, in order to decide upon a given set of alternatives. In such a situation a voting rule is used. Such voting rules may also be employed in systems of artificial intelligence, where different agents have to make a joint decision on some set of alternatives. The study of voting rules from an axiomatic and algorithmic point of view is actively pursued in the field of computational social choice (see e.g., the bookchapters by Zwicker [36], Caragiannis et al. [11], Conitzer

The first author is supported in part by a grant for gender-sensitive universities. The middle author is supported by the Deutsche Forschungsgemeinschaft under Grant GRK 1931 – SocialCars: Cooperative (De-)Centralized Traffic Management.

© Springer International Publishing Switzerland 2015
T. Walsh (Ed.): ADT 2015, LNAI 9346, pp. 469–485, 2015.
DOI: 10.1007/978-3-319-23114-3_28

and Walsh [15], Faliszewski and Rothe [16], Lang and Xia [22], Boutilier and Rosenschein [6]). Most of this work studies the case where the winner of the election is a single candidate, but voting is also used to elect a set of candidates, like in the election of a council or a committee. Consider for example the situation where three friends give a joint party and want to decide what there will be on the buffet. They know that there is only place for a certain number of things on the table, and the three friends will probably have different preferences over the possible choices. In this case a committee election rule may be used. Another application is the design of recommender systems (see [25]), where the task is to choose a number of products for recommendation to a buyer on an online platform.

A widely used rule for committee elections is the minisum approach. Here, the voters decide for each candidate whether they approve or disapprove of her, and the winning committee has a minimum sum of distances to the individual votes, where the distance is measured by the Hamming distance. This corresponds to a utilitarian approach. Another variant of minimizing the voters' dissatisfaction is the minimax approach suggested by Brams et al. [7–9,21]. Here, the maximum distance to an individual vote is minimized using the Hamming distance. Baumeister and Dennisen [2] proposed minisum and minimax committee election rules for trichotomous votes, incomplete linear orders, and complete linear orders. We will study winner determination and manipulation for these newly defined committee election rules.

Closely related to committee elections that minimize the voters' dissatisfaction are systems of proportional representation (see the work of Chamberlin and Courant [13] and Monroe [27]). In these systems the dissatisfaction of a voter is not computed for the whole committee, but only for that candidate from the committee that represents this voter. Computational aspects for problems of proportional representation have been studied in [4,32,34].

2   Definitions and Notations

In this section we introduce the definitions and notations for the voting rules that we analyze in the following sections. A committee election is a triple (C, V, k), where $C = \{c_1, \ldots, c_m\}$ is the set of candidates, $V = (v_1, \ldots, v_n)$ is a list of voters represented by their votes, and k is the size of the committee. We will consider four different types of votes. An approval vote is a $\{0, 1\}^m$ vector, for a fixed order of the candidates. A 1 in the vector stands for an approval of the corresponding candidate, whereas a 0 stands for a disapproval. The second type of votes are trichotomous votes (see Felsenthal [17]). Here, the votes can equally express an approval or disapproval for each candidate, but the voters can also decide to abstain for a candidate. This is realized through $\{-1, 0, 1\}^m$ vectors for a fixed order of the candidates, where again a 1 stands for approval, but now a -1 stands for disapproval, and a 0 stands for an abstention. The last two types of votes we consider are complete and incomplete linear orders. A complete linear order is a total, transitive, and asymmetric binary relation over the set of candidates. We will denote the vote of voter v by $>_v$, where $c >_v d$ means that candidate c is preferred to candidate d by voter v, and $c \not>_v d$ means that voter v does not prefer candidate c to candidate d. Incomplete linear orders are also transitive and asymmetric, but not necessarily a total binary relation over the set of candidates. Note that this allows for general incomplete linear orders and not only linear orders with indifferences. To distinguish between complete and incomplete linear orders, we will denote an incomplete linear order for voter v by $\succ_v$. We will write $a \sim_v b$, if the relation between two candidates a and b is unknown in v, i.e., neither $a \succ_v b$ nor $b \succ_v a$ holds.

To define minisum and minimax voting rules, we need a measure of distance for a single vote with a potential committee. For the sake of readability, we will denote an approval committee as a $\{0,1\}^m$ vector having exactly k ones, or as a set $K \subseteq C$ of candidates. In case of trichotomous votes, we will also denote a committee as a $\{-1,1\}^m$ vector having exactly k ones. And we will say that $weight(v)$ denotes the number of ones in a vector v from $\{0,1\}^m$ or $\{-1,0,1\}^m$. For approval votes we adopt the obvious approach of using the Hamming distance, as proposed by Brams, Kilgour and Sanver [7]. The distance between a vote $v \in \{0,1\}^m$ and a committee $w \in \{0,1\}^m$ is defined through $HD(w,v) = \sum_{1 \leq i \leq m} |w(i) - v(i)|$.

In the case of trichotomous votes, we slightly adapt the Hamming distance, such that a complete disagreement between vote and committee adds two points, and an abstention in the vote and an approval or disapproval in the committee adds only one point. The distance between a vote $v \in \{-1,0,1\}^m$ and a committee $w \in \{-1,1\}^m$ is defined through $\delta(w,v) = \sum_{1 \leq i \leq m} |v(i) - w(i)|$. This distance can be similarly defined for two vectors $v, w \in \{-1,0,1\}^m$.

For complete linear orders, Baumeister and Dennisen [2] use the sum of the ranks of the committee members in a vote to measure the dissatisfaction. This goes back to the Wilcoxon rank-sum test [35]. The normalized ranksum between a vote v and a committee $K \subseteq C$ is $RS(K,v) = \sum_{c \in K} pos(c,v) - k(k+1)/2$, where $pos(c,v)$ denotes the position of candidate c in vote v.

The last type of votes are incomplete linear orders, for which we will use the modified Kemeny distance, as suggested by Baumeister and Dennisen [2]. The distance between a vote v and a committee $K \subseteq C$, is defined as $Dist(K,v) = \sum_{a,b \in C} d_{K,v}(a,b)$, where the distance between two candidates a and b regarding a committee K and a vote v is defined through

$$d_{K,v}(a,b) = \begin{cases} 1 & \text{if } (a \in K, b \notin K \wedge a \sim_v b) \text{ or } (a \notin K, b \in K \wedge a \sim_v b), \\ 2 & \text{if } (a \in K, b \notin K \wedge b \succ_v a) \text{ or } (a \notin K, b \in K \wedge a \succ_v b), \\ 0 & \text{otherwise.} \end{cases}$$

Note that for an increase of dissatisfaction in the definition of $d_{K,v}(a,b)$ it is important to require that one of the candidates a or b is not in the committee. Otherwise the dissatisfaction would already increase if for two candidates from the committee (or outside the committee) the vote specifies an order over them, which may result in different winning committees. Transferring the minisum and minimax approach, that will be defined formally in Sect. 3, to trichotomous votes leads to *minisum-CAV* and *minimax-CAV*, for complete linear orders to *minisum-ranksum* and *minimax-ranksum*, and to *minisum-Kemeny* and *minimax-Kemeny* for incomplete linear orders. In the following sections we will study winner determination and manipulation for these voting rules from a computational point of view. We assume that the reader is familiar with the basic concepts of computational complexity, such as many-one reducibility and the complexity classes P and NP; details and definitions can be found in the textbook by Papadimitriou [30]. Intuitively, the tractable problems are those, which can be solved in polynomial time, whereas problems which are NP-complete can be seen as intractable. Table 1 summarizes the above introduced voting rules and the results concerning winner determination that will be obtained in the following section.

Table 1. Winner determination in minisum and minimax voting rules for different forms of votes.

Votes	Voting rule	Measure	Minisum winner	Minimax winner
Approval votes	Minisum/minimax-approval	Hamming distance (HD)	In P see [8]	NP-hard see [23]
Trichotomous votes	Minisum/minimax-CAV	Mod. Hamming distance (δ)	In P trivial	NP-hard see Theorem 4
Complete linear orders	Minisum/minimax-ranksum	Ranksum (RS)	In P see Theorem 2	
Incomplete linear orders	Minisum/minimax-Kemeny	Mod. Kemeny distance ($Dist$)	In P see Theorem 3	NP-hard see Theorem 7

3   Winner Determination

In a single winner election, the winner is a single winning candidate, whereas in a committee election, the winner is one committee (or several committees) consisting of a fixed number of candidates. Let $F_k(C)$ denote all committees of size k from the set C, where the representation (i.e., approval vote, trichotomous vote, or a set of candidates) will be clear from the context. The number of possible winning committees $|F_k(C)|$ is exponential in the number of participating candidates. Since committee elections may have a huge number of voters and/or candidates, it is desirable that the winning committees may still be determined in a reasonable amount of time. In the following we will present the minisum and minimax approach that can be combined with the above defined measures of disagreement to define committee election rules. These rules will not always return a single winning committee, hence some tie-breaking rule has to be used in order to obtain a single winning committee.

In a minisum rule, the sum of the disagreement values for all individual votes regarding the winning committee is to be minimized. Formally, the set of winning committees in the minisum rule is $\arg\min_{K \in F_k(C)} \sum_{v \in V} \triangle(K, v)$, where $\triangle \in \{HD, \delta, RS, Dist\}$ for approval votes, trichotomous votes, complete linear orders, or incomplete linear orders. In contrast in a minimax rule, the maximum disagreement value for an individual vote is to be minimized. Hence, the set of winning committees in the minimax rule is $\arg\min_{K \in F_k(C)} \max_{v \in V} \triangle(K, v)$, where $\triangle \in \{HD, \delta, RS, Dist\}$ for approval votes, trichotomous votes, complete linear orders, or incomplete linear orders. We will make use of the following theorem from Baumeister and Dennisen [2], which shows that for the case of complete linear orders, the winning committees in a minisum/minimax-Kemeny election and in a minisum/minimax-ranksum election are always equal.

Theorem 1 *(Baumeister and Dennisen [2]). The set of winning committees in a minisum/minimax-Kemeny election with complete linear orders and in the corresponding minisum/minimax-ranksum election are always equal, i.e.,*

$$\arg\min_{K \in F_k(C)} \sum_{v \in V} Dist(K, v) = \arg\min_{K \in F_k(C)} \sum_{v \in V} RS(K, v), \ and$$

$$\arg\min_{K \in F_k(C)} \max_{v \in V} Dist(K, v) = \arg\min_{K \in F_k(C)} \max_{v \in V} RS(K, v).$$

We want to study the complexity of determining a winning committee independently of any tie-breaking rule. Since the number of winners may be exponential, determining all winning committees is obviously not possible in polynomial time, hence we focus on the question whether it is possible to obtain one winning committee. We will see that in the case of minisum elections, this is always possible in polynomial time. For the minimax elections, we show NP-hardness of the corresponding decision problem for trichotomous votes and incomplete linear orders, and present an approximation algorithm for the case of trichotomous votes.

3.1   Minisum Elections

In minisum-approval the total dissatisfaction is minimum if the committee contains k candidates having the highest number of approvals in all votes. Hence, as shown by Brams, Kilgour, and Sanver [8], it is possible to determine a winner in polynomial time. Similar to minisum-approval in minisum-CAV, the total dissatisfaction is minimum if the committee consists of k candidates that have the highest combined approval scores, i.e., the sum of the corresponding values from the votes. Again a winner can obviously be determined in polynomial time.

Theorem 2. *A minisum-ranksum winner can be determined in polynomial time.*

This holds, since a minisum-ranksum winner consists of k candidates with the highest Borda score[1]. The detailed proof is omitted due to space.

---

[1] In an m-candidate Borda election (see [5]) each candidate gets points according to her position in the votes, where a first position gives $m - 1$ points, a second $m - 2$, and so on.

The corresponding single-winner problem for Kemeny elections, i.e., whether there exists a linear order for which the sum of the distances to the votes does not exceed a given bound, was shown to be NP-hard by Bartholdi et al. [20]. In contrast, we show that a winning committee for minisum-Kemeny can be determined in polynomial time, by reducing the determination of a winner committee for an election with incomplete linear orders to an election in which all votes are complete linear orders.

Lemma 1. *For each minisum-Kemeny election (C, V, k) with incomplete linear orders, there is a minisum-Kemeny election (C, V', k) with complete linear orders, so that $\sum_{v \in V'} Dist(K, v) = 2 \sum_{v \in V} Dist(K, v)$ holds for all committees $K \in F_k(C)$ and V' can be constructed in polynomial time.*

Proof. Given a committee election (C, V, k), we will construct a set of voters V' with the following properties: For each vote $v \in V$ there will be two votes v'_1 and v'_2 in V', such that $\forall c, d \in C$

1. if $c \succ_v d$, then $c >_{v'_1} d$ and $c >_{v'_2} d$,
2. if $c \sim_v d$, either $c >_{v'_1} d$ and $d >_{v'_2} c$, or $c >_{v'_2} d$ and $d >_{v'_1} c$ holds.

In general, for a committee K and a voter list U, we have: $\sum_{v \in U} Dist(K, v) = \sum_{c \in K} \sum_{d \notin K} (|\{v \in U | c \sim_v d\}| + 2 \cdot |\{v \in U | d \succ_v c\}|)$. With the above defined properties, it holds $|\{v \in V' | d > c \text{ in } v\}| = |\{v \in V | c \sim_v d\}| + 2 \cdot |\{v \in V | d \succ_v c\}|$ for all $c, d \in C$, and thus

$$\sum_{v \in V'} Dist(K, v) = \sum_{c \in K} \sum_{d \notin K} (|\{v \in V' | c \sim_v d\}| + 2 \cdot |\{v \in V' | d >_v c\}|)$$

$$= 2 \cdot \sum_{c \in K} \sum_{d \notin K} (|\{v \in V | c \sim_v d\}| + 2 \cdot |\{v \in V | d \succ_v c\}|) = 2 \cdot \sum_{v \in V} Dist(K, v).$$

We now describe how V' can be constructed: For each vote $w \in V$ create two directed graphs $G_1 = (V_1, E_1)$ and $G_2 = (V_2, E_2)$.

1. G_1 and G_2 are constructed as follows: There is a node for each candidate $c \in C$, and an edge (c, d) if $c \succ_w d$ holds.
2. Find two nodes $u, v \in V_1$, so that neither $(u, v) \in E_1$ nor $(v, u) \in E_1$ holds. Add (u, v) to E_1. If there is no such pair, go to step 7.
3. Build the transitive closure of G_1.
4. Add for each edge (u, v) which was added in steps 2 and 3 to E_1, the edge (v, u) to E_2.
5. Build the transitive closure of G_2.
6. If in step 5 new edges were added to E_2, add for each edge (u, v) which was added to E_2, the edge (v, u) to E_1 and go to step 3. Otherwise, go to step 2.
7. Determine the complete orders on basis of the indegree of the nodes of G_1 and G_2 and halt. If a node in G_i has indegree j, the corresponding candidate is at position $(j + 1)$ in v'_i.

According to the definition of incomplete linear orders, the original graph in step 1 is acyclic and transitively closed. If we add an edge (u, v) to E_1 in step 2, this cannot create a cycle since there was no path from v to u: If there had been a path from v to u, the edge (v, u) would have been in E_1 because of the transitive closure. This holds analogously for adding an edge (v, u) to E_2 in step 4. If we compute the transitive closure, no cycle can be created if the original graph was acyclic.

If we "mirror" an edge (u, v) from G_1 to G_2, it holds that neither edge (u, v) nor (v, u) existed in G_2 before. Analogously, it holds that neither edge (u, v) nor edge (v, u) existed in G_1 if we "mirror" an edge (u, v) from G_2 to G_1.

A directed graph $G = (V', E')$, which contains for each node pair $u, v \in V'$ either edge (u, v) or edge (v, u), contains exactly $((|V'| - 1) \cdot |V'|) \cdot \frac{1}{2}$ edges. The algorithm runs until both graphs contain $((|V_1| - 1) \cdot |V_1|) \cdot \frac{1}{2} = ((|V_2| - 1) \cdot |V_2|) \cdot \frac{1}{2}$ edges. Step 2 guarantees that this edge number is attained in G_1. As for each edge (u, v) which is added to G_1 in step 2, edge (v, u) is added in step 4 to G_2, this edge number is attained in G_2, too. So, the algorithm terminates, and the properties stated at the beginning of this proof hold.

The initialization of the graphs and the translation of the resulting graphs into two complete linear orders v'_1, v'_2 for a vote $v \in V$ is possible in polynomial time. The transitive closure of a graph G can be computed in $O(|V'|^3)$. For a vote v, the transitive closure has be to computed maximally as often as edges are added to G_1 or G_2. The number of edges is smaller than $|V_1|^2$, i.e., the runtime for the computation of transitive closures for a vote v is in $O(|V_1|^3 \cdot |V_1|^2) = O(m^5)$. So, V' can be constructed in time $O(m^5 \cdot n)$. This completes the proof. $\qquad\square$

Now, we are ready to state our result for winner determination in minisum-Kemeny elections. The proof follows directly from Lemma 1, Theorems 2 and 1.

Theorem 3. *A minisum-Kemeny winning committee can be determined in polynomial time.*

3.2 Minimax Elections

In contrast to the minisum elections, in minimax-approval, minimax-CAV, and minimax Kemeny elections, it is not always possible to determine a winning committee in polynomial time, unless P = NP. This holds, since we show NP-hardness of the following decision problem.

$\triangle$-MINIMAX-SCORE

Given: A committee election (C, V, k) and a positive integer d.
Question: Is there a committee $K \in F_k(C)$ with $\max_{v \in V} \triangle(K, v) \leq d$?

---

where $\triangle \in \{HD, \delta, RS, Dist\}$ for approval votes, trichotomous votes, complete or incomplete linear orders. Since there is a natural upper bound for $\triangle(K, v)$

for every $\triangle \in \{HD, \delta, RS, Dist\}$, it is possible to compute a winning committee in polynomial time, if this decision problem is solvable in polynomial time. LeGrand [23] showed that HD-MINIMAX-SCORE is NP-complete. However, LeGrand et al. [24] showed that the search variant of this problem, where we actually seek a winning committee, can be approximated with a factor of 3, and Caragiannis et al. [12] proposed an LP-based algorithm for this problem where the distance to the optimal solution is at most 2. Recently, Byrka and Sornat [10] provided a PTAS for minimax approval voting.

The above defined score problem for trichotomous votes is also NP-complete.

Theorem 4. δ-MINIMAX-SCORE *is* NP-*complete.*

Proof. Membership in NP is obvious, and to see that δ-MINIMAX-SCORE is NP-hard, it suffices to transform every approval vote v into a trichotomous vote v' by replacing every 0 in v by a -1 in v'. To obtain a committee K' for trichotomous votes, we need to replace every 0 by a -1 in the resulting approval committee K. Then we have $HD(K, v) \le d \Leftrightarrow \delta(K', v') \le 2d$.   □

Similarly to the approximation algorithm for the search variant of HD-MINIMAX-SCORE by LeGrand et al. [24], one can give an approximation algorithm with a factor 3 for the search variant of δ-MINIMAX-SCORE.

The search variant of will be denoted by MINIMAX-CAV, where the input is also a committee election (C, V, k) with votes represented as $\{-1, 0, 1\}^m$ vectors, and the aim is to find a committee $K^* \in \arg \min_{K \in F_k(C)} \max_{u \in V} \delta(K, u)$.

Theorem 5. *There is an approximation algorithm which finds a solution K' in polynomial time, so that for each optimal solution K of the search problem* MINIMAX-CAV *it holds $\delta(K', u) \le 3 \cdot Max\delta(V, K)$ for all votes $u \in V$, where $Max\delta(V, K) = \max_{u \in V} \delta(K, u)$ denotes the maximum distance between K and the votes in V.*

Proof. Denote for a vector $v \in \{-1, 0, 1\}^m$ by $z(v)$ the number of zeros in v, and by $n(v)$ the number of -1 in v. A k-completion of v is a vector v' in $\{-1, 1\}^m$ constructed as follows.

1. If $weight(v) \le k$ and $z(v) + weight(v) < k$, transform all zeros into ones and transform -1 into ones until $weight(v') = k$.
2. If $weight(v) \le k$ and $z(v) + weight(v) \ge k$, transform zeros into ones until $weight(v') = k$, and transform the remaining zeros into -1.
3. If $weight(v) > k$, transform ones into -1 until $weight(v') = k$, and transform all zeros into -1.

Obviously, each k-completion v' for v contains exactly k ones. We will first show, that for an optimal solution K, an initial vector v and a k-completion K' of v, it holds $\delta(K', v) \le \delta(K, v)$. Fix an optimal solution K, an initial vector v and a k-completion K' of v. Consider the possibilities for the i-th position in the vectors, where we can have a -1, a 0, or a 1. For the first case ($weight(v) \le k$

Table 2. Possibilities for the first case

v	1	1	0	0	- 1	- 1	-1	-1
K'	1	1	1	1	1	1	-1	-1
K	1	- 1	1	-1	1	-1	1	-1
	k_1	k_2	k_3	k_4	k_5	k_6	k_7	k_8

and $z(v) + weight(v) < k$), we have the following 8 possibilities. For $1 \leq i \leq 8$, k_i denotes the number of occurrences of the corresponding possibility (Table 2).

We have $weight(K') = k_1 + k_2 + k_3 + k_4 + k_5 + k_6 = k$, and $weight(K) = k_1 + k_3 + k_5 + k_7 = k$. Hence $k_2 + k_4 + k_6 = k_7$, and we have $\delta(K, v) = 2k_2 + k_3 + k_4 + 2k_5 + 2k_7$, and $\delta(K', v) = k_3 + k_4 + 2k_5 + 2k_6$. Now it is easy to verify, that $\delta(K, v) \geq \delta(K', v)$. The other two cases can be handled analogously.

The approximation algorithm APPROX-MINIMAX-CAV proceeds as follows:

1. Select a vote $v \in V$ arbitrarily.
2. Compute a k-completion v' of v.
3. Return $K' = v'$ as solution.

Obviously, this algorithm runs in polynomial time. To estimate the approximation rate of the algorithm, consider the vote $v \in V$ which was chosen, the k-completion K' of v which is returned as solution, and an optimal solution K for MINIMAX-CAV, i.e., a vector K in $\{-1, 1\}^m$ containing exactly k ones so that $Max\delta(V, K)$ is minimum for all vectors in $\{-1, 1\}^m$ with weight k.

We need to show that $\delta(K', u) \leq 3 \cdot Max\delta(V, K)$ holds for all $u \in V$. Since the triangle inequality applies to δ, we have for all $u \in V$: $\delta(K', u) \leq \delta(K', v) + \delta(v, u)$. Repeated application of the triangle inequality leads to:

$$\delta(K', u) \leq \delta(K', v) + \delta(K, v) + \delta(K, u). \tag{1}$$

Since K is an optimal solution, we have $\delta(K, u) \leq Max\delta(V, K)$ for all $u \in V$. Similarly, we have $\delta(K, v) \leq Max\delta(V, K)$. Since K' is a k-completion of v, it also holds that $\delta(K', v) \leq \delta(K, v)$. The three terms on the right hand of inequality (1) are $\leq Max\delta(V, K)$, and we get the desired property: $\delta(K', u) \leq 3 \cdot Max\delta(V, K)$ for all $u \in V$. □

We now turn to the study of the *Dist*-MINIMAX-SCORE for incomplete linear votes. To show that this problem is also NP-hard, we first need to show NP-hardness of RESTRICTED-HD-MINIMAX-SCORE, which corresponds to HD-MINIMAX-SCORE for the case where the number m of candidates is even and the size of the committee is exactly $m/2$.

Theorem 6. RESTRICTED-HD-MINIMAX-SCORE *is* NP-*complete*.

Proof. The NP-hardness of RESTRICTED-HD-MINIMAX-SCORE can be shown via a reduction from HD-MINIMAX-SCORE. Consider a HD-Minimax-Score instance (C, V, k, d), and construct a RESTRICTED-HD-MINIMAX-SCORE instance (C', V', k', d). The set of candidates is $C' = C \cup D$ with the set of dummy candidates $D = \{d_1, ..., d_m\}$, and the list $V' = \{v'_1, ..., v'_n\}$ of votes is constructed as follows: v'_i has the form (v_i, w) where w is a vector that has a 1 at the first $m - k$ positions, and a 0 on the remaining k positions. Denote the set of the $m - k$ candidates from D who receive a 1 in all votes by W. Let the size of the committee be $k' = m$. Now, we show that $(C, V, k, d) \in HD\text{-MINIMAX-SCORE} \Leftrightarrow (C', V', k', d) \in \text{RESTRICTED-}HD\text{-MINIMAX-SCORE}$.

($\Rightarrow$) Assume that (C, V, k, d) is a yes-instance for HD-MINIMAX-SCORE, i.e., there is a committee $K \in F_k(C)$ so that the Hamming distance to all votes in V is at most d. Then, there also exists a committee $K' \in F_{k'}(C') = F_m(C')$ so that the Hamming distance to all votes in V' is at most d, namely the committee $K \cup W$, since for all $i \in \{1, ..., n\}$, it holds that $HD(K', v'_i) = HD(K \cup W, v'_i) = HD(K, v_i) + HD(W, w) = HD(K, v_i) \leq d$.

($\Leftarrow$) Assume that (C', V', k', d) is a yes-instance for RESTRICTED-HD-MINI-MAX-SCORE. Then, there is a committee $K^* \in F_{k'}(C') = F_m(C')$, so that the Hamming distance to all votes in V' is at most d. We will now consider all possible choices of K^* and show that there is always a winning committee with exactly k candidates from C and $m - k$ candidates from D.

Case 1: $K^* = K \cup W$ for a committee $K \in F_k(C)$. If K^* has the form $K \cup W$ for a committee $K \in F_k(C)$, we have with $K = K^* - D$ a committee in $F_k(C)$ so that the Hamming distance to all votes in V is at most d: for all $i \in \{1, ..., n\}$, it holds that $HD(K, v_i) = HD(K^*, v'_i) \leq d$.

Case 2: $K^* \neq K \cup W$ for a committee $K \in F_k(C)$. If K^* does not have the form $K \cup W$ for a committee $K \in F_k(C)$, we can transform K^* into a committee $K' = K \cup W$ so that $K \in F_k(C)$, as follows. If a candidate in W is not elected, "shift" in the vector representation of the committee, if possible, a 1 from a candidate in $D - W$ to this candidate in W. This action reduces the Hamming distance of the committee regarding all votes. Since in all votes the candidates in W are accepted and the candidates in $D - W$ are rejected, the Hamming distance regarding D decreases by 2 via such a shift. Suppose that q such shifts are required and let K^+ denote the resulting committee. Then it holds that $HD(K^+, v'_i) = HD(K^*, v'_i) - 2q \leq HD(K^*, v'_i)$.

Case 2.1: In K^+ more than $m - k$ candidates from D are elected. If in K^+ more than $m - k$ candidates from D are elected, "shift" the surplus ones from candidates in $D - W$ to the candidates in C. Overall, exactly m candidates are elected, i.e., if exactly $m - k$ candidates from D are elected, exactly k candidates from C are elected. The resulting committee K' is a committee of the form $K \cup W$ with $K \in F_k(C)$. If we shift a 1 from $D - W$ to C, the Hamming distance regarding the candidates in D decreases by value 1 since in all votes the candidates in $D - W$ are rejected. The Hamming distance regarding the candidates in C increases at most by 1 in a shift. So, the Hamming distance either decreases or stays the same.

Case 2.2: In K^+ less than $m - k$ candidates from D are elected. We can analogously "shift" the missing ones from C to the candidates in W. With $K = K' - D$, we have a committee from $F_k(C)$, where the Hamming

distance regarding all votes in V is at most d. So, we have for all $i \in \{1, ..., n\}$
$$HD(K, v_i) = HD(K' - D, v_i) = HD(K', v_i) \leq HD(K^+, v_i') \leq HD(K^*, v_i') \leq d.$$
□

With the NP-hardness of RESTRICTED-HD-MINIMAX-SCORE, we can now show NP-hardness of $Dist$-MINIMAX-SCORE for incomplete linear orders.

Theorem 7. $Dist$-MINIMAX-SCORE *is NP-complete.*

Proof. The NP-hardness of $Dist$-MINIMAX-SCORE can be shown via a reduction from RESTRICTED-HD-MINIMAX-SCORE. Consider a RESTRICTED-HD-MINIMAX-SCORE instance (C, V, k, d) and construct a $Dist$-MINIMAX-SCORE instance (C, V', k, d'), The set of votes $V' = \{v_1', ..., v_n'\}$ is constructed as follows: Let W_i denote the set of candidates who received a 1 in v_i, and L_i the set of candidates who received a 0 in v_i. The candidates from W_i and L_i are ordered in v_i' so that $w \succ_{v_i'} l$ holds for all $w \in W_i$ and all $l \in L_i$. The bound on the distance is $d' = d \cdot k$. We now show that due to the property $k = \frac{m}{2}$, $Dist(v_i', K) = HD(K, v_i) \cdot k$ holds for the given construction for all $i \in \{1, ..., n\}$ and for all committees $K \in F_k(C)$. For a vote $v_i' \in V'$ and a committee K define the following sets of candidates:

- $K_i^W = K \cap W_i$,
- $K_i^L = K \cap L_i$,
- $\bar{K}_i^W = (C \setminus K) \cap W_i$, and
- $\bar{K}_i^L = (C \setminus K) \cap L_i$.

The computation of the modified Kemeny distance can then be carried out based on the quantities $|K_i^W|$, $|K_i^L|$, $|\bar{K}_i^W|$, and $|\bar{K}_i^L|$. It suffices to compare all candidates in the committee with the candidates outside of the committee, i.e., we only need to consider the candidate pairs (a, b) with $a \in K$ and $b \in (C \setminus K)$. We have:

- For each pair of candidates $(a, b) \in K_i^W \times \bar{K}_i^W$, the relation between a and b is unknown in v_i, since both of them are contained in W_i. So, we have a distance of 1.
- For each pair of candidates $(a, b) \in K_i^W \times \bar{K}_i^L$, it holds also $a \succ_{v_i} b$, since a is contained in W_i and b in L_i. So, we have a distance of 0.
- For each pair of candidates $(a, b) \in K_i^L \times \bar{K}_i^W$, it holds $b \succ_{v_i} a$ since a is contained in L_i and b in W_i. So, we have a distance of 2.
- For each pair of candidates $(a, b) \in K_i^L \times \bar{K}_i^L$, the relation between a and b is unknown in v_i, since both of them are contained in L_i. So, we have a distance of 1.

Thus, we have:

$$Dist(v_i', K) = |K_i^W| \cdot |\bar{K}_i^W| + 2 \cdot |K_i^L| \cdot |\bar{K}_i^W| + |K_i^L| \cdot |\bar{K}_i^L| \qquad (2)$$

If we determine the Hamming distance between a committee K and a vote v_i, we count the positions where a candidate who is elected in the committee is not accepted in v_i and the positions where a candidate not elected in the committee is accepted in v_i, i.e., we have:

$$HD(K, v_i) = |K_i^L| + |\bar{K}_i^W| \tag{3}$$

For all $i \in \{1, ..., n\}$ and for all committees $K \in F_k(C)$, we have:

$$\begin{aligned}
Dist(v_i', K) &\overset{(2)}{=} |K_i^W| \cdot |\bar{K}_i^W| + 2 \cdot |K_i^L| \cdot |\bar{K}_i^W| + |K_i^L| \cdot |\bar{K}_i^L| \\
&= |\bar{K}_i^W| \cdot (|K_i^W| + |K_i^L|) + |K_i^L| \cdot (|\bar{K}_i^W| + |\bar{K}_i^L|) \\
&= |\bar{K}_i^W| \cdot k + |K_i^L| \cdot (m - k) = (|\bar{K}_i^W| + |K_i^L|) \cdot k + |K_i^L| \cdot (m - 2k) \\
&\overset{(3)}{=} HD(K, v_i) \cdot k + |K_i^L| \cdot (m - 2k).
\end{aligned}$$

With $k = \frac{m}{2}$ it holds for all $i \in \{1, ..., n\}$ and for all committees $K \in F_k(C)$ $Dist(v_i', K) = HD(K, v_i) \cdot k + |K_i^L| \cdot 0 = HD(K, v_i) \cdot k$. Hence, we have that there is a committee for which the Hamming distance to all votes in V is at most d, if and only if there is a committee for which the modified Kemeny distance to all votes in V' is at most dk. This completes the proof. □

The complexity of the corresponding problem for complete linear orders, RS-MINIMAX-SCORE, remains open.

4   Manipulation

The famous Gibbard and Satterthwaite Theorem [18,33] says that, in principle, every preference-based voting rule is manipulable. Bartholdi et al. [1] introduced a decision problem that captures manipulation in elections. They ask whether for a given election and some distinguished candidate, there is a vote of the manipulator that makes this candidate win. Based on manipulation for single winner elections, Meir et al. [26] propose manipulation problems for committee elections. Their definition includes a utility function and its most general form is as follows.

$\mathcal{E}$-UTILITY-COMMITTEE-MANIPULATION

Given: A committee election (C, V, k) with honest voters, a utility function $u : C \to \mathbb{Z}$, and an integer t.

Question: Does there exist a vote s over C, such that in the resulting election with additional vote s under voting rule $\mathcal{E}$ it holds that $\sum_{c \in K} u(c) \geq t$, where K is the winning committee with $|K| = k$?

They consider an adversarial tie-breaking (see [14]), where from several equally performing candidates those with the lower utility for the manipulator win the election. Procaccia et al. [31] state that UTILITY-COMMITTEE-MANIPULATION is in P for committee elections under Approval voting. The proof is given by Meir et al. [26], who also show that if the utility function is mapping to $\{0, 1\}$ rather than $\mathbb{Z}$, UTILITY-COMMITTEE-MANIPULATION is in P for all committee elections held under scoring rules. Since in a minisum-ranksum election the winning committee contains the k candidates with the highest Borda scores (see the remark after Theorem 2), the fact that UTILITY-COMMITTEE-MANIPULATION for minisum-ranksum is in P follows immediately.

Obraztsova et al. [29] follow the approach of Meir et al. [26] by defining manipulation in committee elections through a utility function of the manipulator. They study the complexity of manipulation in committee elections, with a particular focus on the role of tie-breaking. They focus on the case where ties are broken according to a fixed predefined order or by a natural randomized rule.

We will also follow the approach of Meir et al. [26], but here we consider non-unique winners and therefore have to change the definition by asking whether the condition holds for at least one winning committee. Furthermore, we will focus only on the case where the utility function is boolean-valued, and state two special variants of it, that we find most natural. In the first variant, we simply ask whether it is possible for the manipulator to vote such that a given subset of the candidates is in at least one winning committee. In analogy to the manipulation problems in single winner elections, we call this problem $\mathcal{E}$-COMMITTEE-MANIPULATION.

$\mathcal{E}$-COMMITTEE-MANIPULATION (CM)

Given: A committee election (C, V, k) with honest voters and a distinguished set of candidates $L \subseteq C$ with $|L| \leq k$.

Question: Does there exist a vote s over C, such thats $L \subseteq K$ holds for a winning committee $K \in F_k(C)$ in the resulting committee election with additional vote s held under the rule $\mathcal{E}$?

While CM asks whether all candidates in L can become part of the winning committee, one can also ask whether it is possible to make at least t candidates in L part of a winning committee.

$\mathcal{E}$-THRESHOLD-COMMITTEE-MANIPULATION (TCM)

Given: A committee election (C, V, k) with honest voters, a distinguished set of candidates $L \subseteq C$ with $|L| \leq m$, and a non-negative integer $t \leq k$.

Question: Does there exist a vote s over C, such that at least t candidates in L belong to a winning committee $K \in F_k(C)$ in the resulting committee election with additional vote s held under the rule $\mathcal{E}$?

Note that in the case of a committee election rule that always returns a single winner, CM is a special case of UTILITY-COMMITTEE-MANIPULATION with $t = |L|$ and in which the utility function maps all candidates in L to the value 1 and all candidates in $C \setminus L$ to 0. Furthermore, CM is the special case of TCM with $t = |L| \leq k$

Even though we consider a different model regarding tie-breaking, the above mentioned results by Meir et al. [26] can be adapted to CM and TCM.

Theorem 8. CM *and* TCM *for minisum-approval are in* P.

Following Meir et al. [26] it holds that if the manipulators' strategy to approve all candidates in L and to disapprove all candidates in $C \setminus L$ does not succeed, no other does. Thus, both CM and TCM for minisum-approval can be decided in polynomial time. A similar result holds for minisum-CAV.

Theorem 9. CM *and* TCM *for minisum-CAV are in* P.

Similar to minisum-approval we have that if the strategy to give all candidates in L a 1 and all candidates in $C \setminus L$ a -1 does not succeed, no other one does. In the case of complete linear orders, we show, that both manipulation problems can also be solved efficiently.

Theorem 10. CM *and* TCM *for minisum-ranksum are in* P.

Proof. As CM is the special case of TCM where t equals the number of candidates in L, it suffices to give the proof for TCM. Similar to the proofs by Meir et al. [26] and Bartholdi et al. [20] (see also Obraztsova et al. [28]) we show how to decide in polynomial time whether it is possible that at least t members in L belong to a winning committee. We denote the number of candidates in L by l. We define four lists in which the candidates are given in ascending order regarding their Borda scores. List $T = (t_1, \ldots, t_t)$ contains the t candidates in L with the highest Borda scores, list $F = (f_1, \ldots, f_{l-t})$ all other candidates in L, list $G = (g_1, \ldots, g_{\min(k-t,m-l)})$ the $\min(k - t, m - l)$ candidates in $C \setminus L$ with the highest Borda scores and $\hat{C} = (\hat{c}_1, \ldots, \hat{c}_{m-l-\min(k-t,m-l)})$ the remaining candidates. The manipulator's vote is given by $t_1 > \cdots > t_t > f_1 > \cdots > f_{l-t} > g_1 > \cdots > g_{\min(k-t,m-l)} > \hat{c}_1 > \hat{c}_{m-l-\min(k-t,m-l)}$. If at least t members in L are part of a winning committee in the election $(C, (v_1, \ldots, v_n, s), k)$ the manipulation was successful, otherwise it is not possible.   □

The next theorem shows that for the case of incomplete linear orders, CM and TCM are solvable in polynomial time as well.

Theorem 11. CM *and* TCM *for minisum-Kemeny are in* P.

Proof. It again suffices to give the proof for TCM. Given a committee election (C, V, k) with $V = (v_1, \ldots, v_n)$, we construct a set of voters $V' = (v'_1, \ldots, v'_{2n})$ according to Lemma 1.

We introduce a problem 2-TCM which equals TCM with the exception that we ask for the existence of a single vote s over C, such that at least t candidates in L belong to a winning committee K in the committee election $(C, (v'_1, \ldots, v'_n, s, s), k)$. As V' is a list of complete linear orders and according to Theorem 1, the winning committees in the corresponding minisum-Kemeny election with complete linear orders equal the winning committees in a minisum-ranksum-election. The proof that 2-TCM for minisum-ranksum is in P is analogous to the proof of Theorem 10[2]. We now show that it is possible to manipulate (C, V', k) with two manipulators with identical votes if and only if it is possible

---

[2] Note, that this problem does not equal coalitional manipulation which is proved to be NP-hard even for two manipulators and three voters by Betzler et al. [3].

to manipulate (C, V, k) with a single manipulator. It is possible to manipulate the election (C, V, k) according to TCM, i. e. there exists an s such that $L \subseteq K^*$ for a

$$K^* \in \underset{K \in F_k(C)}{\arg\min} \left(\sum_{v \in V} Dist(K, v) + Dist(K, s) \right)$$

$$= \underset{K \in F_k(C)}{\arg\min} \left(2 \left(\sum_{v \in V} Dist(K, v) + Dist(K, s) \right) \right)$$

$$\overset{\text{Lemma 1}}{=} \underset{K \in F_k(C)}{\arg\min} \left(\sum_{v \in V'} Dist(K, v) + 2 \cdot Dist(K, s) \right).$$

Therefore, K^* is a winning committee in $(C, (v_1', \ldots, v_{2n}', s, s), k)$. The other direction can be shown similarly. □

5   Conclusions

We showed that a winning committee under all minisum rules can be determined in polynomial time, whereas the corresponding decision problem for minimax rules is NP-hard for trichotomous votes and incomplete linear orders. In the case of approval votes, NP-hardness was already shown by [23], and the complexity of winner determination for minimax-ranksum remains open. Our analysis focuses on the case where the size of the committee is known in advance. For the case where the size of the committee is not fixed in advance, we can still measure the disagreement of a voter with several committees of different sizes with the Hamming distance in the case of approval or trichotomous votes. Hence for minisum-approval and minisum-CAV it is still possible to determine a winner in polynomial time, since it is enough to compare the disagreement for all possible sizes of the committee. LeGrand [23] argued that the corresponding decision problem for minimax-approval also remains NP-complete. However in the case of complete or incomplete linear orders it is not directly clear how the disagreement of committees of different sizes can be compared with each other. Note that in particular the disagreement of a voter measured by the ranksum or the modified Kemeny distance is always zero for the committee that consists of all candidates and for the empty committee. One task for future research is to compare these rules in terms of their axiomatic properties. Besides winner determination we also studied manipulation in minisum elections, and obtained polynomial-time solvability results in all cases. Another interesting question for future research is the problem of coalitional manipulation, where not a single manipulator, but several voters try to take influence on the outcome of the election.

References

1. Bartholdi III, J., Tovey, C., Trick, M.: The computational difficulty of manipulating an election. Soc. Choice Welfare **6**(3), 227–241 (1989)
2. Baumeister, D., Dennisen, S.: Voter dissatisfaction in committee elections. In: Proceedings of the 14th International Joint Conference on Autonomous Agents and Multiagent Systems, IFAAMAS. Extended abstract (2015)

3. Betzler, N., Niedermeier, R., Woeginger, G.: Unweighted coalitional manipulation under the borda rule is NP-hard. In: Proceedings of the 22nd International Joint Conference on Artificial Intelligence, IJCAI, pp. 55–60 (2011)
4. Betzler, N., Slinko, A., Uhlmann, J.: On the computation of fully proportional representation. J. Artif. Intell. Res. **47**, 475–519 (2013)
5. Borda, J.: Mémoire sur les élections au scrutin. Histoire de L'Académie Royale des Sciences, Paris. English translation appears in [19] (1781)
6. Boutilier, C., Rosenschein, J.: Incomplete information and communication in voting. In: Brandt, F., Conitzer, V., Endriss, U., Lang, J., Procaccia, A. (eds.) Handbook of Computational Social Choice, Chapter 10. Cambridge University Press, Cambridge (2015)
7. Brams, S., Kilgour, D., Sanver, R.: A minimax procedure for negotiating multilateral treaties. In: Wiberg, M. (ed.) Reasoned Choices: Essays in Honor of Hannu Nurmi. Finnish Political Science Association, Turku (2004)
8. Brams, S., Kilgour, D., Sanver, R.: A minimax procedure for electing committees. Public Choice **132**, 401–420 (2007)
9. Brams, S., Kilgour, D., Sanver, R.: A minimax procedure for negotiating multilateral treaties. In: Avenhus, R., Zartmann, I. (eds.) Diplomacy Games: Formal Models and International Negotiations. Springer, Heidelberg (2007)
10. Byrka, J., Sornat, K.: PTAS for minimax approval voting. In: Liu, T.-Y., Qi, Q., Ye, Y. (eds.) WINE 2014. LNCS, vol. 8877, pp. 203–217. Springer, Heidelberg (2014)
11. Caragiannis, I., Hemaspaandra, E., Hemaspaandra, L.: Dodgson's rule and Young's rule. In: Brandt, F., Conitzer, V., Endriss, U., Lang, J., Procaccia, A. (eds.) Handbook of Computational Social Choice, Chapter 5. Cambridge University Press, Cambridge (2015)
12. Caragiannis, I., Kalaitzis, D., Markakis, E.: Approximation algorithms and mechanism design for minimax approval voting. In: Proceedings of the 24th AAAI Conference on Artificial Intelligence, pp. 737–742, AAAI Press (2010)
13. Chamberlin, J., Courant, P.: Representative deliberations and representative decisions: proportional representation and the Borda rule. Am. Polit. Sci. Rev. **77**(3), 718–733 (1983)
14. Conitzer, V., Sandholm, T., Lang, J.: When are elections with few candidates hard to manipulate? J. ACM **54**(3), Article 14 (2007)
15. Conitzer, V., Walsh, T.: Barriers to manipulation in voting. In: Brandt, F., Conitzer, V., Endriss, U., Lang, J., Procaccia, A. (eds.) Handbook of Computational Social Choice, Chapter 6. Cambridge University Press, Cambridge (2015)
16. Faliszewski, P., Rothe, J.: Control and bribery in voting. In: Brandt, F., Conitzer, V., Endriss, U., Lang, J., Procaccia, A. (eds.) Handbook of Computational Social Choice, Chapter 7. Cambridge University Press, Cambridge (2015)
17. Felsenthal, D.: On combining approval with disapproval voting. Behav. Sci. **34**(1), 53–60 (1989)
18. Gibbard, A.: Manipulation of voting schemes: a general result. Econometrica **41**(4), 587–601 (1973)
19. Grazia, A.: Mathematical deviation of an election system. Isis **44**, 41–51 (1953)
20. Bartholdi III, J., Tovey, C., Trick, M.: Voting schemes for which it can be difficult to tell who won the election. Soc. Choice Welf. **6**(2), 157–165 (1989)
21. Kilgour, D., Brams, S., Sanver, R.: How to elect a representative committee using approval balloting. In: Pukelsheim, F., Simeone, B. (eds.) Mathematics and Democracy: Recent Advances in Voting Systems and Collective Choice, pp. 83–95. Springer, Heidelberg (2006)

22. Lang, J., Xia, L.: Voting in combinatorial domains. In: Brandt, F., Conitzer, V., Endriss, U., Lang, J., Procaccia, A. (eds.) Handbook of Computational Social Choice, Chapter 9. Cambridge University Press, Cambridge (2015)
23. LeGrand, R.: Analysis of the minimax procedure. Technical report WUCSE-2004-67, Department of Computer Science and Engineering, Washington University, St. Louis, Missouri, November 2004
24. LeGrand, R., Markakis, E., Mehta, A.: Some results on approximating the minimax solution in approval voting. In: Proceedings of the 6th International Joint Conference on Autonomous Agents and Multiagent Systems, pp. 1193–1195, ACM Press, May 2007
25. Lu, T., Boutilier, C.: Budgeted social choice: from consensus to personalized decision making. In: Proceedings of the 22nd International Joint Conference on Artificial Intelligence, pp. 280–286, AAAI Press (2011)
26. Meir, R., Procaccia, A., Rosenschein, J., Zohar, A.: Complexity of strategic behavior in multi-winner elections. J. Artif. Intell. Res. **33**, 149–178 (2008)
27. Monroe, B.: Fully proportional representation. Am. Polit. Sci. Rev. **89**(4), 925–940 (1995)
28. Obraztsova, S., Elkind, E., Hazon, N.: Ties matter: complexity of voting manipulation revisited. In: Proceedings of the 10th International Joint Conference on Autonomous Agents and Multiagent Systems, pp. 71–78, ACM Press (2011)
29. Obraztsova, S., Zick, Y., Elkind, E.: On manipulation in multiwinner elections based on scoring rules. In: Proceedings of the 12th International Joint Conference on Autonomous Agents and Multiagent Systems, pp. 359–366, IFAAMAS (2013)
30. Papadimitriou, C.: Computational Complexity, 2nd edn. Addison-Wesley, Reading (1995). Reprinted with corrections
31. Procaccia, A., Rosenschein, J., Zohar, A.: Multi-winner elections: complexity of manipulation, control, and winner-determination. In: Proceedings of the 20th International Joint Conference on Artificial Intelligence, pp. 1476–1481, IJCAI (2007)
32. Procaccia, A., Rosenschein, J., Zohar, A.: On the complexity of achieving proportional representation. Soc. Choice Welf. **30**, 353–362 (2008)
33. Satterthwaite, M.: Strategy-proofness and Arrow's conditions: existence and correspondence theorems for voting procedures and social welfare functions. J. Econ. Theor. **10**(2), 187–217 (1975)
34. Skowron, P., Yu, L., Faliszewski, P., Elkind, E.: The complexity of fully proportional representation for single-crossing electorates. In: Vöcking, B. (ed.) SAGT 2013. LNCS, vol. 8146, pp. 1–12. Springer, Heidelberg (2013)
35. Wilcoxon, F.: Individual comparisons by ranking methods. Biom. Bull. **1**(6), 80–83 (1945)
36. Zwicker, W.: Introduction to the theory of voting. In: Brandt, F., Conitzer, V., Endriss, U., Lang, J., Procaccia, A. (eds.) Handbook of Computational Social Choice, Chapter 2. Cambridge University Press, Cambridge (2015)

OWA-Based Extensions
of the Chamberlin–Courant Rule

Edith Elkind[1] and Anisse Ismaili[2]([✉])

[1] Department of Computer Science, University of Oxford, Oxford, UK
elkind@cs.ox.ac.uk
[2] Université Pierre et Marie Curie,
Univ Paris 06, UMR 7606, LIP6, 75005 Paris, France
anisse.ismaili@lip6.fr

Abstract. Given a set of voters V, a set of candidates C, and voters' preferences over the candidates, multiwinner voting rules output a fixed-size subset of candidates (committee). Under the Chamberlin–Courant multiwinner voting rule, one fixes a scoring vector of length $|C|$, and each voter's 'utility' for a given committee is defined to be the score that she assigns to her most preferred candidate in that committee; the goal is then to find a committee that maximizes the joint utility of all voters. The joint utility is typically identified either with the sum of all voters' utilities or with the utility of the least satisfied voter, resulting in, respectively, the *utilitarian* and the *egalitarian* variant of the Chamberlin–Courant's rule. For both of these cases, the problem of computing an optimal committee is NP-hard for general preferences, but becomes polynomial-time solvable if voters' preferences are single-peaked or single-crossing. In this paper, we propose a family of multi-winner voting rules that are based on the concept of *ordered weighted average (OWA)* and smoothly interpolate between the egalitarian and the utilitarian variants of the Chamberlin–Courant rule. We show that under moderate constraints on the weight vector we can recover many of the algorithmic results known for the egalitarian and the utilitarian version of Chamberlin–Courant's rule in this more general setting.

1 Introduction

Local organizers of a conference have to make an important decision: the catering company that was contracted to provide snacks for the coffee break has a menu consisting on 30+ items, but for each coffee break one is allowed to pick at most 6 item types (cookies, pastries, finger sandwiches, etc.). If the organizers have a good estimate of the participants' preferences, they may want to ensure that each participant likes at least one of the items that are served in the coffee break. Note that participants' preferences over items other than their top item (among the ones ordered) are not important here: if Alice likes chocolate cookies, but hates cucumber sandwiches, the organizers can order both, and Alice can simply ignore the sandwiches. However, if the participants' preferences are very diverse, it may be impossible to ensure that everyone's favorite items are

© Springer International Publishing Switzerland 2015
T. Walsh (Ed.): ADT 2015, LNAI 9346, pp. 486–502, 2015.
DOI: 10.1007/978-3-319-23114-3_29

ordered; the organizers may then have to choose between providing 95 % of the attendants with items they rank highly and making alternative arrangements for the remaining picky eaters, or ordering items that noone really likes, but everyone considers acceptable.

A similar issue arises when selecting the board of a large organization, such as, e.g., IFAAMAS (the non-profit foundation that runs the AAMAS conference): it is desirable to ensure that most members of the organization feel that they are represented by the board, but it may be difficult to ensure that this criterion is satisfied for very small minorities.

In their pioneering work, Chamberlin and Courant [10] described a voting procedure for selecting a committee, which is driven by such considerations. Their primary motivation was to provide a method for selecting a representative parliament, but, as illustrated above, this approach has a wider application domain (see, e.g., the recent papers of Elkind et al. [13] and Skowron et al. [25], which discuss a variety of scenarios where the Chamberlin–Courant rule and its variants may be useful). In more detail, Chamberlin and Courant consider the setting where a group of voters has to select a fixed-size committee from the available candidates, and each voter ranks all candidates from best to worst. A voter's satisfaction (utility) from a given committee is determined by the rank of her most preferred candidate in that committee within her preference order, and the goal is to select a committee that maximizes the sum of voters' satisfactions (we provide a formal definition in Sect. 2). A Rawlsian variant of this method was later proposed by Betzler et al. [3]: instead of maximizing the sum of voters' satisfactions, they focus on maximizing the satisfaction of the least happy voter; the subsequent literature refers to the original variant of the Chamberlin–Courant rule as the *utilitarian Chamberlin–Courant rule (U-CC)*, and the Betzler et al.'s variant as the *egalitarian Chamberlin–Courant rule (E-CC)*.

In this paper, we propose a family of committee selection rules that smoothly interpolate between U-CC and E-CC by using the concept of *ordered weighted averages (OWA)*. A rule in this family is determined by a weight vector $w = (w_1, \ldots, w_n)$ where n is the number of voters; the entries of this vector are non-negative reals that sum up to 1. These weights are used when aggregating the voters' satisfaction: the total satisfaction is a weighted sum, where the satisfaction of the least happy voter is taken with weight w_1, the satisfaction of the second least happy voter is taken with weight w_2, etc. This framework captures both U-CC (by setting $w = (\frac{1}{n}, \ldots, \frac{1}{n})$) and E-CC (by setting $w = (1, 0, \ldots, 0)$), and allows us to trade off the utilitarian welfare and egalitarian objectives in a variety of ways. We describe a few such tradeoffs below.

- We can choose to ignore d most unhappy voters (e.g., because we can make alternative provisions for them) and maximize the minimum utility among the remaining voters, by setting the weight vector to be $w = (0, \ldots, 0, 1, 0, \ldots, 0)$, where the only 1 appears in position $d + 1$. We denote this rule by E^{-d}-CC.
- As in the previous case, we can ignore d most unhappy voters and maximize the sum of utilities of the remaining voters; this corresponds to the weight vector $w = (0, \ldots, 0, \frac{1}{n-d}, \ldots, \frac{1}{n-d})$, where 0s appear in the first d positions. We denote this rule by U^{-d}-CC.

– We can take the opposite approach and focus on d least happy voters, aiming to maximize the sum of their utilities. We denote the resulting rule, which corresponds to the weight vector $\boldsymbol{w} = (\frac{1}{d}, \ldots, \frac{1}{d}, 0, \ldots, 0)$, by U^d-CC. Note that U^1-CC is simply E-CC, and U^n-CC is U-CC.
– We can consider a refinement of E-CC, where we first maximize the utility of the least happy voter, then, among all committees that accomplish this, we select the ones that maximize the utility of the second least happy voter, etc.; if voters' utilities take values in the range $[0, K]$, this rule, which we call Lex-E-CC, can be implemented by setting

$$\boldsymbol{w} = \big(\alpha(K + 1)^{n-1}, \alpha(K + 1)^{n-2}, \ldots, \alpha\big),$$

where $\alpha = \frac{K}{(K+1)^n - 1}$ is the normalization factor. The appeal of the resulting rule is that it provides a principled refinement of E-CC, which can easily be seen to be rather indecisive.
– An alternative refinement of E-CC is to break ties based on the utilitarian social welfare, i.e., choose a committee that maximizes the sum of voters' utilities among all committees that maximize the utility of the least satisfied voter. We denote this rule by EU-CC. It is captured by the weight vector

$$\boldsymbol{w} = \left(\frac{nK + 1}{n(K + 1)}, \frac{1}{n(K + 1)}, \ldots, \frac{1}{n(K + 1)} \right);$$

again, we assume that the voters' utilities lie in the range $[0; K]$.

Having defined this family of committee selection rules, we then focus on their computational complexity. Both E-CC and U-CC are known to be computationally hard for general preferences, but admit efficient algorithms for restricted preference domains (see Sect. 1.1 for an overview of the related literature) or if the number of voters/candidates is small. We show that most of these positive results extend to many families of weight vectors, including, in particular, EU-CC, U^{-d}-CC, E^{-d}-CC, and U^d-CC (for some of these results, we also need a mild restriction on the voters' score functions). This means that the additional flexibility obtained by varying the weights does not necessarily imply a substantial penalty in terms of computation time. However, there are families of weight vectors (including, notably, Lex-E-CC) that cannot be captured by our approach. We conclude the paper by discussing the limitations of our techniques and outlining directions for future work.

1.1 Related Work

The committee selection rules considered in this paper build on the original model of Chamberlin and Courant [10], who put forward what we call the utilitarian version of this rule; its egalitarian version was subsequently proposed by Betzler et al. [3]. A modification of the original Chamberlin–Courant rule where each member of the selected committee represents approximately the same number of voters was subsequently suggested by Monroe [21]. Our general framework extends to the Monroe rule; however, this rule appears to be more challenging from the computational perspective [3,27], and therefore we leave the algorithmic analysis of its OWA-based extensions as a topic for future work.

The study of computational aspects of the Chamberlin–Courant rule was initiated by Procaccia et al. [24], who showed that computing U-CC is NP-hard for approval-like utility functions; Lu and Boutilier [20] extend this result to Borda-like utility functions. Betzler et al. [3] demonstrate NP-hardness of E-CC and study fixed-parameter complexity of both E-CC and U-CC; they demonstrate that both of these rules become polynomial-time computable when voters' preferences are single-peaked. Cornaz et al. [11] extend these results to elections with bounded single-peaked width. Skowron et al. [27] show that both E-CC and U-CC remain easy when voters' preferences are single-crossing, or have bounded single-crossing width. Yu et al. [32] analyze the complexity of E-CC and U-CC when voters' preferences are single-peaked on a tree. Lu and Boutilier [20] and Skowron et al. [26] propose approximation algorithms for U-CC and E-CC.

Another extension of the Chamberlin–Courant rule that is based on OWAs was proposed by Skowron et al. [25]. However, their approach is orthogonal to ours: Skowron et al. consider the possibility that the voter cares not just about her most preferred candidate in the committee, but also derives additional utility from her second most preferred committee member, etc. (in our coffee break example, a conference participant may want to try several different snacks). Thus, while we consider weight vectors that have an entry for each *voter*, the weight vectors in the work of Skowron et al. have m entries, where m is the number of *candidates*.

Amanatidis et al. [1] also define a family of voting rules that use OWAs to aggregate voters' preferences; however, in contrast with our work, they consider the setting where voters' preferences are dichotomous. That is, the rules in this family interpolate between Minisum approval voting and Minimax approval voting (see [6,9,19] for the definitions and a discussion of these rules). They prove that most of the rules constructed in this manner are NP-hard, but identify a number of tractable/approximable special cases.

The ordered weighted averaging aggregation operators were introduced in multicriteria decision making by [29] and applied in multiple contexts [23,30,31]. Goldsmith et al. [17] recently proposed using OWAs in the context of voting; however, in contrast with our work, they focus on single-winner settings. OWAs are typically used to accomplish fairness. In the context of multiagent assignment, fairness is formalized as monotonicity with Pareto dominance and Pigou–Dalton transfers [16], and is therefore associated with weight vectors that satisfy $w_i > w_{i+1}$. However, as argued above, in the context of committee elections, weight vectors that do not satisfy these inequalities (such as the ones used by U^{-d}-CC and E^{-d}-CC) may still be useful.

2 Preliminaries

For $a, b \in \mathbb{N}$, let $[\![a, b]\!] = \{u \in \mathbb{N} \mid a \leq u \leq b\}$; $[\![a, b]\!] = \emptyset$ if $b < a$. Similarly, let $]\!]a, b[\![= [\![a, b]\!] \setminus \{a, b\}$. The shorthand for $[\![1, b]\!]$ is $[\![b]\!]$.

An *election* $E = (C, V)$ is defined by a set of *candidates* $C = \{c_1, \ldots, c_m\}$ and a set of *voters* $V = [\![1, n]\!]$. Each voter $i \in V$ is associated with a *preference order* $\succ_i$, which is a linear order over the candidates. Given $c, c' \in C$, we write $c \succ_i c'$ to denote that voter i prefers c to c'. For succinctness, we may sometimes describe the preference order of voter i by listing all candidates from her most preferred one to her least preferred one and suppressing $\succ_i$; for instance abc is an abbreviation for $a \succ_i b \succ_i c$. Let $\text{top}(i) \in C$ denote voter i's most preferred candidate: $\text{top}(i) \succ_i c$ for all $c \in C \setminus \{\text{top}(i)\}$. The *individual score*, or *utility*, of voter i is a function $s_i : C \rightarrow \mathbb{N}$ such that

$$\text{for all } c, c' \in C \quad c \succ_i c' \quad \text{implies} \quad s_i(c) \geq s_i(c').$$

We let $Scr_{C,V} = \{s_i(c) \mid i \in V, c \in C\}$, and set $R_{C,V} = \max_{s \in Scr_{C,V}} s$; we omit (C, V) from the notation when these sets are clear from the context.

We consider the setting where the goal is to select a committee of size κ. Following the ideas of Chamberlin and Courant [10], we assume that the satisfaction that a voter derives from a committee is determined by her score for her most preferred candidate in that committee. Thus, we extend the individual score function to committees by setting

$$s_i(S) = \max_{c \in S}\{s_i(c)\} \quad \text{for each } S \subseteq C \text{ with } |S| = \kappa.$$

Existing work considers two methods of extending individual scores to group scores: the utilitarian score, which was proposed in the original work of Chamberlin and Courant [10], and the egalitarian score, suggested by Betzler et al. [3]. The *utilitarian* and *egalitarian* scores s_U and s_E of a committee $S \subseteq C$, $|S| = \kappa$, are defined by, respectively,

$$s_U(S) = \sum_{i \in V} s_i(S) \quad \text{and} \quad s_E(S) = \min_{i \in V}\{s_i(S)\}.$$

The utilitarian CC rule (U-CC) and the egalitarian CC rule (E-CC) output a κ-sized committee that maximizes the corresponding score (breaking ties arbitrarily).

3   Our Model

We put forward a family of voting rules that aggregate individual scores using *ordered weighted averages*. A rule in this family is associated with a *weight vector* $\boldsymbol{w} = (w_1, \ldots, w_n) \in \mathbb{R}^n$ such that $w_i \geq 0$ for all $i \in V$ and $\sum_{i \in V} w_i = 1$. Given a committee $S \subseteq C$, let σ be a permutation of V that satisfies

$$s_{\sigma(1)}(S) \leq \ldots \leq s_{\sigma(i)}(S) \leq \ldots \leq s_{\sigma(n)}(S),$$

i.e., σ orders the voters according to their score in non-decreasing order. Given an election $E = (C, V)$ and a committee size κ, the rule $\boldsymbol{w}$-CC outputs a committee of size κ with the highest $\boldsymbol{w}$-CC score, which is computed as

$$s_{\boldsymbol{w}}(S) = \sum_{i \in V} w_i s_{\sigma(i)}(S).$$

Note that for $w = (\frac{1}{n}, \ldots, \frac{1}{n})$ we obtain the U-CC rule and for $w = (1, 0, \ldots, 0)$ we obtain the E-CC rule.

Example 1. Suppose that $\kappa = 2$, and consider an election over the set of candidates $C = \{a, b, c, d, e, f\}$ and 6 voters with the following preferences:

$$acebdf: \quad 3 \text{ voters} \qquad bdface: \quad 2 \text{ voters} \qquad efdbca: \quad 1 \text{ voter}.$$

Suppose that each voter uses the Borda score function, i.e., assigns $m - i$ points to the candidate she ranks in position i ($m = 6$ in this example).

It is easy to verify that the only optimal committee with respect to U-CC is $\{a, b\}$, with a total score of 27. Let us now consider E-CC. As no candidate is ranked in top two positions by two different groups of voters, for every committee there is at least one voter who ranks her most preferred committee member in position 3 or lower. On the other hand, there are several committees that ensure a utility of 3 to each voter: some examples are $\{a, f\}$, $\{a, d\}$, $\{e, b\}$, and $\{e, f\}$. E-CC can output any of these committees. In contrast, under EU-CC there is a single optimal committee, namely, $\{a, d\}$. Indeed, $\{a, d\}$ accomplishes a total score of $15 + 8 + 3 = 26$, while providing a utility of at least 3 to each voter. This is also the unique output of Lex-E-CC: there is one voter whose utility is 3, two voters whose utility is 4, and three voters whose utility is 5, and this can easily be seen to be the best possible. The rules E^{-1}-CC and U^{-1}-CC necessarily output $\{a, b\}$, as this is the only committee that ensures the maximum utility to 5 voters; in contrast, E^{-2}-CC and U^{-2}-CC may output either $\{a, b\}$ or $\{a, e\}$, and for E^{-3}-CC and U^{-3}-CC any of the committees $\{a, b\}$, $\{a, c\}$, $\{a, d\}$, $\{a, e\}$, $\{a, f\}$, and $\{b, e\}$ is an acceptable answer.

4   General Observations

We start our algorithmic analysis of w-CC rules by formalizing the respective computational problem and making a few observations about its complexity for general preferences.

To avoid dealing with representation issues for real numbers, from now on we assume that the entries of the weight vector w are non-negative *rational* numbers. For readability, we also make the simplifying assumption that arithmetic operations involving weights and scores can be performed in time $O(1)$; however, this assumption is not essential, and, in particular, all algorithms described in this paper still run in polynomial time if we assume that the running time of arithmetic operations is polynomial in the number of bits in the input.

We observe that, in this model, for any given weight vector w, the w-CC score of a given committee can be computed in polynomial time. We therefore focus on the problem of finding a winning committee, which can be formalized as follows:

WEIGHTED-CC-WINNER:
Input: an election $E = (C, V)$ with $|C| = m$, $|V| = n$, voters' preferences

$(\succ_i)_{i \in V}$ and score functions $(s_i)_{i \in V}$ (where each s_i is described by a list of m integers), a weight vector $\boldsymbol{w} = (w_1, \ldots, w_n)$, and a committee size κ.
Output: Some committee $S \subseteq C$ of size κ with the maximum $\boldsymbol{w}$-CC score.

Clearly, this problem is NP-hard; this follows from the fact that it generalizes the problem of finding a U-CC winning committee, which is known to be NP-hard [20, 24]. Thus, from now on we focus on identifying tractable special cases of this problem by making additional assumption on the properties of the weight vector and/or the voters' preferences.

Few Candidates or Voters We first observe that WEIGHTED-CC-WINNER is easy if the number of candidates m is small: this problem admits an algorithm whose running time is $\text{poly}(n, m)2^m$. The argument is similar to the one in [3]: we can go over all $\binom{m}{\kappa} \leq 2^m$ potential committees, evaluate the $\boldsymbol{w}$-CC score of each committee, and pick a committee with the highest score. This observation implies the following proposition (see, e.g., [22] for an introduction to fixed-parameter tractability).

Proposition 1. WEIGHTED-CC-WINNER *is fixed-parameter tractable with respect to* m.

Betzler et al. [3] also describe an algorithm that finds a U-CC winner in time $\text{poly}(n, m)n^n$, and therefore is useful when the number of voters is small. This algorithm proceeds by considering all ways of partitioning the n voters into at most κ groups. For each partition, it constructs a bipartite graph where the left-hand side corresponds to groups of voters, the right-hand side corresponds to candidates, and the value of an edge is the total utility that the respective group of voters derives from the respective candidate. It then finds a maximum-value matching of size κ in this graph. However, it seems to be difficult to extend this idea to arbitrary weights: the following example illustrates that it is not clear how to define edge values in a meaningful way.

Example 2. Suppose that $\kappa = 2$, and consider an election over the set of candidates $C = \{a, b, c, d\}$ and 4 voters with the following preferences:

$$abdc: \quad 2 \text{ voters} \qquad cdba: \quad 1 \text{ voter} \qquad badc: \quad 1 \text{ voter}$$

Let $\boldsymbol{w} = (.4, .3, .2, .1)$, and suppose that the score function of each voter is the Borda score function (see Example 1).

Suppose we consider a partition where the voters with preferences $abdc$ form one group, and the other two voters form another group, and we would like to evaluate the contribution of the second group of voters to the total utility if we were to match it to candidate d. However, to do so, we would have to know which candidate is matched to the first group: if it is a, then the contribution of the second group is $.4 \cdot 2 + .3 \cdot 1$, whereas if it is c, the contribution of the second group is $.2 \cdot 2 + .1 \cdot 1$.

EU-CC rule and Generalizations. We have argued that the EU-CC rule offers a useful balance between egalitarian and utilitarian welfare. We will now show that it also has computational advantages: computing EU-CC winners is no harder than computing U-CC winners.

Specifically, observe that the E-CC score of any committee is an element of Scr, and note that $|Scr| \leq nm$. Let $\Delta = nR$, where $R = \max_{s \in Scr} s$. We can now consider all elements $z \in Scr$ in decreasing order. For a given value of z, we modify the score functions $s_i(c)$ so that $s_i'(c) = s_i(c) + \Delta$ if $s_i(c) \geq z$ and $s_i'(c) = 0$ otherwise. Note that the modified score functions remain consistent with voters' preferences. If we call an algorithm for U-CC on the resulting instance, we obtain a committee whose U-CC score is at least $\Delta\kappa$ if and only if the original instance admits a committee where the utility of each voter is at least z. Moreover, if the U-CC score of a committee with the highest U-CC score is at least $\Delta\kappa$, then any such committee has the highest U-CC score in the original election among all committees that guarantee a utility of z to each voter.

Thus, by finding the largest value of z for which the modified instance admits a committee with U-CC score of at least $\Delta\kappa$ and outputting a committee with the maximum U-CC score for this instance, we obtain a winning committee under EU-CC for the original instance. As voters' preferences in the modified instance are the same as in the original instance (only the scores change), this implies that EU-CC admits a polynomial-time algorithm when the voters preferences are single-crossing or single-peaked on a line or, more broadly, on a tree with a constant number of leaves, and is fixed-parameter tractable with respect to the number of voters; this follows from the respective results of [3,27,32] for U-CC.

More generally, by considering all values of $z \in Scr$, we obtain the Pareto boundary of the associated bicriteria optimization problem (where the criteria are the egalitarian and the utilitarian social welfare).

Observe that a committee that is optimal with respect to EU-CC is by definition optimal with respect to E-CC (as the former is a refinement of the latter), so in particular the argument above shows that E-CC is tractable whenever U-CC is; however, special-purpose algorithms for E-CC may be considerably faster than the algorithm provided by our reduction (as illustrated by the results of [3]).

5   Algorithms for Single-Peaked Preferences

An election (C, V) described by a collection of preference orders $(\succ_i)_{i \in V}$ is said to be *single-peaked* if there exists an ordering $\sqsubset$ of the candidates (the *left-right axis*) such that for each voter $i \in V$ her preference $\succ_i$ increases from the left of the axis to $\text{top}(i)$ and decreases from $\text{top}(i)$ to the right of the axis: formally,

for all $a, b \in C$ s.t. $a \sqsubset b \sqsubseteq \text{top}(i)$ or $\text{top}(i) \sqsubseteq b \sqsubset a$ it holds that $b \succ_i a$.

Equivalently, for each voter $i \in V$ and each $k \leq m$ the set of candidates that i ranks in top k positions forms a contiguous interval with respect to $\sqsubset$.

Single-peaked elections were introduced by Black [4], and are known to have a number of desirable social-choice properties as well as to admit efficient algorithms for many computational social choice problems (see, e.g., [7,15,28]). Intuitively, such elections arise when the society is aligned along a single axis, and voters rank the candidates according to their position on this axis.

In the rest of this section, we consider two classes of weight vectors, and show how to compute the output of w-CC for weight vectors in these classes, assuming that voters' preferences are single-peaked. It will be convenient to assume that the left-right axis is given explicitly, and, moreover, that this axis is $c_1 \sqsubset \cdots \sqsubset c_m$. This assumption is without loss of generality: given a collection of preferences, we can decide if it is single-peaked, and, if so, find an axis witnessing this, in polynomial time [2,12].

5.1   The U^{-d}-CC rule

In this section, we provide a polynomial-time algorithm for U^{-d}-CC—the rule that maximizes the sum of utilities of all but d voters (and therefore ignores the d least happy voters entirely)—under single-peaked preferences. In contrast with the result of Sect. 5.2, which deals with a larger class of weight vectors, this algorithm works for arbitrary score functions.

Let $\theta = n - d$; recall that the weight vector w^{-d} associated with U^{-d}-CC satisfies

$$w_i^{-d} = \begin{cases} 0 & \text{if } i \leq d \\ \frac{1}{\theta} & \text{if } i > d \end{cases}$$

The key observation is that it is sufficient to maximize the sum of the scores of *some* θ voters, without fixing in advance the voters whose score is taken into account. Indeed, this maximization process naturally selects the θ most happy voters.

Theorem 1. *For single-peaked preferences, Algorithm 1 computes a winning committee under U^{-d}-CC in time $O(\kappa m^2 \theta^2 n \log_2 n) = O(m^3 n^3 \log_2 n)$.*

Proof. We assume without loss of generality that voters' preferences are single-peaked with respect to the axis $c_1 \sqsubset \ldots \sqsubset c_m$. It will be convenient to introduce an additional candidate c_{m+1} who is ranked last by each voter and such that $s_i(c_{m+1}) = 0$ for each $i \in V$; we place this candidate to the right of c_m on the axis so that the election remains single-peaked.

Let $V^{\ell,j} = \{i \in V \mid \text{top}(i) \in \{c_\ell, \ldots, c_j\}\}$, and define

$$s_{(t)}^{\ell,j}(S) = \max \left\{ \sum_{i \in W} s_i(S) \mid W \subseteq V^{\ell,j}, |W| = t \right\},$$

with the convention that $\max \emptyset = -\infty$. When the set $\{W \mid W \subseteq V^{\ell,j}, |W| = t\}$, is not empty, the quantity $s_{(t)}^{\ell,j}(S)$ can be computed by ordering the voters in

---

Algorithm 1. Dynamic program for w^{-d}-CC under single-peaked preferences

---

Input: set of voters $\{1, \ldots, n\}$; set of candidates $\{c_1, \ldots, c_m\}$, voters' preferences $(\succ_i)_{i \in V}$, which are single-peaked with respect to $c_1 \sqsubset \ldots \sqsubset c_m$; score functions $(s_i)_{i \in V}$; target committee size κ; number of represented voters $\theta = n - d$

Output: $\max\{s_{w^{-d}}(S) \mid S \subseteq C, |S| = \kappa\}$

1. INITIALIZATION

for $j = 1, \ldots, m + 1$ *and* $t = 0, \ldots, \theta$ **do**

    $z_1(j, t) \leftarrow s_{(t)}^{1,j}(\{c_j\})$

    for $k = 1, \ldots, \kappa$ **do**

        $z_{k+1}(j, t) = -\infty$

    end

end

2. MAIN LOOP

for $k = 1, \ldots, \kappa$ **do**

    //Predecessors loop:

    for $p = 1, \ldots, m$ *and* $u = 0, \ldots, \theta$

    if $z_k(p, u) \neq -\infty$ **do**

        //Successors sub-loop:

        for $j = p + 1, \ldots, m + 1$ *and* $t = u, \ldots, \theta$ **do**

            $z_{k+1}(j, t) \longleftarrow \max\{z_{k+1}(j, t), z_k(p, u) + s_{(t-u)}^{p+1,j}(\{c_p, c_j\})\}$

        end

    end

end

return $z_{\kappa+1}(m + 1, \theta)$

$V^{\ell, j}$ according to the score they assign to S (from the highest to the lowest), picking the first t voters in this order, and summing their scores for S.

Now, for each $t \in \{0, \ldots, \theta\}$, $k \in \{1, \ldots, \kappa + 1\}$, and $j \in \{k, \ldots, m + 1\}$, let

$$z_k(j, t) = \max\left\{s_{(t)}^{1,j}(S) \mid S \subseteq \{c_1, \ldots, c_j\}, c_j \in S, |S| = k\right\}$$

The quantity $z_k(j, t)$ is the highest total utility that a group of t voters in $V^{1,j}$ can derive from a size-k subset of $\{c_1, \ldots, c_j\}$ that contains c_j. We claim that it can be computed by dynamic programming as follows: for $k = 1$ we have $z_1(j, t) = s_{(t)}^{1,j}(\{c_j\})$ for all $t \in \{0, \ldots, \theta\}$, $j \in \{1, \ldots, m + 1\}$, and for $k > 1$

$$z_k(j, t) = \max_{p \in]\!]k, j[\![} \ \max_{u \in [\![0, t]\!]} \left\{z_{k-1}(p, u) + s_{(t-u)}^{p+1,j}(\{c_p, c_j\})\right\}. \tag{1}$$

Indeed, to find the score of an optimal pair (S, W), where $S \subseteq \{c_1, \ldots, c_j\}$, $c_j \in S$, $|S| = k$, $W \subseteq V^{1,j}$, and $|W| = t$, we guess the last candidate in $S' = S \cap \{c_1, \ldots, c_{j-1}\}$ (let this candidate be c_p) and the number of voters in $W' = W \cap V^{1,p}$ (let this number be u). We then observe that all voters in W'

have their peak at or to the left of c_p and therefore their most preferred candidate in S is not c_j, whereas for each voter $i \in W \setminus W'$ we have $\text{top}(i) \in \{c_p, c_j\}$.

Algorithm 1 implements the calculation in (1) in a forward manner to increase efficiency.

It is easy to see that the U^{-d}-CC score of a winning committee is $z_{\kappa+1}(m, \theta)$; a committee with this score can be found using standard dynamic programming techniques.

The number of variables in our dynamic program is $O(\kappa m\theta)$. To compute the value of variable $z_{k+1}(j, t)$, we need to consider $O(m\theta)$ possibilities for p and u, and, for each of them, compute $s_{(t-u)}^{p+1,j}(\{c_p, c_j\})$; the latter step involves sorting the voters in $V^{p+1,j}$ according to their score for $\{c_p, c_j\}$, and can therefore be implemented in time $O(n \log n)$. This implies our bound on the running time. $\square$

5.2 w-CC for a Bounded Number of Weights

In this section, we show that w-CC admits a polynomial-time algorithm if voters' preferences are single-peaked, their score functions are polynomially bounded (i.e. $R = \text{poly}(m, n)$), and the number of distinct entries in w is small. That is, we consider weight vectors of the form

$$w = (\omega_1, \ldots, \omega_1, \omega_2, \ldots, \omega_2, \ldots, \omega_B, \ldots, \omega_B), \tag{2}$$

where B is assumed to be a constant. Let $\alpha(b)$ and $\beta(b)$ be, respectively, the indices of the first and the last occurrence of ω_b in w, and let $\gamma(b) = \beta(b) - \alpha(b) + 1$. Setting $y_b(S) = \sum_{i=\alpha(b)}^{\beta(b)} s_{\sigma(i)}(S)$, we can rewrite the w-CC score of a committee S as

$$s_w(S) = \sum_{b=1}^{B} \omega_b \sum_{i=\alpha(b)}^{\beta(b)} s_{\sigma(i)}(S) = \sum_{b=1}^{B} \omega_b y_b(S). \tag{3}$$

The main idea of our algorithm is that when B is a constant, we can try to guess the range of scores that will be counted with weight ω_b, for each $b = 1, \ldots, B$; for a given guess, we can use dynamic programming.

Theorem 2. *When voters' preferences are single-peaked and the weight vector w is given by expression (2), a winning committee under w-CC can be computed in time $\text{poly}(n, m, (nm)^B, R^B)$, where $R = \max_{i \in V, c \in C} s_i(c)$.*

Proof. By adding or subtracting a constant to all score functions, we can assume that $\min_{r \in Scr} r = 1$. Let $\mathcal{R}$ be the set of all vectors $(r_0, \ldots, r_B)$ in Scr^{B+1} that satisfy $1 = r_0 \leq r_1 \leq \cdots \leq r_B = R$. Note that $|\mathcal{R}| \leq |Scr|^{B-1} \leq (nm)^{B-1}$; if B is a constant, this quantity is polynomial in the input size.

Each vector $r \in \mathcal{R}$ induces a partition Π^r of Scr into $2B - 1$ sub-ranges, some of which may be empty:

$$\Pi^r = \{ \ [0, r_1[, \{r_1\},]r_1, r_2[, \{r_2\}, \ldots, \{r_{B-1}\},]r_{B-1}, r^*] \ \}.$$

We say that a committee S is r-*compatible* if, when we order the voters according to their score for this committee from lowest to highest, then for each $b = 1, \ldots, B - 1$ it holds that the score of the voter in position β_b is r_b.

Given an instance of WEIGHTED-CC-WINNER where the weight vector w has B distinct entries, our algorithm considers all vectors in $\mathcal{R}$. For each such vector r, it calls the subroutine described in Algorithm 2. It then outputs the maximum of the $|\mathcal{R}| \leq (nm)^B$ numbers obtained in this manner. We will now argue that for a given r, Algorithm 2 returns the maximum w-CC score of an r-compatible committee of size κ. Since an optimal committee of size κ is compatible with some $r \in \mathcal{R}$, this proves that our algorithm computes the maximum w-CC score over all committees of size κ; a committee with this score can then be found using standard techniques.

As in the proof of Theorem 1, we add a dummy candidate c_{m+1} that appears to the right of c_m on the axis and is ranked last by all voters. Also, as in that proof, we denote by $V^{\ell,j}$ the set of voters whose top candidate is in $\{c_\ell, \ldots, c_j\}$.

Fix a vector $r \in \mathcal{R}$ and the respective partition $\Pi^r = \{\rho_1, \ldots, \rho_{2B-1}\}$. For a given committee S, let $V(S, j, d)$ be the set of all voters in $V^{1,j}$ whose score for S lies in ρ_d: $V(S, j, d) = \{i \in V^{1,j} \mid s_i(S) \in \rho_d\}$.

We are now ready to define the variables of our dynamic program. For each $r \in \mathcal{R}$, each $k = 1, \ldots, \kappa + 1$, each $j = k, \ldots, m + 1$, and each $t \in [0, n]^{2B-1}$, we define a variable $Y_k^r(j, t)$. This is a collection of vectors of length $2B - 1$ with entries in $[nR]$; the vectors in $Y_k^r(j, t)$ are 'realizable' combinations of scores. That is, a vector $y \in [nR]^{2B-1}$ is in $Y_k^r(j, t)$ if and only if there exists a committee $S \subseteq \{c_1, \ldots, c_j\}$ with $c_j \in S, |S| = k$, such that for each $\rho_d \in \Pi^r$ we have

$$|V(S, j, d)| = t_d, \qquad \sum_{i \in V(S,j,d)} s_i(S) = y_d.$$

We will now explain how our algorithm computes the sets $Y_k^r(j, t)$. For $k = 1$ and a fixed value of j, we simply compute $s_i(j)$ for all voters in $V^{1,j}$. This score maps each voter in $V^{1,j}$ to some range in Π^r; the number of voters assigned to ρ_d (which is exactly $|V(\{c_j\}, j, d)|$) determines t_d, and the sum of their scores for $\{c_j\}$ determines y_d. We obtain a pair of vectors (t, y) in this manner. We then set $Y_1^r(j, t) = \{y\}$; for other values of t' the set $Y_1^r(j, t')$ remains empty.

For $k > 1$, just as in the proof of Theorem 1, we consider all possibilities for the predecessor-candidate c_p, $p \in [k - 1, j - 1]$. All voters in $V^{1,p}$ prefer c_p to c_j, so we need to focus on voters in $V^{p+1,j}$. Since their preferences are single-peaked with respect to $\sqsubset$, their score for any committee S with $S \cap \{c_p, \ldots, c_j\} = \{c_p, c_j\}$ is equal to their score for $\{c_p, c_j\}$. We compute this score, which maps every such voter to some range in Π^r; the number of voters mapped to ρ_d is given by $|V(\{c_p, c_j\}, j, d) \setminus V(\{c_p, c_j\}, p, d)|$. As the overall number of voters in each range is given by t, we let $t'_d = t_d - |V(\{c_p, c_j\}, j, d) \setminus V(\{c_p, c_j\}, p, d)|$ for each $\rho_d \in \Pi^r$, and consider the set $Y_{k-1}^r(p, t')$. The elements of this set are realizable combinations of scores for p and t'; by adding the scores of voters in $V^{p+1,j}$ in each range, we obtain a realizable combination of scores for j and t. We add the resulting score vector to $Y_k^r(j, t)$. Algorithm 2 performs this computation in a forward manner for added efficiency.

It remains to explain how to use $Y_k^r(j, t)$ to compute the maximum w-CC score of an r-compatible committee of size κ. Recall that γ_b is the number of occurrences of ω_b in w. We say that a vector t in $[\![0, n]\!]^{2B-1}$ is *valid* if $\sum_{d=1}^{2B-1} t_d = n$ and for each $b = 1, \ldots, B - 1$ there exists a non-negative integer $\xi_b \leq t_{2b}$ such that $\gamma(1) = t_1 + \xi_1$, $\gamma(b) = t_{2b-2} - \xi_{b-1} + t_{2b-1} + \xi_b$ for $b > 1$, $\gamma(B) = t_{2B-2} - \xi_{B-1} + t_{2B-1}$. Intuitively, we have to 'distribute' the voters whose score is r_b so that ξ_b of them are counted with weight ω_b and the remaining ones are counted with weight ω_{b+1}. Clearly, one can decide in time $O(B)$ whether a given vector is valid and find the respective integers $(\xi_b)_{b \in [\![1, B-1]\!]}$.

Suppose that t is a valid vector, as witnessed by a collection of integers $\{\xi_b\}_{b \in [\![1, B-1]\!]}$. Then, given a vector $y \in Y_{\kappa+1}^r(m + 1, t)$, let $z_1 = y_1 + \xi_1 r_1$, $z_b = (t_{2b-2} - \xi_{b-1}) r_{b-1} + y_{2b-1} + \xi_b r_b$ for $b = 2, \ldots, B - 1$, $z_B = (t_{2B-2} - \xi_{B-1}) r_{B-1} + y_{2B-1}$, and set $T(t, y) = \sum_{b=1}^{B} \omega_b z_b$. It is not hard to see that for a fixed vector r the maximum w-score of an r-compatible committee of size κ is given by

$$\max\{T(t, y) \mid t \text{ is valid and } y \in Y_{\kappa+1}^r(m + 1, t)\},$$

which is exactly the quantity output by our algorithm for that choice of r.

There are at most $(nm)^{B-1}(\kappa + 1)(m + 1)(n + 1)^{2B-1} = O(\kappa m (n^3 m)^{B-1})$ variables $Y_k^r(j, t)$: at most $(nm)^{B-1}$ choices for r, $\kappa + 1$ choices for k, $m + 1$ choices for j, and at most $(n + 1)^{2B-1}$ choices for t. Moreover, the size of each set $Y_k^r(j, t)$ can be bounded by $(nR)^{2B-1}$; we can improve this bound to $(nR)^B$ by observing that if $d = 2b$ is even, then ρ_d is the singleton $\{r_b\}$, so we have $y_d = r_b \cdot t_d$. This establishes our bound on the running time. □

Remark 1. Theorem 3 relies on the assumption that the ranges of all score functions are polynomially bounded in n and m. This assumption holds for many important score functions, such as the Borda score function; however, it is not without loss of generality. We can exhibit another class of score functions for which w-CC is easy when the number of weights is bounded and the voters' preferences are single-peaked: these are score functions that can only take a constant number of values, i.e., the size of the set *Scr* is bounded by a constant. The algorithm for this setting can be obtained by adapting the algorithm in the proof of Theorem 3; we omit the details due to space constraints.

6   Algorithms for Single-Crossing Preferences

In this section, we focus on elections where voters' preferences are single-crossing [18]. This is another well-known restricted preference domain that models scenarios where voters' preferences are essentially one-dimensional; however, in contrast with single-peaked preferences, which are defined in terms of an ordering of the candidates, single-crossing preferences are defined in terms of an ordering of the voters.

We start by providing a formal definition of this domain.

Algorithm 2. Subroutine for finding an optimal r-compatible committee

---

Input: set of voters $\{1, \ldots, n\}$; set of candidates $\{c_1, \ldots, c_m\}$, voters'
preferences $(\succ_i)_{i \in V}$, which are single-peaked with respect to
$c_1 \sqsubset \ldots \sqsubset c_m$; score functions $(s_i)_{i \in V}$; target committee size κ;
partitioning vector r

Output: The maximum score of an r-compatible committee.

1. INITIALIZATION ($k = 1$)

for $j = 1, \ldots, m$ **do**

   for $\rho_d \in \Pi^r$ **do**

      $t_d \leftarrow |V(\{c_j\}, j, d)|$

      $y_d \leftarrow \sum_{i \in V(\{c_j\}, j, d)} s_i(\{c_j\})$

   end

   $Y_1^r(j, t) = \{y\}$

end

2. MAIN LOOP

for $k = 2, \ldots, \kappa + 1$ **do**

   //Predecessors' loop:

   for $p = 1, \ldots, m$ and all t' such that $Y_{k-1}^r(p, t') \neq \emptyset$ **do**

      //Successors sub-loop:

      for $j = p + 1, \ldots, m + 1$ **do**

         for $\rho_d \in \Pi^r$ **do**

            $t_d \leftarrow t_d' + |V(\{c_p, c_j\}, j, d) \setminus V(\{c_p, c_j\}, p, d)|$

            $z_d \leftarrow \sum_{i \in V(\{c_p, c_j\}, j, d) \setminus V(\{c_p, c_j\}, p, d)} s_i(\{c_p, c_j\})$

         end

         $Y_k^r(j, t) \longleftarrow Y_k^r(j, t) \cup \{y + z \mid y \in Y_{k-1}^r(p, t')\}$

      end

   end

end

return $\max\{T(t, y) \mid t \text{ is valid and } y \in Y_{\kappa+1}^r(m + 1, t)\}$

An election (C, V) described by a collection of preference orders $(\succ_i)_{i \in V}$ is said to be *single-crossing* if there exists an ordering $\sqsubset$ of the voters such that for each pair of candidates $a, b \in C$ such that the first voter in $\sqsubset$ prefers a to b there exists a unique $v \in V$ such that all voters that precede v in $\sqsubset$ prefer a to b, whereas all voters that appear after v in $\sqsubset$ prefer b to a.

In other words, both the set of voters who prefer a to b and the set of voters who prefer b to a are contiguous with respect to $\sqsubset$, i.e., if we graph the positions of a and b in voters' preference orders, the resulting curves intersect at most once. As observed by Bredereck et al. [8], this characterization immediately suggests a polynomial-time algorithm for detecting single-crossing preferences, by reducing this question to the classic consecutive 1 s problem [5] (see also [12,14]). As a consequence, we can assume without loss of generality that an ordering of voters witnessing that a given election is single-crossing is given to us as a part of the input; in fact, it will be convenient to assume that this ordering is $1 \sqsubset \cdots \sqsubset n$.

Skowron et al. [27] have recently shown that given an election with single-crossing preferences, one can solve U-CC and E-CC in polynomial time (specifically, $O(\kappa mn^2)$). The main result of this section is that the algorithm of Skowron et al. extends to w-CC as long as the number of distinct weights in w is bounded by a constant and the score functions are polynomially bounded (these are the same conditions that are used to establish tractability under single-peaked preferences in Sect. 5).

First of all, we observe that a key structural property of optimal solutions established by Skowron et al. for U-CC holds for arbitrary weight vectors.

Given the set of voters $V = [n]$ and a committee $S \subseteq C$, let $\Phi_S : V \to S$ be the function that assigns each voter i to her most preferred candidate in S: $\Phi_S(i) \succ_i c_j$ for all $c_j \in S \setminus \{\Phi_S(i)\}$. Hence, for a candidate $c \in S$, the set of voters she represents is $\Phi_S^{-1}(c)$.

Lemma 1. *Let $E = (C, V)$ be an election with single-crossing preferences, as witnessed by the voter order $1 \sqsubset \ldots \sqsubset n$, where the first voter's preferences are given by $c_1 \succ_1 c_2 \succ_1 \ldots \succ_1 c_m$. Then for any weight vector w and any committee size κ there exists a committee S of size κ that is optimal with respect to w-CC and has the following property: for each $c_j \in S$ the set of voters $\Phi_S^{-1}(c_j)$ is contiguous with respect to $\sqsubset$, and if $c_j, c_\ell \in S$, $\Phi_S^{-1}(c_j) \neq \emptyset$, $\Phi_S^{-1}(c_\ell) \neq \emptyset$ and $j < \ell$ then $\Phi_S^{-1}(c_j)$ precedes $\Phi_S^{-1}(c_\ell)$ in $\sqsubset$.*

Proof. The proof is a simple generalization of the proof of a similar statement in [27], and is omitted due to space constraints.

Using this lemma and a dynamic programming approach similar to the one in the proof of Theorem 3, we obtain the following result (we omit the proof due to space constraints).

Theorem 3. *When voters' preferences are single-crossing and the weight vector w has at most B distinct entries, we can find a winning committee under w-CC in time $\mathrm{poly}(n, m, (nm)^B, R^B)$, where $R = \max_{i \in V, c \in C} s_i(c)$.*

7 Conclusion

We have described a family of voting rules for committee selection that explores a variety of ways to trade off utilitarian and egalitarian objectives. We have developed algorithms for computing the winning committees under these rules for instances where voters' preferences belong to well-known restricted domains, under mild restrictions on the weight vectors and score functions. While in this work we focused on single-peaked and single-crossing preferences, it seems plausible that similar results can be obtained for other domains where U-CC and E-CC are known to be tractable, such as preferences that are single-peaked on trees with a bounded number of leaves, or have bounded single-peaked or single-crossing width.

However, it is not clear if our results can be extended to arbitrary weight vectors. In particular, the complexity of Lex-E-CC for single-peaked or single-crossing preferences remains an intriguing open question, and it appears that obtaining positive results for this rule would require novel algorithmic techniques.

Besides this open problem, our work suggests several other interesting research directions. One of them is considering OWA-based extensions of the Monroe rule and exploring the complexity of finding winning committees under the resulting family of rules. Another one is developing approximation algorithms for WEIGHTED-CC-WINNER for arbitrary weights and voters' preferences; here the work of Skowron et al. [26] provides a useful starting point.

Acknowledgements. This work was partially supported by an STSM Grant from the COST Action IC1205. The authors are grateful to the anonymous ADT referees for their useful suggestions.

References

1. Amanatidis, G., Barrot, N., Lang, J., Markakis, E., Ries, B.: Multiple referenda and multiwinner elections using hamming distances: Complexity and manipulability. In: Proceedings of the 14th International Conference on Autonomous Agents and Multiagent Systems, pp. 715–723 (2015)
2. Bartholdi III, J., Trick, M.: Stable matching with preferences derived from a psychological model. Oper. Res. Lett. **5**(4), 165–169 (1986)
3. Betzler, N., Slinko, A., Uhlmann, J.: On the computation of fully proportional representation. J. Artif. Intell. Res. **47**(1), 475–519 (2013)
4. Black, D.: The Theory of Committees and Elections. Cambridge University Press, Cambridge (1958)
5. Booth, K., Lueker, G.: Testing for the consecutive ones property, interval graphs, and graph planarity using PQ-tree algorithms. J. Comput. Syst. Sci. **13**(3), 335–379 (1976)
6. Brams, S., Kilgour, D.M., Sanver, R.M.: A minimax procedure for electing committees. Public Choice **132**(3-4), 401–420 (2007)
7. Brandt, F., Brill, M., Hemaspaandra, E., Hemaspaandra, L.: Bypassing combinatorial protections: Polynomial-time algorithms for single-peaked electorates. J. Artif. Intell. Res. **53**, 439–496 (2015)
8. Bredereck, R., Chen, J., Woeginger, G.: A characterization of the single-crossing domain. Soc. Choice Welfare **41**(4), 989–998 (2013)
9. Caragiannis, I., Kalaitzis, D., Markakis, E.: Approximation algorithms and mechanism design for minimax approval voting. In: Proceedings of the 25th AAAI Conference on Artificial Intelligence, pp. 737–742 (2010)
10. Chamberlin, B., Courant, P.: Representative deliberations and representative decisions: proportional representation and the borda rule. Am. Polit. Sci. Rev. **77**(3), 718–733 (1983)
11. Cornaz, D., Galand, L., Spanjaard, O.: Bounded single-peaked width and proportional representation. In: Proceedings of the 20th European Conference on Artificial Intelligence, pp. 270–275 (2012)
12. Doignon, J., Falmagne, J.: A polynomial time algorithm for unidimensional unfolding representations. J. Algorithms **16**(2), 218–233 (1994)
13. Elkind, E., Faliszewski, P., Skowron, P., Slinko, A.: Properties of multiwinner voting rules. In: Proceedings of the 13th International Conference on Autonomous Agents and Multiagent Systems, pp. 53–60 (2014)

14. Elkind, E., Faliszewski, P., Slinko, A.: Clone structures in voters' preferences. In: Proceedings of the 13th ACM Conference on Electronic Commerce, pp. 496–513 (2012)
15. Faliszewski, P., Hemaspaandra, E., Hemaspaandra, L., Rothe, J.: The shield that never was: societies with single-peaked preferences are more open to manipulation and control. Inf. Comput. **209**(2), 89–107 (2011)
16. Golden, B., Perny, P.: Infinite order Lorenz dominance for fair multiagent optimization. In: Proceedings of the 9th International Conference on Autonomous Agents and Multiagent Systems, pp. 383–390 (2010)
17. Goldsmith, J., Lang, J., Mattei, N., Perny, P.: Voting with rank dependent scoring rules. In: Proceedings of the 28th AAAI Conference on Artificial Intelligence, pp. 698–704 (2014)
18. Grandmont, J.: Intermediate preferences and the majority rule. Econometrica **46**(2), 317–330 (1978)
19. LeGrand, R., Markakis, E., Mehta, A.: Some results on approximating the minimax solution in approval voting. In: Proceedings of the 6th International Conference on Autonomous Agents and Multiagent Systems, pp. 1185–1187 (2007)
20. Lu, T., Boutilier, C.: Budgeted social choice: from consensus to personalized decision making. In: Proceedings of the Twenty-Second International Joint Conference on Artificial Intelligence, pp. 280–286 (2011)
21. Monroe, B.: Fully proportional representation. Am. Polit. Sci. Rev. **89**(4), 925–940 (1995)
22. Niedermeier, R.: Invitation to Fixed-Parameter Algorithms. Oxford University Press, Oxford (2006)
23. Ogryczak, W.: Multiple criteria linear programming model for portfolio selection. Ann. Oper. Res. **97**(1–4), 143–162 (2000)
24. Procaccia, A., Rosenschein, J., Zohar, A.: On the complexity of achieving proportional representation. Soc. Choice Welfare **30**(3), 353–362 (2008)
25. Skowron, P., Faliszewski, P., Lang, J.: Finding a collective set of items: From proportional multirepresentation to group recommendation. In: Proceedings of the 29th AAAI Conference on Artificial Intelligence, pp. 2131–2137 (2015)
26. Skowron, P., Faliszewski, P., Slinko, A.M.: Achieving fully proportional representation: approximability results. Artif. Intell. **222**, 67–103 (2015)
27. Skowron, P., Yu, L., Faliszewski, P., Elkind, E.: The complexity of fully proportional representation for single-crossing electorates. Theor. Comput. Sci. **569**, 43–57 (2015)
28. Walsh, T.: Uncertainty in preference elicitation and aggregation. In: Proceedings of the 22nd National Conference on Artificial Intelligence, pp. 3–8 (2007)
29. Yager, R.R.: On ordered weighted averaging aggregation operators in multicriteria decisionmaking. IEEE Trans. Syst. Man Cybern. **18**(1), 183–190 (1988)
30. Yager, R.R.: Constrained OWA aggregation. Fuzzy Sets Syst. **81**(1), 89–101 (1996)
31. Yager, R.R., Kacprzyk, J. (eds.): The ordered weighted averaging operators: theory and applications. Springer Publishing Company, Incorporated, New York (2012)
32. Yu, L., Chan, H., Elkind, E.: Multiwinner elections under preferences that are single-peaked on a tree. In: Proceedings of the 23rd International Joint Conference on Artificial Intelligence, pp. 425–431 (2013)

Allocation and Other Problems

Optimal Group Manipulation in Facility Location Problems

Xin Sui[✉] and Craig Boutilier

Department of Computer Science, University of Toronto,
6 King's College Road, Toronto, ON, Canada
{xsui,cebly}@cs.toronto.edu

Abstract. We address optimal group manipulation in multi-dimensional, multi-facility location problems. We focus on two families of mechanisms, *generalized median* and *quantile* mechanisms, evaluating the difficulty of group manipulation of these mechanisms. We show that, in the case of single-facility problems, optimal group manipulation can be formulated as a linear or second-order cone program, under the L_1- and L_2-norms, respectively, and hence can be solved in polynomial time. For multiple facilities, we show that optimal manipulation is NP-hard, but can be formulated as a mixed integer linear or second-order cone program, under the L_1- and L_2-norms, respectively. Despite this hardness result, empirical evaluation shows that multi-facility manipulation can be computed in reasonable time with our formulations.

1 Introduction

Mechanism design deals with the design of protocols to elicit the preferences of self-interested agents to achieve some social objective [22]. An important property in mechanism design is *strategy-proofness*, which requires that there is no incentive for an individual agent to misreport their preferences. While much work in mechanism design deals with settings where monetary transfers can be used to facilitate strategy-proofness [6,18,31], many problems do not admit payments for a variety of reasons [28].

The *Gibbard-Satterthwaite theorem* [16,27] shows that under fairly broad conditions, one cannot construct mechanisms that achieve strategy-proofness in general. However, one can impose restrictions on the preference domain to escape this impossibility result. A widely used restriction is *single-peakedness* [4]. In single-peaked domains, each agent has a single, most-preferred *ideal* point in the outcome space, and (loosely) her preference for outcomes decreases with as the distance of that outcome from the ideal increases. In such settings, strategy-proofness is guaranteed by the classic *median mechanism* and its generalizations for single outcomes [2,24], or *quantile mechanisms* [29] for multiple outcomes. Applications of such models include facility location, voting, product design, customer segmentation, and many others.

While these mechanisms are *individual strategy-proof*, they are not *group strategy-proof*—a group of agents may jointly misreport their preferences to induce a more preferred outcome that makes some group members better off

© Springer International Publishing Switzerland 2015
T. Walsh (Ed.): ADT 2015, LNAI 9346, pp. 505–520, 2015.
DOI: 10.1007/978-3-319-23114-3_30

without harming others. In this paper, we consider the *group manipulation problem* for *facility location problems* (*FLPs*) with multiple facilities in multi-dimensional spaces, with a focus on quantile mechanisms (QMs) (and to some extent *generalized median mechanisms* (*GMMs*)). Since these mechanisms are both transparent and (individual) strategy-proof for general multi-dimensional, multi-FLPs, we seek to understand the difficulty of their group manipulation problems.

Our primary contribution is to formulate the group manipulation problem—for both single- and multi-FLPs under both the L_1- and L_2-norms (where these metrics measure distance/cost between ideal points and facilities)—as convex optimization problems, and study their computational complexity. We show that single-FLPs with L_1 and L_2 costs can be specified as linear programs (LPs) and second-order cone programs (SOCPs), respectively. This means both can be solved in polynomial time (using interior point methods [5]). By contrast, we show that multi-FLPs are NP-hard by reduction from the geometric p-median problem [23] under both norms. Despite this, we provide formulations of problems as mixed integer linear (MILPs) and mixed integer SOCPs (MISOCPs) for L_1 and L_2 costs, respectively. We also test these formulations empirically, with results that suggest commercial solvers can compute group manipulations (or prove that none exists) for multi-FLPs of reasonable size rather effectively, despite the theoretical NP-hardness of the problem.

2   Background and Notation

We begin by defining FLPs, the quantile mechanism, the group manipulation problem we consider, and provide a brief discussion of related work.

2.1   Facility Location and Group Manipulation

A d-dimensional, m-facility *facility location problem (FLP)* involves selecting m facilities in some d-dimensional subspace $\mathbb{S} \subseteq \mathbb{R}^d$ (we omit mention of $\mathbb{S}$ subsequently, assuming all locations fall in $\mathbb{S}$). We assume a set of *agents* $N = \{1, \ldots, n\}$, each with an *ideal location* or *type* $t_i \in \mathbb{R}^d$, which determines her *cost* $s_i(x, t_i)$ for using a facility located at x (we sometimes refer to this as facility x). Given a *location vector* $\mathbf{x} = (x_1, \ldots, x_m)$, $x_j \in \mathbb{R}^d$, of m facilities, we assume each agent uses her most preferred facility, defining $s_i(\mathbf{x}, t_i) = \min_{j \le m} s_i(x_j, t_i)$. Given the ideal points of all agents, our goal is to select an outcome that implements some social choice function (e.g., minimize social cost, ensure Pareto efficiency, etc.). Below we equate cost with L_1 or L_2 distance. A *mechanism* for an FLP is a function f that accepts as input the *reported* ideal points of the n agents and returns a location vector $\mathbf{x}$.

FLPs can be interpreted literally, naturally modeling the placement of homogeneous facilities (e.g., warehouses, public projects) in a geographic space, where agents use the least cost or closest facility. Voting is often modeled this way, where candidates are ordered along each of several dimensions (e.g., stance on environment, fiscal policy, etc.), voters have ideal points in this space, and one elects one or more candidates to a legislative body. Product design, customer segmentation, and other problems can be modeled as FLPs.

Even without explicit distance functions, it is often natural to assume agent (ordinal) preferences are *single-peaked*: an agent's preferences are constrained so that outcomes become less preferred as they are "moved away" from her ideal point (or *peak*). When preferences are single-peaked, the classic *median mechanism* and its generalizations [2,24] guarantee strategy-proofness. Sui *et al.* [29] develop *quantile mechanisms (QMs)* which extend these mechanisms to the multi-facility, multi-dimensional case. We focus here on QMs.

Definition 1. *[29] Let* $\mathbf{q} = \langle \mathbf{q}_1; \ldots; \mathbf{q}_m \rangle$ *be a* $m \times d$ *matrix, where each* $\mathbf{q}_j = \{q_j^1, \ldots, q_j^d\}$ *is a* d-vector in the unit cube. A $\mathbf{q}$-quantile mechanism $f_{\mathbf{q}}$ asks each agent i to report her ideal location (or peak) t_i. The mechanism locates each facility j at the q_j^k th quantile among the n reported peaks in each dimension k independently.*

Example 1. Consider the two-dimensional quantile mechanism in which $\mathbf{q} = \langle 0.25, 0.75; 1.0, 0.5 \rangle$. Given a peak profile of 5 agents $\mathbf{t} = ((1,4), (2,7), (4,2), (7,9), (8,3))$, the $\mathbf{q}$-quantile mechanism will locate the first facility at the intersection of quantile 0.25 in the first dimension and quantile 0.75 in the second dimension, i.e., $(2,7)$ in this example; and the second facility at $(8,4)$.

We note that quantile mechanisms are special case of *generalized median mechanisms (GMMs)* [2,24] when applied to single-FLPs, and can be interpreted as applying a specific form of GMM to the selection of each of the m facilities. As such, QMs are (individual) strategy-proof [29]. However, the characterization of Barberà et al. [2] shows that no (anonymous) mechanism can offer *group* strategy-proofness for multi-dimensional, multi-FLPs in general.[1] The main reason that group manipulation is possible is that a group of manipulators can submit a joint misreport of their ideal locations in which each of them increases her cost in some dimensions but decreases it in others, thereby achieving a lower total cost.

In this paper, we investigate the computational problem of finding just such a group manipulation. Specifically, we consider: (a) the formulation of the *optimal group manipulation problem* as mathematical programs of various types; (b) the computational complexity of this problem; and (c) how much manipulators might gain given optimal manipulations, under different cost functions, when GMMs/QMs are used.[2]

---

[1] Anonymity is critical, as dictatorial mechanisms belong to the class of GMMs and are group strategy-proof.

[2] Barberà et al.'s [2] characterizations do not preclude the existence of group strategy-proof mechanisms when specific cost functions are used. However, it is still meaningful to study the group manipulation of GMMs and QMs due to their simplicity and intuitive nature, their (individual) strategy proofness, and their flexibility. Indeed, these are the only "natural" such mechanisms for multi-dimensional, multi-FLPs of which we are aware.

Informally, the *optimal group manipulation problem* is that of finding a joint misreport for a group of manipulators such that the outcome induced by this misreport is such that: (a) the sum of costs of the manipulators is minimized; and (b) relative to the outcome that would have been induced by truthful reporting, no manipulator is worse-off and at least one is strictly better-off. Our objective of minimizing the sum of costs is the natural one, which represents the social welfare of the manipulators. While one general problem is whether there exists a joint misreport such that no one is worse-off, our optimization version subsumes the former problem.[3] We formalize this as follows:

Definition 2. *Let $N = S \cup M$, where S is a set of sincere agents and M is a set of manipulators with type vectors t_S and t_M. Let $f_{\mathbf{q}}$ be a QM with quantile matrix $\mathbf{q}$. Let $\mathbf{x_q} = f_{\mathbf{q}}(t_M, t_S)$ be the location vector chosen by $f_{\mathbf{q}}$ if all agents report their peaks truthfully, and $\mathbf{x'_q} = f_{\mathbf{q}}(t'_M, t_S)$ be the location vector chosen given some misreport t'_M by the manipulators M. The* optimal group manipulation problem *is to find a joint misreport t'_M for the agents in M satisfying:*

$$t'_M = \arg\min \sum\nolimits_{i \in M} s_i \left(\mathbf{x'_q}, t_i \right) \tag{1}$$

$$\text{s.t. } s_i \left(\mathbf{x'_q}, t_i \right) \leq s_i \left(\mathbf{x_q}, t_i \right), \quad \forall i \in M$$
$$s_i \left(\mathbf{x'_q}, t_i \right) < s_i \left(\mathbf{x_q}, t_i \right), \quad \text{for some } i \in M$$

Given a group of manipulators M, we generally refer to the remaining agents $S = N \backslash M$ as "sincere," though we need not presume that their reports are truthful in general, only that M knows (or can anticipate) their reports.

2.2   Related Work

There has been extensive study of the manipulation problem in other social choice, especially in the contect of voting. While the Gibbard-Satterthwaite theorem shows that strategy-proof mechanisms do not exist in general, Bartholdi et al. [3] demonstrated that manipulation of certain voting rules can be computationally difficult. This spawned an important line of research into the complexity of manipulation for many voting rules—collectively this can be viewed as proposing the use of computational complexity as a barrier to practical manipulation; see, for example, [8,13] for an excellent survey. Recent work has shown that when preferences are single-peaked, the *constructive manipulation problem*—in which a set of manipulators try to find a set of preference rankings (reports) that would make a specific candidate win—becomes polynomial time solvable for many voting rules [12]. Our work is similar in its objective to this approach, with a key difference being that in voting outcomes are discrete and atomic, whereas we deal with a continuous, multi-dimensional space.

---

[3] NP-hardness refers to the corresponding decision problem (as is colloquially understood for optimization problems): is there a misreport that gives the manipulators total cost less than epsilon (for any fixed epsilon). This implies NP-hardness of existence (set cost to truthful cost).

Exploiting computational complexity to prevent (or reduce the odds of) manipulation is somewhat problematic in that it focuses on worst-case scenarios, and usually assumes full knowledge of agent preferences. Recent work has studied average case manipulability (i.e., the probability that a preference profile is "easily" manipulable, assuming some distribution over preferences or preference profiles), and shows that manipulation is often feasible both theoretically and empirically [7,14,19,26,32,33]. The complete information assumption has also been challenged, and manipulation given probabilistic knowledge of other agent's preferences has been studied in equilibrium [1,21] and from an optimization perspective [20].

3   Group Manipulation for Single-FLPs

In this section, we address the problem of group manipulation for single-facility location problems, first describing its general form, then describing a linear programming formulation under the L_1-norm, and finally describing a second-order cone programming formulation under the L_2-norm.

3.1   Group Manipulation Specification

Recall from Definition 2 that a group manipulation is a set of misreports by the manipulating coalition M such that no manipulators is worse off and at least one is better off. The optimization formulation of this problem in Eq. (1) requires that one find the misreport that provides the greatest total benefit to the coalition. This explicit, straightforward formulation considers all possible misreports (i.e., the vector of purported "preferred" locations of each manipulator), which in principle induces a large search space. Fortunately, we can decrease the search space dramatically by only considering *viable* locations for manipulator misreports. We first define *viability*:

Definition 3. *Let $f_\mathbf{q}$ be a QM with quantile matrix $\mathbf{q}$. A location $x \in \mathbb{R}^d$ is viable for a manipulating coalition M if there exists a joint misreport t'_M s.t. $x = f_q(t'_M, t_S)$, where t_S is the report from the sincere agents $S = N \backslash M$. We say t'_M implements x in this case.*

The following critical proposition shows that, in single-FLPs, if a mechanism $f_\mathbf{q}$ selects a location $x'_\mathbf{q} = f_\mathbf{q}(t_S, t'_M)$ under a group manipulation t'_M, then it also selects $x'_\mathbf{q}$ if *each* manipulator misreports $x'_\mathbf{q}$ as her peak.

Proposition 1. *For single-FLPs, let t'_M be a group manipulation and $x'_\mathbf{q}$ be a viable location implemented by t'_M under mechanism $f_\mathbf{q}$. Then $x'_\mathbf{q}$ is also implemented by the group manipulation $t^*_M = \{x'_\mathbf{q}, \ldots, x'_\mathbf{q}\}$.*

Proof (Sketch). We provide a sketch of proof for $d = 2$, but the analysis can be easily generalized. Consider an arbitrary group manipulation t'_M, which implements location $x'_\mathbf{q} = f_\mathbf{q}(t'_M, t_S) \in \mathbb{R}^2$ (as shown in Fig. 1). Let us denote the

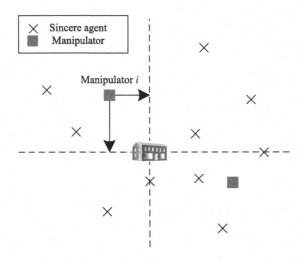

Fig. 1. Each manipulator can move her misreport to $x'_\mathbf{q}$ without changing the outcome.

misreport of each manipulator by $t'_i = (t'^1_i, t'^2_i), \forall i \in M$ and the location by $x'_\mathbf{q} = (x'^1_\mathbf{q}, x'^2_\mathbf{q})$.

Pick an arbitrary manipulator $i \in M$, and assume w.l.o.g. that $t'^1_i \leq x'^1_\mathbf{q}$ and $t'^2_i \geq x'^2_\mathbf{q}$. We construct another group manipulation t''_M by changing the misreport of manipulator i to $x'_\mathbf{q}$. Recall that the mechanism $f_\mathbf{q}$ locates the facility at a specified quantile, so we have:

$$
\begin{aligned}
f_\mathbf{q}(t'_M, t_S) &= f_\mathbf{q}((t'_i, t'_{M \setminus i}), t_S) \\
&= f_\mathbf{q}(((t'^1_i, x'^2_\mathbf{q}), t'_{M \setminus i}), t_S) \\
&= f_\mathbf{q}(((x'^1_\mathbf{q}, x'^2_\mathbf{q}), t'_{M \setminus i}), t_S) \\
&= f_\mathbf{q}((x'_\mathbf{q}, t'_{M \setminus i}), t_S) = f_\mathbf{q}(t''_M, t_S)
\end{aligned}
$$

Repeating this procedure over all manipulators completes our proof.

Proposition 1 demonstrates that we can limit our attention to the "unanimous" reporting of viable locations when searching for optimal misreports, without considering misreports that reveal locations that cannot be implemented or realized by the manipulators. Therefore, we can reformulate the optimal group manipulation problem (Definition 2) as follows:

Definition 4. *Let $f_\mathbf{q}$ be a QM with quantile matrix $\mathbf{q}$. Let $x_\mathbf{q} = f_\mathbf{q}(t_M, t_S)$ and $x'_\mathbf{q} = f_\mathbf{q}(t'_M, t_S)$ be the location chosen by $f_\mathbf{q}$ under truthful reports and misreport t'_M, resp. Optimal group manipulation can be reformulated as:*

$$\min_{x \in \mathbb{R}^d} \sum_{i \in M} s_i(x, t_i) \tag{2}$$

$$\text{s.t.} \quad s_i(x, t_i) \leq s_i(x_{\mathbf{q}}, t_i), \quad \forall i \in M \tag{3}$$

$$s_i(x, t_i) < s_i(x_{\mathbf{q}}, t_i), \quad \text{for some } i \in M \tag{4}$$

$$x \text{ is a viable location under } f_{\mathbf{q}} \tag{5}$$

In the sequel, our specific formulations of the problem will rely on Definition 4. We can also safely omit the constraints in Eq. 4, as they can easily be checked after the fact given the optimized location vector—if no manipulator is strictly better off, then a group manipulation obviously cannot exist.

3.2 LP Formulation Under the L_1-norm

We now consider the formulation of optimal manipulation when the L_1-norm is used as the cost function, i.e., $s_i(x, t_i) = \sum_{k \leq d} |x^k - t_i^k|$ for any location $x \in \mathbb{R}^d$. Let $x = (x^1, \ldots, x^d)$ represent the location to be optimized (i.e., the location induced by the manipulation) in single-FLPs, where each x^k is a continuous variable. Let c_i be a continuous variable denoting the cost of manipulator i given outcome x. We can formulate the objective function Eq. (2), and the constraints Eq. (3), as follows:

$$\min_x \sum_{i \in M} c_i \tag{6}$$

$$\text{s.t.} \quad c_i = \sum_{k \leq d} |x^k - t_i^k|, \quad \forall i \in M \tag{7}$$

$$0 \leq c_i \leq u_i, \quad \forall i \in M \tag{8}$$

where $u_i = s_i(x_{\mathbf{q}}, t_i)$ is the cost of manipulator i under a truthful report t_M by the manipulators.

This formulation contains absolute values in the nonlinear constraints (7). We introduce an additional set of variables to linearize these constraints. Letting D_i^k be an upper bound on the distance between t_i and x in the kth dimension, we linearize the constraints (7) as follows:

$$-D_i^k \leq t_i^k - x^k \leq D_i^k, \quad \forall i \in M, \forall k \leq d \tag{9}$$

$$D_i^k \geq 0, \quad \forall i \in M, \forall k \leq d \tag{10}$$

$$c_i = \sum_{k \leq d} D_i^k, \quad \forall i \in M \tag{11}$$

Finally, we need constraints that guarantee the new location x is viable. Recall that a QM locates the facility at a specified quantile of reported peaks in each dimension independently, and by Proposition 1 we can assume w.l.o.g. that all manipulators use the same misreport. This implies that a viable location for the facility is bounded by the reported coordinates of two sincere agents in each dimension. Formally, let $\mathbf{q} = (q^1, \ldots, q^d)$ be the quantile vector (for single-FLPs, we have a single vector rather than a full matrix), and let

$$\perp^k = \min\{z \in \mathbb{Z}^+ : z + |M| \geq q^k \cdot n\} \text{ and}$$

$$\top^k = \max\{z \in \mathbb{Z}^+ : |S| + |M| - z \geq (1 - q^k) \cdot n\}.$$

If we let $\bar{x}_S^k = \{\bar{x}_1^k, \ldots, \bar{x}_{|S|}^k\}$ denote the ordered coordinates of the reports of agents S in the kth dimension, we have:

Lemma 1. *For single-FLPs, a location* $x = (x^1, \ldots, x^d) \in \mathbb{R}^d$ *is viable if and only if* $\bar{x}^k_{\perp_k} \leq x^k \leq \bar{x}^k_{\top_k}, \forall k \leq d$.

This lemma ensures that we can use the following boundary constraints as to enforce viability (see Eq. (5)):

$$\bar{x}^k_{\perp_k} \leq x^k \leq \bar{x}^k_{\top_k}, \qquad \forall k \leq d \tag{12}$$

To summarize, we can formulate the optimal group manipulation under the L_1-norm as an LP. The objective function (6) minimizes the sum of costs over all manipulators. Constraints (8)–(11) guarantee that no manipulators is worse-off, and constraints (12) ensure that the optimized location induced by the misreport is viable. The LP has $O(d|M|)$ variables. We state this result formally in the following theorem:

Theorem 1. *The optimal group manipulation problem for single facility location under the* L_1-norm *can be formulated as a linear program (LP), with objective function (6) and constraints (8)–(12).*

As such, the optimal manipulation problem can be solved in polynomial time.

3.3 SOCP Formulation Under the L_2-norm

The optimization formulation for the L_1-norm above can be easily modified to account for L_2-costs. Specifically, we need only a minor modification of the constraints (11) to incorporate Euclidean distances as follows:

$$(c_i)^2 \geq \sum_{k \leq d} \left(D_i^k\right)^2, \qquad \forall i \in M \tag{13}$$

Constraint (13), combined with the objective function (6) and constraints (8)–(10) and (12), constitutes a second-order cone program (SOCP) under the L_2-norm:

Theorem 2. *The optimal group manipulation problem for the single facility location under the* L_2-norm *can be formulated as a second-order cone program (SOCP), with objective function (6) and constraints (8)–(10) and (12)–(13).*

Since SOCPs can be solved in polynomial time, we have:

Remark 1. The optimal group manipulation problem for single-facility location under both the L_1- and L_2-norms can be solved in polynomial time.

4   Group Manipulation for Multi-FLPs

In this section, we extend our analysis of group manipulation to multi-facility location problems. Unlike single-FLPs, we show that problem in computationally intractable for multi-FLPs, under both the L_1- and L_2-norms. However, we provide mathematical programming models that are often quite efficient in practice.

4.1   The Complexity of Group Manipulation

We first show group manipulation is NP-hard for multi-FLPs.

Theorem 3. *Optimal group manipulation for multi-facility location under either the L_1- or L_2-norms is NP-hard.*

This hardness result is proved using a reduction from the geometric p-median problem, which is known to be NP-hard under both L_1- and L_2-distance [23]. Given a set of points in the d-dimensional space ($d \geq 2$), the *geometric p-median* is a set P of p points that minimizes the sum of distances between each given point and its closest point in P. A complete proof is provided in a longer version of this paper. The rough intuition is as follows. Considering the optimal group manipulation problem for $(p+1)$ facilities, our proof assumes no sincere agents, and constructs a manipulator location profile and QM $f_{\mathbf{q}}$ such that all $(p+1)$ facilities are located at a single extreme position by $f_{\mathbf{q}}$ given truthful reports. However, the optimal group manipulation induces the mechanism to "spread out" p of the facilities to the benefit of a subset of the manipulators, without harming those who would use the original position. This constitutes an optimal solution to the p-median problem for the p non-extreme locations. As such, an algorithm for optimal group manipulation can be used to solve the p-median problem.

While this implies that worst-case instances may be difficult to solve, it does not mean that instances arising in practice can't be solved efficiently. We now describe formulations of optimal group manipulation for multi-FLPs as integer programs that may support practical solution. Our formulations are quite compact, and combined with the empirical evaluation in Sect. 5, suggest that optimal group manipulations can be found reasonably quickly.

4.2   MILP Formulation Under the L_1-norm

We first describe our mixed integer linear programming (MILP) formulation of optimal group manipulation under the L_1-norm. Due to space limitations, we defer certain technical details and proofs to a longer version of this paper. The following result is analogous to Proposition 1 for single-FLPs.

Proposition 2. *Let t'_M be a group manipulation and $\mathbf{x} = \{(x_1^1, \ldots, x_1^d), \ldots, (x_m^1, \ldots, x_m^d)\}$ be a viable location vector implemented by t'_M. Let $\mathbf{X}^k = \{x_1^k, \ldots, x_m^k\}$ denote the set of coordinates of these facilities in the kth dimension. Then there exists a group manipulation t^*_M that implements $\mathbf{x}$, where $t_i^* \in \prod_{k \leq d} \mathbf{X}^k, \forall i \in M$.*

In other words, we can assume w.l.o.g. that manipulators misreports are drawn from the "intersection positions" in different dimensions induced by the different facilities. The precise misreports at these intersection positions must be coordinated to guarantee a viable location vector (see below).

Let $\mathbf{x} = \{(x_1^1, \ldots, x_1^d), \ldots, (x_m^1, \ldots, x_m^d)\}$ represent the location vector to be optimized. Let c_i be the cost of manipulator i given outcome $\mathbf{x}$, c_{ij} be the cost of manipulator i w.r.t. facility j, and I_{ij} be an indicator variable whose value is 1 iff the closest facility for manipulator i is j. We can formulate the objective Eq. (2), and the constraints Eq. (3), as follows:

$$\min_{\mathbf{x} \in (\mathbb{R}^d)^m} \sum_{i \in M} c_i \tag{14}$$

$$\text{s.t.} \quad c_i = \sum_{j \leq m} I_{ij} \cdot c_{ij}, \qquad \forall i \in M \tag{15}$$

$$\sum_{j \leq m} I_{ij} = 1, \qquad \forall i \in M \tag{16}$$

$$I_{ij} \in \{0, 1\}, \qquad \forall i \in M, \forall j \leq m \tag{17}$$

$$0 \leq c_i \leq u_i, \qquad \forall i \in M, \forall j \leq m \tag{18}$$

$$c_{ij} \geq 0, \qquad \forall i \in M, \forall j \leq m \tag{19}$$

where $u_i = s_i(\mathbf{x_q}, t_i)$ is the cost of manipulator i under a truthful report t_M by the manipulators.

Since both I_{ij} and c_{ij} are variables in constraint (15), we must linearize these quadratic terms by introducing additional variables. Let O_{ij} be some upper bound on the product of I_{ij} and c_{ij}. We can then replace the constraint (15) by

$$c_i = \sum_{j \leq m} O_{ij}, \qquad \forall i \in M \tag{20}$$

$$O_{ij} \geq c_{ij} + (I_{ij} - 1)U, \quad \forall i \in M, \forall j \leq m \tag{21}$$

$$O_{ij} \geq 0, \qquad \forall i \in M, \forall j \leq m \tag{22}$$

where U is any upper bound on manipulator cost.

Let D_{ij}^k be an upper bound on the distance between manipulator i and facility j in the kth dimension. We have:

$$-D_{ij}^k \leq t_i^k - x_j^k \leq D_{ij}^k, \quad \forall i \in M, \forall j \leq m, \forall k \leq d \tag{23}$$

$$D_{ij}^k \geq 0, \qquad \forall i \in M, \forall j \leq m, \forall k \leq d \tag{24}$$

$$c_{ij} = \sum_{k \leq d} D_{ij}^k, \qquad \forall i \in M, \forall j \leq m \tag{25}$$

Finally, we must ensure that $\mathbf{x}$ is viable. Let

$$\perp_j^k = \min\{z \in \mathbb{Z}^+ : z + |M| \geq q_j^k \cdot n\} \text{ and}$$

$$\top_j^k = \max\{z \in \mathbb{Z}^+ : |S| + |M| - z \geq (1 - q_j^k) \cdot n\}$$

and $\bar{x}_S^k = \{\bar{x}_1, \ldots, \bar{x}_{|S|}\}$ be the ordered coordinates of the reports of sincere agents in S in the kth dimension. We break $[\bar{x}_{\perp_j^k}^k, \bar{x}_{\top_j^k}^k]$ into several (ordered) close and open intervals: $[\bar{x}_{\perp_j^k}^k, \bar{x}_{\perp_j^k}^k], (\bar{x}_{\perp_j^k}^k, \bar{x}_{\perp_j^k+1}^k), \ldots, (\bar{x}_{\top_j^k-1}^k, \bar{x}_{\top_j^k}^k), [\bar{x}_{\top_j^k}^k, \bar{x}_{\top_j^k}^k]$ (see Fig. 2 for an illustration). Let Δ_j^k index these intervals ($0 \leq \Delta_j^k < 2|M| + 1$), and let $I_{\Delta_j^k}$ be an indicator variable whose value is 1 iff the coordinate of facility j is contained in the Δ_j^kth interval in the kth dimension. We then have:

$$\sum_{\Delta_j^k} I_{\Delta_j^k} = 1, \qquad \forall j \leq m, \forall k \leq d \tag{26}$$

$$\sum_{\Delta_j^k} I_{\Delta_j^k} \bar{x}_l^k \leq x_j^k \leq \sum_{\Delta_j^k} I_{\Delta_j^k} \bar{x}_r^k, \qquad \forall j \leq m, \forall k \leq d \tag{27}$$

$$I_{\Delta_j^k} \in \{0, 1\}, \qquad \forall j \leq m, \forall k \leq d \tag{28}$$

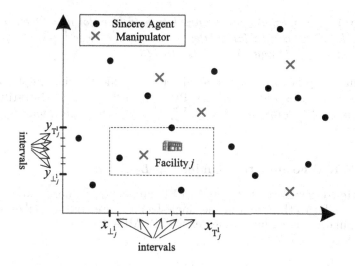

Fig. 2. For each facility in each dimension, the boundaries are split into small intervals, each bounded by one/two sincere agents.

where $l = \perp_j^k + \lfloor \Delta_j^k/2 \rfloor$ and $r = \perp_j^k + \lfloor (\Delta_j^k + 1)/2 \rfloor$.

For each interval, we pre-compute the number of sincere agents that lie to the left of and right of it (including equality) in each dimension k, which we denote by $L_{\Delta_j^k}$ and $R_{\Delta_j^k}$, respectively. We also introduce another indicator variable T_{ij}^k whose value is 1 iff manipulator i misreports the location of facility j in the kth dimension (this binary variable can be relaxed, since all terms in (29) and (30) are integral). Given a quantile matrix $\mathbf{q}$, the location vector $\mathbf{x}$ to be optimized is viable if the following constraints are satisfied:

$$\sum\nolimits_{\Delta_j^k} I_{\Delta_j^k} L_{\Delta_j^k} + \sum\nolimits_{j' \leq_\mathbf{q} j} \sum\nolimits_{i \in M} T_{ij'}^k \geq n q_j^k, \quad \forall j, \forall k \tag{29}$$

$$\sum\nolimits_{\Delta_j^k} I_{\Delta_j^k} R_{\Delta_j^k} + \sum\nolimits_{j' \geq_\mathbf{q} j} \sum\nolimits_i T_{ij'}^k \geq n(1 - q_j^k), \forall j, \forall k \tag{30}$$

$$\sum\nolimits_{j \leq m} T_{ij}^k = 1, \qquad\qquad \forall i \in M, \forall k \leq d \tag{31}$$

$$T_{ij}^k \in [0,1], \qquad\qquad \forall i \in M, \forall j \leq m, \forall k \leq d \tag{32}$$

The LHS of constraint (29) indicates the total number of sincere agents (the first term) and manipulators (the second term) to the left of (or at) facility j in the kth dimension, where $j' \leq_\mathbf{q} j$ denotes the facility j' to the left of j in the kth dimension, (i.e., $q_{j'}^k \leq q_j^k$). According to $f_\mathbf{q}$, this number should be greater than or equal to $n q_j^k$. Constraints (30) are similar, but used to count from the right. Constraints (31) and (32) ensure that each manipulator reports the location of one facility on each dimension.

To summarize, we can formulate optimal group manipulation for multi-FLPs under the L_1-norm as a MILP with $O(dm|M|)$ binary and continuous variables:

Theorem 4. *The optimal group manipulation problem for multi-facility location under the L_1-norm can be formulated as a mixed integer linear program with objective function (14) and constraints (16)–(32).*

The final step is to construct a misreport profile t'_M that implements the location vector optimized above. By Proposition 2, we can arbitrarily choose a set of manipulators of size exactly $\sum_i T_{ij}^k$ for each target facility j in each dimension k.

4.3   MISOCP Formulation Under the L_2-norm

When optimizing misreports for multi-FLPs under the L_2-norm, we can use an approach similar to that used in the single-facility case, and formulate the optimal manipulation as an mixed-integer SOCP (MISOCP). We need only modify constraints (25) as follows:

$$(c_{ij})^2 \geq \sum_{k \leq d} \left(D_{ij}^k \right)^2, \qquad \forall i \in M, \forall j \leq m \tag{33}$$

Using this we obtain the following result:

Theorem 5. *The optimal group manipulation problem for multi-FLPs under the L_2-norm can be formulated as a mixed integer second-order cone program, with objective function (14), and constraints (16)–(24) and (26)–(33).*

5   Empirical Evaluation

In this section, we evaluate the efficiency of the formulations outlined above. We provide empirical results only for multi-facility problems here (since the optimal manipulation problem for single-FLPs is poly-time solvable), testing the efficiency of the MILP/ MISOCP described in Sect. 4.

We test two problems. The first is a two-dimensional, two-facility location problem under the L_2-norm, where the quantile matrix used is $\mathbf{q} = \{0.3, 0.4; 0.8, 0.7\}$. The second is a four-dimensional, three-facility location problem under the L_1-norm, where the quantile matrix used is $\mathbf{q} = \{0.1, 0.6, 0.4, 0.9; 0.4, 0.2, 0.8, 0.6; 0.7, 0.8, 0.3, 0.4\}$. For both problems, we vary the number of sincere agents $|S| \in \{100, 200, 500\}$, and the number of manipulators $|M| \in \{5, 10, 20, 50, 100, 200\}$. We randomly generated 100 problems instances for each parameter setting in which the peaks of both the sincere agents and the manipulators are randomly drawn from the same data set (data sets are explained in detail below). We compute the average execution time of our MILP/MISOCP models, and the probability of manipulation (i.e., the proportion of the 100 instances in which a viable manipulation exists for the randomly chosen manipulators).

For the two-dimensional problem, we use preference data from the Dublin west constituency in the 2002 Irish General Election. Since the data includes only voter *rankings* over the set of candidates, the ideal location of each voter is

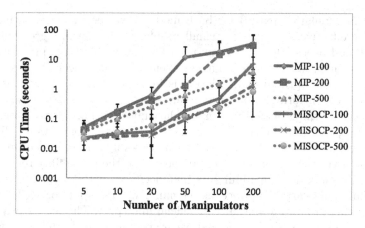

Fig. 3. Time to compute an optimal manipulation (y-axis is log-scale, x-axis is approx. log-scale). Error bars show sample st. dev.

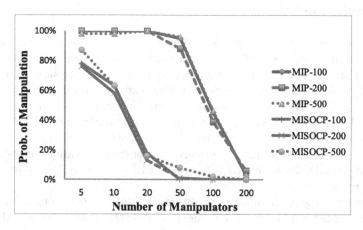

Fig. 4. Probability of a manipulation existing (y-axis is log-scale, x-axis is approx. log-scale).

unknown. Fortunately, recent analysis suggests that this data is approximately single-peaked in two-dimensions [30], and a spatial model using L_2 distance can be used to explain voter preferences [17]. We fit this data to a two-dimensional spatial model, and estimate the voter peaks and candidate positions in the underlying latent space so constructed (Details are provided in a longer version of this paper.) For the four-dimensional problem, we use a synthetic data set in which the peaks of both sincere agents and manipulators are randomly generated from a uniform distribution on the unit cube.

For each instance, the MILP/MISOCP is solved using CPLEX 12.51, on a 2.9GHz, quad-core machine with 8GB memory. Figure 3 shows the average computation time required to find the optimal group manipulation (or show that no group manipulation exists) for both models. We see that our formulations admit very effective solution—for small problems, the optimal group manipulation is found in less than 1 second; even for reasonably large problems, such as the four-dimensional, three-facility problem with 100 sincere agents and 200 manipulators, the optimal manipulation is found in 35.47 s (on average). The performance of our formulations is also very stable (see error bars in the figure).

We illustrate the probability of manipulation for both problems in Fig. 4. For 2D problems, the probability of manipulation decreases from around 80 % to 0 quickly, indicating that it is very hard for a randomly selected set of manipulators to find a viable manipulation; for 4D problems, the probability remains high (close to 1) even with 20 manipulators then decreases with larger sets of manipulator. This is not surprising since, as the number of manipulator get larger, it is harder for them to find a mutually beneficial misreport. The higher probability for 4D problems is due to the fact that we are placing three facilities rather than two, increasing the potential of viable manipulations.

6   Conclusion

In this paper, we addressed the optimal group manipulation problem in multi-dimensional, multi-facility location problems. Specifically, we analyzed the computational problems of manipulating quantile mechanisms. We showed that optimal manipulation for single-facility problems can be formulated as an LP or SOCP, under the L_1- and L_2-norm, respectively, and thus can be solved in polynomial time. By contrast, the optimal manipulation problem for multi-facility problems is NP-hard, but can be formulated as an ILP or MISOCP under the L_1- and L_2-norm, respectively. Our empirical evaluation shows that our MILPs/MISOCPs formulation for multi-FLPs scales well, despite the NP-hardness result.

Our work suggests a number of interesting future directions. First, more empirical results would be helpful in understanding the practical ease or difficulty of group manipulation, as well as the probability of manipulation, the potential gain of manipulators, and the impact on social welfare. Second, other objectives for the manipulating coalition (e.g., minimizing the maximum cost), and mechanisms with other cost functions are also of interest. Finally, some research [17, 25, 30] has shown that agent preferences are often not exactly single-peaked, but may be approximately so under some forms of approximation [9–11, 15]. The theoretical and empirical evaluation of group manipulation in such settings would be extremely valuable.

References

1. Ballester, M.Á., Rey-Biel, P.: Does uncertainty lead to sincerity? simple and complex voting mechanisms. Soc. Choice Welf. **33**(3), 477–494 (2009)
2. Barberà, S., Gul, F., Stacchetti, E.: Generalized median voter schemes and committees. J. Econ. Theory **61**(2), 262–289 (1993)

3. Bartholdi III, J.J., Tovey, C.A., Trick, M.A.: The computational difficulty of manipulating an election. Soc. Choice Welf. **6**(3), 227–241 (1989)

4. Black, D.: On the rationale of group decision-making. J. Polit. Econ. **56**(1), 23–34 (1948)

5. Boyd, S., Vandenberghe, L.: Convex Optimization. Cambridge University Press, Cambridge (2004)

6. Clarke, E.H.: Multipart pricing of public goods. Public Choice **11**(1), 17–33 (1971)

7. Conitzer, V., Sandholm, V.: Nonexistence of voting rules that are usually hard to manipulate. In: Proceedings of the Twenty-first National Conference on Artificial Intelligence (AAAI 2006), pp. 627–634, Boston (2006)

8. Conitzer, V., Sandholm, T., Lang, J.: When are elections with few candidates hard to manipulate? J. ACM **54**(3), 1–33 (2007)

9. Erdélyi, G., Lackner, M., Pfandler, A.: The complexity of nearly single-peaked consistency. In: Proceedings of the Fourth International Workshop on Computational Social Choice (COMSOC 2012), Kraków, Poland (2012)

10. Escoffier, B., Lang, J., Öztürk, M.: Single-peaked consistency and its complexity. In: Proceedings of the Eighteenth European Conference on Artificial Intelligence (ECAI 2008), Patras, Greece, pp. 366–370 (2008)

11. Faliszewski, P., Hemaspaandra, E., Hemaspaandra, L.A.: The complexity of manipulative attacks in nearly single-peaked electorates. In: Proceedings of the Thirteenth Conference on Theoretical Aspects of Rationality and Knowledge (TARK 2011), Groningen, The Netherlands, pp. 228–237 (2011)

12. Faliszewski, P., Hemaspaandra, E., Hemaspaandra, L.A., Rothe, J.: The shield that never was: societies with single-peaked preferences are more open to manipulation and control. In: Proceedings of the 12th Conference on Theoretical Aspects of Rationality and Knowledge, pp. 118–127. ACM (2009)

13. Faliszewski, P., Procaccia, A.D.: AI's war on manipulation: are we winning? AI Mag. **31**(4), 53–64 (2010)

14. Friedgut, E., Kalai, G., Nisan, N.: Elections can be manipulated often. In: Proceedings of the 49th Annual IEEE Symposium on the Foundations of Computer Science (FOCS 2008), Philadelphia, pp. 243–249 (2008)

15. Galand, L., Cornaz, D., Spanjaard, O.: Bounded single-peaked width and proportional representation. In: Proceedings of the Fourth International Workshop on Computational Social Choice (COMSOC 2012), Kraków, Poland (2012)

16. Gibbard, A.: Manipulation of voting schemes: a general result. Econometrica **41**(4), 587–601 (1973)

17. Gormley, I.C., Murphy, T.B.: A latent space model for rank data. In: Airoldi, E.M., Blei, D.M., Fienberg, S.E., Goldenberg, A., Xing, E.P., Zheng, A.X. (eds.) ICML 2006. LNCS, vol. 4503, pp. 90–102. Springer, Heidelberg (2007)

18. Groves, T.: Incentives in teams. Econometrica **41**, 617–631 (1973)

19. Isaksson, M., Kindler, G., Mossel, E.: The geometry of manipulationa quantitative proof of the gibbard-satterthwaite theorem. Combinatorica **32**(2), 221–250 (2012)

20. Lu, T., Tang, P., Procaccia, A.D., Boutilier, C.: Bayesian vote manipulation: optimal strategies and impact on welfare. In: Proceedings of the Twenty-eighth Conference on Uncertainty in Artificial Intelligence (UAI 2012), Catalina, CA, pp. 543–553 (2012)

21. Majumdar, D., Sen, A.: Ordinally Bayesian incentive compatible voting rules. Econometrica **72**(2), 523–540 (2004)

22. Mas-Colell, A., Whinston, M.D., Green, J.R.: Microeconomic Theory. Oxford University Press, New York (1995)

23. Megiddo, N., Supowit, K.J.: On the complexity of some common geometric location problems. SIAM J. Comput. **13**(1), 182–196 (1984)
24. Moulin, H.: On strategy-proofness and single peakedness. Public Choice **35**(4), 437–455 (1980)
25. Poole, K.T., Rosenthal, H.: A spatial model for legislative roll call analysis. Am. J. Polit. Sci. **29**(2), 357–384 (1985)
26. Procaccia, A.D., Rosenschein, J.S.: Junta distributions and the average-case complexity of manipulating elections. J. Artif. Intell. Res. **28**, 157–181 (2007)
27. Satterthwaite, M.A.: Strategy-proofness and Arrow's conditions: Existence and correspondence theorems for voting procedures and social welfare functions. Journal of Economic Theory **10**, 187–217 (1975)
28. Schummer, J., Vohra, R.V.: Mechanism design without money. In: Nisan, N., Roughgarden, T., Tardos, E., Vazirani, V.V. (eds.) Algorithmic Game Theory, pp. 243–265. Cambridge University Press, Cambridge (2007)
29. Sui, X., Boutilier, C., Sandholm, T.: Analysis and optimization of multi-dimensional percentile mechanisms. In: Proceedings of the Twenty-third International Joint Conference on Artificial Intelligence (IJCAI 2013), Beijing (2013)
30. Sui, X., Francois-Nienaber, A., Boutilier, C.: Multi-dimensional single-peaked consistency and its approximations. In: Proceedings of the Twenty-third International Joint Conference on Artificial Intelligence (IJCAI 2013), Beijing (2013)
31. Vickrey, W.: Counterspeculation, auctions, and competitive sealed tenders. J. Finance **16**(1), 8–37 (1961)
32. Walsh, T.: Where are the really hard manipulation problems? the phase transition in manipulating the veto rule. In: Proceedings of the Twenty-first International Joint Conference on Artificial Intelligence (IJCAI 2009), Pasadena, CA, pp. 324–329 (2009)
33. Xia, L., Conitzer, V.: A sufficient condition for voting rules to be frequently manipulable. In: Proceedings of the Ninth ACM Conference on Electronic Commerce (EC 2008), Chicago, pp. 99–108 (2008)

Fairness and Rank-Weighted Utilitarianism in Resource Allocation

Tobias Heinen, Nhan-Tam Nguyen$^{(\boxtimes)}$, and Jörg Rothe

Heinrich-Heine-Universität Düsseldorf, Düsseldorf, Germany
`tobias.heinen@uni-duesseldorf.de,`
`{nguyen,rothe}@cs.uni-duesseldorf.de`

Abstract. In multiagent resource allocation with indivisible goods, boolean fairness criteria and optimization of inequality-reducing collective utility functions (CUFs) are orthogonal approaches to fairness. We investigate the question of whether the proposed scale of criteria by Bouveret and Lemaître [5] applies to nonadditive utility functions and find that only the more demanding part of the scale remains intact for k-additive utility functions. In addition, we show that the min-max fair-share allocation existence problem is NP-hard and that under strict preferences competitive equilibrium from equal incomes does not coincide with envy-freeness and Pareto-optimality. Then we study the approximability of rank-weighted utilitarianism problems. In the special case of rank dictator functions the approximation problem is closely related to the MAXMIN-FAIRNESS problem: Approximation and/or hardness results would immediately transfer to the MAXMIN-FAIRNESS problem. For general inequality-reducing rank-weighted utilitarianism we show (strong) NP-completeness. Experimentally, we answer the question of how often maximizers of rank-weighted utilitarianism satisfy the max-min fair-share criterion, the weakest fairness criterion according to Bouveret and Lemaître's scale. For inequality-reducing weight vectors there is high compatibility. But even for weight vectors that do not imply inequality-reducing CUFs, the Hurwicz weight vectors, we find a high compatibility that decreases as the Hurwicz parameter decreases.

Keywords: Fair division · Indivisible goods · Fairness · Inequality reduction · Computational complexity

1 Introduction

Resource allocation deals with the distribution of goods to agents. We study the case of indivisible goods and cardinal preferences. The quality of allocations can be judged in terms of, e.g., efficiency or fairness. Bouveret and Lemaître [5] proposed a scale (hierarchy) of criteria for fairness. Starting with the weakest, the max-min fair-share criterion, to the strongest, competitive equilibrium from equal incomes, each criterion becomes more demanding towards an allocation.

The approach of optimizing social welfare (see, e.g., the work of Nguyen et al. [19,20]) is orthogonal to the idea of applying fairness criteria, especially

© Springer International Publishing Switzerland 2015
T. Walsh (Ed.): ADT 2015, LNAI 9346, pp. 521–536, 2015.
DOI: 10.1007/978-3-319-23114-3_31

rank-weighted utilitarianism (see, e.g., the book by Moulin [18] and the book chapter by d'Aspremont and Gevers [11]). A collective utility function maps allocations to numerical values. A rank-weighted utilitarian collective utility function is a weighted utilitarian collective utility function, where the weight depends on the sorted rank of utilities. If the weight decreases for agents with more utility, we have an inequality-reducing collective utility function (e.g., [12,16]). Thus maximizers of rank-weighted utilitarianism can also be considered fair allocations. Rank-weighted utilitarianism is also known as ordered weighted averaging [22] and is closely related to so-called Pigou-Dalton transfers (see, e.g., again the book by Moulin [18]).

In our model we make the common assumptions that goods are indivisible, preferences are represented numerically via utility functions, allocations are determined centrally, and agents can trade neither goods nor money. We extend the study of Bouveret and Lemaître [5] by answering the question of whether their proposed scale of criteria holds for nonadditive utility functions. More specifically, by lifting the utility functions to k-additive domains for $k \geq 2$, we find that only the more demanding part of the scale remains intact. The domain of k-additive utility functions is more realistic because substitutabilities can now be expressed (depending on the choice of k). They were first introduced by Grabisch [15] and later by Chevaleyre et al. [10] in resource allocation. In fact, for large enough k we can express any utility function defined over bundles. In addition, we (partially) close two problems left open by Bouveret and Lemaître [5]: We show that it is NP-hard to decide whether a min-max fair-share allocation exists, and we answer (in the negative) the question of whether every envy-free and Pareto-optimal allocation forms a competitive equilibrium from equal incomes under strict preferences.

After that we study the approximability problem of rank-weighted utilitarian CUFs. We explore the relationship between rank dictator functions (that is, rank-weighted utilitarian CUFs where exactly one agent has unit weight) and give first answers to the approximability of rank-weighted utilitarianism.

Then we report on computational experiments that we performed to see how often rank-weighted utilitarianism coincides with the max-min fair-share criterion, the weakest criterion with respect to Bouveret and Lemaître's scale. We find that weakly inequality-reducing rank-weighted utilitarian CUFs are highly compatible with max-min fair-share. For Hurwicz weight vectors, where the worst-off and best-off agents only are assigned positive weights, the number of maximizing allocations that also satisfy the max-min fair-share criterion decreases as the weight distribution is shifted towards the best-off agent. This should be contrasted to Amanatidis et al.'s [1] statement that MaxMin-fairness (egalitarian social welfare) and max-min fair-share "exhibit very different behavior" beyond identical agents. Experiments in a similar spirit were performed by Brams and King [6] on maximin and Borda maximin allocations with respect to envy and by Bouveret and Lemaître [5] with respect to MaxMin-fairness and max-min fair-share, where they find that "approximately only one instance over 3500 is a counter-example," and with respect to their proposed scale.

In addition, Bouveret and Lemaître show that for identical preferences MaxMin-fairness and max-min fair-share coincide, and, similarly, for normalized utilities, if there is an allocation that satisfies proportional fair-share, then every maximizer of MaxMin-fairness satisfies it as well.

In Sect. 2 we introduce the model formally and study the various fairness criteria, continuing in Sect. 3 with approximability results for rank-weighted utilitarianism. In Sect. 4 we study both fairness approaches simultaneously and report on our experimental results. Lastly, we conclude in Sect. 5.

2   Model and Fairness Criteria

Let $A = \{a_1, \ldots, a_n\}$ be the set of *agents*, and let $G = \{g_1, \ldots, g_m\}$ be the set of *goods*. Each agent a_i has a *utility function* u_i : $2^G \to \mathbb{Q}$ that maps a bundle of goods to a utility value. We assume that utility functions are k-additive. A utility function u is k-*additive* if for each $H \subseteq G$, there is a coefficient $w(H)$ with $w(H) = 0$ if $\|H\| > k$ such that $u(H) = \sum_{F \subseteq H, \|F\| \leq k} w(F)$. We denote the utility value for singletons $g \in G$ by $u(g)$. A tuple $(A, G, \{u_i\}_{a_i \in A})$ is called an *allocation setting*. Given such a setting, an *allocation* is a partition $\pi = (\pi_1, \ldots, \pi_n)$ of the set of goods, where π_i denotes the bundle that agent a_i receives. The utility that agent a_i realizes under π is given by $u_i(\pi) = u_i(\pi_i)$.

Bouveret and Lemaître [5] introduced a scale of criteria for 1-additive utility functions and showed the following chain of implications for an allocation π:

$$\pi \models \text{CEEI} \implies \pi \models \text{EF} \implies \pi \models \text{mFS} \implies \pi \models \text{PFS} \implies \pi \models \text{MFS},$$

where "$\pi \models X$" indicates that allocation π satisfies criterion X, CEEI stands for "competitive equilibrium from equal incomes," EF for "envy-freeness," mFS for "min-max fair-share," PFS for "proportional fair-share," and MFS for "max-min fair-share."

We start with the two weakest criteria, max-min and proportional fair-share.

Definition 1. *Let* $(A, G, \{u_i\}_{a_i \in A})$ *be an allocation setting. The* max-min fair-share (MFS) *of agent* a_i *is*

$$u_i^{MFS} = \max_{\pi} \min_{a_j \in A} u_i(\pi_j).$$

An allocation π *satisfies the* max-min fair-share *criterion if for every agent* a_i, *we have* $u_i(\pi) \geq u_i^{MFS}$.

Intuitively, this criterion corresponds to a game where agent a_i chooses an allocation first but picks a bundle of the chosen allocation last, thus maximizing her worst utility over all allocations.

Definition 2. *Let* $(A, G, \{u_i\}_{a_i \in A})$ *be an allocation setting. The* proportional fair-share (PFS) *of agent* a_i *is*

$$u_i^{PFS} = \frac{1}{\|A\|} u_i(G).$$

An allocation π satisfies the proportional fair-share criterion *if for every agent a_i, we have $u_i(\pi) \geq u_i^{PFS}$.*

Intuitively, the proportional fair-share criterion can be explained by pretending that the whole set G of goods were infinitely divisible and allocating to each agent a proportional share of G, from this agent's perspective.

Every allocation that satisfies PFS also satisfies MFS under 1-additive utility functions. If we switch to k-additive utility functions for $k > 1$, though, it becomes obvious that we need additional information. As long as $w_i(G') \geq 0$ for all $G' \subseteq G$ with $\|G'\| \geq 2$, we have

$$\min_{a_j \in A} u_i(\pi_j) \leq \frac{1}{n} \sum_{a_j \in A} u_i(\pi_j) \leq \frac{1}{n} u_i(G) = u_i^{PFS}.$$

and PFS still implies MFS. Without any restrictions on the utilities, however, it is easy to construct counterexamples:

Example 1. Let $A = \{a_1, a_2\}$ and $G = \{g_1, g_2, g_3, g_4\}$. We have utility functions defined by $w_i(g_j) = 1$ for $1 \leq i \leq 2$ and $1 \leq j \leq 4$, and $w_i(\{g_1, g_2\}) = w_i(\{g_3, g_4\}) = w_i(\{g_1, g_3\}) = w_i(\{g_2, g_4\}) = -1$ for $1 \leq i \leq 2$.

Then $u_i^{PFS} = 0$, so every allocation π satisfies PFS. However, $u_i^{MFS} = 2$, thus allocation $(G, \emptyset)$ does not satisfy MFS.

Hence, we can state

Proposition 1. *Under k-additive utility functions with $k \geq 2$,*

1. *for nonnegative weights, every allocation π that satisfies PFS also satisfies MFS;*
2. *for possibly negative weights, there is an allocation setting where an allocation that satisfies PFS does not satisfy MFS.*

Now we look at min-max fair-share in terms of its relationship to the proportional fair-share criterion and its complexity.

Definition 3. *Let $(A, G, \{u_i\}_{a_i \in A})$ be an allocation setting. The* min-max fair-share (mFS) *of agent a_i is*

$$u_i^{mFS} = \min_\pi \max_{a_j \in A} u_i(\pi_j).$$

An allocation π satisfies the min-max fair-share criterion *if for every agent a_i, we have $u_i(\pi) \geq u_i^{mFS}$.*

Intuitively, this criterion corresponds to a game where agent a_i picks her bundle first but does not choose the allocation; thus getting her best bundle in a worst (for her) allocation. Once again, it is easy to see that the implication for 1-additive utility functions that every min-max fair-share allocation satisfies proportional fair-share does not hold for larger k.

Example 2. Let $A = \{a_1, a_2\}$ and $G = \{g_1, g_2, g_3, g_4\}$. We have utility functions defined by $w_i(\{g_1, g_2\}) = w_i(\{g_3, g_4\}) = w_i(\{g_1, g_3\}) = w_i(\{g_2, g_4\}) = 1$ for $1 \leq i \leq 2$.

Note that $u_i^{\text{PFS}} = 2$ for $i \in \{1, 2\}$ and that allocation $\pi = (\{g_1, g_4\}, \{g_2, g_3\})$ does not satisfy PFS. However, it does satisfy mFS:

$$u_i(\pi_j) = 0, 1 \leq i, j \leq 2 \implies u_i^{\text{mFS}} = \min_{\pi'} \max_{a_j \in A} u_i(\pi'_j) \leq 0 \implies u_i^{\text{mFS}} = 0$$

because all coefficients are nonnegative.

Thus we have

Proposition 2. *Under k-additive utility functions with $k \geq 2$, there is an allocation setting where an allocation that satisfies mFS does not satisfy PFS.*

Next, we consider the complexity of deciding whether for a given allocation setting there is an allocation satisfying the min-max fair-share criterion. More precisely, we consider the following decision problem:

MFS-Exist

Given: An allocation setting $(A, G, \{u_i\}_{a_i \in A})$.

Question: Is there an allocation π that satisfies mFS?

Bouveret and Lemaître [5] only state an upper bound by showing that this problem is in Σ_2^p. We now provide a nontrivial lower bound by showing that it is NP-hard.

Proposition 3. MFS-Exist *is NP-hard whenever there are at least three agents.*

Proof. We reduce from the well-known NP-complete problem PARTITION: Given a list $B = (b_1, b_2, \ldots, b_m)$ of positive integer weights that sum up to an even integer, does there a exist a partition of $\{1, 2, \ldots, m\}$ into two sets $X \subseteq \{1, 2, \ldots, m\}$ and $\overline{X} = \{1, 2, \ldots, m\} \setminus X$ such that $\sum_{j \in X} b_j = \sum_{j \in \overline{X}} b_j$? From a given instance $B = (b_1, b_2, \ldots, b_m)$ of PARTITION with $2L = \sum_{1 \leq i \leq m} b_i$, we create the following instance with three agents and $m + 1$ goods:

- $A = \{a_1, a_2, a_3\}$ and $G = \{g_1, \ldots, g_{m+1}\}$.
- $u_i(g_j) = b_j$ for $1 \leq j \leq m$ and $a_i \in A$.
- $u_i(g_{m+1}) = L$ for $a_i \in A$.

Let π be an allocation. Notice that $\max_{a_j \in A} u_1(\pi_j) \geq L$ because some agent has to receive the good g_{m+1}. Because of identical preferences, we have $u_1^{\text{mFS}} = u_2^{\text{mFS}} = u_3^{\text{mFS}} \geq L$. It is easy to see that if π satisfies mFS, one of the shares—say, without loss of generality, π_1—is equal to $\{g_{m+1}\}$ (otherwise there is an agent $a_j \in \{a_2, a_3\}$ such that $u_j(\pi_j) < L \leq u_j^{mFS}$). Hence if $u_i^{\text{mFS}} > L$, there cannot be an allocation satisfying the min-max fair-share criterion.

Suppose there is no equal-sized partition of $\{b_1, \ldots, b_m\}$. Two cases are possible:

(i) There is a share π_i such that $\pi_i \supseteq \{g_{m+1}, g_i\}$, with $b_i \in B$. Then $u_i(\pi_i) > L$.

(ii) One of the shares equals $\{g_{m+1}\}$. Because we cannot split the remaining goods into two subsets of equal utility, there has to be one share such that $u_i(\pi_i) > L$.

In both of these two cases, $\max_{a_j \in A} u_i(\pi_j) > L$ and, therefore, $u_i^{\mathrm{mFS}} > L$, so there is no allocation satisfying the min-max fair-share criterion.

Conversely, let (B_1, B_2) be an equal-sized partition. Consider the allocation $\pi = (\{g_{m+1}\}, B_1, B_2)$. Then $u_i(\pi_i) = L$ for all $a_i \in A$ and thus $u_i^{\mathrm{mFS}} = L$. Hence, π satisfies mFS.   □

The next criterion we consider is envy-freeness. Intuitively, this means that none of the agents would like to swap her share with any of the other agents.

Definition 4. *Let $(A, G, \{u_i\}_{a_i \in A})$ be an allocation setting. An allocation π is envy-free if for all i, j, $1 \leq i, j \leq \|A\|$, we have $u_i(\pi_i) \geq u_i(\pi_j)$.*

Since the proof by Bouveret and Lemaître [5] that an envy-free allocation satisfies the min-max fair-share criterion does not use any information about the utility functions, it is directly applicable to k-additive utility functions. However, extending the scope of this implication turns out to be difficult:

Example 3. Let $A = \{a_1, a_2\}$ and $G = \{g_1, g_2\}$. We have utility functions defined by $w_i(\{g_1\}) = w_i(\{g_2\}) = w_i(\{g_1, g_2\}) = 2$ for $a_i \in A$.

Allocation $\pi = (\{g_1\}, \{g_2\})$ is envy-free, but there is no allocation which gives to each agent her proportional fair-share $u_i^{\mathrm{PFS}} = 3$.

The scale breaks, even if we do not consider the proportional fair-share.[1]

Example 4. Let $A = \{a_1, a_2\}$ and $G = \{g_1, g_2, g_3, g_4\}$. We have utility functions defined by $w_i(\{g\}) = 2$, for $a_i \in A$ and $g \in G$, and $w_1(\{g_1, g_2\}) = w_1(\{g_3, g_4\}) = w_2(\{g_2, g_3\}) = w_2(\{g_1, g_4\}) = 2$.

By evaluating the utilities granted by the allocations $\pi = (\{g_1, g_2\}, \{g_3, g_4\})$ and $\pi' = (\{g_1, g_4\}, \{g_2, g_3\})$ we see that $u_i^{\mathrm{MFS}} \geq 6$. As a consequence, there cannot be an allocation satisfying max-min fair-share. Yet π and π' are both envy-free.

We summarize this in

Proposition 4. *Under k-additive utility functions with $k \geq 2$,*

1. *every envy-free allocation π satisfies mFS;*
2. *there is an allocation setting where an envy-free allocation does not satisfy PFS;*
3. *there is an allocation setting where an envy-free allocation does not satisfy MFS.*

---

[1] Note that proportional fair-share can also be defined differently under k-additive utility functions.

We now consider the last criterion: competitive equilibrium from equal incomes.

Definition 5. *Let $(A, G, \{u_i\}_{a_i \in A})$ be an allocation setting. For a price vector $p \in [0,1]^m$ and an allocation π, the pair (π, p) forms a competitive equilibrium from equal incomes (CEEI) if for every $a_i \in A$, we have*

$$\pi_i \in \arg \max_{G' \subseteq G} \left\{ u_i(G') \mid \sum_{g_i \in G'} p_i \leq 1 \right\}.$$

An allocation π satisfies the competitive equilibrium from equal incomes criterion if there is a price vector p such that (π, p) forms a CEEI.

CEEI can be considered a fairness criterion because prices and budgets are the same for every agent.

Again, proving that an allocation that forms a CEEI is envy-free can be done independently of the utility representation [5]. Thus the implication also holds for k-additive utility functions.

Brânzei et al. [7] prove various hardness results regarding CEEI, such as checking whether a given allocation and price vector form a CEEI. Aziz [2] proves strong NP-hardness for computing an allocation that satisfies CEEI and focuses on a fractional variant of CEEI.

Bouveret and Lemaître asked whether every envy-free and Pareto-optimal allocation is CEEI under strict[2] and 1-additive utility functions. We answer this question in the negative.

Proposition 5. *Under strict and 1-additive utility functions there is an allocation setting that allows for an envy-free, Pareto-optimal allocation that is not CEEI.*

Proof. Let $A = \{a_1, a_2, a_3\}$ and $G = \{g_1, \ldots, g_5\}$. Let the utility matrix where entry $a_{i,j}$ denotes $u_i(g_j) = a_{i,j}$ be as follows:

$$\begin{pmatrix} 88 & 64 & 98 & 4 & 18 \\ 89 & 88 & 70 & 98 & 51 \\ 3 & 19 & 47 & 46 & 58 \end{pmatrix}$$

These utilities are strict. The allocation $\pi = (\{g_3\}, \{g_1, g_4\}, \{g_2, g_5\})$ is the unique allocation that is envy-free and Pareto-optimal. A CEEI price vector p has to satisfy the following inequalities:

$$p_1 + p_4 \leq 1 \tag{1}$$
$$p_2 + p_5 \leq 1 \tag{2}$$
$$p_1 + p_2 > 1 \tag{3}$$
$$p_4 + p_5 > 1 \tag{4}$$

---

[2] A utility function u is *strict* if $u(G) = u(H)$ implies $G = H$.

Allocation π implies inqualities (1) and (2). Agent a_1 realizes a utility of 98, whereas bundle $\{g_1, g_2\}$ would give a utility of 152. Similarly, agent a_3 must not receive bundle $\{g_4, g_5\}$. However, the above inequalities lead to a contradiction. Hence, for every price vector p, the pair (π, p) does not form a CEEI. $\square$

3   Approximability of Rank-Weighted Utilitarianism

In this section, we consider approximation problems of rank-weighted utilitarianism under nonnegative 1-additive utility functions. This approach to fairness is orthogonal to the one in Sect. 2. We first need some additional definitions. Each partition π induces a utility vector $v(\pi) = (u_i(\pi_i))_{1 \leq i \leq n}$. We denote by $v^*(\pi)$ the vector that results from $v(\pi)$ by sorting all entries nondecreasingly. Thus $v_1^*(\pi)$ gives the utility of a worst-off agent (i.e., egalitarian social welfare), $v_{\lceil n/2 \rceil}^*(\pi)$ the utility of a median-off agent[3], and $v_n^*(\pi)$ the utility of a best-off agent (i.e., elitist social welfare). Given a set A of agents, a set G of goods, and for each agent $a_i \in A$ a utility function u_i, we can search for allocations that maximize the *j-rank dictator function* v_j^*. More generally, we have

Definition 6. *Let* $(A, G, \{u_i\}_{a_i \in A})$ *be an allocation setting. Let* $w = (w_1, \ldots, w_n)$ *be a normalized weight vector, i.e.,* $\sum_1^n w_i = 1$. *Define the* rank-weighted utilitarian social welfare *of an allocation* π *as*

$$sw(\pi) = \sum_{i=1}^{n} w_i v_i^*(\pi).$$

We say a rank-weighted utilitarian CUF with weight vector w is *weakly inequality-reducing (inequality reducing)* if w is weakly (strictly) decreasing. Note that, in general, the set of maximizing allocations stays invariant under multiplication of weight vector w by a constant. However, shifting w may change this set. Clearly, the weight vector that has a 1 at position j and a 0 at all remaining positions corresponds to the *j-rank dictator function*. We define the following decision problem:

j-RANK DICTATOR
Given: An allocation setting $(A, G, \{u_i\}_{a_i \in A})$ and a value $k \in \mathbb{Q}$.
Question: Is there an allocation π such that $v_j^*(\pi) \geq k$?

1-RANK DICTATOR is NP-complete for at least two agents by a reduction from PARTITION. This problem is also known as, respectively, the MAXMIN-FAIRNESS problem, the MAXIMUM EGALITARIAN SOCIAL WELFARE problem [19], and (some variant of) the SANTA CLAUS problem [3]. Because $(n-1)$-RANK DICTATOR and $\lceil n/2 \rceil$-RANK DICTATOR are equivalent to 1-RANK DICTATOR for two agents, they are also NP-complete. For intermediate values j,

---

[3] Alternatively, one could define the median for an even number of agents as $\frac{v_{n/2}^*(\pi) + v_{n/2+1}^*(\pi)}{2}$. However, we follow the definition due to Chevaleyre et al. [9].

$1 < j < n-1$, j-RANK DICTATOR is NP-complete as well (see below). However, n-RANK DICTATOR is decidable in polynomial time: For each agent, determine the bundle that maximizes this agent's utility. Assign the required bundle to the agent a_i with maximum utility. If there are at least two agents, all goods with negative utility values for agent a_i are assigned to the other agent. Since for almost all values j (except for $j = n$) j-RANK DICTATOR is NP-complete, we study the approximability of the corresponding optimization problems:

OPTIMAL VALUE j-RANK DICTATOR

Input: An allocation setting $(A, G, \{u_i\}_{a_i \in A})$ with at least j agents.
Output: $\max_\pi v_j^*(\pi)$.

Define the decision problem w-RANK-WEIGHTED UTILITARIANISM analogously, where the weight vector family w is not part of the input, as otherwise we would implicitly study the NP-hard MAXMIN-FAIRNESS problem under a different name.

3.1   Rank Dictator

We first consider j-RANK DICTATOR with constant values j. This case is interesting because of the MAXMIN-FAIRNESS problem.

Proposition 6. *Let $i \in \mathbb{N}$ be a constant. Given an $f(n, m)$-approximation algorithm for* OPTIMAL VALUE $(i + 1)$-RANK DICTATOR, *there is an $f(n + 1, m)$-approximation algorithm for* OPTIMAL VALUE i-RANK DICTATOR.

Proof. Given an allocation setting $P = (A, G, \{u_i\}_{a_i \in A})$ where the ith rank dictator is to be optimized, add a dummy agent d with a utility function that assigns to each good utility 0. Since i is constant, this dummy agent will "shift" the index that is maximized to the left. In more detail, we show $\mathrm{OPT}_{i+1}(P^+) = \mathrm{OPT}_i(P)$, where $\mathrm{OPT}_j(Q)$ denotes the optimal value of the jth rank dictator problem under allocation setting Q and P^+ is allocation setting P extended by the dummy agent. Suppose $\mathrm{OPT}_{i+1}(P^+) > \mathrm{OPT}_i(P)$ where $\mathrm{OPT}_{i+1}(P^+)$ is achieved under allocation π. If $\pi_d = \emptyset$, because the dummy agent is among the worst-off agents, we have $v_{i+1}^*(\pi) = v_i^*(\pi')$, where π' denotes the restriction of π to A. If $\pi_d \neq \emptyset$, reassign the goods in π_d arbitrarily to agents in A. Call the resulting allocation π''. Because utilities are nonnegative, it holds that $v_{i+1}^*(\pi) \leq v_i^*(\pi'')$ (contradiction). Thus we have $\mathrm{OPT}_{i+1}(P^+) \leq \mathrm{OPT}_i(P)$. Suppose $\mathrm{OPT}_{i+1}(P^+) < \mathrm{OPT}_i(P)$, where $\mathrm{OPT}_i(P)$ is achieved under allocation π. Then consider the extension $\widehat{\pi}$ of π to $A \cup \{d\}$ by assigning the empty bundle to d. Since utilities are nonnegative, the dummy agent is among the worst-off agents. Thus $v_{i+1}^*(\widehat{\pi}) = v_i^*(\pi)$ (contradiction). To summarize, we have $\mathrm{OPT}_{i+1}(P^+) = \mathrm{OPT}_i(P)$. $\square$

This result shows that a good approximation algorithm for OPTIMAL VALUE j-RANK DICTATOR for constant j would imply a good approximation for the MAXMIN-FAIRNESS problem. Thus we cannot hope for a good approximation algorithm for, e.g., OPTIMAL VALUE 2-RANK DICTATOR without new ideas.

Corollary 1. *For each constant $i \in \mathbb{N}$, $i \geq 2$, i-RANK DICTATOR is NP-complete.*

Proof. Membership in NP is clear. Given i and an instance of 1-RANK DICTATOR, add $i - 1$ dummy agents as in the proof of Proposition 6. Because $\text{OPT}_{i+1}(P^+) = \text{OPT}_i(P)$ for each constant $i \in \mathbb{N}$, the result follows. □

Proposition 7. *Let $i \in \mathbb{N}$ be a constant. Given an $f(n,m)$-approximation algorithm for OPTIMAL VALUE $(n-i)$-RANK DICTATOR, there is an $f(n+1, m+1)$-approximation algorithm for OPTIMAL VALUE $(n - i + 1)$-RANK DICTATOR.*

Proof. The proof is analogous to the proof of Proposition 6: Add a dummy agent that values every good zero except for a new good that is valued at least OPT (e.g., $m \max_{a_s \in A, g_t \in G} u_s(g_t)$) only by this dummy agent. □

Corollary 2. *For each constant $i \in \mathbb{N}$, $i \geq 1$, $(n - i)$-RANK DICTATOR is NP-complete.*

Corollary 3. *Let $i \in \mathbb{N}$ be a constant with $i < \lceil n/2 \rceil$. Given an $f(n,m)$-approximation algorithm for OPTIMAL VALUE $\lceil n/2 \rceil$-RANK DICTATOR, there is an $f(2n - 2i, m)$-approximation algorithm for OPTIMAL VALUE i-RANK DICTATOR.*

Proof. Let ℓ be the smallest number such that $\lceil \frac{n+\ell}{2} \rceil = \ell + i$. Apply the proof of Proposition 6 ℓ times. □

Similarly, Corollary 3 says that a good approximation algorithm for the median rank dictator problem would yield an approximation algorithm for the MAXMIN-FAIRNESS problem where the approximation guarantee is worse by a factor of 2.

Proposition 8. *Let $i \in \mathbb{N}$ be a constant. Given an $f(n,m)$-approximation algorithm for OPTIMAL VALUE 1-RANK DICTATOR, there is an $f(n,m)$-approximation algorithm for OPTIMAL VALUE i-RANK DICTATOR.*

Proof. Let I be an instance of OPTIMAL VALUE i-RANK DICTATOR. All agents a_j that get less than $v_i^*(\pi)$ utility realize in fact zero utility. Consider all subsets of the agents of size $n - i + 1$ and the corresponding restriction of I to this subset of agents. We have restrictions $I_1, \ldots, I_{\binom{n}{n-i+1}}$. Denote by OPT the maximum value of I with respect to OPTIMAL VALUE i-RANK DICTATOR and denote by OPT_k the maximum value of I_k with respect to OPTIMAL VALUE 1-RANK DICTATOR. Then we have $\text{OPT} = \max_k \text{OPT}_k$ (otherwise, OPT cannot be optimal). There are $\binom{n}{n-i+1} = \binom{n}{i-1} \leq n^{i-1}$ such subsets, i.e., polynomially many in n. Choose the maximum value that the approximation algorithm returns when applying it to each restricted instance. This gives an $f(n,m)$-approximation for OPTIMAL VALUE i-RANK DICTATOR because we have $\text{OPT}_k > \text{OPT}_\ell \iff f(n,m)\text{OPT}_k > f(n,m)\text{OPT}_\ell$. □

Thus hardness results for, e.g., OPTIMAL VALUE 2-RANK DICTATOR translate to hardness results for the MAXMIN-FAIRNESS problem. On the other hand, from [8] we get the following approximation algorithm for OPTIMAL VALUE j-RANK DICTATOR, $j \in \mathbb{N}$.

Corollary 4. *For any $j \in \mathbb{N}$, there is an $\widetilde{\mathscr{O}}(1/m^\varepsilon)$- approximation algorithm in time $m^{\mathscr{O}(1/\varepsilon)}$ for $\varepsilon \in \Omega(\frac{\log \log m}{\log m})$ for* OPTIMAL VALUE j-RANK DICTATOR.

For superconstant f, the $f(n)$-RANK DICTATOR is at least as hard as the MAXMIN-FAIRNESS problem.

Proposition 9. *Let f be function over the natural numbers with $f(n) < n$. $f(n)$-RANK DICTATOR is NP-complete.*

Proof. For $n = 2$, we have $f(2) = 1$. Hence, $f(n)$-RANK DICTATOR is equivalent to 1-RANK DICTATOR for two agents. $\qquad\square$

Corollary 5. *For every $\varepsilon > 0$, there is no $(1/2 + \varepsilon)$-approximation algorithm for* OPTIMAL VALUE $\lceil n/2 \rceil$-RANK DICTATOR, *unless* P = NP.

Proof. It is NP-hard to approximate OPTIMAL VALUE 1-RANK DICTATOR better than $1/2$ (see [4]). $\qquad\square$

3.2 Rank-Weighted Utilitarianism

Now we turn to general inequality-reducing weight vectors and study the computational hardness with regard to rank-weighted utilitarianism. Note that even for two agents maximizing rank-weighted utilitarianism is NP-complete. This follows from a proof by Golden and Perny [13].

Proposition 10. w-RANK-WEIGHTED UTILITARIANISM *is NP-complete, even for two agents with identical utility functions.*

If the number of agents is not fixed, there is no FPTAS unless P = NP. To prove this, we use basic notions from the theory of majorization (see, e.g., the book by Marshall et al. [17]). Given two vectors $x, y \in \mathbb{R}^n$, x is said to be *majorized by* y if

$$\sum_{i=1}^{k} x_{(i)} \geq \sum_{i=1}^{k} y_{(i)}, \quad 1 \leq k \leq n - 1, \text{ and}$$

$$\sum_{i=1}^{n} x_{(i)} = \sum_{i=1}^{n} y_{(i)}$$

hold, where $v_{(i)}$ denotes the ith smallest entry of a vector v. A real-valued function f over $\mathbb{R}^n$ is *strictly Schur-concave* if x being majorized by y implies $f(x) > f(y)$, unless x is a permutation of y. Note that the rank-weighted utilitarian CUF with a strictly decreasing weight vector is a strictly Schur-concave function.

Proposition 11. *If the number of agents is not fixed, w-RANK-WEIGHTED UTILITARIANISM is strongly NP-complete, even for identical utility functions.*

Proof. Membership in NP follows from the fact that guessing and checking can be done in polynomial time. For proving hardness, we reduce from 3-PARTITION, which is defined as follows: Given $3m$ numbers $b_1, \ldots, b_{3m}$ with cumulative sum mB and $B/4 < b_i < B/2$ for $1 \leq i \leq 3m$, is it possible to partition them into m subsets such that each subset sums up to B? Given such an instance of 3-PARTITION, we generate m agents and $3m$ goods. All utility functions are identical, with $u(g_i) = b_i$, $1 \leq i \leq 3m$. Set the bound to $\left(\sum_i w_i\right) B$. For proving completeness, a partition gives each agent B utility. Hence, social welfare is $\left(\sum_i w_i\right) B$. For proving soundness, we use the fact that the CUF is strictly Schur-concave. If utility vector x is majorized by y, then the social welfare with utility vector x is strictly greater than the social welfare with utility vector y. Let π be an allocation and let y be the utility vector that is induced by π. We denote by x the utility vector for which every agent gets B utility. We show that x is majorized by y, and hence the social welfare induced by vector y is strictly less than B. Let $y_{(k)}$ be the utility of the k-poorest agent. We perform an induction over k. For $k = 1$, we have $B \geq y_{(1)}$ because for every allocation π', there is an agent i with $u_i(\pi') < B$; otherwise, it would not be a no-instance. For the inductive step, assume for the sake of contradiction that we have $\sum_{i=1}^{k} y_{(i)} + y_{(k+1)} > (k+1)B$. Thus $y_{(k+1)} > B$ by the inductive hypothesis $\sum_{i=1}^{k} y_{(i)} \leq kB$. Then all agents after the $(k+1)$st agent get more than B utility:

$$\sum_{i=1}^{k} y_{(i)} + \sum_{i=k+1}^{m-1} y_{(i)} > kB + (m-1-k-1+1)B$$

$$= kB + mB - 2B - kB + B$$

$$= (m-1)B$$

$$= \sum_{i=1}^{m} y_{(i)} - B.$$

This gives the desired contradiction $y_{(m)} < B$. □

4   Computational Experiments

We start this section with a result that states that a maximizing allocation of a rank-weighted utilitarian CUF with a weight vector whose first entry is not positive never satisfies the max-min fair-share criterion. Hence, k-rank dictator functions for $k > 1$ are less interesting.

Proposition 12. *Let $m \geq n$, let utility values be positive and $w = (w_1, \ldots, w_n)$ be a weight vector with $w_1 \leq 0$. Then every allocation π that maximizes the rank-weighted utilitarian social welfare with respect to w does not satisfy MFS.*

Proof. Since all utilities are positive, π assigns no goods to a worst-off agent. Because of $m \geq n$ there exists an allocation that assigns a good with positive value to every agent. Hence, MFS is positive for every agent, whereas the worst-off agent realizes zero utility under π. □

Before explaining our experimental setup we introduce a set of weight vector families (see [14,21]):

- **k-worst:** The first k entries are ones and we have a zero everywhere else. The weight vector has an egalitarian flavor. We only care about the utilities of the k worst agents.
- **Borda:** Position i has weight $n-i+1$. Hence, every agent's utility is accounted for with decreasing priority.
- **Gini:** This is a well-known weight vector (neglecting normalization) that is similar to Borda. Position i has weight $2(n-i)+1$. Every agent has a positive weight, but agents with less utility have more weight (see, e.g., the papers by Endriss [12] and Lesca and Perny [16] for more information on Gini social welfare and the Gini index in the context of indivisible goods).
- **Lex:** Position i has weight 2^{n-i}. The weights decrease exponentially: Poor agents get the most attention but the best-off agent is not neglected.
- **Reverse-Borda:** Position i has weight i (the reverse of the Borda weight vector).
- **λ-Hurwicz:** Position 1 has weight λ, position n has weight $1-\lambda$, and we have zero weight everywhere else. This weight vector bridges both extreme ends of the utility vector.

Recall that the set of maximizing allocations is invariant under multiplication of the weight vector by a constant. Hence we do not normalize weight vectors for our experiments. This also holds for max-min fair-share [5].

We conducted experiments to determine to what extent rank-weighted utilitarianism coincides with the weakest fairness criterion, max-min fair-share. We tested for the number of agents being $n = 2, \ldots, 5$ and the number of goods being $m = n, \ldots, 10$. For $m < n$, every allocation satisfies MFS for nonnegative utility functions. Utility functions were drawn uniformly (similarly to the impartial culture assumption in voting theory) from $[0, 1)$ by uniformly drawing utility values for each good. For each data point we performed 1000 iterations. We checked whether all maximizing allocations satisfied the max-min fair-share criterion, under the condition that there is an allocation that satisfies MFS (which was the case for all 490,000 sampled profiles). The number of profiles where not all maximizing allocations but at least one maximizer satisfied MFS were negligible.

The figures depict the relation between number of goods and the number of profiles where all maximizing allocations satisfy MFS. For (weakly) decreasing weight vectors (Fig. 1(a)–(g)), we see a very high compatibility between maximizing allocations of weakly inequality-reducing rank-weighted utilitarianism and the max-min fair-share criterion. On the other hand, Fig. 1(h) shows that the Reverse-Borda-based social welfare function is highly unfair (with respect to

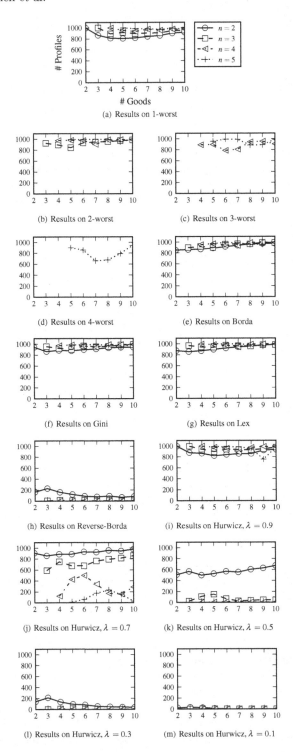

Fig. 1. Results of the computational experiments

max-min fair-share). Hurwicz weight vectors (Fig. 1(i)–(m)) show a more interesting behavior. Although they are not decreasing weight vectors, maximizing allocations also satisfy MFS for high values of λ. In addition, as λ decreases, the number of profiles that admit maximizers that satisfy MFS decreases as well.

5   Conclusions and Open Questions

Continuing the work of Bouveret and Lemaître [5] we have investigated the question of whether their scale of fairness criteria extends also to nonadditive utility functions. For k-additive utility functions, we have seen that this scale remains intact only for the more restrictive criteria. We have also shown that the min-max fair-share allocation existence problem is NP-hard and that under strict preferences competitive equilibrium from equal incomes does not coincide with envy-freeness and Pareto-optimality. Then, we have proved that k-rank dictator functions are closely connected to MaxMin fairness with respect to approximability. We have studied the computational hardness of general inequality-reducing rank-weighted utilitarianism problems. Finally, we have presented computational experiments that answer the question how often maximizers of weakly inequality-reducing rank-weighted utilitarianism satisfy the max-min fair-share criterion, the weakest among the fairness criteria due to Bouveret and Lemaître [5]. We have seen that rank-weighted utilitarianism with weakly inequality-reducing weight vectors and the Hurwicz weight vectors for larger values of λ are highly compatible with MFS.

As open problems we propose resolving the (exact) complexity of the max-min fair-share and min-max fair-share existence problems. In terms of approximability, we conjecture that there is a PTAS for decreasing weight vectors and identical utilities.

Acknowledgments. We thank the anonymous reviewers for their helpful comments. This work was supported in part by DFG grant RO 1202/14-1.

References

1. Amanatidis, G., Markakis, E., Nikzad, A., Saberi, A.: Approximation algorithms for computing maximin share allocations. Technical Report arXiv:1503.00941 [cs.GT], ACM Computing Research Repository (CoRR), March 2015
2. Aziz, H.: Competitive equilibrium with equal incomes for allocation of indivisible objects. Technical Report arXiv:1501.06627 [cs.GT], ACM Computing Research Repository (CoRR), March 2015
3. Bansal, N., Sviridenko, M.: The Santa Claus problem. In Proceedings of the 38th annual ACM Symposium on Theory of Computing, pp. 31–40. ACM Press (2006)
4. Bezáková, I., Dani, V.: Allocating indivisible goods. ACM SIGecom Exchanges 5(3), 11–18 (2005)
5. Bouveret, S., Lemaître, M.: Characterizing conflicts in fair division of indivisible goods using a scale of criteria. Journal of Autonomous Agents and Multi-Agent Systems, 1–32, 2015. To appear. A preliminary version appeared in the proceedings of AAMAS-2014

6. Brams, S.J., King, D.L.: Efficient fair division - help the worst off or avoid envy? Rationality and Society **17**(4), 387–421 (2005)
7. Brânzei, S., Hosseini, H., Miltersen, P.: Characterization and computation of equilibria for indivisible goods. Technical Report arXiv:1503.06855 [cs.GT], ACM Computing Research Repository (CoRR), March 2015
8. Chakrabarty, D., Chuzhoy, J., Khanna, S.: On allocating goods to maximize fairness. In: Proceedings of the 50th IEEE Symposium on Foundations of Computer Science, pp. 107–116. IEEE Computer Society Press (2009)
9. Chevaleyre, Y., Dunne, P., Endriss, U., Lang, J., Lemaître, M., Maudet, N., Padget, J., Phelps, S., Rodríguez-Aguilar, J., Sousa, P.: Issues in multiagent resource allocation. Informatica **30**(1), 3–31 (2006)
10. Chevaleyre, Y., Endriss, U., Estivie, S., Maudet, N.: Multiagent resource allocation in k-additive domains: preference representation and complexity. Ann. Oper. Res. **163**(1), 49–62 (2008)
11. d'Aspremont, C., Gevers, L.: Social welfare functionals and interpersonal comparability. In: Arrow, K.J., Sen, A.K., Suzumura, K. (eds) Handbook of Social Choice and Welfare, vol. 1, pp. 459–541. Elsevier (2002)
12. Endriss, U.: Reduction of economic inequality in combinatorial domains. In: Proceedings of the 12th International Joint Conference on Autonomous Agents and Multiagent Systems, pp. 175–182. IFAAMAS (2013)
13. Golden, B., Perny, P.: Infinite order Lorenz dominance for fair multiagent optimization. In: Proceedings of the 9th International Conference on Autonomous Agents and Multiagent Systems, pp. 383–390. IFAAMAS (2010)
14. Goldsmith, J., Lang, J., Mattei, N., Perny, P.: Voting with rank dependent scoring rules. In: Proceedings of the 28th AAAI Conference on Artificial Intelligence, pp. 698–704. AAAI Press (2014)
15. Grabisch, M.: k-order additive discrete fuzzy measures and their representation. Fuzzy Sets Syst. **92**(2), 167–189 (1997)
16. Lesca, J., Perny, P.: LP Solvable Models for Multiagent Fair Allocation Problems. In: Proceedings of the 19th European Conference on Artificial Intelligence, pp. 393–398 (2010)
17. Marshall, A., Olkin, I., Arnold, B.: Inequalities: Theory of Majorization and its Applications. Springer (2010)
18. Moulin, H.: Axioms of Cooperative Decision Making. Cambridge University Press (1991)
19. Nguyen, N., Nguyen, T., Roos, M., Rothe, J.: Computational complexity and approximability of social welfare optimization in multiagent resource allocation. Journal of Autonomous Agents and Multi-Agent Systems **28**(2), 256–289 (2014)
20. Nguyen, T., Roos, M., Rothe, J.: A survey of approximability and inapproximability results for social welfare optimization in multiagent resource allocation. Ann. Math. Artif. Intell. **68**(1–3), 65–90 (2013)
21. Skowron, P., Faliszewski, P., Lang, J.: Finding a collective set of items: from proportional multirepresentation to group recommendation. In: Proceedings of the 29th AAAI Conference on Artificial Intelligence, pp. 2131–2137 (2015)
22. Yager, R.R.: On ordered weighted averaging aggregation operators in multicriteria decisionmaking. IEEE Transactions on Systems, Man and Cybernetics **18**(1), 183–190 (1988)

Randomized Assignments for Barter Exchanges: Fairness vs. Efficiency

Wenyi Fang[1]([✉]), Aris Filos-Ratsikas[2], Søren Kristoffer Stiil Frederiksen[2], Pingzhong Tang[1], and Song Zuo[1]

[1] Institute for Interdisciplinary Information Sciences, Tsinghua University, Beijing, China
{fwy13,zuos13}@mails.tsinghua.edu.cn, kenshin@tsinghua.edu.cn
[2] Department of Computer Science, Aarhus University, Aarhus, Denmark
{filosra,ssf}@cs.au.dk

Abstract. We study fairness and efficiency properties of randomized algorithms for barter exchanges with direct applications to kidney exchange problems. It is well documented that randomization can serve as a tool to ensure fairness among participants. However, in many applications, practical constraints often restrict the maximum allowed cycle-length of the exchange and for randomized algorithms, this imposes constraints of the cycle-length of every realized exchange in their decomposition. We prove that standard fairness properties such as envy-freeness or symmetry are incompatible with even the weakest notion of economic efficiency in this setting. On the plus side, we adapt some well-known matching mechanisms to incorporate the restricted cycle constraint and evaluate their performance experimentally on instances of the kidney exchange problem, showing tradeoffs between fairness and efficiency.

1 Introduction

Over the past years, barter exchanges, with kidney exchange as a representative example, have become a topic of intensive research at the intersection of AI and economics. In a barter exchange, participants enter the system with some endowment and then exchange their endowments in order to obtain better allocations. Such exchanges are very popular in settings such as exchanges of used books or DVDs where agents do not have very high values for their endowments and would rather trade them with others.

This work was supported by the National Basic Research Program of China Grant 2011CBA00300, 2011CBA00301, the Natural Science Foundation of China Grant 61033001, 61361136003, 61303077, a Tsinghua Initiative Scientific Research Grant and a China Youth 1000-talent program. Aris Filos-Ratsikas and Søren Kristoffer Stiil Frederiksen acknowledge support from the Danish National Research Foundation and The National Science Foundation of China (under the grant 61061130540) for the Sino-Danish Center for the Theory of Interactive Computation, within which this work was performed. The authors also acknowledge support from the Center for Research in Foundations of Electronic Markets (CFEM), supported by the Danish Strategic Research Council.

© Springer International Publishing Switzerland 2015
T. Walsh (Ed.): ADT 2015, LNAI 9346, pp. 537–552, 2015.
DOI: 10.1007/978-3-319-23114-3_32

The current literature on barter exchanges presents two major challenges, *fairness* and *implementability* [2,3,6–9,13]. Most would agree that a fair procedure should guarantee at least properties like *symmetry* and *envy-freeness*. In practical applications, barter exchanges tend to be carried out *deterministically*; on the other hand as argued in a number of papers [5,11,14], central economic notions of fairness such as the ones stated above, require randomization.

Implementability has to do with whether the designed exchanges can be carried out in practice. The simplest exchanges are *pairwise* exchanges, involving only two agents [12,14]. Exchanges can also be more complicated, involving multiple participants, exchanging endowments in a cycle. A vital constraint of most such exchange systems (kidney exchanges, room exchanges[1]) is that the number of agents involved in a cycle must be bounded [2,14]. Such constraints may be imposed for a number of reasons; a real-life motivating example comes from perhaps the most widespread applications of barter exchanges, the *kidney exchange problem*.

In a kidney exchange market, pairs consisting of incompatible donors and patients enter the market, in search for other pairs to exchange kidneys with. In case of inter-pair compatibility, i.e. when the donor of the first pair is compatible with the patient of the second pair and vice-versa, an exchange is carried out. In many countries, such exchange systems have been in effect for several years.

There are several constraints on the length of such exchanges however, imposed for both practical and ethical reasons. First of all, participants involved in an exchange cycle must conduct surgery simultaneously at the same hospital[2], making it logistically infeasible for any hospital to host a large cycle. Secondly, donors can not be contractually obligated to donate their kidneys, since it is illegal in most countries. If an "offline" exchange were to take place, there is no guarantee that donors would not opt out after their counterparts receive their transplants [8].

For this reason, the length of the exchange cycles is constrained to be a small number (three in most cases). This however, makes the problem much more challenging. It is known that, under such constraints on the cycle-length, the problem of finding an efficient exchange is NP-hard [2]. Despite the theoretical hardness results, Abraham et al. [2] designed an algorithm that through several optimization techniques produces an optimal exchange on typical instances of the problem in reasonable running time. The algorithm, while efficient, is deterministic and not tailored to incorporate fairness criteria. On the other hand, fairness is a key property here; between patients with similar compatibility characteristics and similar needs, no deterministic choice can be justified, especially if it results in loss of human life. In fact, Dickerson et. al. [8] observe that, among the exchanges in the major organ exchange system in the United States, UNOS,[3]

---

[1] http://reslife.umd.edu/housing/reassignments/roomexchange/.

[2] In this paper, we do not consider the use of altruistic chains, which may circumvent this requirement.

[3] www.unos.org.

only 7 percent of them finally make it to surgery. The lack of fairness guarantees might be a possible explanation for this phenomenon.

As mentioned earlier, randomization can be used as a way of achieving fairness and there are many candidate mechanisms in matching and exchange literature to choose from. Perhaps the two best-studied are the Probabilistic Serial mechanism [5] and Random Serial Dictatorship [1]. Both achieve fairness in the sense of symmetry but the former is also envy-free. However, the implementability constraint introduces complications to the use of randomized mechanisms as well. Given the natural interpretation of a randomized exchange as a probability mixture over deterministic exchanges, the constraint requires that *every* such exchange does not contain long cycles. If we restrict the possible outcomes to those assignments only, it is unclear whether the mechanisms maintain any of their fairness properties. This is one of the questions that we address in this paper.

We investigate the problem of designing barter exchanges that meet the two desiderata above. In particular, we explore the use of randomized mechanisms for achieving tradeoffs between efficiency and fairness, under the added constraint of restricted cycle-length in their decomposition. We make the following two-fold contribution.

- First, we consider the tradeoffs between economic efficiency and fairness from a theoretical point of view and prove that even the weakest form of economic efficiency (a relaxation of ex-post Pareto efficiency that is suitable for the problem) is incompatible with both *envy-freeness* and *symmetry*. On the other hand, we show that it is possible to satisfy each property independently, together with the restricted cycle-length.
- Next, we adapt two well-known mechanisms, *Random Serial Dictatorship* and *Probabilistic Serial* to incorporate the cycle-length constraint and evaluate their performance on instances of the kidney exchange problem. We show tradeoffs between the efficiency (in the sense of social welfare) and quantified envy-freeness of the exchange for those adaptations of the mechanisms and compare them to the mechanism that produces an optimal assignment.

Most relevant to the current paper is the work by Balbuzanov [4], where the author considers deterministic and randomized mechanisms for barter exchanges, under the restriction of the cycle length in the components of the decomposition, very similarly to what we do here. Interestingly, he presents an adaptation of the Probabilistic Serial mechanism (named the "2-cycle Probabilistic Serial") that always produces components with cycles of length two and satisfies two desired properties: *ordinal efficiency* (a stronger notion of efficiency than the one we consider here) and anonymity, i.e. a guarantee that the outcome is imprevious to renaming agent/item pairs. It is well-known that ordinal efficiency implies ex-post Pareto efficiency; in particular this is also true for the relaxed versions of efficiency that we consider in this paper. Furthermore, anonymity (together with neutrality) is known to imply symmetry. The existence of 2-cycle Probabilistic Serial however does not contradict our negative result on compatibility between ex-post efficiency and symmetry; crucially, the anonymity notion used in [4] does not imply symmetry in our setting.

2 Background

Let $N = \{1, \ldots, n\}$ be a set of agents and let $M = \{1, \ldots, n\}$ be a set of items. We assume that each agent is associated with exactly one item and without loss of generality, let agent i be associated with item i. Let item i be the *endowment* of agent i. Each agent has *valuations* over the items., i.e. numerical values that denote her levels of satisfaction. Let $\mathbf{v_i} = (v_{i1}, \ldots, v_{in})$ be the *valuation vector* of agent i and let $\mathbf{V} = (\mathbf{v_1} \ldots \mathbf{v_n})$ be a *valuation matrix*.

An *assignment* D is a matching of agents to items, such that each agent receives exactly one item. This is precisely a permutation matrix, where entry $d_{ij} = 1$ if agent i receives item j and 0 otherwise. Alternatively, one can view D as a directed graph $D = (N, E)$, where a vertex v_i corresponds to both agent and item i and an edge (i, j) means that agent i is matched with item j in D. Given this interpretation, an assignment is a set of disjoint cycles, where agents exchange endowments along a cycle. If the maximum length of any cycle in such an assignment is k, we will say that the assignment is k-*restricted*.[4]

Since each agent receives exactly one item in expectation, a *probabilistic* or *randomized* assignment is a bistochastic matrix P where entry p_{ij} denotes the probability that agent i is matched with item j. We will call $p_i = (p_{i1}, \ldots, p_{in})$ an *assignment vector*. A *mechanism* is a function that on input a valuation matrix V outputs an assignment P.

A probabilistic assignment P can be viewed as a probability mixture over deterministic assignments. This is due to the *Birkhoff-von Neumann* theorem that states that each bistochastic matrix of size n can be written as a convex combination of at most n^2 permutation matrices. Since it is particularly relevant to the design of our mechanisms, we will describe the decomposition process in more detail in Sect. 4. We will say that a randomized assignment P is k-restricted if it can be written as a probability mixture of k-restricted determnistic assignments.

The two standard notions of fairness that we consider in this paper are *envy-freeness* and *symmetry*. An assignment is (ex-ante) envy-free if no agent would prefer to swap assignment vectors with any other agent. An assignment is *symmetric* if all agents that have identical valuation vectors receive identical assignment vectors. Note that envy-freeness does not imply symmetry; two agents with identical valuations could be equally satisfied with an assignment without having the same probabilities of receiving each item[5].

A deterministic assignment D is *Pareto efficient* if there is no other assignment D' that is not less preferable for any agent and strictly more preferable for at least one agent. If such an assignment D' exists, we will say that D' *Pareto dominates* D.

It is not hard to see that standard efficiency is not compatible with k-restricted assignments. To see this, let $n > k$ and let V be such that for

---

[4] Balbuzanov [4] uses the term k-*constrained* to describe such assignments.

[5] This is true in particular because we consider all mechanisms, including *cardinal* mechanisms, i.e. mechanisms that can use the numerical values when outputting assignments.

$i = 1, \ldots, n - 1$, $\max_{j \in M} v_{ij} = i + 1$ and $\max_{j \in M} v_{nj} = 1$. Any ex-post Pareto efficient assignment must consist single permutation matrix D consisting of only one cycle of length n, which is of course not a k-restricted assignment. For that reason, it makes sense to only define Pareto efficiency in terms of k-restricted assignments.

Definition 1. *A deterministic assignment D is Pareto efficient if there does not exist any k-restricted assignment that Pareto dominates it.*

Remark 1. Note that this is the same definition of *k-constrained Pareto efficiency* used in [4]. For simplicity and since the only notion of efficiency in this paper is with respect to k-restricted assignments, we simply use the term Pareto-efficiency.

For randomized mechanisms, an assignment is *ex-ante* Pareto efficient, if there is no other assignment that satisfies the condition above in expectation. An assignment is *ex-post* Pareto efficient if for every realization of randomness, there is no other assignment that Pareto dominates it. In other words, a randomized assignment is ex-post Pareto efficient if it can be written as a probability mixture of Pareto efficient deterministic assignments. Note that ex-post Pareto efficiency is the weakest efficiency notion for randomized mechanisms in literature.

3   Fairness and Economic Efficiency

In this section, we explore the compatibility and incompatibility between fairness and efficiency properties given the constraint of small cycles. We show that even the weakest form of efficiency is incompatible with two well-known fairness criteria, even in the case when $k = 3$. Note that $k = 3$ is the common choice for the maximum allowed cycle length in the most important applications of the problem, that of kidney exchange that we discuss in Sect. 5.

Theorem 1. *There is no mechanism that always outputs an assignment which is envy-free, ex-post Pareto efficient and 3-restricted.*

Proof. First, we prove the theorem for $n = 4$. First we construct the valuation profile $\mathbf{v}$ with $\mathbf{v_1} = (22, 27, 81, 79)$, $\mathbf{v_2} = (14, 67, 36, 16)$, $\mathbf{v_3} = (48, 6, 33, 88)$ and $\mathbf{v_4} = (36, 87, 91, 90)$. Next, we generate all possible permutation matrices with four elements and eliminate those that contain cycles of length more than 3. Then also remove those that are not Pareto efficient according to the preference orderings profile induced by $\mathbf{v}$. Let $D = \{D_1, D_2, \ldots, D_{|D|}\}$ be the resulting set of permutation matrices. Then, we need to solve the following constraint satisfaction problem:

Find $P, \alpha_1, \ldots, \alpha_{|D|}$ such that

$$\text{(i) } \sum_j p_{kj} v_{kj} \geq \sum_j p_{lj} v_{kj} \ \forall k, l, \qquad \textit{(envy-freeness)}$$

(ii) $P = \sum_{i=1}^{|D|} \alpha_i D_i,$ (decomposition)

(iii) $\sum_{i=1}^{n} p_{ij} = \sum_{j=1}^{n} p_{ij} = 1,$ (valid assignment)

(iv) $\alpha_i \geq 0 \ \forall i,$ (valid coefficients)

(v) $0 \leq p_{ij} \leq 1 \ \forall i,j.$ (valid probabilities)

For the valuation profile that we created, the constraint satisfaction problem is infeasible. This proves the theorem for $n = 4$. We will use that as the base case to prove that the theorem is true for any n using induction on the size of the valuation matrix.

For size $k - 1$, there is no envy-free, ex-post Pareto efficient mechanism that produces a 3-restricted assignment by the induction hypothesis. Let V_{k-1} be the valuation matrix for size $k - 1$ and let U_k be the matrix obtained by V_{k-1} by adding agent k and item k such that for all agents $i \neq k$, $v_{ij} > v_{ik}$ for all items $j \neq k$ and $v_{kk} > v_{jk}$ for all items $j \neq k$. Then, any ex-post Pareto efficient mechanism on input V_k must allocate item k to agent k with probability 1, otherwise there would be some Pareto-dominated permutation matrix in the decomposition. If such a mechanism existed, the assignment of the remaining $k - 1$ items to the remaining $k - 1$ agents would imply the existence of an envy free, ex-post Pareto efficient mechanism for input V_{k-1}, contradicting the induction hypothesis.   □

Next, we prove a similar impossibility theorem for the case when the notions of fairness is symmetry.

Theorem 2. *There is no mechanism that always outputs an assignment which is symmetric, ex-post Pareto efficient and 3-restricted.*

Proof. The proof idea is similar to that of the proof of Theorem 1, the main difference being that in the constraint satisfaction problem, the constraint for envy-freeness is replaced by:

$$P_{ik} = P_{jk}, \forall k \in M, \forall i,j \in N \text{ such that } v_{il} = v_{jl} \ \forall l \in M,$$

which is the constraint for symmetry. For the new constraint satisfaction problem, for $n = 5$, if we let $\mathbf{v_1} = \mathbf{v_2} = (4,3,5,1,2)$, $\mathbf{v_3} = \mathbf{v_4} = \mathbf{v_5} = (2,5,1,4,3)$, then the problem is infeasible. To extend the theorem for any n, we can use exactly the same inductive argument we use in the proof of Theorem 1.   □

Remark 2. As we mention in the introduction, Balbuzanov proposes the *2-cycle Probabilistic Serial* mechanism, which is ex-ante Pareto efficient and anonymous.

This does not contradict Theorem 2 because his definition of anonymity is with respect to pairs of agents and endowed items and does not imply symmetry.[6]

Next, we prove that both ex-post Pareto efficiency and envy-freeness or symmetry are needed for the impossibilities. If we remove ex-post efficiency, the simple mechanism that allocates all items uniformly at random, which is trivially both envy-free and symmetric, also produces k-restricted assignments.

Theorem 3. *The mechanism U_n that always outputs a uniform random assignment always produces a k-restricted assignment for $k \geq 2$. Furthermore, if n is odd, then the decomposition of the assignment consists of n permutation matrices, each one of which contains a self-loop and $\frac{n-1}{2}$ pairs. If n is even, then the decomposition consists of n permutation matrices, $n-1$ of which contain $\frac{n}{2}$ pairs and one which contains n self-loops.*

Proof. We will consider the cases when n is odd and n is even separately. Since all entries p_{ij} of the assignment matrix are $1/n$, the coefficients of the decomposition will be $1/n$ and for any i and j, d_{ij} will be 1 in exactly one component D and 0 in all others. Assume first that n is odd. Recall the graph interpretation of D and observe that D is a regular $n-1$-sided polygon. For vertex i, let e_i be the opposite side of G to vertex i and let D_i be the permutation matrix consisting of the self-loop (i) and pairs (kl) where k and l are adjacent to e_i or adjacent to a diagonal parallel to e_i. To get the decomposition, we iterate over all $i = 1, \ldots, n$ and obtain the permutation matrices D_i. Note that each permutation matrix consists of one self-loop and $\frac{n-1}{2}$ pairs. Next, assume that n is even, which means that $n-1$ is odd. Let U_{n-1} be the assignment matrix of size $n-1$ and U_n be the assignment matrix after we add agent n and item n. Since $n-1$ is odd, U_{n-1} can be decomposed into permutation matrices that contain one self-loop and $\frac{n-2}{2}$ pairs. The decomposition of U_n will be exactly the same, except that for each self-loop of each permutation matrix, we create a pair with item n, and we add an additional permutation matrix consisting only of self-loops. Again, it is not hard to see that the decomposition consists of n components, $n-1$ of which contain $\frac{n}{2}$ pairs and one that contains n self-loops. □

Finally, if we only require Pareto efficiency without any regard to fairness, it is trivial to obtain a deterministic Pareto efficient mechanism. The mechanism is the following simple one. Given an input valuation matrix V, generate all possible permutation matrices and find a feasible one that is Pareto efficient (with respect to the set of k-restricted components).

4 Randomized Mechanisms

In this section, we design mechanisms that output k-restricted assignments. Recall that a randomized mechanism inputs a valuation profile (or a preference profile) for n agents and outputs a bistochastic assignment matrix P. The

---

[6] A simple example with two agents $1, 2$ that have the same preference over items $1, 2$ is sufficient to see this.

assignment P can then be decomposed into at most n^2 permutation matrices using the Birkhoff-von Neumann decomposition.

The Birkhoff-von Neumann Decomposition

The decomposition works as follows. First, from P, construct a binary matrix P^{bin} by setting $p_{ij}^{\text{bin}} = 1$ if $p_{ij} > 0$ and 0 otherwise. From P^{bin}, construct a bipartite graph G with vertices corresponding to the rows and the columns of P^{bin} and with edges corresponding to the non-zero entries of P^{bin}. In other words, edge (i, j) exists in G if and only if $p_{ij}^{\text{bin}} = 1$. Using Hall's theorem, one can easily prove that G has a perfect matching. Note that this matching corresponds to some permutation matrix Π. Find such a D in G and find the smallest entry (i, j) in P such that $\Pi_{ij} = 1$ and let a be the value of that entry. For every entry (i, j) in P such that $\Pi_{ij} = 1$, subtract a from (i, j) to obtain a substochastic matrix P'. Then apply the same procedure again on P'. Note that a will be the coefficient of the first component of the decomposition and D will be that component. Also note that since P' has at least one more zero entry than P, the procedure will terminate in at most n^2 steps.

In our case, we are interested only in k-restricted components of the decomposition. One way to handle components with longer cycles is to remove them from the decomposition and redistribute their probabilities (given by their coefficients) to k-restricted components. It is conceivable that some of the properties of the assignment that are satisfied *in expectation* might be lost during the process; on the other hand, properties satisfied ex-post are preserved. To evaluate the ex-ante properties of the new assignment, we can re-construct the bistochastic matrix based on the components that survived the previous step. We will call this process the *recomposition* of the assignment matrix.

The process described above can be used to transform any mechanism to one that produces k-restricted assignments, assuming that the original decomposition had at least one k-restricted component. In general, this is not always the case however; it could be that some other decomposition (with possibly more than n^2 components) is needed in order to find such a component. Even worse, it could be the case that such a decomposition does not exist. We observe that in general, it is hard to decide whether this is the case or not.

Theorem 4. *Let P be an assignment. Deciding whether any decomposition of P has a 3-restricted component is NP-hard.*

Proof. Abraham et al. [2] proved that finding a cycle-cover consisting of cycles of length at most 3 is NP-hard. In their reduction, they use the gadget shown in Fig. 1, also known as a *clamp*. They construct a graph where clamps only intersect with other clamps on vertices, x_a, y_b and z_c. To get some intuition about the construction, one can think as x_a, y_b and z_c as elements in sets X, Y and Z respectively. Let $T \subseteq X \cup Y \cup Z$ be a set of triples. Two clamps intersect at a vertex x_a if x_a is part of two different triples (x_a, y_b, z_c) and (x_a, y_b', z_c').[7]

---

[7] This interpretation is very natural given that the proof in [2] uses a reduction from 3D-Matching.

We will refer to the subgraph consisting of vertices $x_a, 1, \ldots, L-1$ (on the left in Fig. 1), x_a^i and the edges indicident to them as "the x part of the clamp".

Recall the definition of graph D corresponding to the binary matrix P^{bin} at the Birkhoff-von Neumann decomposition described earlier; the graph has edges (i, j) only between vertices satisfying $P_{ij}^{\mathrm{bin}} = 1$. We claim that there exists a decomposition of P with at least one 3-restricted component if and only if graph D has a cycle cover consisting of cycles of length at most 3. It is not hard to see that there exists a decomposition of P with at least one 3-restricted component if and only if Graph D has a cycle cover consisting of cycles of length at most 3. It then suffices to prove that the graph used in [2] corresponds to some binary matrix P^{bin} associated with a bistochastic matrix P.

To find such a matrix, it is enough to find an assignment of weights $p_1, \ldots, p_{|E|}$ to the edges of the graph, such that for every vertex, the total weight of incoming edges and the total weight of outgoing edges is 1; such a weight assignment corresponds directly to a bistochastic matrix. We will only specify the weights for edges in the x part of the clamp; the rest are defined symmetrically. Let s be the in-degree of x_a.

First, for edges $e = (1, 2), \ldots, (L-2, L-1)$, let $p_e = 1$.

Then let:

$$p_{(L-1, x_a)} = \frac{1}{s}, \qquad\qquad p_{(L-1, x_a^i)} = \frac{s-1}{s},$$

$$p_{(x_a, 1)} = \frac{1}{s}, \qquad\qquad p_{(z_c^i, x_a^i)} = \frac{1}{s},$$

$$p_{(x_a^i, 1)} = \frac{s-1}{s}, \qquad\qquad p_{(x_a^i, y_b^i)} = \frac{s-1}{s}.$$

It is not hard to see that the assigned weights satisfy the constraint above and hence correspond to some bistochastic matrix P. $\qquad\qquad\square$

In the following, we describe a general method to generate k-restricted assignments based on some original assignment P. We will call this method *small-cycle projection*. Note that the decomposition-recomposition procedure that we described earlier can be viewed as such a projection.

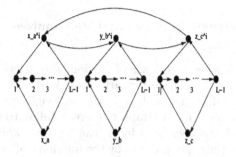

Fig. 1. The gadget used in the proof of Theorem 4 as it appears in Abraham et al. [2].

SMALL-CYCLE PROJECTION: Given a bistochastic matrix P and a "distance" measure d, the small-cycle projection P^* of P with respect to d is the solution of the following program,

$$\text{minimize} \quad d(P^*, P) \tag{1}$$
$$\text{subject to } P^* \in \text{Conv}(\mathcal{D}_k)$$

where $\mathcal{D}_k$ is the set of all k-restricted deterministic assignments and $\text{Conv}(\mathcal{D}_k)$ is the convex hull of $\mathcal{D}_k$.

A key observation here is that if d is a linear function, Program (1) is a linear program. Unfortunately, even if k is chosen to be 3, the set $\mathcal{D}_k$ is exponentially large in n. For this reason, we present two ways to approximate $\text{Conv}(\mathcal{D}_k)$ via a small subset of $\mathcal{D}_k$.

Small-Cycle Projection via Randomized Birkhoff-von Neumann Decomposition

Recall that the Birkhoff-von Neumann decomposition operates by finding a perfect matching on a bipartite graph in each step. Using the decomposition-recomposition procedure, we can approximate P by the matrix composed of the k-restricted components of the decomposition. It is conceivable however that different decompositions yield different k-restricted components and the choice of decomposition plays a central role to the quality of the approximation. For this reason, we need to have some freedom to choose between decompositions. On the other hand, iterating over all decompositions is computationally intractable. To balance the need for flexibility and the computational burdens, we employ the algorithm proposed by Goel, Kapralov, and Khanna [10]. Their algorithm computes random perfect matchings which can then be used to obtain random decompositions. The mechanism is then simply:

(i) decompose P as $\sum_i \lambda_i D_i$;
(ii) recompose $P^* = \sum_{D_i \in \mathcal{D}_k} \lambda_i^* D_i$.

The redistribution of probabilities can be done in various ways; the simplest being equally among k-restricted components.

Small Cycle Projection via Sequential Randomized Small-Cycle Cover

The second approach generates a set of random k-restricted permutations to approximate $\mathcal{D}_k$ in program (1). Particularly, these k-restricted permutations are generated by an algorithm that finds a maximum weight cycle cover consisting of cycles of length at most k, on Graph G corresponding to assignment P. For example, when $k = 3$, the algorithm by [2] can be used, where the weights are randomly assigned to the edges induced by the fractional allocation P. Formally, we describe our generating method in Algorithm 1.

Algorithm 1. Generating Small-Cycle Permutations.

---

input : P
output: $\hat{\mathcal{D}}_k$
$\hat{\mathcal{D}}_k \leftarrow \emptyset$;
while $|\hat{\mathcal{D}}_k| < \alpha(n)$ **do**
   | **for** i, j in $[n]$ **do**
   | | **if** $P_{ij} = 0$ **then**
   | | $\lfloor$ $G_{ij} \leftarrow 0$;
   | | **else**
   | | $\lfloor$ $G_{ij} \leftarrow$ rand();
   | $\lfloor$ $\hat{\mathcal{D}}_k \leftarrow \hat{\mathcal{D}}_k \cup \{\text{MaxWeightSmallCycleCover}(G, k)\}$;
return $\hat{\mathcal{S}}_k$

In the algorithm, $\alpha(n)$ is the desired size of set $\hat{\mathcal{D}}_k$, rand() generates a random number in $[1, 1 + \epsilon]$, and finally MaxWeightSmallCycleCover(G, k) returns the maximum weighted cycle cover of length at most k in graph G.

Unlike the randomized Birkhoff-von Neumann decomposition, the k-restricted permutations generated here are not a decomposition of the input P. Hence we need to solve Program (1) with some properly chosen "distance measure" d to approximate the assignment P. For linear distance measures, the program can be easily solved.

In the remainder of the section, we adapt two well-known mechanisms, *Probabilistic Serial* and *Random Serial Dictatorship* to make them compatible with 3-restricted allocations, in order to use them in our experiments in Sect. 5. As mentioned earlier, the reason for the choice of $k = 3$ is because this is the standard maximum allowed cycle length in kidney exchange operations. For Probabilistic Serial, we apply the small-cycle projection method; for Random Serial Dictatorship, we apply a different construction that always admits a decomposition with small cycles.

4.1 Probabilistic Serial with Restricted Cycles

The Probabilistic Serial mechanism works as follows. Each item is interpreted as an infinitely divisible good that the agents consume over the unit interval $[0, 1]$ at the same fixed speed. Each agent starts consuming her favorite item until the item is entirely consumed. Then, she moves to the next item on her preference list that has not been entirely consumed and starts consuming it. The procedure terminates when all items are entirely consumed. The fraction p_{ij} of item j consumed by agent i is then interpreted as the probability of assigning item j to agent i.

Using the small-cycle projection methods described above, we can construct two variants of the mechanism. For the variants generated using the randomized small-cycle cover method, we choose two appropriate distance metrics as follows to generate the approximate assignment P^*.

- *Social welfare distance*: $d_{\text{welfare}}(P, P^*) = \langle \mathbf{V}, P - P^* \rangle$ measuring the difference of social welfares induced by P and P^*, where $\langle \cdot, \cdot \rangle$ is the pointwise product of matrices.
- l_∞ *distance*: $d_{\text{norm}}(P, P^*) = \max_{i,j} |P_{ij} - P^*_{ij}|$ measuring the l_∞ error of P^* approaching P.

4.2   Random Serial Cycle

Random Serial Cycle, or *RSC* for short is a straightforward adaptation of Random Serial Dictatorship to incorporate the short cycle constraint. Specifically, the mechanism first uniformly at random fixes an ordering of agents and then matches them serially with their favorite items from the set of available items, just like Random Serial Dictatorship does. The difference is that whenever the length of an exchange is $k - 1$ and the exchange is not a cycle, the next agent to be picked is matched with the item that "closes" the cycle, regardless of her preferences. For example, for $k = 3$, if some agent i is matched with item j and agent j is matched with item l and agent l is next to pick an item, she will be matched with item i.

By construction, any run of the mechanism outputs a k-restricted assignment. To evaluate its properties however, we need to compose the assignment matrix from the probability mixture of the outcomes, which if done naively would require us to generate all possible $n!$ orderings of agents. In fact, it has been shown [15] that it is #P-hard to compute the assignment matrix of RSD given an input valuation matrix.

To sidestep this complication, we modify the mechanism to instead generate n^2 orderings at random using a Monte-Carlo process. The details follow. First we explain how to generate the orderings that we use and then we describe the mechanism.

Orderings Generation

- Fix a permutation π of $\{1, \ldots, n\}$ uniformly at random from the set of all permutations with n elements. Initialize position i of order π to be 1.
- For each position i, swap the agent in position i, denoted by π_i, with randomly chosen from π_i to π_n (could be agent π_i as well). Increase i by 1 at each iteration.
- Repeat until $i = n$.

Random Serial Cycle Mechanism

- Generate n^2 orders uniformly at random by Monte-Carlo process. Initialize the assignment matrix to be n-by-n zero matrix.
- For each order π, Initialize position i of order π to be 1. Agent π_i tries to be matched with the most preferred available item, denoted by $\mu(\pi_i)$, until reaches one of the following conditions:

- Agent $\mu(\pi_i)$ ranks item π_i on top of her preference list and $\pi_i, \mu(\pi_i)$ are matched together.
- Agent $\mu(\pi_i)$ finds an item j such that agent j has positive value for π_i after searching for the most preferred available item. Then the exchange cycle is completed.

Remove these matched pairs from the ordering. If agent π_i does not find a qualified item $\mu(\pi_i)$, then she is matched to item π_i. Increase i by 1 at each iteration.

- When position i is bigger than n, the k-restricted permutation matrix for ordering π can be produced. Multiply it by $1/n^2$ and add it to allocation matrix.
- After processing all the orderings, the assignment matrix P is produced.

5   Experimental Results

In this section, we design experiments to evaluate the performance of our mechanisms in terms of efficiency and fairness. Our experiments are conducted on the kidney exchange domain with input from realistic data. Our data generator is carefully designed based on statistics of the population in the United States. The generator incorporates patients and donors' physiological characteristics such as age, gender, blood type, HLA antigen, panel reactive antibodies (PRA) and number of waiting years prior to transplantation. It then produces a matching score for each patient-donor pair to quantify the transplantation quality. The input is then a weighted, directed, bipartite graph where the weight of an edge (i, j) is the utility (transplantation quality) of the patient of a donor-patient pair i when being matched with the donor of pair j. The transplantation quality can be interpreted as the probability that an exchange between the two participating pairs will be successful. For all of the experiments, k will be equal to 3.

Our measure of efficiency will be the *social welfare*, i.e. the sum of the transplantation qualities of the expected matching over all participating pairs. For fairness, we will try to minimize the fraction of envious agents, where the envy is calculated *in expectation*, i.e. an agent is envious if she would prefer another agents expected assignment to hers. We run experiments for different input sizes, ranging from a few agents to a hundred agents. Real-life input sizes can be larger but exchanges involving no more than a hundred agents are often carried out as well in practice. The bottleneck for the running time is the decomposition; as the number of agents grows larger, the harder it gets to achieve a decomposition with 3-restricted components.

We compare three mechanisms in terms of their social welfare and fraction of envy. The first one is the modification of Probabilistic Serial (PS) that we obtain by applying the randomized small-cycle cover method for small-cycle projection. We implement two variants of modified Probabilistic Serial, namely *PS-welfare* and *PS-norm*, based on the *social welfare distance* and the l_∞ *distance* respectively. The second mechanism that we use is Random Serial Cycle (RSC) with Monte-Carlo random generation of n^2 orderings of agents. Finally, we consider the optimal mechanism (denoted OPT in Figs. 2 and 3) that computes the optimal social welfare of the exchange, by Abraham et al. [2]. The variant of Probabilistic Serial that we obtain from applying the randomized Birkhoff-von Neumann small-cycle projection turned out to be ineffective for our goals, the problem being that k-restricted components were not present in most decompositions, but it is conceivable that it could be effective using some other mechanism as the "basis" of the decomposition.

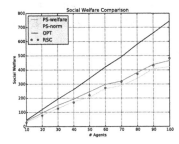 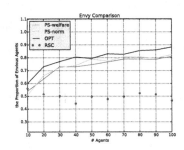

Fig. 2. Social welfare comparison (Color figure online).

Fig. 3. Envy comparison (Color figure online).

5.1   Fairness/Efficiency Tradeoffs

In Figs. 2 and 3, we show the results of the experiments for data from the U.S. population. As expected, the optimal mechanism performs better in terms of social welfare. The performance of our mechanisms (the two variants of PS and RSC) is very similar and not too far from the performance of the optimal mechanism, at least for smaller input sizes. Among the three, the best social welfare is achieved PS-welfare, since it was designed to extract higher levels of welfare.

In terms of fairness, the optimal mechanism fairs worse in comparison to our mechanisms. From the two versions of PS, PS-welfare is also slightly more fair, which suggests that it is a better choice than its norm-counterpart for this particular problem. Interestingly, RSC outperforms all mechanisms in terms of fairness by a big margin. In fact, for some input sizes, the proportion of envious agents is less than 40 % whereas for the optimal mechanism it is close to 80 %. This suggests that among the mechanisms we consider, RSC is the one that achieves the best fairness/efficiency guarantees.

The final result seems to be in contrast to the theoretical superiority of Probabilistic Serial over Random Serial Dictatorship in terms of fairness but it can be attributed to two factors: the inputs are not worst-case inputs and more importantly, it seems that the assignment produced by RSC is "closer" to the one outputted by Random Serial dictatorship, when compared to the outputs of Probabilistic Serial and its 3-restricted counterparts.

6 Conclusion

We considered the problem of random assignments in barter exchanges under the additional constraint on the cycle-length of the decomposition. We proposed two new mechanisms for the problem and a general method for designing mechanisms that produce assignments with small cycles. We evaluated our mechanisms on instances of the kidney exchange problem and found that they are better in terms of fairness and not much worse in terms of efficiency, when compared with the optimal exchange. An interesting future direction is to consider other notions of fairness and design different mechanisms for achieving better tradeoffs between fairness and efficiency. The 2-cycle Probabilistic Serial mechanism of Balbuzanov [4] seems like an obvious choice, given that it is designed to produce 3-restricted assignments.

References

1. Abdulkadiroğlu, A., Sönmez, T.: Random serial dictatorship and the core from random endowments in house allocation problems. Econometrica **66**(3), 689–701 (1998)
2. Abraham, D.J., Blum, A., Sandholm, T.: Clearing algorithms for barter exchange markets: enabling nationwide kidney exchanges. In: Proceedings of the 8th ACM Conference on Electronic Commerce, pp. 295–304. ACM (2007)
3. Ashlagi, I., Roth, A.: Individual rationality and participation in large scale, multi-hospital kidney exchange. In: Proceedings of the 12th ACM Conference on Electronic Commerce, pp. 321–322. ACM (2011)
4. Balbuzanov, I.: Short trading cycles: kidney exchange with strict ordinal preferences (2014)
5. Bogomolnaia, A., Moulin, H.: A new solution to the random assignment problem. J. Econ. Theor. **100**(2), 295–328 (2001)
6. Dickerson, J.P., Goldman, J.R., Karp, J., Procaccia, A.D., Sandholm, T.: The computational rise and fall of fairness. In: Proceedings of the Twenty-eighth AAAI Conference on Artificial Intelligence, Québec City, Québec, Canada, 27–31 July 2014, pp. 1405–1411 (2014)
7. Dickerson, J.P., Procaccia, A.D., Sandholm, T.: Dynamic matching via weighted myopia with application to kidney exchange. In: AAAI (2012)
8. Dickerson, J.P., Procaccia, A.D., Sandholm, T.: Failure-aware kidney exchange. In: Proceedings of the Fourteenth ACM Conference on Electronic Commerce, pp. 323–340. ACM (2013)
9. Dickerson, J.P., Procaccia, A.D., Sandholm, T.: Price of fairness in kidney exchange. In: International Conference on Autonomous Agents and Multi-agent Systems, AAMAS 2014, Paris, France, 5–9 May 2014, pp. 1013–1020 (2014)
10. Goel, A., Kapralov, M., Khanna, S.: Perfect matchings in $o(n \log n)$ time in regular bipartite graphs. SIAM J. Comput. **42**(3), 1392–1404 (2013)
11. Katta, A.-K., Sethuraman, J.: A solution to the random assignment problem on the full preference domain. J. Econ. Theor. **131**(1), 231–250 (2006)
12. Li, J., Liu, Y., Huang, L., Tang, P.: Egalitarian pairwise kidney exchange: fast algorithms vialinear programming and parametric flow. In: International Conference on Autonomous Agents and Multi-agent Systems, AAMAS 2014, Paris, France, 5–9 May 2014, pp. 445–452 (2014)

13. Roth, A.E., Sönmez, T., Ünver, M.: Kidney exchange. Q. J. Econ. **119**(2), 457–488 (2004)
14. Roth, A.E., Sönmez, T., Ünver, M.U.: Pairwise kidney exchange. J. Econ. Theor. **125**(2), 151–188 (2005)
15. Saban, D., Sethuraman, J.: The complexity of computing the random priority allocation matrix. In: Chen, Y., Immorlica, N. (eds.) WINE 2013. LNCS, vol. 8289, pp. 421–421. Springer, Heidelberg (2013)

Doctoral Consortium

CP-nets: From Theory to Practice

Thomas E. Allen$^{(\boxtimes)}$

University of Kentucky, Lexington, KY, USA
thomas.allen@uky.edu

1 Introduction

Consider the problem of buying an automobile. The vehicle that a customer prefers may depend on many factors including the customer's life commitments, hobbies, income level, concern about the environment, and so on. For example, a customer may prefer a minivan to a sports car if he has young children. Another customer may prefer a pick-up truck with a towing package to one that lacks such a feature if she enjoys hiking and kayaking. Numerous other factors, such as whether the paint shows pollen or whether the spare tire is accessible could also prove important.

Many decision domains, like this one, have a combinatorial structure. If I am asked to choose from a very small set of alternatives $\mathcal{O} = \{O_1, \ldots, O_4\}$, I can usually provide at least a partial ordering. I may be unsure how to compare alternatives O_1 and O_3, but know that I like O_4 better than both, and O_2 less than O_1 and O_3. However, if the preference involves a large number of features (e.g., make, model, dealership, warranty, color, etc.), then this approach is no longer practical, or even possible, given time constraints. Indeed, the best alternative may not yet exist in the physical world. Perhaps I could describe my most preferred alternative in terms of its features, but it does not exist unless I have it built to my custom specifications.

Formally, a preference relation $\succsim$ is a partial preorder on a set of alternatives (or outcomes) $\mathcal{O}$. The expression $o \succ o'$ means that o is preferred to o'. If neither outcome is preferred to the other, they are said to be incomparable, written $o \bowtie o'$. In this case, $O_4 \succ O_1$, $O_4 \succ O_3$, $O_1 \succ O_2$, and $O_3 \succ O_2$. Since the relationship is assumed to be transitive, one can further reason that $O_4 \succ O_2$. However, since the relationship between O_1 and O_3 is left unspecified, we say that these alternatives are incomparable ($O_1 \bowtie O_3$). In the discussion that follows, let us assume that $\mathcal{O}$ is finite and can be factored into features (variables) $V = \{X_1, \ldots, X_n\}$ with associated domains, s.t. $\mathcal{O} = X_1 \times \cdots \times X_n$. For example, the binary valued feature BRAKES $= \{antilock, conventional\}$ would be one of many features that collectively model all conceivable alternatives in the set **Automobiles**.

Conditional preference networks (CP-nets) have been proposed for such problems [4]. Rather than comparing alternatives as atoms, the decision maker considers the interplay of the features—how the preference over one feature depends on the values of others in the decision domain. For example, if I am purchasing a vintage sports car, I may be willing to forgo modern safety features such as

© Springer International Publishing Switzerland 2015
T. Walsh (Ed.): ADT 2015, LNAI 9346, pp. 555–560, 2015.
DOI: 10.1007/978-3-319-23114-3_33

anti-lock brakes, but if I am purchasing a new minivan, then I definitely prefer anti-lock brakes. One can formalize this as

$$minivan \wedge new : antilock \succ conventional.$$

Such statements are known as *ceteris paribus rules*, from the Latin "as long as everything else stays the same." That is, if I am comparing two new minivans, one with anti-lock brakes and one with conventional brakes, I will always prefer the alternative with antilock brakes, provided the values of other features (price, color, make, model, etc.) are the same for both alternatives. The relationship among the features is further specified through a *dependency graph*, a directed graph in which the nodes represent the relevant features and a directed edge from one feature to another indicates that the latter (known as the child) depends on the former (the parent). The rules themselves are stored in *conditional preference tables* of the feature to which they apply (Fig. 1).

Definition 1. *A CP-net $\mathcal{N}$ is a directed graph. Each node represents a variable $X_i \in V$ and is annotated with a* conditional preference table (CPT) *describing the subject's preferences over the domain of X_i given its dependencies. An edge (X_h, X_i) indicates that the preferences over X_i depend directly on the value of X_h.*

Alternatives that differ in just one feature (e.g., two **Automobile**s that are identical except for their AGE) can be compared directly, provided CPT(AGE) contains an applicable rule. Alternatives that differ in more than one feature can be compared only if a transitive sequence of rules exists between the two alternatives. For example, $\langle children, sportscar, vintage, conventional \rangle \prec \langle children, minivan, vintage, conventional \rangle \prec \langle children, minivan, new, conventional \rangle \prec \langle children, minivan, new, antilock \rangle$. Such sequences are known as (improving) flipping sequences, since each rule "flips" the value of just one variable.

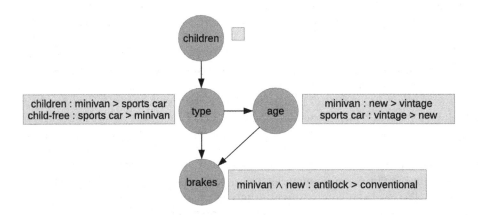

Fig. 1. A simple CP-net

There is much to like about CP-nets. They let us model preferences over factored domains with exponentially many conceivable alternatives. They capture visually the if-then rules that many of us think we employ when we reason about such alternatives. They are qualitative; that is, they only ask us to specify whether one thing is better than another, without assigning a numeric weight as to precisely *how much* we prefer it. Finally, the problem of determining the optimal (most preferred) outcome with respect to a CP-net can be answered in linear time in the number of features if CPTs are complete.

On the other hand, while many academic papers discuss CP-nets (at last count, the seminal journal paper had over 750 citations, according to Google Scholar) and many applications have been proposed, we are not yet aware of their use in real-world applications. There are several reasons for this. First, determining dominance—whether one arbitrary outcome is better than another with respect to a CP-net—is known to be very hard (PSPACE-complete) in the general case [5]. An application for automobile shoppers, to continue our example, is not particularly useful if it requires several days in the worst case to determine whether one vehicle is better than some other vehicle that I am considering! Additionally, much of the research on CP-nets makes strong, often unrealistic assumptions, such as that the features must be binary, CPTs must be complete (contain a rule for every combination of parent features), that indifference is not allowed, or that the graph must conform to a particular structure (e.g., rooted tree, acyclic digraph, etc.). My research involves addressing these issues to make CP-nets more useful for complex engineering applications.

In the sections that follow, I briefly discuss three areas of ongoing research: how to generate CP-nets i.i.d., practical approaches to the problem of dominance testing (DT), and new methods for learning CP-nets.

2   Generating CP-nets

Often, one needs to generate CP-nets in an i.i.d. manner. For example, consider that we wish to compare the *expected* running time of two algorithms that perform dominance testing. Both algorithms are designed for the same sort of input—e.g., binary valued features and complete, acyclic CPTs with an assumed bound on in-degree (number of parents). To compare the two algorithms, we wish to generate a set of dominance testing problems that are representative of the DT problems in this set. We are not aware of real-world datasets of DT problems; moreover, even if such datasets were readily available, one would like to compare the algorithms' expected performance on *any* allowable input. Generating the outcome pairs for such an experiment is easy since one can just assign the value of each feature as an independent coin-flip. However, it is not so clear how to generate the CP-nets such that each instance is valid and equally likely with respect to the set of all CP-nets under consideration. Often researchers simply permute the nodes and insert edges at random so as to avoid cycles and then randomly assign CPT entries. However, it is easy to show that this approach leads to statistical bias, with high-indegree CP-nets grossly undersampled at the expense of low-indegree CP-nets.

Other motivations for generating CP-nets i.i.d. include Monte Carlo algorithms (e.g., for learning, reasoning with, or aggregating CP-nets), statistical analyses of the properties of CP-nets (such as how expected flipping sequence length varies as the number of edges in the dependency graph increases), voting profiles for social choice experiments, black-box testing of algorithms for correctness, and finding hard problem instances. In each case, an unbiased generation algorithm is needed.

Many asymptotic results are known for problems that involve learning or reasoning with CP-nets. However, the combinatoric properties of CP-nets are less understood. Since these properties are central to understanding how to generate unbiased random instances, I began by studying how to count the number of CP-nets for various types of graphs (directed trees, polytrees, directed acyclic graphs with bounded indegree, etc.) [2]. For this I built upon results such as Prüfer codes, which can be used to represent and generate labeled trees [8], and the so-called DAG codes for directed acyclic graphs [11], introducing novel recurrences for counting and generating CP-nets efficiently. As part of this process, I studied the problem of assuring that a CPT generated at random was consistent with the graph, showing that this is equivalent to the set of *nondegenerate functions* of k Boolean inputs where k is the node's indegree. I have also developed and implemented (in C++ and x86 assembly language) a novel method for generating random binary, complete CP-nets i.i.d. for various graphs, including directed trees, polytrees, DAGs with bounded indegree, and unbounded DAGs. My implementation can generate thousands of CP-nets with up to 100 features uniformly at random in just a few seconds. I am presently extending this method to accommodate more general classes of CP-nets, such as those with multi-valued variables, incomplete CPTs, CP-nets that can model indifference as well as strict preferences, and CP-nets that are compatible with a partially specified (incomplete) CP-net.

3   Reasoning with CP-nets

While the problem of finding the most preferred outcome can be conducted in linear time in the number of nodes, the problem of dominance testing is known to be hard in general, requiring worst-case exponential time in practice. A significant aspect of this complexity is the possibility of exponentially long flipping sequences [4]. In earlier work, I observed that long sequences appeared to be rare [1]. I also reasoned that very long transitive sequences were unlikely to provide useful information about human preferences if the possibility of noise was taken into account. That is, if even some small percentage of the rules could be represented incorrectly—for example, due to an entry error during the elicitation process, then the probability that a given sequence correctly represents the decision maker's preferences diminishes to what would be expected from chance as the length of the transitive sequence increases. However, my earlier experiments relied on generation methods that did not provide i.i.d. guarantees. More recently, I have been performing additional experiments using the method of i.i.d. generation discussed above. These show that in most cases the expected flipping length is relatively close to the Hamming distance, the number of features in which the two alternatives differ. I have also shown that the average path length of the dependency graph is a good predictor of flipping sequence length.

I have also shown how to reduce general dominance testing problems to one of Boolean satisfiability so that the heuristic methods employed by modern SAT solver can be leveraged. Others have elsewhere shown how to reduce the problem to one of planning [4], model checking [10], or (indirectly) constraint satisfaction [9]. In more recent work, I have suggested the possibility of limiting search depth based on the expected length of flipping sequence given parameters that are easy to compute.

In future work, I hope to show the effect of indifference and incomparability on flipping sequence lengths. I am also comparing the performance of different algorithms that have been proposed and considering how to combine such algorithms into a portfolio for dominance testing. Finally, I am interested in discovering better heuristic methods for dominance testing.

4 Eliciting and Learning CP-nets

CP-nets can be elicited directly or indirectly from a user or learned from observational data. Direct elicitation assumes human subjects can introspect on the cause-and-effect processes that underlie their preferences. This is a particularly strong assumption that seems unrealistic in many settings [3].

Indirect elicitation relies on the weaker assumption that subjects can answer whether one alternative is preferred to another ("I prefer the blue minivan to the red one") without providing explanations for such preferences. At the ADT-2013 conference [7], I presented a heuristic algorithm, earlier proposed by Guerin [6], for indirectly eliciting CP-nets through user queries. In that paper we assumed strict, complete preferences over binary variables and that the user could answer queries consistently.

In later work [1], I considered the problem of learning CP-nets from choice data. There I relaxed some of the customary modeling assumptions in favor of models with multi-valued variables over which the subject may be indifferent and/or inconsistent. I presented the case that such CP-nets are necessary for many simple, real-world problems. I then showed how to leverage the power of SAT solvers to learn such CP-nets from choice data. More recently, I have proposed a novel encoding for tree-shaped CP-nets that enables local search. This method allows searching for a suitable tree-shaped CP-net even if the comparison set is inconsistent due to noise. Moreover, since it is a learning- rather than an elicitation-based method, it does not depend on the subject's capacity to introspect or to respond or choose consistently. In future work, I hope to improve on this method and extend it to a richer class of CP-nets.

References

1. Allen, T.E.: CP-nets with indifference. In: 2013 51st Annual Allerton Conference on Communication, Control, and Computing (Allerton), pp. 1488–1495. IEEE (2013)
2. Allen, T.E., Goldsmith, J., Mattei, N.: Counting, ranking, and randomly generating CP-nets. In: MPREF 2014 (AAAI-14 Workshop) (2014)
3. Allen, T.E., Chen, M., Goldsmith, J., Mattei, N., Popova, A., Regenwetter, M., Rossi, F., Zwilling, C.: Beyond theory and data in preference modeling: bringing humans into the loop. In: Proceedings of ADT. LNAI. Springer, Heidelberg (2015, In print)
4. Boutilier, C., Brafman, R., Domshlak, C., Hoos, H., Poole, D.: CP-nets: a tool for representing and reasoning with conditional ceteris paribus preference statements. J. Artif. Intell. Res. **21**, 135–191 (2004)
5. Goldsmith, J., Lang, J., Truszczynski, M., Wilson, N.: The computational complexity of dominance and consistency in CP-nets. J. Artif. Intell. Res. **33**(1), 403–432 (2008)
6. Guerin, J.T.: Graphical Models for Decision Support in Academic Advising. Ph.D. thesis. University of Kentucky (2012)
7. Guerin, J.T., Allen, T.E., Goldsmith, J.: Learning CP-net preferences online from user queries. In: Perny, P., Pirlot, M., Tsoukiàs, A. (eds.) ADT 2013. LNCS, vol. 8176, pp. 208–220. Springer, Heidelberg (2013)
8. Kreher, D.L., Stinson, D.: Combinatorial Algorithms: Generation, Enumeration, and Search. CRC Press, Boca Raton (1999). ISBN 9780849339882
9. Li, M., Vo, Q.B., Kowalczyk, R.: Efficient heuristic approach to dominance testing in CP-nets. In: Proceedings of AAMAS, pp. 353–360 (2011)
10. Santhanam, G.R., Basu, S., Honavar, V.: Dominance testing via model checking. In: AAAI (2010)
11. Steinsky, B.: Efficient coding of labeled directed acyclic graphs. Soft Comput. 7(5), 350–356 (2003). doi:10.1007/s00500-002-0223-5. http://dx.doi.org/10.1007/s00500-002-0223-5. ISSN 1432-7643

Possible Optimality and Preference Elicitation for Decision Making

Nawal Benabbou[✉]

CNRS, LIP6 UMR 7606, Sorbonne Universités, UPMC Univ Paris 06,
4 Place Jussieu, 75005 Paris, France
nawal.benabbou@lip6.fr

Decision support systems often rely on a mathematical decision model allowing the comparison of alternatives and the selection of a proper solution. In the field of Multicriteria Decision Making, an aggregation function is often used to synthesize the different evaluations of each alternative into an aggregated value representing its overall utility. The aggregation function must be sufficiently expressive to efficiently approximate the decision maker's preferences in human decision support, or simulate a prescribed decision behavior in automated decision systems. This explains the diversity of decision models available in the literature but also the increasing interest for sophisticated parameterized models such as the Choquet integral [14,32] which enables the representation of complex preferences and includes many other models as special cases (e.g. leximin and lexicographic aggregators [11], the Ordered Weighted Average operator [38], and Weighted Ordered Weighted Average [34]).

To make use of such models, one needs to assess the model parameters in order to fit to the decision maker's preferences. Most of the previous work on the elicitation of Choquet integral parameters consider a static database of preference statements, and focus on the determination of the parameters that best fit to the available database (e.g. [13,15,16,26,27]) for instance by minimizing a quadratic error. However, these approaches require a relatively large number of preference statements to model the decision maker's behaviour accurately which are not always possible to obtain. Preference elicitation with limited available information is a crucial task in many application domains, including recommender systems and interface customization [28]. Departing from these standard approaches, we consider incremental elicitation methods based on the minimax regret which is a decision criterion that has been advocated as a means for robust optimization in the presence of data uncertainty [21] and has been used for decision making with utility function uncertainty [5,6,31]. The general principle of this approach is to iteratively ask questions to the decision maker so as to reduce the set of possible parameters until the preferred alternative can be detected with some guarantees (as given by the minimax regret). This elicitation approach enables limiting the decision maker's burden as preference information is only required to discriminate between alternatives (not to assess the model parameters). Incremental elicitation methods have been proposed for the simple case of linear utilities but have never been studied for Choquet Integrals. This constitutes the first challenging issue considered during my Ph.D.

© Springer International Publishing Switzerland 2015
T. Walsh (Ed.): ADT 2015, LNAI 9346, pp. 561–566, 2015.
DOI: 10.1007/978-3-319-23114-3_34

Incremental Elicitation of Choquet Integrals. Adapting the incremental elicitation approach used for linear functions to the case of Choquet integrals is not straightforward. In the linear case, the selection of new preference queries requires multiple linear programs to be solved, but these problems become significantly more difficult when considering a Choquet integral. More precisely, the Choquet integral's parameters take the form of a capacity, i.e. a monotonic function defined on the power set of criteria which enables the control of the importance attached to all subsets of criteria, and possibly positive or negative synergies between criteria. Thus the number of parameters to assess is exponential in the number of criteria. As a consequence, the nature of the Choquet integral's parameters induces an exponential number of optimization variables and an exponential number of constraints over these variables in the linear programs considered in this elicitation scheme. First, we have proved that, by focusing on a specific type of queries involving binary profiles versus constant profiles, the optimization problems to be solved can be simplified to problems admitting only a linear number of variables and constraints. Then, using the constraint graph associated with these simplified linear programs, we have proposed an iterative procedure to solve these optimization problems in polynomial time (instead of using linear programming); our iterative procedure reduces computation times by around five orders of magnitude. Finally, we tested "the query selection strategy" which consists of selecting the question that is the most informative in the worst-case scenario of answers and we observed that it enables the detection of the preferred option without asking too many questions. This work has been accepted for publication in the last European Conference on Artificial Intelligence and honored to receive the ECAI'14 Best Student Paper award [4].

The next step was to consider decision situations where the decision space is very large, which is often the case in recommender systems where the possible alternatives can be thousands. Elicitation on combinatorial domains is a challenging issue that recently motivated several contributions in various contexts, e.g. in constraint satisfaction problems [6], in Markov Decision Processes [30,36], in stable matching problems [10] and in multiattribute spaces [7,12,20]. We consider decision spaces defined implicitly as the set of possible solutions of a multiobjective combinatorial optimization problem. In standard interactive methods for multicriteria decision support, preference elicitation methods consists in iterating the generation of a feasible instance of the parameters and the computation of the corresponding optimal solution, until the decision maker is satisfied with the latter solution (e.g. [35,39]). However, this elicitation approach does not guarantee that the final solution is actually the best alternative for the decision maker since the final instance of the model parameters does not necessarily fit to the decision maker's preferences. We proposed instead to combine search for possibly optimal solutions and elicitation in order to reduce the uncertainty over the model parameters during the resolution so as to more focus the search while determining a necessary optimal solution. A possibly (resp. necessary) optimal solution is an option that is optimal for some (resp. all) parameters compatible with our knowledge about the decision maker's preferences. We need first to propose efficient search procedures for the determination of all possibly optimal solutions given a set of feasible parameters, and then to design elicitation methods reducing the set of feasible parameters during the search while ensuring the determination of a necessary optimal alternative at the end of the resolution. So far, we have considered the two following multiobjective combinatorial optimization problems: multiobjective state space search and multicriteria spanning tree problem. In order to address one difficulty at a time, we have first considered the case of linear utilities and then Choquet Integrals.

Multiobjective State Space Search. Consider a state space graph endowed with q evaluation criteria, i.e. q cost functions to be minimized (e.g. time, distance, energy, risk). Each path is therefore valued by a cost vector and preferences over paths are inherited from the preference over their cost vectors. Preference over cost vectors are defined using a linear aggregation function f_ω defining the overall cost (or disutility) $f_\omega(x)$ attached to any cost vector x, where ω is a vector of preference parameters representing the relative importance of criteria. We want to find a path from an initial node to a goal node that minimizes the overall cost function f_ω that represents the DM's preferences, but the vector of weights ω is imprecisely known. Since all preference models considered in multicriteria analysis are compatible with Pareto dominance, preference-based search methods in multiobjective optimization are often based on the exploration of the set of Pareto-optimal solutions. The so-called MOA* algorithm [25,33] is a multiobjective extension of A^* [17] that determines the set of Pareto non-dominated cost vectors attached to solution paths and returns one path for each element. In the multiobjective case, there possibly exists several optimal paths with different cost vectors to reach a given node. Therefore, the basic graph exploration procedure consists in iteratively expanding subpaths rather than nodes. In order to compute the set of possibly optimal solution paths, we need to define new pruning rules that use sharper conditions than those based on Pareto-dominance tests. In other words, we need to be able to detect subpaths that cannot lead to possibly optimal solution given our knowledge about the decision maker's preferences. To do so, we introduced a dominance relation between sets of cost vectors and given a set of cost vectors, we proved that it enables to detect vectors that cannot be possibly optimal. Then, we proposed a filtering algorithm based on this dominance relation, enabling us to compute the set of possibly optimal cost vectors in a polynomial time. Finally, we proposed two pruning rules based on this filtering algorithm which enable us to discard subpaths that cannot lead to a possibly optimal solution path and we proved that the corresponding graph exploration procedure returns exactly the set of possibly optimal solution paths. Finally, to detect a necessary optimal solution paths, we proposed two incremental elicitation strategies based on the minimax regret criterion so as to reduce the uncertainty during the search. This work has been accepted for publication in the last AAAI Conference on Artificial Intelligence [2]. The next step was to extend our approach to work with a non-linear f_ω function such as a Choquet integral so as to obtain better fitting capacities to the decision maker's preferences. However, this extension raises challenging algorithmic questions since the use of a Choquet integral complicates the definition of pruning rules by not satisfying the Bellman principle (i.e. discarding supbaths by comparison with other subpaths is no longer possible). In order to design an efficient algorithm, we proposed to work on near optimal cost vectors using an approximate version of minimax regret. This work has been accepted for publication in the last IJCAI conference [1].

Multicriteria Spanning Tree Problem. Consider a connected graph G where each edge is valued by a cost vector corresponding to its evalutation with respect to different criteria. Every criterion is assumed to be additive over the edges; therefore the cost of any subgraph is the sum of the cost of its constituent edges. A spanning tree of G is a connected subgraph of G which contains no cycle while including every node of G. We assume that preferences over cost vectors are defined using a linear aggregation function f_ω but the weighting vector ω is imprecisely known. Here again, given a feasible set of weights, we first consider the problem of determining all possibly optimal spanning trees. In the single objective case, the minimum spanning tree problem can be solved in polynomial time using standard greedy algorithms due to Kruskal [22] and Prim [29]. Unfortunately, as soon as the number of criteria is greater than 2, the problem becomes intractable because the number of Pareto-optimal cost vectors associated to spanning trees is, in the worst case, exponential in the number nodes. In the paper "On Possibly Optimal Tradeoffs in Multicriteria Spanning Tree Problems" that has been accepted for publication in ADT 2015 [3], we proposed a multiobjective extension of Prim's algorithm which can compute the exact set of possibly optimal cost vectors associated to spanning trees; this algorithm is a greedy search based on a specific decomposition of the feasible set of parameters. Then, we proposed to interweave incremental elicitation and search to determine a necessary optimal spanning tree. This algorithm consists in selecting, at each iteration step, an edge in the cocycle of the current subgraph that is necessarily (or almost) optimal in the cocycle; if no such edge exists, the procedure asks questions to the decision maker to reduce the set of feasible weights until such edge can be detected. We are now studying the case of non-linear utility functions, which seems to be a challenging issue because our multiobjective greedy search is no longer valid.

Perspectives. As future work, we plan to study combinatorial voting with partial preference profiles. When individual preferences are incomplete, one can indeed study possible and necessary winners (e.g., [9, 19, 23, 37]). In this setting, incremental elicitation methods are used to progressively reduce the set of possible winners until a winner can be determined with some guarantee [8, 18, 24]). As a next step, we can study the potential of incremental elicitation methods in combinatorial voting (i.e. the set of alternatives has a combinatorial structure) with partial preference profiles. We also plan to adapt these approaches for utility elicitation in the context of decision making under risk, not only for expected utility models but also for rank-dependent utility model (elicitation of the probability distortion combined with the elicitation of utilities).

References

1. Benabbou, N., Perny, P.: Combining preference elicitation and search in multiobjective state-space graphs. In: Proceedings of IJCAI 2015 (2015)
2. Benabbou, N., Perny, P.: Incremental weight elicitation for multiobjective state space search. In: Proceedings of AAAI 2015, pp. 1093–1098 (2015)
3. Benabbou, N., Perny, P.: On possibly optimal tradeoffs in multicriteria spanning tree problems. In: Proceedings of ADT 2015 (2015)
4. Benabbou, N., Perny, P., Viappiani, P.: Incremental elicitation of Choquet capacities for multicriteria decision making. In: Proceedings of ECAI 2014, pp. 87–92 (2014)
5. Boutilier, C., Bacchus, F., Brafman, R.I.: UCP-Networks: a directed graphical representation of conditional utilities. In: Proceedings of UAI 2001, pp. 56–64 (2001)
6. Boutilier, C., Patrascu, R., Poupart, P., Schuurmans, D.: Constraint-based optimization and utility elicitation using the minimax decision criterion. Artif. Intell. **170**(8–9), 686–713 (2006)
7. Braziunas, D., Boutilier, C.: Minimax regret based elicitation of generalized additive utilities. In: Proceedings of UAI 2007, pp. 25–32 (2007)
8. NaamaniDery, L., Kalech, M., Rokach, L., Shapira, B.: Reaching a joint decision with minimal elicitation of voter preferences. Inf. Sci. **278**, 466–487 (2014)
9. Ding, N., Lin, F.: Voting with partial information: what questions to ask? In: Proceedings of AAMAS 2013, pp. 1237–1238 (2013)
10. Drummond, J., Boutilier, C.: Preference elicitation and interview minimization in stable matchings. In: Proceedings of AAAI 2014, pp. 645–653 (2014)
11. Fishburn, P.C.: Axioms for lexicographic preferences. Rev. Econ. Stud. **42**, 415–419 (1975)
12. Gonzales, C., Perny, P.: GAI networks for utility elicitation. In: Proceedings of the 9th International Conference on the Principles of Knowledge Representation and Reasoning, pp. 224–234 (2004)
13. Grabisch, M., Kojadinovic, I., Meyer, P.: A review of methods for capacity identification in Choquet integral based multi-attribute utility theory. Eur. J. Oper. Res. **186**(2), 766–785 (2008)
14. Grabisch, M., Labreuche, C.: A decade of application of the Choquet and Sugeno integrals in multi-criteria decision aid. Ann. Oper. Res. **175**(1), 247–286 (2010)
15. Grabisch, M., Marichal, J.-L., Mesiar, R., Pap, E.: Aggregation Functions. Encyclopedia of Mathematics and its Applications. Cambridge University Press, New-York (2009)
16. Grabisch, M., Nguyen, H.T., Walker, E.A.: Fundamentals of Uncertainty Calculi, with Applications. Encyclopedia of Mathematics and its Applications. Kluwer Academic Publishers, dordrecht (1995)
17. Hart, P.E., Nilsson, N.J., Raphael, B.: A formal basis for the heuristic determination of minimum cost paths. IEEE Trans. Syst. Man Cybern. **4**(2), 100–107 (1968)
18. Kalech, M., Kraus, S., Kaminka, G.A.: Practical voting rules with partial information. Auton. Agent. Multi-Agent Syst. **22**(1), 151–182 (2010)
19. Konczak, K., Lang, J.: Voting procedures with incomplete preferences. In: Proceedings of IJCAI-2005 Multidisciplinary Workshop on Advances in Preference Handling, vol. 20 (2005)
20. Koriche, F., Zanuttini, B.: Learning conditional preference networks. Artif. Intell. **174**(11), 685–703 (2010)

21. Kouvelis, P., Yu, G.: Robust Discrete Optimization and Its Applications. Kluwer, Dordrecht (1997)
22. Kruskal, J.B.: On the shortest spanning subtree of a graph and the traveling salesman problem. Proc. Am. Math. Soc. **7**, 48–50 (1956)
23. Lang, J., Pini, M.S., Rossi, F., Salvagnin, D., Brent, K., Venable, K.B., Walsh, T.: Winner determination in voting trees with incomplete preferences and weighted votes. Auton. Agent. Multi-Agent Syst. **25**(1), 130–157 (2012)
24. Lu, T., Boutilier, C.: Robust approximation and incremental elicitation in voting protocols. In: Proceedings of IJCAI 2011, pp. 287–293 (2011)
25. Mandow, L., Pérez De la Cruz, J.L.: A new approach to multiobjective A* search. In: Proceedings of IJCAI 2005, pp. 218–223 (2005)
26. Marichal, J.-L., Roubens, M.: Determination of weights of interacting criteria from a reference set. EJOR **124**(3), 641–650 (2000)
27. Meyer, P., Roubens, M.: On the use of the Choquet integral with fuzzy numbers in multiple criteria decision support. Fuzzy Sets and Syst. **157**(7), 927–938 (2006)
28. Peintner, B., Viappiani, P., Yorke-Smith, N.: Preferences in interactive systems: technical challenges and case studies. AI Mag. **29**(4), 13–24 (2008)
29. Prim, R.C.: Shortest connection networks and some generalizations. Bell Syst. Tech. J. **36**, 1389–1401 (1957)
30. Regan, K., Boutilier, C.: Eliciting additive reward functions for markov decision processes. In: Proceedings of IJCAI 2011, pp. 2159–2164 (2011)
31. Salo, A., Hämäläinen, R.P.: Preference ratios in multiattribute evaluation (PRIME)-elicitation and decision procedures under incomplete information. IEEE Trans. Syst. Man Cybern. **31**(6), 533–545 (2001)
32. Schmeidler, D.: Integral representation without additivity. Proc. Am. Math. Soc. **97**(2), 255–261 (1986)
33. Stewart, B.S., White III, C.C.: Multiobjective A*. J. ACM **38**(4), 775–814 (1991)
34. Torra, V.: The weighted OWA operator. Int. J. Intell. Syst. **12**(2), 153–166 (1997)
35. Vanderpooten, D., Vincke, P.: Description and analysis of some representative interactive multicriteria procedures. Appl. Math. Comp. **83**(2–3), 261–280 (1997)
36. Weng, P., Zanuttini, B.: Interactive value iteration for markov decision processes with unknown rewards. In: Proceedings of IJCAI 2013, pp. 2415–2421 (2013)
37. Xia, L., Conitzer, V.: Determining possible and necessary winners given partial orders. J. Artif. Intell. Res.(JAIR) **41**, 25–67 (2011)
38. Yager, R.R.: On ordered weighted averaging aggregation operators in multicriteria decision making. IEEE Trans. Syst. Man Cybern. **18**(1), 183–190 (1998)
39. Zionts, S., Wallenius, J.: An interactive programming method for solving the multiple criteria problem. Manage. Sci. **22**(6), 652–663 (1976)

Measuring Intrinsic Quality of Human Decisions

Tamal T. Biswas[✉]

Department of CSE, University at Buffalo, Amherst, NY 14260, USA
tamaltan@buffalo.edu

1 Research Problem

Decision making is an integral part of artificial intelligence. Humans often make sub-optimal decisions in a bounded rational environment. This raises the issue of getting a measure of the quality of decisions made by the person. Most of the time, the evaluation of decisions, considers only a few parameters. For example, in test-taking one might consider only the final score; for a competition, the results of the game; for the stock market, profit and loss, as the only parameters used when evaluating the quality of the decision. We regard these as *extrinsic* factors. If an artificial intelligent agent is available which is superior than any human for any domain, it is possible to evaluate the quality of the decision by analyzing the decisions made with such entities and thus move from bounded toward strict rationality. This approach gives a measure of the *intrinsic* quality of the decision taken. Ideally, this removes all dependence on factors beyond the agent's control, such as, performance by other agents (on tests or in games) or accidental circumstances (which may affect profit or loss). Decisions taken by humans are often effectively governed by *satisficing*, a cognitive heuristic that looks for an acceptable sub-optimal solution among possible alternatives. We aim to measure the loss in quality and opportunity from satisficing and express the bounded-rational issues in terms of *depth* of thinking. Any aptitude test allows multiple participants to answer the same problem, and based on their responses, the difficulty of the problem is measured. The desired measure of difficulty is used when calculating the relative importance of the question on their overall scores. The first issue is how to distinguish the *intrinsic* difficulty of a question, problem or chess position from a simple poor performance by respondents? A second issue is how to judge whether the problem is hard because it requires specialized knowledge, requires deep reasoning, or is "tricky"—with plausible wrong answers. Classical test theory approaches are less able to address these issues owing to design limitations such as in test questions, with only one answer receiving credit. Our work aims to address these issues.

2 Outline of Objectives

We have identified three research goals:

1. Find an intrinsic way to judge the difficulty of decision problems, such as test questions,

© Springer International Publishing Switzerland 2015
T. Walsh (Ed.): ADT 2015, LNAI 9346, pp. 567–572, 2015.
DOI: 10.1007/978-3-319-23114-3_35

2. Quantify *level-k* thinking [8] by which to identify satisficing and measure the degree of boundedness in rational behavior.
3. Use an application context (namely, chess) in which data is large and standards are well known so as to calibrate extrinsic measures of performance reflecting difficulty and depth. Then transfer the results to validate goals 1 and 2 in applications where conditions are less regular.

Putting together all these aspects, we have developed a model that can segregate agents by their skill level via rankings based on their decisions and the difficulty of the problems faced, rather than being based only on total test scores and/or outcomes of games. Moreover, it is possible to predict an actor's future performance based on the past decisions made by similar agents. In our setting, we have chosen chess games played by thousands of players spanning a wide range of ranking. The moves played in the games are analyzed with chess programs, called *engines*, which are known to play stronger than any human player.

This approach can be extended to other fields of bounded rationality, for example stock market trading and multiple choice questions, for several reasons, one being that the model itself does not depend on any game-specific properties. The only inputs are numerical values for each option, values that have authoritative hindsight and/or depth beyond a human actor's immediate perception. Another is the simplicity and generality of the mathematical components governing its operation, which are used in other areas.

3   State of the Art

Various descriptive theories of decision models have been proposed to date. *Prospect theory*, handles a few fundamental requirements for dealing with decision measures, such as eliminating clearly inferior choices and simplifying and ordering outcomes. Sequential sampling/accumulation based models are the most influential type of decision models to date. *Decision field theory* (DFT) applies sequential sampling for decision making under risk and uncertainty [4]. One important feature of DFT is 'deliberation', i.e., the time taken to reach a decision.

Although item response theory (IRT) models do not involve any decision making models directly, they provide tools to measure the skill of a decision-maker. IRT models are used extensively in designing questionnaires which judge the ability or knowledge of the respondent. Morris and Branum et al. have demonstrated the application of IRT models to verify the ability of the respondents with a particular test case [10].

On the chess side, a reference chess engine $E \equiv E(d, mv)$ was postulated by DiFatta, Haworth, and Regan [6]. The parameter d indicates the maximum depth the engine can compute, where mv represents the number of alternative variants the engine used. In their model, the fallibility of human players is associated to a likelihood function L with engine E to generate a stochastic chess engine $E(c)$, where $E(c)$ can choose any move among at max mv alternatives with non zero probability defined by the likelihood function L.

In relation to test-taking and related item-response theories [16], our work is an extension of Rasch modeling [1,11] for *polytomous* items, and has similar mathematical ingredients (cf. [9]). Rasch models have two main kinds of parameters, *person* and *item* parameters. These are often abstracted into the single parameters of actor *location* (or "ability") and item *difficulty*. It is desirable and standard to map them onto the same scale in such a way that '*location* > *difficulty*' is equivalent to the actor having a greater than even chance of getting the right answer, or of scoring a prescribed norm in an item with partial credit. For instance, the familiar 4.0 to 0.0 (A to F) grading scale may be employed to say that a question has exactly B level difficulty if half of the B-level students get it right. The formulas in Rasch modeling enable predicting distributions of responses to items based on differences in these parameters.

4 Methodology

Our work builds on the original model of Regan and Haworth [13], which has two person parameters called s for *sensitivity* and c for *consistency*. The main schematic function $E(s, c)$ is determined by regression from training data to yield an estimation of *Elo rating*, which is the standard measure of player quality or strength in the chess world. Our main departure from Rasch modeling is that the engine's "authoritative" utility values are used to infer probabilities for each available response, without recourse to a measure of difficulty on the Elo scale itself. That is, we have no prior notion of "a position of Grandmaster-level difficulty", or "expert difficulty", or "beginning difficulty" per-se. Chess problems, such as White to checkmate in two moves or Black to move and win, are commonly rated for difficulty of solution, but the criteria for these do not extend to the vast majority of positions faced in games, let alone their reference to chess-specific notions (such as "sacrifices are harder to see"). Instead, we aim to infer difficulty from the expected loss of utility from that of the optimal move, and separately from other features of the computer-analysis data itself.

5 Research Accomplished

We have demonstrated the connection between game play and psychometric modeling in other decision making domain, such as test taking [15]. Later we have shown how humans act and play differently than computers [12]. The main observations found in the paper can be listed as:

- Humans often are prone to blunder. For any position, for humans, it is often beneficial to play as an opponent and let the player make the first blunder. This phenomenon is absent in computer games.
- Humans perceive differences in value in proportion to the total value involved.
- When humans are assisted with computer to play they often play more forcing moves than computer's tendency to 'wait-and-see' and play more defensively.

These findings are important in cheating detection where players illegally use computers to consult their moves and also in player modeling. We revisited the fundamentals of skill assessment and have done a survey of existing models and assessed their performance, prediction and/or profiling of players [7]. I have shown how an intelligent agent can be designed which can act as a judge for decisions made by human [2]. The input to the system is the move played at any particular position and the intelligent agent applies chess engine to analyze the move to find its difficulty and discriminatory power and finally gives its verdict regarding the quality of the decision made. We have quantified the notion of depth of thinking, *swing*, *difficulty* and *complexity* of positions and how these metrics behave across players of various rankings [3]. We found swing moves are often tricky, even for the best chess player to find.

6   Research Plan

In our yet unpublished work, we have proposed the concept of *satisficing depth* and found a measure of satisficing depth which increases linearly with respect to the rating of the respective player. We also have found the impact of time pressure on satisficing depth. Currently we are working on finding the notion of 'gamble', i.e., a sub optimal decision which produces better outcomes than the optimal decision if the opponent falls into the trap of the decision. In some applications, such as multiple-choice tests, we would like to establish an isomorphism of the underlying mathematical quantities, which induces a correspondence between various measurement theories and the chess model. We are trying to find results toward the objective of applying the correspondence in reverse to obtain and quantify the measure of depth and difficulty for multiple-choice tests, stock market trading, and other real-world applications and utilizing this knowledge to design intelligent and automated systems to judge the quality of human or artificial agents. The two central contributions of the proposed work are the marriage of IRT models to traditional decision making processes, as quantified in chess, and the integration of *depth* as a concept.

7   Expected Outcome

Our model can be used for predictive analysis and data mining. This model also gives an insight about performances of an agent in time constrained environments. The result can be used in the following domains.

7.1   Speed-Accuracy Trade-Off

The model can be applied to verify the impact on accuracy if faster decisions are taken. The effect is well known in chess tournaments [5], almost all of which use a time control at move 40. Players often use up almost all of the allotted time before move 30 or so, thus incurring 'time pressure' for 10 or more moves.

Regan, Macieja, and Haworth [14] show a steep monotonic increase in errors by move number up to 40, then a sudden drop off as players have more time. Our yet-unpublished work has quantified the drop of intrinsic rating at 'rapid' and 'blitz' chess played at faster controls.

7.2   Agent Modeling

This model can be extended to model decision-makers, which would be advantageous to plan strategy for or against him/her. For player modeling, we need to find various characteristics unique to the player. The performance rating from our model is a strong indicator of the aptitude level. It can be used to find out any specific trend followed by the player while taking decisions. Measures of blunders and the proclivity for procrastination also could contribute in the player modeling. This may be relevant for player profiling in other online battle games.

7.3   Cheating Detection and Verification

Proved and alleged instances of cheating with computers at chess have increased many-fold in recent years. If a successful technique for detection of cheating is possible, the same idea can be applied to other fields of online gaming or online test-taking. We aim to compare this model with other predictive analytic models used in fraud detection.

7.4   Multiple-Criteria Decision Analysis

Our model can be applied for multiple-criteria decision analysis and verifying the rationality of the intrinsic quality measured with respect to multi-criteria decision rules. In a setting, where an examinee cannot return to previous questions, he often needs to split his time for each question keeping in mind the difficulty of future questions. Prior articulation of preferences in multiple-criteria decision problems plays a key role in agent modeling.

7.5   Decision Making in Multi-agent Environment

Does the quality of the decisions of any agent get affected based on the presence of other agents? How does a player play against a weaker versus a stronger opponent? How does an examinee response when he knows the other examinees are either far superior or far inferior than him? Our model tries to answer these questions and measures the displacement from the mean in these either extreme cases.

References

1. Andrich, D.: Rasch Models for Measurement. Sage Publications, Beverly Hills (1988)
2. Biswas, T.: Designing intelligent agents to judge intrinsic quality of human decisions. In: Proceedings of International Conference on Agents and Artificial Intelligence (ICAART) (2015)
3. Biswas, T., Regan, K.: Quantifying depth and complexity of thinking and knowledge. In: Proceedings of International Conference on Agents and Artificial Intelligence (ICAART) (2015)
4. Busemeyer, J.R., Townsend, J.T.: Decision field theory: a dynamic-cognitive approach to decision making in an uncertain environment. Psychol. Rev. **100**(3), 432 (1993)
5. Chabris, C., Hearst, E.: Visualization, pattern recognition, and forward search: effects of playing speed and sight of the position on grandmaster chess errors. Cogn. Sci. **27**, 637–648 (2003)
6. DiFatta, G., Haworth, G.M^{c}C., Regan, K.: Skill rating by Bayesian inference. In: Proceedings of 2009 IEEE Symposium on Computational Intelligence and Data Mining (CIDM 2009), pp. 89–94, Nashville, TN, 30 March – 2 April 2009 (2009)
7. Haworth, G., Biswas, T., Regan, K.: A comparative review of skill assessment: performance, prediction and profiling. In: Proceedings of International Conference on Advances in Computer Games (ACG 2014) (2015)
8. Stahl II, D.O., Wilson, P.W.: Experimental evidence on players' models of other players. J. Econ. Behav. Organ. **25**(3), 309–327 (1994)
9. Van Der Maas, H.L.J., Wagenmakers, E.J.: A psychometric analysis of chess expertise. Am. J. Psychol. **118**, 29–60 (2005)
10. Morris, G.A., Branum-Martin, L., Harshman, N., Baker, S.D., Mazur, E., Dutta, S., Mzoughi, T., McCauley, V.: Testing the test: item response curves and test quality. Am. J. Phys. **74**(5), 449–453 (2006)
11. Rasch, G.: Probabilistic Models for Some Intelligence and Attainment Tests. Danish Institute for Educational Research, Copenhagen (1960)
12. Regan, K., Biswas, T., Zhou, J.: Human and computer preferences at chess. In: Proceedings of the 8th Multidisciplinary Workshop on Advances in Preference Handling (MPref 2014) (2014)
13. Regan, K., Haworth, G.M^{c}C.: Haworth intrinsic chess ratings. In: Proceedings of AAAI 2011, San Francisco (2011)
14. Regan, K.W., Macieja, B., Haworth, G.M.C.: Understanding distributions of chess performances. In: van den Herik, H.J., Plaat, A. (eds.) ACG 2011. LNCS, vol. 7168, pp. 230–243. Springer, Heidelberg (2012)
15. Regan, K., Biswas, T.: Psychometric modeling of decision making via game play. In: Proceedings of IEEE Conference on Computational Intelligence in Games (2013)
16. Thorpe, G.L., Favia, A.: Data analysis using item response theory methodology: an introduction to selected programs and applications. Psychol. Fac. Sch. (2012). Paper 20. http://digitalcommons.library.umaine.edu/psy_facpub/20

Sequential Decision Making Under Uncertainty Using Ordinal Preferential Information

Hugo Gilbert[✉]

Sorbonne Universités, UPMC Univ Paris 06, CNRS, LIP6 UMR 7606, Paris, France
hugo.gilbert@lip6.fr

1 Introduction

The research work undertaken in my thesis aims at facilitating the conception of autonomous agents able to solve complex problems in sequential decision problems (e.g., planning problems in robotics). In such problems, an agent has to choose autonomously the actions to perform according to the state of the world in order to accomplish a task. Yet, for a given task, describing and modeling appropriate behaviors reveal to be a major difficulty in the conception of autonomous agents. Standard approaches (e.g., Markov Decision Processes [11], reinforcement learning [13]) require a precise numeric evaluation of action values (e.g., rewards, utilities) to compute preferred behaviors. In practice, it is observed that those values are not always available. Indeed, for real life problems, thousands or millions of values may have to be specified manually.

Thus, even when simplifying assumptions are made, setting those parameters is hard, and even impossible to realize when dealing with most large size problems. Moreover, in some situations, even the precise evaluation of an action can be problematic and costly to determine. For instance, in a medical treatment problem, how can one evaluate the value of a patient's life or well-being? To make conventional algorithms applicable, those unknown values are often arbitrarily set. Employing such a method amounts to adding some preference information that was not present in the original problem. As the optimal policies (prescription of actions to execute) can be highly sensitive to those values, we observe that a slight change in the setting could lead to completely different solutions as shown by Example 1 [14].

We recall that a Markov Decision Problem (MDP) is defined by a tuple $\mathcal{M} = (S, A, P, R, \gamma)$ where S is a finite set of states, A is a finite set of actions, $P : S \times A \to \mathcal{P}(S)$ is a transition function with $\mathcal{P}(S)$ being the set of probability distributions over states, $R : S \times A \to \mathbb{R}$ is a reward function and $\gamma \in [0, 1]$ is a discount factor. A solution to an MDP is a policy π^* (that prescribes the action to be used according to the current state) which is optimal according to a decision criterion. For instance, one of the most classic decision criterion is to maximize the expectation of discounted future cumulated rewards ($\pi^* = \max_\pi \mathbb{E}^\pi(\sum_{t=0}^\infty \gamma^t R(s_t, a_t))$).

Example 1. *Consider an MDP consisting of two states $\{s_1, s_2\}$ and where the available actions are a and b. In s_1 both actions are available while in s_2 only a*

© Springer International Publishing Switzerland 2015
T. Walsh (Ed.): ADT 2015, LNAI 9346, pp. 573–577, 2015.
DOI: 10.1007/978-3-319-23114-3_36

is. A discount factor is set at $\beta = 0.5$. The transition function $P(s_t|a_{t-1}, s_{t-1})$ (giving the probability of the new state s_t resulting from the previous state-action pair (s_{t-1}, a_{t-1})) is defined as follows:

$$P(s_1|a, s_1) = 1 \quad P(s_1|b, s_1) = 0.5 \quad P(s_1|a, s_2) = 1$$

In this case, there are only two possible policies depending on the choice of the action in state s_1. Let R denote the reward function where $R(s_1, a)$ is the value of the immediate reward obtained when performing action a in state s_1. Assume that we only know the following constraints:

$$R(s_1, b) > R(s_1, a) > R(s_2, a)$$

$R(s_2, a)$ represents no reward, $R(s_1, a)$ a "small"reward and $R(s_1, b)$ a "big" reward. If the reward function is arbitrarily defined as follows:

$$R(s_1, b) = 2 \quad R(s_1, a) = 1 \quad R(s_2, a) = 0$$

then we can easily check that the best policy (using the expectation of discounted future cumulated rewards as criterion) choses action b in s_1. Now, if the reward function were defined as follows:

$$R(s_1, b) = 10 \quad R(s_1, a) = 9 \quad R(s_2, a) = 0$$

the best policy would have been the one choosing action a.

Although the two functions respect the order imposed on rewards, we observe a preference reversal.

Thus, this method is highly questionable and the objective of this thesis is to construct a theoretically grounded framework to solve this problem.

2   Related Works

A first alternative, when a reward function is hard to specify, is to proceed by letting an expert demonstrate quasi-optimal behaviors. The agent must then try to reproduce the behavior of the expert either by finding a policy that mimics the observed demonstrations (behavioral cloning [3]) or by finding a reward function that explains the behavior of the expert as much as possible (inverse reinforcement learning [1]).

However, these solutions are not always adapted either because the expert can be unable to perform an optimal behavior [2] or because in some cases, experts do not provide any objective guidance, for the concepts to be learnt are subjective. Another solution, called robust MDPs [9], represents imperfectly the reward function (e.g., unknown parameters represented by intervals). However, as this approach often considers the worst scenario, it often yields overly pessimistic behaviors. Furthermore, finding an optimal robust policy can be very computationally cumbersome.

3   Preference-Based Framework

The approach I consider in my thesis consists in working in a less demanding framework using preferential pieces of information such as "this element is preferred to this other element". The elements to be qualitatively compared are numerous: states, actions, histories, policies, etc. and the designed processes can be interactive or not.

Preferences over bags of rewards. Weng and Zanuttini [16] developed an algorithm called interactive value iteration that interleaves the elicitation and resolution processes necessary to solve an MDP. At the beginning of the algorithm, only an order is assumed to be known on the different rewards. The algorithm performs a value iteration like procedure and queries an expert when needed about his preference over two bags of rewards. Not only does the acquired ordinal information enable to continue the solving procedure but it also increases the knowledge about the underlying reward function reducing the number of future queries necessary to find the optimal policy. A similar procedure was defined by Weng *et al.* [15] to adapt the q-learning algorithm to this framework in the more general context of reinforcement learning.

The authors proved that the number of queries issued by interactive value iteration before finding an optimal policy is upper-bounded by a polynomial in the size of the problem. However, the original procedure as presented by Weng and Zanuttini does not explicitly attempt to minimize the number of queries issued by the algorithm and a lot of work remains to be done in order to increase the efficiency of this approach. Moreover the question of the form of queries that can reasonably be asked to the expert (for cognitive reasons) needs to be investigated.

Preferences over histories. In this framework, specifying a problem becomes a much simpler task as illustrated by the cancer treatment problem from Cheng *et al* [5]. In this problem, a state is a pair consisting of the tumor size and toxicity resulting from the treatment (representing the health and well-being of a patient with cancer). The agent should learn how to control the treatment dosage for different time steps in order to cure the patient. In the authors' framework, two behaviors are compared in the following way. A behavior where the patient stays alive is always preferred to a behavior where the patient dies. If a behavior yields both lower tumor size and toxicity level versus a different behavior, it will be preferred to that other behavior. Not only is this model simple but it does not add any fictive numerical information.

However, current models, criteria and algorithms need to be radically modified to cope with ordinal feedbacks. Indeed, in the standard framework, expectation of cumulative rewards is used as criterion to maximize. This criterion can not be used in a qualitative framework and alternative models from decision theory must be considered to tackle this issue. Some criteria that suits this qualitative framework are for instance: the probability threshold criterion [17] (where one tries to maximise the probability of exceeding a given threshold), reference point criterion [14] (where one tries to maximise the probability of doing better than a given probability distribution over possible outcomes), and quantile criteria [12]. The development of new algorithmic tools is required to optimize such criteria.

Preferential-based approaches belong to this stream of research, building bridges between several fields such as reinforcement learning, preference learning, elicitation, etc. This topic is becoming a very active research field with researchers developing methods for ordinal MDPs [14], preference-based reinforcement learning problems [2,4,5], preferential-based multi-armed bandit problems [18], preferential-based case-based reasoning problems [10], etc.

4   Contributions, Current Advances and Perspectives

During the first year of my PhD, we considered both the framework with ordinal preferences over bags of rewards and the framework with ordinal preferences over histories.

In the former, a modified version of interactive value iteration was designed [7], using several strategies in order to reduce as much as possible the number of queries issued to the expert. The key insights are that, in interactive value iteration, (1) the queries should be delayed as much as possible, avoiding asking queries that might not be necessary to find the optimal policy, (2) queries should be asked following a priority order because the answers to some queries can enable to resolve some other queries, (3) queries can be avoided using heuristique information to guide the process.

In the latter, an automated decision algorithm was designed [8] to handle a non-linear model for decision under uncertainty [6] in a sequential setting. The family of utility functions induced by this model encompasses many decision criteria, among which expected utility, threshold criterion, and likely dominance. Those criteria are particularly well-suited for our preference-based framework as it proceeds by comparing pairs of alternatives.

We are currently undertaking a study of reinforcement learning with such non-classic criteria; we are also investigating the problem tackled by interactive value iteration in order to design a similar algorithm but with stronger theoretical guarantees on the number of queries issued; another interesting research direction is the study of direct elicitation protocols that interactively elicit the reward function during the solution procedure (as in Akrour [2], but with non-classic criteria). Finally a topic of interest is the design of scalable algorithms for real problems with very large state/action spaces, as for instance robot navigation.

References

1. Abbeel, P., Ng, A.: Apprenticeship Learning via Inverse Reinforcement Learning. In: Proceedings of Twenty-first International Conference on Machine Learning. ICML 2004, ACM, New York, NY, USA (2004). http://doi.acm.org/10.1145/1015330.1015430

2. Akrour, R., Schoenauer, M., Sebag, M.: APRIL: Active preference-learning based reinforcement learning. In: CoRR (2012). http://arxiv.org/abs/1208.0984
3. Bain, M., Sammut, C.: A Framework for Behavioural Cloning. In: Machine Intelligence vol. 15, pp. 103–129. Oxford University Press (1996)
4. Busa-fekete, R., Sznyi, B., Weng, P., Cheng, W., Hullermeier, E.: Preference-based Evolutionary Direct Policy Search (2014)
5. Cheng, W., Fürnkranz, J., Hüllermeier, E., Park, S.-H.: Preference-based policy iteration: leveraging preference learning for reinforcement learning. In: Gunopulos, D., Hofmann, T., Malerba, D., Vazirgiannis, M. (eds.) ECML PKDD 2011, Part I. LNCS, vol. 6911, pp. 312–327. Springer, Heidelberg (2011)
6. Fishburn, P., LaValle, I.: A nonlinear, nontransitive and additive-probability model for decisions under uncertainty. Ann. Statist. **15**(2), 830–844 (1987)
7. Gilbert, H., Spanjaard, O., Viappiani, P., Weng, P.: Reducing the number of queries in interactive value iteration. In: ADT (2015)
8. Gilbert, H., Spanjaard, O., Viappiani, P., Weng, P.: Solving MDPs with skew symmetric bilinear utility functions. In: IJCAI (2015)
9. Givan, R., Leach, S., Dean, T.: Bounded-parameter Markov decision processes. Artif. Intell. **122**(1–2), 71–109 (2000)
10. Hüllermeier, E., Schlegel, P.: Preference-Based CBR: first steps toward a methodological framework. In: Ram, A., Wiratunga, N. (eds.) ICCBR 2011. LNCS, vol. 6880, pp. 77–91. Springer, Heidelberg (2011)
11. Puterman, M.: Markov Decision Processes: Discrete Stochastic Dynamic Programming, 1st edn. John Wiley & Sons Inc, New York (1994)
12. Rostek, M.: Quantile Maximization in Decision Theory. Rev. Econ. Stud. **77**(1), 339–371 (2010)
13. Sutton, R., Barto, A.: Reinforcement learning: An introduction, vol. 116. Cambridge University Press, Cambridge (1998)
14. Weng, P.: Ordinal decision models for Markov decision processes. In: ECAI 2012-20th European Conference on Artificial Intelligence. Including Prestigious Applications of Artificial Intelligence (PAIS-2012) System Demonstrations Track, Montpellier, France, 27–31 August, 2012. pp. 828–833 (2012)
15. Weng, P., Busa-Fekete, R., Hüllermeier, E.: Interactive Q-learning with ordinal rewards and unreliable tutor. In: ECML/PKDD Workshop Reinforcement Learning with Generalized Feedback (September 2013). http://www-desir.lip6.fr/weng/pub/ecml2013-ws.pdf
16. Weng, P., Zanuttini, B.: Interactive Value Iteration for Markov Decision Processes with Unknown Rewards. In: Rossi, F. (ed.) IJCAI. IJCAI/AAAI (2013)
17. Yu, S., Lin, Y., Yan, P.: Optimization models for the first arrival target distribution function in discrete time. J. Math. Anal. Appl. **225**(1), 193–223 (1998). http://www.sciencedirect.com/science/article/pii/S0022247X98960152
18. Yue, Y., Broder, J., Kleinberg, R., Joachims, T.: The K-armed Dueling Bandits Problem. Journal of Computer and System Sciences (2012). (in press)

Efficiency in Multi-objective Games

Anisse Ismaili[(✉)]

Université Pierre et Marie Curie, University of Paris 06, UMR 7606, LIP6,
75005 Paris, France
anisse.ismaili@lip6.fr

Abstract. In a multi-objective game, each agent individually evaluates each overall action-profile by a vector. I generalize the price of anarchy to multi-objective games and provide a polynomial-time algorithm to assess this new generalization. A working paper is on: http://arxiv.org/abs/1506.04251.

1 Introduction

In multi-agent economic systems, for each individual decision, there are multiple objectives hanging in the balance, like costs, resources, commodities, goods, financial income, sustainability, happiness and life expectancy, motivating the introduction of a super-class of normal form games: multi-objective (MO) games [5,22], where each agent evaluates each overall action profile by a *vector*. Each agent's individual preference is modelled by the Pareto-dominance (to introduce a bit of rationality), inducing Pareto-equilibria (PE) as the concept of overall selfish stability. Concerning game-theoretic and economic models, vectorial evaluations can be seen as a humble backtrack from the intrinsic and subjective theories of value, towards a non-theory of value where the evaluations are maintained vectorial, in order to enable partial (or bounded) rationalities and to avoid critical losses of information. Therefore, in this more realistic framework, thoroughly measuring efficiency is a tremendous necessity.

Let $N = \{1, \ldots, n\}$ denote the *set* of agents. Let A^i denote each agent i's *action-set* (discrete, finite). Each agent i decides an *action* $a^i \in A^i$. Given a subset of agents $M \subseteq N$, let A^M denote $\times_{i \in M} A^i$ and let $A = A^N$ denote the *set* of overall action-profiles. Let $\mathcal{O} = \{1, \ldots, d\}$ denote the *set* of all the objectives, with d fixed. Let $v^i : A \to \mathbb{R}_+^d$ denote an agent i's *individual MO evaluation function*, which maps each overall action-profile $a = (a^1, \ldots, a^n) \in A$ to a MO evaluation $v^i(a) \in \mathbb{R}_+^d$. Hence, agent i's evaluation for objective k is $v_k^i(a) \in \mathbb{R}_+$.

Definition 1. *A* Multi-objective Game (MOG) is a tuple $\left(N, \{A^i\}_{i \in N}, \mathcal{O}, \{v^i\}_{i \in N}\right)$.

MO games encompass single-objective (discrete) optimization problems, MO optimization problems and non-cooperative games. Let us assume $\alpha = |A^i| \in \mathbb{N}$, for each agent. Then, the representation of an MOG requires $n\alpha^n$ d-dimensional vectors and the complexity of an algorithm taking an MOG as an input, must be

© Springer International Publishing Switzerland 2015
T. Walsh (Ed.): ADT 2015, LNAI 9346, pp. 578–586, 2015.
DOI: 10.1007/978-3-319-23114-3_37

established with respect to the length $L = n\alpha^n d$. Let us now supply the vectors with a preference relation. Assuming a *maximization* setting, given $x, y \in \mathbb{R}_+^d$, the following relations state respectively that y (1) weakly-Pareto-dominates and (2) Pareto-dominates x:

$$y \succsim x \quad \Leftrightarrow \quad \forall k \in \mathcal{O}, \ y_k \geq x_k \tag{1}$$

$$y \succ x \quad \Leftrightarrow \quad \forall k \in \mathcal{O}, \ y_k \geq x_k \text{ and } \exists k \in \mathcal{O}, \ y_k > x_k \tag{2}$$

The Pareto-dominance is a *partial* order, inducing a multiplicity of Pareto-efficient outcomes. Formally, the set of efficient vectors is defined as follows:

Definition 2 (Pareto-efficiency). *For $Y \subseteq \mathbb{R}_+^d$, the efficient vectors $EFF[Y] \subseteq Y$ are:*

$$EFF[Y] = \{y^* \in Y \ | \ \forall y \in Y, \ not \ (y \succ y^*)\}$$

The Pareto-dominance enables to define as efficient all the trade-offs that cannot be improved on one objective without being downgraded on another one, that is: the best compromises between objectives. At the individual scale, Pareto-efficiency defines a *partial* rationality, enabling to model behaviours that single-objective (SO) games would not model consistently. Similarly, I denote the subset of worst vectors by $\text{WST}[Y] = \{y^- \in Y \text{ s.t. } \forall y \in Y, \ not \ (y^- \succ y)\}$. Given an overall action-profile $a \in A$, a^M is the restriction of a to A^M, and a^{-i} to $A^{N \setminus \{i\}}$.

Definition 3 (Pareto-equilibrium [22]). *Given an MOG, an action-profile $a \in A$ is a Pareto equilibrium (denoted by $a \in PE$), if and only if, for each agent $i \in N$, we have:*

$$v^i(a^i, a^{-i}) \quad \in \quad EFF\left[\{v^i(b^i, a^{-i}) \ | \ b^i \in A^i\}\right]$$

Pareto-equilibria encompass many behaviourally relevant action-profiles. For instance, whatever the subjective positive weighted combination of the objectives applied by an agent, the decision is included in Pareto-efficiency. I will distinguish the objectives on which I focus the efficiency study and the purely behavioural ones.

Equilibrium Existence. In many sound probabilistic settings, Pareto-efficiency is not demanding on the conditions of individual rationality, and there are multiple Pareto-efficient responses. Consequently, I strongly suspect the number of pure PE to be numerous in average: $|PE| \sim \alpha^{\frac{d-1}{d}n}$, justifying their existence in a probabilistic manner. Furthermore, in MO games with MO potentials [16,18,19], the existence is guaranteed.

Example 1 (A didactic toy-example in Ocean Shores). Five shops in Ocean Shores (the nodes) can decide upon two activities: renting bikes or buggies, selling clams or fruits, etc. Each agent evaluates his local action-profile depending on the actions of his inner-neighbours and according to two objectives: financial revenue and sustainability.

For instance, we have $(b^1, b^2, a^3, b^4, b^5) \in$ PE, since each of these individual actions, given the adversary local action profile (column), is Pareto-efficient among the two actions of the agent (row). Even if the relative values of the objectives cannot be certainly ascertained, all the subjectively efficient vectors are encompassed by Pareto-efficiency. In this MO game, there are 13 such Pareto-equilibria. Their utilitarian evaluations are in Fig. 1 (Sect. 2).

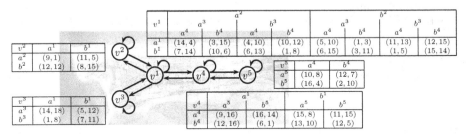

2 The Multi-objective Price of Anarchy

It is well known in game theory that an equilibrium can be overall inefficient with regard to the sum of the individual evaluations. This loss of efficiency is measured by the *price of anarchy* [3,4,6,9,12,20,21] (PoA) $\min[u(PE)]/\max[u(A)]$. Regrettably, when focusing on one sole objective (e.g. making money or a higher GDP), there are losses of efficiency that are not observed: non-sustainability of productions, poor quality of life for workers abroad, production of carcinogens, and others caused by bounded rationality. This appeals for a more thorough analysis of the loss of efficiency at equilibrium and the definition of a *multi-objective* price of anarchy. The higher generality of our model makes it fall outside of the dogma of Smoothness-analysis [20], on which most known PoA results rely [3,4,6,9]. Indeed, with Pareto-equilibria, regardless of the efficiency measurement chosen, best-response inequalities cannot but summed. Instead of analytical results, I will rather show that this MO-PoA can be computed.

The utilitarian social welfare $u : A \rightarrow \mathbb{R}_+^d$ is a vector-valued function measuring social welfare with respect to the d objectives: $u(a) = \sum_{i \in N} v^i(a)$. Given a function $f : A \rightarrow Z$, the *image set* $f(E)$ of a subset $E \subseteq A$ is defined by $f(E) = \{f(a) | a \in E\} \subseteq Z$. Given $\rho, y, z \in \mathbb{R}_+^d$, the vector $\rho \star y \in \mathbb{R}_+^d$ is defined by $\forall k \in \mathcal{O}, (\rho \star y)_k = \rho_k y_k$ and the vector $y/z \in \mathbb{R}_+^d$ is defined by $\forall k \in \mathcal{O}, (y/z)_k = y_k/z_k$. For $x \in \mathbb{R}_+^d$, I denote $x \star Y = \{x \star y \in \mathbb{R}_+^d \mid y \in Y\}$. Given $x \in \mathbb{R}_+^d$, I denote $\mathcal{C}(x) = \{y \in \mathbb{R}_+^d \mid x \succsim y\}$.

I also introduce the notations $\mathcal{E}$ and $\mathcal{F}$, illustrated in Figs. 1 and 2:

- $\mathcal{E} = u(\text{PE})$ the set of *e*quilibria outcomes.
- $\mathcal{F} = \text{EFF}[u(A)]$ the set of *e*fficient outcomes.

For SO games, the worst-case efficiency of equilibria is measured by the PoA $\min[u(PE)]/\max[u(A)]$. However, for MO games, there are many equilibria and optima, and a ratio of the (green) set $\mathcal{E}$ over the (red) set $\mathcal{F}$ is not defined yet and ought to maintain each critical information without introducing arbitrary choices.

Firstly, the efficiency of one equilibrium $y \in \mathcal{E}$ is quantified without taking side with any efficient outcome, by defining with flexibility and no dictatorship, a disjunctive set of guaranteed ratios of efficiency $R[y, \mathcal{F}] = \bigcup_{z \in \mathcal{F}} \mathcal{C}(y/z)$.

Secondly, in MOGs, in average, there are many Pareto-equilibria. An efficiency guarantee $\rho \in \mathbb{R}_+^d$, must hold for each equilibrium-outcome, inducing the conjunctive definition of the set of guaranteed ratios $R[\mathcal{E}, \mathcal{F}] = \bigcap_{y \in \mathcal{E}} R[y, \mathcal{F}]$.

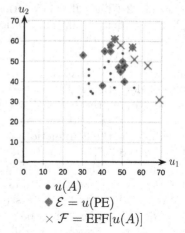

● $u(A)$
◆ $\mathcal{E} = u(\text{PE})$
× $\mathcal{F} = \text{EFF}[u(A)]$

Fig. 1. The utilitarian vectors of Ocean Shores

Technically, $R[\mathcal{E}, \mathcal{F}]$ only depends on $WST[\mathcal{E}]$ and $\mathcal{F}$. Moreover, if two bounds on the efficiency ρ and ρ' are such that $\rho \succ \rho'$, then ρ' brings no more information. Consequently, MO-PoA is defined by using EFF on the guaranteed efficiency ratios.

Definition 4 (MO Price of Anarchy). *Given a MOG, a vector $\rho \in \mathbb{R}_+^d$ bounds the MOG's inefficiency if and only if it holds that: $\forall y \in \mathcal{E}, \quad \exists z \in \mathcal{F}, \quad y \succsim \rho \star z$. Consequently, the set of guaranteed ratios is defined by:*

$$R[\mathcal{E}, \mathcal{F}] = \bigcap_{y \in \mathcal{E}} \bigcup_{z \in \mathcal{F}} \mathcal{C}(y/z) = R[WST[\mathcal{E}], \mathcal{F}]$$

and the MO-PoA is defined by: $MO\text{-}PoA = EFF[R[WST[\mathcal{E}], \mathcal{F}]]$

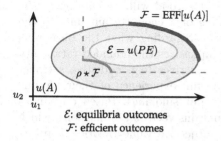

$\mathcal{E}$: equilibria outcomes
$\mathcal{F}$: efficient outcomes

Fig. 2. $\rho \in$ MO-PoA bounds below $\mathcal{E}$'s inefficiency: $\mathcal{E} \subseteq (\rho \star \mathcal{F}) + \mathbb{R}_+^d$

Having ρ in MO-PoA means that for each $y \in \mathcal{E}$, there is an efficient outcome $z^{(y)} \in \mathcal{F}$ such that y dominates $\rho \star z^{(y)}$. In other words, if $\rho \in R[\mathcal{E}, \mathcal{F}]$, then each equilibrium satisfies the ratio of efficiency ρ. This means that equilibria-outcomes are at least as good as $\rho \star \mathcal{F}$. That is: $\mathcal{E} \subseteq (\rho \star \mathcal{F}) + \mathbb{R}_+^d$ (see Fig. 2). Moreover, since ρ is tight, $\mathcal{E}$ sticks to $\rho \star \mathcal{F}$.

Example 2 (*The* Efficiency ratios of Example 1). I depict the efficiency ratios of Ocean Shores (intersected with $[0,1]^d$) which only depend of $\mathrm{WST}[\mathcal{E}] = \{(30,53),(40,38)\}$ and $\mathcal{F} = \{(46,61),\dots,(69,31)\}$. The part below the red line corresponds to $R[(30,53),\mathcal{F}]$, the part below the blue line to $R[(40,38),\mathcal{F}]$ and the yellow part below both lines is the conjunction on both equilibria $R[\mathrm{WST}[\mathcal{E}],\mathcal{F}]$. The freedom degree of deciding what the overall efficiency should be is left free which results in several ratios in the MO-PoA. Firstly, for each $\rho \in R[\mathcal{E},\mathcal{F}]$, we have $\rho_1 \le 65\,\%$. Hence,

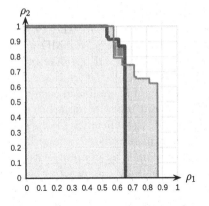

Fig. 3. The MO-PoA of Ocean Shores

whatever the choices of overall efficiency, one cannot guarantee more than 65 % of efficiency on objective 1. Secondly, for each equilibrium, there will always exist some subjectivities for which the efficiency on objective 2 is already total (100 %, if not more) while only 50 % can be obtained on objective 1. Thirdly, from 50 % to 65 % of guaranteed efficiency on objective 1, the various subjectivities trade the guaranteed efficiency on objective 2 from 100 % to 75 % (Fig. 3).

3 Application to the Tobacco Economy

According to the World Health Organisation [24], 17.000 humans die each day of smoking related diseases, but meanwhile, the financial revenue of this industry fosters the politics of tobacco businesses and industries. Let us issue the thorough economic value of a cigarette. According to the *intrinsic theory of value* [2,14], the value of a cigarette amounts to the quantities of raw materials used for its production, or is the combination of the labour times put into it. However, each economic agent needs to keep the freedom to evaluate and act how he pleases, in order to keep his good will.. According to the *subjective theory of value* [2,15,23], the value of a cigarette amounts to the price an agent is willing to pay for it, but disregarding what the disastrous consequence is on his life expectancy. This emphasizes the bounded rationality of behaviours: Agents behave according to objectives (e.g. addictive satisfaction) that they would avoid if they had the full experience of their lifetime (e.g. a lung cancer) and a strong will (e.g. quit smoking). Based on a non-theory of value, the theory of MO games maintains vectorial evaluations, which is a backtrack from both the intrinsic and subjective value theories. **I modelled** the tobacco industry and its consumers [1,13] by a succinct MOG, with the help of (..) the association "Alliance contre le tabac". The set of agents is $N = \{\mathrm{industry}, \nu \text{ consumers}\}$, where there are about $\nu = 6.10^9$ prospective consumers. Each consumer decides in $A^{\mathrm{consumer}} = \{\mathrm{not\text{-}smoking}, \mathrm{smoking}\}$ and cares about money, his addictive satisfaction, and living. The industry only cares about money and decides in $A^{\mathrm{industry}} = \{\mathrm{not\text{-}active}, \mathrm{active}, \mathrm{advertise\,\&\,active}\}$.

We have $\mathcal{O} = \{$money, reward, life-expectancy$\}$. The tables below depict the evaluation vectors (over a life-time and ordered as in $\mathcal{O}$) of one prospective consumer and the evaluations of the industry with respect to the number $\theta \in \{0, \ldots, \nu\}$ of consumers who decide to smoke. Money (already an aggregation) is expressed in kilo-dollars; the addictive reward is on an ordinal scale $\{1, 2, 3, 4\}$; life-expectancy is in years.

v^{consumer}	Not-active	Active	Advertise&active
not-smoking	$(48, 1, 75)$	$(48, 1, 75)$	$(48, 1, 75)$
smoking	$(48, 1, 75)$	$(12, 3, 65)$	$(0, 4, 55)$

$v^{\text{industry}}(\theta)$	Not-active	Active	Advertise&active
$(\nu - \theta) \times$	$(0, -, -)$	$(0, -, -)$	$(0, -, -)$
$+ \theta \times$	$(0, -, -)$	$(26, -, -)$	$(36, -, -)$

Pareto Equilibria. If the industry is active, then for the consumer, deciding to smoke or not depends on how the consumer subjectively values/weighs money, addictive satisfaction and life expectancy: both decisions are encompassed by Pareto-efficiency. For the industry, advertise&active is a dominant strategy. Consequently, Pareto-equilibria are all the action-profiles in which the industry decides advertise&active.

Efficiency. I focus on money and life-expectancy, since addiction is only a behavioural objective. We have $\mathcal{E} = \{\theta \cdot (36, -, 55) + (\nu - \theta) \cdot (48, -, 75) \mid 0 \leq \theta \leq \nu\}$ and $\mathcal{F} = \{\nu \cdot (48, -, 75)\}$, where ν is the world's population, and θ the number of smokers. Since $\text{WST}[\mathcal{E}] = \{(36, -, 55)\}$, the MO-PoA is the singleton $\{(75\%, -, 73\%)\}$: we can lose up to 12k\$ and 20 years of life-expectancy per-consumer. Some economists would say: "Since consumers value the product, then the industry creates value." However, these Pareto-equilibria are the worst action-profiles for money and life-expectancy. This critical information was not lost by the MO-PoA.

Lessons. Advertising tobacco fosters consumption. Independently from this work, the association "Alliance contre le tabac" passed a law for standardized neutral packets (April 3rd 2015), in order to annihilate the benefits of branding, but only in France. My model says that *this law will promote a higher efficiency*.

4 Computation of the MO-PoA

In this section, I provide a polynomial-time algorithm for the computation of MO-PoA, for MOGs given in MO normal form, which representation length is $L = n\alpha^n d$.

Theorem 1 (PTIME Computation of the MO-PoA). *Given a MO normal form, one can compute the MO-PoA in time $O(L^{4d-2})$. If $d = 2$, it lowers to $O(L^4 \log_2(L))$.*

The proof of Theorem 1 relies on a very general procedure based on **two phases**:

1. Given a MOG, compute the worst equilibria $\text{WST}[\mathcal{E}]$ and the efficient outcomes $\mathcal{F}$.
2. Given $\text{WST}[\mathcal{E}]$ and $\mathcal{F}$, compute MO-PoA = EFF[$R[$ WST$[\mathcal{E}]$, $\mathcal{F}$]].

Phase 1 is easy, if the MOG is given in normal form. For Phase 2, using the shorthand $q = |\text{WST}[\mathcal{E}]|$ and $m = |\mathcal{F}|$, at first glance, the development of the intersection of unions $R[\text{WST}[\mathcal{E}], \mathcal{F}] = \cap_{y \in \text{WST}[\mathcal{E}]} \cup_{z \in \mathcal{F}} \mathcal{C}(y/z)$ causes an exponential m^q. But fortunately, one can compute the MO-PoA in polynomial time. Below, D^t is a set of vectors. Given two vectors $x, y \in \mathbb{R}^d_+$, let $x \wedge y$ denote the vector defined by $\forall k \in \mathcal{O}$, $(x \wedge y)_k = \min\{x_k, y_k\}$ and recall that $\forall k \in \mathcal{O}$, $(x/y)_k = x_k/y_k$. Algorithm 1 is the development of $\cap_{y \in \text{WST}[\mathcal{E}]} \cup_{z \in \mathcal{F}} \mathcal{C}(y/z)$, on a set-algebra of cone-unions.

Algorithm 1. Computing MO-PoA in polynomial-time

Input: $\text{WST}[\mathcal{E}] = \{y^1, \ldots, y^q\}$ and $\mathcal{F} = \{z^1, \ldots, z^m\}$
Output: MO-PoA = EFF$[R[\text{WST}[\mathcal{E}], \mathcal{F}]]$

create $D^1 \leftarrow \{y^1/z \in \mathbb{R}^d_+ \mid z \in \mathcal{F}\}$
for $t = 2, \ldots, q$ **do**
$\quad \mid \quad D^t \leftarrow \text{EFF}[\{\rho \wedge (y^t/z) \mid \rho \in D^{t-1}, \ z \in \mathcal{F}\}]$
end
return D^q

For compact representations of massively multi-agent games, one can approximate:

Theorem 2 (Approximation for MO-PoA in Compact MOGs). *Given approximates E of $\mathcal{E}$ and F of $\mathcal{F}$, Algorithm 1 outputs a covering R of MO-PoA in the sense that:*

if	$\forall y \in \mathcal{E}, \exists y' \in E, \quad y \succsim y'$	and	$\forall y' \in E, \exists y \in \mathcal{E}, \quad (1+\varepsilon_1)y' \succsim y$
and	$\forall z' \in F, \exists z \in \mathcal{F}, \quad z' \succsim z$	and	$\forall z \in \mathcal{F}, \exists z' \in F, \quad (1+\varepsilon_2)z \succsim z'$
then	$\forall \rho \in \text{MO-PoA}, \quad \exists \rho' \in R,$		$(1+\varepsilon_1)(1+\varepsilon_2)\rho' \succsim \rho$

5   Prospects

This algorithmic work could be extended to massively multi-agent compact game representations, like for instance MO extensions of graphical games [8] or action-graph games [7,11]. Studying the efficiency of MO generalization of auctions or Cournot-competitions [9,10,17] could provide insights on critical economic malfunctions.

Acknowledgements. I am grateful to anonymous ADT referees, to (..) *"Alliance contre le tabac"* and particularly to Clémence Cagnat-Lardeau for her help on the model.

References

1. Global Issues: Tobacco (2014). http://www.globalissues.org/article/533/tobacco
2. Adam, S.: An Inquiry into the Nature and Causes of the Wealth of Nations, Edwin Cannan's annotated edition edn. Random House, New York (1776)
3. Aland, S., Dumrauf, D., Gairing, M., Monien, B., Schoppmann, F.: Exact price of anarchy for polynomial congestion games. In: Durand, B., Thomas, W. (eds.) STACS 2006. LNCS, vol. 3884, pp. 218–229. Springer, Heidelberg (2006)
4. Awerbuch, B., Azar, Y., Epstein, A.: The price of routing unsplittable flow. In: Proceedings of the Thirty-seventh Annual ACM Symposium on Theory of Computing, pp. 57–66. ACM (2005)
5. Blackwell, D., et al.: An analog of the minimax theorem for vector payoffs. Pac. J. Math. **6**(1), 1–8 (1956)
6. Christodoulou, G., Koutsoupias, E.: The price of anarchy of finite congestion games. In: Proceedings of the Thirty-seventh Annual ACM Symposium on Theory of Computing, pp. 67–73. ACM (2005)
7. Daskalakis, C., Goldberg, P.W., Papadimitriou, C.H.: The complexity of computing a Nash equilibrium. SIAM J. Comput. **39**(1), 195–259 (2009)
8. Gottlob, G., Greco, G., Scarcello, F.: Pure nash equilibria: Hard and easy games. J. Artif. Intell. Res. **24**, 357–406 (2005)
9. Guo, X., Yang, H.: The price of anarchy of cournot oligopoly. In: Deng, X., Ye, Y. (eds.) WINE 2005. LNCS, vol. 3828, pp. 246–257. Springer, Heidelberg (2005)
10. Immorlica, N., Markakis, E., Piliouras, G.: Coalition formation and price of anarchy in cournot oligopolies. In: Saberi, A. (ed.) WINE 2010. LNCS, vol. 6484, pp. 270–281. Springer, Heidelberg (2010)
11. Jiang, A.X., Leyton-Brown, K., Bhat, N.A.: Action-graph games. Games Econ. Behav. **71**(1), 141–173 (2011)
12. Koutsoupias, E., Papadimitriou, C.: Worst-case equilibria. In: Meinel, C., Tison, S. (eds.) STACS 1999. LNCS, vol. 1563, pp. 404–413. Springer, Heidelberg (1999)
13. Madeley, J.: Big Business, Poor Peoples: The Impact of Transnational Corporations on the World's Poor. Palgrave Macmillan, New York (1999)
14. Marx, K.: Das Kapital. Verlag von Otto Meisner, Hamburg (1867)
15. Menger, C.: Principles of economics. Ludwig von Mises Institute, Auburn (1871)
16. Monderer, D., Shapley, L.S.: Potential games. Games Econ. Behav. **14**(1), 124–143 (1996)
17. Nadav, U., Piliouras, G.: No regret learning in oligopolies: cournot vs. bertrand. In: Kontogiannis, S., Koutsoupias, E., Spirakis, P.G. (eds.) SAGT 2010. LNCS, vol. 6386, pp. 300–311. Springer, Heidelberg (2010)
18. Patrone, F., Pusillo, L., Tijs, S.: Multicriteria games and potentials. Top **15**(1), 138–145 (2007)
19. Rosenthal, R.W.: A class of games possessing pure-strategy Nash equilibria. Int. J. Game Theor. **2**(1), 65–67 (1973)
20. Roughgarden, T.: Intrinsic robustness of the price of anarchy. In: Proceedings of the Forty-first Annual ACM Symposium on Theory of Computing, pp. 513–522. ACM (2009)

21. Roughgarden, T., Tardos, E.: Introduction to the inefficiency of equilibria. Algorithmic Game Theor. **17**, 443–459 (2007)
22. Shapley, L.S.: Equilibrium points in games with vector payoffs. Naval Res. Logistics Q. **6**(1), 57–61 (1959)
23. Walras, L.: Éléments d'économie politique pure, ou, Théorie de la richesse sociale. F Rouge, Lausanne (1896)
24. WHO: WHO report on the global tobacco epidemic, 2011: warning about the dangers of tobacco: executive summary (2011)

Modeling, Learning and Reasoning
with Qualitative Preferences

Xudong Liu$^{(\boxtimes)}$

Department of Computer Science, University of Kentucky, Lexington, KY, USA
liu@cs.uky.edu

1 Research Overview

My research is focused on knowledge representation and reasoning, especially, preference modeling, learning and reasoning, and computational social choice. Preference modeling, learning and reasoning is a major research area in artificial intelligence (AI) and decision theory, and is closely related to the social choice theory considered by economists and political scientists. In my research I explore emerging connections between preferences in AI and social choice theory. My main focus is on qualitative preference representation languages extending and combining formalisms such as lexicographic preference trees (LP-trees) [1], answer-set optimization theories (ASO-theories) [3], possibilistic logic [4]; and conditional preference networks (CP-nets) [2], on learning problems that aim at discovering qualitative preference models and predictive preference information from empirical data; and on qualitative preference reasoning problems centered around preference optimization and strategy-proofness of preference aggregation methods. Applications of my research include recommendation systems, decision support tools, multi-agent systems, and Internet trading and marketing platforms.

2 Preliminaries

My research focuses on problems involving *qualitative preferences*, that is, simple and intuitive qualitative statements about *preferred properties* of alternatives. These alternatives are described in terms of *attributes* or *issues*, each assuming values from some finite domain. For instance, vacations can be described in terms of issues such as *activity* (A), *destination* (D), *time* (T), and *transportation* (R), where *activity* has values *water-sports* (ws) and *hiking* (h), *destination* has values *Florida* (fl) and *Colorado* (co), *time* has values *summer* (s) and *winter* (w), and *transportation* has values *car* (c) and *plane* (p). Thus, a sequence of attribute values, for example, $\langle ws, fl, s, c \rangle$ describes a specific *summer* vacation involving *water-sports* in *Florida* to which we travel by *car*. Spaces of alternatives of this type are referred to as *combinatorial domains*.

The exponential size of the combinatorial domain leads to the infeasibility of correctly putting precise numbers on the utility of specific choices; thus, we turn

© Springer International Publishing Switzerland 2015
T. Walsh (Ed.): ADT 2015, LNAI 9346, pp. 587–592, 2015.
DOI: 10.1007/978-3-319-23114-3_38

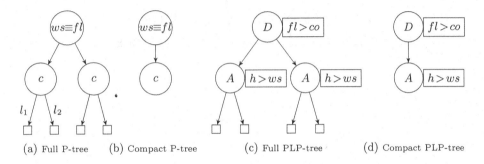

(a) Full P-tree   (b) Compact P-tree   (c) Full PLP-tree   (d) Compact PLP-tree

Fig. 1. P-trees and PLP-trees on vacations

to languages specifying preferences qualitatively. The sheer number of alternatives in the combinatorial domain also makes it impossible to enumerate from the most preferred to the least. Consequently, my research focuses on designing *concise* formalisms in which qualitative preferences over such domains could be expressed compactly and intuitively, and solving problems in preference learning and reasoning in the context of these formalisms.

One of such preference systems, *preference trees* (P-trees), was introduced by Fraser [5,6], and further discussed in my work [9]. Let us illustrate the formalism with preferences over the vacation domain. The most important property for our agent involves activity and destination. She prefers vacations with water sports in Florida or hiking in Colorado over the other options. This preference is described as an equivalence formula $ws \equiv fl$. Within each of the two groups of vacations (satisfying the formula and not satisfying the formula), driving (c) is the preferred transportation mode. These preference statements are described in Fig. 1a. Clearly, the P-tree partitions the vacations into four clusters, denoted by the leaves, with the leftmost representing the set of most preferred vacations satisfying the formulas $ws \equiv fl$ and c. Thus, the alternative $\langle h, co, s, c \rangle$ is better than vacation $\langle ws, fl, s, p \rangle$, because the former descends to leaf l_1 and the latter l_2, and l_1 precedes l_2 in the order of leaves. Since the subtrees of the root are the same and leaves can be omitted in Fig. 1a, we can collapse the full tree to its *compact* version in Fig. 1b. Compactness of preference models is crucial in studying problems in preference learning and reasoning.

I introduced the preference formalism of *partial lexicographic preference trees*, or PLP-trees [10], where nodes in the tree are labeled not by a formula but by an attribute and a total ordering of the values of the attribute. To illustrate PLP-trees, let us consider again preferences over vacations. As shown in Fig. 1c, our agent puts *destination* as the most important attribute on which she prefers *Florida*. Similarly as before, she next considers *activity* and prefers *hiking* for both *Florida* and *Colorado* vacations. Like the above P-tree, this full PLP-tree induces a total preorder of four clusters of equivalent alternatives as the box-labeled leaves, and it is collapsed to a much more compact one in Fig. 1d, where preferences on all attributes are *unconditional*. I have shown that

PLP-trees are special cases of P-trees but more general than the restrictive LP-trees. Studying learning and reasoning problems for PLP-trees will contribute to the most general setting of P-trees.

3 Research Experience

Working on my research I collaborated with professors and colleagues in our department. My work on preferences has led to publications on preference modeling [9], learning [10] and reasoning [7,8,12].

3.1 Preference Modeling

In joint work with my Ph.D. advisor Dr. Miroslaw Truszczynski, we focused on the language of P-trees, studied the relationship between P-trees and other existing preference languages, and showed that P-trees extend possibilistic logic, LP-trees and ASO-rules [9]. Moreover, we established computational complexity results of commonly considered decision problems in the setting of P-trees, such as *dominance testing* (asking if an alternative is preferred to another given the preferences), *optimality testing* (deciding if an alternative is optimal given the preferences), and *optimality testing w.r.t a property* (determining if there exists an optimal alternative satisfying a given property).

3.2 Preference Learning

Another joint work with my Ph.D. advisor introduced the formalism of PLP-trees, a novel formalism for lexicographic preference models, also a subclass of P-trees, over combinatorial domains of alternatives [10]. For PLP-trees we investigated the problem of *passive learning*, that is, the problem of learning preference models given a set of pairwise preferences between alternatives, called *training examples*, provided by the user upfront. Specifically, for several classes of PLP-trees, we studied how to learn (i) a PLP-tree, preferably of a small size, consistent with a dataset of examples, and (ii) a PLP-tree correctly ordering as many of the examples as possible in case of inconsistency. We established complexity results of these problems and, in each case where the problem is in the class P, proposed a polynomial time algorithm.

3.3 Preference Reasoning

In the work with Dr. Truszczynski [8,11], we investigated two preference-aggregation problems, the *winner* problem, computing the winning alternative in an election, and the *evaluation* problem, computing an alternative scoring at least above some threshold in an election, based on *positional scoring rules* (such as k-approval and Borda) when preferences are represented as LP-trees. We obtained new computational complexity results of these two problems and provided computational methods to model and solve the problems in two programming formalisms, answer set programming (ASP) and weighted partial maximum satisfiability (WPM). To support experimentation, we designed methods to generate LP-tree votes randomly and presented experimental results with ASP and WPM solvers. In a joint work [12] with Dr. Judy Goldsmith and other fellow graduate students, we introduced a new variant of hedonic coalition formation games in which agents have two levels of preference on their own coalitions: preference on the set of "roles" that make up the coalition, and preference on their own role within the coalition. We defined and studied several stability notions and optimization problems for this model.

4   Current and Future Research

Moving forward I plan to continue working on problems on preferences in AI and social choice theory, particularly when the preferences concern alternatives ranging over combinatorial domains.

4.1   Preference Learning and Approximation

I will generalize my results on learning PLP-trees to the case of P-trees. I will also design and implement algorithms to learn, from both synthetic and real-world datasets, preferences described in formalisms of LP-trees, PLP-trees and P-trees for both passive learning and active learning, where, unlike passive learning, training examples are elicited from the user interactively. To support evaluation of these learning algorithms, I will design and implement algorithms to randomly generate instances of LP-trees, PLP-trees and P-trees. To facilitate the preference learning process, I will develop datasets of examples from existing learning datasets, and apply machine learning methods to obtain preferences from these developed datasets.

Some models of preference orders do not support effective reasoning. For instance, if a preference order is represented by a CP-net, the *dominance testing* problem is known to be NP-hard even for the simple case where the dependency among the nodes is acyclic, and it is PSPACE-complete in general. Learning can provide a way to circumvent the difficulty. Compared to the formalism of CP-nets, P-trees are more practical, more intuitive and more transparent for representing preferences over combinatorial domains. Since reasoning with P-trees is easier (e.g., dominance is straightforward), approximating (or exactly representing) CP-nets using P-trees learned from examples consistent with the CP-net might open a way to more effective approximate, or even exact, reasoning with CP-nets. I plan to design algorithms to find a *small* set of P-trees that can best approximate the given CP-net.

4.2   Preference Aggregation

Provided that we have obtained preferences from the agents as P-trees, I will apply two approaches to aggregate P-trees to compute the collective decision: the Pareto method and voting rules. Using the Pareto method is similar to a previous work on ASO. As for the voting rules that could be applied, I will investigate positional scoring rules (e.g., Plurality, k-Approval and Borda), comparison-based rules (e.g., the Copeland, Simpsons and Maximin rules), and distance-based rules (e.g., the Kemeny and Dodgson rules). To compare the two approaches of aggregating P-trees, I will perform experiments on both randomly generated and learned P-trees using the two methods separately, and analyze the winning alternatives computed by both of them.

4.3 Misrepresentation of Preferences

I will study problems relating to vulnerability of collective decisions under misrepresentation of preferences specified over combinatorial domains. Take the *coalitional manipulation problem* as an example. This problem asks to decide if the small coalition set of manipulative voters can make some candidate a winner. I have already obtained preliminary complexity results for LP-trees when the voting rules are Plurality and half-Approval where each voter approves her top half candidates. I will examine other positional scoring rules, as well as some comparison-based and distance-based voting systems, for LP-trees, and extend these results to elections over complicated domains when votes are specified as P-trees.

5 Conclusion

My research concerns problems in the fields pertaining to preferences and social choice, and exploits emerging connections between preferences in AI and social choice theory. My main focus is on qualitative preference representation languages extending and combining existing formalisms such as lexicographic preference trees (LP-trees) and answer-set optimization theories (ASO-theories), on learning problems that aim at discovering predictive preference models from empirical data, and on qualitative preference reasoning problems centered around preference optimization and strategy-proofness of preference aggregation methods.

References

1. Booth, R., Chevaleyre, Y., Lang, J., Mengin, J., Sombattheera, C.: Learning conditionally lexicographic preference relations. In: European Conference on Artificial Intelligence (ECAI) (2010)
2. Boutilier, C., Brafman, R.I., Domshlak, C., Hoos, H.H., Poole, D.: CP-nets: a tool for representing and reasoning with conditional ceteris paribus preference statements. J. Artif. Intell. Res. **21**, 135–191 (2004)
3. Brewka, G., Niemelä, I., Truszczynski, M.: Answer set optimization. In: International Joint Conference on Artificial Intelligence (IJCAI) (2003)
4. Dubois, D., Lang, J., Prade, H.: Towards possibilistic logic programming. In: International Conference on Logic Programming (ICLP) (1991)

5. Fraser, N.M.: Applications of preference trees. In: IEEE Systems Man and Cybernetics Conference (IEEE SMC), pp. 132–136. IEEE (1993)
6. Fraser, N.M.: Ordinal preference representations. Theor. Decis. **36**(1), 45–67 (1994)
7. Liu, X.: Aggregating lexicographic preference trees using answer set programming: Extended abstract. In: IJCAI-13 Doctoral Consortium (2013)
8. Liu, X., Truszczynski, M.: Aggregating conditionally lexicographic preferences using answer set programming solvers. In: Perny, P., Pirlot, M., Tsoukiàs, A. (eds.) ADT 2013. LNCS, vol. 8176, pp. 244–258. Springer, Heidelberg (2013)
9. Liu, X., Truszczynski, M.: Preference trees: a language for representing and reasoning about qualitative preferences. In: Multidisciplinary Workshop on Advances in Preference Handling (MPREF), pp. 55–60. AAAI Press (2014)
10. Liu, X., Truszczynski, M.: Learning partial lexicographic preference trees over combinatorial domains. In: AAAI Conference on Artificial Intelligence (AAAI) (2015)
11. Liu, X., Truszczynski, M.: Reasoning about lexicographic preferences over combinatorial domains. In: Algorithmic Decision Theory (ADT) (2015)
12. Spradling, M., Goldsmith, J., Liu, X., Dadi, C., Li, Z.: Roles and teams hedonic game. In: Perny, P., Pirlot, M., Tsoukiàs, A. (eds.) ADT 2013. LNCS, vol. 8176, pp. 351–362. Springer, Heidelberg (2013)

Author Index